Methods in Cell Biology

VOLUME 85

Fluorescent Proteins

Series Editors

Leslie Wilson

Department of Molecular, Cellular and Developmental Biology
University of California
Santa Barbara, California

Paul Matsudaira

Whitehead Institute for Biomedical Research
Department of Biology
Division of Biological Engineering
Massachusetts Institute of Technology
Cambridge, Massachusetts

Methods in Cell Biology

VOLUME 85

Fluorescent Proteins

Edited by

Kevin F. Sullivan

Department of Biochemistry
National University of Ireland
Galway

AMSTERDAM • BOSTON • HEIDELBERG • LONDON
NEW YORK • OXFORD • PARIS • SAN DIEGO
SAN FRANCISCO • SINGAPORE • SYDNEY • TOKYO
Academic Press is an imprint of Elsevier

Cover photo: The structure of GFP is shown with a spectrum of colors representing the diverse family of fluorescent proteins developed and discovered since 1995. Surrounding it are stylized illustrations based on the structure with iconographic representations of various techniques and approaches that incorporate fluorescent proteins in biological research.

Academic Press is an imprint of Elsevier
84 Theobald's Road, London WC1X 8RR, UK
Radarweg 29, PO Box 211, 1000 AE Amsterdam, The Netherlands
The Boulevard, Langford Lane, Kidlington, Oxford OX5 1GB, UK
30 Corporate Drive, Suite 400, Burlington, MA 01803, USA
525 B Street, Suite 1900, San Diego, CA 92101-4495, USA

First edition 2008

ISBN: 978-0-12-372558-5

ISSN: 0091-679X

For information on all Academic Press publications
visit our website at books.elsevier.com

Printed and bound in USA

08 09 10 11 12 10 9 8 7 6 5 4 3 2 1

CONTENTS

CONTRIBUTORS

Numbers in parentheses indicate the pages on which the authors' contributions begin.

Chris Allan (555) Division of Gene Regulation and Expression, College of Life Sciences, Wellcome Trust Biocentre, University of Dundee, Dundee, Scotland DD1 5EH, United Kingdom

Anjon Audhya (179) Ludwig Institute for Cancer Research, Department of Cellular and Molecular Medicine, University of California, San Diego, La Jolla, California 92093

Roger N. Beachy (153) Donald Danforth Plant Science Center, St. Louis, Missouri 63132

Kerry S. Bloom (127) Department of Biology, University of North Carolina, Chapel Hill, North Carolina 27599

R. Howard Berg (153) Integrated Microscopy Facility, Donald Danforth Plant Science Center, St. Louis, Missouri 63132

Jean-Marie Burel (555) Division of Gene Regulation and Expression, College of Life Sciences, Wellcome Trust Biocentre, University of Dundee, Dundee, Scotland DD1 5EH, United Kingdom

Heather Butler (23) Department of Cell Biology and Neuroscience, Montana State University, Bozeman, Montana 59717

Pascal Chartrand (273) Département de Biochimie, Université de Montréal, 2900 Edouard-Montpetit, Montréal, Québec H3C 3J7, Canada

Maïté Coppey-Moisan (395) Institut Jacques Monod, UMR 7592 CNRS, University Paris 6/University Paris 7, 2 Place Jussieu, 75251 Paris Cedex 05, France

F. P. Cordelières (83) Institut Curie, Section de Recherche/CNRS UMR 146, Plateforme d'Imagerie Cellulaire et Tissulaire, Centre Universitaire, 91405 Orsay Cedex, France

Ralf Dahm (293) Center for Brain Research, Division of Neuronal Cell Biology, Medical University of Vienna, Spitalgasse 4, A-1090 Vienna, Austria

Alexander Dammermann (179) Ludwig Institute for Cancer Research, Department of Cellular and Molecular Medicine, University of California, San Diego, La Jolla, California 92093

Gaudenz Danuser (497) Laboratory for Computational Cell Biology, Department of Cell Biology, CB167, The Scripps Research Institute La Jolla, California 92037

Arshad Desai (179) Ludwig Institute for Cancer Research, Department of Cellular and Molecular Medicine, University of California, San Diego, La Jolla, California 92093

J. R. De Mey (83) École Supérieure de Biotechnologie de Strasbourg, UMR-7175 CNRS/Université Louis Pasteur (Strasbourg I), BP10413, 67412 IllKIRCH Cedex, France

H. William Detrich, III (219) Department of Biology, Northeastern University, Boston, Massachusetts 02115

A. Dieterlen (83) Laboratoire MIPS, Groupe LAB.EL, Université de Haute-Alsace, IUT Génie Electrique II, 61, rue Albert Camus, 68093 Mulhouse Cedex, France

Ian M. Dobbie (1) Department of Biochemistry, National University of Ireland, Galway, Ireland

J. Dompierre (83) Institut Curie, Section de Recherche/CNRS UMR 146, Saudou Group, Centre Universitaire, 91405 Orsay Cedex, France

Jonas F. Dorn (497) Laboratory for Computational Cell Biology, Department of Cell Biology, CB167, The Scripps Research Institute, La Jolla, California 92037

Adam D. Douglass (113) Department of Cellular and Molecular Pharmacology, University of California, The Howard Hughes Medical Institute, San Francisco, California 94107

R. Eils (539) Department of Bioinformatics and Functional Genomics, German Cancer Research Center (DKFZ), University of Heidelberg, IPMB, Im Neuenheimer Feld 267, D-69120 Heidelberg, Germany

Grigori Enikolopov (243) Cold Spring Harbor Laboratory, Cold Spring Harbor, New York 11724

Mark Fricker (353) Department of Plant Sciences, University of Oxford, Oxford OX1 3RB, United Kingdom

Bernhard Götze* (293) Center for Brain Research, Division of Neuronal Cell Biology, Medical University of Vienna, Spitalgasse 4, A-1090 Vienna, Austria *Present address: Carl Zeiss MicroImaging GmbH, Carl Zeiss Promenade 10, D-07745 Jena, Germany

Rebecca A. Green (179) Ludwig Institute for Cancer Research, Department of Cellular and Molecular Medicine, University of California, San Diego, La Jolla, California 92093

Klaus M. Hahn (63) Department of Pharmacology, Lineberger Cancer Center, University of North Carolina at Chapel Hill, Chapel Hill, North Carolina 27599

N. Harder (539) Department of Bioinformatics and Functional Genomics, German Cancer Research Center (DKFZ), University of Heidelberg, IPMB, Im Neuenheimer Feld 267, D-69120 Heidelberg, Germany

Robert M. Hoffman (485) AntiCancer, Inc., San Diego, California 92111, Department of Surgery, University of California, San Diego, California 92103

Louis Hodgson (63) Department of Pharmacology, Lineberger Cancer Center, University of North Carolina at Chapel Hill, Chapel Hill, North Carolina 27599

Thomas Hughes (23) Department of Cell Biology and Neuroscience, Montana State University, Bozeman, Montana 59717

Anthony Hyman (179) Max-Planck Institute of Molecular and Cellular Biology and Genetics, Dresden 01307, Germany

Ajit P. Joglekar (127) Department of Biology, University of North Carolina, Chapel Hill, North Carolina 27599

Tom K. Kerppola (431) Department of Biological Chemistry, Howard Hughes Medical Institute, University of Michigan Medical School, Ann Arbor, Michigan 48109

P. Kessler (83) Imaging Center of the IGBMC, 1, rue Laurent Fries, 67404 Illkirch Cedex, France

Michael A. Kiebler (293) Center for Brain Research, Division of Neuronal Cell Biology, Medical University of Vienna, Spitalgasse 4, A-1090 Vienna, Austria

J. Langowski (471) German Cancer Research Center, Division Biophysics of Macromolecules, Im Neuenheimer Feld 580, D-69120 Heidelberg, Germany

Jennifer Lippincott-Schwartz (45) Cell Biology and Metabolism Branch, National Institute of Child Health and Human Development, National Institutes of Health, Bethesda, Maryland 20892

Brian Loranger (555) Division of Gene Regulation and Expression, College of Life Sciences, Wellcome Trust Biocentre, University of Dundee, Dundee, Scotland DD1 5EH, United Kingdom

Noel F. Lowndes (1) Department of Biochemistry, National University of Ireland, Galway, Ireland

Paolo Macchi* (293) *Present address: Center for Integrative Biology, Laboratory of Molecular and Cellular Neurobiology, University of Trento, Via delle Regole 101, 38060 Mattarello, Trento, Italy. Center for Brain Research, Division of Neuronal Cell Biology, Medical University of Vienna, Spitalgasse 4, A-1090 Vienna, Austria

Donald MacDonald (555) Division of Gene Regulation and Expression, College of Life Sciences, Wellcome Trust Biocentre, University of Dundee, Dundee, Scotland DD1 5EH, United Kingdom

Juan Manuel Encinas (243) Cold Spring Harbor Laboratory, Cold Spring Harbor, New York 11724

James G. McNally (329) Laboratory of Receptor Biology and Gene Expression, National Cancer Institute, Bethesda, Maryland 20892

Robert Mealer (23) Department of Cell Biology and Neuroscience, Montana State University, Bozeman, Montana 59717

Atsushi Miyawaki (381) Laboratory for Cell Function Dynamics, Brain Science Institute, The Institute of Physical and Chemical Research (RIKEN), 2-1 Hirosawa, Wako-city, Saitama 351-0198, Japan

Joost Monen (179) Ludwig Institute for Cancer Research, Department of Cellular and Molecular Medicine, University of California, San Diego, La Jolla, California 92093

Jonathan Monk (555) Division of Gene Regulation and Expression, College of Life Sciences, Wellcome Trust Biocentre, University of Dundee, Dundee, Scotland DD1 5EH, United Kingdom

Ian Moore (353) Department of Plant Sciences, University of Oxford, Oxford OX1 3RB, United Kingdom

Joshua Moore (555) Division of Gene Regulation and Expression, College of Life Sciences, Wellcome Trust Biocentre, University of Dundee, Dundee, Scotland DD1 5EH, United Kingdom

Karen Oegema (179) Ludwig Institute for Cancer Research, Department of Cellular and Molecular Medicine, University of California, San Diego, La Jolla, California 92093

George H. Patterson (45) Cell Biology and Metabolism Branch, National Institute of Child Health and Human Development, National Institutes of Health, Bethesda, Maryland 20892

Hayley Pemble (179) Ludwig Institute for Cancer Research, Department of Cellular and Molecular Medicine, University of California, San Diego, La Jolla, California 92093

Olivier Pertz (63) Institute of Biochemistry and Genetics, Department of Clinical-Biological Sciences (DKBW), University of Basel, Center for Biomedicine, CH-4058 Basel, Switzerland

David W. Piston (415) Department of Molecular Physiology and Biophysics, Vanderbilt University Medical Center, Nashville, Tennessee 37232

Nathan Portier (179) Ludwig Institute for Cancer Research, Department of Cellular and Molecular Medicine, University of California, San Diego, La Jolla, California 92093

Andrei Pozniakovsky (179) Max-Planck Institute of Molecular and Cellular Biology and Genetics, Dresden 01307, Germany

Emmanuelle Querido (273) Département de Biochimie, Université de Montréal, 2900 Edouard-Montpetit, Montréal, Québec H3C 3J7, Canada

Mark A. Rizzo (415) Department of Physiology, University of Maryland School of Medicine, Baltimore, Maryland 21201

K. Rohr (539) Department of Bioinformatics and Functional Genomics, German Cancer Research Center (DKFZ), University of Heidelberg, IPMB, Im Neuenheimer Feld 267, D-69120 Heidelberg, Germany

E. D. Salmon (127) Department of Biology, University of North Carolina, Chapel Hill, North Carolina 27599

Marketa Samalova (353) Department of Plant Sciences, University of Oxford, Oxford OX1 3RB, United Kingdom

Satoshi Shimozono (381) Laboratory for Cell Function Dynamics, Brain Science Institute, The Institute of Physical and Chemical Research (RIKEN), 2-1 Hirosawa, Wako-city, Saitama 351-0198, Japan

J.-B. Sibarita (83) Institut Curie, Section de Recherche/CNRS 144, Compartimentation et Dynamique Cellulaires, 26 rue d'Ulm, 75248 Paris Cedex 05, France

Kevin F. Sullivan (1) Department of Biochemistry, National University of Ireland, Galway, Ireland

Jason R. Swedlow (555) Division of Gene Regulation and Expression, College of Life Sciences, Wellcome Trust Biocentre, University of Dundee, Dundee, Scotland DD1 5EH, United Kingdom

Marc Tramier (395) Institut Jacques Monod, UMR 7592 CNRS, University Paris 6/ University Paris 7, 2 Place Jussieu, 75251 Paris Cedex 05, France

Ronald D. Vale (113) Department of Cellular and Molecular Pharmacology, University of California, The Howard Hughes Medical Institute, San Francisco, California 94107

J.-L. Vonesch (83) Imaging Center of the IGBMC, 1, rue Laurent Fries, 67404 Illkirch Cedex, France

Ge Yang (497) Laboratory for Computational Cell Biology, Department of Cell Biology, CB167, The Scripps Research Institute, La Jolla, California 92037

Manuel Zeitelhofer (293) Center for Brain Research, Division of Neuronal Cell Biology, Medical University of Vienna, Spitalgasse 4, A-1090 Vienna, Austria

PREFACE

Microscopy has played a central, in fact defining, role in understanding life at the level of the cell, the level at which the molecular components of living systems give rise to the actual mechanisms of life. The development of an unusual autofluorescent protein from the jellyfish, *A. victoria*, has had a tremendous impact on the use of microscopy and related techniques on dissecting cellular mechanisms at the level of individual molecules. Green fluorescent protein (GFP) was originally shown in 1994 to exhibit autofluorescent properties when expressed in exogenous systems and as fusions to other proteins. Since those discoveries, led by Prasher and Chalfie and their colleagues, GFP has revolutionized our concept of mechanistic investigation of cellular processes and become as routine a tool in the cell biology laboratory as PCR is in a molecular biology laboratory. The insights gained by incorporating a genetically encoded fluorochrome into experimental thinking have been unprecedented, and they continue to advance. Many of the concepts, discoveries, developments, and techniques that will fuel this continued advance are presented in this edition of *Fluorescent Proteins*, derived from a previous volume entitled *Green Fluorescent Proteins*. Indeed, one of the key advances in autofluorescent protein technologies has been a tremendous proliferation of spectral derivatives, engineered from *A. victoria* GFP and isolated *de novo* from other species, notably Anthozoans. The rich palette of autofluorescent proteins now spans the spectral range from blue to deep red and this is discussed in multiple chapters in this volume.

Methodologies play out at multiple levels and the authors of this volume have very impressively developed all of those levels. From presentation of the ideas and concepts that provide the foundation for methods through discussing the factual knowledge and sources required to design experiments to the detailed exposition of actual experimental protocols, the chapters in this book combine to provide an essential tool for thinking about using genetically encoded fluorescent molecules. The experimental goals and systems presented range from biophysical interrogation of individual molecules to analysis of the behavior of cell populations in whole animals. A book like this works at two levels. On the one hand, it serves as a source of factual information and procedural guidelines to support specific experimental goals, to facilitate asking specific mechanistic questions in biology. On the other hand, by illustrating the range of techniques, the depth of resources available for molecular fluorescence, it serves as a resource for helping actually pose those questions. For the critical step from hypothesis to experiment requires knowing what can be done, and knowing what can be done facilitates this translation from idea to approach.

Work in biological imaging is founded on the biology of the systems in which the experiments take place, and several model systems are represented in this volume, including fungi, plants, invertebrates, and vertebrates, from cellular to whole organism strategies. The range of autofluorescent proteins available is presented and discussed in several chapters and, rather than being redundant, illustrates how different experimentalists organize and present this crucial and rapidly developing resource. Construction of FP fusions is discussed in several contexts, from developing biosensors and optimizing FRET to constructing intramolecular fusions and hemi-FP chimeras used for detecting protien–protein interactions. Specialized photodynamic techniques, such as photoactivation, photobleaching analysis, and fluorescence polarization techniques are presented that, with the allied approach of fluorescence correlations spectroscopy, allow analysis of protein diffusion and binding in cells. Microscopy of single molecules and approaches to quantitating molecule numbers extend these techniques right to the level of individual protein molecules and complexes, observable in their functional state. A wide variety of biological systems for analysis of specific cellular processes are discussed, ranging from chromosome segregation to mRNA transport, from membrane protein trafficking to cell population dynamics. These provide both the specific resources necessary to carry out experiments in those systems and also models to consider when tackling a novel problem. Microscopy itself is a critical issue, as well as management and analysis of the data produced by live cell imaging and these are addressed in a number of chapters. A critical element in moving biology forward at a quantitative level is computational modeling and analysis of systems and this is well represented in this volume.

I wish to thank the authors of this volume for their outstanding efforts to consider and to write the chapters presented here. This is their work and they have done a superb job of providing a dynamic view of a rapidly moving area. It is appropriate to recall that a very small number of pioneers in photobiology worked to make the essential breakthroughs in fluorescent protein technologies, reminding us of the value of research in far flung areas of biology. This book is dedicated to their efforts to literally bring light forth into scientific understanding, and to your efforts, reader, in shining that lamp into your own domain of ideas and intent, to discover the molecular mechanisms of living systems.

Kevin F. Sullivan
Galway, October 2007

CHAPTER 1

Autofluorescent Proteins

Ian M. Dobbie, Noel F. Lowndes, and Kevin F. Sullivan

Department of Biochemistry
National University of Ireland
Galway, Ireland

Abstract

Autofluorescent proteins (AFPs) have revolutionized molecular cell biology, and applications continue to harness the power of these genetically encoded fluorescent tags. Here, we review the discovery and physical properties of AFPs as well as their development through mutational optimization for several functional parameters. A practical guide to selection and use of major AFPs is provided as well as an overview of techniques for experimental applications.

0091-679X/08 $35.00
DOI: 10.1016/S0091-679X(08)85001-7

I. History

Autofluorescent proteins (AFPs) are now a major tool for biochemistry and cell biology with a huge number available and a dizzying range of properties. It is difficult to imagine that the first of these proteins was cloned only in 1992 (Prasher *et al.*, 1992). Since then PubMed entries related to fluorescent proteins have gone from 0 to ~2000 per year (Fig. 1). The growth is now leveling off, but probably just because the use of green fluorescent protein (GFP) and its variants in biology are so ubiquitous that it is no longer mentioned in titles or abstracts.

Green fluorescence was discovered in *Aequorea* as early as 1955 (Davenport and Nichol, 1955), and the fluorescent properties of GFP from the jellyfish *Aequorea victoria* were probed in more detail by Shimomura *et al.* (1962) as a companion protein to aequorin, the chemiluminescent protein from *A. victoria*. The function of GFP in *Aequorea* is to convert the blue light emitted by aequorin into the green bioluminescence observed in the jellyfish, giving it the property of absorption in the UV/blue region and fluorescent emission in the green. Exactly why *A. victoria*, and other similar light-emitting creatures, should emit in the green rather than blue region is not clear. This explanation is further brought into doubt by the existence of nonfluorescent homologues in related *Aequorea coerulescens* (Gurskaya *et al.*, 2003) and the fact that the wild-type protein absorbs more readily at 395 nm emission of aequorin than the wild-type protein at 475 nm emission of aequorin does.

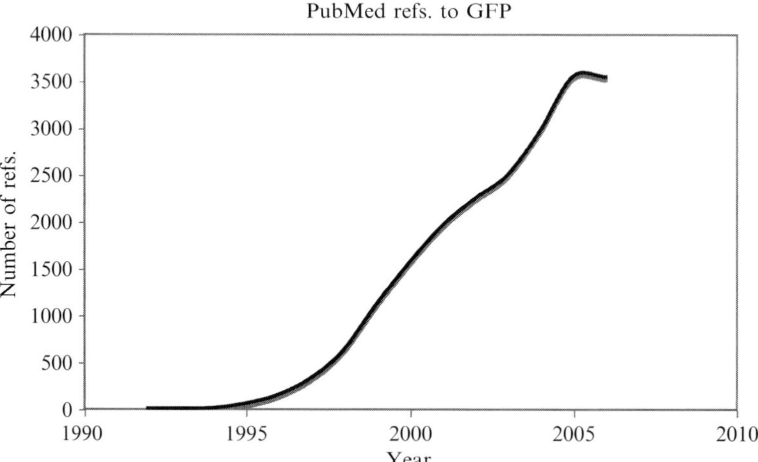

Fig. 1 The number of references quoting "GFP," "green fluorescent protein" or "fluorescent protein," in their PubMed entry per year. Entries included titles, abstracts, and keywords. The years 1992–1995 were hand checked to remove to 1–4 references where GFP did not refer to GFP or where fluorescent protein referred to fluorescence not from GFP-derived sources. After this date, it was assumed that these references were trivial compared with those correctly selected by the search.

Some progress was made between 1955 and 1992 such as the fluorescent properties were measured (Shimomura *et al.*, 1962), the structure of the fluorescent region was solved (Shimomura, 1979), and the protein was crystallized (Morise *et al.*, 1974). However, the field only exploded once the gene was cloned in 1992 (Prasher *et al.*, 1992). Apparently only two requests were made for the cloned gene after this paper was published; fortunately, one of these was by Roger Tsien, who has since become a leader in the fluorescent protein field. Following the cloning of *A. victoria* GFP, it was shown that the protein would rapidly produce fluorescence when expressed in other organisms (Chalfie *et al.*, 1994), demonstrating that there are no additional cofactors needed for the posttranslational modification or folding to produce fluorescent emission. The final important feature of GFP is that it preserves its fluorescence when fused with other proteins (Wang and Hazelrigg, 1994). These discoveries were a major breakthrough, as they enabled fluorescent tagging of specific proteins in live cells by simple techniques of molecular biology.

The fact that GFP was functional in other organisms made a number of groups start working on creating mutants with different excitation and emission profiles and optimizing their other properties. This has resulted in a large number of GFP variants with a range of emission wavelengths from 442 to 529 nm (Zacharias and Tsien, 2006). Other biophosphorescent organisms, such as *Renilla* (Morin and Hastings, 1971; Ward and Cormier, 1979), contained similar fluorophores with the same restrictions. Unfortunately mutagenesis, both random and directed, has been unable to move the peak beyond that magic 529 nm. With this limitation, effort was devoted to find similar fluorescent proteins from other organisms. Matz *et al.* (1999) were able to clone a number of homologues from fluorescent, but not bioluminescent, organisms of the species *Anthozoa*. Their strategy involved degenerate primers to regions they thought were structurally important. This was extremely successful and managed to produce proteins with only 25–30% homology but virtually identical structures (see alignment in Fig. 2; Matz *et al.*, 1999). A similar strategy has also been used to clone nonfluorescent chromoproteins from *Anthozoa* species that were then mutated to produce new fluorescent proteins (Gurskaya *et al.*, 2001). Since this discovery, several other organisms have provided fluorescent proteins with a range of emission spectra, though none of these has yet had the same impact as GFP or DsRed and their variants.

By far the most widely studied and used of these new fluorescent proteins was originally labeled drFP583 (Matz *et al.*, 1999) but has since been released as a commercial product known as DsRed. This had excitation and emission peaks of 558 and 583 nm, respectively (Matz *et al.*, 1999). The native protein, DsRed, exists as an obligate tetramer (Baird *et al.*, 2000), which can often lead to spurious interactions and the formation of insoluble aggregates. Although early efforts were unable to reduce the tendency for the protein to multimerize, improvements were made to the other properties of wild-type DsRed (Baird *et al.*, 2000). Recent work, using extensive mutagenesis, has led to a monomer with broadly similar fluorescent properties (Campbell *et al.*, 2002), which has then been further improved (Shaner *et al.*, 2004).

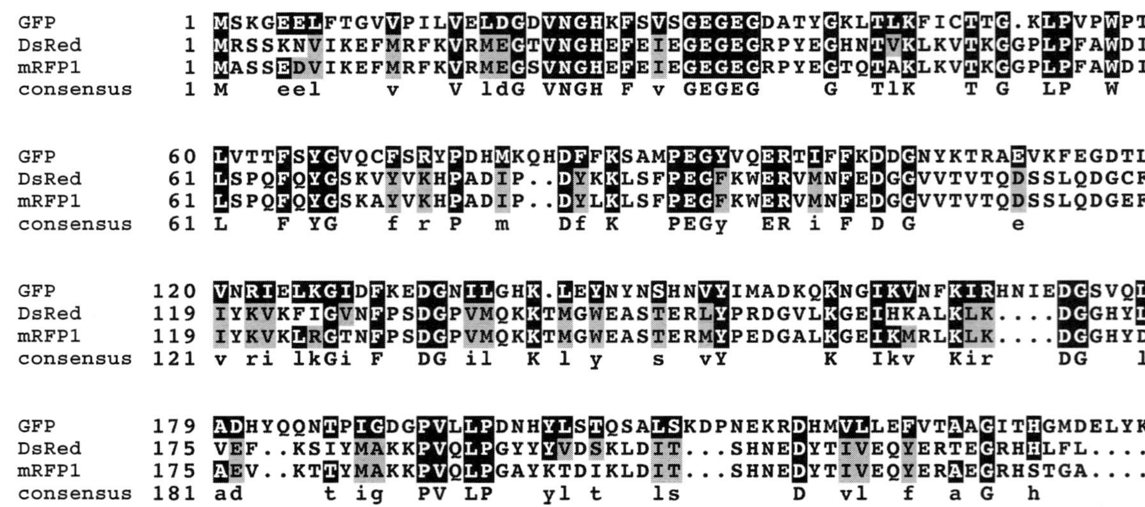

Fig. 2 Sequence alignment of *Aequorea victoria* green fluorescent protein (GFP) with *Discosoma* sp. DsRed and its extensively mutated derivative mRFP1.

Currently, a whole palette of AFPs with a wide range of properties are available. These span the visible spectrum from blue variants with emission peaks ~440 nm (Yang *et al.*, 1998; Zacharias and Tsien, 2006) to red and far red with emissions ranging to >640 nm (Gurskaya *et al.*, 2001; Shkrob *et al.*, 2005). There is also continual development with proteins cloned from other organisms and mutagenesis, leading to new AFPs. Additionally, there are now a large number of biosensors based on fluorescent proteins. These enable monitoring of cellular concentrations, such as Ca^{2+} or Cl^{-}, or cellular processes, such as caspase or kinase activation (Evanko and Haydon, 2005; Guerrero and Isacoff, 2001; Harpur *et al.*, 2001; Heim and Tsien, 1996; Miyawaki *et al.*, 1997; Takemoto *et al.*, 2003; Ting *et al.*, 2001; Zhang *et al.*, 2002).

II. Variants

The properties of wild-type GFP from *A. victoria* [referred to as wild-type green fluorescent protein (wtGFP) below], though tremendously useful, are far from ideal. It has two distinct excitation peaks, with a photoconversion pathway. Its brightness is rather low, though its quantum yield is relatively high ~0.8 (Heim *et al.*, 1995). Since it was first cloned, a large amount of work has gone into improving various features by creating variants, produced by single and multiple changes in the primary sequence. In addition to improving the fluorescent properties of these proteins, mutations have been made in order to shift the excitation and emission peaks, producing fluorescent emissions that are separable by color filters (Yang *et al.*, 1996, 1998). Variants of wtGFP have produced a limited range of wavelengths, from the blue to yellow-green, from 440 to 529 nm (Tsien, 1998; Zacharias and Tsien, 2006). Because of this limited spectral range available from variants of wtGFP, a lot of work has been devoted to discover similar proteins from other organisms with different emission ranges (Labas *et al.*, 2002; Matz *et al.*, 1999; Verkhusha and Lukyanov, 2004). These new proteins have then been subjected to similar efforts in order to optimize their properties and to produce a wider range of emission wavelengths. An exhaustive list of these proteins is beyond this introduction, but a selection of the variants produced to date are discussed below. A very comprehensive list can be found on George McNamara's web site (McNamara, 2006). One mutation to note is Q80R that existed in the originally distributed plasmids probably due to a PCR error. This mutation does not seem to affect the properties of GFP (Tsien, 1998).

A. Structure

The structure of all the AFPs consists of an 11-stranded β barrel, with both ends capped by hairpins between β strands (Fig. 3; Ormo *et al.*, 1996). This structure is conserved across all fluorescent proteins, though at the sequence level there is a relatively low level of identity (ca. 26–30% between *A. Victoria* and anthozoan

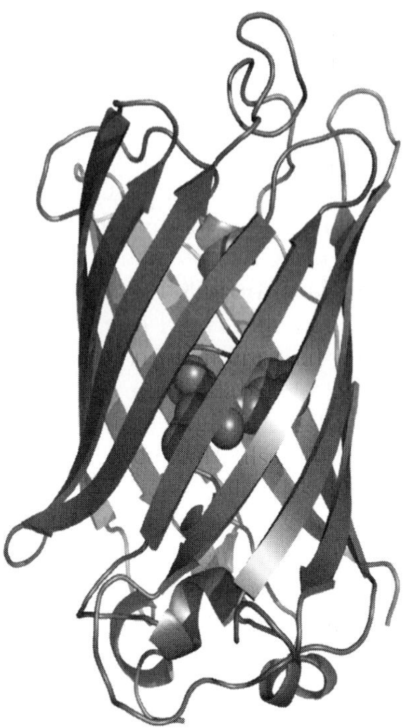

Fig. 3 Crystal structure of green fluorescent protein (GFP) (Q80R) with the β barrel shown in structural cartoons and the core chromophore in a space-filling representation (Ormo *et al.*, 1996).

AFPs; Matz *et al.*, 1999). The fluorophore is buried as a distorted α helix in the interior of the "β can" structure (Ormo *et al.*, 1996), largely protected from solvent, leading to relatively low-environmental sensitivity and its ability to form AFPs when fused with other proteins or peptides (Tsien, 1998). The minimal fluorescent region appears to be 2–232 (AFP sequences are usually numbered based on wtGFP; Remington, 2000), but many variants have insertion either before or after this minimal fluorescent region. Both the N- and C-termini are exposed on the surface of the structure (Ormo *et al.*, 1996) and are therefore available for fusing with other proteins, and these regions are often engineered to provide DNA polylinkers for cloning or to optimize linkage with fusion partners. The termini are also physically close to each other and can be fused with a linker, producing cyclic permutations with the sequence start and end at other residues (Baird *et al.*, 1999; Zacharias and Tsien, 2006), or allow insertion into the middle of other proteins without compromising folding and fluorophore formation (Hughes *et al.*, 2001, Chapter 2 by Mealer *et al.*, this volume).

Protein sequence alignments show that there are few essential residues in the AFPs. The fluorophore is formed from residues 65–67, with the only strictly required residue being Gly67 (Matz *et al.*, 1999). The residues Arg96 and Glu222 are also conserved and are involved in stabilizing the hydrogen-bonding network for the autocatalytic fluorophore formation. Other residues facing into the center of the β barrel influence the exact fluorescence properties such as spectra, maturation time, and photostability.

The fluorophore is formed by an intramolecular cyclization of the core amino acids, Ser65, Tyr66, Gly67 producing a *p*-hydroxybenzylideneimidazolinone in the center of the interior α helix (Fig. 4; Tsien, 1998). This structure is required for both absorption and fluorescence. The rate at which this process occurs is determined not only by the central fluorophore sequence, residues 65–67, but also by the surrounding sequence and the external environment. In DsRed, and other red-shifted AFPs, the peptide bond before the X65/Y66/G67 fluorophore is oxidized, leading to delocalized electron density over the polypeptide bond and the longer wavelength excitation and emission (Shu *et al.*, 2006). Variations on this structure produce the range emissions from blue to far red (Labas *et al.*, 2002; Remington *et al.*, 2005; Shu *et al.*, 2006).

B. Stability, Folding, and Multimerization

Environmental stability of the AFP β can structure is surprisingly high, contributing to its utility and versatility in experimental applications. wtGFP will produce normal fluorescence at 70°C, in a crystal or when frozen (Chattoraj *et al.*, 1996). However, despite the overall robustness of fluorescence in respect to external conditions, fluorescence may be quantitatively affected by factors such as pH depending on which variant is used. AFPs are quite stable in the folded state,

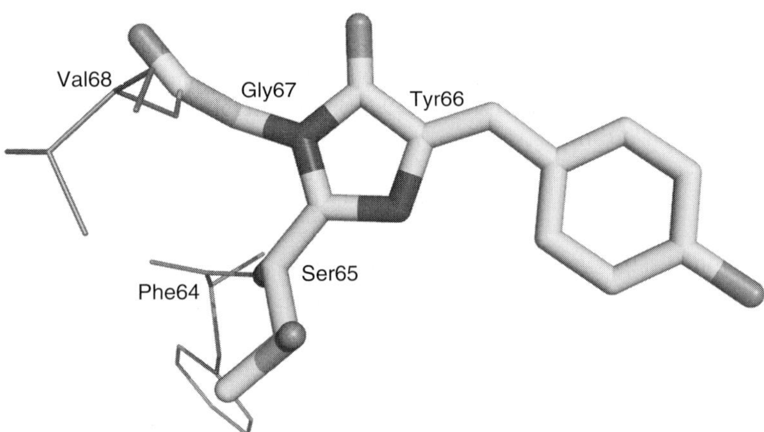

Fig. 4 The central chromophore, a *p*-hydroxybenzylideneimidazolinone, formed by the cyclization of the core amino acids, Ser65, Tyr66, and Gly67 (see Section II.A for details; Tsien, 1998). The α carbons are labeled, with the preceding and following amino acids shown in a line representation to demonstrate the path of the protein backbone.

but the rate of folding and fluorophore formation and the propensity for multimerization are factors that have been aggressively engineered to provide improved experimental tools.

The rate of AFP synthesis, folding, and fluorochrome formation potentially limit expression in exogenous systems. Mutagenesis has been applied to modify the codon bias for improved translation, and *A. victoria* GFP derivatives have been optimized for expression in a variety of organisms, such as yeasts, *C. elegans*, mammals, and plants (Crameri *et al.*, 1996; Dixit *et al.*, 2006). For use in plants, a variant must be used as the wtGFP sequence leads to unwanted splicing, hence, a noncomplete protein and no fluorescence. The rate of protein folding is limited in wtGFP, particularly at 37°C necessary for mammalian cells. The standard mutations, F64L/M153T/V163A/S175G (Nagai *et al.*, 2002), greatly increase this rate. Such enhanced GFPs improve fluorescent signal intensities when expressed in cells by improving the proportion of properly folded protein, though they do not significantly affect the fluorescent properties or stability of the fully mature protein.

The extra catalytic step involved in the longer wavelength redshifted variants can lead to even slower maturation, with $\tau_{1/2}$ times of >24 h (Zhang *et al.*, 2002). Several residues structurally close to the fluorophore can significantly influence this process. In proteins such as DsRed, this is a key issue as the immature protein emits in the green and full maturation takes up to 30 h, limiting its usefulness (Baird *et al.*, 2000). One major use of red fluorescent proteins is to label multiple proteins, in conjunction with shorter wavelength variants. The slow maturation can limit the usefulness of DsRed in combination with such shorter wavelength AFPs.

wtGFP has a slight tendency to dimerise, which can be a significant problem at high concentrations. Following determination of the structure of wtGFP, mutagenesis demonstrated that the mutations F223R, L221K, and A206L all reduce this tendency, with A206L being the most effective (Zhang *et al.*, 2002). A major limiting factor for DsRed is its tendency to tetramise. Campbell and co-workers (2002) performed a random and directed mutation series to first remove its multimerization and then to re-establish its fluorescence. The final product of this massive endeavor was mRFP1, with 33 separate mutations from the original DsRed sequence (Campbell *et al.*, 2002). The mRFP1 protein has since become a new starting point for a wide range of mutations to increase its brightness, photostability, and shift its fluorescent emission to both longer and shorter wavelengths (Shaner *et al.*, 2004).

C. Spectra and Photophysical Dynamics

Modification of the spectral properties of AFPs has been a major effort, in order to provide a palette of proteins with a range of excitation and emission spectra, for multiple labeling and other applications *in vivo*. This has resulted in a large number of AFP derivatives, reviewed in detail in Shaner *et al.* (2005).

The spectra of wtGFP are far from simple, with two excitation but a single emission peak. Further, in low oxygen conditions it can be photoexcited to produce red fluorescent emissions (Sawin and Nurse, 1997; van Thor *et al.*, 2002). The two emission peaks present in wtGFP are due to two different protonation states of the fluorophore (see Heim and Tsien, 1996; Tsien, 1998, for a more detailed discussion). Mutations within the fluorophore that affect this protonation, the most common being the S65T mutation, effectively produce a single excitation peak at ~480 nm with emission at ~505 nm, leading to a 6 times brightness increase without any other mutations (Heim *et al.*, 1995). However, the dual excitation has been engineered to produce photoactivatable derivatives of wtGFP that are able to switch their fluorescent emission properties on illumination with a specific wavelength of light, either permanently (Ando *et al.*, 2002; Chudakov *et al.*, 2003; Patterson and Lippincott-Schwartz, 2002) or switchably (Ando *et al.*, 2002, 2004; see Chapter 3 by Lippincott-Schwartz and Patterson, this volume). Other photoactivatable fluorescent proteins come from different organisms, see Lukyanoy *et al.* (2005) for a review. These molecules are not strictly photoactivatable, as they are fluorescent even before photoactivation, though possibly very weakly. Illumination with a specific wavelength shifts the fluorescent absorption and emission to longer wavelengths, effectively turning on fluorescence at these wavelengths.

Modification of excitation and emission wavelengths has been achieved by mutagenesis of wtGFP (Heim *et al.*, 1995; Shaner *et al.*, 2005). Substitutions at residue 65 lead to variants with a range of excitation and emission wavelengths. The substitution S65G produces yellow, Y66W produces cyan, and Y66H produces blue variants (Zacharias and Tsien, 2006). The final range of emission wavelengths available from wtGFP mutants is now from ~440 to ~530 nm. In order to obtain a wider range of emission spectra fluorescent proteins from other organisms must be used. After wtGFP from *A. victoria*, the second most widely exploited AFP is DsRed from *Anthozoa* (originally called drFP583) with an emission peak ~583 nm (Matz *et al.*, 1999). The main breakthrough in its discovery was the fact that it came from a fluorescent, but nonbioluminescent organism, greatly widening the possible sources of AFPs. The fact that it came from a coral reef, one of the most diverse and colorful ecosystems on the planet hints that many more fluorescent proteins are awaiting discovery (Field *et al.*, 2006). A practical guide to spectral derivatives of the AFP family is presented below (Table I) and is covered in detail in Shaner *et al.* (2005).

Relatively early in the development of AFPs, wtGFP was mutated to produce blue, cyan, and yellow emissions (Tsien, 1998). These proteins were shown to exhibit fluorescence resonance energy transfer (FRET) when the complimentary fluorescent domains were in close proximity, <10 nm, a scale ideal for looking at protein–protein interactions. FRET is a process that transfers excitation energy from a donor molecule to an acceptor molecule (Jares-Erijman and Jovin, 2003; Pollok and Heim, 1999). A number of chapters in this volume address the use and measurement of AFP FRET techniques *in vivo* (Chapter 4 by Hodgson and Hahn, Chapter 16 by

Table I
A Selection of Useful Autofluorescent Proteins and Some Basic Information

Name	Excitation max (nm)	Emission max (nm)	Relative brightness	Relative photostability	Laser line (nm)
ECFP	433	475	13	64	405/440 Diode
Cerulean	433	475	27	36	405/440 Diode
EGFP	487	507	34	174	488 Argon
Emerald	487	509	39	.69	488 Argon
EYFP	514	527	51	60	514 Argon
mCitrean	516	529	59	49	514 Argon
mOrange?	548	562	49	9.0	532 Diode/543 HeNe
m/dTomato?	554	581	95	98	543/568 HeNe
mCherry	587	610	16	96	568/594 HeNe

Source: Data taken from Shaner *et al.* (2005).

Shimozono and Miyawaki, Chapter 17 by Tramier and Coppey-Moisan, and Chapter18 by Piston and Rizzo, this volume).

After excitation, the fluorescent core can lose its excitation energy via several routes. One of these is photobleaching, where the energy is used to break a covalent bond, creating reactive oxygen species and permanently destroying fluorescence. This reactive oxygen species can be toxic to cells. Maximizing photostability of AFPs not only improves long-term imaging but also reduces phototoxicity. However, the variant KillerRed has been optimized to enhance phototoxicity and enables site-specific elimination of targeted cells (Bulina *et al.*, 2006). Experimental exploitation of photobleaching in fluorescence recover after photobleaching (FRAP) experiments is addressed in Chapter 14 by McNally, this volume.

The complexity of the fluorescent core, the interactions that create and stabilize it, leads to differing photophysical properties in different variants. This has been exploited in order to differentiate between spectrally similar AFPs based on their excited state lifetimes (Pepperkok *et al.*, 1999).

The development of photoactivatable AFPs has allowed a whole range of novel experiments to be performed. The ability to locally activate a selection of your tagged protein allows tracking of dynamics, particularly steady-state dynamics, untraceable using standard AFP tags. This is detailed in Chapter 3 by Lippincott-Schwartz and Patterson, this volume.

In summary, the spectral and photophysical properties of the AFPs have turned out to be remarkably plastic, and it has been possible to harness and modify these to develop a variety of experimental strategies by using photophysical techniques to interrogate the behavior of proteins in living cells and organisms. These techniques have revolutionized cell biology and promise to continue, through refinement and further modifications, to extend the power of molecular microscopy *in vivo*.

III. Practical Considerations

The range of fluorescent proteins, their photophysical properties and potential applications, is large and the choices as to which to employ for a given experimental aim can be a daunting task. Which fluorescent protein to use largely depends on the specific experimental requirements, but can also be influenced by the equipment available. It is often easier to clone another AFP than to purchase a new laser or filter set for a microscope. An excellent guide to the latest variants has been published by Shaner *et al.* (2005), and in this section we consider a number of practical aspects of AFP choice and application.

An important aspect of all AFPs is their brightness, the product of their extinction coefficient and quantum efficiency (see Table I). This is the product of the probability of them absorbing an excitation photon, and the probability of them then emitting the energy as fluorescence rather than through other processes such as photobleaching or triplet state formation. Both these properties are not simply related to any single mutation, but a complex function of the whole sequence such that the only reliable way to determine it is direct measurement (Shaner *et al.*, 2005). Comparison between variants in different spectral regions is also very difficult, as the apparent brightness may be limited by excitation/emission filters and detector sensitivity in the relevant spectral range rather than the true relative fluorescent emission.

The increase in sensitivity of light detectors, such as avalanche photo-diodes (APDs) and charge coupled detectors (CCD cameras), has revolutionized microscopy. Coupled with techniques such as confocal microscopy or total internal reflection fluorescence (TIRF) to suppress background and bright photostable fluorescent proteins, it is possible to detect single AFP molecules in live or fixed cells (Chapter 6 by Douglass and Vale, this volume; Betzig *et al.*, 2006; Leake *et al.*, 2006; Steinmeyer *et al.*, 2005).

The simplest experimental situation is using an AFP alone or as a fusion protein as a marker for cellular or subcellular behavior (Lippincott-Schwartz and Patterson, 2003). A green wtGFP variant is ideal based on the criteria of brightness and suitability in most fluorescence-based devices, as it effectively mimics the widely used dye fluorescein. Currently, the best green variants are EGFP or Emerald (Tsien, 1998). Commercially distributed by Invitrogen, Emerald is relatively photostable, has a high-quantum yield, simple excitation and emission spectra close to those of fluorescein. The green emission, around 505 nm, is close to the peak sensitivity of the human eye; many CCD cameras and photo multiplier tubes (PMTs) are used in analytical equipment. The excitation wavelength is also very close to the commonly used 488 nm Ar laser line used in most confocal laser scanning microscopes (CLSM) and fluorescence activated cell sorters (FACS). One drawback of Emerald is that it is a weak dimer. If monomeric behavior is essential, or very high-expression levels are expected, the recent discovery of the A206K mutation should be exploited to eliminate the possibility of dimerization.

While AFPs are ideal for visualization in living cells, experiments using fixed specimens are readily accomplished as AFP fluorescence is largely preserved on fixation by paraformaldehyde or methanol.

For any reason if Emerald or EGFP is not suitable for your situation, then the next best choice is one of the monomeric red emitting AFPs such as mCherry (Shaner *et al.*, 2004). While not as bright as Emerald or GFP (relative brightnesses are ~35 and 16 for Emerald or EGFP and mCherry, respectively; Shaner *et al.*, 2005), red AFP spectra are closely matched the commonly used rhodamine derivatives, providing the advantages discussed above for green AFPs. Major drawbacks reduced brightness, due to lower quantum yield, and reduced sensitivity of detectors, including the human eye, to these longer wavelength emissions.

Although green variants are the best choice for many simple experiments, the relatively high cellular autofluorescence in the green region can make visualizing very low-expression levels extremely difficult. By using red emitting AFPs such as mCherry, dTomato, or mPlum (Shaner *et al.*, 2004) where the cellular autofluorescence is greatly reduced, greater sensitivity to low-expression levels can be achieved. While dTomato has a high-quantum yield and good photostability, its dimeric nature may limit its usefulness.

Less useful alternatives for single wavelength experiments are the wavelength-shifted wtGFP variants, such as mCFP and Cerulean in the cyan range or EYFP and mCitrine in the yellow range (Shaner *et al.*, 2005). The cyan variants are less bright than Emerald, and the yellow variants have significantly lower photostability. These AFPs are readily visualized using arc illumination with specific filter sets, with the cyan variants peak excitation at ~440 nm, emitting ~480 nm, and the yellow ones exciting at 515 nm and emitting at up to 530 nm. Some more expensive laser-based instruments such as multiline Kr/Ar lasers have wavelengths suitable for these AFPs. A list of useful variants with some basic data is shown in Table I and some common laser lines and which variants to use with them. The use of multiple colors simultaneously and other more advanced techniques are discussed in the next section.

Fluorescent proteins have been found to be compatible with multiphoton excitation. In fact, EGFP fusion constructs and 2-photon excitation are often used for *in-vivo* imaging tasks due to the enhanced depth penetration of the 2-photon excitation. The 2-photon excitation spectra tend to be broader, often allowing multiple AFPs to be excited simultaneously in a 2-photon setup (Yuste, 2005).

AFPs can also have significantly different photostabilities, the rate at which they photobleach. This rate is difficult to quantify, though direct comparisons between different AFPs with measures that correct for fluorescent emission rate, such as those in Shaner *et al.* (2005), are helpful. Mutations that affect photostability are not simply transferable from one variant to another, so it must be reoptimized in each variant (Nagai *et al.*, 2002). Photobleaching is an important practical limitation of experiments with AFPs. All AFPs are much more susceptible to photobleaching than modern fluorescent dyes such as the Cy or Alexa dyes. Photobleaching is especially prominent while doing high-resolution 3D stacks or time-series experiments due to the extended light exposure. Using 2-photon excitation can ease this light exposure to an extent, as areas outside the excitation plane are not exposed. The best method of reducing photobleaching is to minimize light exposure with short exposures at low-light intensities. These types of experiment are increasingly possible with the advent of better detectors. The presence of photobleaching must be taken into account when designing and interpreting experiments.

IV. Advanced FP Applications

AFPs provide simple and robust methods for creating chimeric fusion proteins for visualization in cells. In addition, AFPs are also useful in a wide range of advanced fluorescent techniques. Key techniques are discussed here, and some are covered in more detail in subsequent chapters.

A. Multiple Labeling

Dual and multiple labeling experiments with AFPs require careful selection of fluorochromes, resolving individual tags by either excitation or emission spectra differentiation (Cowan *et al.*, 2000; Yang *et al.*, 1996, 1998). The simplest situation is using two AFPs with different excitation wavelengths and emission spectra can be separated with filters (Anderson *et al.*, 2006; Cowan *et al.*, 2000; Ellenberg *et al.*, 1998). This allows cross talk between the two signals to be minimized by exciting one fluorophore and recording its emission and then exciting the other and recording its emission. This is necessary, as the AFPs have long tails in their emission spectra meaning the shorter wavelength protein will be detected in the longer wavelength channel unless they are excited separately. An alternative is to use a spectral detector, such as fluorescent spectrometer or a spectrally resolving CLSM, in order to allow the separation of the signals.

The use of a spectral detector allows the selection of AFPs with less separation in emission wavelengths, so long as suitable single labeled controls are used to provide the calibration spectra (Zimmerman *et al.*, 2003). One major drawback of linear unmixing to separate closely spaced emission peaks is the requirement that both signals be present in roughly equal amounts. Intensity differences of more than a factor of 5 can lead to unreliable linear unmixing.

For two channel detection, the best combination of brightness, separation between excitation and emission spectra and commonly available filters, is achieved using Emerald and mCherry. An alternative choice is to use cyan and the yellow variants of wtGFP, such as ECFP or Cerulean and EYFP or mCitrine. These proteins have less spectral separation and so more likelihood of cross talk between signals (Anderson *et al.*, 2006; Cowan *et al.*, 2000; Yang *et al.*, 1996, 1998). It is also possible to separate fluorescent protein emissions via differences in fluorescence lifetime; however, this is much more technically demanding and requires more expensive and complicated equipment (Pepperkok *et al.*, 1999). The recent development of Keima with the extremely large stokes shift, with excitation peaks at 440 nm and emission at 620 nm, allows a third option. This AFP can be used in combination with ECFP, using a single source to excite both proteins and then dichroic filters to separate their respective emissions. The stokes shift of 180 nm means that even the long emission tail of ECFP is almost entirely avoided (Kogure *et al.*, 2006).

For three color imaging, the best options are found by using a combination such as ECFP, EYFP, and mCherry, or equivalent cyan, yellow, and red emitting AFPs. More than three color imaging is certainly feasible with careful selection of the fluorescent proteins and microscope equipment, but the technical approaches are beyond the scope of this chapter. Multiple labeling is not limited to using AFPs alone; they can easily be combined with the whole range of existing fluorescent techniques (Giepmans *et al.*, 2006; Miyawaki *et al.*, 2003). Multiple labels are often used to assess colocalization, and it should be noted that colocalization is not a simple matter to determine, particularly in multicolor microscopy. Care should be taken while performing such experiments to ensure that the observed distributions actually reflect real colocalizations (Bolte and Cordelieres, 2006).

B. Dynamic Imaging

One of the major advantages of AFPs is the ability to observe specific proteins in live cells, following the localization of the protein of interest through time and using photodynamic techniques (Lippincott-Schwartz *et al.*, 2001). Time-lapse fluorescence microscopy allows cellular behavior to be followed over a wide range of time scales from milliseconds to hours or days (Ellenberg *et al.*, 1998; Higashida *et al.*, 2004; Niell and Smith, 2004). Using multiple AFP fusions allows the response of two or more proteins to be followed simultaneously, highlighting similarities and differences in the behavior of proteins and cellular compartments. Questions such as "How quickly does the protein diffuse about the cell?," "What is the turnover rate between cellular compartments?," can be of vital importance in understanding cellular functions. The complimentary techniques of FRAP and fluorescence loss in photobleaching (FLIP) allow study of these types of questions (Dunn *et al.*, 2002; Lippincott-Schwartz *et al.*, 2000, 2003; Reits and Neefjes, 2001; Zicha *et al.*, 2003, Chapter 14 by McNally, this volume). AFP fusion proteins are particularly effective for these types of study, as they enable fluorescent labeling of specific cellular proteins in minimally perturbed live cells.

FRAP involves using a bright light source, usually a laser, to photobleach a region and then time-lapse observation of recovery of fluorescence within this region or structure, and is detailed in Chapter 14 by McNally, this volume. FLIP is similar but rather than observing the return of fluorescence to the bleach region it measures the loss of fluorescence from other regions, allowing direct measurement of the rate of turnover between the observed and the bleach regions. These two techniques are usually performed using a CLSM because they contain a very bright, easily controlled light source and can be configured to interleave bleaching with imaging (Lippincott-Schwartz *et al.*, 2001, 2003). Using Emerald, or similar variant, and the 488 Ar line on a CLSM is the most common combination for photobleaching techniques. The relatively bright signal from green wtGFP variants means that it is easy to image, and the high-bleaching power available from the 488 nm line of most Ar lasers makes this a good choice for these techniques. Nevertheless, all the AFPs are relatively easy to photobleach and can be used for photobleaching experiments given appropriate bleaching illumination.

Another technique for measuring protein dynamics is fluorescence correlation spectroscopy (FCS; Kohl and Schwille, 2005), and is detailed in Chapter 20 by Langowski, this volume. By observing signal fluctuation in a confocal volume, autocorrelation analysis can yield a measure of concentration and diffusion rate for the fluorescent species, and hence molecular mass (Braet *et al.*, 2007). By cross-correlation of multiple fluorophores, molecular interactions may be observed *in vivo*. Three main drawbacks of FCS circumscribe its usefulness for protein dynamics. First, concentrations must be $<1\ \mu M$. Second, it is only effective on diffusing species, so it will not work in fixed samples or stable structures. Third, the diffusion rate is not a sensitive measure of molecular mass, as it is proportional to the cube root of the mass, and is also affected by changes in shape and viscosity. With appropriate control experiments FCS can certainly be used to get order of magnitude mass estimates, and is one of the few ways to get absolute concentrations from within live cells (Braet *et al.*, 2007).

C. Protein–Protein Interactions

Cellular processes fundamentally rely on protein–protein interactions. A number of techniques able to measure protein–protein interactions by using AFPs have been developed and are detailed in this volume (Chapter 16 by Shimozono and Miyawaki, Chapter 17 by Tramier and Coppey-Moisan, Chapter18 by Piston and Rizzo, Chapter 19 by Kerppola, and Chapter 20 by Langowski, this volume). Importantly, these techniques enable such interactions to be measured in live cells. One elegant approach relies on intermolecular complementation of split AFP molecules (Chapter 19 by Kerppola, this volume). The AFP sequence is split at one of its turn regions between the β-sheets making up the barrel section. The two segments are fused with different proteins, and when the proteins interact a functioning fluorescent protein is produced (Hu *et al.*, 2002; Walter *et al.*, 2004). The use of fragments from multiple AFPs, which produce measurably different emission when combined allow, several different interactions to be simultaneously monitored (Hu and Kerppola, 2003).

A more advanced, but more technically challenging, method of measuring protein–protein interactions relies on FRET (Chapter 16 by Shimozono and Miyawaki, Chapter 17 by Tramier and Coppey-Moisan, and Chapter18 by Piston and Rizzo, this volume; Jares-Erijman and Jovin, 2003; Kenworthy, 2001). In fact, the biological role of wtGFP is to act as a FRET acceptor from a blue bioluminescent protein, aquorin, to produce green emission (Morise *et al.*, 1974). This property can easily be exploited to measure protein–protein interactions. In their simplest form, FRET-based experiments tag two different proteins with a donor and an acceptor AFP, respectively (Zhang *et al.*, 2002). This process has two effects on the fluorescent signal, first the fluorescent emission intensity from the donor species is reduced and that from the acceptor is increased (Gordon *et al.*, 1998), and second the fluorescence lifetime of the donor is reduced (Pepperkok *et al.*, 1999). The FRET signal, the fraction of energy that is transferred from donor to acceptor, is then used as a measure of the interacting fraction between these two proteins. This can be further extended to a cascade of three interacting proteins (Galperin *et al.*, 2004). However, this approach is not without its difficulties. The FRET efficiency is highly dependent on the separation distance of the donor and acceptor fluorophores and their geometric orientation. Thus, FRET sensors must be developed and optimized empirically, guided by protein structure where appropriate.

In practice, most FRET experiments take a few simple precautions to minimize systematic errors. First, they measure difference in FRET between states, such as before and after stimulation, rather than the absolute FRET efficiency. Second, by using flexible linkers, between fusion proteins and fluorescent tags, orientation artifacts can be minimized, though maximal signal is also reduced. Third, careful choice of AFP to use as the donor and the acceptor can maximize signal and ease detection. In terms of achieving maximal FRET efficiency then ECFP and EYFP, or equivalents such as Cerulean and mCitrean, are the best choice for donor and acceptor. An ECFP–EYFP construct has been made that achieves a FRET efficiency of over 90% (Shimozono *et al.*, 2006 and Chapter 16 by Shimozono and Miyawaki, this volume). For protein–protein interaction FRET experiments an AFP pair with significantly enhanced FRET efficiencies has been developed, CyPet and Ypet (Nguyen and Daugherty, 2005); however, CyPet folds poorly at 37 °C (Shaner *et al.*, 2005) and so is not suitable for experiments in mammalian cells. An alternative choice is Emerald and mCherry, or similar green and red AFPs, which provide lower maximum efficiency but are more easily separated by wavelength.

For detection of FRET signals there are several techniques that can be used, each with their own advantages and disadvantages. The simplest technique is to use the relative intensity of the two emission channels, when excited with either the donor excitation or the acceptor excitation (Gordon *et al.*, 1998). A technique less susceptible to bleed through or intensity-based artifacts is to observe FRET interactions using the fluorescence lifetime of the donor species. This is usually done by fluorescence lifetime imaging (FLIM), referred to as FLIM–FRET.

The ability of fluorescent fusions of AFPs to FRET on close interaction has been exploited in order to create a number of AFP-based FRET sensors. These sensors rely on binding, rearrangement, or cleavage of the proteins the AFPs are fused with to measure some specific variable. A range of sensors have been developed, such as Ca^{2+} (Evanko and Haydon, 2005; Miyawaki *et al.*, 1997), caspase activity

(Takemoto *et al.*, 2003), kinase activity (Ting *et al.*, 2001; Zhang *et al.*, 2001), and others (Guerrero and Isacoff, 2001; Zhang *et al.*, 2002).

A final method for measuring protein–protein interactions relies on an extension of FCS to two colors, fluorescence cross-correlation spectroscopy (FCCS). In this technique, the cross-correlation between the intensity fluctuations of two fluorescent species within a confocal volume is measured (Baudendistel *et al.*, 2005; Kogure *et al.*, 2006). From this, the fraction of each species that are traveling together can be deduced, and hence the interacting fraction. This technique has a significant advantage over FRET measurements in that it does not rely on the separation distance between the two fluorophores, hence can be used to monitor two components of a large complex. It does however suffer from similar drawbacks to FCS in that it only works at concentrations <1 μM and is only effective on diffusing species.

The study dynamics of cytoskeletal filaments has been transformed by the use of fluorescence speckle microscopy (FSM) developed by Waterman-Storer and colleagues (Adams *et al.*, 2004; Waterman-Storer *et al.*, 1998). FSM works by introducing a small fraction of tagged molecules ($\sim$0.5%). Some of these molecules are then bound into stable positions when incorporated into cytoskeletal filaments. The freely diffusing background can be removed by comparing images from a time series. The movement of the speckles is then used to track movement of the filaments (Danuser and Waterman-Storer, 2006). This technique would also work with other proteins with similar dynamics to cytoskeletal components.

V. Future Directions

With such a diverse palate of fluorescent proteins already available, it is difficult to predict where the field will go in the future. It is likely that spectral mutations of wtGFP have been fairly exhaustively covered by a number of directed and random mutation strategies; however, this is not true of the AFPs derived from other sources such as *Anthozoa*. The relatively large number of mutational studies of wtGFP and derivatives has probably also maximized, or nearly so, the benefits that are gained in terms of stability, folding speed, maturation rates, suppressing dimerization, and increasing brightness. Most progress in these features will probably occur in AFPs derived from other species. There are a number of multimeric proteins derived from a range of species that could possibly be made monomeric by a similar strategy to that used on DsRed (Campbell *et al.*, 2002). These proteins could then benefit from the kind of extensive development that wtGFP has undergone over the last 15 years. It also seems likely that the future imaging developments, such as using photoactivatable fluorescent proteins to achieve nanometer resolution in fluorescence microscopy (Betzig *et al.*, 2006) and controlled light exposure microscopy (Hoebe *et al.*, 2007) to reduce phototoxicity, will enhance the usefulness of existing AFPs.

The main advantage of fluorescent proteins over other types of fluorescent tag is that they are usable in minimally perturbed live cells. This allows monitoring of dynamic processes occurring within these cells. It seems likely that a substantial amount of future progress in the field will be related to utilizing new and existing AFPs as biosensors to measure these processes, discussed in Chapter 4 by Hodgson and Hahn, this volume. There already exist sensors to measure a wide range of factors, including protein–protein interactions, ion concentrations, and posttranslational modifications such as phosphorylation or peptide cleavage. Both the range and types of process that can be monitored will increase, and new techniques will be developed to measure other dynamic processes occurring within cells.

References

Adams, M. C., Matov, A., Yarar, D., Gupton, S. L., Danuser, G., and Waterman-Storer, C. M. (2004). Signal analysis of total internal reflection fluorescent speckle microscopy (TIR-FSM) and wide-field epi-fluorescence FSM of the actin cytoskeleton and focal adhesions in living cells. *J. Microsc.* **216,** 138–152.

Anderson, K. I., Sanderson, J., Gerwig, S., and Peychl, J. (2006). A new configuration of the Zeiss LSM 510 for simultaneous optical separation of green and red fluorescent protein pairs. *Cytometry A* **69,** 920–929.

Ando, R., Hama, H., Yamamoto-Hino, M., Mizuno, H., and Miyawaki, A. (2002). An optical marker based on the UV-induced green-to-red photoconversion of a fluorescent protein. *Proc. Natl. Acad. Sci. USA* **99,** 12651–12656.

Ando, R., Mizuno, H., and Miyawaki, A. (2004). Regulated fast nucleocytoplasmic shuttling observed by reversible protein highlighting. *Science* **306,** 1370–1373.

Baird, G. S., Zacharias, D. A., and Tsien, R. Y. (1999). Circular permutation and receptor insertion within green fluorescent proteins. *Proc. Natl. Acad. Sci. USA* **96,** 11241–11246.

Baird, G. S., Zacharias, D. A., and Tsien, R. Y. (2000). Biochemistry, mutagenesis, and oligomerization of DsRed, a red fluorescent protein from coral. *Proc. Natl. Acad. Sci. USA* **97,** 11984–11989.

Baudendistel, N., Muller, G., Waldeck, W., Angel, P., and Langowski, J. (2005). Two-hybrid fluorescence cross-correlation spectroscopy detects protein–protein interactions *in vivo*. *Chemphyschem* **6,** 984–990.

Betzig, E., Patterson, G. H., Sougrat, R., Lindwasser, O. W., Olenych, S., Bonifacino, J. S., Davidson, M. W., Lippincott-Schwartz, J., and Hess, H. F. (2006). Imaging intracellular fluorescent proteins at nanometer resolution. *Science* **313,** 1642–1645.

Bolte, S., and Cordelieres, F. P. (2006). A guided tour into subcellular colocalization analysis in light microscopy. *J. Microsc.* **224,** 213–232.

Braet, C., Stephan, H., Dobbie, I. M., Togashi, D. M., Ryder, A. G., Foldes-Papp, Z., Lowndes, N., and Nasheuer, H. P. (2007). Mobility and distribution of replication protein A in living cells using fluorescence correlation spectroscopy. *Exp. Mol. Pathol.* **82**(2), 156–162.

Bulina, M. E., Chudakov, D. M., Britanova, O. V., Yanushevich, Y. G., Staroverov, D. B., Chepurnykh, T. V., Merzlyak, E. M., Shkrob, M. A., Lukyanov, S., and Lukyanov, K. A. (2006). A genetically encoded photosensitizer. *Nat. Biotechnol.* **24,** 95–99.

Campbell, R. E., Tour, O., Palmer, A. E., Steinbach, P. A., Baird, G. S., Zacharias, D. A., and Tsien, R. Y. (2002). A monomeric red fluorescent protein. *Proc. Natl. Acad. Sci. USA* **99,** 7877–7882.

Chalfie, M., Tu, Y., Euskirchen, G., Ward, W. W., and Prasher, D. C. (1994). Green fluorescent protein as a marker for gene expression. *Science* **263,** 802–805.

Chattoraj, M., King, B. A., Bublitz, G. U., and Boxer, S. G. (1996). Ultra-fast excited state dynamics in green fluorescent protein: Multiple states and proton transfer. *Proc. Natl. Acad. Sci. USA* **93,** 8362–8367.

Chudakov, D. M., Belousov, V. V., Zaraisky, A. G., Novoselov, V. V., Staroverov, D. B., Zorov, D. B., Lukyanov, S., and Lukyanov, K. A. (2003). Kindling fluorescent proteins for precise *in vivo* photolabeling. *Nat. Biotechnol.* **21,** 191–194.

Cowan, S. E., Gilbert, E., Khlebnikov, A., and Keasling, J. D. (2000). Dual labeling with green fluorescent proteins for confocal microscopy. *Appl. Environ. Microbiol.* **66,** 413–418.

Crameri, A., Whitehorn, E. A., Tate, E., and Stemmer, W. P. (1996). Improved green fluorescent protein by molecular evolution using DNA shuffling. *Nat. Biotechnol.* **14,** 315–319.

Danuser, G., and Waterman-Storer, C. M. (2006). Quantitative fluorescent speckle microscopy of cytoskeleton dynamics. *Annu. Rev. Biophys. Biomol. Struct.* **35,** 361–387.

Davenport, D., and Nichol, J. A. C. (1955). Luminescence in Hydromedusae. *Proc. R. Soc. London, Ser. B* **144,** 399–411.

Dixit, R., Cyr, R., and Gilroy, S. (2006). Using intrinsically fluorescent proteins for plant cell imaging. *Plant J.* **45,** 599–615.

Dunn, G. A., Dobbie, I. M., Monypenny, J., Holt, M. R., and Zicha, D. (2002). Fluorescence localization after photobleaching (FLAP): A new method for studying protein dynamics in living cells. *J. Microsc.* **205,** 109–112.

Ellenberg, J., Lippincott-Schwartz, J., and Presley, J. F. (1998). Two-color green fluorescent protein time-lapse imaging. *Biotechniques* **25,** 838–842, 844–846.

Evanko, D. S., and Haydon, P. G. (2005). Elimination of environmental sensitivity in a cameleon FRET-based calcium sensor via replacement of the acceptor with Venus. *Cell Calcium* **37,** 341–348.

Field, S. F., Bulina, M. Y., Kelmanson, I. V., Bielawski, J. P., and Matz, M. V. (2006). Adaptive evolution of multicolored fluorescent proteins in reef-building corals. *J. Mol. Evol.* **62,** 332–339.

Galperin, E., Verkhusha, V. V., and Sorkin, A. (2004). Three-chromophore FRET microscopy to analyze multiprotein interactions in living cells. *Nat. Methods* **1,** 209–217.

Giepmans, B. N., Adams, S. R., Ellisman, M. H., and Tsien, R. Y. (2006). The fluorescent toolbox for assessing protein location and function. *Science* **312,** 217–224.

Gordon, G. W., Berry, G., Liang, X. H., Levine, B., and Herman, B. (1998). Quantitative fluorescence resonance energy transfer measurements using fluorescence microscopy. *Biophys. J.* **74,** 2702–2713.

Guerrero, G., and Isacoff, E. Y. (2001). Genetically encoded optical sensors of neuronal activity and cellular function. *Curr. Opin. Neurobiol.* **11,** 601–607.

Gurskaya, N. G., Fradkov, A. F., Pounkova, N. I., Staroverov, D. B., Bulina, M. E., Yanushevich, Y. G., Labas, Y. A., Lukyanov, S., and Lukyanov, K. A. (2003). A colourless green fluorescent protein homologue from the non-fluorescent hydromedusa Aequorea coerulescens and its fluorescent mutants. *Biochem. J.* **373,** 403–408.

Gurskaya, N. G., Fradkov, A. F., Terskikh, A., Matz, M. V., Labas, Y. A., Martynov, V. I., Yanushevich, Y. G., Lukyanov, K. A., and Lukyanov, S. A. (2001). GFP-like chromoproteins as a source of far-red fluorescent proteins. *FEBS Lett.* **507,** 16–20.

Harpur, A. G., Wouters, F. S., and Bastiaens, P. I. (2001). Imaging FRET between spectrally similar GFP molecules in single cells. *Nat. Biotechnol.* **19,** 167–169.

Heim, R., Cubitt, A. B., and Tsien, R. Y. (1995). Improved green fluorescence. *Nature* **373,** 663–664.

Heim, R., and Tsien, R. Y. (1996). Engineering green fluorescent protein for improved brightness, longer wavelengths and fluorescence resonance energy transfer. *Curr. Biol.* **6,** 178–182.

Higashida, C., Miyoshi, T., Fujita, A., Oceguera-Yanez, F., Monypenny, J., Andou, Y., Narumiya, S., and Watanabe, N. (2004). Actin polymerization-driven molecular movement of mDia1 in living cells. *Science* **303,** 2007–2010.

Hoebe, R. A., Van Oven, C. H., Gadella, T. W., Jr., Dhonukshe, P. B., Van Noorden, C. J., and Manders, E. M. (2007). Controlled light-exposure microscopy reduces photobleaching and phototoxicity in fluorescence live-cell imaging. *Nat. Biotechnol.* **25,** 249–253.

Hu, C. D., Chinenov, Y., and Kerppola, T. K. (2002). Visualization of interactions among bZIP and Rel family proteins in living cells using bimolecular fluorescence complementation. *Mol. Cell* **9,** 789–798.

Hu, C. D., and Kerppola, T. K. (2003). Simultaneous visualization of multiple protein interactions in living cells using multicolor fluorescence complementation analysis. *Nat. Biotechnol.* **21,** 539–545.

Hughes, T. E., Zhang, H., Logothetis, D. E., and Berlot, C. H. (2001). Visualization of a functional Galpha q-green fluorescent protein fusion in living cells. Association with the plasma membrane is disrupted by mutational activation and by elimination of palmitoylation sites, but not be activation mediated by receptors or AlF4. *J. Biol. Chem.* **276,** 4227–4235.

Jares-Erijman, E. A., and Jovin, T. M. (2003). FRET imaging. *Nat. Biotechnol.* **21,** 1387–1395.

Kenworthy, A. K. (2001). Imaging protein–protein interactions using fluorescence resonance energy transfer microscopy. *Methods* **24,** 289–296.

Kogure, T., Karasawa, S., Araki, T., Saito, K., Kinjo, M., and Miyawaki, A. (2006). A fluorescent variant of a protein from the stony coral Montipora facilitates dual-color single-laser fluorescence cross-correlation spectroscopy. *Nat. Biotechnol.* **24,** 577–581.

Kohl, T., and Schwille, P. (2005). Fluorescence correlation spectroscopy with autofluorescent proteins. *Adv. Biochem. Eng. Biotechnol.* **95,** 107–142.

Labas, Y. A., Gurskaya, N. G., Yanushevich, Y. G., Fradkov, A. F., Lukyanov, K. A., Lukyanov, S. A., and Matz, M. V. (2002). Diversity and evolution of the green fluorescent protein family. *Proc. Natl. Acad. Sci. USA* **99,** 4256–4261.

Leake, M. C., Chandler, J. H., Wadhams, G. H., Bai, F., Berry, R. M., and Armitage, J. P. (2006). Stoichiometry and turnover in single, functioning membrane protein complexes. *Nature* **443,** 355–358.

Lippincott-Schwartz, J., and Patterson, G. H. (2003). Development and use of fluorescent protein markers in living cells. *Science* **300,** 87–91.

Lippincott-Schwartz, J., Altan-Bonnet, N., and Patterson, G. H. (2003). Photobleaching and photoactivation: Following protein dynamics in living cells. *Nat. Cell Biol.* (Suppl.), S7–S14.

Lippincott-Schwartz, J., Roberts, T. H., and Hirschberg, K. (2000). Secretory protein trafficking and organelle dynamics in living cells. *Annu. Rev. Cell Dev. Biol.* **16,** 557–589.

Lippincott-Schwartz, J., Snapp, E., and Kenworthy, A. (2001). Studying protein dynamics in living cells. *Nat. Rev. Mol. Cell Biol.* **2,** 444–456.

Lukyanov, K. A., Chudakov, D. M., Lukyanov, S., and Verkhusha, V. V. (2005). Innovation: Photoactivatable fluorescent proteins. *Nat. Rev. Mol. Cell Biol.* **6,** 885–891.

Matz, M. V., Fradkov, A. F., Labas, Y. A., Savitsky, A. P., Zaraisky, A. G., Markelov, M. L., and Lukyanov, S. A. (1999). Fluorescent proteins from nonbioluminescent Anthozoa species. *Nat. Biotechnol.* **17,** 969–973.

McNamara, G. (2006). Multi-Probe Microscopy http://home.earthlink.net/~mpmicro/.

Miyawaki, A., Llopis, J., Heim, R., McCaffery, J. M., Adams, J. A., Ikura, M., and Tsien, R. Y. (1997). Fluorescent indicators for Ca^{2+} based on green fluorescent proteins and calmodulin. *Nature* **388,** 882–887.

Miyawaki, A., Sawano, A., and Kogure, T. (2003). Lighting up cells: Labelling proteins with fluorophores. *Nat. Cell Biol.* **5**(Suppl.), S1–S7.

Morin, J. G., and Hastings, J. W. (1971). Energy transfer in a bioluminescent system. *J. Cell Physiol.* **77,** 313–318.

Morise, H., Shimomura, O., Johnson, F. H., and Winant, J. (1974). Intermolecular energy transfer in the bioluminescent system of *Aequorea. Biochemistry* **13,** 2656–2662.

Nagai, T., Ibata, K., Park, E. S., Kubota, M., Mikoshiba, K., and Miyawaki, A. (2002). A variant of yellow fluorescent protein with fast and efficient maturation for cell-biological applications. *Nat. Biotechnol.* **20,** 87–90.

Nguyen, A. W., and Daugherty, P. S. (2005). Evolutionary optimization of fluorescent proteins for intracellular FRET. *Nat. Biotechnol.* **23,** 355–360.

Niell, C. M., and Smith, S. J. (2004). Live optical imaging of nervous system development. *Annu. Rev. Physiol.* **66,** 771–798.

Ormo, M., Cubitt, A. B., Kallio, K., Gross, L. A., Tsien, R. Y., and Remington, S. J. (1996). Crystal structure of the *Aequorea victoria* green fluorescent protein. *Science* **273,** 1392–1395.

Patterson, G. H., and Lippincott-Schwartz, J. (2002). A photoactivatable GFP for selective photolabeling of proteins and cells. *Science* **297,** 1873–1877.

Pepperkok, R., Squire, A., Geley, S., and Bastiaens, P. I. (1999). Simultaneous detection of multiple green fluorescent proteins in live cells by fluorescence lifetime imaging microscopy. *Curr. Biol.* **9,** 269–272.

Pollok, B. A., and Heim, R. (1999). Using GFP in FRET-based applications. *Trends Cell Biol.* **9,** 57–60.

Prasher, D. C., Eckenrode, V. K., Ward, W. W., Prendergast, F. G., and Cormier, M. J. (1992). Primary structure of the *Aequorea victoria* green-fluorescent protein. *Gene* **111,** 229–233.

Reits, E. A., and Neefjes, J. J. (2001). From fixed to FRAP: Measuring protein mobility and activity in living cells. *Nat. Cell Biol.* **3,** E145–E147.

Remington, S. J. (2000). Structural basis for understanding spectral variations in green fluorescent protein. *Methods Enzymol.* **305,** 196–211.

Remington, S. J., Wachter, R. M., Yarbrough, D. K., Branchaud, B., Anderson, D. C., Kallio, K., and Lukyanov, K. A. (2005). zFP538, a yellow-fluorescent protein from Zoanthus, contains a novel three-ring chromophore. *Biochemistry* **44,** 202–212.

Sawin, K. E., and Nurse, P. (1997). Photoactivation of green fluorescent protein. *Curr. Biol.* **7,** R606–R607.

Shaner, N. C., Campbell, R. E., Steinbach, P. A., Giepmans, B. N., Palmer, A. E., and Tsien, R. Y. (2004). Improved monomeric red, orange and yellow fluorescent proteins derived from *Discosoma* sp. red fluorescent protein. *Nat. Biotechnol.* **22,** 1567–1572.

Shaner, N. C., Steinbach, P. A., and Tsien, R. Y. (2005). A guide to choosing fluorescent proteins. *Nat. Methods* **2,** 905–909.

Shimomura, O. (1979). Structure of the chromophore of Aequorea green fluorescenct protein. *FEBS Lett.* **104,** 220–222.

Shimomura, O., Johnson, F. H., and Saiga, Y. (1962). Extraction, purification and properties of aequorin, a bioluminescent protein from the luminous hydromedusan, Aequorea. *J. Cell Comp. Physiol.* **59,** 223–239.

Shimozono, S., Hosoi, H., Mizuno, H., Fukano, T., Tahara, T., and Miyawaki, A. (2006). Concatenation of cyan and yellow fluorescent proteins for efficient resonance energy transfer. *Biochemistry* **45,** 6267–6271.

Shkrob, M. A., Yanushevich, Y. G., Chudakov, D. M., Gurskaya, N. G., Labas, Y. A., Poponov, S. Y., Mudrik, N. N., Lukyanov, S., and Lukyanov, K. A. (2005). Far-red fluorescent proteins evolved from a blue chromoprotein from Actinia equina. *Biochem. J.* **392,** 649–654.

Shu, X., Shaner, N. C., Yarbrough, C. A., Tsien, R. Y., and Remington, S. J. (2006). Novel chromophores and buried charges control color in mFruits. *Biochemistry* **45,** 9639–9647.

Steinmeyer, R., Noskov, A., Krasel, C., Weber, I., Dees, C., and Harms, G. S. (2005). Improved fluorescent proteins for single-molecule research in molecular tracking and co-localization. *J. Fluoresc.* **15,** 707–721.

Takemoto, K., Nagai, T., Miyawaki, A., and Miura, M. (2003). Spatio-temporal activation of caspase revealed by indicator that is insensitive to environmental effects. *J. Cell Biol.* **160,** 235–243.

Ting, A. Y., Kain, K. H., Klemke, R. L., and Tsien, R. Y. (2001). Genetically encoded fluorescent reporters of protein tyrosine kinase activities in living cells. *Proc. Natl. Acad. Sci. USA* **98,** 15003–15008.

Tsien, R. Y. (1998). The green fluorescent protein. *Annu. Rev. Biochem.* **67,** 509–544.

van Thor, J. J., Gensch, T., Hellingwerf, K. J., and Johnson, L. N. (2002). Phototransformation of green fluorescent protein with UV and visible light leads to decarboxylation of glutamate 222. *Nat. Struct. Biol.* **9,** 37–41.

Verkhusha, V. V., and Lukyanov, K. A. (2004). The molecular properties and applications of Anthozoa fluorescent proteins and chromoproteins. *Nat. Biotechnol.* **22,** 289–296.

Walter, M., Chaban, C., Schutze, K., Batistic, O., Weckermann, K., Nake, C., Blazevic, D., Grefen, C., Schumacher, K., Oecking, C., Harter, K., and Kudla, J. (2004). Visualization of protein interactions in living plant cells using bimolecular fluorescence complementation. *Plant J.* **40,** 428–438.

Wang, S., and Hazelrigg, T. (1994). Implications for bcd mRNA localization from spatial distribution of exu protein in Drosophila oogenesis. *Nature* **369,** 400–403.

Ward, W. W., and Cormier, M. J. (1979). An energy transfer protein in coelenterate bioluminescence. Characterization of the Renilla green-fluorescent protein. *J. Biol. Chem.* **254,** 781–788.

Waterman-Storer, C. M., Desai, A., Bulinski, J. C., and Salmon, E. D. (1998). Fluorescent speckle microscopy, a method to visualize the dynamics of protein assemblies in living cells. *Curr. Biol.* **8,** 1227–1230.

Yang, T. T., Kain, S. R., Kitts, P., Kondepudi, A., Yang, M. M., and Youvan, D. C. (1996). Dual color microscopic imagery of cells expressing the green fluorescent protein and a red-shifted variant. *Gene* **173,** 19–23.

Yang, T. T., Sinai, P., Green, G., Kitts, P. A., Chen, Y. T., Lybarger, L., Chervenak, R., Patterson, G. H., Piston, D. W., and Kain, S. R. (1998). Improved fluorescence and dual color detection with enhanced blue and green variants of the green fluorescent protein. *J. Biol. Chem.* **273,** 8212–8216.

Yuste, R. (2005). Fluorescence microscopy today. *Nat. Methods* **2,** 902–904.

Zacharias, D. A., and Tsien, R. Y. (2006). Molecular biology and mutation of green fluorescent protein. *Methods Biochem. Anal.* **47,** 83–120.

Zhang, J., Campbell, R. E., Ting, A. Y., and Tsien, R. Y. (2002). Creating new fluorescent probes for cell biology. *Nat. Rev. Mol. Cell Biol.* **3,** 906–918.

Zhang, J., Ma, Y., Taylor, S. S., and Tsien, R. Y. (2001). Genetically encoded reporters of protein kinase A activity reveal impact of substrate tethering. *Proc. Natl. Acad. Sci. USA* **98,** 14997–15002.

Zicha, D., Dobbie, I. M., Holt, M. R., Monypenny, J., Soong, D. Y., Gray, C., and Dunn, G. A. (2003). Rapid actin transport during cell protrusion. *Science* **300,** 142–145.

Zimmerman, T., Rietdorf, J., and Pepperkok, R. (2003). Spectral imaging and its applications in live cell microscopy. *FEBS Lett.* **546,** 87–92.

CHAPTER 2

Functional Fusion Proteins by Random Transposon-Based GFP Insertion

Robert Mealer, Heather Butler, and Thomas Hughes

Department of Cell Biology and Neuroscience
Montana State University
Bozeman, Montana 59717

0091-679X/08 $35.00
DOI: 10.1016/S0091-679X(08)85002-9

Abstract

Fusions with fluorescent proteins are usually created by fusing the ends of two coding sequences. Appending the coding region of a fluorescent protein to the N- or C-terminus of another protein is typically the easiest way of creating a functional, fluorescent fusion protein. Another strategy involves placing the fluorescent protein in the middle of another protein. Such sandwich fusions are feasible, and there are many reasons for creating these fusion proteins. For example, sandwich fusions can be used to place two fluorescent proteins close to one another for optimization of a biosensor based on fluorescence resonance energy transfer, or they can be used to place the fluorescent protein in a region that moves during conformational changes of the host protein. Designing a sandwich fusion that produces a functional, fluorescent fusion protein is often challenging. This protocol describes an alternative approach. A simple, *in vitro*, transposon reaction is used to randomly insert the sequence encoding a fluorescent fusion protein into a target protein. This random labeling strategy makes it possible to create a small library of sandwich fusion proteins that can then be screened for activity. The approach makes it possible to test many possible solutions to the complex problem of building new biosensors.

I. Introduction

The initial descriptions of fluorescent fusion proteins involved placing the GFP sequence at the end of a variety of other coding sequences (Marshall *et al.*, 1995; Wang and Hazelrigg, 1994). Since then, thousands of fluorescent fusion proteins have been created in this way, and a wide variety of cloning vectors and strategies are available to easily and quickly create the fluorescent fusion protein of choice. Indeed, there are even systems available that are suitable for large-scale tagging efforts (Huh *et al.*, 2003). Given this record of success, it is reasonable to ask why would you insert fluorescent proteins into the middle of other fusion partners?

The process of creating fluorescent fusion proteins that retain biological function involves adding the relatively large fluorescent protein to a surface of the fusion protein that is not involved in interactions with other proteins. This is often a matter of guesswork. Usually, fusing the fluorescent protein to either the N- or the C-terminus of the other protein maintains function, but there are proteins where both ends are near a surface critical to protein interactions. For example, the ends of the G-protein subunit αq are at the surface critical to interactions with both the receptor and many effectors. Placing a fluorescent protein, which is nearly as large as the α subunit, at either the N- or the C-terminus destroys biological function, while inserting it into a loop on the other side of the structure produces a fluorescent, functional protein (Hughes *et al.*, 2001). One good reason for inserting a fluorescent protein internally often involves the desperation that follows the realization that end labeling will not work.

Fluorescent fusion proteins give us new views of the dynamics of protein or organelle movement within the cell. The next step, a more difficult step, is to create fusion proteins that signal changes in fluorescence that reflect changes in the conformational state of the protein. The earliest such constructs relied on fluorescence resonance energy transfer (FRET, see Chapter 7 by Ajit Joglekar, E. D. Salmon, and Kerry Bloom, this volume). Two different fluorescent proteins were positioned in different domains, and the relative distance between the two proteins, the conformational state of the protein, was readily measured by the change in emission ratios. This was feasible because the efficiency of FRET changes with the sixth power of the distance between the fluorophores. FRET is a remarkably sensitive measurement of molecular distances (Stryer and Haugland, 1967). While FRET constructs are useful tools, the critical dependence on the distance between the donor and acceptor fluorophores means that success depends upon two variables. First, the donor and acceptor have to be positioned in domains that move relative to one another during the conformational change. Second, the two fluorophores have to be close enough to one another for readily detectable FRET changes to occur. Another good reason to insert fluorescent proteins in the internal domains of other proteins is that this placement can be critical for creating good FRET pairs.

Internally inserted fluorescent proteins can also produce signals when the conformation of the fusion partner changes. In 1997, Siegel and Isacoff showed that a GFP placed just beneath the S6 domain of the Shaker potassium channel produces a truly novel, genetically encoded biosensor (Guerrero *et al.*, 2002; Siegel and Isacoff, 1997). Voltage-driven rearrangements of the channel protein cause a change in the fluorescence intensity produced by the fusion protein. It is unclear what the mechanism is that couples channel movement to fluorescence changes, but it appears that this approach can be generalized. Ataka and Pieribone (2002) placed a GFP in an intracellular loop of a voltage-gated sodium channel and also observed a voltage-dependent change in fluorescence. Without better structural information on these channels and without mechanistic insights about the coupling of conformational changes and fluorescence, the optimization of these fluorescent voltage sensors remains a matter of guesswork. One way of circumventing an arduous process of iteratively guessing and testing is to use the process described here to quickly generate a library of different proteins with the fluorescent protein inserted randomly throughout the sequence. This moves the discovery effort from designing and creating fusion proteins to a process of screening them for the desired biosensor properties (Fig. 1).

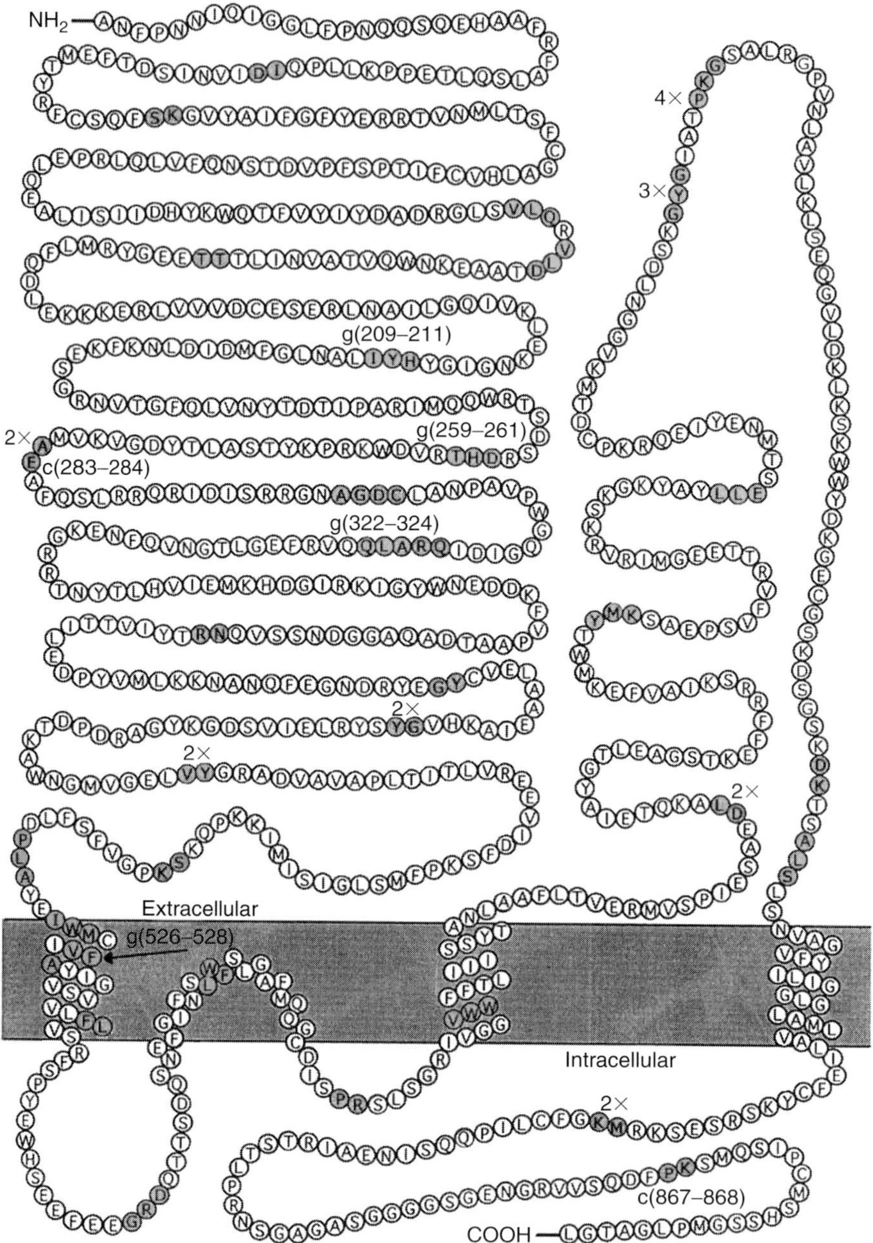

Fig. 1 The process described in this protocol makes it possible to quickly insert fluorescent proteins into many different portions of another protein. This figure illustrates the primary sequence of the glutamate-gated ion channel GluR 1. The colored residues are where either a cyan or a green fluorescent

Finding internal insertions sites that can tolerate a fluorescent protein, optimizing FRET pairs, or finding fusions that signal conformational changes through fluorescence changes is challenging. The potential space that needs to be explored is quite large and multidimensional. Which fluorescent protein is best? What are the optimal linkers between the fusion partners? Where exactly should the fluorescent protein(s) be inserted? The techniques presented here simplify one part of this approach by making it possible to quickly, and relatively randomly, create a library of GFP insertions in a target.

II. Rationale

The earliest examples of sandwich fusions involved placing alkaline phosphatase in many different regions of the multispanning membrane protein, MalF. Surprisingly, the reporter was enzymatically active even when it was inserted into the middle of another protein (Ehrmann *et al.*, 1990). One can imagine that constraining both the N- and the C-termini of the reporter, by attaching it to other protein domains, could inhibit the correct folding of the reporter. Similarly, it is surprising that fluorescent proteins can be inserted deep into the structure of other proteins and still fold correctly. There are now many reports of sandwich fusions with GFP, cyan fluorescent protein (CFP), and yellow fluorescent protein (YFP), indicating that this is quite feasible. Indeed, in our work with transposon insertions and hundreds of fusion proteins, the results have been consistent with a model in which a fluorescent GFP is always formed, often at the expense of its fusion partner (Giraldez *et al.*, 2005; Sheridan and Hughes, 2004; Sheridan *et al.*, 2002, 2006). However, the extent to which these results can be generalized to the other fluorescent fusion proteins or to circularly permuted forms of GFP remains to be determined.

Since the original report of sandwich fusions (reviewed in Doi and Yanagawa, 1999), transposons have been the method of choice for inserting reporters into the middle of other proteins (e.g., Hoekstra *et al.*, 1991; Merkulov and Boeke, 1998; Ross-Macdonald *et al.*, 1999a, b). They provide a remarkably fast and efficient way to randomly insert reporters or tags. In designing a transposon system for tagging, there are several constraints that must be considered. First, at the DNA level, there are specific sequences flanking the transposon that are recognized by the transposase. These define the portion of DNA that will be excised by the transposase enzyme and then inserted into another sequence. The sequence of these ends must

---

protein was randomly inserted into the GluR 1 coding sequence by a transposon. All of these insertions produced a fluorescent fusion protein, but only a subset of these continue to function as a ligand-gated channel (functional sites are numbered). The results of such tagging experiments often reveal improbably insertions sites that continue to produce channel function (figure from Sheridan *et al.*, 2002). (See Plate no. 1 in the Color Plate Section.)

contain at least one open reading frame such that a fusion protein can be produced when the transposon lands in another coding sequence. A second constraint is that the sequences encoded by these transposon ends will inevitably add linkers between the target protein and the reporter, which will be difficult to remove. Finally, even the most efficient transposon systems still require antibiotic selection, so the transposon must initially carry a selectable marker.

Several transposon systems have been used to tag with the GFP coding sequence. The approach described here uses a modified version of the Tn5 transposon (reviewed in Reznikoff, 2003, 2006). A variety of improvements and modifications of the tranposase, and the DNA ends that it recognizes, has produced a remarkably efficient *in vitro* reaction in which entire libraries of fluorescent fusion proteins can be created in a few hours. The recombinant transposase is commercially available (Epicentre) and the sequences recognized by the transposase are 19-bp inverted repeats that are quite easy to add with PCR primers or simple subcloning steps (Fig. 2).

The Tn5 *in vitro* reaction is remarkably efficient. The transposon DNA is simply mixed with the target plasmid, and a 2-h reaction creates thousands of plasmids carrying the transposon. The reaction involves inserting the transposon into a 9-bp nick in the target sequence. This is subsequently filled in, so three codons of the target sequence are duplicated. This leads to the same three codons being present before and after the transposon (Fig. 3). The reaction is remarkably efficient, but maximum yields are only ~1%. This means that the transposon has to carry a selectable marker. In our designs, the selectable marker is flanked by a rare 8-bp restriction site. Initially, when the transposon lands in another coding sequence, in the correct orientation and relative reading frame, a fluorescent fusion protein is produced in which the target protein is truncated and the fluorescent protein is at

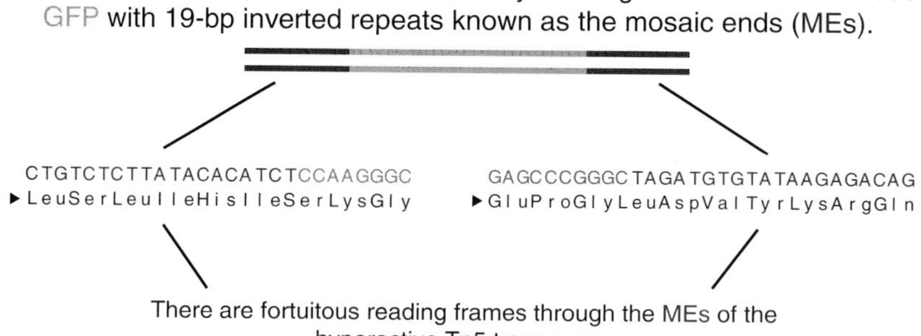

Fig. 2 To create a synthetic Tn5 transposon, 19-bp sequences need to be added to the ends of the green fluorescent protein (GFP). The 19 bp are inverted sequences recognized by the Tn5 transposase. There is a reading frame that passes through both mosaic ends (MEs) such that sequence encoding GFP can be placed in between the MEs to create one continuous reading frame with no stop codons. (See Plate no. 2 in the Color Plate Section.)

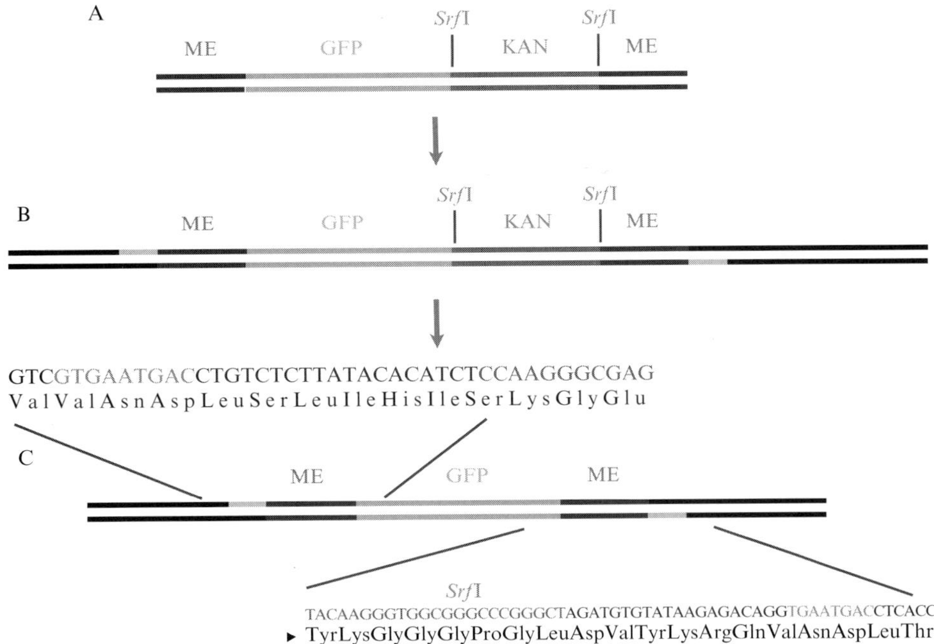

Fig. 3 The transposition process is only ~1% efficient, so a selectable marker has to be included. The simplest transposon we have built is illustrated in (A). This is a transposon containing the mosaic ends, the sequence encoding the green fluorescent protein (GFP), and a kanamycin resistance gene for selection. *Srf*I restriction sites are positioned on either side of the kanamycin gene so that it can be removed by digestion and re-ligation. When the transposon lands in another coding region, it is inserted into a 9-bp staggered nick. This is subsequently repaired to produce a 9-bp duplication of the target sequence. If the transposon lands in the target sequence in the correct orientation and relative reading frame, a truncated protein is created with GFP at the C-terminus (B). This is due to a stop codon in the kanamycin cassette. *Srf*I digestion and re-ligation removes the kanamycin cassette and produces a continuous open reading frame that crosses the entire transposon (C). This is how the sandwich fusion is created. (See Plate no. 3 in the Color Plate Section.)

the C-terminus. This is due to a termination codon in the selection cassette. The clones isolated after the transposition reaction can be screened at this point in a transient expression system. The presence of fluorescence indicates that the transposon landed in the target sequence, and that it is in the correct orientation and relative reading frame. The next step (Fig. 3B) involves restriction digestion, re-ligation, and retransformation steps to remove the selection cassette and produce a full-length sandwich fusion with the fluorescent protein in the middle of the protein (Fig. 3C).

There are several ways in which the simple tagging transposon illustrated in Figs. 1 and 3 can be modified to produce a variety of different tagging systems.

For example, one design places two sequences encoding fluorescent proteins in opposite orientations in the transposon (Fig. 4). This improves the efficiency of the process because a fusion protein is produced when the transposon lands in either orientation (Sheridan and Hughes, 2004). This process depends on the judicious use of alternating restriction sites that make it possible to drop the selectable marker and one of the other fluorescent coding regions. A similar use of staggered sites can be used to create a transposon that carries two different fluorescent proteins, a FRET pair for example, in such way that either the donor or the acceptor fluorescent protein can be left at each site (Sheridan and Hughes, 2004). Or an entire FRET pair of fluorescent proteins, with a hinge in between, can be carried on the transposon (Butler and Hughes, 2006). Quite recently our group has discovered that this approach can be used to leave fragments of the fluorescent protein at each insertion site as well. This makes it possible to screen for fluorescent protein complementation that can occur between fluorescent protein fragments inserted into interacting surfaces of large proteins (Mealer *et al.*, 2006; see Chapter 19 by Kerppola, this volume for more on fluorescent protein complementation).

The final step before beginning the transposition process involves optimizing the target plasmid. If, for example, the target is a 3-kb sequence encoding a 1000 amino acid protein, this sequence has to be moved into a suitable expression plasmid with the following characteristics. The plasmid should be small. Because the transposon insertion process is nearly random (see Ason and Reznikoff, 2004 for a description of insertion preference), the probability that the transposon will land in a 3-kb coding sequence can be predicted quite well. In a small ~4.5-kb CMV expression plasmid, there is about 3 kb of sequence that the transposon can land in and still

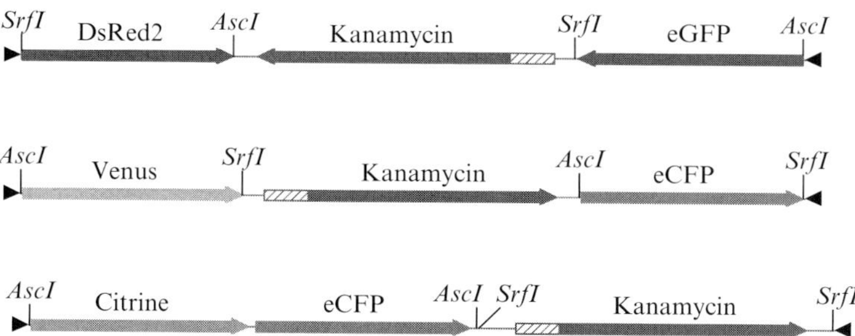

Fig. 4 Transposons may be designed in a variety of ways to create useful molecular tools. Top: The *either orientation works* transposon produces either a green fluorescent protein or a red fluorescent protein depending on how it lands in another coding sequence. *Asc*I or *Srf*I digestion can then be used to remove the selection cassette and the other fluorescent protein sequence. This doubles the number of recovered insertion locations. Middle: The *either yellow or cyan* transposon makes it possible to leave either a yellow or a cyan fluorescent protein at each transposon insertion site. Bottom: The *FRET cassette* transposon makes it possible to scan a protein with a conformationally sensitive fluorescent resonance energy transfer (FRET) probe. (See Plate no. 4 in the Color Plate Section.)

produce a viable plasmid (transposon insertions in the origin of replication or selectable marker are usually not recovered). This means that there will be a 50% chance of the transposon landing in the sequence encoding your protein. Next, there is only a 50% chance of the transposon landing in the correct orientation, and finally, there is only a 33% chance of it being in the right relative reading frame. Typically, only 5–8% of the plasmids carrying a transposon insertion produce a fluorescent protein. In addition to these design considerations, it is important to make sure that the target plasmid does not contain the antibiotic resistance that is carried in the transposon, and it is crucial that plasmid does not carry restriction sites for the enzyme(s) used to ultimately remove the selection cassette from the transposon.

III. Methods

Assuming either you have created a Tn5 transposon of your own or you are using pBNJ24.6, the first step is to create enough transposon for a workable reaction. You will also need a target plasmid for the reaction.

A. PCR Amplification of the Transposon

The first step is to amplify the transposon. This can be done by amplifying from a plasmid containing a transposon with PCR. This reaction requires only a single primer that is complimentary to Tn5 sequences at either end of the transposon, which the transposase recognizes, known as mosaic ends (MEs). If a template that carries the same antibiotic resistance as either the transposon or the target plasmid is being used, it is a good idea to cut the plasmid first to linearize it and reduce the possibility that it will transform *Escherichia coli* at a later step.

1.1. Digest 4 μl of the pBNJ24.6 mini-prep with *Hin*dIII. Mix the following, in order, in a 0.5-ml tube on ice. Remember to mix the final reaction mixture by pipetting up and down a few times.

dDH_2O	4 μl
Mini-Prep DNA	4 μl
*Hin*dIII	1 μl
10× NEB buffer #2	1 μl
Total volume	10 μl

1.2. Incubate at 37 °C for 1 h.

1.3. Heat inactivate the reaction at 65 °C for 20 min (store the excess reaction at −20 °C).

1.4. Use a single primer complementary to the transposon MEs to amplify the transposon with the high fidelity polymerase PfuUltra (or a similar high fidelity polymerase). This works because the MEs of the transposon are an inverted pair of identical 19-bp sequences on either end of the transposon. Note that the annealing temperature of the ME primer is remarkably low (~45 °C).

1.5. Mix the following on ice, in this order, in a thin walled tube suitable for PCR cycling.

dDH_2O	40.75 μl
10× Pfu buffer	5 μl
dNTPs (10 mM)	1 μl
Digestion reaction (*Hin*dIII cut pBNJ24)	1 μl
ME primer (100 ng/μl)	1.25 μl
PfuUltra polymerase (2.5 U/μl)	1 μl
Total volume	50 μl

1.6. When the reaction is mixed well, by pipetting up and down, amplify the transposon with the following temperature protocol:

94 °C	47 °C	72 °C	Cycles
4 min	0 min	0 min	1
1 min	30 sec	5.30 min	24
0	0	10 min	1

2.1. Resolve 1 μl of the target plasmid on an agarose gel (1%) along with a lane of markers and 1, 3, and 6 μl of the PCR reaction. The PCR reaction should generate one band around 1900 bp for the pBNJ24.6 transposon. Use a comparison of the PCR sample with the standards to estimate the concentration of the PCR product.

2.2. A 1:1 mole ratio of target plasmid to transposon PCR product is optimal for the transposition reaction. Deviation from this ratio can lower yield, but in practice you can visualize the relative amounts on a gel and get a good reaction.

B. The Transposition Reaction

1.1. Using the estimate from the gel, set up a transposition reaction with the amplified transposon and the target plasmid. The reaction should contain ~100 ng of target plasmid and 300 ng of transposon in a reaction tube with the transposase and buffer (based on the assumption that the transposon is one-third the size of your target).

1.2. It is critical that the DNA used in the following reaction, both the target plasmid and the transposon PCR product, is not exposed to ethidium bromide and a transilluminator. This will ensure undamaged DNA is used in the transposition reaction. Visualizing the DNA in an agarose gel before using it is the most common mistake made in this protocol.

dDH_2O	–
10× EZ::TN™ reaction buffer	1 μl
Target plasmid	200 ng
Transposon (PCR product)	300 ng
EZ::TN5™ transposase	1 μl
Total volume	10 μl

1.3. Incubate at 37°C for 2 h.
1.4. Add 1 μl of 10× stop solution.
1.5. Heat inactivate at 70°C for 10 min (reaction can be stored at −20°C for >6 months).

It is very important to add the stop solution (1% SDS) and heat inactivate the transposon reaction. The transposon/transposase complex is very stable and will readily transform the *E. coli* genome during transformation. This can result in many kanamycin resistant colonies of bacteria that do not contain a plasmid carrying a transposon.

C. Transformation Requirements and Troubleshooting

It is critical to use electroporation to transform the transposition reaction. Chemical transformations do not work for this reaction. After transformation and an hour of growth in SOC media, plate the reaction on LB agar plates with 100 μg/ml ampicillin and 50 μg/ml kanamycin as well as on one plate that contains only 100 μg/ml ampicillin. A comparison of the two types of plates the following day will provide a way of calculating the efficiency of the transposition reaction.

1.1. In preparation for transformation, prechill a 0.1-cm gap cuvette on wet ice, prewarm SOC media to 37°C, and prewarm 1 ampicillin and 8 ampicillin + kanamycin plates.
1.2. Add 40 μl of dDH_2O to the transposition reaction. This dilution enables pipetting of small amounts of the reaction more accurately and controls exactly how much of the reaction mixes with the cells.
1.3. Carefully pipette 1 μl of the transposition reaction onto the back wall of a prechilled 0.1-cm-gapped cuvette. Next add 100 μl (or the manufacturer's recommended amount) of electrocompetent *E. coli* and pipette gently to

mix the DNA and cells. Shock the cuvette contents with a 1.7-kV pulse and immediately add 1 ml of prewarmed SOC. Pipette up and down to mix, transfer to a Falcon 5029 tube, and incubate at 37°C, shaking at 250 RPM, for 40 min.

1.4. In the event of arcing, dilute the reaction with dDH_2O.
1.5. The transposition reaction only inserts a transposon into ~1% of the target plasmids in the reaction tube, so highly competent cells are needed for a reasonable yield of colonies (10^9 or greater CFU).
1.6. Plate 100 μl of the transformed *E. coli* that has been shaking for an hour, on each of the 8-LB agar plates with 100 μg/ml ampicillin and 50 μg/ml kanamycin.
1.7. Plate 10 μl (dilute 10 μl of the transformation with 90 μl of fresh SOC broth) on a 100 μg/ml ampicillin only plate as a control.
1.8. Put all nine plates in a 37°C incubator for ~12 h.

D. *E. coli* Colony Selection and Growth in a 96-Well Format

1.1. Remove the plates and compare the amount of colonies on the ampicillin only plate to the other plates containing ampicillin and kanamycin. When you adjust for the different dilutions during plating, you should see that the number of ampicillin + kanamycine resistant clones is only ~1% of the clones that are just ampicillin resistant.
1.2. Plates may be stored for a few days time at 4°C with parafilm around each plate.
1.3. Add 200 μl of LB broth to each well of a deep well 96-well plate. Use autoclaved wooden toothpicks to pick individual *E. coli* colonies from the ampicillin and kanamycin plates and add one colony to each well. It is advantageous to leave the toothpicks in the wells until all 96 wells are inoculated. This decreases the chance of inoculating a well twice. Add a lid to each 96-well plate and place in a shaker incubator for 12–16 h (preferably on a slanted holder in the shaker, 250 RPM, 37°C).

E. Backing Up the Experiment: Making 10% Glycerol Stocks

At this point it is worth making a backup of your clones. Using either a 96-pin replicator tool or a multichannel pipette, you will create a replica plate of the clones in 10% glycerol and freeze the clones for future use.

1.1. Using a multichannel pipette, aliquot 50 μl of 20% filter sterilized glycerol into each well of a standard 96-well plate.

1.2. Use the multichannel pipette to transfer 50 μl of each culture from the deep well growing plate to this plate. Put the plate in a −70 °C freezer for 20 min to freeze, and then put it in a ziplock bag in the freezer for storage.

F. 96-Well Mini-Preparation Purification of Plasmid DNA

Qiagen produces a high-quality kit for preparing mini-prep DNA. There are alternatives, but this is the protocol we use, which is only slightly different from the one recommended by the manufacturer.

1.1. Pellet overnight cultures, with the lid on, in the centrifuge at 1900 × *g* for 5 min at 4 °C. Repeat if pellets have not formed. Quickly invert the culture plate to drain media into the sink, and then place the plate upside down on a paper towel to drain any excess media.
1.2. Resuspend pelleted bacterial cells in 250-μl Buffer P1 (make sure that RNase A has been added to Buffer P1). Place a piece of sealing tape on the top of the plate, let it sit for 5 min. Shake plate vigorously on a platform shaker. Shake until there are no longer cell clumps visible in the bottom of the wells. Good resuspension is critical for a good yield, and it often takes up to 30 min.
1.3. Remove tape and add 250-μl Buffer P2 to each sample, then seal the plate with a new piece of tape. Gently invert plate four to six times to mix, and then incubate at room temperature for 5 min. It is important to mix gently by inverting the plate. Do not shake vigorously, as this will shear the genomic DNA.
1.4. Remove the tape from the plate, add 350-μl Buffer N3 to each sample and seal the block with a new tape sheet, and gently invert the plate four to six times. To avoid localized precipitation, mix the samples gently but thoroughly, immediately after addition of Buffer N3. The solutions should become cloudy.
1.5. Centrifuge the plate at 1900 × *g* for 5 min to pellet as much of the white precipitate as possible.
1.6. During Step 5, prepare QIAvac 96 by placing the TurboFilter 96 plate (white plate) in the QIAvac top plate. Make sure that the plate is seated securely. Seal unused wells of the TurboFilter with tape.
1.7. Place the plate holder inside the QIAvac base. Place QIAprep 96 plate (teal plate) into the plate holder.
1.8. Place QIAvac 96 top plate squarely over base. The QIAprep plate should now be positioned under the TurboFilter plate. Attach QIAvac to a vacuum source.
1.9. Pipette the supernatant (850 μl per well) into the wells of the TurboFilter plate. Try to avoid the white precipitate, but make sure to get as much of the supernatant as possible. Unused wells of the TurboFilter plate should

be sealed with tape. Apply vacuum until all samples have passed through. The optimal flow rate is ~1–2 drops per sec, which can be regulated by using a three-way valve or vacuum regulator between the QIAvac and the vacuum source.

1.10. Ventilate the QIAvac 96 slowly, and switch off vacuum. Discard the Turbo-Filter plate. Transfer the QIAprep plate containing the cleared lysates to the top plate of the manifold. Seal any unused wells of the QIAprep plate with tape. Replace plate holder in the base with waste tray. Place the top plate squarely over the base, making sure that the QIAprep plate is seated securely. Apply vacuum. The flow-through is collected in the waste tray.

1.11. Switch off vacuum, and wash QIAprep plate by adding 0.9-ml Buffer PB to each well and applying vacuum.

1.12. Switch off vacuum, and wash QIAprep plate by adding 0.9 ml of Buffer PE to each well and applying vacuum.

1.13. After Buffer PE has been drawn through all wells, apply maximum vacuum for an additional minute to dry the membrane. Then remove the QIAprep plate and place it on top of the 96-well collection plate, secure with a few pieces of tape, and spin in the centrifuge for 2 min. Discard the flow-through (keep the QIAprep plate which has your DNA). It is important to use this extra spin in the centrifuge to remove all traces of the PE solution. The solution has ethanol, and any residual carried over in the next step will destroy the transfection of the mammalian cells used in the visual screen for fluorescent fusion proteins.

1.14. For elution into provided collection microtubes: Replace waste tray from vacuum with the blue collection microtube rack containing 1.2-ml collection microtubes. Place the top plate back on the base, and place the QIAprep plate securely on top.

1.15. To elute DNA, add 100 μl of Buffer EB (10-mM Tris–HCl, pH 8.5) or dDH_2O to the center of each well of the QIAprep plate, let it stand for ~1 min, and apply maximum vacuum for ~5 min. Ventilate QIAvac 96 slowly and switch off vacuum. A smaller elution volume can be used to increase DNA concentration. Seal with provided lids and label this plate with appropriate information.

G. Preparation of HEK 293 Cells

1.1. Prior to transient transfection, if cells are not already growing, follow ATCC guidelines for growth of new HEK 293 cells in a 75 cm^2 sterile flask. One flask of 95% confluent cells is necessary for three to four 96-well glass bottom plates.

1.2. Approximately 24 h prior to transient expression, plate cells in 100 μl of media per well of a 96-well glass bottom plate such that the cells will be 80–95% confluent at the time of transfection. The recommended media is

Dulbececco's Modified Eagle Medium with 10% Fetal Bovine Serum and 1% L-glutamine and nonessential amino acids. Store in an incubator with 5% CO_2 at 37°C.

H. Transient Transfection of HEK 293 Cells in a 96-Well Format

Having prepared plates of plasmid DNA, the next step is to transiently transfect HEK 293 cells to screen for fluorescent fusion proteins. This is done with a Lipofectamine™ 2000 protocol in 96-well plates that are fitted with glass bottoms for microscopy. This step is to determine which clones encode a truncated, fluorescent fusion protein.

1.1. The goal is to create a mix of DNA, Lipofectamine™ 2000, and Opti-MEM I in a new 96-well plate and then transfer this mix to a plate with HEK 293 cells, or a different cell line in which expression is desired. All of this is performed in a sterile tissue culture hood with sterile equipment for best results. Contamination will not yield ideal results.
1.2. Use the 96-well mini-preps previously prepared.

The following is for one 96-well plate, but may be increased for additional plates.

1.3. Add 60 μl of Lipofectamine™ 2000 to a 3 ml aliquot of Opti-MEM I and mix by pipetting up and down (120 × 25 μl for each well = 3 ml of Opti-MEM; 120 × 0.5 μl for each well = 60 μl of Lipofectamine).
1.4. Incubate for 5 min.
1.5. While waiting for the above, grab a new sterile 96-well V-bottom plate to use as a mixing plate. Use a multichannel pipette to *dispense* 25 μl of plain Opti-MEM I to each of the 96 wells in the mixing plate.
1.6. Using the multichannel pipette, transfer 20 μl of the mini-prep DNA from the mini-prep 96-well plate to each of the wells in the mixing plate (use new sterile tips for each well to avoid cross-contamination of the DNA).
1.7. Pour the Lipofectamine™ 2000/Opti-MEM I mix that was set aside into a new, sterile V-bottom boat.
1.8. Use a multichannel pipette to distribute 25 μl to each well of the mixing plate. (You may mix once or twice by pipetting up and down.)
1.9. Incubate for 20 min.
1.10. The 96-well mini-prep DNA may be stored at −20°C for future use.
1.11. The Lipofectamine™ 2000 is forming complexes with the mini-prep DNA during this interval.
1.12. Remove the glass bottom 96-well plate containing cells for transient expression from the incubator.
1.13. Use a multichannel pipette, with new sterile tips, to add the Lipofectamine™ 2000/Opti-MEM/DNA complexes (this is everything in the mixing plate) to the plated cells.

1.14. No mixing is needed in this step except maybe a gentle rocking of the plate after everything has been added.

1.15. Place the cell plate in an incubator with 5% CO_2 for 24–48 h before screening for fluorescence with a fluorescence microscope.

I. Screening Live HEK 293 Cells for GFP Fluorescence

1.1. The screen for fluorescent colonies involves systematically going through the 96-well plates from one well to the next to look for fluorescence. Use DIC optics or Phase contrast to make sure that the actual correct focal plane is being utilized, and then switch to fluorescence illumination (any FITC compatible filter set will work for this; to order your own set, see http://chroma.com/ or http://www.omegafilters.com). If a clone does not express well, then use objectives with a high numerical aperture (1.2–1.4) or positive clones can be missed. Systematically work through the plate looking for fluorescent cells. If there are doubts about a signal, and autofluorescence at times generate a dim yellow background that can be confusing, switch to the RITC optics. Autofluorescence will still be visible, the GFP signal will not. Keep careful records of which clones produced fluorescent cells at this point to avoid later confusion.

J. Removing the Selection Cassette with Restriction Digestion and Re-Ligation

When clones that encode fluorescent proteins are located, use a plate map to return to the glycerol stock plates made earlier. Use these stocks to inoculate new tubes of LB with ampicillin and kanamycin to grow up more cells. The following day, mini-prep the cultures using a Qiagen kit. Reserve half of the mini-prep DNA for sequencing. A primer complementary to the transposon, positioned ~50 bp from the end, can be used to sequence each clone and identify where the transposon landed in the sequence encoding your protein.

1.1. The kanamycin cassette in the transposon of pBNJ24.6 is flanked by *Srf*I sites. *Srf*I digestion will remove this cassette so that a full-length fluorescent protein can be created.

1.2. Use the mini-prep DNA of the identified fluorescent fusion proteins for kanamycin cassette removal. Set up the necessary digests to remove the kanamycin cassette within the transposon. When processing more than one clone, make up a reaction mixture and then add aliquots of the reaction mixture to each tube containing a unique clone.

	Single reaction	10 Reaction master mixture
dDH_2O	4.5 μl	45 μl
10× Stratagene universal buffer	1 μl	10 μl
*Srf*I	0.5 μl	5 μl
Plasmid DNA	4 μl	–
Total volume	10 μl	Aliquot at 6 μl

1.3. Digest for 1 h at 37°C.

1.4. Resolve the restriction digest on a 1% agarose gel. There should be one large band, the plasmid and coding region, as well as a smaller band, the kanamycin cassette, present. Cut out the large band containing the plasmid and coding region and purify the DNA with a Qiagen min elute column. Elute the DNA with 10 μl of elution buffer or water, then set up the ligation.

1.5. Mix on ice, in order, the following ligation reaction.

dDH_2O	7.75 μl
10× fast link buffer	1.5 μl
10-mM ATP	0.75 μl
*Srf*I fragment you isolated from the gel	4 μl
Fast link ligase	1 μl
Total volume	15 μl

1.6. The efficiency of ligating two different pieces of DNA together is directly related to the insert/vector ratio *and* the molar concentration of free ends. If the concentration of free ends is high enough, recombinants in which two different pieces are joined to one another is likely. The converse of this rule is true too. If the sample is dilute enough, the plasmid closing on itself without an insert is favored. This is desired, so now simply dilute the restriction digestion from above and set up a ligation reaction that favors the plasmid closing on itself without the kanamycin cassette.

1.7. Incubate at room temperature for 15 min.

1.8. Heat inactivate at 70°C for 15 min. Failure to inactivate can cause decreased transformation efficiencies.

1.9. Transform electrocompetent *E. coli* with 0.5 μl of the ligation reaction using the transformation protocol provided earlier. Plate 100 μl of the transformation on an LB ampicillin plate and 100 μl on an LB ampicillin + kanamycin plate.

1.10. Clones that no longer contain the kanamycin resistance (the colonies growing on the ampicillin plate) are to now be grown up and mini-prepped. The simplest way to check each clone for the loss of the kanamycin cassette is to grow each clone in duplicate tubes of broth with ampicillin *or* with ampicillin + kanamycin.

1.11. Comparing the plates, calculate how many of the colonies on the ampicillin only plate still contain the kanamycin cassette. Using this calculation, estimate how many clones will need to be grown and checked to ensure the kanamycin cassette has been successfully removed.

1.12. Set up pairs of culture tubes with 1 ml of LB ampicillin and 1 ml of LB ampicillin + kanamycin.

1.13. Use a sterile inoculation loop to pick individual colonies from the correct plate and inoculate first the tube with ampicillin and then the tube with ampicillin + kanamycin. Grow up the clones in the shaking incubator for at least 12 h.

1.14. Find the pair of tubes in which the clone has ampicillin resistance and does not have kanamycin resistance.

1.15. Use a Qiagen mini-prep kit to mini-prep the clone.

1.16. Use a *Sma*I digest to check clones and verify the loss of the kanamycin cassette. This works because *Sma*I recognizes the inner 6 bp of the *Srf*I site. Including a digestion of the parent plasmid as a control.

1.17. The next step is to transiently express your protein in more HEK 293 cells. This can be done in a new 96-well format or expressed in individual 35-mm dishes with glass coverslips. The same Lipofectamine™ 2000 protocol stated earlier can be used for the 96-well format or modified for larger 35-mm glass bottom dishes.

IV. Materials

A. PCR Amplification of the Transposon

1.1. pBNJ24.6 in Top 10f′ cells (available from the author) or a transposon that you have created.

1.2. Qiagen mini-prep spin column kit (Qiagen QIAprep Spin Mini-Prep Kit Cat. # 27104).

1.3. *Hin*dIII enzyme and 10× buffer (New England Biolabs R0104S).

1.4. PfuUltra enzyme and 10× buffer (Stratagene Cat. # 600380).

1.5. Nucleotides (New England Biolabs Cat. # N0446S).

1.6. ME primer # 63288 (5′ CTGTCTCTTATACACATCT 3′).

B. The Transposition Reaction

1.1. EZ-Tn5 Transposase Kit (Epicentre Cat. # TNP92110).

C. Transformation Requirements and Troubleshooting

1.1. Electrocompetent *E. coli* (10^9 or greater CFU). This is not a very efficient reaction, so you need to use very competent cells.

1.2. SOC medium (Invitrogen Cat. # 15544-034).
1.3. 0.1-cm gap width cuvette for electroporation (Bio-Rad Cat. # 1652089).
1.4. Culture tubes (Falcon 35-5029).
1.5. 1-LB agar, 100 μg/ml ampicillin plate.
1.6. 8-LB agar, 100 μg/ml ampicillin + 50 μg/ml kanamycin plates.

D. *E. coli* Colony Selection and Growth in a 96-Well Format

1.1. Deep 96-well growth plates (provided in the QIAprep mini-prep kit listed below).
1.2. LB broth containing 100 μg/ml ampicillin + 50 μg/ml kanamycin.
1.3. Autoclaved toothpicks or pipette tips.

E. Backing Up the Experiment: Making 10% Glycerol Stocks

1.1. Multichannel pipette (Eppendorf Research pro 50–1250 μl).
1.2. 20% sterile filtered glycerol.

F. 96-Well Mini-Preparation Purification of Plasmid DNA

1.1. Qiagen QIAprep 96 Turbo Mini-Prep Kit (Qiagen Cat. # 27191).

G. Preparation of HEK 293 Cells

1.1. HEK 293 cells (ATCC, Cat. # CRL-1573).
1.2. 75 cm^2 Sterile Flasks (Becton Dickinson BD 353136).
1.3. 5- and 10-ml sterile serological pipettes (Falcon 357543 and 357551).
1.4. Dulbececco's Modified Eagle Medium High Glucose 1× (Gibco 11965 from Invitrogen Cat. # 11965-084).
1.5. Fetal Bovine Serum (FBS; Invitrogen Cat. #16000-044).
1.6. MEM NonEssential Amino Acids Solution (Invitrogen Cat. # 11140-50).
1.7. L-Glutamine 200 mM (100×) (Invitrogen Cat. # 25030-081).
1.8. Trypsin–EDTA (Invitrogen Cat. # 25300-112).
1.9. Multichannel pipette (Eppendorf Research pro 5–100 and 50–1250 μl).
1.10. V-boats (Labcor Cat. # 730-004).

H. Transient Transfection of HEK 293 Cells in a 96-Well Format

1.1. 96-well U-bottom (or V-bottom) mixing plates (Falcon Cat. # 353918).
1.2. Lipofectamine™ 2000 Transfection Reagent (Invitrogen Cat. # 11668-019).
1.3. Opti-MEM I Reduced Serum Medium (Invitrogen Cat. # 31985-070).

I. Screening Live HEK 293 Cells for GFP Fluorescence

1.1. Epifluorescence microscope fitted with FITC optics (480-nm excitation, 530-nm long pass emission) and a 40× immersion lens with at least a 0.9 NA.

1.2. 96-well Optical CVG Sterile, w/lid plates with coverglass bottom (Nalge Nunc International, Cat. # 164588).

J. Removing the Selection Cassette with Restriction Digestion and Re-Ligation

1.1. *Srf*I enzyme and universal buffer (Stratagene Cat. # 501064).

1.2. 1% agarose gel (Cambrex Cat. # 50000) and gel electrophoresis boat.

1.3. λDNA *Hin*dIII digest lane marker (New England Biolabs N3012L).

1.4. Qiagen MinElute Gel Extraction Kit (Qiagen Cat. # 28604).

1.5. Fast Link Ligation Kit (Epicenter Cat. # LK6201H).

1.6. Electrocompetent *E. coli* (10^9 or greater CFU).

1.7. 0.1-cm gap width cuvette for electroporation (Bio-Rad Cat. # 1652089).

1.8. SOC medium (Invitrogen Cat. # 15544-034).

1.9. Culture Tubes (Falcon Cat. # 35-5029).

1.10. LB agar 100 μg/ml ampicillin plates.

1.11. LB agar 100 μg/ml ampicillin + 50 μg/ml kanamycin plates.

1.12. LB broth with 100 μg/ml ampicillin + 50 μg/ml kanamycin.

1.13. LB broth with 100 μg/ml ampicillin.

1.14. Inoculation loop.

1.15. Qiagen QIAprep mini-prep spin kit (Qiagen Cat. # 27106).

V. Discussion

Building functional, fluorescent fusion proteins involves a combination of guesswork, strategic decisions, and a bit of luck. One approach is to use structural and functional information about the proteins to rationally design a series of fusions. This has certainly been a successful strategy in designing important fusion proteins currently used as biosensors in the field (see, e.g., Chapter 16 by Shimozono and Miyawaki, this volume). In some ways, this is analogous to the process of reverse genetics; models of what a gene is, and how it works, are used to develop perturbation schemes that should better test gene function.

All too often the process of reverse genetics reveals that our models of gene function are less than complete. Forward genetics on the other hand, the process of using random mutation and functional screens, makes few assumptions and simply searches for interesting explanations. Transposons have been powerful tools for forward genetics because they can readily create and mark mutations in the genomes of many different genetic organisms (Bushman, 2002). In the process described here, however, the transposable element is used not to disrupt genes or proteins, but rather to find just the right place to insert a fluorescent protein. The same efficiency that makes them powerful tools for mutagenesis can be harnessed for exploring large potential spaces for the few feasible solutions.

What are the limits of this process, what is the resolution you can obtain? These are the questions that arise after screening the first 1000 clones. How many is enough? The limit in the process usually involves labor and expense. The probability of encountering the same transposon insertion site rises as an almost exponential function, and this drives the cost of the screen up to the point that it is no longer worth searching for new proteins.

Acknowledgments

The authors thank all of the participants of the Cold Spring Harbor Lab summer course *Advanced Techniques in Molecular Neuroscience* for their creative feedback and ideas on how to improve the protocols in this chapter.

References

Ason, B., and Reznikoff, W. S. (2004). DNA sequence bias during Tn5 transposition. *J. Mol. Biol.* **335**(5), 1213–1225.

Ataka, K., and Pieribone, V. A. (2002). A genetically targetable fluorescent probe of channel gating with rapid kinetics. *Biophys. J.* **82**(1), 509–516.

Bushman, F. (2002). "Lateral DNA Transfer: Mechanisms and Consequences." Cold Spring Harbor Laboratory Press, Cold Spring Harbor, NY.

Butler, H., and Hughes, T. E. (2006). A FRET based strain gauge: a new Tn5 transposon for randomly inserting CFP::Venus pairs in proteins. *Biophys. J.* (Supplement) Abstract #137.

Doi, N., and Yanagawa, H. (1999). Insertional gene fusion technology. *FEBS Lett.* **457**(1), 1–4.

Ehrmann, M., Boyd, D., and Beckwith, J. (1990). Genetic analysis of membrane protein topology by a sandwich gene fusion approach. *Proc. Natl. Acad. Sci. USA* **87**(19), 7574–7578.

Giraldez, T., Hughes, T. E., and Sigworth, F. J. (2005). Generation of functional fluorescent BK channels by random insertion of GFP variants. *J. Gen. Physiol.* **126**(5), 429–438.

Guerrero, G., Siegel, M. S., Roska, B., Loots, E., and Isacoff, E. Y. (2002). Tuning FlaSh: Redesign of the dynamics, voltage range, and color of the genetically encoded optical sensor of membrane potential. *Biophys. J.* **83**(6), 3607–3618.

Hoekstra, M. F., Burbee, D., Singer, J., Mull, E., Chiao, E., and Heffron, F. (1991). A Tn3 derivative that can be used to make short in-frame insertions within genes. *Proc. Natl. Acad. Sci. USA* **88**(12), 5457–5461.

Hughes, T. E., Zhang, H., Logothetis, D. E., and Berlot, C. H. (2001). Visualization of a functional Galpha q-green fluorescent protein fusion in living cells. Association with the plasma membrane is disrupted by mutational activation and by elimination of palmitoylation sites, but not be activation mediated by receptors or $AlF4^-$. *J. Biol. Chem.* **276**(6), 4227–4235.

Huh, Won-Ki., James, V. Falvo., Luke, C. Gerke., Adam, S. Carroll., Russell, W. Howson., Jonathan, S. Weissman., and Erin, K. O'Shea. (2003). Global analysis of protein localization in budding yeast. *Nature* **425**(2003), 686–91.

Marshall, J., Molloy, R., Moss, G. W., Howe, J. R., and Hughes, T. E. (1995). The jellyfish green fluorescent protein: A new tool for studying ion channel expression and function. *Neuron* **14**(2), 211–215.

Merkulov, G. V., and Boeke, J. D. (1998). Libraries of green fluorescent protein fusions generated by transposition *in vitro*. *Gene* **222**(2), 213–222.

Reznikoff, W. S. (2003). Tn5 as a model for understanding DNA transposition. *Mol. Microbiol.* **47**(5), 1199–1206.

Reznikoff, W. S. (2006). Tn5 transposition: A molecular tool for studying protein structure-function. *Biochem. Soc. Trans.* **34**(Pt. 2), 320–323.

Robbie, Mealer., Arnd Pralle., Ehud Isacoff., and Thomas Hughes. (2006). Scanning the Shaker channel subunit with fragments of GFP: Can complementing fragments on adjacent subunits produce a fluorophore? *Biophys. J.* (Supplement) Abstract # 2266.

Ross-Macdonald, P., Coelho, P. S., Roemer, T., Agarwal, S., Kumar, A., Jansen, R., Kei-Hoi, C., Sheehan, A., Symoniatis, D., Umansky, L., Heidtman, M., Nelson, F. K., *et al.* (1999a). Large-scale analysis of the yeast genome by transposon tagging and gene disruption. *Nature* **402**(6760), 413–418.

Ross-Macdonald, P., Sheehan, A., Friddle, C., Roeder, G. S., and Snyder, M. (1999b). Transposon mutagenesis for the analysis of protein production, function, and localization. *Meth. Enzymol.* **303,** 512–532.

Sheridan, D. L., Berlot, C. H., Robert, A., Inglis, F. M., Jakobsdottir, K. B.,, *et al.* (2002). A new way to rapidly create functional, fluorescent fusion proteins: Random insertion of GFP with an *in vitro* transposition reaction. *BMC Neurosci.* **3**(1), 7.

Sheridan, D. L., and Hughes, T. E. (2004). A faster way to make GFP-based biosensors: Two new transposons for creating multicolored libraries of fluorescent fusion proteins. *BMC Biotechnol.* **4**(1), 17.

Sheridan, D. L., Robert, A., Cho, C. H., Howe, J. R., and Hughes, T. E. (2006). Regions of alpha-amino-5-methyl-3-hydroxy-4-isoxazole propionic acid receptor subunits that are permissive for the insertion of green fluorescent protein. *Neuroscience* **141**(2), 837–849.

Siegel, M. S., and Isacoff, E. Y. (1997). A genetically encoded optical probe of membrane voltage. *Neuron* **19**(4), 735–741.

Stryer, L., and Haugland, R. P. (1967). Energy transfer: A spectroscopic ruler. *Proc. Natl. Acad. Sci. USA* **58**(2), 719–726.

Wang, S., and Hazelrigg, T. (1994). Implications for bcd mRNA localization from spatial distribution of exu protein in *Drosophila* oogenesis. *Nature* **369**(6479), 400–403.

CHAPTER 3

Fluorescent Proteins for Photoactivation Experiments

Jennifer Lippincott-Schwartz and George H. Patterson

Cell Biology and Metabolism Branch
National Institute of Child Health and Human Development
National Institutes of Health
Bethesda, Maryland 20892

Abstract

The discoveries, improvements, and alterations of fluorescent protein (FP) variants are having a profound impact on the ability of investigators to observe and quantify the behavior of proteins within cells and organisms. Among the most promising of FPs are photoactivatable fluorescent proteins (PA-FPs). Invisible at the imaging wavelength until activated by irradiation at a different wavelength,

0091-679X/08 $35.00
DOI: 10.1016/S0091-679X(08)85003-0

PA-FPs allow the controlled highlighting of distinct molecular populations within the cell. This chapter introduces the different types of PA-FPs and discusses their use for monitoring protein movement, protein turnover, protein interactions, and high-resolution protein localization.

I. Why Use a Fluorescent Protein?

Originally, light microscopy techniques employing fluorescence relied mainly on fluorescence from small organic dyes attached by means of antibodies to proteins, but this required cell permeabilization and fixation. Later, fluorophores capable of directly binding to organelles, nucleic acids, and particular ions in living cells were used, but these were largely nonspecific. More recently, the discovery, gene cloning (Prasher *et al.*, 1992), and heterologous expression of fluorescent proteins (FPs) have provided a means for studying proteins of interest (Chalfie *et al.*, 1994). In this approach, straightforward methods of molecular cloning are used to attach an FP to a protein of interest. Expression of the genetic fusion protein within cells then results in visible fluorescence without requiring any cofactors except for oxygen. Because fluorescence from the construct can be observed within cells without fixation or addition of a secondary label or reactant, it becomes possible both to optimally localize proteins within the cell and to visualize their dynamics over time.

II. Why Use a Photoactivatable Fluorescent Protein?

The reasons for using FPs in general also hold for using photoactivatable fluorescent proteins (PA-FPs), but the "activation" characteristic of PA-FPs offers many advantages over normal FPs and has led to the development of novel imaging strategies. One advantage is that PA-FPs are observed as fluorescent signals over dark backgrounds. This allows subpopulations of proteins, organelles, or cells that have been highlighted to be monitored temporally and/or spatially. This approach is distinct from usual highlighting techniques involving photobleaching. PA-FP tagged molecules are highlighted directly rather than indirectly as occurs by photobleaching all the molecules surrounding a particular population. Photoactivation can often be done more rapidly and results in higher contrast over the nonhighlighted pool compared with photobleaching. By varying the amount of light used for photoactivation, it becomes possible to control the number of fluorescent molecules used for a given experiment. In addition, protein synthesis or protein folding during an experiment does not factor into the observed results when using PA-FPs. Photobleaching techniques performed on non–PA-FPs, on the other hand, must contend with recovery signals contaminated by newly synthesized or newly folded proteins.

III. Survey of Photoactivatable Fluorescent Proteins

Proteins from nine species have been reported with several variations or approaches for their use as PA-FPs (Lukyanov *et al.*, 2005) (Table I). For their description, these molecules have been categorized based on the species of origin, on the spectral characteristics of their activated and nonactivated states (green vs red fluorescence), or on the reversibility of the activated state (irreversible vs reversible).

Table I
Selected Photoactivatable Fluorescent Proteins

Proteins	Wavelengths (nm)		Organisms	References
	λ_{ex}	λ_{em}		
PA-GFP	400 (Pre)	515	*Aequorea victoria*	Patterson and Lippincott-Schwartz (2002)
	504 (Post)	517		
Kaede	508 (Pre)	518	*Trachyphyllia geoffroyi*	Ando *et al.* (2002)
	572 (Post)	582		
KiKGR	507 (Pre)	517	*Favia favus*	Tsutsui *et al.* (2005)
	583 (Post)	593		
PAmRFP1-1	578 (Post)	605	*Discosoma* sp.	Verkhusha and Sorkin (2005)
PAmRFP1-2	578 (Post)	605	*Discosoma* sp.	Verkhusha and Sorkin (2005)
PAmRFP1-3	578 (Post)	605	*Discosoma* sp.	Verkhusha and Sorkin (2005)
EosFP	506 (Pre)	516	*Lobophyllia hemprichii*	Wiedenmann *et al.* (2004)
	571 (Post)	581		
d1EosFP	505 (Pre)	516	*Lobophyllia hemprichii*	Wiedenmann *et al.* (2004)
	571 (Post)	581		
d2EosFP	506 (Pre)	516	*Lobophyllia hemprichii*	Wiedenmann *et al.* (2004)
	569 (Post)	581		
mEosFP	505 (Pre)	516	*Lobophyllia hemprichii*	Wiedenmann *et al.* (2004)
	569 (Post)	581		
DRONPA	503 (Post)	518	*Pectiniidae* spp.	Ando *et al.* (2004)
Dendra	486 (Pre)	505	*Dendronephthya* sp.	Gurskaya *et al.* (2006)
	558 (Post)	575		
Dendra2	486 (Pre)	505	*Dendronephthya* sp.	
	558 (Post)	575		
asFP595	572 (Post)	595	*Anemonia sulcata*	Lukyanov *et al.* (2000)
KFP1	580 (Post)	600	*Anemonia sulcata*	Chudakov *et al.* (2003)
PS-CFP	402 (Pre)	468	*Aequorea coerulescens*	Chudakov *et al.* (2004)
	490 (Post)	511		
PS-CFP2	400 (Pre)	470	*Aequorea coerulescens*	
	490 (Post)	511		

(Pre) represents the major peak before photoactivation.
(Post) represents the major peak after photoactivation.

A. Photoactivatable Fluorescent Proteins: *Aequorea victoria* GFP

One early approach to develop a PA-FP relied on the phenomenon that several *A. victoria* variants, wild-type green fluorescent protein (wtGFP), GFPmut1, 2, and 3 (Cormack *et al.*, 1996), S65T (Heim *et al.*, 1995), I167T (Heim *et al.*, 1994), and GFPuv (Crameri *et al.*, 1996), convert into red fluorescent species upon irradiation with 488 nm light (Elowitz *et al.*, 1997; Sawin and Nurse, 1997). This "photoactivation" produced high contrast with background fluorescence and worked for monitoring protein diffusion in bacteria (Elowitz *et al.*, 1997) and mitochondria matrix (Jakobs *et al.*, 2003). However, because low oxygen conditions were required for the photoconversion in this approach, its use among cell biologists has been limited.

Another early approach exploited the photoinduced conversion (photoconversion) that occurs within the wtGFP under aerobic conditions (Yokoe and Meyer, 1996). The wtGFP chromophore population exists as neutral phenols when Y66 is protonated and as anionic phenolates when Y66 is deprotonated, and these different populations produce a major absorbance peak at ~397 nm and a minor absorbance peak at 475 nm, respectively. The chromophore undergoes a proton transfer and converts into the anionic form after irradiation (Chattoraj *et al.*, 1996; Creemers *et al.*, 1999; Tsien, 1998). The resulting absorbance increase at the minor peak leads to an increase in the fluorescence when excited at this wavelength. Unfortunately, owing to the high initial background, photoconversion of the wtGFP results in modest, approximately threefold, fluorescence increase (Patterson and Lippincott-Schwartz, 2002; Tsien, 1998; Yokoe and Meyer, 1996), making it impractical as a highlighter.

An advance in the development of a PA-FP came when wtGFP was mutated in an effort to obtain a GFP variant with decreased absorbance at 488 nm that could undergo photoconversion (Patterson and Lippincott-Schwartz, 2002). The T203I mutant of GFP has a mostly neutral phenol chromophore population, giving a major peak at ~400 nm and little absorbance at ~488 nm (Heim *et al.*, 1994; Ehrig *et al.*, 1995) that provided the starting point for this strategy. Substitution of the threonine at the 203 position leads to several proteins with spectral characteristics similar to that of the T203I mutant but with the additional capability to undergo photoconversion (Patterson and Lippincott-Schwartz, 2002). The T203H mutant (Fig. 1A) gave the highest fluorescence contrast compared with its nonphotoactivated protein after photoactivation (Patterson and Lippincott-Schwartz, 2002). Named photoactivatable green fluorescent protein (PA-GFP) it displays similar overall protein properties as the original *A. victoria* GFP and thus remains widely used as green PA-FP.

B. Photoactivatable Fluorescent Proteins: DsRed Fluorescent Protein

DsRed proteins from *Discosoma* coral initially form green fluorescent molecules, which mature in to red fluorescent proteins (RFPs). Since it is an obligate tetramer, the proteins exist as mixed populations of immature green fluorescent molecules

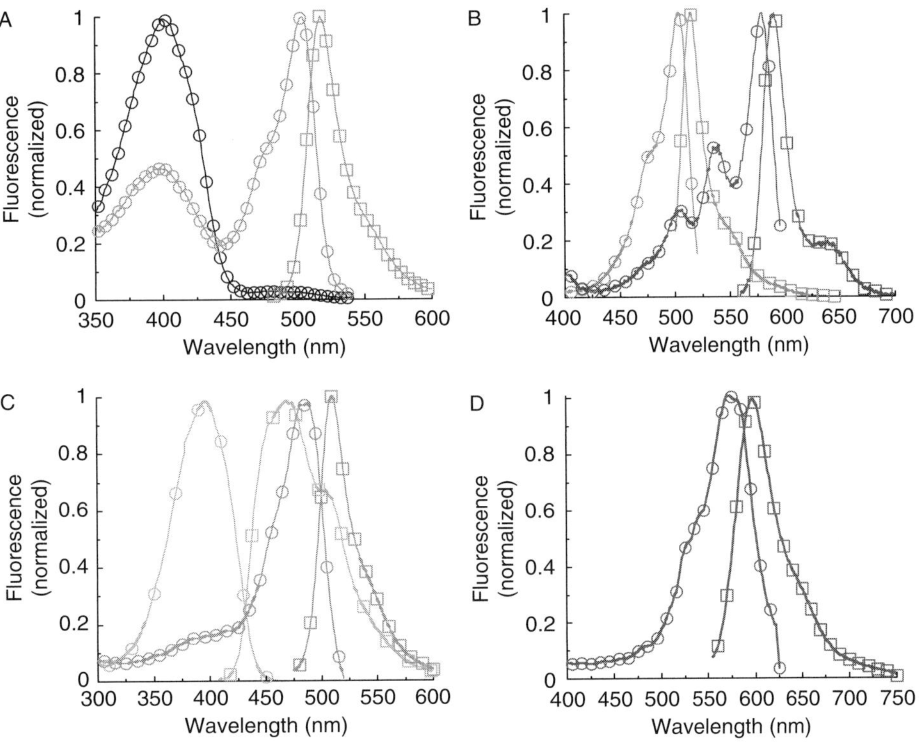

Fig. 1 Fluorescence excitation and emission spectra of selected photoactivatable fluorescent proteins (PA-FPs) show the diversity that is now available from this class of fluorescent proteins. (A) The excitation spectra (open circles) for photoactivatable green fluorescent protein (PA-GFP) are shown in blue before photoactivation and in green after photoactivation. The emission spectrum (green, open squares) was acquired with excitation at 475 nm on photoactivated PA-GFP. (B) The spectra for KiKGR are shown in green before photoactivation and in red after photoactivation. Open circles represent excitation spectra and open squares indicate emission spectra. (C) The spectra for photo-switchable cyan fluorescent protein 2 (PS-CFP2) are shown in cyan before photoactivation and in green after photoactivation. Open circles represent excitation spectra and open squares indicate emission spectra. (D) The excitation spectra (open circles) and emission spectrum (open squares) are shown for photoactivated kindling fluorescent protein (KFP1). (See Plate no. 5 in the Color Plate Section.)

and mature red fluorescent molecules. The close proximity of the two forms and the overlap of the green emission with the excitation of the red absorbance lead to Förster resonance energy transfer (FRET) between the two. Photobleaching of the red molecules dequenches the green form and leads to an increase in its fluorescence (Marchant *et al.*, 2001). After some efforts, the tetrameric DsRed was eventually converted into a monomeric red fluorescent protein, mRFP1 (Campbell *et al.*, 2002), which was then converted into a series of PA-FPs, PAmRFP1-1, PAmRFP1-2, and PAmRFP1-3 (Verkhusha and Sorkin, 2005). The brightest, PAmRFP1-1, gives ~70-fold increase in red fluorescence upon ultraviolet light excitation (Verkhusha and Sorkin, 2005). Since it has little fluorescence signal at any wavelength before activation, it offers much promise for use in multifluorescent protein experiments, including double photoactivation studies with green PA-FPs.

C. Photoactivatable Fluorescent Proteins: Green-to-Red Photoconversions

Many of the naturally occurring and engineered PA-FPs exhibit a spectral shift from a GFP to a RFP (Fig. 1B). The first of these to be discovered was Kaede, from a stony coral, *Trachyphyllia geoffroyi* (Ando *et al.*, 2002). Kaede absorbs maximally at 508 nm and emits at 518 nm. When photoactivated by irradiation at ~400 nm, however, it exhibits absorbance at 572 nm and emission at 582 nm afterward. Since both the excitation and emission peaks are shifted, ratio imaging results in a >2000-fold increase in the red-to-green ratio. Despite its impressive signal, Kaede is limited as a protein tag because it forms tetramers (Ando *et al.*, 2002). Hopefully this characteristic will be eventually alleviated by developing a monomeric form of Kaede as was developed with the DsRed protein (Campbell *et al.*, 2002).

KiKGR is another coral fluorescent protein that undergoes green-to-red photoactivation (Tsutsui *et al.*, 2005) (Fig. 1B). The original KikG from *Favia favus* did not exhibit photoactivatable properties. To obtain a photoactivatable form, Miyawaki and colleagues engineered KikG into the KikGR variant based on structural characterizations of Kaede. KikGR, which is also a tetramer, lacks the brightness of Kaede *in vitro*, but exhibits several fold more fluorescence than Kaede when expressed in cells (Tsutsui *et al.*, 2005).

EosFP from another stony coral, *Lobophyllia hemprichii*, also exhibits a green-to-red fluorescence photoconversion after exposure to ultraviolet or near ultraviolet light (Wiedenmann *et al.*, 2004). EosFP has a preactivated excitation maximum at 506 nm with emission at 516 nm, which shift to 571 and 581 nm, respectively, after photoactivation. EosFP was engineered into two dimeric forms, d1EosFP and d2EosFP, from its original tetrameric form and the combination of the mutations produced monomeric mEosFP. Because mEosFP has reduced tendency to form intermolecular associations or aggregates, it is preferable to other FPs in this subclass in this respect. However, mEosFP forms a fluorescent molecule inefficiently when expressed at 37 °C and requires a lower temperature incubation for proper folding (Wiedenmann *et al.*, 2004).

The latest addition to this category is Dendra from *Dendronephthya* sp., which gives up to 4500-fold increase in the red-to-green ratio after photoactivation (Gurskaya *et al.*, 2006). Uniquely, Dendra can be activated with potentially less phototoxic wavelengths (~488 nm) in addition to the ~400 nm light required by the other green-to-red PA-FPs. Although Dendra can efficiently develop fluorescence when expressed at 37 °C, a mutated version, Dendra2, offered by Evrogen reportedly has improved folding efficiency compared to the original.

D. Photoactivatable Fluorescent Proteins: Cyan-to-Green Photoconversion

Photoswitchable-cyan fluorescent protein (PS-CFP) similarly displays a spectral shift of both absorbance and emission, but this is a change from a cyan-to-green fluorescent protein (Chudakov *et al.*, 2004) (Fig. 1C). It was originally derived from nonfluorescent acGFPL (Gurskaya *et al.*, 2003) and now exhibits initial excitation at 402 nm and emission at 468 nm (Chudakov *et al.*, 2004) which shift to 490 and 511 nm, respectively, and give ~1500-fold increase in the green-to-cyan fluorescence ratio. As a monomeric protein, PS-CFP has been imaged when tagged to actin and dopamine transporter (Chudakov *et al.*, 2004). An improved version, PS-CFP2, which develops fluorescence more efficiently and is thus brighter, is also available (Table I).

E. Photoactivatable Fluorescent Proteins: Reversible

Remarkably, some PA-FPs display the fascinating property of switching "on" and "off" by excitation at different wavelengths. While the on–off switching was observed earlier in single molecules of GFP T203 mutants (Dickson *et al.*, 1997), reversible PA-FPs show this behavior across the entire population. Less prominent is the red fluorescence (λ_{max} ~595 nm) of asFP595, a protein isolated from the sea anemone, *Anemonia sulcata*. It can be enhanced by exposure to green light and quenched by exposure to blue light (Lukyanov *et al.*, 2000). Initially having little practical use as a marker because of its slow maturation and low quantum yield, the reversible capabilities of asFP595 have been employed in a new fluorescence microscopy technique called reversible saturable optical fluorescence transitions (RESOLFT) to break the diffraction barrier (Hofmann *et al.*, 2005).

Kindling fluorescent protein (KFP1) is a variant of asFP595 that retains the reversible photoactivation capability but can also be irreversibly photoactivated (Chudakov *et al.*, 2003). Activation with 532 nm laser light increases its red fluorescence ~30-fold and the reversion to the inactive state depends on the intensity of the photoactivation light (Fig. 1D). Lower level excitation (~1 W/cm^2 for 2 min) produced red KFP1 fluorescence that relaxed to the nonfluorescent state with a halftime of ~50 s whereas 200 times this excitation (~20 W/cm^2 for 20 min) irreversibly photoactivated ~50% of the population. Its oligomerization into a tetramer and slow maturation time ($t_{1/2}$ ~5 h) are certainly pitfalls, but the use of potentially less phototoxic green activation light (532 nm) makes it attractive for use in live cell imaging experiments and one of the few proteins that can be activated by something other than ~400 nm light (Chudakov *et al.*, 2003).

Perhaps the most intriguing and well known of the reversible PA-FPs is the one derived from *Pectimiidae* named Dronpa (Ando *et al.*, 2004). Its spectra resemble those of PA-GFP (Fig. 1A), except it has a much higher extinction coefficient and is reversible. Initially, Dronpa displays green fluorescence with excitation at 503 nm and emission at 518 nm. Intense excitation at 490 nm (0.4 W/cm^2) results in a loss of the absorbance at 503 nm. But, upon irradiation at 400 nm (0.14 W/cm^2), the green fluorescence is restored. Dronpa can undergo this on–off cycling 100 times with only a loss of 25% of the original fluorescence. The molecule is monomeric and has an extinction coefficient (95,000 M^{-1} cm^{-1}) and a quantum yield (0.85) that make it one of the brightest FPs available. Its photoactivation capabilities may make it useful as a marker for any photoactivation experiment; the reversible nature of the fluorescence allows the same photoactivation experiment to be repeated multiple times within the same region of interest (Ando *et al.*, 2004). More intriguingly, the brightness and ease of photoswitching of Dronpa make it suitable for high-resolution microscopy techniques, such as RESOLFT (Hofmann *et al.*, 2005).

IV. Uses of Photoactivatable Fluorescent Proteins

A. Protein Dynamics

The common characteristic of all PA-FPs is that they are observed as fluorescent signals over darker backgrounds. As such, subpopulations of proteins, organelles, or cells can be highlighted and their dynamics monitored temporally and spatially within the whole population. Thus, molecules can be activated in one region and monitored as they move to another (Fig. 2A). Often, a disadvantage in photoactivation is that the signal generated in one region dissipates as it moves throughout surrounding regions. To compensate for this signal dilution, it is possible to repeatedly photoactivate the same region while monitoring the fluorescence increases in another region (Kim *et al.*, 2006). This is a variation of the photobleaching technique called fluorescence loss in photobleaching (FLIP) that relies on repeatedly photobleaching within the same region and monitoring fluorescence loss in another region.

Photoactivation can also be used to monitor diffusion of molecules in a manner similar to another photobleaching technique, fluorescence recovery after photobleaching (FRAP). However, a distinct difference between FRAP and photoactivation is that rather than monitoring the recovery of fluorescence within a region, the decrease in fluorescence is monitored (Fig. 2B). This potentially provides two advantages compared with FRAP when monitoring rapid diffusion. First, photoactivation can often produce higher contrast between highlighted and nonhighlighted molecules more rapidly and efficiently than photobleaching. (Of course, this is dependent on the fluorophore being photobleached and the laser power available for photobleaching.) Second, the highest signal-to-background ratio is found in the earliest points of the photoactivation experiment, whereas the early time points of a photobleaching experiment have the lowest signal-to-background ratio. Thus, the photoactivation experiment may provide new insight into the dynamics of a rapidly diffusing protein of interest.

B. Fluorescence Pulse-Labeling

During photobleaching techniques, newly synthesized FPs can contaminate a slow recovery of the fluorescence unless synthesis is inhibited, whereas photoactivation produces signal from a synthesized and properly folded FP. Thus, proteins

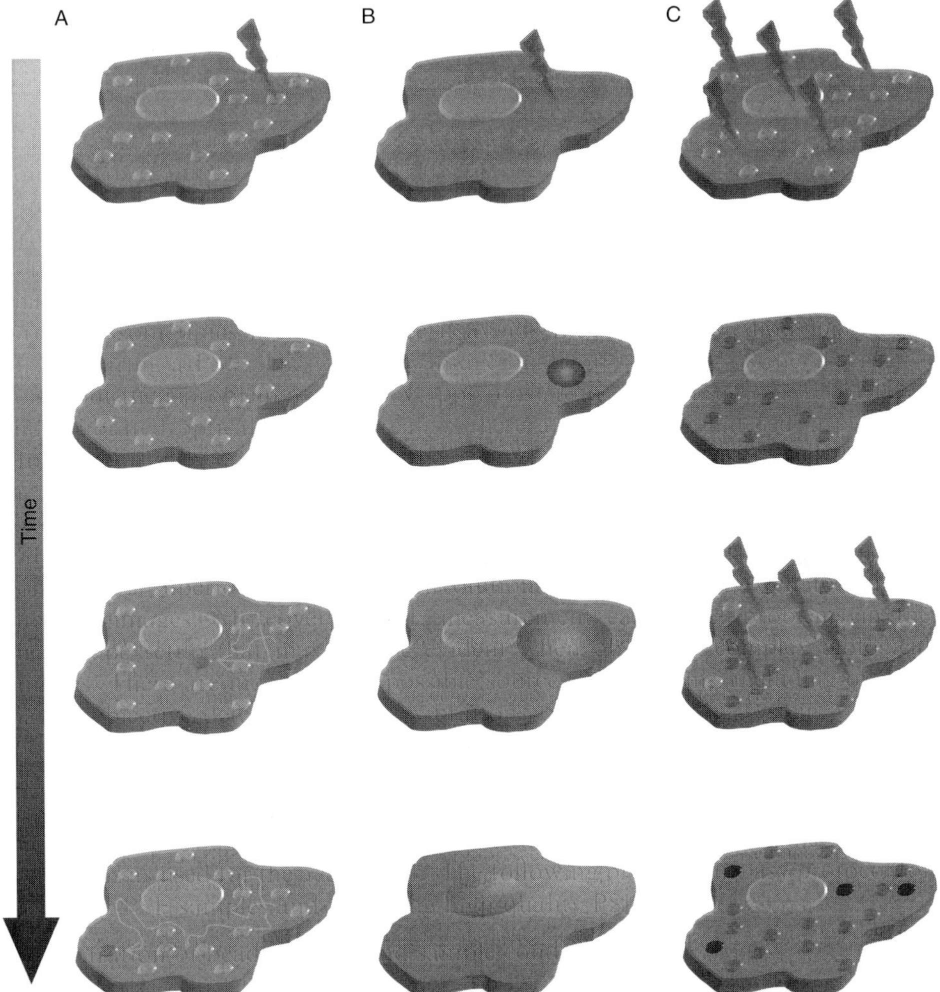

Fig. 2 These cartoons describe some of the imaging techniques that are capable using photoactivatable fluorescent proteins. (A) Proteins or organelles can be tracked by photoactivating a subpopulation and monitoring its movement with time. (B) Protein diffusion can be monitored by photoactivating a small region and monitoring either the decrease in fluorescence within this region or the area encompassed by the photoactivated fluorescence with time. (C) Synthesis of organelles, such as peroxisomes, can be monitored by photoactivating the population at an early time point to label the existing structures and then rephotoactivating at a later time point to highlight newly formed structures (shown in red). (See Plate no. 6 in the Color Plate Section.)

that are subsequently synthesized fold into the inactive form and do not fluoresce at the wavelengths indicative of photoactivated proteins (Fig. 2C). This characteristic of PA-FPs has been used to monitor the fate of cells during embryogenesis with KFP1 (Chudakov *et al.*, 2003), the movements of chromatin loci in

Drosophila embryos (Post *et al.*, 2005), and the formation of new peroxisomes in cell culture (Kim *et al.*, 2006). In the latter studies, cells expressing a PA-GFP tagged peroxisome protein were photoactivated, cultured for a number of hours, and photoactivated again. Since peroxisomes that had not been previously photoactivated were evident after the second photoactivation, these experiments indicated that new peroxisomes can form *de novo* rather than from existing peroxisomes.

This characteristic simplifies interpretation of many trafficking experiments, and also introduces an approach to monitor protein turnover. The basis for a fluorescence pulse label is similar to a radioactive pulse label. A brief activation with the activation light of a tagged protein of interest labels a population of molecules similar to the incubation of cells with S^{35}-labeled methionine and the degradation of the protein is monitored by imaging the loss of fluorescence as the loss of radioactivity is monitored in biochemical pulse-labeling. It is important to recognize that the population of fluorescent molecules labeled under these conditions includes only those molecules that are fully synthesized and properly folded, whereas radioactive labeling can highlight either the entire population or just the newly synthesized molecules. In addition, the FP must be degraded along with the protein of interest for proper readout. Biochemical pulse-labeling thus should be carried out in parallel with fluorescent pulse-label experiments to verify results. An advantage of fluorescence pulse-labeling for the study of protein turnover is that questions related to the path and location of degradation can be visualized directly in a single living cell. The temporal resolution of fluorescence pulse-labeling is essentially limited by the instrument parameters (usually requiring milliseconds–seconds for activation) and has subcellular spatial resolution (dependent on the optics used for imaging).

C. Photoquenching Fluorescence Resonance Energy Transfer

Fluorescent proteins have contributed to the study of protein–protein interactions within living cells by FRET (Day *et al.*, 2001). Since the FRET requires that the distance between the donor and acceptor fluorophores be <10 nm, the energy transfer can be interpreted as an interaction of the tagged proteins of interest. The readout for energy transfer in many such experiments is enhanced fluorescence of the acceptor molecule during excitation of the donor. However, this method requires numerous control experiments to account for donor fluorescence in the acceptor channel as well as direct excitation of the acceptor (Day *et al.*, 2001). In addition, discerning dynamic interactions when the molecules of interest have reached a steady state is difficult. The method referred to as acceptor photobleaching relies on the destruction of the acceptor molecule while monitoring increases in donor fluorescence (Day *et al.*, 2001). The advantage is that each experiment contains its own control but lacks the capability for kinetic measurements since the acceptor can usually be photobleached only once.

An interesting approach to the study of protein–protein interactions in cells has been termed photoquenching FRET (PQ-FRET) and relies on the use of photoactivated PA-GFP as an acceptor to quench the fluorescence of a donor fluorophore, cyan fluorescence protein (CFP) (Demarco *et al.*, 2006). In the inactive state, the absorbance spectrum of PA-GFP has little overlap with the emission spectrum of CFP, but after photoactivation exhibits spectral overlap that makes FRET possible (Fig. 3). Using PQ-FRET, Demarco *et al.* were able to monitor both the movement of the proteins and the interactions of the heterochromatin protein (HP1α) and the transcription factor CCAAT/enhancer binding protein alpha (C/EBPα) within distinct domains of the nuclei of living cells (Demarco *et al.*, 2006). Thus, PQ-FRET offers the capability to monitor the dynamic interactions of molecules at steady state within different regions of the cell. Although the use of other PA-FPs in PQ-FRET has not been reported, this technique should be expandable by using them in combination with the plethora of FPs that are now available.

D. Photoactivated Localization Microscopy

Fluorescent proteins tagged to a protein of interest offer spatial information that is accurate within a few nanometers. Yet the limit of resolution for conventional optical techniques is $\sim$100 $\times$ more than the size of the FPs. Recently, a new optical technique, termed photoactivated localization microscopy (PALM), was introduced that provides near molecular resolution of proteins (Betzig *et al.*, 2006). PALM involves imaging of molecules individually and then localizing them to high precision by determining their centers of fluorescence emission. This is achieved by performing a statistical fit of their measured photon distributions with ideal point spread functions (PSFs). When the background noise is negligible compared with the molecular signal, the error in the fitted position is $\sigma_{xy} = s/\sqrt{N}$, where s is the standard deviation of a Gaussian approximating the PSF of the objective ($\approx$200 nm for light of $\lambda = 500$ nm wavelength), and N is the total number of detected photons (Cheezum *et al.*, 2001; Thompson *et al.*, 2002). The potential for precisely determining a molecule's position is remarkable. For example, if 10,000 photons can be collected from a single fluorophore molecule before it bleaches, its localization can be determined to nearly 1 nm precision (Churchman *et al.*, 2005; Shaner *et al.*, 2004; Yildiz *et al.*, 2003).

Determination of protein localization using PALM is usually limited to molecules that are separated by the distance required of conventional optics ($\sim$250 nm). However, within most fluorescent specimens, hundreds or thousands of molecules may be present within one point spread function. PALM approaches this problem using both spectral and temporal means to isolate individual molecules. To construct an image involving thousands of molecules within the diffraction limit of conventional optics, PALM relies on the serial photoactivation of a small

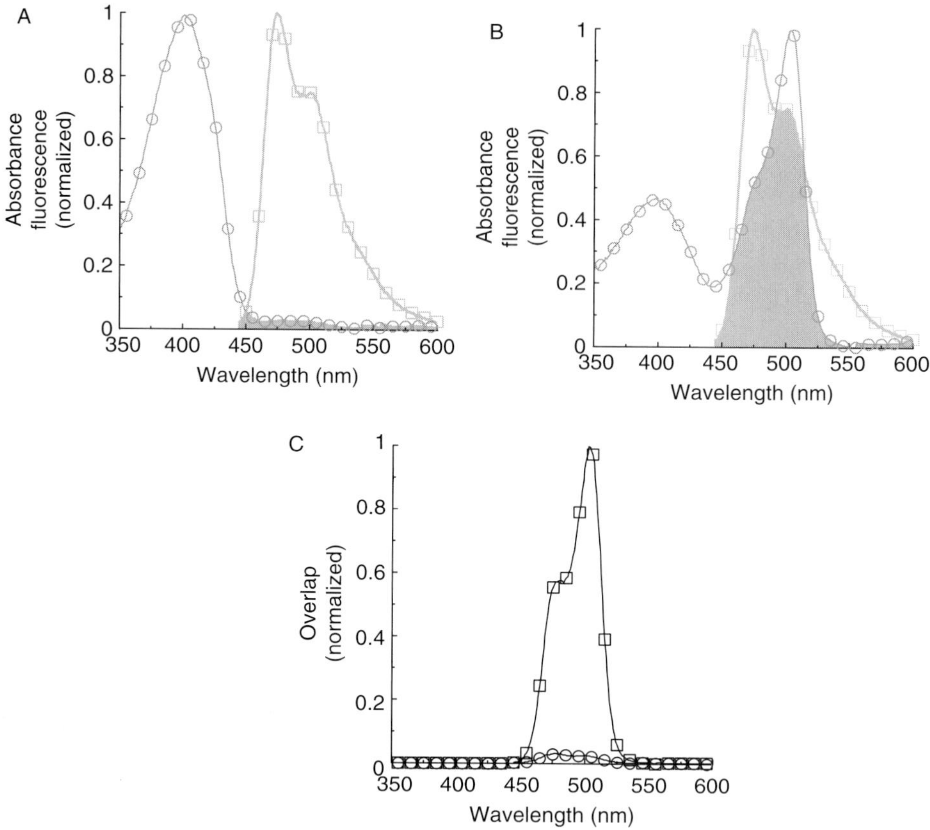

Fig. 3 The principle behind photoquenching fluorescence resonance energy transfer relies on the overlap of the acceptor, photoactivatable green fluorescent protein (PA-GFP), absorbance with the emission of the donor, enhanced cyan fluorescent protein (ECFP). (A) The absorbance spectrum of PA-GFP before photoactivation is shown in green (open circles) with the emission spectrum of ECFP (open blue squares). The gray-shaded area indicates the spectral overlap. (B) The absorbance spectrum of PA-GFP after photoactivation is shown in green (open circles) with the emission spectrum of ECFP (open blue squares). The gray-shaded area again indicates the spectral overlap. (C) The calculated overlap integrands of PA-GFP and ECFP before (open circles) and after (open squares) photoactivation. The integrands are the product of the normalized ECFP fluorescence, the absorbance (in units of extinction coefficient, M^{-1} cm^{-1}), and the wavelength (λ^4). This graph was normalized to the post-photoactivation integrand, which has >20× the overlap of the prephotoactivation integrand. (See Plate no. 7 in the Color Plate Section.)

population of PA-FPs and their subsequent imaging until photodestruction (Fig. 4A–C). The photoactivation of small populations ensures that the density of imaged molecules remains much less than one molecule per normal resolution limit and their subsequent photodestruction allows a new subset of molecules to be imaged. By fitting the fluorescent signal from each molecule to a Gaussian PSF of free center coordinates x_0, y_0, the coordinates x_m, y_m for the location of the

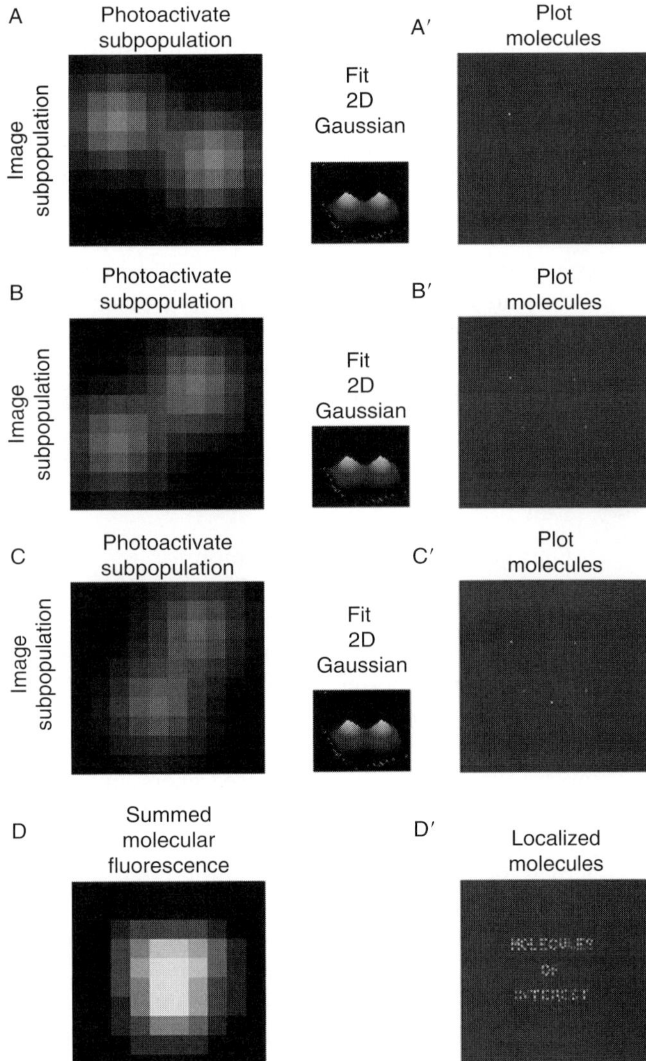

Fig. 4 The principle behind PALM requires serial photoactivation and imaging of small subpopulations of tagged molecules of interest in concert with determination of each molecule's localization by 2D Gaussian fit of the photon distribution. This procedure is simulated here on images representing an area of 1 μm^2 containing ~150 molecules giving enough photons to be localized to 10 nm certainty. (A) By photoactivating a sparse subpopulation (<1 molecule per resolution limit), the molecules can be imaged individually by conventional optics (resolution ~250 nm) and localized more accurately by fitting the collected photons to a 2D Gaussian distribution. The center of the distribution indicates the location of the molecules on the *x*- and *y*-coordinates, and these molecules can be rendered in a new image (A′) as 2D Gaussian distribution with the centers located at the determined *x*- and *y*-coordinates and a width indicated by the uncertainty of the 2D Gaussian fit of the original photon distribution. By repeating this

molecule and its uncertainty $(\sigma_{x,\ y})_m$ are determined. Each molecule is then rendered in a new image as a Gaussian distribution of standard deviation $(\sigma_{x,\ y})_m$ centered at x_m, y_m normalized to integrated intensity. The superresolution image is obtained by rendering the Gaussian distributions of each fitted molecule into one frame (Fig. 4D). Thus, the photoactivation of the PA-FPs provides the necessary capability to isolate single molecules even when present at high densities (up to $\sim 10^5/\mu m^2$) and the localization provides spatial information 10- to 100-fold better than conventional optics allow. PALM imaging of intracellular structures in cryo-prepared thin sections has been demonstrated (Betzig *et al.*, 2006). When these thin sections were examined by transmission electron microscopy, a correlative fluorescence-EM image was obtained. It is thus possible using a correlative PALM–EM approach to relate nanometer scale distribution of a specified protein at high density to cellular ultrastructure.

V. Future Directions of Photoactivatable Fluorescent Proteins

One goal in imaging is to monitor the behavior of individual molecules of interest within the living cells of an organism. Developments in imaging techniques using PA-FPs have advanced light microscopy beyond the diffraction limit to the point that we are able to image at resolutions approaching the molecular scale (Betzig and Chichester, 1993; Dyba and Hell, 2002; Gustafsson, 2005; Hofmann *et al.*, 2005) in both fixed (Betzig *et al.*, 2006) and living cells (Willig *et al.*, 2006). The techniques to achieve these goals have differed in their approach and owe much to the ingenuity of the researchers, but in all cases, the developments would have benefited from the availability of brighter, more stable FPs. Thus, these techniques will benefit from the discovery or development of FPs, both photoactivatable and nonphotoactivatable, that have larger, more defined absorbance cross sections, more narrow emission spectra, higher fluorescence quantum efficiencies, more photostability, and less blinking and/or flickering. Given that FPs and PA-FPs, specifically, have undergone much development and improvement within a fairly short period, we should expect molecules with the aforementioned list of desired characteristics to be developed in the near future. At the very least, en route to developing these molecules, a better understanding of the photochemical characteristics of the molecules currently in use will expedite attaining the goals of research involving cell imaging.

---

sequence of photoactivation and imaging (B and B′–C and C′), all of the molecules within a specimen can be imaged. (D) Summing the fluorescence from all of the molecules results is an image that is equivalent to a conventional optical image (resolution ~250 nm) with limited observable details. (D′) Rendering all of the localized molecules on the same image gives a superresolution image (~10 nm) for these simulated molecules of interest. (See Plate no. 8 in the Color Plate Section.)

References

Ando, R., Hama, H., Yamamoto-Hino, M., Mizuno, H., and Miyawaki, A. (2002). An optical marker based on the UV-induced green-to-red photoconversion of a fluorescent protein. *Proc. Natl. Acad. Sci. USA* **99**(20), 12651–12656.

Ando, R., Mizuno, H., and Miyawaki, A. (2004). Regulated fast nucleocytoplasmic shuttling observed by reversible protein highlighting. *Science* **306**(5700), 1370–1373.

Betzig, E., and Chichester, R. J. (1993). Single molecules observed by near-field scanning optical microscopy. *Science* **262,** 1422–1425.

Betzig, E., Patterson, G. H., Sougrat, R., Lindwasser, O. W., Olenych, S., Bonifacino, J. S., Davidson, M. W., Lippincott-Schwartz, J., and Hess, H. F. (2006). Imaging intracellular fluorescent proteins at nanometer resolution. *Science* **313**(5793), 1642–1645.

Campbell, R. E., Tour, O., Palmer, A. E., Steinbach, P. A., Baird, G. S., Zacharias, D. A., and Tsien, R. Y. (2002). A monomeric red fluorescent protein. *Proc. Natl. Acad. Sci. USA* **99**(12), 7877–7882.

Chalfie, M., Tu, Y., Euskirchen, G., Ward, W. W., and Prasher, D. C. (1994). Green fluorescent protein as a marker for gene expression. *Science* **263**(5148), 802–805.

Chattoraj, M., King, B. A., Bublitz, G. U., and Boxer, S. G. (1996). Ultra-fast excited state dynamics in green fluorescent protein: Multiple states and proton transfer. *Proc. Natl. Acad. Sci. USA* **93**(16), 8362–8367.

Cheezum, M. K., Walker, W. F., and Guilford, W. H. (2001). Quantitative comparison of algorithms for tracking single fluorescent particles. *Biophys. J.* **81**(4), 2378–2388.

Chudakov, D. M., Belousov, V. V., Zaraisky, A. G., Novoselov, V. V., Staroverov, D. B., Zorov, D. B., Lukyanov, S., and Lukyanov, K. A. (2003). Kindling fluorescent proteins for precise *in vivo* photolabeling. *Nat. Biotechnol.* **21**(2), 191–194.

Chudakov, D. M., Verkhusha, V. V., Staroverov, D. B., Souslova, E. A., Lukyanov, S., and Lukyanov, K. A. (2004). Photoswitchable cyan fluorescent protein for protein tracking. *Nat. Biotechnol.* **22**(11), 1435–1439.

Churchman, L. S., Okten, Z., Rock, R. S., Dawson, J. F., and Spudich, J. A. (2005). Single molecule high-resolution colocalization of Cy3 and Cy5 attached to macromolecules measures intramolecular distances through time. *Proc. Natl. Acad. Sci. USA* **102**(5), 1419–1423.

Cormack, B. P., Valdivia, R. H., and Falkow, S. (1996). FACS-optimized mutants of the green fluorescent protein (GFP). *Gene* **173**(1), 33–38.

Crameri, A., Whitehorn, E. A., Tate, E., and Stemmer, W. P. (1996). Improved green fluorescent protein by molecular evolution using DNA shuffling. *Nat. Biotechnol.* **14**(3), 315–319.

Creemers, T. M., Lock, A. J., Subramaniam, V., Jovin, T. M., and Volker, S. (1999). Three photoconvertible forms of green fluorescent protein identified by spectral hole-burning. *Nat. Struct. Biol.* **6**(6), 557–560.

Day, R. N., Periasamy, A., and Schaufele, F. (2001). Fluorescence resonance energy transfer microscopy of localized protein interactions in the living cell nucleus. *Methods* **25**(1), 4–18.

Demarco, I. A., Periasamy, A., Booker, C. F., and Day, R. N. (2006). Monitoring dynamic protein interactions with photoquenching FRET. *Nat. Methods* **3**(7), 519–524.

Dickson, R. M., Cubitt, A. B., Tsien, R. Y., and Moerner, W. E. (1997). On/off blinking and switching behaviour of single molecules of green fluorescent protein. *Nature* **388**(6640), 355–358.

Dyba, M., and Hell, S. W. (2002). Focal spots of size lambda/23 open up far-field fluorescence microscopy at 33 nm axial resolution. *Phys. Rev. Lett.* **88**(16), 163901.

Ehrig, T., O'Kane, D. J., and Prendergast, F. G. (1995). Green-fluorescent protein mutants with altered fluorescence excitation spectra. *FEBS Lett.* **367**(2), 163–166.

Elowitz, M. B., Surette, M. G., Wolf, P. E., Stock, J., and Leibler, S. (1997). Photoactivation turns green fluorescent protein red. *Curr. Biol.* **7**(10), 809–812.

Gurskaya, N. G., Fradkov, A. F., Pounkova, N. I., Staroverov, D. B., Bulina, M. E., Yanushevich, Y. G., Labas, Y. A., Lukyanov, S., and Lukyanov, K. A. (2003). A colourless green fluorescent protein homologue from the non-fluorescent hydromedusa Aequorea coerulescens and its fluorescent mutants. *Biochem. J.* **373**(Pt. 2), 403–408.

Gurskaya, N. G., Verkhusha, V. V., Shcheglov, A. S., Staroverov, D. B., Chepurnykh, T. V., Fradkov, A. F., Lukyanov, S., and Lukyanov, K. A. (2006). Engineering of a monomeric green-to-red photoactivatable fluorescent protein induced by blue light. *Nat. Biotechnol.* **24**(4), 461–465.

Gustafsson, M. G. (2005). Nonlinear structured-illumination microscopy: Wide-field fluorescence imaging with theoretically unlimited resolution. *Proc. Natl. Acad. Sci. USA* **102**(37), 13081–13086.

Heim, R., Cubitt, A. B., and Tsien, R. Y. (1995). Improved green fluorescence. *Nature* **373**(6516), 663–664.

Heim, R., Prasher, D. C., and Tsien, R. Y. (1994). Wavelength mutations and posttranslational autoxidation of green fluorescent protein. *Proc. Natl. Acad. Sci. USA* **91**(26), 12501–12504.

Hofmann, M., Eggeling, C., Jakobs, S., and Hell, S. W. (2005). Breaking the diffraction barrier in fluorescence microscopy at low light intensities by using reversibly photoswitchable proteins. *Proc. Natl. Acad. Sci. USA* **102**(49), 17565–17569.

Jakobs, S., Schauss, A. C., and Hell, S. W. (2003). Photoconversion of matrix targeted GFP enables analysis of continuity and intermixing of the mitochondrial lumen. *FEBS Lett.* **554**(1–2), 194–200.

Kim, P. K., Mullen, R. T., Schumann, U., and Lippincott-Schwartz, J. (2006). The origin and maintenance of mammalian peroxisomes involves a de novo PEX16-dependent pathway from the ER. *J. Cell Biol.* **173**(4), 521–532.

Lukyanov, K. A., Chudakov, D. M., Lukyanov, S., and Verkhusha, V. V. (2005). Innovation: Photoactivatable fluorescent proteins. *Nat. Rev. Mol. Cell Biol.* **6**(11), 885–891.

Lukyanov, K. A., Fradkov, A. F., Gurskaya, N. G., Matz, M. V., Labas, Y. A., Savitsky, A. P., Markelov, M. L., Zaraisky, A. G., Zhao, X., Fang, Y., Tan, W., and Lukyanov, S. A. (2000). Natural animal coloration can be determined by a nonfluorescent green fluorescent protein homolog. *J. Biol. Chem.* **275**(34), 25879–25882.

Marchant, J. S., Stutzmann, G. E., Leissring, M. A., LaFerla, F. M., and Parker, I. (2001). Multiphoton-evoked color change of DsRed as an optical highlighter for cellular and subcellular labeling. *Nat. Biotechnol.* **19**(7), 645–649.

Patterson, G. H., and Lippincott-Schwartz, J. (2002). A photoactivatable GFP for selective photolabeling of proteins and cells. *Science* **297**(5588), 1873–1877.

Post, J. N., Lidke, K. A., Rieger, B., and Arndt-Jovin, D. J. (2005). One- and two-photon photoactivation of a paGFP-fusion protein in live Drosophila embryos. *FEBS Lett.* **579**(2), 325–330.

Prasher, D. C., Eckenrode, V. K., Ward, W. W., Prendergast, F. G., and Cormier, M. J. (1992). Primary structure of the Aequorea victoria green-fluorescent protein. *Gene* **111**(2), 229–233.

Sawin, K. E., and Nurse, P. (1997). Photoactivation of green fluorescent protein. *Curr. Biol.* **7**(10), R606–R607.

Shaner, N. C., Campbell, R. E., Steinbach, P. A., Giepmans, B. N., Palmer, A. E., and Tsien, R. Y. (2004). Improved monomeric red, orange and yellow fluorescent proteins derived from Discosoma sp. red fluorescent protein. *Nat. Biotechnol.* **22**(12), 1567–1572.

Thompson, R. E., Larson, D. R., and Webb, W. W. (2002). Precise nanometer localization analysis for individual fluorescent probes. *Biophys. J.* **82**(5), 2775–2783.

Tsien, R. Y. (1998). The green fluorescent protein. *Annu. Rev. Biochem.* **67,** 509–544.

Tsutsui, H., Karasawa, S., Shimizu, H., Nukina, N., and Miyawaki, A. (2005). Semi-rational engineering of a coral fluorescent protein into an efficient highlighter. *EMBO Rep.* **6**(3), 233–238.

Verkhusha, V. V., and Sorkin, A. (2005). Conversion of the monomeric red fluorescent protein into a photoactivatable probe. *Chem. Biol.* **12**(3), 279–285.

Wiedenmann, J., Ivanchenko, S., Oswald, F., Schmitt, F., Rocker, C., Salih, A., Spindler, K. D., and Nienhaus, G. U. (2004). EosFP, a fluorescent marker protein with UV-inducible green-to-red fluorescence conversion. *Proc. Natl. Acad. Sci. USA* **101**(45), 15905–15910.

Willig, K. I., Rizzoli, S. O., Westphal, V., Jahn, R., and Hell, S. W. (2006). STED microscopy reveals that synaptotagmin remains clustered after synaptic vesicle exocytosis. *Nature* **440**(7086), 935–939.

Yildiz, A., Forkey, J. N., McKinney, S. A., Ha, T., Goldman, Y. E., and Selvin, P. R. (2003). Myosin V walks hand-over-hand: Single fluorophore imaging with 1.5-nm localization. *Science* **300**(5628), 2061–2065.

Yokoe, H., and Meyer, T. (1996). Spatial dynamics of GFP-tagged proteins investigated by local fluorescence enhancement. *Nat. Biotechnol.* **14**(10), 1252–1256.

CHAPTER 4

Design and Optimization of Genetically Encoded Fluorescent Biosensors: GTPase Biosensors

Louis Hodgson,* Olivier Pertz,† and Klaus M. Hahn*

*Department of Pharmacology
Lineberger Cancer Center
University of North Carolina at Chapel Hill
Chapel Hill, North Carolina 27599

†Institute of Biochemistry and Genetics
Department of Clinical-Biological Sciences (DKBW)
University of Basel, Center for Biomedicine,
CH- 4058 Basel, Switzerland

0091-679X/08 $35.00
DOI: 10.1016/S0091-679X(08)85004-2

Abstract

This chapter details the design and optimization of biosensors based on a design used successfully to study nucleotide loading of small GTPase proteins in living cells. This design can be generalized to study many other protein activities, using a single, genetically encoded chain incorporating the protein to be studied, an "affinity reagent" which binds only to the activated form of the targeted protein, and mutants of the green fluorescent protein (GFP) that undergo fluorescence resonance energy transfer (FRET). Specific topics include procedures and caveats in the design and cloning of single-chain FRET sensors, *in vitro* and *in vivo* validation, expression in living cell systems for biological studies, and some general considerations in quantitative fluorescence imaging.

I. Introduction

Direct visualization of proteins in their native environment has been a powerful tool in cell biological studies for over two decades (Taylor and Wang, 1980). Early approaches were limited in scope to proteins that could be purified *in vitro*, chemically labeled, and reintroduced into cells via methods such as microinjection and electroporation. The applicability of biosensors was greatly expanded by the discovery of the green fluorescent protein (GFP) from the jellyfish *Aequorea victoria* (Chalfie *et al.*, 1994; Heim and Tsien, 1996) and very importantly by the development of mutants with enhanced photophysical properties and the ability to undergo fluorescence resonance energy transfer (FRET). Nonradiative FRET between different spectral variants of fluorescent proteins is strongly dependent on the distance and orientation between the GFP mutants. This property can be used to design genetically encoded biosensors that report posttranslational modifications and conformational changes, rather than simply tagging proteins to follow changes in protein localization. There are now multiple proven biosensors based on FRET between GFP mutants (Adams *et al.*, 1991; Hahn *et al.*, 1992; Haj *et al.*, 2002; Llopis *et al.*, 2000; Miyawaki *et al.*, 1997; Ting *et al.*, 2001). Figure 1 illustrates several of these biosensors, including the RhoA activation sensor from our laboratory that is used as an example in this chapter.

FRET signals are often weak relative to fluorescence background, easily leading to false interpretations, or alternately requiring biosensor expression levels that perturb normal cell physiology. Here, we will examine how to optimize biosensor design characteristics that impact fluorescence properties, and discuss controls that validate the biological information obtained from living cells. We will attempt to lay out a straightforward procedure to develop biosensors, based on our experience with the p21 Rho family of small GTPases. Fluorescence microscopy will not be covered in detail in this chapter; readers are referred to previous publications in this area (Gordon *et al.*, 1998; Hodgson *et al.*, 2006; Kenworthy, 2001; Periasamy, 2001; Periasamy and Day, 1999; Xia and Liu, 2001).

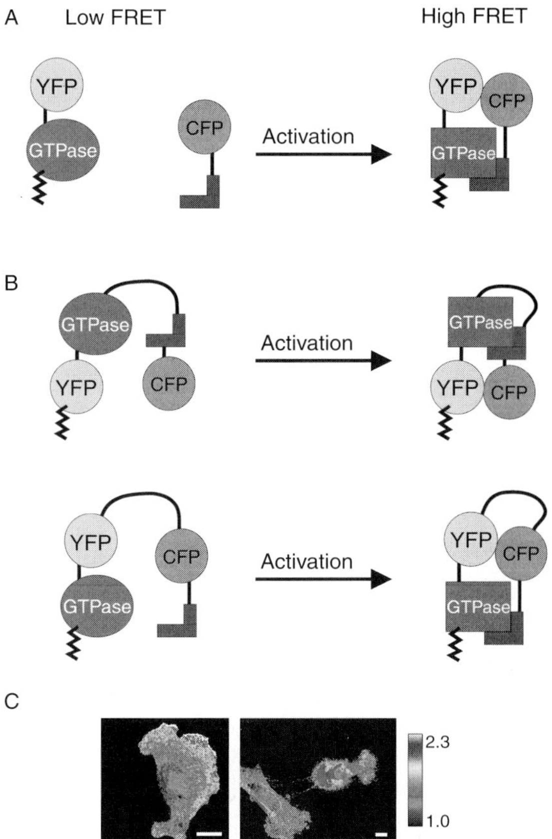

Fig. 1 Examples of biosensors based on fluorescence resonance energy transfer (FRET) between the green fluorescent protein (GFP) mutants. (A) Intermolecular FRET. The donor and the acceptor fluorophores are attached to two separate molecules, the targeted protein and an "affinity reagent" that interacts only with the activated state of the target. Activation of the target causes the two to bind, generating FRET. (B) Intramolecular FRET. Fluorescent proteins are placed on the N- and C-termini, so that the separation between the fluorophores is strongly dependent on protein activation. Activation results in the interaction of the affinity reagent with the target protein, leading to increased FRET. In some cases, the C-terminal end of the biosensor is modified to attach a lipid modification motif from K-Ras, producing constitutive membrane localization. In other cases, proteins are placed in the internal portion of the biosensor. This design is advantageous when the C-terminus of the protein must remain intact for the normal regulation of subcellular localization. (C) Images of the RhoA biosensor in living mouse embryonic fibroblasts during motility. Ratiometric images in pseudo-color show RhoA activation during tail retraction (right) and in extending protrusions at the cell's leading edge (left). Scale bar, 20 μm. Images reproduced from Pertz *et al.* (2006). (See Plate no. 9 in the Color Plate Section.)

II. Background: Factors Influencing FRET Efficiency

FRET is sensitive to both the distance and orientation of the two fluorophores in the biosensor—when the fluorophores are sufficiently far apart or have orthogonal dipole orientations, excitation of the donor simply leads to donor emission rather than to FRET. However, when the distance is sufficiently small, and the orientation enables sufficient dipole coupling, excitation energy is transferred from the donor to the acceptor, leading to decreased donor emission and increased emission from the acceptor. This produces a characteristic FRET excitation/emission spectrum, different from that of the donor or the acceptor alone.

There has been a continuing evolution of useful GFP mutants suitable for FRET in living cells. Mutants incorporate different trade-offs between brightness, FRET efficiency, photostability, and the pH dependence of fluorescence characteristics (Heikal *et al.*, 2000; Miyawaki and Tsien, 2000). Enhanced brightness improves the overall signal-to-noise ratio in cells, but is not always an improvement if it comes at the cost of FRET efficiency (which affects the difference between the activated and inactivated states of the biosensor) (Nguyen and Daugherty, 2005). The cyan and yellow fluorescent proteins (CFP and YFP) (Miyawaki and Tsien, 2000) are relatively fast maturing, bright GFP mutants that have proven useful in many FRET biosensors. More recent mutants with improved FRET efficiency (Nguyen and Daugherty, 2005) include CyPet and YPet, which exhibit 6.7-fold greater FRET efficiency than the original CFP–YFP pair does (Nguyen and Daugherty, 2005). In this chapter, we will refer to the GFP mutants simply as CFP and YFP.

FRET efficiency is quantified by the Förster equation:

$$R_0 = [8.8 \times 10^{23} K^2 n^{-4} Q_d J]^{1/6}$$

R_0 is the Förster distance, where energy transfer is 50% efficient. K^2 is the dipole orientation factor, a function of the donor emission transition moment and the acceptor absorption transition moment. $K^2 = 2/3$ is generally assumed when fluorophore rotation can occur about the bond attaching the fluorophore to the protein. Q_d is the fluorescence quantum yield of the donor in the absence of the acceptor, n is the refractive index of the medium, generally assumed to be 1.4 for proteins, and J is the spectral overlap integral, indicating the extent of overlap between the donor fluorescence emission spectrum and the acceptor excitation spectrum (Lakowicz, 1999).

In biosensors, the activation of the targeted protein affects FRET efficiency by altering the distance and/or the orientation of the fluorophores. FRET efficiency, E, is exquisitely sensitive to the distance between the fluorophores:

$$E = \frac{R_0^6}{(R_0^6 + R^6)}$$

In this equation, R_0 is the Förster distance and R is the actual distance (Lakowicz, 1999). FRET varies as the power of 6 of the distance between the fluorophores. R_0 is dependent on the dipole orientation factor K^2. The K^2 factor can be assumed to be 2/3 only when both fluorophores are free to rotate isotropically during the excited state lifetime. A change in the rotational mobility or fixed angle of the fluorophores in different biosensor states can also affect FRET (K^2 can change between 0 and 4) (dos Remedios and Moens, 1995; Lakowicz, 1999). The effects of fluorophore separation (linear displacement) versus angular reorientation cannot be readily separated in live cell studies. Therefore, FRET changes are not used to precisely determine distances between proteins. Rather, the extent of FRET produced by fully activated versus inactive target protein is determined, and these endpoints are used to interpret FRET signals in cells.

III. Design and Cloning of Biosensors

In single-chain FRET biosensors, the target protein is linked to the affinity reagent and to two fluorescent proteins (Fig. 1). This can affect the interactions of the target protein with upstream regulatory proteins, downstream effectors, and scaffolding proteins. Preservation of upstream regulatory interactions is most important, as these affect the activation being monitored. Competing with effector interactions need not invalidate biosensor readouts, provided these interactions do not impact biosensor localization, and provided the biosensor can be expressed at low concentrations that minimize dominant negative effects. In fact, effects of biosensor overexpression can be reduced when the affinity reagent competes with native effectors. Pull down experiments demonstrated that the affinity reagent in the RhoA biosensor competed effectively against native effectors (Pertz *et al.*, 2006).

Validation of biosensors must include experiments to determine if the reported protein localization has been perturbed by the modification of the target protein, to determine which ligands are being affected, and to determine intracellular biosensor concentrations below which normal cell function is not affected. Point mutations in the biosensor that block interaction between the affinity reagent and the target protein should knock out activation signals, while activating mutations should lead to maximal activation. The effects of such mutations on pull down of endogenous ligands can reveal how the target protein's interaction with various ligands is affected by competition with the affinity reagent in the biosensor. The ability to pull down competing ligands should be increased when point mutations knock out the affinity reagent interactions (see Pertz *et al.*, 2006, supplementary data). The distribution of the fluorescent biosensor should mimic that of the native protein visualized via antibody staining, or at least the GFP fusions of the target protein. Finally, the biosensor should not perturb normal cell behaviors known to be mediated by the target protein. Usually, this is a function of intracellular biosensor concentration. Although precise concentrations are very difficult to measure, it is not so challenging to determine a concentration cutoff above which experiments should not be performed. In a population of randomly loaded cells, a plot of brightness/area versus inhibitor effects usually reveals clear cutoffs where biological perturbation occurs (Chamberlain *et al.*, 2000; Kraynov *et al.*, 2000; Nalbant *et al.*, 2004).

The placement of the GFP mutants in the biosensor chain can be varied, to affect which portions of the target protein are exposed to biological interactions. In our RhoA GTPase biosensor, we were forced to alter the usual arrangement of components. Previous biosensors had placed the GFP mutants at the termini of the chain, so that the separation of the GFP mutants would be maximally affected when the affinity reagent changed between the bound and unbound states. However the Rho family GTPases require an intact C-terminus to interact with guanine nucleotide dissociation inhibitors (GDI). GDI control reversible membrane localization of the GTPases (Bokoch *et al.*, 1994; Chuang *et al.*, 1993; DerMardirossian and Bokoch, 2005). Through optimization of linker length, we were able to place the GFP mutants on the interior of the chain, between the target protein and the affinity reagent (Fig. 1). Here changes in the orientation of two GFP mutants may have played a larger role.

To maximize FRET changes, the length of the linker connecting the intrachain fluorescent proteins had to be optimized. We used a flexible linker of 18 amino acids, encoding "GSTSGSGKPGSGEGSTKG" (Whitlow *et al.*, 1993) in tandem cassettes. The linker encoded a *Bam*HI restriction site on the 5′ side and *Bgl*II site on the 3′ side. These two restriction sites form compatible termini when cut, but ligated product cannot be cut by either of the enzymes. The sense and antisense oligonucleotides were produced with 5′-phosphorylation modification (Invitrogen) and were annealed and ligated for 1 h using T4 DNA ligase in the presence of *Bam*HI and *Bgl*II enzymes. This produced a ladder of multiple linker lengths containing different cassette units. We cloned this mixture into the biosensor backbone to obtain 1–4 linker cassette versions whose FRET efficiency was compared.

The choice of the affinity reagent is one of the most important factors determining the sensitivity of the biosensor. Ideally, there should be a large difference in affinity for the active versus inactive target protein. Low-affinity binding is greatly increased when the affinity reagent is incorporated in the same chain as the target protein. This can reduce the FRET change between the bound and unbound states, because the affinity reagent shows substantial binding even to inactive target. In such cases, the biosensor produces an elevated FRET:CFP ratio even when dominant negative GTPase mutants are used. We have found that subtle differences in affinity can strongly impact the functioning of the biosensor.

In order to facilitate the optimization and the cloning of various GTPases and binding domains, we developed a "master construct" based on the pTriEX-4 backbone (EMD Chemicals Inc, San Diego) that contains multiple, unique restriction sites between every component of the single-chain biosensor (Fig. 2; sequence

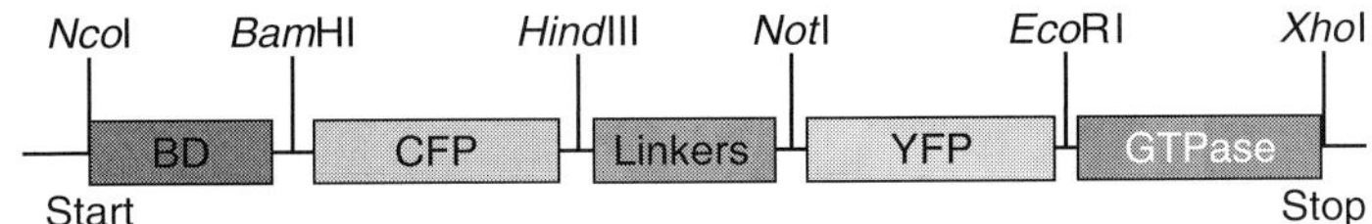

Fig. 2 Schematic representation of the biosensor DNA construct in pTriEX4 vector. The indicated restriction sites are unique sites. (See Plate no. 10 in the Color Plate Section.)

information in Appendix I). We used this construct to develop and optimize several GTPase biosensors including the published RhoA biosensor (Pertz *et al.*, 2006).

IV. Validation of the Biosensor in Cell Suspensions

The single-chain biosensors are quite large (>90 kDa) and so cannot readily be expressed and isolated for *in vitro* characterization (i.e., of FRET, target response, and effect of ligands). However, they can be characterized by expressing biosensors in cells suspended in a fluorometer cuvette. We use HEK293T cells because of their transfection efficiency: expression levels must be high to detect the biosensor changes in cell suspension. When we test response by coexpressing ligands, coexpression must occur in a large percentage of cells. The constitutively active and dominant negative mutants of the biosensor are compared, and the sensitivity to ligands that affect activation is examined by coexpressing the biosensor and the ligand [litration of DNA concentrations is critically important. Typically one- to fourfold excess of regulators, for RhoA including GDI, GEF, and GTPase activating protein (GAP)]. Adherent cells are transfected, then detached and measured as a suspension in the fluorormeter cuvette without lysis. Lysis in some cases releases factors that strongly affect the activation of the target protein, overwhelming the overexpressed regulatory protein.

Despite variable transfection efficiency and cell health, we have found that this assay provides robust and repeatable readouts of the biosensor behavior. Importantly, the overexpression of GTPases in HEK293T cells saturates endogenous cellular regulators, so that the overexpressed biosensor containing wild-type target protein will be "all-on" and will give a signal like that of the overexpressed constitutively active mutants. For RhoA, the overexpressed wild-type protein overwhelms the capacity of native GDI to maintain the biosensor in an inactive state. Usually, a fourfold excess cotransfection of GDI results in an "all-off" readout, similar to that of the dominant negative biosensor, biosensor that cannot bind the affinity reagent, or the coexpression of GAP.

Biosensors can efficiently be developed through a series of optimization steps. The optimization of the linker length is best carried out first, using relatively straightforward tests of biosensor response (i.e., for GTPases, the effect of GDI coexpression, as described above). The procedure described above can be used to test linkers between 1 and 4 of repeating units. Depending on the trend of the response, it may be necessary to further shorten the linker to GSGSGS or lengthen to 5–6 linkers. The biosensors with the best response and brightness are further validated using point mutations and more complex tests of interaction with regulatory ligands (i.e., biosensors containing the activated and inactivated GTPases, and effects of GEFs and GAPs) (Pertz *et al.*, 2006). Finally, the best biosensor is introduced into viral transduction vectors (see Section IV.B).

In many cases, cells are very sensitive to biosensor expression levels; viral transfection of inducible vectors, and stable cell lines with low-expression levels must be used. This is often important not only to avoid toxicity but also because biosensor response is seen only at lower expression levels. As described above, our RhoA sensor is constitutively "on" when overexpressed, because it overwhelms endogenous GDI. An uncautious experimentalist using higher levels of sensor to overcome typically weak FRET fluorescence might simply assume that the sensor did not work.

It may be possible to alter the relative orientations of the components to effect the changes in the FRET efficiency using circularly permutated fluorescent proteins (Baird *et al.*, 1999; Nagai *et al.*, 2004). We have implemented this to greatly enhance the FRET ratio changes of some biosensors (approximately threefold increase in ratio change, unpublished data). With circularly permutated fluorescent protein built into the biosensor, the intrachain linker lengths need to be reoptimized.

The following sections describe more detailed procedures for some of the steps outlined above.

A. Expression in HEK293T Cells for Assay of Biosensors in Cell Suspension

This is a protocol for testing the effects of protein coexpression on the RhoA biosensor. It can provide a template for the development of other GTPase biosensors.

1. *Day 1, 3 days before the assay*: Prepare 6-well plates by coating the wells with a 1:10 dilution of polylysine (P4832–50 ML; Sigma Co, St. Louis, MO) in 1 ml PBS per well for 1 h at room temperature on a rocker. Prepare enough wells to test each condition in triplicate, plus 1 extra well for mock transfection (this will be used to obtain the background fluorescence on the fluorometer). HEK293T cells should be detached and plated onto polylysine coated 6-well plates at $(5–6) \times 10^5$ cells per well. This should be optimized in the given range, depending on cell health and the doubling rate.
2. *Day 2*: On the following day, transfect typically 500–750 ng DNA per well, using the following ratios and the Lipofectamine Plus (Invitrogen, Carlsbad, CA) protocol:
    a. Biosensor and GDI: 100 ng biosensor DNA and 400 ng GDI DNA. In those wells containing only biosensor DNA (GDI negative), make sure to equalize the total DNA quantity by cotransfecting control empty vector.

b. Biosensor mutants: 100 ng wild-type biosensor plus 400 ng empty vector, and 100 ng biosensor mutants plus 400 ng empty vector. Also prepare biosensor plus GDI (100:400 ng) as a control.

c. GEF-rescue of GDI binding: 100 ng biosensor DNA, 400 ng GDI DNA, and 100 ng GEF. Variable amounts of GEFs have been needed to rescue the GDI binding, depending on the relative efficiency of the GEF DH-PH expression constructs. In some cases such as Dbl, Dbs, and Intersectin, we have had to use 100 ng to see effects without affecting cell viability. Other GEFs, including Tiam-1, can be used at higher concentrations. We have used ratios form 1:4:1 to 1:4:10, depending on the particular GEF/biosensor combination.

d. GAP inhibition of biosensor: 100 ng biosensor DNA and 100–400 ng GAP DNA. Here, excess GAP will be toxic and each GAP DNA should be titrated.

Suspend the DNA into 100 μl of serum and antibiotic-free DMEM. Add 16 μl of Plus reagent (Invitrogen). Vortex and let stand for 15 min at room temperature. Add to this mixture 100 μl of serum and antibiotic-free DMEM containing 5 μl of Lipofectamine reagent (Invitrogen). Vortex and let stand for 15 min at room temperature. While the transfection mixtures are incubating, wash the cells on 6-well plates once with PBS and add 0.8 ml of serum and antibiotic-free DMEM into each well. Apply the transfection mixture into the wells, swirl to mix fully, and incubate under standard tissue culture conditions for 6 h. At the end of this incubation, exchange medium with standard culture medium, adding 3 ml of medium per well.

3. *Day 3*: Check fluorescence—it is particularly important to check for proper localization of the biosensor, especially if GDI and GEFs are coexpressed. There should be distinct membrane or cytosol localization, depending on the conditions used (Pertz *et al.*, 2006). For this purpose, a tissue culture microscope equipped with FITC/GFP epifluorescence filters should suffice.

4. *Day 4*: On the day of the assay, cells are washed once with PBS, and 1 ml per well of trypsin/EDTA is added and immediately aspirated. Using 0.5 ml of chilled PBS, cells are detached by repeated pipetting and placed on ice. The samples are then read on the flourometer by placing 0.4 ml of the cell suspension into the fluorometer cuvette (18/9F-Q-10; Starna Cells, Inc., Atascadero, CA). Fluorescence emission scans are performed by excitation at 433 nm and scanning between 450 and 600 nm, stepping every 3 nm. Mock-transfected (empty vector control) cells should be used to obtain the background spectra, which should be subtracted from all subsequent measurements. The spectra should be normalized to the maximum CFP emission at 477 nm to standardize the data analysis. The ratio of FRET emission at 525 nm to the CFP emission at 477 nm is compared among various mutants and transfection conditions.

In Fig. 3, sample validation results for the single-chain RhoA GTPase biosensor are shown.

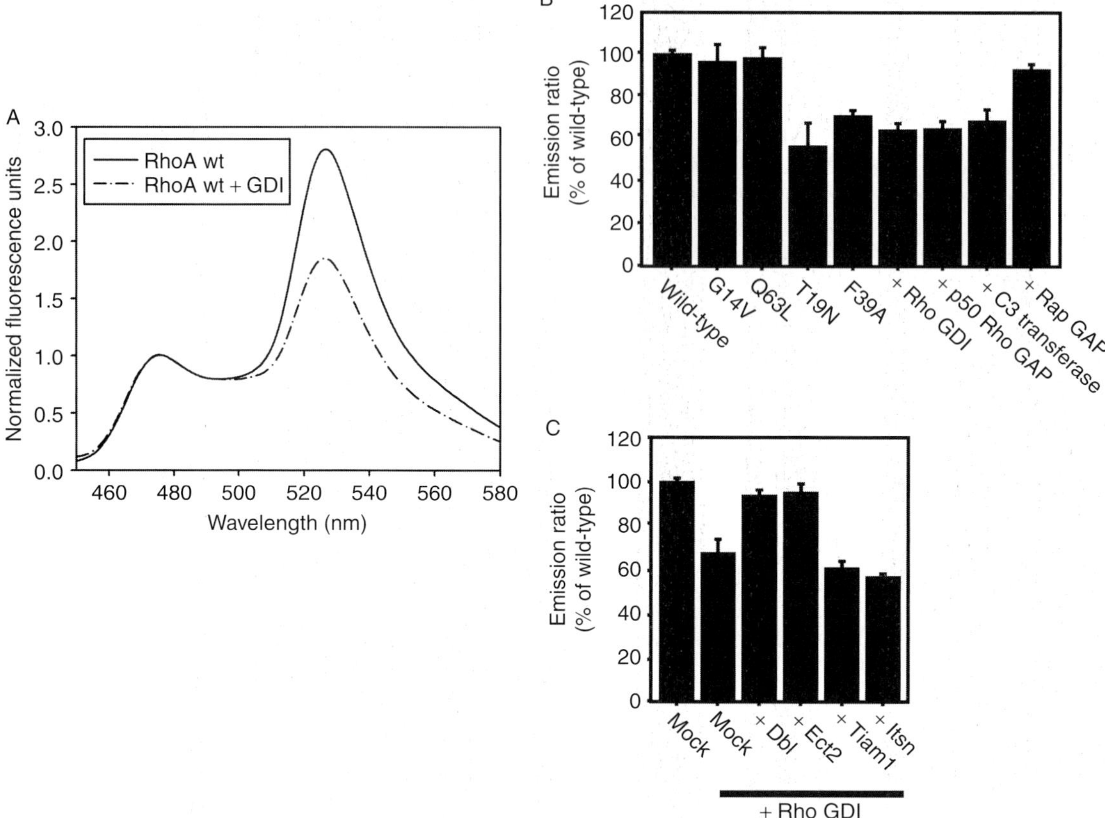

Fig. 3 *In vitro* validation of the single-chain cyan and yellow fluorescent proteins fluorescence resonance energy transfer (CFP–YFP FRET) RhoA biosensor. (A) Emission spectra for the wild-type RhoA biosensor alone and the biosensor plus guanine nucleotide dissociation inhibitors (GDI) coexpression. (B) Versions of the biosensor incorporating the indicated mutations in the GTPase. GDI coexpression shows interaction with G14V constitutively active mutant, but not with Q63L constitutively active mutant. (C) The reduced ratio produced by the expression of excess GDI is rescued by the coexpression of the GEFs Dbs or Dbl, but not by Intersectin (a Cdc42 GEF) or Tiam-1 (a Rac GEF). p50RhoGAP expression results in a low ratio. Data panels reproduced from Pertz *et al.* (2006).

B. Expression of the Biosensor in Cells

The pTriEX backbone used for the cloning of biosensors allows overexpression in mammalian cells driven by the *Cytomegalovirus* (CMV) promoter. However, this expression strategy requires that one is aware of the potential toxicity of GTPases upon overexpression (Pertz *et al.*, 2006) and the need for a near-stoichiometric relationship between endogenous GTPases and endogenous regulators in cells (Del Pozo *et al.*, 2002; Michaelson *et al.*, 2001).

We have addressed these issues by using retroviral transduction to stably incorporate low copy numbers of the biosensor DNA (Pertz *et al.*, 2006). Transfected cells were sorted by fluorescence activated cell sorting (FACS) to obtain low expressors only, so that the biosensor expression level was comparable to endogenous GTPase. We have shown that the competition with endogenous effectors of the GTPase is not a significant problem (Pertz *et al.*, 2006).

In order to address the toxicity of the biosensor expression, we used a tetracycline-inducible retroviral construct based on a pBabe-Sin-Puro-tet-CMV backbone. We use the tet-off stable MEF/3T3 cell system (Clontech Mountain View, CA), and the infected cells are repressed using 1 μg/ml Doxycyclin. Cells are selected with Puromycin up to 10 μg/ml, increased gradually in concentration following the infection. At the end of the selection, cells are induced for the biosensor expression by the removal of Doxycyclin and reseeding at a sparse concentration (1×10^4 cells per 10 cm TC dish) for 48 h. These cells are then sorted by FACS to produce near 100% cells positive for the biosensor expression, which are then repressed again with the application of Doxycyclin.

For experiments, it is important to use sparse reseeding following the removal of Doxycyclin from the culture media. We routinely maintain repressed cells in 1 μg/ml Doxycyclin in standard tissue culture conditions. For induction, cells are detached by a brief trypsinization and spun at 1200 rpm for 5 min. Supernatant media are suctioned out carefully and completely, and cells are resuspended in fresh culture medium without Doxycyclin. Cells are replated at 1×10^4 cells per 10 cm TC dish without Doxycyclin and checked for fluorescence 24 h after the induction. The cells are kept in this condition for an additional 24 h before the assay. This additional 24 h will ensure that overexpressors will die off, and cells only with a sustainable expression level will survive for the experiment. Cells are detached and replated on coverslips coated with fibronectin (10 μg/ml; Sigma) on the morning of the experiment and allowed to adhere for 5 h before imaging.

V. Microscopy and Imaging Considerations

The basic aspects of the fluorescence imaging using FRET biosensors have been covered elsewhere (Gordon *et al.*, 1998; Kenworthy, 2001; Periasamy, 2001; Periasamy and Day, 1999; Xia and Liu, 2001). It is worthwhile to mention a few key points that apply specifically to the use of the single-chain biosensors at low-intracellular concentrations.

Unlike single-chain FRET biosensors, intermolecular FRET biosensors (Chamberlain *et al.*, 2000; Kraynov *et al.*, 2000; Tzima *et al.*, 2003) require bleed-through corrections due to the varying subcellular distribution of the two components. They do produce enhanced dynamic range. For a single-chain design, it is sufficient to simply take a ratio of the FRET emission over the donor emission

(Pertz *et al.*, 2006). For both designs, it is important to realize that the two fluorophores will bleach at different rates, so that the bleaching corrections are required to counter a bias in the signal over time. The computational steps involved in the calculation of ratios and photobleach correction are covered elsewhere (Hodgson *et al.*, 2006).

For the relatively dim fluorescence we use to maintain biosensor concentrations that do not overwhelm endogenous regulatory molecules, we must be careful to maximize the efficiency of light collection. It is not desirable to compensate for low biosensor concentrations simply by increasing irradiation, as this bleaches the biosensor and increases the phototoxicity. We routinely use an oil immersion 40× DIC (differential interference contrast) objective with a numerical aperture of 1.3, together with 2 × 2 binning on our charge coupled device (CCD) camera. Using 60× or higher magnification objective lens cuts down greatly on the brightness of the transmitted signal. It is important to use a DIC rather than phase contrast objective lens, as the phase objective contains a phase ring that substantially reduces the transmittance of light through the objective lens. Neutral density filters of 0.6–1.0 (54.9–36.8% transmittance) are used to cut the brightness of the excitation light. Longer exposure using dimmer excitation light is preferable to shorter, more intense irradiation; this reduces both bleaching and phototoxicity. We routinely use methods to remove oxygen from the medium to further decrease photobleaching and phototoxicity [an oxygen scavenger reagent, OxyFluor (Oxyrase, Inc.), antioxidants including vitamin C at 1 mM concentration, and/or purging the assay medium with Argon; Fig. 4].

We currently use a Sony ICX285-based interline transfer cooled CCD camera, the Roper/Photometrics CoolSnapESII, cooled to 0 °C. This camera can be

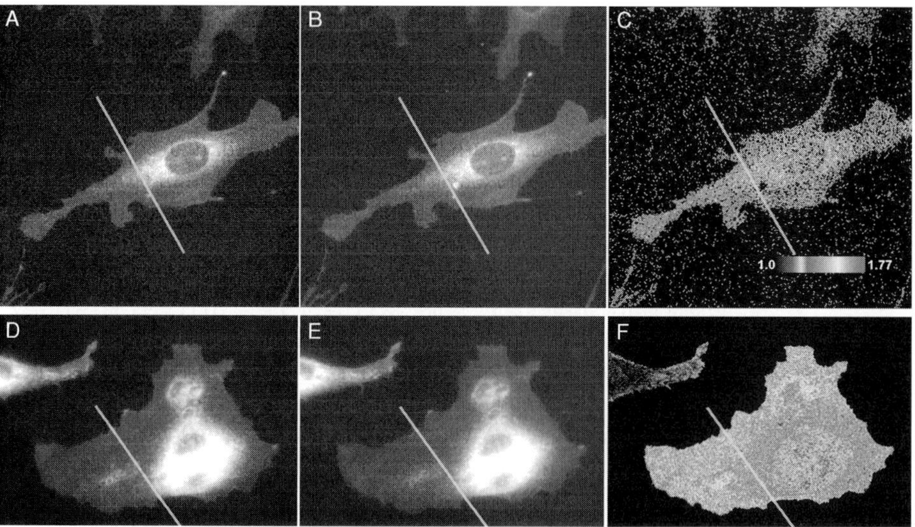

Fig. 4 (*continues*)

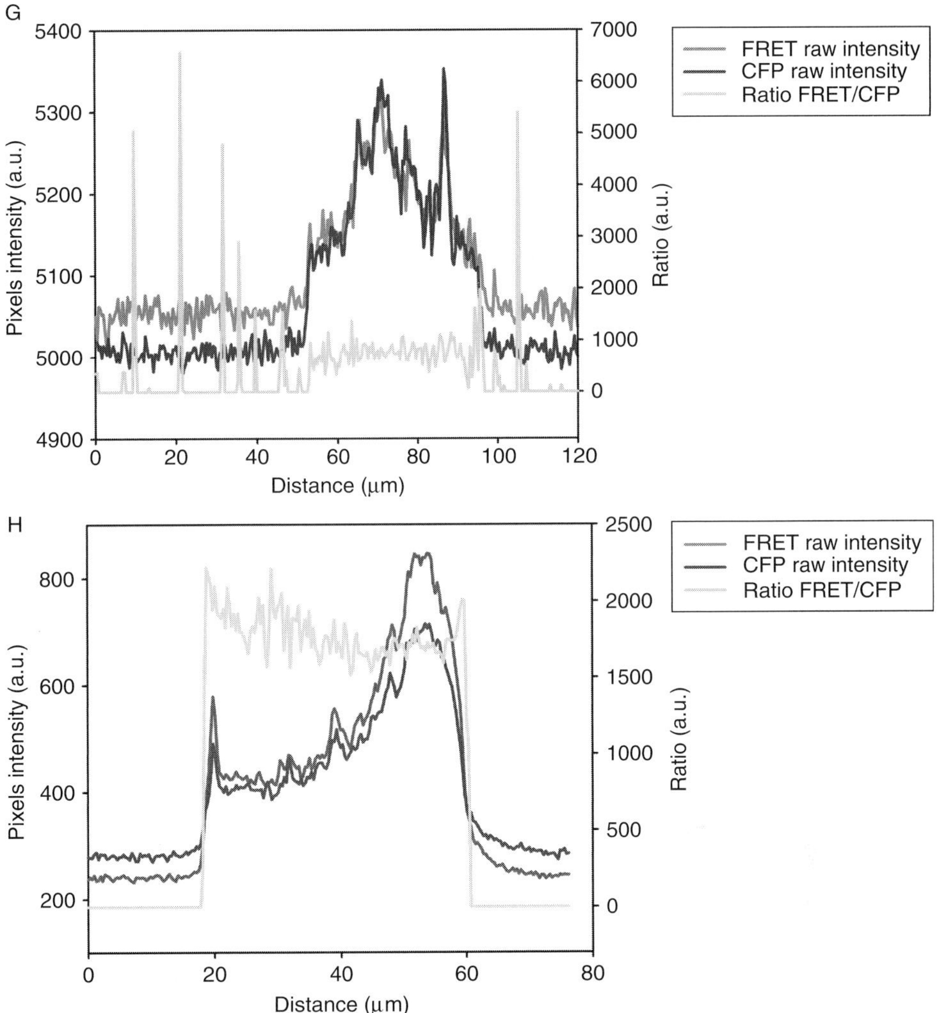

Fig. 4 Comparison between an electron multiplying charge coupled detector (EMCCD) camera and a normal cooled charge coupled detector (CCD) camera for wide-field imaging of a cyan and yellow fluorescent protein fluorescence resonance energy transfer (CFP–YFP FRET) biosensor. (A) FRET and (B) CFP channels taken using a Roper Photometrics Cascade II 512 camera at 1 × 1 binning, 3× gain, 3300 EM-gain, 50 and 100 ms exposures per frame for FRET and CFP, respectively. Both channels were averaged for 10 frames. (C) Ratiometric result (FRET–CFP) from the images shown in A and B. The color bar is scaled from 1.0 to 1.77. (D) FRET and (E) CFP channels taken using a Roper Photometrics CoolsnapES camera at 2 × 2 binning, 2× gain, 400 and 800 ms exposures for FRET and CFP, respectively. (F) Ratiometric result (FRET–CFP) from the images shown in D and E. Scaling is identical to that shown in (C). All images were taken using a Zeiss EC-Plan Neofluor 40× NA = 1.4 oil immersion DIC objective lens, with similar intensities. (G) Line scans from the green lines shown on panels A–C. (H) Line scans from the green lines shown on panels D–F. The raw data from the two cameras (A and B vs C and D) appear similar by visual observation, but the ratio images (C and F) and the line scans reveal higher noise associated with the EMCCD camera. This is due to both the greater stochastic noise in the EMCCD camera raw images and the propagation of this noise when the two noisier images are divided by one another. Thus this issue is more acute where division of image for ratio imaging is required. (See Plate no. 11 in the Color Plate Section.)

obtained with 5 e-read noise, and 0.01e-/pixel/s dark current. The quantum efficiency of the chip is 60% for 450–625 nm light, and it has a small pixel size (6.45 × 6.45 μm) for good spatial resolution. We routinely use 2 × 2 binning and expose 800 ms for CFP and 400 ms for FRET when imaging the RhoA biosensor. These conditions usually result in gray values filling 75% of the full 12-bit range of camera digitization. We do not recommend the current on-chip gain amplification electron multiplying CCD (EMCCD) cameras for biosensor imaging. While these cameras can capture images under extremely dim illumination conditions, the gain-circuitry introduces too much stochastic noise for ratiometric imaging (Fig. 4). The stochastic noise is greatly increased when one image is divided by another. Other viable options may include back-thinned, back-illuminated cooled CCD cameras with high-quantum efficiency. These cameras offer ultra-high quantum efficiency (over 90%), but the pixel size is usually large (16 × 16 μm), reducing spatial resolution.

The choice of imaging medium is also an important consideration when imaging FRET biosensors. Background fluorescence from the media is a significant issue at the wavelengths used. We have performed a quantitative comparison of various media available commercially, and have concluded that Ham's F-12 K medium without phenol red (Kaighn, 1973; Robey and Termine, 1985) is a good choice. Unfortunately, this medium is no longer commercially available. The formulation can be found in Appendix II. Commercially available phenol red-free medium 199 (Mediatech Inc, Herndon, VA) has slightly worse background fluorescence than does Ham's F-12K medium without phenol red.

VI. Conclusion

Here, we present some basic methods for designing, building, and validating single-chain biosensors, with procedures specifically adapted to study small GTPase proteins. The techniques and approaches are potentially applicable to a much wider group of proteins, those where an appropriate binding-domain/target can be identified. We hope that this description of building a single-chain CFP–YFP FRET biosensor, including common pitfalls, will be valuable for those wishing to develop biosensors for their favorite molecule.

VII. Appendix I

A. DNA Sequence for the pTriEX-4-Biosensor Construct

Start (NcoI site) into pTriEX NcoI site, and then 6×His tag plus GSG linker

*CCATGG*CACACCATCACCACCATCACGGTAGTGGC

Rhotekin RBD

ATCCTGGAGGACCTCAATATGCTCTACATCCGGCAGATGGCACT
CAGCCTGGAGGACACAGAGCTGCAGAGGAAACTAGATCATGAG
ATCCGGATGAGGGATGGGGCCTGCAAGCTGCTGGCAGCCTGCT
CCCAGCGAGAGCAGGCTCTGGAAGCCACCAAGAGCCTGCTGGT
GTGCAACAGCCGTATTCTCAGCTACATGGGTGAGCTGCAGCGG
CGAAAGGAGGCCCAGGTGCTGGAGAAGACA

GSG linker (BamHI)

*GGATCC*GGA

CFP (HindIII)

ATGGTGAGCAAGGGCGAGGAGCTGTTCACCGGGGTGGTGCCC
ATCCTGGTCGAGCTGGACGGCGACGTAAACGGCCACAAGTTCA
GCGTGTCCGGCGAGGGCGAGGGCGATGCCACCTACGGCAAGC
TGACCCTGAAGTTCATCTGCACCACCGGCAAGCTGCCCGTGCCC
TGGCCCACCCTCGTGACCACCCTGACCTGGGGCGTGCAGTGCTT
CAGCCGCTACCCCGACCACATGAAGCAGCACGACTTCTTCAAGT
CCGCCATGCCCGAAGGCTACGTCCAGGAGCGCACCATCTTCTTC
AAGGACGACGGCAACTACAAGACCCGCGCCGAGGTGAAGTTCG
AGGGCGACACCCTGGTGAACCGCATCGAGCTGAAGGGCATCGA
CTTCAAGGAGGACGGCAACATCCTGGGGCACAAGCTGGAGTAC
AACTACATCAGCCACAACGTCTATATCACCGCCGACAAGCAGAA
GAACGGCATCAAGGCCAACTTCAAGATCCGCCACAACATCGAG
GACGGCAGCGTGCAGCTCGCCGACCACTACCAGCAGAACACCC
CCATCGGCGACGGCCCCGTGCTGCTGCCCGACAACCACTACCT
GAGCACCCAGTCCGCCCTGAGCAAAGACCCCAACGAGAAGCGC
GATCACATGGTCCTGCTGGAGTTCGTGACCGCCGCCGGGATCA
CTCTCGGCATGGACGAGCTGTACAA*A AGC TT*A

Linker Cassette

ACTTCTGGTTCTGGTAAACCTGGTTCTGGTGAAGGTTCTAC
TAAAGGTGGATCT

Link into YFP (NotI)

GGATCTGCGGCCGCA

YFP (EcoRI)

ATGGTGAGCAAGGGCGAGGAGCTGTTCACCGGGGTGGTGCCC
ATCCTGGTCGAGCTGGACGGCGACGTAAACGGCCACAAGTTCA
GCGTGTCCGGCGAGGGCGAGGGCGATGCCACCTACGGCAAGC
TGACCCTGAAGTTCATCTGCACCACCGGCAAGCTGCCCGTGCCC

TGGCCCACCCTCGTGACCACCTTCGGCTACGGCCTGATGTGCTT
CGCCCGCTACCCCGACCACATGAAGCAGCACGACTTCTTCAAGT
CCGCCATGCCCGAAGGCTACGTCCAGGAGCGCACCATCTTCTTC
AAGGACGACGGCAACTACAAGACCCGCGCCGAGGTGAAGTTCG
AGGGCGACACCCTGGTGAACCGCATCGAGCTGAAGGGCATCGA
CTTCAAGGAGGACGGCAACATCCTGGGGCACAAGCTGGAGTAC
AACTACAACAGCCACAACGTCTATATCATGGCCGACAAGCAGA
AGAACGGCATCAAGGTGAACTTCAAGATCCGCCACAACATCGA
GGACGGCAGCGTGCAGCTCGCCGACCACTACCAGCAGAACACC
CCCATCGGCGACGGCCCCGTGCTGCTGCCCGACAACCACTACCT
GAGCTACCAGTCCGCCCTGAGCAAAGACCCCAACGAGAAGCGC
GATCACATGGTCCTGCTGGAGTTCGTGACCGCCGCCGGGATCAC
TCTCGGCATGGACGAGCTGTACAAG*GAATTC*

RhoA wild type (XhoI)

ATGGCTGCCATCCGGAAGAAACTGGTGATTGTTGGTGATGGAG
CCTGTGGAAAGACATGCTTGCTCATAGTCTTCAGCAAGGACCAG
TTCCCAGAGGTGTATGTGCCCACAGTGTTTGAGAACTATGTGG
CAGATATCGAGGTGGATGGAAAGCAGGTAGAGTTGGCTTTGTG
GGACACAGCTGGGCAGGAAGATTATGATCGCCTGAGGCCCCTC
TCCTACCCAGATACCGATGTTATACTGATGTGTTTTTCCATCGAC
AGCCCTGATAGTTTAGAAAACATCCCAGAAAAGTGGACCCCAG
AAGTCAAGCATTTCTGTCCCAACGTGCCCATCATCCTGGTTGGG
AATAAGAAGGATCTTCGGAATGATGAGCACACAAGGCGGGAGC
TAGCCAAGATGAAGCAGGAGCCGGTGAAACCTGAAGAAGGCAG
AGATATGGCAAACAGGATTGGCGCTTTTGGGTACATGGAGTGTT
CAGCAAAGACCAAAGATGGAGTGAGAGAGGTTTTTGAAATGGC
TACGAGAGCTGCTCTGCAAGCTAGACGTGGGAAGAAAAAATCT
GGGTGCCTTGTCTTGTGAAACTAA*CTCGAG*

VIII. Appendix II

A. Media Formulation for Ham's F-12K Phenol Red-Free (Kaighn, 1973; Robey and Termine, 1985)

Volume: 500 mL; without glutamine, pH 7.4 to 7.5

Formulation:

Inorganic salts	mg/liter	Other compounds	mg/liter
NaCl	7530.00	Glucose	1260.00
KCl	285	Linoleic acid	0
$MgCl_2$•$6H_2O$	106	Hypoxanthine•Na	4
$MgSO_4$•$7H_2O$	393	Phenol red	0
$CaCl_2$	135	Putresine•HCl	0.3

$Na_2HPO_4•7H_2O$	218	Sodium pyruvate	220
KH_2PO_4	59	Thymidine	0.7
$NaHCO_3$	2500.00		
$FeSO_4•7H_2O$	0.8		
$CuSO_4•5H_2O$	2 μg		
$ZnSO_4•7H_2O$	0.14		
Amino acids	mg/liter	Vitamins	mg/liter
L-Alanine	17.8	L-Ascorbic acid	0
L-Arginine•HCl	421.3	Biotin	0.07
L-Asparagine•H_2O	30	D-Calcium Pantothenate	0.48
L-Aspartic acid	26.6	Choline chloride	13.96
L-Cysteine•HCl•H_2O	70.04	Cyanocobalamin	1.36
L-Cystine	0	Folic acid	1.32
L-Glutamic acid	29.4	Inositol	18
Glycine	15	Nicotinamide	0.04
L-Histidine•HCl•H_2O	41.9	Pyridoxine•HCl	0.06
L-Isoleucine	7.9	Riboflavin	0.04
L-Leucine	26.2	Thiamine•HCl	0.21
L-Lysine•HCl	73.1	DL-Thioctic acid	0.21
L-Methionine	8.9		
L-Phenylalanine	9.9		
L-Proline	69.1		
L-Serine	21		
L-Threonine	23.8		
L-Tryptophan	4.1		
L-Tyrosine	10.9		
L-Valine	23.4		

References

Adams, S. R., Harootunian, A. T., Buechler, Y. J., Taylor, S. S., and Tsien, R. Y. (1991). Fluorescence ratio imaging of cyclic AMP in single cells. *Nature* **349,** 694–697.

Baird, G. S., Zacharias, D. A., and Tsien, R. Y. (1999). Circular permutation and receptor insertion within green fluorescent proteins. *Proc. Natl. Acad. Sci. USA* **96,** 11241–11246.

Bokoch, G. M., Bohl, B. P., and Chuang, T. H. (1994). Guanine nucleotide exchange regulates membrane translocation of Rac/Rho GTP-binding proteins. *J. Biol. Chem.* **269,** 31674–31679.

Chalfie, M., Tu, Y., Euskirchen, G., Ward, W. W., and Prasher, D. C. (1994). Green fluorescent protein as a marker for gene expression. *Science* **263,** 802–805.

Chamberlain, C. E., Kraynov, V., and Hahn, K. M. (2000). Imaging spatiotemporal dynamics of Rac activation *in vivo* with FLAIR. *Meth. Enzymol.* **325,** 389–400.

Chuang, T.-H., Xu, X., Knaus, U. G., Hart, M. J., and Bokoch, G. M. (1993). GDP dissociation inhibitor (GDI) prevents intrinsic and GTPase activating protein (GAP)-stimulated GTP hydrolysis by the Rac1 and Rac2 GTP-binding proteins. *J. Biol. Chem.* **268,** 775–778.

Del Pozo, M. A., Kiosses, W. B., Alderson, N. B., Meller, N., Hahn, K. M., and Schwartz, M. A. (2002). Integrins regulate GTP-Rac localized effector interactions through dissociation of Rho-GDI. *Nat. Cell Biol.* **4,** 232–239.

DerMardirossian, C., and Bokoch, G. M. (2005). GDIs: Central regulatory molecules in Rho GTPase activation. *Trends Cell Biol.* **15,** 356–363.

dos Remedios, C. G., and Moens, P. D. (1995). Fluorescence resonance energy transfer spectroscopy is a reliable "ruler" for measuring structural changes in proteins. Dispelling the problem of the unknown orientation factor. *J. Struct. Biol.* **115,** 175–185.

Gordon, G. W., Berry, G., Liang, X. H., Levine, B., and Herman, B. (1998). Quantitative fluorescence resonance energy transfer measurements using fluorescence microscopy. *Biophys. J.* **74,** 2702–2713.

Hahn, K., DeBiasio, R., and Taylor, D. L. (1992). Patterns of elevated free calcium and calmodulin activation in living cells. *Nature* **359,** 736–738.

Haj, F. G., Verveer, P. J., Squire, A., Neel, B. G., and Bastiaens, P. I. (2002). Imaging sites of receptor dephosphorylation by PTP1B on the surface of the endoplasmic reticulum. *Science* **295,** 1708–1711.

Heikal, A. A., Hess, S. T., Baird, G. S., Tsien, R. Y., and Webb, W. W. (2000). Molecular spectroscopy and dynamics of intrinsically fluorescent proteins: Coral red (dsRed) and yellow (Citrine). *Proc. Natl. Acad. Sci. USA* **97,** 11996–12001.

Heim, R., and Tsien, R. Y. (1996). Engineering green fluorescent protein for improved brightness, longer wavelengths and fluorescence resonance energy transfer. *Curr. Biol.* **6,** 178–182.

Hodgson, L., Nalbant, P., Shen, F., and Hahn, K. (2006). Imaging and photobleach correction of Mero-CBD, sensor of endogenous Cdc42 activation. *Meth. Enzymol.* **406,** 140–156.

Kaighn, M. E. (1973). "Tissue Culture Methods and Applications," pp. 54–56. Academic Press, New York.

Kenworthy, A. K. (2001). Imaging protein–protein interactions using fluorescence resonance energy transfer microscopy. *Methods* **24,** 289–296.

Kraynov, V. S., Chamberlain, C., Bokoch, G. M., Schwartz, M. A., Slabaugh, S., and Hahn, K. M. (2000). Localized Rac activation dynamics visualized in living cells. *Science* **290,** 333–337.

Lakowicz, J. R. (1999). "Principles of Fluorescence Spectroscopy," pp. 368–377. Kluwer Academic/Plenum, New York.

Llopis, J., Westin, S., Ricote, M., Wang, Z., Cho, C. Y., Kurokawa, R., Mullen, T. M., Rose, D. W., Rosenfeld, M. G., Tsien, R. Y., and Glass, C. K. (2000). Ligand-dependent interactions of coactivators steroid receptor coactivator-1 and peroxisome proliferator-activated receptor binding protein with nuclear hormone receptors can be imaged in live cells and are required for transcription. *Proc. Natl. Acad. Sci. USA* **97,** 4363–4368.

Michaelson, D., Silletti, J., Murphy, G., D'Eustachio, P., Rush, M., and Philips, M. R. (2001). Differential localization of Rho GTPases in live cells: Regulation by hypervariable regions and RhoGDI binding. *J. Cell Biol.* **152,** 111–126.

Miyawaki, A., Llopis, J., Heim, R., McCaffery, J. M., Adams, J. A., Ikura, M., and Tsien, R. Y. (1997). Fluorescent indicators for Ca^{2+} based on green fluorescent proteins and calmodulin [see comments]. *Nature* **388,** 882–887.

Miyawaki, A., and Tsien, R. Y. (2000). Monitoring protein conformations and interactions by fluorescence resonance energy transfer between mutants of green fluorescent protein. *Meth. Enzymol.* **327,** 472–500.

Nagai, T., Yamada, S., Tominaga, T., Ichikawa, M., and Miyawaki, A. (2004). Expanded dynamic range of fluorescent indicators for $Ca(^{2+})$ by circularly permuted yellow fluorescent proteins. *Proc. Natl. Acad. Sci. USA* **101,** 10554–10559.

Nalbant, P., Hodgson, L., Kraynov, V., Toutchkine, A., and Hahn, K. M. (2004). Activation of endogenous Cdc42 visualized in living cells. *Science* **305,** 1615–1619.

Nguyen, A. W., and Daugherty, P. S. (2005). Evolutionary optimization of fluorescent proteins for intracellular FRET. *Nat. Biotechnol.* **23,** 355–360.

Periasamy, A. (2001). Fluorescence resonance energy transfer microscopy: A mini review. *J. Biomed. Opt.* **6,** 287–291.

Periasamy, A., and Day, R. N. (1999). Visualizing protein interactions in living cells using digitized GFP imaging and FRET microscopy. *Methods Cell Biol.* **58,** 293–314.

Pertz, O., Hodgson, L., Klemke, R. L., and Hahn, K. M. (2006). Spatiotemporal dynamics of RhoA activity in migrating cells. *Nature* **440,** 1069–1072.

Robey, P. G., and Termine, J. D. (1985). Human bone cells *in vitro*. *Calcif. Tissue Int.* **37,** 453–460.

Taylor, D. L., and Wang, Y. L. (1980). Fluorescently labeled molecules as probes of the structure and function of living cells. *Nature* **284,** 405–410.

Ting, A. Y., Kain, K. H., Klemke, R. L., and Tsien, R. Y. (2001). Genetically encoded fluorescent reporters of protein tyrosine kinase activities in living cells. *Proc. Natl. Acad. Sci. USA* **98,** 15003–15008.

Tzima, E., Kiosses, W. B., del Pozo, M. A., and Schwartz, M. A. (2003). Localized cdc42 activation, detected using a novel assay, mediates microtubule organizing center positioning in endothelial cells in response to fluid shear stress. *J. Biol. Chem.* **278,** 31020–31023.

Whitlow, M., Bell, B. A., Feng, S. L., Filpula, D., Hardman, K. D., Hubert, S. L., Rollence, M. L., Wood, J. F., Schott, M. E., Milenic, D. E., Takashi, Y., and Schlom, J. (1993). An improved linker for single-chain Fv with reduced aggregation and enhanced proteolytic stability. *Protein Eng.* **6,** 989–995.

Xia, Z., and Liu, Y. (2001). Reliable and global measurement of fluorescence resonance energy transfer using fluorescence microscopes. *Biophys. J.* **81,** 2395–2402.

CHAPTER 5

Fast 4D Microscopy

J. R. De Mey,* P. Kessler,† J. Dompierre,‡ F. P. Cordelières,§ A. Dieterlen,¶ J.-L. Vonesch,† and J.-B. Sibarita#

*École Supérieure de Biotechnologie de Strasbourg
UMR-7175 CNRS/Université Louis Pasteur (Strasbourg I)
BP10413, 67412 IllKIRCH Cedex, France

†Imaging Center of the IGBMC
1, rue Laurent Fries, 67404 Illkirch Cedex, France

‡Institut Curie, Section de Recherche/CNRS UMR 146
Saudou Group, Centre Universitaire
91405 Orsay Cedex, France

§Institut Curie, Section de Recherche/CNRS UMR 146
Plateforme d'Imagerie Cellulaire et Tissulaire, Centre Universitaire
91405 Orsay Cedex, France

¶Laboratoire MIPS, Groupe LAB.EL, Université de Haute-Alsace
IUT Génie Electrique II, 61, rue Albert Camus
68093 Mulhouse Cedex, France

#Institut Curie, Section de Recherche/CNRS 144
Compartimentation et Dynamique Cellulaires, 26 rue d'Ulm
75248 Paris Cedex 05, France

METHODS IN CELL BIOLOGY, VOL. 85

0091-679X/08 $35.00
DOI: 10.1016/S0091-679X(08)85005-4

Abstract

Many cellular processes involve fast movements of weakly labeled cellular structures in all directions, which should be recorded in 3D time-lapse microscopy (4D microscopy). This chapter introduces fast 4D imaging, which is used for sampling the cell's volume by collecting focal planes in time-lapse mode as rapidly as possible, without perturbing the sample by strong illumination. The final images should contain sufficient contrast allowing for the isolation of structures of interest by segmentation and the analysis of their intracellular movements by tracking. Because they are the most sensitive, systems using wide-field microscopy and deconvolution techniques are discussed in greater depth. We discuss important points to consider, including system components and multifunctionality, spatial resolution and sampling conditions, and mechanical and optical stability and how to test for it. We consider image formation using high numerical aperture optics and discuss the influence of optical blur and noise on image formation of living cells. Spherical aberrations, their consequences for axial image quality, and their impact on the success of deconvolution of low intensity image stacks are explained in detail. Simple protocols for acquiring and treating point spread functions (PSFs) and live cells are provided. A compromise for counteracting spherical aberration involving the use of a kit of immersion oils for PSF and cell acquisition is illustrated. Recommendations for evaluating acquisition conditions and deconvolution parameters are given. Finally, we discuss future developments based on the use of adaptive optics which will push back many of today's limits.

I. Introduction

Macromolecular cellular components operate in large "protein machineries" of varying composition and cellular localizations (Alberts, 1998). Elucidating how the concerted activities of several such machines drive cellular processes is often very complex (Alberts and Klug, 2000). Microscopic imaging techniques, visualizing fluorescent fusion proteins in the context of living cells, embryos, and tissues, are providing ways to study the dynamics of structures they associate with (Giepmans *et al.*, 2006; Lippincott-Schwartz and Patterson, 2003; Shaner *et al.*, 2005). This provides quantitative descriptions of key processes and, in several cases, has lead to the discovery of essential short-lived protein accumulations at cellular structures (Draviam *et al.*, 2006; Howell *et al.*, 2004; Lippincott-Schwartz *et al.*, 2001; Mora-Bermudez and Ellenberg, 2007; Presley *et al.*, 2002; Stephens *et al.*, 2000). This quantitative descriptive knowledge is gaining in interest as it provides valuable phenotypic assays for probing gene perturbation consequences in experimental models. The combination of imaging routines with other microscopy-based techniques, such as laser ablation (Skibbens *et al.*, 1995), fluorescence recovery after photobleaching (FRAP) (Reits and Neefjes, 2001), and microscale-fluorophore-assisted laser inactivation (FALI) (Ilag, 2003), allow assaying of gene product residence time and function in a subcellular context. In addition, techniques such as Förster resonance energy transfer (FRET) can allow visualization of *in vivo* interactions between proteins (Chen and Periasamy, 2006; Tramier *et al.*, 2002)or reveal functional states (biosensors) (Ballestrem *et al.*, 2006; Sato *et al.*, 2002; Sekar and Periasamy, 2003). Together, all these approaches make up the discipline of functional cell imaging, the results of which provide significant progress in understanding molecular mechanisms. This recognition continuously stimulates efforts aiming at reaching beyond their limits, and fast 4D imaging is an application for which such attempts continue to be made.

Many processes involve linked events taking place in the cell's volume. In addition, many cellular structures and transport intermediates move very fast (up to 2 μm/sec) and in all directions. These need to be analyzed by collecting, as rapidly as possible, an appropriate number of focal planes with adapted narrow z-spacing and without perturbing the sample by strong illumination. Image acquisition speed and high quantum efficiency of fluorophores then become a major limit. This chapter will discuss exclusively this fast 4D imaging. A fast 4D microscopy system is capable of recording a given volume by optical sectioning over time at high spatial and temporal resolution. It must do this over a sufficiently long time and without perturbing too much the cell activity. Rapid 4D systems include wide-field and multipoint-confocal microscopes, which take advantage of both wide area illumination of the sample and high-efficiency charge coupled detector (CCD) detectors.

4D recording of fast cellular processes has been accomplished by the pioneering work of Fay and colleagues (Carrington *et al.*, 1995; Fay *et al.*, 1989; Rizzuto *et al.*, 1998b). They used wide-field optical sectioning fluorescence microscopy followed by a novel computational image restoration approach, a technique initiated by J. Sedat and D. Agard for high-resolution optical sectioning microscopy (Agard *et al.*, 1989; Hiraoka *et al.*, 1989, 1991). Their imaging system involved software for programming a piezoelectric driver to change the focus by moving the objective relative to the sample in concert with successive exposures of a high-speed CCD. Their results clearly demonstrated the potential of fast 4D imaging (Paemeleire *et al.*, 2000; Patki *et al.*, 2001; Rizzuto *et al.*, 1998a). In the late 1990s, we and others have assembled fast 4D systems using easily accessible components, taking advantage of progress in the fields of CCD cameras, illuminators, observation chambers, and steering software (Gerlich *et al.*, 2001; Savino *et al.*, 2001) (Fig. 1). While working on mitosis, which is controlled by stress sensitive checkpoints and is perturbed very easily, we have faced the necessity of limiting sample irradiation at the same time as speeding up sample acquisition. In our system, the acquisition software was set to operate the camera at frame rate in the overlapped mode and at minimal exposure time (typically 50 msec). The camera drove Z-stepping by controlling a precise Piezo stepper displaying a response time shorter than 2 msec. In this way, a stack acquisition cycle consisted of greater than 90% image collection, and the need for shuttering during (but not between) stack acquisition was avoided. We also wished to monitor two differently tagged

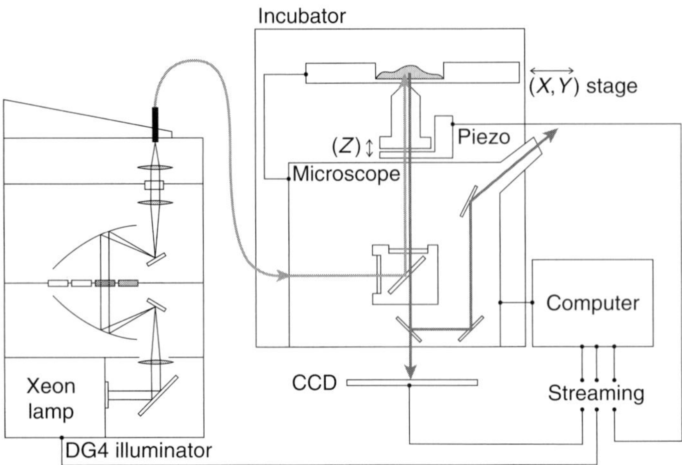

Fig. 1 Scheme of a rapid 4D microscopy system based on the system developed in 1998: The system is assembled on a bottom port inverted microscope (Leica DM IRBE, Leica Microsystems, Rueil-Malmaison, France) mounted on an anti-vibration table with a circular perforation of 25 cm (TMC, Photonetics, Marly-le-Roi, France) and placed into an incubator for temperature control (Life Imaging Services, Olten, Switzerland). A motorized stage (Märzhauser/Leica) holds a perfusable observation chamber mounted into an adapter plate (Ludin chamber, LIS, Olten, Switzerland). It adapts a round 18 mm No. 1 coverslip onto which the cells have been growing or attached and is closed by a round 24 mm coverslip, mounted above the bottom one, creating a chamber holding 0.75 ml of growth medium. The fluorescence is imaged using a 100× PL APO HC NA 1.4 oil immersion objective and captured by a cooled charge coupled detector (CCD) (Roper Coolsnap HQ, pixel size 6.45 μm × 6.45 μm, 1024 × 582 pixels, offering 10 or 20 MHz readout speed and 12 bits dynamics) (Roper Scientific, Trenton, NJ 08619, USA). *Z*-positioning is accomplished by a precision piezo-electric driver (PIFOC P-721.17 with LVDT feed back sensor and E-662.LR Driver, 10 nm precision and 40 nm repetitiveness, Physik Instrumente, Waldbronn, Germany) mounted underneath the objective lens with the help of an intercalating ring, 5 mm thick. In order to accommodate the piezo-driver, the microscope stage is raised with the help of spacers (20 mm high, Leica, Wetzlar, Germany). Illumination is assured by a DG4 instrument (Sutter Instr. Comp., Novato, CA 94949, USA). It houses a 175 W Xenon lamp, the output of which is focused onto one end of a liquid light guide which is coupled to the microscope using the adapter for coupling laser sources to Leica confocal scanning laser systems. It also functions as a modulator for the intensity of the excitation light and as a quick wavelength changer. Filter cubes are mounted in the motorized microscope turret. For single fluorochrome experiments, standard narrow band excitation and emission filter sets are used. For two-color 4D imaging, appropriate high-transmittance dual narrow pass band filter block are used. In addition, in the DG-4 illuminator, adapted short- and long-pass filters are mounted into positions 1 and 2. The light for phase contrast microscopy is shuttered by a Uniblitz shutter, mounted beneath the field diaphragm by an adaptor (Roper Scientific France, Evry, France). The microscope, its stage and filter cubes, the Uniblitz shutter, camera, Piezo driver, and DG-4 illuminator are steered by Metamorph Software (Universal Imaging, West Chester, PA 19380, USA). Installed on a standard computer (PIV 3 GHz, 1 Go RAM, SCSI U2W hard disk) operating under Windows XP Pro, this software drives the acquisition process. It is also possible to use a monochromator instead of the DG-4 illuminator. Using a beam splitter for the emitted fluorescence allows capturing light by two synchronized cameras or side by side on the same camera. In that case, one can use white light illumination.

proteins for studying the spatial and temporal relationships between visualized structures or the dynamic association of proteins with labeled structures. This was achieved by using a fast wavelength changer based on use of a xenon lamp, galvanometric mirrors and appropriate excitation filters (Horrigan and Bookman, 1995). The acquisition software was set to trigger a rapid wavelength change to acquire two images at each *Z* step. This assured that the two fluorescent tags were recorded sequentially at maximal speed, with no moving filters in the microscope. In the late 1990s, existing deconvolution methods required heavy computing, and this was problematic when the very large data sets (1–3 Gb/h) 4D microscopy produces had to be processed. We have developed a solution which enabled batch-wise deconvolution of hundreds of two-color stacks on a normal PC in reasonable amount of time (Sibarita, 2005; Sibarita *et al.*, 2002). Our system has been used in several applications involving a variety of experimental models (Angelier *et al.*, 2005; Gauthier *et al.*, 2004; Jacquet *et al.*, 2003; Piel *et al.*, 2000, 2001; Savino *et al.*, 2001; Thery *et al.*, 2005). It has contributed to the proof that 2C 4D acquisition and subsequent computationally deblurring is feasible within certain limits that we will discuss below. Today, this approach has become standard, and has been implemented and improved on several commercially available imaging systems.

In this chapter, we will first address questions relevant to any type of 4D imaging system and then address those pertaining to using wide-field optical sectioning microscopy and image restoration by deconvolution. We will focus on the requirements for recording subcellular dynamics at the best possible spatial and temporal resolution, enabling subsequent image analysis using image segmentation. This practice necessitates insights into some optical principles and limits imposed by diffraction and optical aberrations.

II. Fast 4D Imaging: Definition, Interest, and Limits

4D microscopy is the combination of optical sectioning and time-lapse microscopy. Its simplest form consists of collecting a few optical planes and producing an appropriate data representation such as maximal intensity projections (MIPs). This can significantly improve monitoring and measuring correlated or linked events. Observing combined XY and XZ MIPs, anaglyphs, or stereo projections often produces interesting additional qualitative insights (Fig. 2). The additional burden for the sample consists in the extra illumination energy it must endure. Within the framework of quantitative biology, an important goal of a 4D experiment is obtaining data sets allowing for the isolation of structures of interest by segmentation, and analysis of their intracellular movements by tracking (Fig. 2) (Eils and Athale, 2003; Schiffmann *et al.*, 2006). This yields quantitative information on parameters such as speed, directionality, and mean square displacement of such structures. How to adapt acquisition settings is a question that

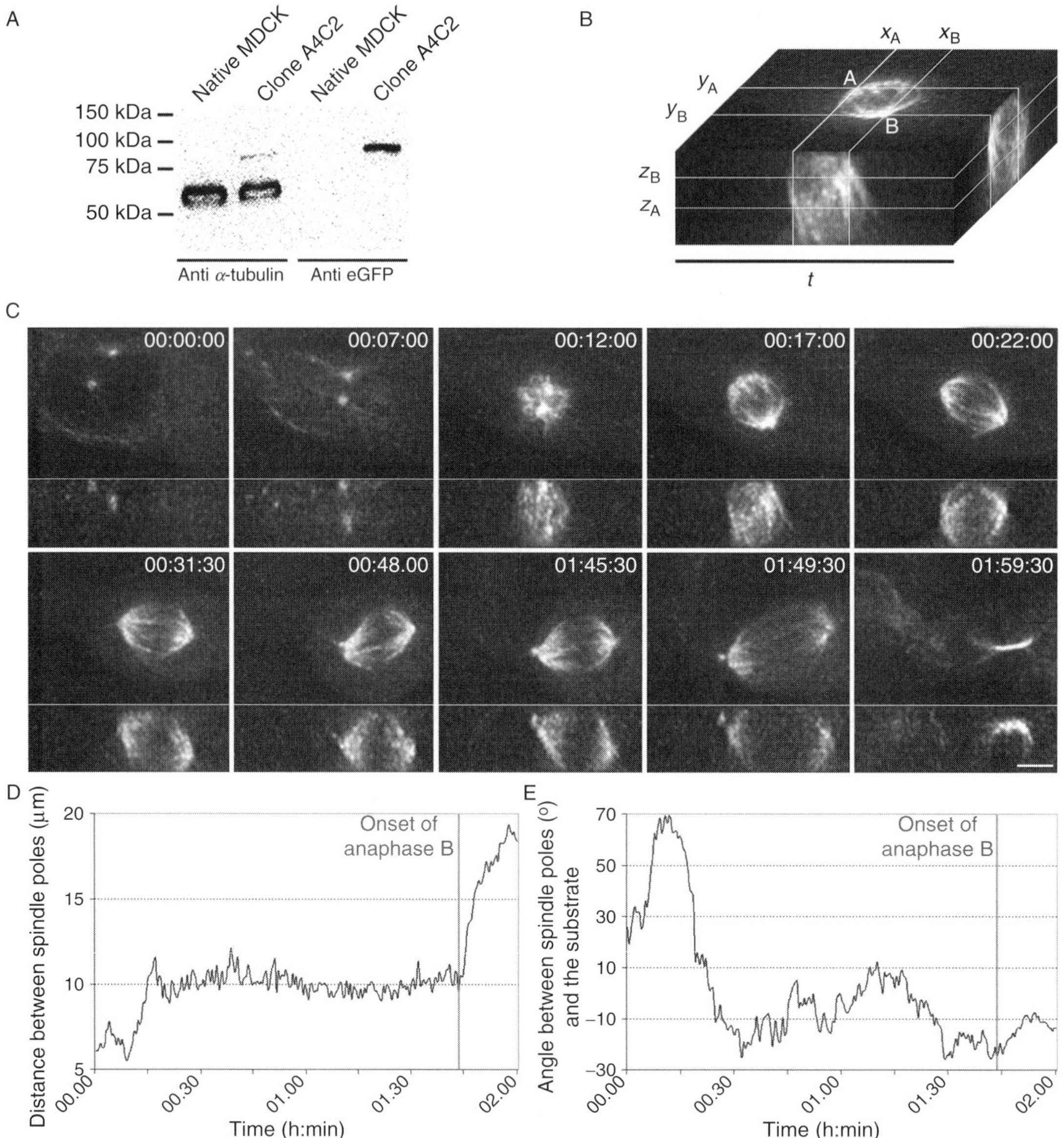

Fig. 2 4D tracking of spindle positioning: 4D analysis of spindle orientation in living Madin Darbey Canine Kidney (MDCK) cells. A stable clone expressing enhanced Green Fluorescent Protein (eGFP)-tagged α-tubulin expressing low levels (A: 11% of endogenous tubulin) was used. The cells were grown to confluency on glass coverslips. Image stacks comprising 34 planes were collected with an interval of 20 sec. The mean duration of prometaphase was 38 min ± 7 ($n = 44$), indicating no spindle defects or problems with spindle assembly or chromosome segregation occurred. In order to extract the 3D coordinates of the pole, the structure was tracked five times on two orthogonal maximal intensity projections (MIPs) (XY MIP and XZ MIP) using the "track point" tool of Metamorph (B). The mean

will be discussed below. Adequate data sets can be obtained using simple 3D systems, provided sufficient signal is generated without perturbing the process under study. Volume of acquisition, acquisition speed, frequency, and duration as well as the intensity of the illumination are closely linked parameters. Depending on the dynamics of the biological process of interest, some compromises have to be made. We consider that as soon as the study implies movements exceeding 0.2 μm/sec and necessitates collecting more than six planes, 4D recording requires more advanced equipment and precautions. For example, this is the case when the structures of interest contain low copy numbers of proteins and move very rapidly (faster than 1 μm/sec). Working with low signal levels can easily put a system to the limits imposed by the physical parameters of its components and of the fluorescent tag. In many cases, increasing the signal by overexpressing the fluorescent protein leads to undesired phenotypes. The case of carrier intermediates in endocytosis, transporting a limited copy number of a given protein in any direction and at speeds exceeding 2.0 μm/sec, is an example of a phenomenon approaching today's limits. Nevertheless, combination of such high-speed 4D acquisition with laser-assisted techniques like FRAP or fluorescence correlation spectroscopy (FCS) can help in getting quantitative measurements even for very high-speed biological processes.

III. Points to Consider Before Working with Fast 4D Imaging Systems

A. Imaging Modes and Combination with Other Functionalities into a Multifunctional System

Until recently, fast 4D imaging was thought to necessitate the wide-field mode of illumination because it could collect images faster and with greater sensitivity. Recent developments on confocal microscopy systems, including the combination of optimized spinning discs and electron multiplying (EM) CCDs on the one hand and much faster "conventional" confocal instruments on the other, have revived the debate on the question whether to use a confocal or a wide-field microscope. Nevertheless, since the confocal process consists in rejecting a considerable part of the emitted photons collected by the objective, the wide-field microscopy will remain the most efficient system, and can make the difference in the case of extreme conditions.

Conventional laser scanning confocal systems collect data on punctual detectors [photo multiplier tubes (PMTs)], and have the important advantage of inherently

values were used to calculate the distance between poles (D), the angle between the spindle axis (E), and horizontal plane and their velocity (not shown). Images shown in (C) are MIPs of deconvolved stacks, viewed in top (XY MIPs) and side (XZ MIPs) views. At nuclear envelope breakdown, the spindle axis is perpendicular to the horizontal plane (C, times 00:07:00–00:12:00). The spindle axis rapidly aligns to within 20° of the substrate. It undergoes a series of smaller reorientations resulting in its final orientation (C, time 00:12:00–01:45:00). Deformations of the spindle pole region followed by reorientations can be observed (C, time 00:48:00). (See Plate no. 12 in the Color Plate Section.)

displaying other functionalities such as using the laser beam for illuminating regions of interest with increased power for FRAP or photoactivation (PA) applications for example, or for simultaneous multicolor imaging. They can also benefit from biphotonic excitation using femtosecond pulsed laser illumination. Two laser sources, one for bleaching or PA and another for visualization, can also be combined. Nevertheless, the price to pay to make high-speed acquisitions is to excite the sample with a very high intensity during a very short amount of time.

On the other hand, confocal spinning disk and wide-field systems collect data on a CCD which consists of a checkerboard-like array of light sensors, which will correspond to the pixels in the digital image. Modern CCDs combine small pixel sizes (6.45–8 μm on a side), excellent quantum efficiencies (>90%) and high speed (>20 MHz), and virtually noise-free readout. It is also recently feasible to equip CCD-based systems (both multipoint confocal and wide field) with additional illuminations for FRAP/PA, but also for total internal reflection fluorescence (TIRF) applications (Toomre and Manstein, 2001). As optical paths for imaging and laser perturbation are completely separated, there is no delay between the two modalities, increasing the capabilities of the system for measuring highly dynamic processes. Acquiring two wavelengths simultaneously is also possible through the use of beam splitters at the emission side. This can be achieved using either a dual-view system displaying the two colors on the same chip at the expense of decreasing the field of view and increasing the readout time, or a dual-cam system sending the two colors on two different CCDs. The discussion below will reveal that both confocal and wide-field systems display advantages and disadvantages, and that the combination of various modalities under the same instrument is certainly the best solution to monitor complex biological processes.

We also strongly recommend any newcomer to carefully assess not only the type of applications that will be performed in the short term with a given system, but also others which may be more demanding. As a general rule, a system offering "advanced" possibilities will be optimal for "normal" applications.

B. Keeping the Sampled Volume Immobile and Test for It

1. The Components of Fast Multidimensional Systems

a. Minimal Requirements

An essential requirement of multidimensional imaging, and even more so for fast versions, is that the volume sampled by optical sectioning microscopy remains immobile within well-defined limits, and that the light source produces an even and stable illumination. These requirements are particularly stringent when data sets need restoration by deconvolution. Practically, all high-end microscopes have excellent inherent mechanical stability, but differences exist and their implementation in an imaging system may further impact overall performance. In the following, we list the essential components and comment points to survey.

b. The Microscope Table and Microscope setup

A vibration dampening table is an absolute necessity. All the elements susceptible to produce vibrations should be mounted without contacting parts of the microscope. Microscopes equipped with a CCD camera are best equipped with a bottom port and positioned above a hole in the table. This allows the most efficient light path, keeping the image capturing device outside the heating chamber and letting more accessibility to the microscope (see below). If the system has to be equipped with accessory equipment such as a microinjector, or a perfusion system, careful planning is a necessity. For example, for combining microinjection with CO_2 control, we are using the LIS Brick CO_2 control system (Life Imaging Services, Switzerland), a LIS heating cabinet and observation chamber in combination with a microinjection system. This allows microinjection without interrupting CO_2 control. Equipping the system with a motorized high-precision stage enables both routines, facilitating finding cells of interest and multipositioning acquisitions (for screening purpose for example).

c. Temperature and CO_2 Control

We recommend using a heating cabinet into which the whole microscope, and its peripheral components such as a microinjection system, is fitting. This allows much higher focus stability, indispensable for time-lapse experiments. As heated air movement is provided by fans, the system should be unable to transfer vibrations it generates onto the microscope. When selecting commercial materials, the access to the microscope table and other microscope parts must also be considered carefully.

d. Z Movement Control

Standard equipment is provided by a piezo stepper for moving the objective(s), or by a galvanometric sample holder, mounted on the microscope stage. The first allows faster movements and leaves the sample immobile by moving the objectives, while the second has the advantage to allow the displacement for all the objectives. *Z*-step precision and repeatability (the error made when repositioning at the original *Z* position) is important, and should be well within the optical resolution of the system. We have obtained good results with a precision of 10 nm and repeatability of 50 nm, well within 10% of the system's optical resolution.

e. Wavelength Selection and Changers

Conventional confocal microscopes use a combination of lasers and wavelength selectors and several PMTs. Spectral detection makes it easy to define on-the-go so-called "digital fluorescence filters," allowing separation of close-emitting fluorophores. Combined with spectral unmixing, they are very suitable for multiwavelength acquisition. Spinning disc confocal microscopes are somewhat less flexible in this respect and use either slower filter wheels for wavelength selection or dedicated excitation/emission combinations for faster imaging. Laser illumination imposes the selection of fixed excitation wavelengths, which are not always optimal for the selected fluorochromes.

For wide-field systems, a Xenon lamp is preferred since it provides an even spectrum and excellent photon stability which is crucial for image restoration. Common practice today is to use a monochromator which allows selecting a narrow band excitation. Combined with appropriate excitation filters, each fluorophore can be excited optimally. Selecting an offset with respect to the excitation pass band or narrowing the pass band width allows for precise and reproducible regulation of excitation energy. A monochromator changes the excitation wavelength in less than 2 msec, and is therefore compatible with the overlapped mode of functioning of interline or frame-shift CCD captors (see below). A multipass band filter set thus allows recording two or more colors sequentially per *Z* position, without loss of time and the need for shuttering. For splitting two colors on a single CCD chip, or using two CCD working in parallel, a multiwavelength excitation is required. In this case, a multibands excitation filter with white xenon illumination will be preferred. It is therefore possible to make simultaneous double color acquisitions at high frame rate. In combination with rapid readout CCDs, these options make dual color, rapid 4D imaging feasible.

f. Integration of Z-Stepper and Wavelength Changer with Overlapped Mode of Functioning of the CCD

Interline and frame-transfer CCD shift the electrons integrated into the capture wells into a readout frame. Provided integration time is equal or larger than the readout time, a new illumination/integration cycle can be started within 2 msec. The signal produced at the moment of the shift allows steering a fast-reacting *Z*-stepper (piezo stepper) or/and a fast wavelength changer (monochromator or DG4). Mechanical shuttering and the use of filter wheels are avoided during streaming acquisition, thus contributing to the system's overall stability. In this mode, more than 90% of the recording time is concerned with data collection.

g. Type of Observation Chamber

A last component which can contribute a lot to the system's performance is the observation chamber. In our hands, coverslip bottomed plastic dishes are not suitable. We use an LIS Ludin chamber (Savino *et al.*, 2001). It is made of stainless steel and is assembled around a No. 1 18 mm diameter glass coverslip. It can be used in the closed or open configuration and can also be perfused. In the open configuration, gas exchange is allowed while limiting evaporation by covering the medium with paraffin oil. It is screwed into an adaptor plate fitting in the microscope table.

h. Controlling Focus Drift

Fast 4D experiments often do not last very long and, therefore, the mechanical stability is often not tested for. Our experience, however, has shown that considerable *x*, *y*, and *z* drifting can occur during the time following mounting the observations chamber on the stage (see below). The slightest temperature shift also induces notable drifting. If the system must also be used for TIRF experiments or for long duration recordings, it is recommended to extensively test for drifting.

Two kinds of devices exist to monitor the *z* drift. The first one is based on a mechanical sensor allowing the precise position of the stage to be recorded. While correcting for positional drift this does not correct for misfocusing due to thermal perturbation. The second is based on the use of an LED, the light beam of which is sent the coverslip. Analyzing the reflected light allows determining the front lens/coverslip distance and further correcting by sending the information to a feedback loop. As, however, both introduce a monitoring step, an additional time delay is added to the acquisition sequence.

2. Testing for System Performance with Respect to Mechanical and Optical Stability

We use point spread function (PSF)-Speck microspheres (Microscope Point Source Kit P-7220, Invitrogen, Molecular Probes) displaying a diameter of 0.175 ± 0.005 μm and emitting blue, green, or red wavelengths which we also use for measuring PSFs (see below). Coverslips with attached beads (see below) are mounted with water in the observation chamber, which is subsequently fixed into the chamber holder. In order to test the system's multiple color conditions, a mix of different color beads is used. Prewarmed water is used, and upon focusing on a few microspheres, a time-lapse sequence (1 image every 10 sec) on one focal level is started. We monitor a microsphere on the screen with a strong digital zoom factor until it no longer drifts. This usually takes 15–20 min! We then start a 3D time-lapse sequence using conditions identical to those of the experiment. The low light conditions assure that no bleaching occurs. The result is analyzed visually using a 400% zoom factor by checking that a microsphere stays focused on the initial *Z* position and does not vibrate or drift. Plotting signal intensity over time at the best focus of the microspheres is used for evaluating the system's optical and mechanical performance (Fig. 3). Comparing the Z-level of the best focus for each color also characterizes the system's chromatic aberration (see below), and allows correcting for this factor. This can be done optimally using multicolor beads (Tetraspec, Molecular Probes Inc.), which contain four fluorophores inside the same 175-nm diameter bead, allowing to measure chromatic aberrations. A mix of single color beads can also be used to characterize the cross-talk between colors.

C. The Impact of Optical Blur, Noise, Aberrations, and Calibration Defects

We recommend reading some reviews and book chapters which describe in detail the principles of image formation in a fluorescent microscope and the problems caused by aberrations, noise, and calibration defects (Bolte and Cordelieres, 2006; Conchello, 1998; Dieterlen *et al.*, 2002; Hanser *et al.*, 2004; Holmes, 1992; Scalettar *et al.*, 1996; Sibarita, 2005; Swedlow and Platani, 2002; Wallace *et al.*, 2001). Here, we will summarize these only briefly and discuss their consequences when using fast 4D imaging of living samples.

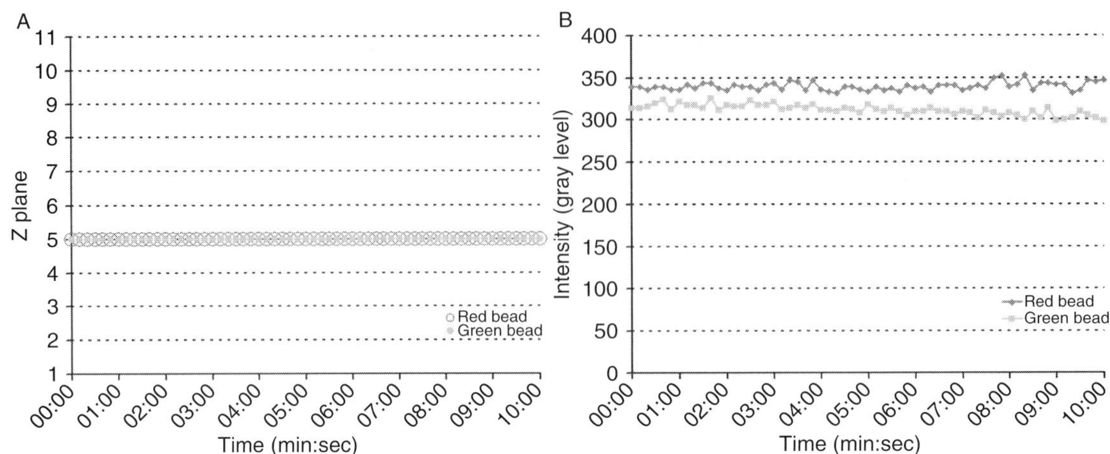

Fig. 3 Stability testing using 175 nm fluorescent beads: A mixed population of green and red 175 nm fluorescent beads were attached to a coverslip, mounted in the observation chamber with cell growth medium, and recorded in two-color 4D mode. Sixty stacks each consisting of 2×17 images, taken at 50 msec exposure times, were recorded at 10 sec intervals. Using Metamorph, the center of individual green and red beads extracted from the same stack was detected automatically and their x, y, and z coordinates determined in each of the stacks. The x, y (not shown), and z values (A) never changed and (B) the maximum intensity of the beads remained stable within 10%. This demonstrates that the Piezo positioner, observation chamber, and microscope incubator form a solid assembly which is compatible with fast two-color 4D imaging. (See Plate no. 13 in the Color Plate Section.)

1. Optical Blur and Noise

A fluorescent sample is a collection of spatially distributed point source objects which can only be imaged imperfectly. Each image of the sample is formed by the optical system and projected on an electronic device composed of a discrete detector that will spatially sample the signal and translate the light signal into an electronic signal for further processing by the computer. Most importantly, however, for both confocal and wide-field microscopy, the image formation process is the convolution between the object and the PSF. Therefore, any microscope produces a distorted image of the object, mostly due to optical blur and photon noise. Optical blur is inherent to any optical system. It is reproducible and a consequence of light diffraction. The image of a single point source object formed by a microscope is called the PSF. It consists of a bright disc in the center and a series of weaker concentric discs in 2D called Airy discs (spheres in 3D). In fact, the image of the point source is a 3D diffraction pattern centered on the best focus image plane. The intensity distribution and shape along the z-axis appears elongated, like a rugby ball, and in the absence of aberrations, displays an axial symmetry along the z-axis as well as symmetry with respect to the center of the pattern with an hourglass-like shape.

From the above, it is clear that even in the "ideal" case, what we see through a microscope, and on the screen of a computer, is only a distorted representation of the reality. Optical blur indeed not only limits optical resolution, but also strongly diminishes the intensity and therefore the original contrast of the object. Out-of-focus haze reduces the contrast in wide-field images leading to big bright fluorescent sources hiding well-focused small weak objects.

Noise also strongly contributes to image degradation, in particular, in the case of low intensity objects. It is a random process which mainly arises from the statistical distribution of the photons following a Poisson distribution. In practice, image quality is a function of the number of detected photons, which depends on the sample, opacity of the specimen to the photons, excitation energy, the fluorescence filter set, the integration time, and the quantum efficiency of both the fluorophore and the detector. Optimizing each step in the optical pathway, including optimizing fluorescence-tagged protein expression levels in the sample, choice of the fluorescent protein or fluorochrome, the pH sensibility of the latter, etc., and choosing the right detection device can therefore greatly influence the success of a rapid 4D recording sequence.

2. Image Restoration by Deconvolution

Confocal microscopy partially solves the problem with out-of-focus haze by placing a pinhole before the detector, through which only the main peak of the Airy disc is transmitted. The consequence is a spectacular reduction of haze and a gain in axial resolution. Nevertheless, the image at the focal plane remains distorted by diffraction and the pinhole rejects almost 90% of the emitted signal while the excitation process remains the same. This requires the expense of more excitation energy, in order to get images with sufficient signal-to-noise ratio for further analysis of the labeled structures, therefore leading to a greater bleaching rate, as we observed by direct comparisons.

When the PSF of a microscope is known, image restoration algorithms can be used to reassign the optical blur to its original location. Upon each iteration, a new estimate of the object, closer to the real object, is obtained. At the end of the iterative process, the blur is reduced and the signal considerably increased in the regions of the structures of interest (Fig. 4).

Despite the problems due to aberrations discussed below, deconvolution represents therefore a great interest for fast wide-field 4D imaging. Wide-field fluorescence microscopy collects much more of the available signal on the detection device so that excitation energy can be limited. Structures can be acquired at a contrast level too low for segmentation, but upon deconvolution, they get these properties. 3D confocal images can also be improved by deconvolution. Because of the axial resolution improvement caused by the pinhole, image sampling along the z-axis must be finer, a requirement very often incompatible with rapid 4D imaging.

The noise present in the image considerably limits the efficiency of deconvolution algorithms. It must be taken into account, either directly in the algorithms

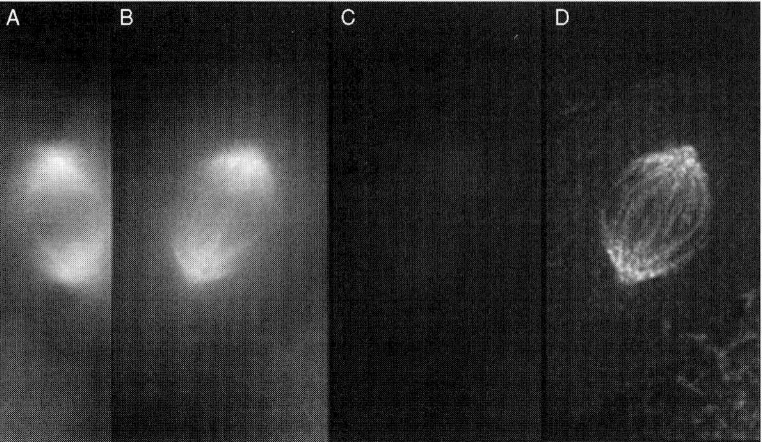

Fig. 4 Image restoration by deconvolution: A living mitotic MDCK cell expressing α-tubulin-eGFP recorded with a refractive index of 1.520. This is the index yielding the best PSF at the coverslip (see Fig. 5). (A) A subregion in (B) was tilted 90°. The bottom is to the left. Note the asymmetric halo of light diffusing from the spindle towards the bottom and top. (B and C) XY MIPs of the same image stack shown in different look up tables (LUTs): in (B) matched to that in (A), in (C) matched to that of the deconvolved stack in (D). See also Fig. 8. Deconvolution was done using a synthetic PSF, maximum likelihood estimation (MLE) algorithm and 60 iterations.

themselves or by adapted prefiltering. It follows that a minimum signal-to-noise ratio must be achieved by using proper acquisition conditions on living cells (see below). In practice, this comes as always down to finding the best compromise between cell viability, spatial and temporal resolutions, and final image quality.

3. Spatial Resolution and Sampling Conditions Compatible with Deconvolution

For studying intracellular processes and improving image quality by deconvolution, the spatial sampling should occur at conditions approaching the Nyquist theorem: sampling must be less or equal half the optical resolution. In this respect, the size of the CCD's pixels is important and must be compatible with the objective's magnification and numerical aperture. In wide-field microscopy, lateral sampling Δ_{xy} is achieved by the CCD chip and is a function of the CCD pixel size P_S, camera binning b, and objective magnification M, as follows:

$$\Delta_{xy} = \frac{P_S b}{M}$$

For example, using a 100× objective with high numerical apperture (NA) (1.25–1.4), CCDs with a pixel size of 6.5 μm width yield an image pixel size of 0.65 μm and can therefore be used with a binning of 2 × 2. Indeed, the image pixel size will be 0.129 μm, which very close to the Nyquist theorem of 0.100 μm. Binning increases the photon sensitivity (sensor surface × 4) and readout speed (number of pixels to read out 4× less) and is essential when using a nonamplified CCD. The *Z*-step size between focal planes is also following the same sampling rules, and a 300 nm plane to plane distance is well suited for high NA objectives.

We will describe below how wide-field 4D imaging can be realized using low energy illumination (<1 mW at the fiber optic outlet) by selecting empirically conditions able to generate the required image quality after the deconvolution process. Before that, however, we need to discuss an important limitation to the technique, when using oil immersion objectives.

4. Spherical Aberrations and Their Consequences for Axial Image Quality and the Success of Deconvolution of Low Intensity Image Stacks

The image formation as described above, assumes that the 3D image formation process of a fluorescence microscope is invariant over the *Z*-distance, and to a lesser extent also in the *xy* plane. For live cell recording, however, this situation is never met with, because the emitted light passes through stratified media with different refraction indices. The refractive index *n* within a living cell is variable ($n =$ ca. 1.41) (Kam *et al.*, 2007), but in general closer to that of water ($n = 1.33$) than to that of immersion oil (usually $n = 1.518$ at 22°C). The *n* of a cell observed at 37°C is far from the *n* of oil at 22°C, for which optics are designed. In addition, few objective lenses are corrected for work at 37°C. Consequently, when light is emitted from a distance well above the coverslip, it passes through various layers with different refractive indices, resulting in the generation of spherical aberrations (Hanser *et al.*, 2004; Scalettar *et al.*, 1996). Spherical aberrations are characterized by an asymmetry in the shape of the PSF along the optical axis, an increase in the flare of the PSF, and a decrease in its maximal intensity. The phenomenon gradually worsens as the focal plane moves away from the coverslip, deeper into the sample (Kam *et al.*, 2007). Empirically, one observes image quality deteriorating notably beyond 5 μm and severely beyond 15 μm. This can be observed on isolated small structures located far from the coverslip. It is useful to remind here that both wide-field and confocal microscopy are affected. The problem is that wide-field image stacks must be restored by deconvolution, and that current methods work with a spatially invariant PSF. It is our frustrating experience that this leads to additional artifacts, sometimes even leading to the suppression, instead of enhancement of structures in deeper regions within the sample. It also makes segmentation and tracking, nearly impossible, and quantification of signals on structures moving through the stack illusionary.

These problems are well known by specialists, but not widely recognized by biologists, and today, clearly impose strong limitations to the use of rapid 4D imaging on thick samples. For these purposes, it may be recommendable to use a rapid confocal system, even if the image quality along the *Z*-axis also decrease notably.

Objective lenses used at 37°C should be kept at all moments at 37°C, since temperature changes cause "fatigue" which impacts negatively on the lenses performance as reflected by its PSF.

In practice, in wide-field microscopy, losses by deconvolution can be limited by using higher illumination energy, changing optics (Hiraoka *et al.*, 1990), or by computational approaches (Kam *et al.*, 2001).

In principle, it would be better to use a water immersion objective lens, but these are limited to a 60× magnification and are less sensitive, especially when combined with a 1.5× zoom lens, in order to keep the image pixel size small enough. The nearby development of adaptive optics compatible with rapid fast 4D microscopy and optimized objective lenses will offer a significant improvement and allow full exploitation of the advantages of wide-field deconvolution microscopy (Kam *et al.*, 2007). One of the key difficulties remains the measurement of precise PSFs which faithfully represent the spatial variations of living cells (Hanser *et al.*, 2004).

5. *Z*-Distance Calibration

Focusing light through layers with different refractive indices introduces geometric distortions along the optical axis. These factors are strongly linked to the numerical aperture of the objective (Zwier *et al.*, 2004). It is therefore necessary to calibrate the acquisition system to avoid wrong measurements and for optimal sampling. This can be done by imaging fluorescent 6 μm calibration beads, attached to the coverslip. In ZX cross-sections through their center, they appear elongated and pear shaped. Calibration is achieved by introducing in the acquisition software an axial correction factor (usually ranging from 1.5 to 1.8) (Fig. 5). An easy additional check is the number of sections needed to pass through such beads (should be 20 for a step size of 300 nm). This correction is crucial since axial sampling errors cause major problems in deconvolution processes using a theoretical PSF. In addition, without correction, *Z*-intervals are too small and the sample

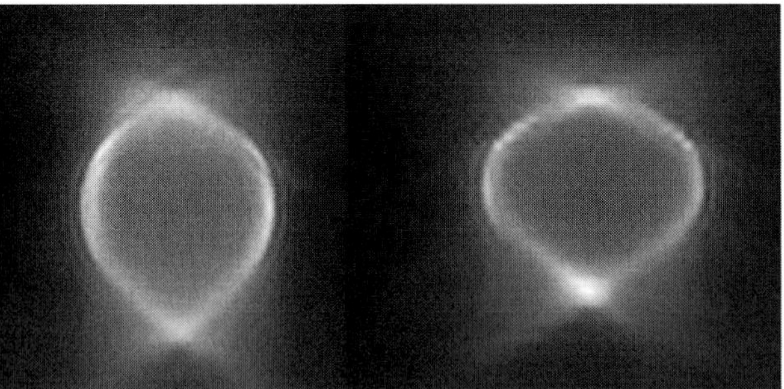

Fig. 5 *Z* calibration using 6 μm beads: XZ sections through a 6 μm calibration bead. Without correction (left), the bead is elongated. Application of a 1.6 correction factor corrects this aberration in part. With this factor, the bead is sectioned in 20 steps.

is oversampled, leading to unnecessary photobleaching, cell damage, and decreasing speed acquisition.

D. Setting Up a Rapid 4D Acquisition

1. Sample Preparation and Mounting onto the Microscope Stage

Rapid 4D imaging is not much more difficult to perform than any other time-lapse imaging experiment. For use with Ludin's chamber, cultured cells are seeded on No. 1 18 mm diameter glass coverslips, placed into six-well plates, and cultivated for 36–72 h, during which time various kinds of treatments can be applied. When using other samples, it is important that at least one face of the specimen is touching the coverslip, since only here, image formation is optimal. The coverslip is mounted in the observation chamber which is then placed in the stage adaptor and the CO_2 incubator. When using the chamber in the open configuration, it is carefully overlaid with paraffin oil. The cells are kept for a further 20 min during which pH regulation and stabilization of the assembly in the x, y, and z dimensions takes place (see above).

2. Selection of the Object and Region of Interest

The sample is screened visually or automatically until a suitable object is found. Automatic screening can also be performed, involving incrementally moving the motorized stage, while taking a full chip image at each position and recording the stage coordinates. These images are taken using low light level illumination, 3×3 binning, and short exposure times in order to limit photon damage and photobleaching. Once the most suitable cell is selected, the stage is positioned to put it in the field of observation. A region of interest (ROI) surrounding the cell is set to minimize the number of pixels to read and increase acquisition speed if necessary. Choosing a square ROI size taking the form of 2^X will facilitate subsequent image processing such as deconvolution as general transforms such as fast fourier transform (FFT) have been optimized for such pixel matrix sizes.

3. Selection of the Volume and Camera Settings

While displaying images live on the screen, the piezo controller is moved up and down to mark the top and bottom limits. Ideally, these limits lay a few micrometers below and above the cell's limits. It should be reminded that the number of pixels within the ROI determines the readout speed of the camera and the minimum exposure possible for overlapped mode compatibility. It is therefore important to select the exposure time long enough to allow for full speed (see below). Whenever the recording must be performed with a minimal time lapse between time points, the data of a recording should either be temporarily stored in the RAM or streamed directly on the hard disk. With a typical 20 MHz CCD device set at binning 2 × 2, and for a 512 × 512 pixels ROI (ca. 33 μm^2 for a 6.45 μm pixel and a 100× objective), this minimum is around 50 msec. Whenever possible, conditions close to the minimum exposure time, allowing the maximum speed, should be selected. The *Z*-interval is usually fixed at 300 nm for 1.4 NA objectives. For example, using these settings, a 10 μm thick cell will be recorded in 50 optical sections (covering a *Z* distance of 15 μm with the cell included in the 33 center sections of the stack) per fluorophore. This takes slightly more than 2.5 sec per stack. For recording movements faster than 0.5 μm/sec, the acquisition frequency must be increased by recording a smaller thickness (fewer optical sections), decreasing the size of the ROI, or both. Eight sections still provide quite a lot of image improvement upon deconvolution. Using 35 msec exposure time and eight *Z* positions, four stacks per second can be recorded. Faster and more sensitive CCDs, in principle, allow further speed improvements; even with such a high frame rate, the number of photons becomes very low, leading to very noisy images. With very strong illumination, this is still feasible, since very fast movements only need short recordings.

4. Selecting Illumination and Camera Parameters

This step and the next one are crucial for obtaining the best possible data sets. The maximal output from a 75 W Xenon lamp housed in a monochromator and guided into the system through an optical fiber is about 4–8 mW with 10 nm bandwidth. As 1 mW is often enough, intensity attenuation can be performed by using neutral density filters or/and by changing the excitation wavelength relative to absorption spectrum of the fluorophore. For other fast illuminations devices, like the DG4, it is possible to precisely attenuate the excitation energy by computer control. The operator should know which fluorescent structures are important for the subsequent analysis. Suitable acquisition conditions for visualization, segmentation, and tracking can be adjusted by performing preliminary deconvolution tests for various acquisition conditions. This involves collecting various stacks using a range of illumination settings with the shortest possible recording time. These stacks are then restored by deconvolution (see below) and scored for compatibility with quantification techniques. An important point to consider is the degree of photobleaching and photodamaging of the sample, which are often linked phenomena. If the acquisition conditions, for example, cause notable bleaching, slow down the kinetics, or prevent the cell from entering into mitosis, the amount of total light must be reduced. This can be done by decreasing the sampling frequency, the total duration of the recording, and the excitation energy. For living samples enclosing rapidly moving structures, it is not a good idea to increase the acquisition time per time point, since this will generate artifacts in the deconvolution process. Since the goal of deconvolution is to "reassign" the light to its original place, an object must not move more than 0.25 μm (about 2 pixels) during the optical sectioning acquisition process of a 3 μm thick volume.

For example, for structures moving at 0.5 μm/sec, this is achieved by recording 10 sections in 0.5 sec.

5. Recording a PSF

The PSF describes the way in which the objective distorts the image of each point of the sample imaged during the acquisition process. This information is required for most deconvolution algorithms. The accuracy and quality of the PSF strongly impacts the performance of any deconvolution algorithm. Noise, aberrations, and incorrect scaling of the PSF may cause major artifacts in the restored image. The PSF may be calculated theoretically or measured empirically. The former is sometimes preferred, since it avoids the somewhat tedious procedure for measuring a PSF. The main advantage of theoretical PSFs is the absence of noise, but their problem is that they apply only to perfect lenses and well defined and calibrated optical paths. In reality, these are never encountered in practice. Most importantly, any discrepancy between the theoretically estimated PSF and the actual PSF of a microscope may result in faulty results, even close to the coverslip. After extensive quantitative testing, it is our experience that a measured PSF is always outperforming a theoretically estimated PSF. Nonblind deconvolution is also superior to blind deconvolution, which performs well for higher intensity images. Moreover, a PSF measurement can detect problems in the microscope setup that may not be evident when observing complex biological samples. There are indeed many possible sources of problems, including oil mismatch, objective defect, drifts and vibrations from the heating system, or CCD cooling. In the face of these potential sources of signal degradation, it is strongly recommended that the microscope's optical properties, stability and reproducibility, should be checked with microbead samples. For accurate determination of the microscope's PSF, it is essential to measure the PSF under the same sampling conditions as used for the acquisition. The following provides an easy protocol for preparing PSF samples and recording high-quality PSFs.

Preparation of bead samples and sample conditioning

1. PSF-Speck microspheres (Microscope Point Source Kit P-7220, Invitrogen, Molecular Probes) with a diameter of 0.175 ± 0.005 μm, are used as point sources for PSF recording. Kits with multiple wavelengths covering the full visible spectrum range are available.
2. For attachment to coverslips, coverslips are placed in six-well plates and covered by a 100 μl drop of 0.025% poly-L-lysine (Sigma), incubated for 10 min, washed with water, and air dried. Microspheres are appropriately diluted (around 1000×) and a 100 μl drop is placed on each coverslip. A suitable dilution yields preparations in which the beads are sufficiently widely spaced, so that the signal of an individual bead covers a region of 600 × 600 image pixels without the presence of signals from neighboring beads. After 10 min, excess liquid is drained off, the coverslip briefly rinsed and mounted into the observation chamber.

3. Before recording a PSF, a microsphere is focused and imaged periodically on the screen with a strong digital zoom factor (400%), until it no longer drifts. As mentioned above, this can take up to 15–20 min! With the help of the zoom factor, the bead is carefully focused on the preview window to yield its maximum intensity in such a way that a symmetric pixel pattern is visible. A short time-lapse movie, using 2-sec intervals, is recorded to assure that the pixel pattern is stable, reflecting the lack of drifts and vibrations.

Most reports, including our own, recommend using measured PSFs having the same sampling than the acquired images of the sample. In our case, this involves setting the camera at binning 2 × 2 and the plane spacing to 300 nm for 100× 1.4 NA objectives.

1. PSF acquisition protocol
   - A suitable bead is focused and surrounded by a 512 × 512 ROI.
   - *Z*-step is set at 300 nm, starting at 5.1 μm below the best focus, and the number of *Z* planes adjusted to 34 planes.
   - The exposure time is chosen to yield a value inside the central pixel close to 4000 grey levels (for 12-bit cameras) to reach the maximum dynamics of the CCD camera.
2. PSF treatments
   - The stack is cropped in a 256 × 256 × 30 planes sub-volume around the centered bead, with the goal to entirely contain the most out-of-focus rings.
   - The arithmetic background subtraction sets the lowest value pixels to zero. In case of variable background, which can arise when using some detectors in specific acquisition modes, adapted background subtraction methods have to be applied to perform a better PSF extraction.
   - To increase the signal-to-noise ratio (SNR) of the measured PSF, especially in the out-of-focus regions, in which signal intensity is weak, it is recommended to average the acquisitions. To simplify the acquisition process, the same bead can be acquired many times. It is our experience that when using illumination energy lower than 3 mW, bleaching of beads is negligible. Knowing that the SNR is proportional to the square root of the number of photons, the number of images needs to be high (16 to improve by a factor of 4) to subsequently improve the image quality in the out-of-focus regions.
   - The SNR can be further increased by applying circular averaging with respect to the optical axis, provided the original bead acquisitions show the symmetry predicted by the theory.
3. Multiwavelength PSFs
   - The PSF is a function of emission wavelength and the objective must therefore be characterized with different color beads. Each image stack corresponding to a given fluorophore must be deconvolved with the corresponding PSF.

- Multicolor beads (Tetraspec, Molecular Probes Inc.), conjugated with four different fluorescent probes, provide a range of emitted wavelengths ranging from blue, green, red to infrared. This enables spectral characterization of the objective. Importantly, these measurements can be used to check, and even to correct after acquisition, the nonapochromatism of the lens as discussed previously. We indeed observed *Z*-shift ranging from 0.2 to 0.3 μm between colors, especially between blue and green and between red and IR, even for a set of selected objectives!

6. Counteracting Spherical Aberration

Whenever the structures of interest have incorporated enough fluorescence and the volume to be sampled is within 5 μm of the coverslip, no special precautions are needed. Standard, low intrinsic fluorescence immersion oil can be used. For even better results, it is recommended to

1. Use a series of immersion oils (1.516, 1.518, 1.520, 1.522, and 1.524, from Gargill, series A) for recording PSFs of point sources at the temperature of the experiment (37 °C in our study) (Fig. 6).
2. Choose the refractive index which provides the most symmetric PSF along the optical axis.
3. Use this refractive index also for recording image stacks.

When the structures of interest are also occurring further away than 5 μm from the coverslip and yield signals which cannot be distinguished clearly because of their weak intensity, it may happen that in the deeper regions of the sample, these will be artifactually disappearing upon deconvolution. This can, in part, be avoided by

1. Acquiring optical sections of the sample using refractive index of 0.002–0.004 points higher than that providing a symmetric PSF. This will displace the zone of optimal image collection upward (Fig. 7).
2. Using a PSF recorded with a refractive index of 0.002–0.004 points lower than that used for the sample to best restore these stacks.

For example, when a 1.520 index of refraction provides a symmetrical PSF from a bead attached to the coverslip, images will be collected using a refractive index of 1.522 or 1.524 and deconvolution will be performed using a PSF recorded at 1.518 or 1.520. Below, we provide a procedure allowing choosing the best compromise (Fig. 8).

7. Deconvolution

Deconvolution is used to reassign the intensities, distributed due to diffraction, to their originated place. It locally increases the contrast of the image to allow structures of interest to be better resolved and segmented. Different deconvolution algorithms and strategies for noise control yield different results, but each result is

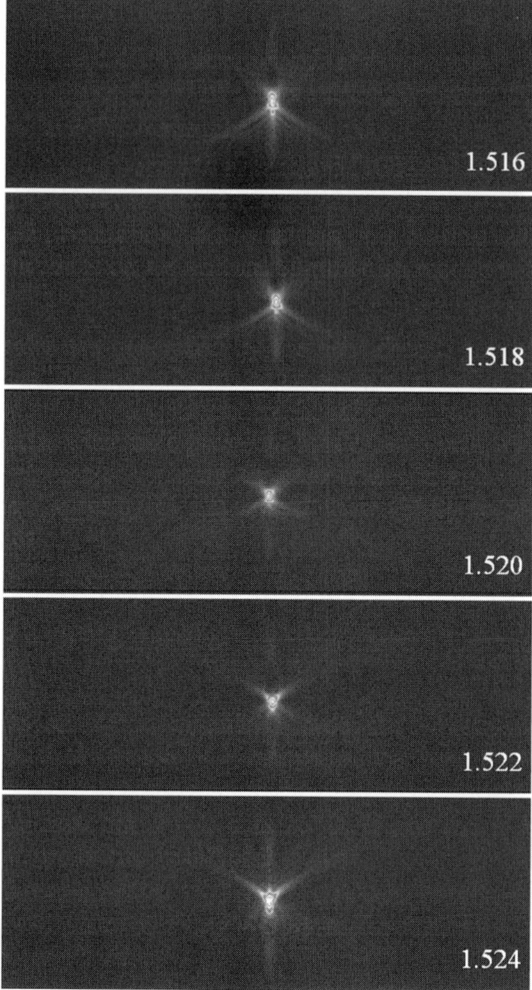

Fig. 6 PSFs recorded at different refractive indices: XZ views of PSFs recorded at 37 °C using a Leica CS PlanAPO objective 100×, NA 1.4 and a series of Cargill immersion series A oils with refractive indices ranging from 1.516 to 1.524. (See Plate no. 14 in the Color Plate Section.)

an estimate of the object that is closer to the observed object than the acquired image, and therefore more suitable for analysis. The decision of which is the "best" estimate can only be based on practical criteria: absence of obvious artefacts and getting the necessary contrast enabling the required analysis.

There are various families of deconvolution algorithms, each with its own advantages and limitations. The maximum likelihood estimation (MLE), with or without regularization, and the iterative constrained algorithms are the most

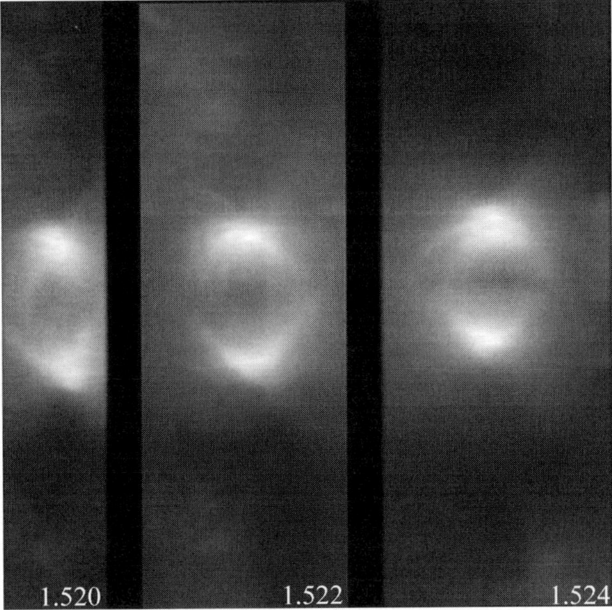

Fig. 7 The effect of varying the refractive index for recording living cells: XZ maximal intensity projections (MIPs) of stacks recorded with immersion oils with the indicated refractive indices. One can see how the asymmetric light cones get more symmetric.

commonly used ones in cellular imaging. The iterative constrained algorithms give poor results if the original data are noisy. This can be partially solved by adding a noise controlling step throughout the iterative process. The stronger the filtering is, the better the noise control is likely to be, but the less sharp is the result. The user must take the decision concerning the acceptability of such a compromise, depending on the subsequent analyses to be performed. The MLE algorithms imply a mathematical strategy for noise minimization, and are therefore well adapted to noisy data produced by rapid 4D microscopy. Their major disadvantage is that they display very slow convergence (between 50 and 1000 iterations) and require more computing time per iteration. First-guess smoothing and regularization functions can be applied to increase the convergence speed.

Terminating iterative algorithms before complete convergence also contributes to avoiding artefacts and permit to compare data together, a key point for quantitative 4D experiment. Artefacts may arise for various reasons. For example, a given algorithm may not be efficient enough to process the data in the presence of very poor SNR. In addition, the choice of algorithm parameters may not be optimal, PSF estimation may be inaccurate, or the data may have been undersampled. There is little point in trying to illustrate all these artefacts on real samples because their appearance depends on object morphology and noise level.

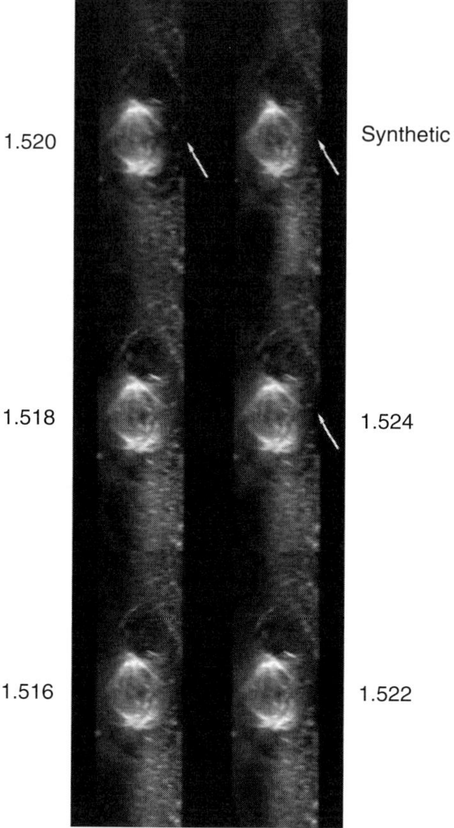

Fig. 8 The effect on image restoration of selecting refractive indices for both the point spread functions (PSF) and the images: maximal intensity projections (MIPs) of the stack recorded with a refractive index of 1.522 (see Fig. 7) deconvolved with PSFs (see Fig. 6) recorded using immersion oils with the indicated refractive indices and one synthetic PSF. The cell bottom is to the right. The arrows indicate regions where the deconvolution led to signal losses near the bottom. This illustrates the importance of inspecting stacks in XZ MIPs (and XY Z-series, not shown).

Image quality is very difficult to assess for biological objects, and is not the only parameter in practice. In some cases, it is preferable to use a given algorithm simply because it is 10 times faster than another "more rigorous" algorithm. Comparing speeds of different algorithms is difficult, because they depend on the way the algorithms are implemented. Nevertheless, for a user point of view, this is not the main purpose. For a routine use of the deconvolution, batch processing is perhaps the most important feature to consider. Effectively, and thanks to the evolution of computer speed, deconvolution reached a point where the processing time per stack becomes very short, between a few seconds to a few minutes depending on the algorithm and the file size. It should be therefore easy to process all the acquired data in the most efficient and integrated way, automatically taking into account the best parameters (PSF, algorithm parameters). At the end, for a given algorithm, the most important criterion will be software interface and its capability to easily process thousands of files acquired from various microscopes and organised in a complex file system tree. Despite all of this, we recommend using some of the guidelines below in order to try to obtain the best possible results with the solutions commercially available. In all cases, the *z*-calibration procedure described above must always be carried out first.

As predicted, the robust but time-consuming MLE algorithm provides the best compromise for very low SNR images. It requires at least 60 iterations, and seems optimal after 100 (Fig. 9). The modified iterative constrained Gold algorithm is much faster but performs less well with very low SNR images. It yields acceptable results whenever the initial contrast is visible above the background. It requires less than 10 iterations and is particularly useful for the deconvolution of very large data sets.

We recommend that deconvolution tests must be part of the strategy deployed for determining optimal acquisition parameters. We usually perform several deconvolutions until the best compromise is found. Evaluation must be done both visually, plane by plane, and by measurement of SNR and contrast through a ROI. Here are some examples to help to qualitatively compare the results.

1. For visual inspections, we display the restored stacks in a gallery. It is useful to use a spectrum look-up table. This allows one to visually compare the results plane by plane, which is important since the quality of deconvolution may vary in *Z* because of the PSF variance along the optical axis.
2. Intensity profiles along a given line are also very helpful tools for more precisely comparing enhancements and possible artifacts in regions of interests, in the three dimensions. They can be made on (x, y) optical planes, on (x, z) or (y, z) axial cross-sections (Fig. 10), and finally on maximum intensity projections made at various angles.

This practice cannot overcome all of the problems caused by spherical aberrations, but it can help to improve results and better handle the image quality and the possible artifacts of the processed data.

IV. Conclusions

Rapid 4D imaging microscopy with deconvolution is the method of choice when maximal sensitivity and speed are needed. In this chapter, we have summarized parameters to consider when using this technique, and discussed the limitations mainly coming from the detectors and spherical aberrations introduced by using high numerical aperture optics. We have put emphasis on the importance of optimizing the acquisition protocol in order to obtain the best results. Whenever the structures of interest are occurring further away than 15 μm from the coverslip, it may become necessary to use water immersion optics. It remains possible to

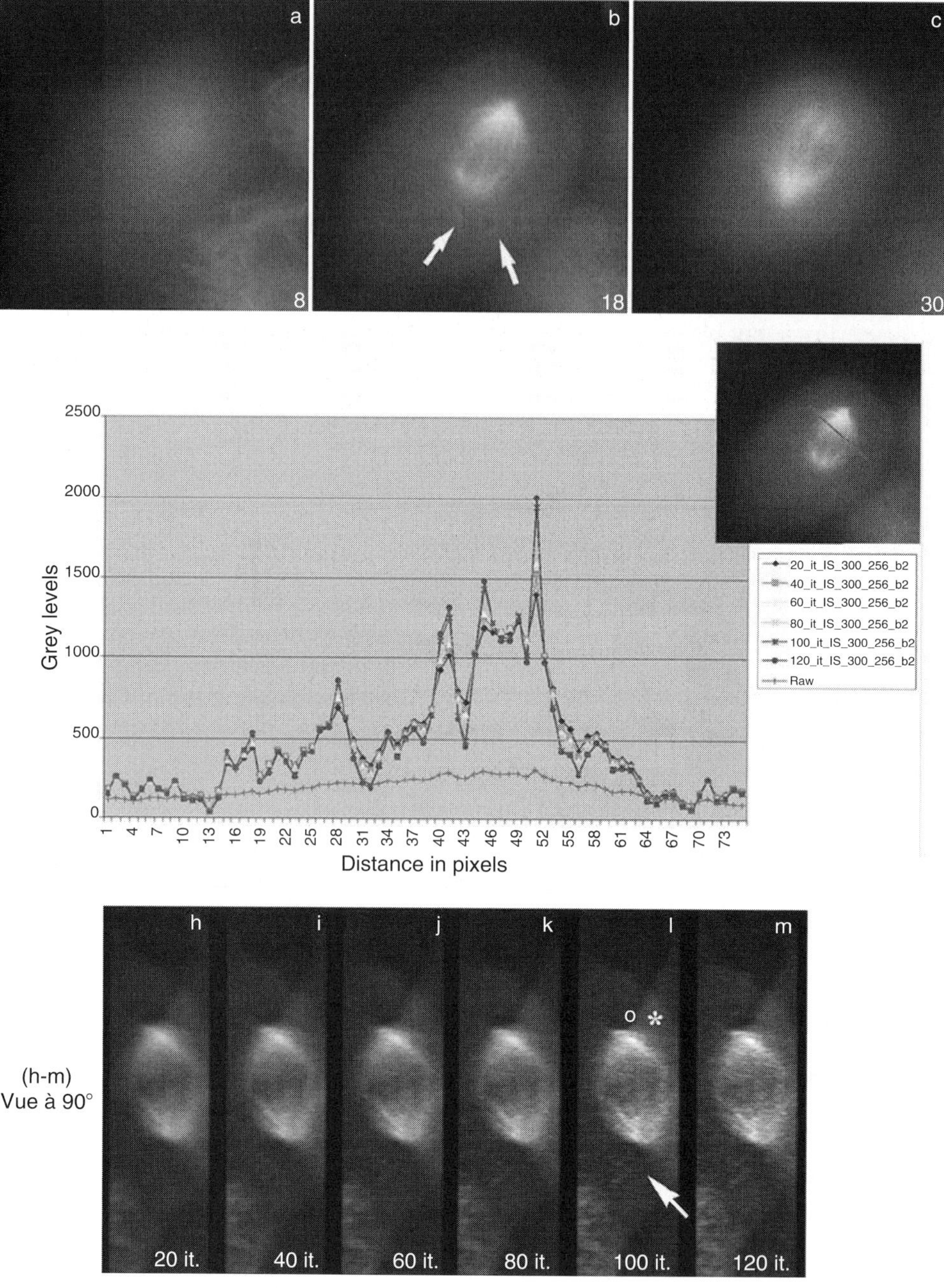

Fig. 9 Determining the number of iterations. Top panel: Three optical sections of the stack of Fig. 3, the numbers indicating the plane number. The arrows indicate transverse sections through very weakly labeled astral microtubules. Middle panel: intensity tracings along a line drawn over the half spindle as indicated in the insert, showing the very significant increase in contrast already upon 40 iterations (maximum likelihood estimation, MLE) and reaching a plateau after 100 iterations. A good compromise is 60 iterations (See Plate no. 15 in the Color Plate Section.)

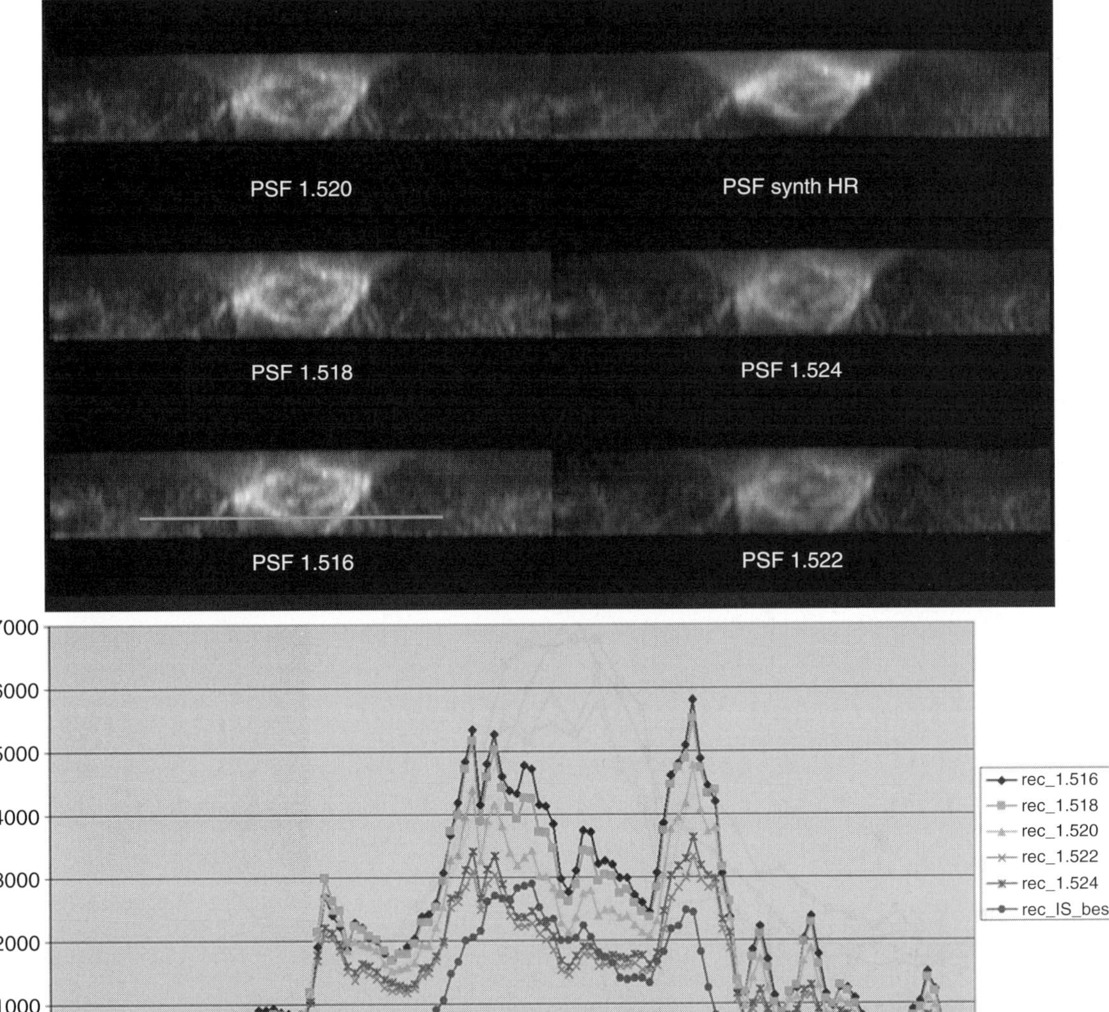

Fig. 10 Quantification of deconvolution by intensity tracing: Deconvolution (maximum likelihood estimation (MLE), 60 iterations) of the stack of Fig. 4 with point spread functions (PSFs) recorded with a range of refractive indices as indicated, and a synthetic PSF. Lower panel: Intensity tracing along the indicated line shows the improvement of the visualization of the astral microtubules, but also differences between the PSFs. Inferior quality was obtained using a synthetic PSF of the same dimensions. The light cone towards the top of the cell (to the right) is not completely disappearing. Calculations not shown here have shown that the 1.516 PSF performs somewhat better than the 1.518 one, but it still masks the astral microtubules in that region (See Plate no. 16 in the Color Plate Section.)

analyze larger structures, but fine structures will be lost. It is clear that we need fluorescent tags that are brighter and more sensitive cameras, so that in most cases, water immersion optics can be used. Until then, the practical hints and solutions described here can help obtaining valuable data sets compatible with the appropriate analysis techniques. New developments in the field of deconvolution algorithms are expected to bring better solutions in the coming years. Distributed deconvolution is even offering the perspective of online deconvolution during the acquisition phase. In certain cases, it can also be useful exploring the use of fast multipoints confocal imaging, knowing that these techniques are much more expensive and need higher excitation energies for generating suitable data sets.

Acknowledgments

We gratefully acknowledge D. Louvard (Curie Institute, Director of the Research Division) for his enthusiastic support, and the important contributions of J. Stuckey (Universal Imaging), M. Pontoizeau (Roper Scientific France), Beat Ludin (LIS), and H. Rühl, M. Ganser, and H. Martel (Leica Microsystems, Wetzlar, Germany), for advice and help with assembling the Curie Institute's fast 4D system in 1997–1998.

References

Agard, D. A., Hiraoka, Y., Shaw, P., and Sedat, J. W. (1989). Fluorescence microscopy in three dimensions. *Meth. Cell Biol.* **30,** 353–377.

Alberts, B. (1998). The cell as a collection of protein machines: Preparing the next generation of molecular biologists. *Cell* **92,** 291–294.

Alberts, B., and Klug, A. (2000). Genes: We can't expect full understanding yet. *Nature* **404,** 542.

Angelier, N., Tramier, M., Louvet, E., Coppey-Moisan, M., Savino, T. M., De Mey, J. R., and Hernandez-Verdun, D. (2005). Tracking the interactions of rRNA processing proteins during nucleolar assembly in living cells. *Mol. Biol. Cell* **16,** 2862–2871.

Ballestrem, C., Erez, N., Kirchner, J., Kam, Z., Bershadsky, A., and Geiger, B. (2006). Molecular mapping of tyrosine-phosphorylated proteins in focal adhesions using fluorescence resonance energy transfer. *J. Cell Sci.* **119,** 866–875.

Bolte, S., and Cordelieres, F. P. (2006). A guided tour into subcellular colocalization analysis in light microscopy. *J. Microsc.* **224,** 213–232.

Carrington, W. A., Lynch, R. M., Moore, E. D., Isenberg, G., Fogarty, K. E., and Fay, F. S. (1995). Superresolution three-dimensional images of fluorescence in cells with minimal light exposure. *Science* **268,** 1483–1487.

Chen, Y., and Periasamy, A. (2006). Intensity range based quantitative FRET data analysis to localize protein molecules in live cell nuclei. *J. Fluoresc.* **16,** 95–104.

Conchello, J. A. (1998). Superresolution and convergence properties of the expectation-maximization algorithm for maximum-likelihood deconvolution of incoherent images. *J. Opt. Soc. Am. A Opt. Image Sci. Vis.* **15,** 2609–2619.

Dieterlen, A., Xu, C., Gramain, M. P., Haeberle, O., Colicchio, B., Cudel, C., Jacquey, S., Ginglinger, E., Jung, G., and Jeandidier, E. (2002). Validation of image processing tools for 3-D fluorescence microscopy. *C. R. Biol.* **325,** 327–334.

Draviam, V. M., Shapiro, I., Aldridge, B., and Sorger, P. K. (2006). Misorientation and reduced stretching of aligned sister kinetochores promote chromosome missegregation in EB1- or APC-depleted cells. *EMBO J.* **25,** 2814–2827.

Eils, R., and Athale, C. (2003). Computational imaging in cell biology. *J. Cell Biol.* **161,** 477–481.

Fay, F. S., Carrington, W., and Fogarty, K. E. (1989). Three-dimensional molecular distribution in single cells analysed using the digital imaging microscope. *J. Microsc.* **153,** 133–149.

Gauthier, L. R., Charrin, B. C., Borrell-Pages, M., Dompierre, J. P., Rangone, H., Cordelieres, F. P., De Mey, J., MacDonald, M. E., Lessmann, V., Humbert, S., and Saudou, F (2004). Huntingtin controls neurotrophic support and survival of neurons by enhancing BDNF vesicular transport along microtubules. *Cell* **118,** 127–138.

Gerlich, D., Beaudouin, J., Gebhard, M., Ellenberg, J., and Eils, R. (2001). Four-dimensional imaging and quantitative reconstruction to analyse complex spatiotemporal processes in live cells. *Nat. Cell Biol.* **3,** 852–855.

Giepmans, B. N., Adams, S. R., Ellisman, M. H., and Tsien, R. Y. (2006). The fluorescent toolbox for assessing protein location and function. *Science* **312,** 217–224.

Hanser, B. M., Gustafsson, M. G., Agard, D. A., and Sedat, J. W. (2004). Phase-retrieved pupil functions in wide-field fluorescence microscopy. *J. Microsc.* **216,** 32–48.

Hiraoka, Y., Minden, J. S., Swedlow, J. R., Sedat, J. W., and Agard, D. A. (1989). Focal points for chromosome condensation and decondensation revealed by three-dimensional in vivo time-lapse microscopy. *Nature* **342,** 293–296.

Hiraoka, Y., Sedat, J. W., and Agard, D. A. (1990). Determination of three-dimensional imaging properties of a light microscope system. Partial confocal behavior in epifluorescence microscopy. *Biophys. J.* **57,** 325–333.

Hiraoka, Y., Swedlow, J. R., Paddy, M. R., Agard, D. A., and Sedat, J. W. (1991). Three-dimensional multiple-wavelength fluorescence microscopy for the structural analysis of biological phenomena. *Semin. Cell Biol.* **2,** 153–165.

Holmes, T. J. (1992). Blind deconvolution of quantum-limited incoherent imagery: Maximum-likelihood approach. *J. Opt. Soc. Am. A* **9,** 1052–1061.

Horrigan, F., and Bookman, R. J. (1995). Construction and performance of a rapidly switching, filter-based fluorescence excitation system. *Biophys. J.* **68,** A289.

Howell, B. J., Moree, B., Farrar, E. M., Stewart, S., Fang, G., and Salmon, E. D. (2004). Spindle checkpoint protein dynamics at kinetochores in living cells. *Curr. Biol.* **14,** 953–964.

Ilag, L. L. (2003). Direct methods for modulating protein function. *Curr. Opin. Drug Discov. Develop.* **6,** 262–268.

Jacquet, M., Renault, G., Lallet, S., De Mey, J., and Goldbeter, A. (2003). Oscillatory nucleocytoplasmic shuttling of the general stress response transcriptional activators Msn2 and Msn4 in *Saccharomyces cerevisiae*. *J. Cell Biol.* **161,** 497–505.

Kam, Z., Kner, P., Agard, D., and Sedat, J. W. (2007). Modelling the application of adaptive optics to wide-field microscope live imaging. *J. Microsc.* **226,** 33–42.

Kam, Z., Zamir, E., and Geiger, B. (2001). Probing molecular processes in live cells by quantitative multidimensional microscopy. *Trends Cell Biol.* **11,** 329–334.

Lippincott-Schwartz, J., and Patterson, G. H. (2003). Development and use of fluorescent protein markers in living cells. *Science* **300,** 87–91.

Lippincott-Schwartz, J., Snapp, E., and Kenworthy, A. (2001). Studying protein dynamics in living cells. *Nat. Rev. Mol. Cell Biol.* **2,** 444–456.

Mora-Bermudez, F., and Ellenberg, J. (2007). Measuring structural dynamics of chromosomes in living cells by fluorescence microscopy. *Methods* **41,** 158–167.

Paemeleire, K., Martin, P. E., Coleman, S. L., Fogarty, K. E., Carrington, W. A., Leybaert, L., Tuft, R. A., Evans, W. H., and Sanderson, M. J. (2000). Intercellular calcium waves in HeLa cells expressing GFP-labeled connexin 43, 32, or 26. *Mol. Biol. Cell* **11,** 1815–1827.

Patki, V., Buxton, J., Chawla, A., Lifshitz, L., Fogarty, K., Carrington, W., Tuft, R., and Corvera, S. (2001). Insulin action on GLUT4 traffic visualized in single 3T3-l1 adipocytes by using ultra-fast microscopy. *Mol. Biol. Cell* **12,** 129–141.

Piel, M., Meyer, P., Khodjakov, A., Rieder, C. L., and Bornens, M. (2000). The respective contributions of the mother and daughter centrioles to centrosome activity and behavior in vertebrate cells. *J. Cell Biol.* **149,** 317–330.

Piel, M., Nordberg, J., Euteneuer, U., and Bornens, M. (2001). Centrosome-dependent exit of cytokinesis in animal cells. *Science* **291,** 1550–1553.

Presley, J. F., Ward, T. H., Pfeifer, A. C., Siggia, E. D., Phair, R. D., and Lippincott-Schwartz, J. (2002). Dissection of COPI and Arf1 dynamics *in vivo* and role in Golgi membrane transport. *Nature* **417,** 187–193.

Reits, E. A., and Neefjes, J. J. (2001). From fixed to FRAP: Measuring protein mobility and activity in living cells. *Nat. Cell Biol.* **3,** E145–E147.

Rizzuto, R., Carrington, W., and Tuft, R. A. (1998a). Digital imaging microscopy of living cells. *Trends Cell Biol.* **8,** 288–292.

Rizzuto, R., Pinton, P., Carrington, W., Fay, F. S., Fogarty, K. E., Lifshitz, L. M., Tuft, R. A., and Pozzan, T. (1998b). Close contacts with the endoplasmic reticulum as determinants of mitochondrial Ca2+ responses. *Science* **280,** 1763–1766.

Sato, M., Ozawa, T., Inukai, K., Asano, T., and Umezawa, Y. (2002). Fluorescent indicators for imaging protein phosphorylation in single living cells. *Nat. Biotechnol.* **20,** 287–294.

Savino, T. M., Gebrane-Younes, J., Mey, J. D., Sibarita, J.-B., and Hernandez-Verdun, D (2001). Nucleolar assembly of the rRNA processing machinery in living cells. *J. Cell Biol.* **153,** 1097–1110.

Scalettar, B. A., Swedlow, J. R., Sedat, J. W., and Agard, D. A. (1996). Dispersion, aberration and deconvolution in multi-wavelength fluorescence images. *J. Microsc.* **182,** 50–60.

Schiffmann, D. A., Dikovskaya, D., Appleton, P. L., Newton, I. P., Creager, D. A., Allan, C., Nathke, I. S., and Goldberg, I. G. (2006). Open microscopy environment and findspots: Integrating image informatics with quantitative multidimensional image analysis. *Biotechniques* **41,** 199–208.

Sekar, R. B., and Periasamy, A. (2003). Fluorescence resonance energy transfer (FRET) microscopy imaging of live cell protein localizations. *J. Cell Biol.* **160,** 629–633.

Shaner, N. C., Steinbach, P. A., and Tsien, R. Y. (2005). A guide to choosing fluorescent proteins. *Nat. Methods* **2,** 905–909.

Sibarita, J. B. (2005). Deconvolution microscopy. *Adv. Biochem. Eng. Biotechnol.* **95,** 201–243.

Sibarita, J.-B., Magnin, H., and De Mey, J. R. (2002). *In* "Ultra-fast 4D Microscopy and High Throughput Distributed Deconvolution." *Proceedings of 2002 IEEE InternSymp Biomed Imaging, Publ by IEEE* Washington, pp. 669–772.

Skibbens, R. V., Rieder, C. L., and Salmon, E. D. (1995). Kinetochore motility after severing between sister centromeres using laser microsurgery: Evidence that kinetochore directional instability and position is regulated by tension. *J. Cell Sci.* **108,** 2537–2548.

Stephens, D. J., Lin-Marq, N., Pagano, A., Pepperkok, R., and Paccaud, J. P. (2000). COPI-coated ER-to-Golgi transport complexes segregate from COPII in close proximity to ER exit sites. *J. Cell Sci.* **113,** 2177–2185.

Swedlow, J. R., and Platani, M. (2002). Live cell imaging using wide-field microscopy and deconvolution. *Cell Struct. Funct.* **27,** 335–341.

Thery, M., Racine, V., Pepin, A., Piel, M., Chen, Y., Sibarita, J. B., and Bornens, M. (2005). The extracellular matrix guides the orientation of the cell division axis. *Nat. Cell Biol.* **7,** 947–953.

Toomre, D., and Manstein, D. J. (2001). Lighting up the cell surface with evanescent wave microscopy. *Trends Cell Biol.* **11,** 298–303.

Tramier, M., Gautier, I., Piolot, T., Ravalet, S., Kemnitz, K., Coppey, J., Durieux, C., Mignotte, V., and Coppey-Moisan, M. (2002). Picosecond-hetero-FRET microscopy to probe protein–protein interactions in live cells. *Biophys. J.* **83,** 3570–3577.

Wallace, W., Schaefer, L. H., and Swedlow, J. R. (2001). A workingperson's guide to deconvolution in light microscopy. *Biotechniques* **31,** 1076–1078, 1080, 1082 passim.

Zwier, J. M., Van Rooij, G. J., Hofstraat, J. W., and Brakenhoff, G. J. (2004). Image calibration in fluorescence microscopy. *J. Microsc.* **216,** 15–24.

CHAPTER 6

Single-Molecule Imaging of Fluorescent Proteins

Adam D. Douglass and Ronald D. Vale

Department of Cellular and Molecular Pharmacology
University of California
The Howard Hughes Medical Institute
San Francisco, California 94107

Abstract

Single molecule imaging techniques overcome the averaging effects inherent in ensemble measurements and enable characterization of the enormous heterogeneity that exists in biomolecular systems. Though long the domain of a few highly specialized labs, optical imaging of single molecules in living cells is becoming a widely accessible technique. The development of commercially available microscopes, robust analysis tools, and sensitive, low-noise detectors has contributed to this dissemination, as has the ever-growing array of fluorescent proteins. The relative ease with which genetically-tagged proteins can be created and introduced into a cell has largely eliminated more cumbersome and less precise means of particle labeling. A number of special considerations apply when using genetically encoded fluorophores for single molecule experiments, however. We discuss the

0091-679X/08 $35.00
DOI: 10.1016/S0091-679X(08)85006-6

means by which fluorescent proteins can be transfected into living cells to obtain the low particle densities required for single molecule imaging, and consider the limitations that are placed on single molecule analysis by the fluorophore's photophysical properties. We also discuss the types of morphology and subcellular localization that make certain preparations more amenable to single particle imaging than others. Last, we discuss some common pitfalls involved in analyzing single molecule datasets, and consider some of the unique information that can be obtained using these techniques.

I. Introduction

The ability to image single molecules offers substantially new biological information when compared to traditional forms of microscopy. In contrast to ensemble measurements, which average underlying heterogeneity in a population and mask interesting, time-variant behavior, single-molecule visualization provides a direct window into molecular motions, time courses of reactions, and the different properties of subpopulations of molecules. The statistical advantages afforded by these techniques are enabling completely new types of questions to be asked in biochemistry and cell biology.

The first single-molecule measurement occurred over three decades ago with the invention of the patch-clamp technique (Neher and Sakmann, 1976), which allowed researchers to monitor the electrical activity of individual ion channels. While this technique had enormous impact and gave rise to entirely new subdisciplines in the neurosciences, optical measurements of single molecules were relatively slow to follow. The first steps in this direction came from studies of membrane dynamics, where observations of fluorescent tags (Barak and Webb, 1981) or antibody-coated gold particles (Edidin *et al.*, 1991) allowed the movements of individual, cell surface receptors to be tracked by video microscopy. However, while these early attempts were informative, the labeling methods involved made it impossible to unambiguously show that single-molecule sensitivity had been achieved; in many cases, the observed behaviors might have been attributed to small clusters of receptors acting in tandem.

Because of the ambiguities involved in imaging single molecules in living cells, *in vitro* studies came to dominate the single-molecule landscape, and these were largely centered initially around observations of cytoskeletal motor proteins. Howard *et al.* (1989) indirectly measured the processivity of single molecules of the microtubule-associated motor, kinesin, by tracking the movement of fluorescently tagged microtubules that were being translocated by glass-adsorbed motor proteins. Optical trapping was later proven to be a very powerful approach for measuring the forces and steps taken by individual motors (Finer *et al.*, 1994; Svoboda *et al.*, 1993). A direct means of observing single molecules unambiguously was subsequently developed by the Yanagida laboratory (Funatsu *et al.*, 1995), who found that single fluorophores could be observed by total internal reflection fluorescence (TIRF) microscopy in an aqueous environment, with rapid acquisition speeds (30 fps), and wide-field observation. Although TIRF microscopy was developed many years earlier (Axelrod, 1981), the refinement of TIRF for single-molecule imaging dramatically changed the scope of single-molecule studies. Early studies demonstrated binding and dissociation events of nucleotides to single myosin motors (Funatsu *et al.*, 1995) and tracking of single kinesin motors moving along microtubules (Vale *et al.*, 1996). Shortly thereafter, it was shown that single green fluorescent protein (GFP) molecules could be imaged by TIRF microscopy (Dickson *et al.*, 1997; Pierce *et al.*, 1997), opening up subsequent studies of imaging single GFP molecules in living cells. Because of the precisely defined stoichiometry between a molecule of GFP and the protein to which it is attached, it became relatively simple to show that single-molecule resolution had been obtained. Since that time, a number of groups have used GFP labeling strategies to study single molecules in many different cellular contexts, including signal transduction (Vazquez *et al.*, 2006), cell adhesion (Iino *et al.*, 2001), cytoskeletal dynamics (Watanabe and Mitchison, 2002), and supramolecular patterning at the cell surface (Douglass and Vale, 2005).

In this chapter, we will discuss some of the applications of imaging single fluorescent proteins in living cells and describe the experimental requirements for doing so. Technological improvements over the last decade have made such measurements much less difficult, although there are still a number of nontrivial technical issues that must be addressed. We will focus in particular on the properties of GFP and its variants that pertain to single-molecule imaging, and on methods for delivering fluorescent proteins to cells so as to achieve the low protein densities needed for resolving single molecules. Finally, we will discuss how single-molecule imaging of fluorescent proteins might be expanded in the future.

II. Instrumentation

While it is possible to detect single GFP molecules using other fluorescence microscopy techniques (e.g., confocal imaging), the overwhelming majority of studies have employed TIRF imaging. In TIRF imaging, the illuminating light (usually a laser beam) is steered toward the sample at an angle such that the light totally internally reflects at the interface between the glass substrate and the aqueous sample. In doing so, it creates an exponentially decaying evanescent wave that penetrates a short distance beyond the point of reflection (Fig. 1A). This wave excites fluorophores lying near the substrate, but not those out of the focal plane. Indeed, the utility of this technique stems from the shallowness of the TIRF illumination field, which is typically on the order of 100–200 nm, approximately one-fifth the axial dimension of a typical confocal illumination volume and less than one-tenth that of standard wide-field, epifluorescence illumination. The resulting reduction in out-of-focus fluorescence increases the overall signal-to-noise ratio and greatly facilitates detection of single molecules (Fig. 1B

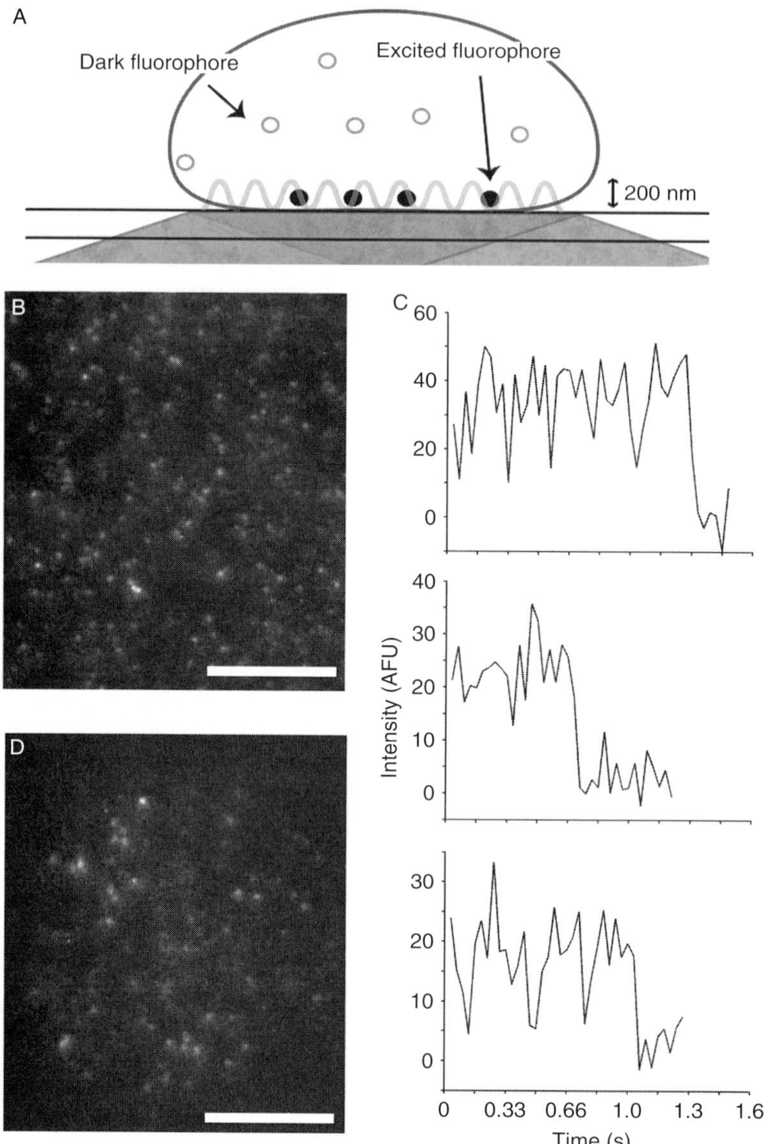

Fig. 1 Single-molecule imaging of GFP by total internal reflection fluorescence microscopy. (A) Reduction in background signal from out-of-focus fluorophores by TIRF illumination. (B) Field of single, purified GFP adhering to a glass coverslip, as visualized by TIRF. Scale bar = 10 μm. (C) Intensity versus time profiles for three representative spots in the field shown in panel B demonstrate single-step photobleaching. (D) Single molecules of CD2-GFP in the surface of a Jurkat T cell adhering to an antibody-coated coverslip, as visualized by TIRF. Scale bar = 10 μm. TIRF images were acquired at 30 fps, using 33-ms exposures.

and D). The depth of penetration of the evanescent wave can be tuned to some extent by varying the angle of incidence of the illuminating light source on the sample.

Not long ago, the difficulty in building a TIRF microscope constituted a barrier for many researchers hoping to conduct single-molecule work. This is no longer the case, as several microscope vendors have marketed their own "turnkey" systems for TIRF imaging. These systems can easily be mounted on conventional, epifluorescence microscope bodies without requiring an optical bench to route the laser beam through the microscope objective at the desired angle. Microscope companies now package several small, powerful diode lasers in housings that are coupled by fiber optics to the microscope. In addition, the major microscope vendors have all developed 100× objectives with numerical apertures of 1.45–1.49, which has allowed through-the-objective TIRF illumination to become standard practice [a substantial improvement in terms of ease-of-use compared with prism-type TIRF microscopy (Axelrod, 2003)]. Microscope companies have also made alignment of lasers for through-the-objective TIRF illumination a relatively trivial operation. Thus, commercial systems now enable any laboratory to perform single-molecule fluorescence imaging. However, for certain custom applications, an optical bench TIRF microscope might be preferable due to their greater flexibility.

In addition to the illumination mode, the sensitivity of the detector is also a critical factor in achieving single-molecule fluorescence imaging. TIRF microscopy, being a wide-field technique, normally uses a CCD camera rather than a point detector such as a photomultiplier tube. If the process being observed occurs on a relatively slow timescale (>500 ms), then lower-sensitivity CCD cameras can be used and the images can be integrated over longer periods to increase the fluorescence signal. This approach has been used to make very precise (a few nanometer resolution) positional measurements of single motor proteins traveling along cytoskeletal filaments *in vitro* (Yildiz *et al.*, 2003). However, many cellular processes cannot be adequately captured at such sampling rates. For instance, the lateral diffusion coefficients of many membrane proteins can approach 1 $\mu m^2/s$, in which case a protein might move 2 μm in a 100-ms time interval. To capture such rapid events, the camera must be able to acquire data at video rate or higher, and must do so with very high sensitivity. There are two types of camera that meet these needs: intensified CCDs (ICCDs) and electron multiplying CCDs (EMCCDs).

ICCD technology has been used in single-molecule imaging for a number of years and derives its high (up to single-photon) sensitivity from the placement of a multichannel plate (MCP) intensifier before a CCD detector. Using such a camera, we have been able to acquire single-molecule images of very rapidly moving particles at frame rates of over 100 fps. While intensifier cooling (to reduce dark noise) and fiber optic coupling between the MCP and CCD have continued to improve the capabilities of ICCDs, these cameras do suffer from several limitations, including a relatively small linear range and image artifacts arising from the intensifier. While changing the intensifier gain settings allows one to tune the camera's sensitivity over a wide range, at higher gain, the relationship between photon number and pixel intensity becomes nonlinear. This makes quantitative measurements of particle intensity relatively difficult to make, particularly at high frame rates. Image quality also suffers due to imperfections in the coupling of the MCP to the CCD, as small deviations in alignment between these components produce artifacts in the image. This problem becomes more significant as chip resolution increases, and imposes an upper limit on the true resolution attainable with ICCDs.

EMCCDs represent the second camera choice for single-molecule imaging and have become increasingly popular. In these cameras, the signal is multiplied on the CCD chip itself, as the pixel array is read out through a gain register. The sensitivity of these cameras is close to that of an ICCD, but with fewer restrictions on potential resolution and none of the artifacts that result from coupling between the MCP and the CCD. The sensitivity offered by current-generation EMCCDs is sufficient for most applications, and this technology is likely to further develop to allow faster temporal acquisition. These sensors also have very high quantum efficiencies (QEs) of detection—approaching 95%—and dramatically reduce the negative contribution of readout noise.

III. Fluorophores

Genetically encoded fluorescent proteins are widely used for live cell imaging. Genetic fusions of fluorescent proteins have the advantage of a precisely defined stoichiometry; the number of fluorophores attached to a given protein is invariant and directly determined by the engineered DNA sequence. Adverse effects on protein activity can occur in some cases, however, and thus the activity of GFP-fusion proteins is best tested in a genetically tractable system where reconstitution-of-function experiments can be performed.

Whether a process can be visualized or not is determined directly by the intrinsic photophysical properties of the fluorescent tag. In this area, the fluorescent proteins fall somewhat short of the brightest organic dyes. A fluorophore's brightness is determined primarily by its extinction coefficient and quantum yield. EGFP (enhanced green fluorescence protein), with an extinction coefficient of roughly 40,000 and a fluorescence quantum yield of ~0.6, is not as bright as Alexa 488, with an extinction coefficient of 71,000 and a quantum yield near that of EGFP. That said, it has still been possible to image single GFP molecules at or near the plasma membrane. In our experience, EGFP has been too dim only when imaging certain extremely rapidly diffusing membrane proteins, which required acquisition speeds in excess of 100 fps and a concomitant decrease in the number of photons available in a given frame. Other fluorescent proteins have also been employed successfully—EYFP, which is even brighter than EGFP, is a useful tag, as is mCherry and the tandem dimeric version of DsRed. In general, any fluorophore that is approximately as bright as EGFP will be sufficient for most applications, though the continued development of brighter fluorescent proteins will improve single-molecule imaging and allow faster temporal acquisition.

One serious limitation in using EGFP is its tendency to enter transient "dark" states, during which time the protein does not emit fluorescent light (Dickson *et al.*, 1997). These states can last for hundreds of milliseconds and produce gaps in single-molecule trajectories that can lead to premature truncation of a trace. While conversion to the dark state is not frequent enough to prevent most types of single-molecule measurement, it presents a significant analytical problem that is less problematic for synthetic fluorophores. In the absence of a means of eliminating these states, one must design a particle tracking procedure that can compensate for them by tracking through transient dark periods or simply settle for shorter contiguous traces.

Photostability also is of major importance in single-molecule work. In order to optimize signal to noise, single-molecule images are often acquired at very high illumination intensities. It is sometimes necessary to increase laser power to the point where the fluorophore becomes photosaturated, which will provide the highest possible signal-to-noise characteristics. Under these conditions, photobleaching tends to occur more quickly than in typical ensemble imaging experiments. While EGFP's photostability is reasonable, improvements in this area would enable the position and activity of a single molecule to be tracked over longer timescales. *In vitro* experiments have followed single molecules of the organic dye Cy3 for over 1 min at video rate, whereas comparable measurements of GFP rarely exceed 5–10 s. The development of quantum dots as fluorescent tracers has also led some researchers away from the use of fluorescent proteins, as these dyes are effectively unbleachable (Alivisatos, 2004). On the positive side, photobleaching can actually be useful as a means of demonstrating that single-molecule resolution has been attained. Despite some variability in fluorescence intensity over time, bleaching is an all-or-none process. If a particle's intensity profile suddenly drops to zero, it can be safely inferred that it represented a single fluorophore (Fig. 1C), whereas bleaching in multiple discrete steps or in a slow, continuous fashion would indicate the presence of multiple fluorophores within a diffraction-limited volume.

IV. Reducing Protein Expression Levels

It should be emphasized that single-molecule imaging, when performed using TIRF and without any additional optical or computational manipulation, does not entail an increase in lateral resolution as compared to other wide-field microscopies. This technique simply provides a much greater sensitivity than conventional methods. The lateral dimensions of a single particle will still be subject to the diffraction limit of light, such that molecules lying closer together than about 300 nm will not be resolvable as discrete objects. One ramification of this is that high GFP densities will obscure or completely prevent one from achieving single-molecule resolution, since the diffraction-limited, ~300-nm spots will overlap. Thus, obtaining a low particle density is a critical factor in single-molecule live cell imaging. Since most gene expression reagents are designed to achieve the highest levels of transgene expression possible, one typically has to make special efforts to achieve low protein expression in order to image single molecules in cells.

There are a number of ways to reduce fluorescence to achieve single-molecule imaging in cells. A simple method is to employ photobleaching. By increasing the power of the excitation source for a short time, it is usually possible to bleach most of the fluorescent particles lying in the evanescent wave. By then turning off the excitation light and waiting for diffusion or active transport to redistribute the remaining particles from the unbleached regions of the cell, imaging at a single-molecule density can then be performed. We have used this technique to achieve the necessary number of adequately spaced GFP molecules for single-molecule tracking in already low-expressing cells (Douglass and Vale, 2005). As with most imaging experiments, however, photobleaching should be avoided if possible. Generation of reactive oxygen species during the bleaching process can harm or, in extreme cases, kill the cell (although this is less of a problem with TIRF than with epifluorescence microscopy). Another factor to consider is that imaging a region very soon after bleaching can obscure the equilibrium behavior of the population being studied. For instance, if a population of molecules equilibrates between slow- and fast-diffusing states, the fast-diffusing class is likely to reenter the bleached region sooner than the slowly diffusing one. Imaging too soon after bleaching would then bias the results to the fast-diffusing state. Thus for some experiments, bleaching should not be used to reduce particle density.

The first step for achieving low fluorophore levels involves the proper choice for transcriptional and translational regulation of the transgene. There are many modifications that can be made to expression constructs to make them produce less protein. In our laboratory, we have seen that the expression of certain genes can be reduced to very low levels by simply omitting a consensus Kozak sequence (or any components of the gene's 5′ untranslated region) in the expression plasmid. Others have demonstrated that deleting large regions of the promoter sequence can produce the desired result (Watanabe and Mitchison, 2002). Another strategy involves the use of inducible promoters, such as those that turn on gene expression in response to the drug tetracycline. By titrating the amount of inducer, it should be possible to find a concentration of drug that drives transgene expression at the desired level. Indeed, basal expression in the absence of the inducer often produces sufficient (or even too much) expression of protein for single-molecule imaging.

Finally, it is also possible to achieve low fluorophore densities through a modified form of transient expression. We have transiently transfected mammalian cell lines with GFP-fused proteins regulated by a standard *Cytomegalovirus* promoter. While expression is very high during the first 24 h, the transgene is not integrated into the host genome in most cells and is steadily lost from the population as the cells divide. By ~72 h after transfection, there are a large number of cells that have presumably lost the plasmid, but retain low residual levels of tagged protein; these cells are suitable for single-molecule imaging. Prior to imaging, this population of cells can also be enriched by fluorescence-activated cell sorting, though we have not typically found it necessary to do so.

Photoactivation (the converse of photobleaching) holds promise for single-molecule work, although it has not been extensively employed to date. Because the amount of visible fluorophore can be precisely controlled by varying the duration and intensity of the activating light, a wide range of particle densities can be achieved through photoactivation. A number of photoactivatable fluorescent proteins now exist and have fluorescence properties that should make them suitable for single-molecule detection. Among these are PA-GFP, Kaede, and PA-mRFP. Recently, single-molecule photoactivation has been used as the basis of PALM microscopy (Betzig, 2006), a specialized application in which the distribution of fluorescence can be determined with nanometer-scale precision in fixed cells.

V. Biological Preparations

Whether a single-molecule approach is an appropriate avenue of investigation or not depends on both the experimental question and the suitability of the biological preparation for single-molecule imaging. The geometry of TIRF illumination imposes significant restrictions on the types of molecules and structures that can be investigated. In general, because of the short penetration depth of the evanescent wave, only molecules lying in or near the plasma membrane are observable by TIRF. This makes certain systems, such as the cortical cytoskeleton and membrane-proximal signaling networks, very amenable to single-molecule study while other systems are completely inaccessible. Cellular geometry is also quite important; in general, the more the cell can be induced to flatten out against its substrate, the easier it will be to image by TIRF. For this reason, the utility of TIRF in studying plant cells, yeast, and bacteria is relatively limited, though other approaches may still allow single-molecule imaging to work. For instance, increasing the number of GFP molecules attached to a single target molecule through aptamer binding can dramatically increase the signal available and makes it possible to image single molecules using other imaging techniques (Shav-Tal *et al.*, 2004).

We have performed the majority of our single-molecule work in cultured mammalian T cells, adhering to glass coverslips that have been coated with either antibodies directed against cell surface receptors or planar lipid bilayers containing mixtures of receptor ligands. In both cases, the cells adhere to and flatten on the substrate, thereby providing an excellent sample for TIRF imaging (Fig. 1D). The planar bilayer substrate has the additional advantage of being laterally fluid, such that proteins incorporated into the bilayer can diffuse freely in two dimensions. By including an adhesion molecule and a ligand for the T cell receptor in the bilayer, we and others (Varma *et al.*, 2006) have stimulated immune synapse formation in these preparations and have been able to study single-molecule behavior in a context that approximates a real cell–cell contact. Such bilayer preparations are being adapted to a number of other systems, and all of these should be accessible to TIRF imaging.

VI. Data Analysis and Interpretation

The number of parameters that can be measured in a single-molecule data set is potentially large and diverse. Perhaps the simplest, and often most informative, is measuring the position of a molecule over time. For linearly moving particles, such as many cytoskeletal motor proteins, the position provides a simple measure of a molecule's velocity over time and thus an indicator of the mechanisms driving its movement. The molecular trajectories of membrane proteins also can be used to determine the diffusion coefficients of these particles. We and others have observed dramatic changes in single-molecule diffusion coefficients over time, with periods of free diffusion alternating with periods of complete immobilization (Douglass and Vale, 2005; Suzuki, 2005). By imaging the population distribution of a second protein labeled with a different fluorophore in the same cell, we have been able to assess the single-molecule behavior relative to defined subcellular structures. This relatively simple modification has allowed us to show a strong correlation between a protein's diffusion behavior and its local environment.

Fluorescence intensity can potentially provide information about a particle's oligomerization state. Because fluorescence emission is quantized, the normalized intensity of a diffraction-limited spot can be used to infer the number of fluorophores lying within that spot. While such measurements are relatively simple in certain *in vitro* preparations (Collins *et al.*, 2004), it can be very difficult to make the equivalent measurement in a living cell, for several reasons. For one, the cell membrane is never completely flattened against the substrate, and since the evanescent wave decays in the *z*-direction, such membrane undulations will give rise to variability in the intensities of single fluorophores. In addition, because these experiments require very low fluorophore densities, it is rare that all molecules of a particular type can be labeled and still provide single-molecule resolution. This means that the majority of oligomerization events will go undetected, as such events will typically involve a single labeled molecule at most. Despite these limitations, a few groups have managed to obtain information on protein clustering from intensity data (Iino *et al.*, 2001; Sako *et al.*, 2000). In general, these measurements are easiest when differentiating between large protein assemblies (i.e., containing tens or hundreds of molecules) and single particles, but it is possible to make finer assessments as well. In an impressive example of this, Leake *et al.* (2006) used intensity and photobleaching measurements to determine the stoichiometry of the MotB protein at the flagellar motor assembly of *Escherichia coli*.

Recently, single-molecule imaging has also been used to measure the duration of associations between cytoplasmic signaling proteins and the cell surface (Vazquez *et al.*, 2006). This is simply another example of measuring position over time, but with the difference that photobleaching must be absolutely avoided or carefully corrected for. Since particles are effectively invisible when diffusing in the cytoplasm, cell surface residence times can be measured by counting the number of contiguous frames during which a particle is visible. If bleaching is properly controlled for, such measurements can reveal the half-life for dissociation from the membrane, a value that might provide insight into its interactions with other proteins in the membrane. Vasquez *et al.* (2006) used this approach to show that single molecules of the tumor suppressor PTEN spend up to a few hundred milliseconds at a time bound to the plasma membrane, and mapped the relevant interactions to a specific protein domain.

One of the most the most difficult aspects of single-molecule imaging is being able to track single particles in an image sequence. Manual tracking is possible but extremely tedious, and usually less accurate than automated methods. Due to the low signal-to-noise levels inherent in these experiments, image segmentation and centroid tracking are nontrivial. As described earlier, GFP's blinking behavior also can be problematic as transient disappearance of a particle will truncate a trajectory prematurely. Progress continues to be made, however, and a number of computational methods have been developed for single particle tracking. We primarily use a suite of tools written in the IDL language by Weeks and colleagues for tracking colloidal particles in suspension, which we have optimized for single-molecule analysis. One of the useful features of these routines is the inclusion of a "memory" function, which allows one to define a maximal number of image frames over which to continue searching for a particle that has disappeared before terminating the trace. Though artificial truncations do still occur, particularly when analyzing rapidly diffusing particles, the number of particles that can be analyzed in a given data set is often very large. Hundreds of molecules can be tracked in a single cell in just 1 min of acquired data. Because of GFP's relatively low brightness it cannot be localized with particularly high precision, however, and the current generation of fluorescent proteins will need to be dramatically improved to allow the nanometer-scale precision that is currently possible with other dyes (Yildiz *et al.*, 2003).

VII. Future Prospects

As the refinement of imaging technologies has enabled many laboratories to perform single-molecule imaging in living cells, the types of information that can be gained from single-molecule imaging also will undoubtedly expand. For example, the growing list of fluorescent proteins will enable one to track multiple proteins simultaneously. We have used a simple beam splitter to observe single molecules of GFP and tandem dimeric DsRed in the same sample. We also have used a linear unmixing strategy to visualize single GFP and YFP simultaneously. Such approaches enable the spatial-temporal tracking of multiple species at the single-molecule level. Perhaps most exciting, single-molecule FRET—which is relatively straightforward in *in vitro* systems (Tomishige *et al.*, 2006)—can be used to measure binding events, conformational changes, and enzymatic activity in live cells, taking traditional biochemical measurements out of the test tube and into the complex environment of the cell.

References

Alivisatos, P. (2004). The use of nanocrystals in biological detection. *Nat. Biotechnol.* **22,** 47–52.

Axelrod, D. (1981). Cell-substrate contacts illuminated by total internal reflection fluorescence. *J. Cell Biol.* **89,** 141–145.

Axelrod, D. (2003). Total internal reflection fluorescence microscopy in cell biology. *Methods Enzymol.* **361,** 1–33.

Barak, L. S., and Webb, W. W. (1981). Fluorescent low density lipoprotein for observation of dynamics of individual receptor complexes on cultured human fibroblasts. *J. Cell Biol.* **90,** 595–604.

Betzig, E., Patterson, G. H., Sougrat, R., Lindwasser, O. W., Olenych, S., Bonifacino, J. S., Davidson, M. W., Lippincott-Schwartz, J., and Hess, H. F. (2006). Imaging intracellular fluorescent proteins at nanometer resolution. *Science* **313,** 1642–1645.

Collins, S. R., Douglass, A., Vale, R. D., and Weissman, J. S. (2004). Mechanism of prion propagation: Amyloid growth occurs by monomer addition. *PLoS Biol.* **2,** e321.

Dickson, R. M., Cubitt, A. B., Tsien, R. Y., and Moerner, W. E. (1997). On/off blinking and switching behaviour of single molecules of green fluorescent protein. *Nature* **388,** 355–358.

Douglass, A. D., and Vale, R. D. (2005). Single-molecule microscopy reveals plasma membrane microdomains created by protein-protein networks that exclude or trap signaling molecules in T cells. *Cell* **121,** 937–950.

Edidin, M., Kuo, S. C., and Sheetz, M. P. (1991). Lateral movements of membrane glycoproteins restricted by dynamic cytoplasmic barriers. *Science* **254,** 1379–1382.

Finer, J. T., Simmons, R. M., and Spudich, J. A. (1994). Single myosin molecule mechanics: Piconewton forces and nanometre steps. *Nature* **368,** 113–119.

Funatsu, T., Harada, Y., Tokunaga, M., Saito, K., and Yanagida, T. (1995). Imaging of single fluorescent molecules and individual ATP turnovers by single myosin molecules in aqueous solution. *Nature* **374,** 555–559.

Howard, J., Hudspeth, A. J., and Vale, R. D. (1989). Movement of microtubules by single kinesin molecules. *Nature* **342,** 154–158.

Iino, R., Koyama, I., and Kusumi, A. (2001). Single molecule imaging of green fluorescent proteins in living cells: E-cadherin forms oligomers on the free cell surface. *Biophys. J.* **80,** 2667–2677.

Leake, M. C., Chandler, J. H., Wadhams, G. H., Fan, B., Berry, R. M., and Armitage, J. P. (2006). Stoichiometry and turnover in single, functioning membrane protein complexes. *Nature* **443,** 355–358.

Neher, E., and Sakmann, B. (1976). Single-channel currents recorded from membrane of denervated frog muscle fibres. *Nature* **260,** 799–802.

Pierce, D. W., Hom-Booher, N., and Vale, R. D. (1997). Imaging individual green fluorescent proteins. *Nature* **388,** 338.

Sako, Y., Minoghchi, S., and Yanagida, T. (2000). Single-molecule imaging of EGFR signalling on the surface of living cells. *Nat. Cell Biol.* **2,** 168–172.

Shav-Tal, Y., Darzacq, X., Shenoy, S. M., Fusco, D., Janicki, S. M., Spector, D. L., and Singer, R. H. (2004). Dynamics of single mRNPs in nuclei of living cells. *Science* **304,** 1797–1800.

Suzuki, K., Ritchie, K., Kajikawa, E., Fujiwara, T., and Kusumi, A. (2005). Rapid hop diffusion of a G-protein-coupled receptor in the plasma membrane as revealed by single-molecule techniques. *Biophys. J.* **88,** 3659–3680.

Svoboda, K., Schmidt, C. F., Schnapp, B. J., and Block, S. M. (1993). Direct observation of kinesin stepping by optical trapping interferometry. *Nature* **365,** 721–727.

Tomishige, M., Stuurman, N., and Vale, R. D. (2006). Single-molecule observations of neck linker conformational changes in the kinesin motor protein. *Nat. Struct. Mol. Biol.* **13,** 887–894.

Vale, R. D., Funatsu, T., Pierce, D. W., Romberg, L., Harada, Y., and Yanagida, T. (1996). Direct observation of single kinesin molecules moving along microtubules. *Nature* **380,** 451–453.

Varma, R., Campi, G., Yokosuka, T., Saito, T., and Dustin, M. L. (2006). T cell receptor-proximal signals are sustained in peripheral microclusters and terminated in the central supramolecular activation cluster. *Immunity* **25,** 117–127.

Vazquez, F., Matsuoka, S., Sellers, W. R., Yanagida, T., Ueda, M., and Devreotes, P. N. (2006). Tumor suppressor PTEN acts through dynamic interaction with the plasma membrane. *Proc. Natl. Acad. Sci. USA* **103,** 3633–3638.

Watanabe, N., and Mitchison, T. J. (2002). Single-molecule speckle analysis of actin filament turnover in lamellipodia. *Science* **295,** 1083–1086.

Yildiz, A., Forkey, J. N., McKinney, S. A., Ha, T., Goldman, Y. E., and Selvin, P. R. (2003). Myosin V walks hand-over-hand: Single fluorophore imaging with 1.5-nm localization. *Science* **300,** 2061–2065.

CHAPTER 7

Counting Kinetochore Protein Numbers in Budding Yeast Using Genetically Encoded Fluorescent Proteins

Ajit P. Joglekar, E. D. Salmon, and Kerry S. Bloom

Department of Biology
University of North Carolina
Chapel Hill, North Carolina 27599

METHODS IN CELL BIOLOGY, VOL. 85

0091-679X/08 $35.00
DOI: 10.1016/S0091-679X(08)85007-8

Abstract

Genetically encoded fluorescent proteins are an essential tool in cell biology, widely used for investigating cellular processes with molecular specificity. Direct uses of fluorescent proteins include studies of the *in vivo* cellular localization and dynamics of a protein, as well as measurement of its *in vivo* concentration. In this chapter, we focus on the use of genetically encoded fluorescent protein as an accurate reporter of *in vivo* protein numbers. Using the challenge of counting the number of copies of kinetochore proteins in budding yeast as a case study, we discuss the basic considerations in developing a technique for the accurate evaluation of intracellular fluorescence signal. This discussion includes criteria for the selection of a fluorescent protein with optimal characteristics, selection of microscope and image acquisition system components, the design of a fluorescence signal quantification technique, and possible sources of measurement errors. We also include a brief survey of available calibration standards for converting the fluorescence measurements into a number of molecules, since the availability of such a standard usually determines the design of the signal measurement technique as well as the accuracy of final measurements. Finally, we show that, as in the case of budding yeast kinetochore proteins, the *in vivo* intracellular protein numbers determined from fluorescence measurements can also be employed to elucidate details of cellular structures.

I. Introduction

Fluorescence microscopy is finding increasing usage in studies of diverse aspects of cell biology at the cellular as well as the molecular level, with a variety of fluorescent probes available for studying molecular function (Giepmans *et al.*, 2006). Genetically encoded fluorescent proteins have become the most widely used fluorophores by cell biologists (Chudakov *et al.*, 2005). The green fluorescent protein (GFP) was the first to be optimized as a genetically encoded fluorescent marker for use *in vivo* (Tsien, 1998). The exploding number of available fluorescent proteins with characteristics tailor-made to suit experimental needs (Miyawaki, 2004; Miyawaki *et al.*, 2005; Sawano and Miyawaki, 2000; Zhang *et al.*, 2002) has allowed researchers to devise elegant ways of employing them to reveal cellular processes. Their typical uses range from *in vivo* protein localization and dynamics (Lippincott-Schwartz and Patterson, 2003), characterizing intracellular chemistry (Kohl and Schwille, 2005; Lippincott-Schwartz *et al.*, 2001; Schwille, 2001), to studying gene expression and regulation patterns (Bar-Even *et al.*, 2006; Colman-Lerner *et al.*, 2005; Raser and O'Shea, 2005; Rosenfeld *et al.*, 2005). Besides their utility in studying spatiotemporal protein localization patterns within a cell, quantification of intracellular protein concentration is intuitively the simplest application of genetically encoded fluorescent proteins in cell biology.

Estimation of intracellular concentration of fluorophores is commonly employed using flow cytometry (Huh *et al.*, 2003; Newman *et al.*, 2006). This technique measures whole-cell fluorescence signal to deduce the concentration of a protein or DNA within the cell. Flow cytometry works well for bright signals in samples that have relatively low autofluorescence and has the advantage of rapidly accumulating very large datasets. However, it may not provide the desired flexibility necessary for the accurate measurement of proteins that have low cellular abundance, with protein numbers per cell ranging from a few to a few hundred protein molecules (Newman *et al.*, 2006). This is commonly the case for studies involving prokaryotes and lower eukaryotes such as budding and fission yeast. In recent years, a number of microscopic imaging-based methods have been devised for this purpose. These methods typically involve high numerical aperture (NA) objectives for imaging, and sensitive, low-noise cooled CCD cameras for image acquisition. Current technology allows the observation of protein expression one molecule at a time *in vivo* (Xie *et al.*, 2006; Yu *et al.*, 2006).

There are two important considerations that must be addressed while developing a quantitative fluorescence microscopy method for counting protein numbers. The first involves the development of an accurate technique for the quantification of the fluorescence signal that is based on features of the biological structure being studied, and the characteristics of the imaging system. The second factor is to obtain a calibration standard that allows for the accurate conversion of measured fluorescence signal into the corresponding number of fluorophores. While the details of the first factor are largely dictated by the imaging optics, diverse methodologies have been developed over the years to address the need of a good calibration standard. The choice of a particular calibration standard can play a significant role in determining the accuracy of the protein counts obtained. A fluorescence signal measurement and conversion method designed with these considerations can yield valuable data on intracellular protein numbers.

In this chapter, we describe a technique for accurately counting the numbers of kinetochore proteins in budding yeast by measuring the fluorescence signal from fluorescently tagged kinetochore proteins. First, the criteria for choosing optimal fluorescent proteins for counting intracellular protein numbers are discussed briefly. We then describe the sample preparation method and the microscope setup used. The next section discusses characterization of the microscope performance, which is critical in developing an appropriate signal measurement scheme. Fluorescence signal quantification method for the budding yeast kinetochore cluster, along with the obtained results, is then described. The last section contains a discussion of a more generalized fluorescence signal measurement method that may be used for proteins that are diffusely distributed within a cell, as well as the sources of errors in fluorescence signal measurements. This section also includes a survey of the varied fluorescence signal calibration standards that have been developed to convert the fluorescence signal into the corresponding number of proteins.

II. Counting Kinetochore Protein Numbers in Budding Yeast

The eukaryotic kinetochore is a highly complex protein structure composed of more than 60 different proteins (McAinsh *et al.*, 2003; Meraldi *et al.*, 2006). It establishes attachment of microtubule plus-ends with the centromeric DNA during mitosis, and generates force necessary to move and segregate chromosomes. The end-on attachment of a microtubule plus-end with centromeric DNA requires at least eight different proteins and protein complexes. To understand how these proteins assemble together to make a functional kinetochore-microtubule attachment, it is necessary to understand their arrangement within the microtubule attachment site at the kinetochore. Although electron microscopy has revealed the overall structure of the kinetochore, the arrangement of the protein complexes within this structure remains unknown. The number of copies of each protein or protein complex involved in making a functional kinetochore-microtubule attachment is critical for understanding the molecular architecture of the kinetochore. The budding yeast (*Saccharomyces cerevisiae*) as a model organism provides some unique advantages as an experimental system for counting kinetochore protein numbers *in vivo* using quantitative fluorescence microscopy.

A wealth of information about the composition of the budding yeast kinetochore is now available. The budding yeast kinetochore is a relatively simple structure with only one microtubule attachment site as compared with vertebrate kinetochores that have multiple microtubule attachments. Each kinetochore is based on ~150 base pair long DNA sequence wrapped around one centromeric nucleosome containing centromere-specific histone Cse4p (human homolog CENP-A). In metaphase, the centromeric DNA is stably attached to the plus-end of one microtubule by eight other linker proteins or protein complexes (Fig. 1). Since each yeast kinetochore supports only one stable microtubule attachment in metaphase, the copy number of each protein complex per kinetochore can be directly useful in understanding the molecular architecture of the microtubule attachment site. The stability of microtubule attachment raises the possibility that the protein complexes that make up the microtubule attachment site may also be stably associated with the kinetochore. Most of the proteins in this linkage are conserved in all eukaryotes including humans (McAinsh *et al.*, 2003; Meraldi *et al.*, 2006). Therefore, protein architecture within a kinetochore-microtubule attachment can also be expected to be conserved from budding yeast to humans.

The versatile molecular biology and genetics of budding yeast provides a critical advantage for protein number evaluation through the measurement of fluorescence signal from fluorescently tagged proteins. Most proteins can be easily tagged at the C-terminus with a fluorescent protein by insertion of the gene sequence at the endogenous locus. Thus, the fusion protein is the only species of protein produced within the cell. The protein level can be expected to be similar to the native strain as the fusion protein expression is controlled by the native promoter. Furthermore, the fusion protein can be considered as functionally equivalent, since it replaces the

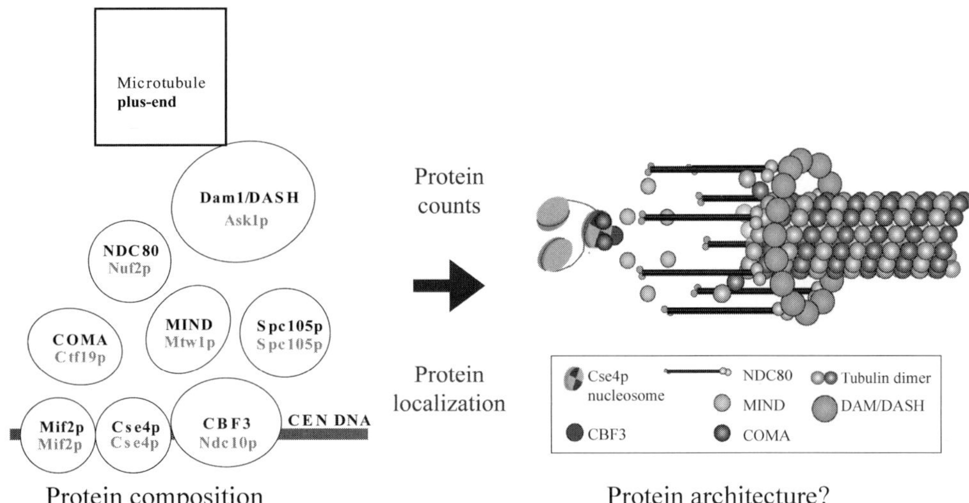

Fig. 1 Use of quantitative fluorescence microscopy for understanding protein structure. The yeast kinetochore is a complex protein structure composed of 8 different proteins and protein complexes. From each complex, the constituent protein that was selected as the representative is indicated in green. Understanding protein architecture requires protein numbers and their relative localization within the kinetochore. Quantitative fluorescence microscopy can provide protein counts necessary for visualizing the three-dimensional (3-D) arrangement of different protein complexes. (See Plate no. 17 in the Color Plate Section.)

native protein without an observable effect on chromosome segregation or gross cell growth. Therefore, the protein counts obtained with fluorescence microscopy can be expected to accurately reflect the functional needs in the cell. This critical advantage is absent in vertebrate systems, since expression of a fusion proteins from the native promoter has proven difficult. Therefore, the fluorescently tagged protein is typically expressed with the help of an extra copy of the gene fused to the gene for a fluorescent protein from an artificial promoter. This results in nonnative protein expression of the fluorescently tagged protein along with the native untagged protein. Therefore, suitable control experiments must be designed to account for the presence of two species of proteins within the cells, and possibly preferential recruitment of these proteins at the site of action. The use of budding yeast as a model system avoids these complications.

The geometry of the budding yeast spindle in metaphase and anaphase/telophase facilitates accurate evaluation of the fluorescence signal from fluorescently labeled kinetochore proteins (Fig. 2A and B). The metaphase spindle is ~1500 nm in length, with two clusters of sister kinetochores separated by 800 nm and positioned on either side of the spindle equator (Fig. 2C). Each cluster consists of 16 kinetochores that are distributed over a 200–300 nm region. Each kinetochore is stably connected to the plus-end of one microtubule that is anchored at its minus

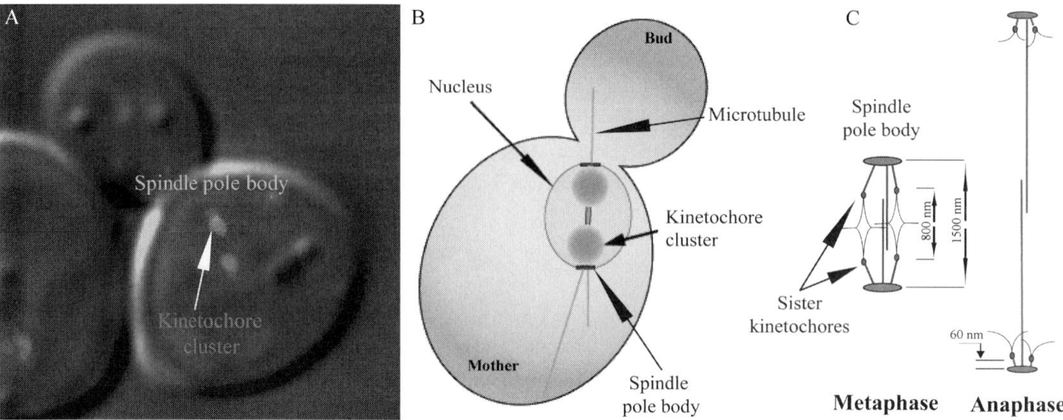

Fig. 2 Distribution of kinetochores in a budding yeast cell in metaphase. (A) An image overlay of a metaphase cell expressing Cse4p-GFP to visualize the kinetochores and Spc29-RFP, which is a spindle pole body protein. (B) Cartoon of a metaphase budding yeast cell. Sixteen sister kinetochores cluster on either side of the spindle equator in a subdiffraction volume. Each kinetochore stably binds the plus-end of only one microtubule. (C) Dimensions of a typical metaphase and telophase spindle measured from the fluorescence intensity distribution agree with direct measurements of average kinetochore microtubule lengths using serial section electron microscopy (O'Toole *et al.*, 1999; Winey *et al.*, 1995). (See Plate no. 18 in the Color Plate Section.)

end within a spindle pole body. In anaphase, the kinetochore microtubules shrink to very short lengths (~50 nm), pulling their kinetochores very close to a spindle pole body. The clusters of sister kinetochores become highly separated by spindle elongation to a length of 8–10 μm.

Budding yeast also provides an excellent calibration standard, which is critical for converting the fluorescence signal into the number of fluorophores accurately. As mentioned earlier, the yeast kinetochore is built around a single centromeric nucleosome that contains two Cse4p molecules (Collins *et al.*, 2004; Joglekar *et al.*, 2006; Meluh *et al.*, 1998). Thus, the signal from a kinetochore cluster expressing Cse4p-GFP represents the fluorescence of 32 GFP molecules (16 kinetochores with 2 Cse4p-GFP molecules per kinetochore). The spatial protein distribution within the cluster for all the kinetochore proteins is nearly identical to that of Cse4p, which avoids some of the potential complications in fluorescence signal measurement. We have developed a ratiometric method for counting the number of GFP-tagged kinetochore proteins. This method evaluates the fluorescence signal for a GFP-tagged kinetochore protein, and compares it with the signal for Cse4p-GFP by obtaining a ratio of the two fluorescence signals. The conversion of this ratio into protein number is then a straightforward task, since the exact number of Cse4p per kinetochore is known.

Figure 3 compares the fluorescence intensity from kinetochore clusters in metaphase and anaphase/telophase cells expressing the different GFP-tagged kinetochore proteins (shown in green in Fig. 1). The images in both the panels were obtained with nearly identical imaging conditions. The intensity from the

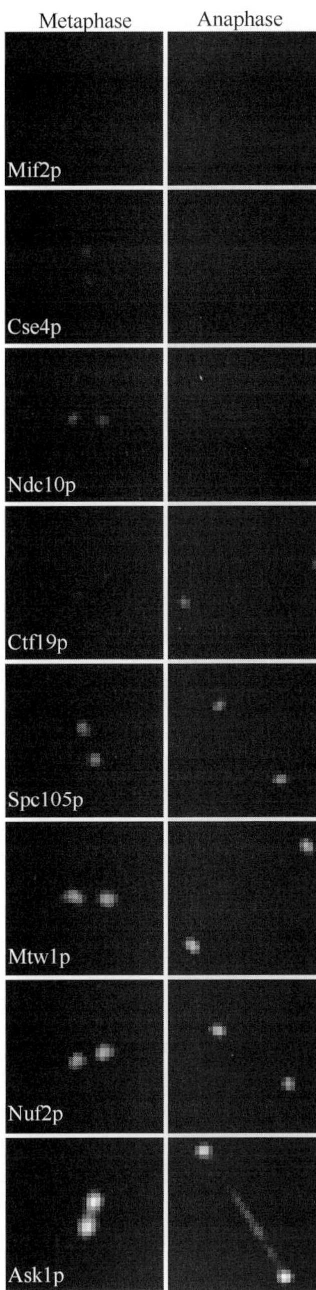

Fig. 3 Fluorescence signal for different kinetochore proteins in metaphase and anaphase cells. A visual examination of the apparent intensity for different kinetochore proteins indicates that these proteins must be present in different numbers within the kinetochore cluster. The fluorescence signal from Cse4p-GFP provides an excellent calibration standard, as the number of Cse4p molecules per kinetochore, and hence the total number of GFP molecules in a cluster is known.

representative protein can be expected to accurately reflect the number of copies of the complex with which it is associated (Meraldi *et al.*, 2006). Even a cursory examination of the images clearly shows that different proteins are incorporated within the kinetochore in different numbers. An image analysis method is necessary to accurately evaluate the fluorescence signals, and convert them into absolute numbers of corresponding proteins.

A. Optimal Fluorescent Proteins

Selection of the appropriate fluorescent protein is critical for its use as a faithful reporter of intracellular protein concentration (Shaner *et al.*, 2005). The following characteristics of the fluorescent protein are highly desirable: (1) a high and constant maturation efficiency at the typical physiological temperature, (2) a maturation time faster than the temporal dynamics of the protein under investigation, (3) a high molecular brightness to produce a robust signal that is also insensitive to its microenvironment, and (4) a low bleaching rate. EGFP was the first fluorescent protein to be optimized as a suitable genetically encoded fluorescent protein for quantitative microscopy (Patterson *et al.*, 1997; Piston *et al.*, 1999). It is an ideal reporter protein in budding yeast because of its fast maturation time, high folding efficiency, and satisfactory molecular brightness. Its color variants YFP and CFP provide suitable alternatives, although properties such as bleaching rates are not as good as the properties of EGFP. YFP can be useful because of a higher molecular brightness, and because it allows the use of filters that enable a greater rejection of autofluorescence background in certain cases (Wu and Pollard, 2005). Low bleaching rate is especially important when counting low protein numbers, and when temporal changes in protein numbers are being studied. Bleaching also becomes an issue if multiple image slices are necessary to cover the entire thickness of a cell. Table I lists relevant properties of fluorescent proteins that are particularly suited for counting experiments.

III. Sample Preparation

The budding yeast strains used in our study were created by tagging the endogenous copy of the kinetochore protein gene with EGFP at the C-terminus (Longtine *et al.*, 1998). A strain made in this manner shows wild-type growth characteristics at 25 °C in complete media (YPD), indicating that the EGFP-tagged kinetochore protein can functionally substitute the native, untagged protein. Standard methodologies have now been established for *in vivo* imaging of fluorescent proteins in budding yeast (Bloom *et al.*, 1999). In brief, the cells are grown to mid-log phase in YPD at 25 °C. For microscopy, the cells are spun down, washed in water, and then resuspended in filter-sterile synthetic media (SD). Coverslips are immersed in 1 M NaOH overnight. After washing them thoroughly with distilled water, a thin layer of 0.5 mg/ml concanavalin A (cat # C7275, Sigma, St. Louis, MO) solution (10 mM phosphate buffer, 1 mM $CaCl_2$, pH 6.0) is then applied on the coverslip for ~20 min. The

Table I
Characteristics Selection of Optimal Fluorescent Proteins for Fluorescence Quantification Experiments

Protein	Excitation wavelength[a] (nm)	Emission wavelength[a] (nm)	Brightness[a]	$t_{1/2}$ maturation time (min)[a]	Photostability[a]
Cerulean	433	475	27	<5[b]	36
EGFP	488	507	34	<5	174
EYFP	514	527	51	<5[c]	60
mCherry	587	610	16	15[d]	96

[a]Shaner *et al.* (2005) provides a comprehensive compilation of all the important characteristics of the available fluorescent proteins. $t_{1/2}$ as estimated by completely denaturing the protein *in vitro*, and then measuring the fluorescence recovery half-life as the protein renatures.
[b]Kremers *et al.* (2006).
[c]Nagai *et al.* (2002).
[d]Shaner *et al.* (2004).

coverslips are washed again with distilled water and allowed to dry. In each experiment, cells belonging to two strains—one expressing the protein of interest tagged with GFP and the other expressing Cse4p-GFP—are mixed together in approximately equal concentration. Five microliters of this mixture is then spread over a concanavalin A coated coverslip. The coverslip edges are then sealed with VALAP (a mixture of equal amounts of vaseline, lanoline, and paraffin) to prevent media evaporation.

Sample preparation protocols used in live cell microscopy strive to maintain optimal conditions for cells under the microscope. For optimal fluorescence microscopy however, a low but uniform background fluorescence due to growth media is also desirable. Immobilizing yeast cells with concanavalin A avoids the use of thin gelatin slabs that are commonly used to hold the yeast cells in place during live-cell microscopy (Bloom *et al.*, 1999). We found that the autofluorescence from gelatin slabs increases with depth, which makes accurate evaluation of fluorescence signals from kinetochore clusters difficult.

IV. Microscope and Image Acquisition System

We used wide-field epifluorescence microscopy and digital imaging for counting kinetochore proteins in budding yeast. The common considerations in selecting the critical components for a wide-field epifluorescence microscope have been previously described (Salmon *et al.*, 2003). For fluorescence signal measurements in budding yeast, we used a Nikon TE 2000-U microscope (Nikon, Melville, NY) equipped with a 1.4 NA, 100× DIC oil immersion objective. We used the HQ filter cube for EGFP from Chroma (Chroma, Rockingham, VT; exciter: HQ470/40x, dichroic: Q495LP, and emission: HQ525/50m). The microscope is equipped with an *XY* translation stage (LEP, Hawthorne, NY). The objective can be translated along the optical axis with a servo stepper motor (LEP). Images were acquired with an Orca ER cooled CCD camera (pixel size of 6.47 μm, Hamamatsu). The camera was operated in the 2 × 2 binning mode. Only a 300 × 300 pixel region at the center of the field of view was acquired. For each selected field, 21 *Z* sections were taken, with a distance of 200 nm separating successive image planes. Image acquisition was carried out with Metamorph (Molecular Devices, Downington, PA). Image analysis was done by using a custom written graphical user interface in MatLAB (MatLAB, Natick, MD).

The selection of appropriate imaging conditions is carried out to maximize the signal-to-noise ratio of the images. It depends on the characteristic of the fluorescent protein as well as the imaging system. The response of the fluorescent protein to the excitation intensity is the first important characteristic. GFP signal increases nonlinearly with the excitation intensity, and saturates beyond a critical intensity, behavior that is typical of three-state systems (Kues *et al.*, 2001). This nonlinear response becomes especially important when comparing fluorescent protein signals acquired at two different intensities. Another characteristic of a typical fluorophore is fluorescence or concentration quenching. Fluorescence quenching is the effect of nonradiative depletion of fluorophores in the excited state that becomes pronounced at high fluorophore concentrations. Fluorescence quenching thus distorts the expected linear relationship between fluorescent protein concentration and fluorescent protein signal. As discussed later, we found that local EGFP concentrations (over the range considered in our study) do not distort the fluorescence signal. The excitation intensity and exposure time were optimized to minimize photobleaching, and to allow for optimal usage of the dynamic range of the camera. An exposure of 400 ms with the Orca ER camera allowed us to image kinetochore clusters with the lowest signal with a satisfactory signal-to-noise ratio, while allowing the recording of fluorescence signal of the most abundant proteins without pixel saturation. This exposure was used for all the measurements. For a large field of view, it is important to evaluate the variation in the excitation intensity as a function of position in the object plane in wide-field microscopy. These variations can be characterized by imaging a uniformly labeled surface and applying a "shading correction" to the acquired images. Depending upon the amount of variation over the necessary imaging area and the desired accuracy, advanced image calibration methods may be used (Ghauharali and Brakenhoff, 2000; Zwier *et al.*, 2004). In our experiments, we acquired images only from a square in the center of the image plane representing less than 10% of the total area of the image plane. The excitation intensity can be assumed to be spatially invariant over this small region.

The choice of a digital CCD camera plays an important role in deciding both the accuracy and sensitivity of fluorescence signal measurements. The two critical characteristics of a CCD camera to consider are: (1) the photon conversion efficiency of the CCD chip and (2) camera noise. Modern electronics makes electronic noise less significant in comparison with the background noise from the sample or microscope optics. The linearity of CCD response to the incident intensity is excellent, thus making its use in signal quantification straightforward.

An optimal balance between the spatial resolution in an image and the signal-to-noise ratio of that image is important for carrying out fluorescence signal measurements. The process of recording a microscope image with a CCD pixel array results in a convolution of the image by a top-hat function corresponding to the dimensions of each CCD pixel. The pixel size of the CCD camera used thus is an important parameter in determining the magnification needed in the image plane (Piston, 1998; Salmon *et al.*, 2003). With 100× magnification of the objective and a pixel size of 6.47 μm, each pixel corresponds to 65 nm in the object plane. Objective magnification and camera pixel size thus together decide the sampling frequency for the objective. The sampling frequency can be varied within limits by either changing the objective magnification or focal length of the tube lens or by electronic binning of neighboring pixels on the camera. The sampling frequency is important especially when measuring the fluorescence signal from subdiffraction objects. For imaging kinetochore clusters in budding yeast, we found that 2 × 2 binning of the camera chip provided sufficient spatial resolution to clearly distinguish metaphase clusters, while maintaining a sufficiently high signal-to-noise ratio for accurate signal measurement at moderate excitation intensity.

V. Measurement of Fluorescence Signal

A. Characterization of the Point Spread Function of the Objective

An objective produces a diffraction pattern in the form of an Airy disk in the image plane of a point source of light from the specimen such as a single fluorophore (Agard *et al.*, 1989; Inoue and Springer, 1997; Taylor and Salmon, 1989). The intensity distribution in the XY plane at focus along the Z axis is given by:

$$I = \left(\frac{J_1(R \times r)}{R \times r} \right)^2$$
$$R = \frac{0.61\lambda}{\mathrm{NA}}, \quad r = \sqrt{x^2 + y^2} \tag{1}$$

where J_1 is the Bessel function of the first kind, λ is the wavelength, and NA is the numerical aperture.

The dimensions of this diffraction pattern, known as the Point Spread Function (PSF), are determined by the NA of the objective and the wavelength of fluorescent light. In the image plane, the Airy disk intensity distribution can be closely approximated by a two-dimensional (2-D) Gaussian function given by:

$$I = A\mathrm{e}^{-(x^2/2\sigma_x^2)}\mathrm{e}^{-(y^2/2\sigma_y^2)} \tag{2}$$

where σ_x, σ_y are the standard deviation in X and Y direction.

The spread of the Gaussian function is given by its standard deviation (σ_x, σ_y). The total signal from a point source (or from a subdiffraction object) is thus spread over the area of the PSF, and thus can be evaluated by integrating the intensity over this entire region. The $\pm 2\sigma$ limits include ~96% of the area under a 1-D Gaussian curve (~90% in 2-D), and is commonly used as the characteristic dimension for the distribution. These limits correspond to a circle with 444 nm diameter for an image of a point source of light emitting at 510 nm and recorded with a 1.4 NA objective. It should be noted that for high NA oil immersion objectives, the PSF approximates the theoretical PSF only very close to the coverslip surface. Spherical aberrations due to the mismatch in the refractive indices of glass and aqueous media rapidly distort the PSF away from the coverslip surface (Gibson and Lanni, 1992) and reduces the overall intensity.

It is important to accurately characterize the PSF for the objective to be used in the image plane (XY) and along the optical axis (Z) to verify the expected performance from the objective. This was done by imaging fluorescent beads of subdiffraction size (100 nm Transfluobeads, Invitrogen Carlsbad, CA) immobilized on the coverslip. It is important to open up the field diaphragm for these measurements. In epifluorescence microscopy, the same objective is used to excite the fluorophores and to collect the emission light. Therefore, the focus change required to image out-of-focus light will also change the illumination intensity. This effect depends inversely on the size of the field diaphragm (Hiraoka *et al.*, 1990) and can be minimized by opening up the field diaphragm for PSF characterization. A line-scan through the experimentally obtained PSF in the XY plane of focus can be fitted with a 1-D Gaussian function to obtain the standard deviation (σ) for the Gaussian function as the fit parameter. The PSF along the optical axis can also be similarly approximated with a Gaussian curve.

B. Characterization of Fluorescence Intensity Distribution for a Kinetochore Cluster and Signal Measurement

The next step in developing the fluorescence measurement technique for a kinetochore cluster is to accurately determine its fluorescence intensity distribution in the image plane (the in-focus XY plane) and along the optical axis of the microscope (Z axis, Fig. 4B). This was done by recording the intensity distribution along a line through the maximum intensity pixel in an in-focus image of kinetochore clusters in cells expressing Nuf2p-GFP (Fig. 4C, left graph). The intensity distribution along the line was then fitted with a Gaussian function to obtain the standard deviation for the Gaussian as the fitting parameter. The standard deviation ($\sigma = 155$ nm) obtained from fitting the curve was used to determine the number of camera pixels to be included in the calculation of the integrated intensity from a kinetochore cluster (4×155 nm $< 5 \times 133$ nm, where 133 nm is the effective pixel size due to 2×2 binning). For metaphase kinetochore clusters, intensity had to be integrated over a larger area to account for the larger space occupied by the kinetochore clusters.

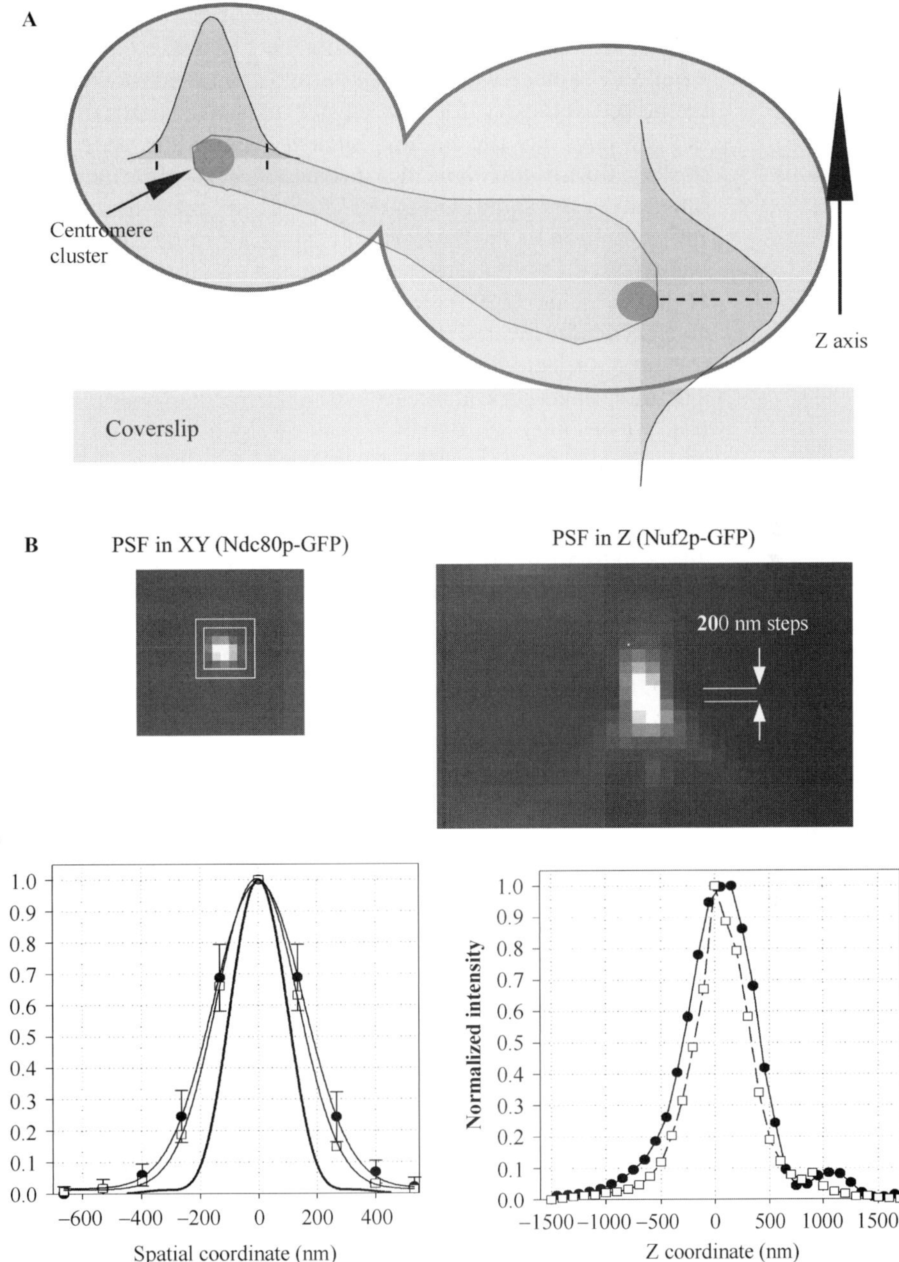

Fig. 4 Characterization of the intensity distribution from a kinetochore cluster. (A) The kinetochore clusters are positioned at random depths from the coverslip surface. Therefore, *Z*-stacks must be obtained by stepping the stage with an appropriate step size with respect to the objective. (B) The intensity distribution for a kinetochore clusters from cells expressing GFP-tagged kinetochore proteins

For accurate comparison, we obtained the average fluorescence signal for cells expressing Cse4p-GFP along with the average signal for the protein of interest from the same coverslip. Signal for a kinetochore cluster was defined as the integrated intensity for the cluster in the in-focus image plane. The in-focus image plane has the highest intensity, and thus provides the highest signal-to-noise ratio for intensity measurements. This approach also avoids the need for integrating the signal along the *Z* axis. An important step in signal calculation is the evaluation of the background signal. For an isolated fluorescent spot, accurate background evaluation can be carried out by using the method described by Hoffman *et al.* (2001). In brief, a box of the appropriate dimension is drawn concentrically around the box used for signal measurement. The dimension of this box is selected so as to equalize the area of the background and signal region. The integrated pixel intensity within the signal region after background subtraction was defined as the fluorescence signal from the kinetochore cluster. This method avoids errors due to inhomogeneities in the intracellular background levels. In cases where this method cannot be adopted, for example, budding yeast cells in metaphase, the background region may be manually chosen to obtain an estimate of the true background.

The kinetochore clusters are situated at unknown distances from the coverslip surface in a yeast cell. Thus, a stack of optical sections through each cell along the optical axis must be acquired to capture an in-focus image of the kinetochore clusters within the cell. The precision with which the *XY* plane at the center/peak of the PSF along the optical axis can be captured depends on the separation between successive images along the *Z* axis (Fig. 4C, right graph). For example, the worst case situation with a 200-nm step size would be images captured 100 nm apart on either side of the actual peak in the PSF. This corresponds to a worst-case error of 8% of the maximum intensity. On average, random sampling of the PSF along the *Z* axis will result in an average underestimation of the measured signal by 4% of the maximum intensity. A 200-nm step size thus allows imaging through the entire thickness of the cell with minimal photobleaching during acquisition at the cost of a reasonably small measurement error. Depending on the signal-to-noise ratio and photobleaching rate for the fluorophore used, a smaller step size may be used.

VI. Validation of Measurement Method

To calibrate the linear response of the system, we took advantage of the known stoichiometry for constituent proteins within the NDC80 complex. Biochemical analysis of purified NDC80 complex from budding yeast shows that the four

(Nuf2p-GFP and Ndc80p-GFP) in the *XY* and *XZ* plane are shown. (C) The characteristic dimensions for the intensity distribution for a kinetochore cluster in telophase (●) are obtained by fitting it with a Gaussian function. The standard deviation obtained from the fit is similar to that for 200 nm beads (□). The theoretical PSF based on the objective NA (1.4) and wavelength (510 nm) is also plotted (solid curve). The *Z* intensity distribution allowed us to determine the optimal step size along the *Z* axis (● for a kinetochore cluster and □ for a 200 nm bead).

proteins within this complex have a 1:1:1:1 stoichiometry (Wei *et al.*, 2005). We constructed three yeast strains expressing either Nuf2p-GFP or Ndc80-GFP or both Nuf2p-GFP and Ndc80p-GFP. Using the method described earlier, we compared the fluorescence signal from kinetochore clusters in each case with the signal from a strain expressing Cse4p-GFP (Fig. 5A). As expected, the

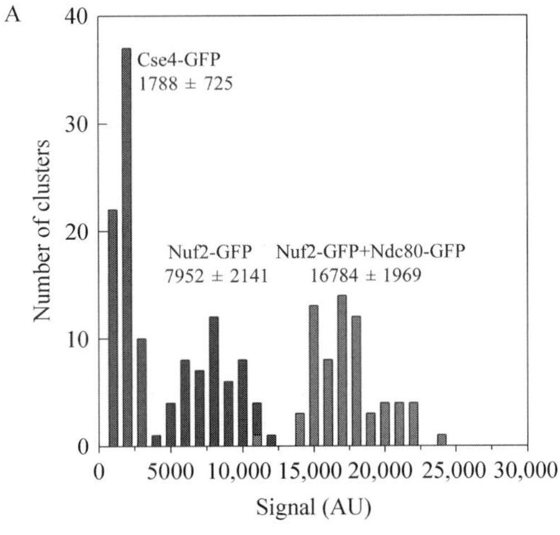

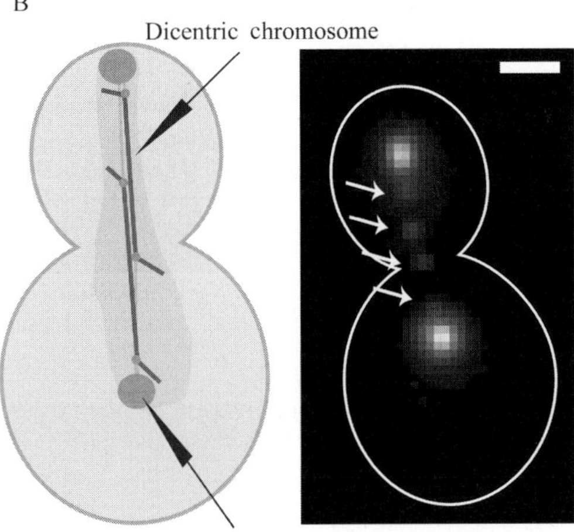

Fig. 5 Linear response of the technique for measuring the fluorescence signal from diffraction-limited images of kinetochore clusters in telophase. (A) The doubling of the kinetochore proteins that have been tagged with GFP is accurately reflected by a doubling of the measured fluorescence signal. This validates our ratiometric method for counting the number of molecules within a kinetochore cluster. (B) Single kinetochores can be imaged in a cell expressing Nuf2p-GFP after the activation of the dicentric chromosome. (See Plate no. 19 in the Color Plate Section.)

ratio of the average signal from Nuf2p-GFP was found to be identical to that for Ndc80p-GFP. The ratio for the strain expressing both Nuf2p-GFP and Ndc80p-GFP on the other hand was exactly twice in magnitude, thus validating our method.

These measurements also demonstrate that at least at these concentrations, there is no detectable quenching of the fluorescence signal. The GFP molecules are tethered by the kinetochore proteins in a stable protein structure. This insolubility may make the nonradiative transfer of energy from one fluorophore to another unlikely.

VII. Results

Table II lists the ratios for representative kinetochore proteins obtained by using the method described earlier. As mentioned earlier, we obtained the average fluorescence intensity for a kinetochore cluster for the reference (Cse4p-GFP) strain and the strain with the GFP-tagged protein of interest. Three experiments were performed for each protein in both metaphase and anaphase/telophase to obtain three ratios. The number of observations in each experiment for each strain was more than 20. We found that the coefficient of variation (standard deviation/mean) was sufficiently small (<0.25) for each data set. The values reported are the average of the ratios from the three experiments for each protein.

It must be demonstrated that the cumulative signal from a kinetochore cluster can be divided by 16 (the number of kinetochores in a cluster) to obtain the number

Table II
Kinetochore Protein Counts

Complex	Protein	Vertebrate homolog	Ratio		Number	
			Metaphase	Anaphase	Metaphase	Anaphase
Nucleosome	Cse4p	CENP-A	1	1	2	2
	Mif2p*	CENP-C	5.4 ± 0.4	5.5 ± 0.1	1–2	1–2
	Ndc10p	–	1.9 ± 0.2	1.3 ± 0.01	4	2–3
CBF3	Cep3p	–	0.9 ± 0.2	0.6 ± 0.01	2	1–2
COMA	Ctf19p*	CENP-F	3.4 ± 0.3	3.4 ± 0.2	3	2
	Spc105p	KNL-1, AF15q14	2.4 ± 0.01	2.4 ± 0.01	5	5
MIND	Mtw1p	Mis12, CENP-H	3.3 ± 0.2	2.4 ± 0.1	6–7	4–5
NDC80	Nuf2p	Nuf2	4.0 ± 0.2	3.4 ± 0.05	8	7
DAM/DASH	Ask1p	xDam1p?	9.0 ± 1.0	5.3 ± 0.3	16–20	10–12

The table lists the average ratio of the average signal from the protein of interest with that for Cse4p-GFP from three experiments. Cells from two strains in each experiment were distinguished based on the magnitude of the fluorescence signal. This approach could not be used for Mif2p and Ctf19p, as the intensities for these proteins were very similar to that for Cse4p-GFP making manual identification of cells impossible. For these proteins, we used the fluorescence signal from cells expressing Nuf2p-GFP as the reference. The ratios reported for these proteins are Nuf2p-GFP:protein of interest.

*Cells expressing Mif2p-GFP or Ctf19p-GFP could not be visually distinguished from cells belonging to the standard strain expressing Cse4p-GFP based on the image intensity of the kinetochore clusters. Therefore, budding yeast strain expressing Nuf2p-GFP was used as the standard strain for comparing the average fluorescence intensity of these two proteins. The ratio reported in the table represents (Average signal for Nuf2p-GFP/Average signal for Mif2p-GFP or Ctf19p-GFP).

of proteins per kinetochore. We verified this assumption by making use of a conditional dicentric chromosome. Inserted into chromosome III is an additional copy of the centromeric DNA sequence that is placed under the control of a GAL1 promoter. When the cells are grown on media with galactose as the carbon source, the transcription activity at the GAL1 promoter silences the additional centromere. In cells grown on glucose-containing media however, this additional centromere becomes active, and builds a functional kinetochore on the same chromosome. Two active kinetochores on the same chromosome prevent the sister chromatids from segregating to their respective poles in anaphase in a fraction of cells. The lagging kinetochores, at positions away from the monocentric kinetochore clusters near the spindle poles, can be clearly imaged in cells expressing high copy number kinetochore protein such as Nuf2p-GFP (Fig. 5B). A comparison of the fluorescence signal from these single kinetochores with the cluster of 15 monocentric chromosomes close to the spindle pole bodies yielded an average ratio of 16 ± 2, as expected. This experiment validates the assumption that each kinetochore contributes equally to the cumulative fluorescence signal from a cluster of 16 kinetochores.

The observed standard deviation of the signal about the mean value from metaphase or telophase kinetochore clusters contains important information about the variation in protein numbers from cell to cell. The observed standard deviation may also contain contributions from experimental errors in addition to the biological variation. We therefore analyzed the data to look for potential experimental sources of error to find that the distance of the kinetochore cluster away from the coverslip strongly affects the signal magnitude due to spherical aberrations that increase with depth as discussed above. As shown in Fig. 6, this effect does not depend on the absolute magnitude of the signal, and thus does not distort the ratio of two fluorescence signals. However, the resultant variation of the signal about the mean signal masks information about the variation in the protein number. To avoid this experimental source of error, we compared the fluorescence signals for kinetochore cluster pairs that had a relative separation along the optical axis of 600 nm or less. Table III lists the mean and standard deviation for three different strains spanning the range of signals measured in this study. As can be seen from the table, the difference in measured intensity values for these kinetochore clusters is small as compared with the total signal. The standard error of the mean fluorescence value based on this difference is also very small. It can be stated in terms of the number of GFP molecules, by using the average Cse4p-GFP signal (1945 counts for 32 GFP molecules at 16 kinetochores $\rightarrow$ 60 counts per GFP molecule). Thus, the difference between two kinetochore clusters in the same cell for Cse4p is $\sim$4 GFP molecules out of 32, while that for Ndc80p-GFP + Nuf2p-GFP is 20 GFP molecules out of 256. This translates into a variation of less than one molecule per kinetochore for each protein. It should also be noted that the standard deviation roughly scales with the mean.

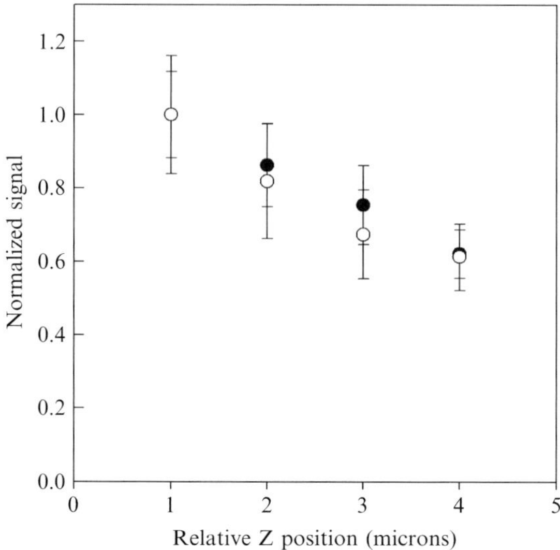

Fig. 6 Decrease in signal with increasing depth from the coverslip surface. Because of spherical aberrations, the signal measured from a kinetochore cluster decreases as a function of the distance from the coverslip surface. This decrease is independent of the magnitude of the signal (●: Cse4p-GFP; ○: Nuf2p-GFP+Ndc80p-GFP).

VIII. Discussion

Counting protein copy numbers *in vivo* using genetically encoded fluorescent proteins is a relatively simple but powerful approach. Its judicious use can potentially reveal critical information about subcellular structures as well as protein stoichiometry. This is especially true for many lower eukaryotes such as fungi and prokaryotes that are amenable to easy genetic manipulation for expressing genetically encoded fluorescent proteins. We have demonstrated the use of this method for understanding a key aspect of the molecular architecture of the kinetochore-microtubule attachment by counting the number of copies of protein complexes involved in linking centromeric DNA to a microtubule plus-end. Similar quantitative fluorescence microscopy assays can be extremely useful in establishing the lower eukaryotes as a platform for studying basic biological machinery that is conserved in all eukaryotes.

The use of similar techniques in studying vertebrate cell lines faces two challenges. The expression of fluorescently labeled proteins in vertebrate cells is commonly driven by an artificial promoter, in addition to the endogenous protein. As a result, the cells contain two protein species. More importantly, the fluorescently labeled protein is expressed at levels that can be significantly higher than the endogenous protein levels. With two species of the same protein within the cell, one protein type may get incorporated into the cellular process preferentially over the fusion protein. The interpretation of the experimental measurements thus can

Table III
Analysis of the Observed Variance

	Cse4p-GFP	Nuf2p-GFP	Nuf2p-GFP + Ndc80p-GFP
N (clusters)	90	104	104
Average	1945	6796	15523
Standard deviation	428	1987	2773
Clusters separated by less than 400 nm			
N (cells)	21	13	23
Average difference	240	840	1195
SEM based on average difference	45	232	249

Decrease in the measured signal as a function of the depth at which signal measurement is carried out contributes significantly to the observed variance in the signal. By minimizing its contribution, we find that the biological variance in the number of kinetochore proteins from cell to cell is very low.

be a complicated task in such cells. More importantly, these complications can deteriorate the accuracy achievable in counting protein numbers. The lack of suitable calibration standards is the second challenge if absolute protein numbers are desirable. Our work with the budding yeast centromere-specific histone protein Cse4p demonstrates that yeast cells expressing Cse4p-GFP can serve as a good fluorescent signal standard. Many other types of standards have been used in cell biology, and these are discussed below.

A. The Choice of Calibration Standards for Quantitative Fluorescence Microscopy

The most intuitive method of converting the observed GFP signal into a number of molecules is to divide the total fluorescence signal from a cell or cell organelle by the single molecule GFP fluorescence. Accurate determination of the fluorescence signal from a single GFP molecule, however, can be difficult. There are technical as well as practical issues with this approach. Evaluation of single molecule GFP fluorescence typically requires a highly sensitive imaging and image acquisition setup. While the fluorescence signal from single fluorophores very close to the coverslip can be accurately evaluated, it cannot be directly used as the reference signal for converting intracellular fluorescence into a number of molecules. More importantly, *in vitro* single molecule fluorescence properties such as quantum efficiency may differ from those *in vivo*.

A number of alternative methods have been developed to obtain an accurate calibration standard. The simplest standard is reconstituted GFP in solutions of known concentrations (Hirschberg *et al.*, 1998). The fluorescence signal obtained from such a standard is used as the reference for comparing the measured intracellular signal. The error in the measured number of molecules scales as the square of the average number of proteins in solution. Furthermore, the *in vitro* signal may differ from the intracellular GFP signal because of different excitation strength and quantum efficiency.

Another method was the use of virus capsids incorporating a GFP-fusion capsid protein (Dundr *et al.*, 2002). Because of the crystal-like structure of the shell, each individual shell incorporates a precise number of each of the constituent proteins including the GFP-fusion protein. Individual virus particles introduced/expressed in the cell can therefore be used to determine the single molecule GFP fluorescence intracellular environments.

Wu and Pollard (2005) used quantitative immunoblotting of GFP-fusion constructs in conjunction with quantitative fluorescence microscopy to construct a standard curve that can be used for converting the fluorescence signal from the entire cell for any other protein into a molecular count. This method directly links the measured fluorescence signal for a protein with biochemical determination of the concentration of the same protein via quantitative immunoblotting.

Recently, Rosenfeld *et al.* (2005, 2006) developed an ingenious method for converting the fluorescence signal into the number of fluorophores by making use of transiently expressed fluorescent proteins in dividing bacteria. A fixed number of proteins in a small dividing cell will be randomly partitioned into the daughter cells with a binomial distribution for the number of molecules partitioned in each cell.

Yet another possible approach is to use continuous photobleaching of a small number of closely clustered GFP molecules to resolve a step-wise decrease in the total fluorescence signal from the cluster (Leake *et al.*, 2006; Watanabe and Mitchison, 2002). These steps correspond to individual GFP molecules lapsing into a nonfluorescent state. This technique, however, requires a highly sensitive, low-noise image acquisition system, a small number of GFP molecules in a diffraction limited volume, and a careful selection of excitation and image acquisition conditions to be successful.

Advanced methods such as fluorescence correlation spectroscopy are particularly geared for measuring intracellular protein concentration of soluble proteins (Schwille, 2001). These methods, however, require specialized setups as well as expertise to acquire and analyze images to obtain concentration measurements.

B. Counting Protein Numbers from Volumes Larger than the Diffraction Limit

The kinetochore proteins in budding yeast are distributed over a subdiffraction volume. The technique described in this chapter focuses on ratiometric measurement of protein numbers by using the fluorescence from kinetochore clusters with Cse4p-GFP as the reference. More general instances of protein distribution in a cell include objects that are larger than the diffraction limit of the objective or proteins that are diffusely distributed over the volume of the cell. The nature of protein distribution along with the intended reference standard being used must both be considered in devising a suitable methodology for signal measurement. In either case, the extent of the spatial distribution of fluorescence signal intensity is a function of the actual dimensions of the object and the PSF of the objective. Especially for objects that are much larger than the diffraction-limit along the optical axis, it is important to use either a confocal microscope or a wide-field microscope with suitable deconvolution to assign the out-of-focus light to the correct plane. This has been described previously in Swedlow *et al.* (2002) using 6 μm fluorescent beads. Alternatively, confocal microscopy may also be useful as discussed below.

Accurate quantification of diffusely distributed proteins in thick specimens is best carried out using a confocal microscope. Measurements carried out with high NA objectives become susceptible to spherical and chromatic aberrations making it difficult to collect all the emitted light from molecules dispersed throughout the cell. Wu and Pollard (2005) made simultaneous use of fluorescence microscopy and quantitative immunoblotting of a set of candidate proteins to obtain a "standard curve" that relates measured fluorescence signal to a protein concentration value. Fluorescence signal was measured integrating the signal over the entire volume of the cell from a stack of images for each cell. To avoid deconvolution, Wu and Pollard used a Z step size that is equal to the full width at half maximum (FWHM) of the PSF of the objective along the optical axis. The assumption thus made is that the illumination intensity is uniform over this volume along the optical axis and over the entire thickness of the specimen. After establishing the cell boundaries in differential interference contrast (DIC) images, the fluorescence within these boundaries was integrated through all the image planes. The mean integrated signal for a given protein was then converted into an intracellular concentration by using the standard curve. The accuracy of this technique depends on the errors arising from immunoblotting, the errors in estimating protein content per cell from the total protein extracted for immunoblotting. These errors must be carefully minimized and evaluated in order to impose limits on the accuracy of the molecular counts.

C. Sources of Error in Fluorescence Signal Measurement

Variations in the number of protein molecules obtained from different cells arise from a combination of experimental errors and inherent biological variability. It is therefore important to characterize the contribution of experimental errors so that the nature and reasons for biological variance can be studied. This is especially relevant in the study of the stochasticity in gene expression that is the major goal of fluorescence quantification experiments performed with prokaryotes and simple eukaryotes such as budding yeast (Raser and O'Shea, 2004; Rosenfeld *et al.*, 2005). As discussed earlier, the variance in the kinetochore protein number in each cluster is very low (less than 1 molecule per kinetochore). We also carried out fluorescence recovery after photobleaching (FRAP) experiments to demonstrate that there is no measurable turnover of kinetochore proteins (Joglekar *et al.*, 2006). Together with the low variance of protein numbers from cell to cell, the absence of protein turnover strongly suggests that the protein assemblage at the kinetochore is a built from a specific number of proteins that have a specific arrangement within the kinetochore structure.

Each step in the imaging, image acquisition, and data analysis contributes to the observed errors in the fluorescence signal. The relative contribution of each error must therefore be characterized and minimized. Possible sources in signal variation start from the excitation intensity variations. Mercury arc lamps generally emit a steady intensity, but it may vary over long periods of time. Also, misalignment of the lamp can result in significant changes in the excitation intensity. Commonly used laser sources in confocal microscopy show short- and long-term variations (Swedlow *et al.*, 2002). Periodic alignment of the excitation source and calibration of the excitation intensity is therefore important. Comparative fluorescence measurements are not sensitive to long-term changes in intensity variation. Autofluorescent background in cells is another common source of error. The relative magnitude and spatial variation in the background within the cell both limit the accuracy of measurements. The background light is usually the limiting factor that decides the lowest number of molecules that can be counted within the cell.

There are two major sources of noise that originate in the process of image acquisition: (1) shot noise arising from the photon counting statistics and (2) electronic noise due to the camera electronics. With the low-noise electronics used in modern CCD cameras, the contribution of electronic noise is typically miniscule as compared to other sources of error. The low-noise, back-thinned CCD cameras are also finding more and more use in cell biology. These cameras allow up to 90% photon conversion efficiency, thus significantly improving the signal for the same excitation intensity. The added efficiency is also useful in live-cell microscopy, since it allows the use of lower exposure times for the same sample reducing photobleaching and phototoxicity.

Finally, approximation of the extent of the Gaussian intensity spread, and measuring it with an array of square pixels also introduces errors in signal measurement (Joglekar *et al.*, 2006). The simplest way of measuring a signal is to draw a box centered on a pixel with the maximum signal value (which approximates the centroid of a symmetric spot). This procedure assumes, however, that the object centroid is approximately aligned with the center of a pixel in the CCD array. In reality, positioning of the object centroid is random over any given pixel. As a result, a box drawn with the brightest pixel as the center will on average result in clipping of the signal area. This error can be avoided for diffraction-limited spots by using Gaussian curve-fitting or cross-correlation algorithms to determine the exact centroid of the spot, and using the centroid location to determine the boundaries of the spot. It should be noted that ratio measurements are relatively insensitive to this error.

IX. Conclusions

Determination of fluorophore numbers from the total fluorescence signal measurement is one of the most immediate uses of fluorescence microscopy. With the focus of cell biology experiments shifting to a quantitative, mechanistic characterization of cellular functions, quantitative fluorescence microscopy makes it important to understand the organization of proteins within a protein assemblage. Quantitative fluorescence microscopy provides a new approach to studying protein organization *in vivo* by accurately counting the number of copies of each protein within the assemblage. Accurate protein counts along with nanometer-scale localization data can provide a level of detail that is otherwise not accessible with conventional methods.

Acknowledgments

A.P.J. holds a Career Award at the Scientific Interface from Burroughs-Wellcome Fund. This work was supported by NIH GM32238 to K.S.B. and NIH GM60678, NIH GM24364 to E.D.S.

References

Agard, D. A., Hiraoka, Y., Shaw, P., and Sedat, J. W. (1989). Fluorescence microscopy in three dimensions. *Methods Cell Biol.* **30,** 353–377.

Bar-Even, A., Paulsson, J., Maheshri, N., Carmi, M., O'Shea, E., Pilpel, Y., and Barkai, N. (2006). Noise in protein expression scales with natural protein abundance. *Nat. Genet.* **38,** 636–643.

Bloom, K. S., Beach, D. L., Maddox, P., Shaw, S. L., Yeh, E., and Salmon, E. D. (1999). Using green fluorescent protein fusion proteins to quantitate microtubule and spindle dynamics in budding yeast. *Methods Cell Biol.* **61,** 369–383.

Chudakov, D. M., Lukyanov, S., and Lukyanov, K. A. (2005). Fluorescent proteins as a toolkit for *in vivo* imaging. *Trends Biotechnol.* **23,** 605–613.

Collins, K. A., Furuyama, S., and Biggins, S. (2004). Proteolysis contributes to the exclusive centromere localization of the yeast Cse4/CENP-A histone H3 variant. *Curr. Biol.* **14,** 1968–1972.

Colman-Lerner, A., Gordon, A., Serra, E., Chin, T., Resnekov, O., Endy, D., Pesce, C. G., and Brent, R. (2005). Regulated cell-to-cell variation in a cell-fate decision system. *Nature* **437,** 699–706.

Dundr, M., McNally, J. G., Cohen, J., and Misteli, T. (2002). Quantitation of GFP-fusion proteins in single living cells. *J. Struct. Biol.* **140,** 92–99.

Ghauharali, R. I., and Brakenhoff, G. J. (2000). Fluorescence photobleaching-based image standardization for fluorescence microscopy. *J. Microsc.* **198**(Pt 2), 88–100.

Gibson, S. F., and Lanni, F. (1992). Experimental test of an analytical model of aberration in an oil-immersion objective lens used in three-dimensional light microscopy. *J. Opt. Soc. Am. A* **9,** 154–166.

Giepmans, B. N., Adams, S. R., Ellisman, M. H., and Tsien, R. Y. (2006). The fluorescent toolbox for assessing protein location and function. *Science* **312,** 217–224.

Hiraoka, Y., Sedat, J. W., and Agard, D. A. (1990). Determination of three-dimensional imaging properties of a light microscope system. Partial confocal behavior in epifluorescence microscopy. *Biophys. J.* **57,** 325–333.

Hirschberg, K., Miller, C. M., Ellenberg, J., Presley, J. F., Siggia, E. D., Phair, R. D., and Lippincott-Schwartz, J. (1998). Kinetic analysis of secretory protein traffic and characterization of golgi to plasma membrane transport intermediates in living cells. *J. Cell Biol.* **143,** 1485–1503.

Hoffman, D. B., Pearson, C. G., Yen, T. J., Howell, B. J., and Salmon, E. D. (2001). Microtubule-dependent changes in assembly of microtubule motor proteins and mitotic spindle checkpoint proteins at PtK1 kinetochores. *Mol. Biol. Cell* **12,** 1995–2009.

Huh, W. K., Falvo, J. V., Gerke, L. C., Carroll, A. S., Howson, R. W., Weissman, J. S., and O'Shea, E. K. (2003). Global analysis of protein localization in budding yeast. *Nature* **425,** 686–691.

Inoue, S., and Springer, K. R. (1997). "Video Microscopy: The Fundamentals." Plenum, New York.

Joglekar, A. P., Bouck, D. C., Molk, J. N., Bloom, K. S., and Salmon, E. D. (2006). Molecular architecture of a kinetochore-microtubule attachment site. *Nat. Cell Biol.* **8,** 581–585.

Kohl, T., and Schwille, P. (2005). Fluorescence correlation spectroscopy with autofluorescent proteins. *Adv. Biochem. Eng. Biotechnol.* **95,** 107–142.

Kremers, G. J., Goedhart, J., van Munster, E. B., and Gadella, T. W., Jr. (2006). Cyan and yellow super fluorescent proteins with improved brightness, protein folding, and FRET Forster radius. *Biochemistry* **45,** 6570–6580.

Kues, T., Peters, R., and Kubitscheck, U. (2001). Visualization and tracking of single protein molecules in the cell nucleus. *Biophys. J.* **80,** 2954–2967.

Leake, M. C., Chandler, J. H., Wadhams, G. H., Bai, F., Berry, R. M., and Armitage, J. P. (2006). Stoichiometry and turnover in single, functioning membrane protein complexes. *Nature* **443,** 355–358.

Lippincott-Schwartz, J., and Patterson, G. H. (2003). Development and use of fluorescent protein markers in living cells. *Science* **300,** 87–91.

Lippincott-Schwartz, J., Snapp, E., and Kenworthy, A. (2001). Studying protein dynamics in living cells. *Nat. Rev. Mol. Cell Biol.* **2,** 444–456.

Longtine, M. S., McKenzie, A., 3rd, Demarini, D. J., Shah, N. G., Wach, A., Brachat, A., Philippsen, P., and Pringle, J. R. (1998). Additional modules for versatile and economical PCR-based gene deletion and modification in *Saccharomyces cerevisiae*. *Yeast* **14,** 953–961.

McAinsh, A. D., Tytell, J. D., and Sorger, P. K. (2003). Structure, function, and regulation of budding yeast kinetochores. *Annu. Rev. Cell Dev. Biol.* **19,** 519–539.

Meluh, P. B., Yang, P., Glowczewski, L., Koshland, D., and Smith, M. M. (1998). Cse4p is a component of the core centromere of *Saccharomyces cerevisiae*. *Cell* **94,** 607–613.

Meraldi, P., McAinsh, A. D., Rheinbay, E., and Sorger, P. K. (2006). Phylogenetic and structural analysis of centromeric DNA and kinetochore proteins. *Genome Biol.* **7,** R23.

Miyawaki, A. (2004). Fluorescent proteins in a new light. *Nat. Biotechnol.* **22,** 1374–1376.

Miyawaki, A., Nagai, T., and Mizuno, H. (2005). Engineering fluorescent proteins. *Adv. Biochem. Eng. Biotechnol.* **95,** 1–15.

Nagai, T., Ibata, K., Park, E. S., Kubota, M., Mikoshiba, K., and Miyawaki, A. (2002). A variant of yellow fluorescent protein with fast and efficient maturation for cell-biological applications. *Nat. Biotechnol.* **20,** 87–90.

Newman, J. R., Ghaemmaghami, S., Ihmels, J., Breslow, D. K., Noble, M., DeRisi, J. L., and Weissman, J. S. (2006). Single-cell proteomic analysis of *S. cerevisiae* reveals the architecture of biological noise. *Nature* **441,** 840–846.

O'Toole, E. T., Winey, M., and McIntosh, J. R. (1999). High-voltage electron tomography of spindle pole bodies and early mitotic spindles in the yeast *Saccharomyces cerevisiae*. *Mol. Biol. Cell* **10,** 2017–2031.

Patterson, G. H., Knobel, S. M., Sharif, W. D., Kain, S. R., and Piston, D. W. (1997). Use of the green fluorescent protein and its mutants in quantitative fluorescence microscopy. *Biophys. J.* **73,** 2782–2790.

Piston, D. W. (1998). Choosing objective lenses: The importance of numerical aperture and magnification in digital optical microscopy. *Biol. Bull.* **195,** 1–4.

Piston, D. W., Patterson, G. H., and Knobel, S. M. (1999). Quantitative imaging of the green fluorescent protein (GFP). *Methods Cell Biol.* **58,** 31–48.

Raser, J. M., and O'Shea, E. K. (2004). Control of stochasticity in eukaryotic gene expression. *Science* **304,** 1811–1814.

Raser, J. M., and O'Shea, E. K. (2005). Noise in gene expression: Origins, consequences, and control. *Science* **309,** 2010–2013.

Rosenfeld, N., Perkins, T. J., Alon, U., Elowitz, M. B., and Swain, P. S. (2006). A fluctuation method to quantify *in vivo* fluorescence data. *Biophys. J.* **91,** 759–766.

Rosenfeld, N., Young, J. W., Alon, U., Swain, P. S., and Elowitz, M. B. (2005). Gene regulation at the single-cell level. *Science* **307,** 1962–1965.

Salmon, E. D., Shaw, S. L., Waters, J., Waterman-Storer, C. M., Maddox, P. S., Yeh, E., and Bloom, K. (2003). A high-resolution multimode digital microscope system. *Methods Cell Biol.* **72,** 185–216.

Sawano, A., and Miyawaki, A. (2000). Directed evolution of green fluorescent protein by a new versatile PCR strategy for site-directed and semi-random mutagenesis. *Nucleic Acids Res.* **28,** E78.

Schwille, P. (2001). Fluorescence correlation spectroscopy and its potential for intracellular applications. *Cell Biochem. Biophys.* **34,** 383–408.

Shaner, N. C., Campbell, R. E., Steinbach, P. A., Giepmans, B. N., Palmer, A. E., and Tsien, R. Y. (2004). Improved monomeric red, orange and yellow fluorescent proteins derived from *Discosoma* sp. red fluorescent protein. *Nat. Biotechnol.* **22,** 1567–1572.

Shaner, N. C., Steinbach, P. A., and Tsien, R. Y. (2005). A guide to choosing fluorescent proteins. *Nat. Methods* **2,** 905–909.

Swedlow, J. R., Hu, K., Andrews, P. D., Roos, D. S., and Murray, J. M. (2002). Measuring tubulin content in *Toxoplasma gondii*: A comparison of laser-scanning confocal and wide-field fluorescence microscopy. *Proc. Natl. Acad. Sci. USA* **99,** 2014–2019.

Taylor, D. L., and Salmon, E. D. (1989). Basic fluorescence microscopy. *Methods Cell Biol.* **29,** 207–237.

Tsien, R. Y. (1998). The green fluorescent protein. *Annu. Rev. Biochem.* **67,** 509–544.

Watanabe, N., and Mitchison, T. J. (2002). Single-molecule speckle analysis of actin filament turnover in lamellipodia. *Science* **295,** 1083–1086.

Wei, R. R., Sorger, P. K., and Harrison, S. C. (2005). Molecular organization of the Ndc80 complex, an essential kinetochore component. *Proc. Natl. Acad. Sci. USA* **102,** 5363–5367.

Winey, M., Mamay, C. L., O'Toole, E. T., Mastronarde, D. N., Giddings, T. H., Jr., McDonald, K. L., and McIntosh, J. R. (1995). Three-dimensional ultrastructural analysis of the *Saccharomyces cerevisiae* mitotic spindle. *J. Cell Biol.* **129,** 1601–1615.

Wu, J. Q., and Pollard, T. D. (2005). Counting cytokinesis proteins globally and locally in fission yeast. *Science* **310,** 310–314.

Xie, X. S., Yu, J., and Yang, W. Y. (2006). Living cells as test tubes. *Science* **312,** 228–230.

Yu, J., Xiao, J., Ren, X., Lao, K., and Xie, X. S. (2006). Probing gene expression in live cells, one protein molecule at a time. *Science* **311,** 1600–1603.

Zhang, J., Campbell, R. E., Ting, A. Y., and Tsien, R. Y. (2002). Creating new fluorescent probes for cell biology. *Nat. Rev. Mol. Cell Biol.* **3,** 906–918.

Zwier, J. M., Van Rooij, G. J., Hofstraat, J. W., and Brakenhoff, G. J. (2004). Image calibration in fluorescence microscopy. *J. Microsc.* **216,** 15–24.

CHAPTER 8

Fluorescent Protein Applications in Plants

R. Howard Berg* and Roger N. Beachy†

*Integrated Microscopy Facility
Donald Danforth Plant Science Center
St. Louis, Missouri 63132

†Donald Danforth Plant Science Center
St. Louis, Missouri 63132

Abstract

Study of plant cell biology has benefited tremendously from the use of fluorescent proteins (FPs). Development of well-established techniques in genetics, by transient expression or by *Agrobacterium*-mediated plant cell transformation, makes it possible to readily create material for imaging molecules tagged with FPs. Confocal microscopy of FPs is routine and, in highly scattering tissues, multiphoton microscopy improves deep imaging. The abundance of autofluorescent compounds in plants in some cases potentially interferes with FP signals, but spectral imaging is

0091-679X/08 $35.00
DOI: 10.1016/S0091-679X(08)85008-X

an effective tool in unmixing overlapping signals. This approach allows separate detection of DsRed and chlorophyll, DsRed and GFP, and green fluorescent protein (GFP) and yellow fluorescent protein (YFP).

FPs have been targeted to most plant organelles. Free (untargeted) FPs in plant cells are not only cytoplasmic, but also go into the nucleus due to their small size. FP fluorescence is potentially unstable in acidic vacuoles. FPs have been targeted to novel compartments, including protein storage vacuoles in seeds. Endoplasmic reticulum (ER)-targeted GFP has identified novel inclusion bodies that are surprisingly dynamic. FP-tagged Rab GTPases have allowed documentation of the dynamics of membrane trafficking. Investigation of virus infections has progressed significantly with the aid of FP-tagged virus proteins.

Advanced techniques are giving plant scientists the ability to quantitatively analyze the behavior of FP-tagged proteins. Fluorescence lifetime microscopy is becoming the method of choice for fluorescence resonance energy transfer (FRET) analysis of FP-tagged proteins. Fluorescence correlation spectroscopy (FCS) of FPs provides information on molecular diffusion and intermolecular interactions.

Use of FPs in elucidating the behavior of plant cells has a bright future.

I. Introduction

Fluorescent proteins (FPs) have changed the way we think of looking at living plant cells. They allow the behavior of specific proteins to be followed at high resolution and for relatively long periods of time. These probes have come at a fortuitous time when light microscopy offers powerful imaging technology, including a variety of optical sectioning techniques, sophisticated detectors, and effective image reconstruction software (Moreno *et al.*, 2006).

Looking past the efforts required to generate expressing cells, FP imaging does not require staining, and allows analysis of cells in a relatively undisturbed, living state. FPs are the most important fluorophores in investigation of plant cell biology, judging from the extensive literature reporting their use (recent reviews include Dixit *et al.*, 2006; Ehrhardt, 2003; Fricker *et al.*, 2006; Haseloff and Siemering, 2006). In addition to their use in localizing specific proteins in the cytoplasm, FPs have been targeted to all plant organelles (Tian *et al.*, 2004). They also have been adapted to use as biosensors for pH and calcium. FP-tagged viral proteins have shed light on the cell biology of the interaction of viral proteins with plant organelles (Lazarowitz and Beachy, 1999).

The original demonstration of the utility of GFP by Chalfie *et al.* (1994) was followed by extensive molecular analysis of this protein and derivation of many different spectral mutants (Zacharias and Tsien, 2006). Additional FPs have been discovered in coral organisms (Lukyanov *et al.*, 2006), providing the plant scientist with potentially a rainbow of different biological fluorophores to work with.

High-throughput technology is helping to bring to reality the promise of FPs to reveal the "workings" of plant cells. Carefully considering the effects of the position of insertion of the FP open reading frame (ORF) within the target gene sequence, and the physiological conditions of the cellular compartment that is targeted, Tian *et al.* (2004) designed a high-throughput technique for tagging full-length proteins in a way that minimizes disruption of their targeting or function. This approach may aid in revealing the cellular location and dynamics of the over one-third of the *Arabidopsis* proteins that have no predicted function (Wortman *et al.*, 2003). Methods for high-throughput microscopy to screen mutant cell lines or whole organisms have been developed by Avila *et al.* (2003). Real time scanning microscopes, available in the form of slit- or Nipkow disk-scanning designs (Pawley, 2006a), allow direct ocular viewing of samples and rapid screening by eye. Avila *et al.* also developed a method for using an automated confocal microscope to screen mutant pools of *Arabidopsis* seedlings grown in modified lids of microtiter plates that can be robotically controlled by the microscope. The system is accompanied by software capable of analyzing expression patterns. A novel transient virus infection system uses cDNA fusions with GFP to screen for cDNAs that localize to specific cell sites (Escobar *et al.*, 2003). Each infection site carries a specific gene construct and makes it possible for the investigator to screen hundreds of cell phenotypes per day. Twelve plasmodesmata-localized signals were found using this technique, proving more successful than other methods.

It is not our intent to present a comprehensive review of applications of FPs in plants or plant cells. Rather, we address the primary considerations one undertakes when deriving plant cells that express genes that encode FPs, the methods for microscopy of FP-expressing cells, and briefly examine advanced techniques for the quantitative analysis of FP-expressing cells in plants.

II. Expression and Function of FPs in Plants

A. Gene Expression

The original GFP isolated from the jellyfish *Aequorea victoria* has an excitation peak at 395 nm and a weaker excitation at 475 nm, and emits at 508 nm (FP molecular biology is reviewed in Zacharias and Tsien, 2006). Early attempts to express wild-type GFP in *Arabidopsis* were unsuccessful, requiring modification of codons and other sequences in order to yield a functional protein in plants (Chiu *et al.*, 1996; Haseloff *et al.*, 1997). Subsequent genetic modifications have produced a number of GFP mutants, with the enhanced green fluorescence protein (EGFP) mutant being one of the most commonly used forms. EGFP is relatively bright, has a single excitation maximum at 488 nm and an emission at 509 nm, closely matching those for fluorescein (commonly configured for detection on fluorescence microscopes), and matures about four times faster than wild-type GFP. FPs from a variety of other marine organisms are now in use, including DsRed and mRFP1.

GFP spectral mutants are divided into seven types based on the type of chromophore (Zacharias and Tsien, 2006). The cyan form enhanced cyan fluorescent proteins (ECFP) has excitation/emission maxima at 434/476 nm, and the yellow form enhanced yellow fluorescent proteins (EYFP) has excitation/emission maxima at 514/527 nm. Choice of FP for a given study depends on whether there will be several FPs under observation, requiring selection of pairs that can be spectrally separated. Conventional, filter-based separation cannot separate GFP and YFP, but these can be separated using spectral imaging and unmixing (as discussed below, Section IV—Advanced Techniques). Two common pairs are ECFP and EYFP, which can be used to study protein–protein interactions by fluorescence resonance energy transfer (FRET, Periasamy and Day, 2005), which signals proximity between two fluorophores (of appropriate spectral overlap) when they are within ~9 nm.

Brightness of the FP depends on a combination of its extinction coefficient (efficiency in absorbing excitation light) and quantum yield (photons emitted/photons absorbed). A high extinction coefficient allows the use of lower excitation light levels, desirable because this reduces specimen photooxidation damage that results as a consequence of free radical formation that photobleaches the FP and other nearby molecules. A high quantum yield also allows use of lower levels of excitation wavelength light, resulting in a higher brightness relative to an FP with lower quantum yield when using the same intensity of excitation light. EGFP and EYFP have similar quantum yields, but EYFP has a higher extinction coefficient, making it more prone to photobleaching due to excessive free radical formation with those excited state electrons not used for emission (Dixit *et al.*, 2006). ECFP is less prone to photobleaching but is not as bright as EGFP or EYFP. Rizzo *et al.* (2004) modified the barrel structure for the ECFP chromophore, increasing its brightness ~2.5-fold. Termed "Cerulean," this form of CFP also is improved for use in CFP-YFP FRET. In addition to an increased quantum yield, Cerulean CFP has mono-exponential fluorescence lifetime decay that simplifies FRET analysis by fluorescence lifetime imaging, discussed below (Section IV—Advanced Techniques). Cerulean exhibits a brighter signal than ECFP when expressed in plant cells (unpublished results, Danforth Center investigators).

The destination compartment can affect FP fluorescence. Fluorescence of GFP targeted to the lumen of lytic vacuoles is sensitive to light, acidic pH, and cysteine proteinases (Tamura *et al.*, 2003). YFPs are sensitive to anions and acidic pH, leading Tian *et al.* (2004) to use the Citrine YFP mutant for their studies, as it is less sensitive to anions and pH than other YFPs. Choosing an FP with a pK_a less than the pH of the destination compartment will avoid loss of fluorescence due to acidic pH. Dixit *et al.* (2006) summarized the data as follows: the pK_a of EGFP and EYFP is 6.1, ECFP is 6.4, Citrine YFP is 5.7, and Cerulean CFP is 4.7.

DsRed, isolated from the coral *Discosoma* sp., is an FP useful in the red spectral region (excitation/emission maxima at 558/583 nm, Campbell *et al.*, 2002). It is bright (high extinction coefficient and quantum yield) and resistant to photobleaching, but has a relatively long maturation period (~24 h). A major drawback to its use is that the native protein forms tetramers. Campbell *et al.* (2002) derived a mutant of DsRed, mRFP1, that is monomeric and matures 10 times faster than DsRed.

An additional improvement of mRFP1 over DsRed is a ~25 nm red spectral shift, yielding an excitation/emission maxima of 584/607. Thus, mRFP1 is readily separated from GFP in coexpression studies. A major drawback to mRFP1 is its low brightness (low quantum yield) and high rate of photobleaching. Additional mutations of this protein have improved quantum yield and photostability of this class of FP, producing a useful library of red-shifted FPs (Shaner *et al.*, 2004).

Plants contain a variety of autofluorescent compounds (Rost, 1995). Figure 1 shows the emission spectra of some common plant autofluorescence compounds, with two photon excitation at 790 nm. The most commonly encountered autofluorescence is from lignin, a cell wall polymer that emits in the green range (~440 to ~540 nm), and chlorophyll, which emits at 680 nm. Therefore, detection of red-emitting FPs potentially could be interfered with by chlorophyll autofluorescence in above-ground plant tissues, and GFP detection could be hindered by cells containing lignified cell walls. Cutin- and suberin-containing walls have a lignin-like matrix as well. Spectral imaging can be used to separate autofluorescence from FP signals, as discussed below (Section IV—Advanced Techniques).

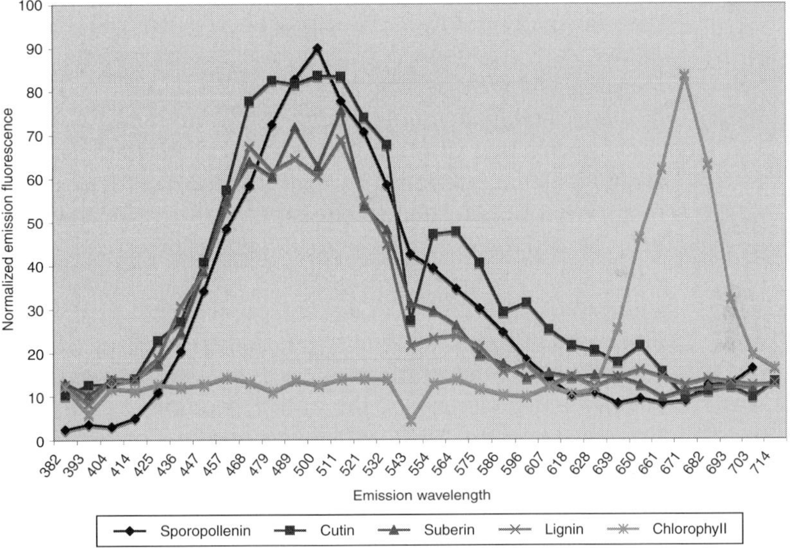

Fig. 1 Emission spectra of common plant autofluorescent compounds. Most of these emit in the blue-green region (~440 to ~540 nm), potentially overlapping with cyan fluorescent protein (CFP), green fluorescent protein (GFP), and yellow fluorescent (YFP). However, this depends on whether they are in the same cells as the FP fusion product. Sporopollenin is a component of the pollen wall. Cutin and suberin autofluoresce due to their lignin matrix, and are found in shoot and leaf epidermal cell walls (cutin) or in the root endodermis, hypodermis, and other specialized cells (suberin). Chlorophyll emits at 680 nm and only occurs in chloroplast-containing tissues of the shoot and leaf. In these tissues, there is the potential for overlapping emission with red-emitting FPs. Spectra obtained from *Arabidopsis* tissues excited with two photon light (790 nm) and collected from spectral images using a Zeiss LSM 510 META microscope. From Berg (2004) with permission.

Design of fusion proteins should consider the potential effect of the fusion on function of the native protein. Computer-assisted designs can help to identify potential targeting sequences at either the C- or N-terminus of the native protein (Tian *et al.*, 2004). N-terminal fusions have the potential to disrupt targeting to the ER, mitochondria, and plastids. C-terminal fusions may cause mislocalization. C-terminal fusions have failed to target the plasma membrane, and also could mask posttranslational modification sites, for example, myristylation or farnesylation sites that target membranes (Tian *et al.*, 2004). These problems have been avoided by either adding short linker peptides at the termini, which help maintain protein function, or by using internal insertion sites to create the fusion.

Transcriptional promoters have a dominant effect on the type of organ and tissue in which the fluorescent signal is located. Constitutive promoters such as the 35S Cauliflower mosaic virus, actin, and ubiquitin promoters can lead to high levels of gene expression. Beware of transgenic cells and plants that produce very high levels of FPs. Overexpression is to be avoided because excess protein can saturate the target compartment and excess product can mislocalize to other compartments (Dixit *et al.*, 2006). Furthermore, high levels of the gene product can be harmful to the cell (Hashimoto, 2002). Ectopic expression and genes that cause dominant negative phenotypes can be avoided by selecting transformants that have lower expression levels. In selecting stably transformed plants, there is a chance that lines in which the transgene causes a negative phenotype can be eliminated from among the pool of transgenic lines. However, transient expression systems that result in bright, overexpression in cells are especially suspect for errant protein targeting. The ideal is to use native promoters, which are more likely to appropriately regulate expression on the tissue and cell level than are nonnative promoters. However, low levels of gene expression from native promoters may be difficult to detect. One approach to increase signal in native promoter-driven studies is to use multiple, tandem copies of the FP in the fusion protein. This is illustrated in Fig. 2, showing tip localization of the actin-related protein Arp2/3 in caulonemal filaments of the moss *Physcomitrella patens* (Perroud and Quatrano, 2006). Under control of the native promoter, this protein fusion contains two tandem copies of YFP, allowing detection of a protein that is present in low amounts.

Inducible promoters (Moore *et al.*, 2006) can provide additional control in timing the expression of transgenes and generally reduce or eliminate negative phenotypes that can be associated with gene overexpression. We have had success using the chemically inducible gene switch based on the ecdysone receptor, induced with the insecticide methoxyfenozide (Padidam *et al.*, 2003).

1. Protocol for Using the Ecdysone Gene Switch System in *Arabidopsis*

The plant-based ecdysone gene switch is based on activation of the insect steroid ecdysone receptor (VGE) with methoxyfenozide, and subsequent activation of a target gene by binding of activated VGE to an upstream target sequence that

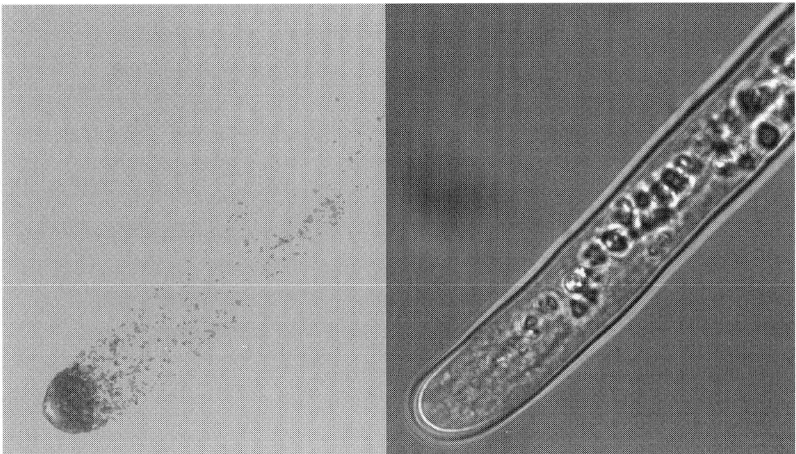

Fig. 2 Caulonemal filament of *Physcomitrella patens* expressing a yellow fluorescent protein (YFP): ARPC4 fusion protein. Localized to the caulonema tip, detection of this low-abundance protein is enhanced by using two (tandemly inserted) YFPs fused with the protein. (Cell Motility and the Cytoskeleton, Vol. 63, No. 3, 2006, pp. 162–171. © 2006 Wiley-Liss, Inc. Reprinted with permission of Wiley-Liss, Inc., a subsidiary of John Wiley & Sons, Inc.)

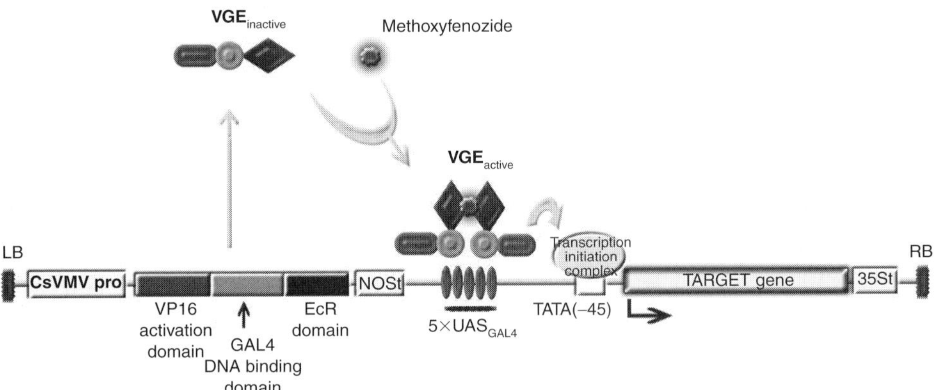

Fig. 3 Gene construct for the plant-based ecdysone gene switch. The ecdysone receptor is activated by methoxyfenozide, subsequently activating the target gene via binding to the GAL4 binding sequence (see text).

contains the GAL4 binding sequence (Fig. 3). The gene construct is introduced to the target tissue by transformation; in the case of transformation of *Arabidopsis*, genes are inserted into an *Agrobacterium*-based vector and plants are transformed by a simple floral dip method (Lorence and Verpoorte, 2004). Transgenic plants of other types are produced by similar or different mechanisms. The inducer is available commercially as the insecticide Intrepid™, which is diluted 1:10,000 or more to reach methoxyfenozide concentrations of 61 μM and less, according to the target plant and transgenic plant line (Koo *et al.*, 2004). The target gene is induced by soil drench or, in the case shown in Fig. 4, by brushing leaves with a suspension containing the inducer. Using this system, expression of a gene encoding the tobacco mosaic virus (TMV) movement protein (MP) in *Arabidopsis* was developed. It is possible to show that viral spread is dependent on induction of MP in leaves infected with TMV that lacks the movement protein ORF (Fig. 4).

Expression of multiple FPs can lead to cosuppression of gene expression, and gene silencing of either or both genes (Baulcombe, 2005); this especially when the same promoter is used for the coexpressed genes. Inducible promoters help to avoid this because gene expression levels are absent until the ligand is added to induce gene expression with both temporal and spatial control.

Agrobacterium-based binary vectors are commonly used when constructing genes that encode protein fusions (Lorence and Verpoorte, 2004). Transient expression can be induced by a variety of methods, including by biolistic bombardment, electroporation (Asurmendi *et al.*, 2004), or by infiltrating leaf tissue with *Agrobacterium* containing the plasmid.

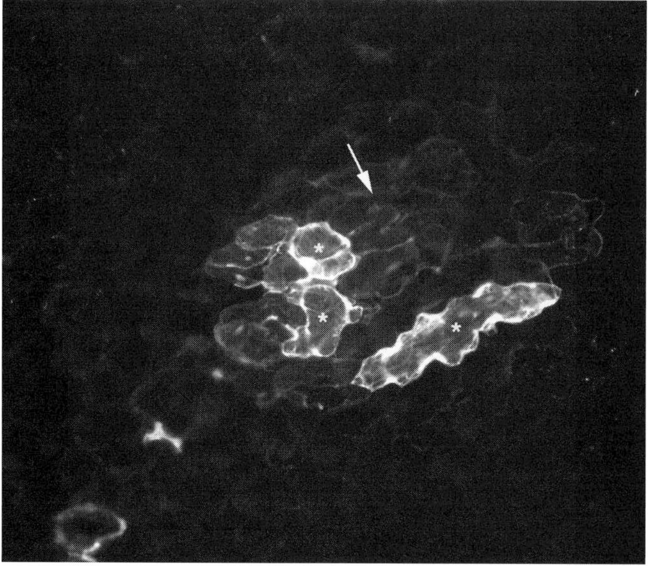

Fig. 4 Induction of tobacco mosaic virus (TMV) cell–cell movement using the ecdysone gene switch. This leaf tissue is from a transgenic *Arabidopsis* plant expressing TMV movement protein (MP) regulated by the ecdysone gene switch. The leaf was infected with TMV that expresses free green fluorescent protein (GFP) and lacks MP. Virus was limited to the initially infected cells (asterisks) until the gene switch was induced. Shown here 3 days after induction, viral movement to adjacent cells (e.g., arrow) was only possible after induction of MP expression by the plant cell (M. Soto-Aguilar and R.N. Beachy, unpublished results).

Recent technology simplifies construction of multiple forms of expression vectors, including constructing multiple gene fusions for expression in a variety of host systems. Gateway cloning vectors (Earley *et al.*, 2006) includes examples of vectors that facilitate coexpression of multiple FPs as C- or N-terminal fusions. pSAT is another novel modular vector system (Tzfira *et al.*, 2005) that provides C- or N-terminal fusions for up to five different FPs that are controlled by a variety of promoters. This has the potential for reducing gene silencing effects. Imaging of multiple FPs is possible using spectral imaging and algorithms that "unmix" the signals (Berg, 2004, see Section IV—Advanced Techniques).

B. Biosensors

FP mutants that are sensitive to ions or that shift emission spectra in response to specific wavelengths of light have proven useful as probes in plant cell biology. Mutants of GFP have been constructed that have sensitivity to calcium, pH, and various signaling compounds (reviewed in Lalonde *et al.*, 2005). Other biosensors have been based on FRET between two FPs. FRET biosensors have been developed for calcium, pH, sugars, signaling compounds, and enzymes, including caspases and kinases (reviewed in Lalonde *et al.*, 2005). Photoswitchable FPs change excitation and/or emission maxima upon photoactivation (Chapman *et al.*, 2005). In this case, the photoactivatable FP, termed an optical highlighter (Chapman *et al.*, 2005), is excited with a specific wavelength that changes its spectral characteristics. These changes can be monitored with the appropriate excitation and emission detection. Their use in plants includes detecting photoactivation of an optical highlighter fused with TMV movement protein (Chapman *et al.*, 2005), which provides a means for studying the role of this protein in viral movement.

C. Assessing Function

Proteins of unknown function and cellular location present a challenge to researchers because of problems in identifying possible artifacts that may be associated with their localization. As discussed previously, overexpressing proteins that produce very bright signals can lead to ectopic localization. If the fusion protein misfolds hydrophobic residues which are exposed can lead to aggregation and punctate fluorescent signals that are artifacts. Computer aided analysis of protein sequence may identify a targeting sequence, which in turn should be consistent with the pattern of accumulation of the protein. The ideal method for confirming that the fusion protein is functional in the target compartment is to complement a mutant with the fusion protein. This is uncomplicated in moss plants because they exhibit homologous recombination, allowing exchange of mutant gene with fusion gene. The signal in caulonema in Fig. 2 was derived from complementation of the mutant gene by homologous recombination (Perroud and Quatrano, 2006).

If there is no known targeting sequence, distribution of fluorescent signal can be verified by immunofluorescence with antibodies against the native protein. If localization of the FP appears to be organellar, the result can be confirmed by colocalizing fluorescence in plants that express an organelle-targeted FP with an emission spectrum that can be separated from that of the fusion protein. In shoot and leaf tissue, colocalization of the fusion protein with chloroplasts can be confirmed using chlorophyll autofluorescence (excitation with blue or red wavelengths, emission at 680 nm). Virtually all plant organelles have been targeted with FPs (Dixit *et al.*, 2006), and dual expression might be accomplished by transient expression of the protein–FP fusion, or by crossing two parental lines. It is possible to use spectral unmixing to separate two (or more) spectrally similar FPs (see Section IV—Advance Techniques). As previously discussed, some FPs are sensitive to pH or anions and this may lead to reduced or no fluorescence in lytic vacuoles (acidic) or the cell wall (acidic, anionic). This may be avoided by choosing an FP having an appropriately low pK_a.

Counterstaining tissues samples with fluorescent dyes can aid in identifying cell compartments. Stains are available for cell walls, mitochondria, nuclei, and other organelles (Hepler and Gunning, 1998; Ruzin, 1999). The cell wall stain Congo Red is convenient as a counterstain for GFP, as it is excited by 488 nm and its emission has a large Stoke's shift (595 nm) that is easily separated from GFP. The dye MDY-64 (Inselman *et al.*, 1999) has been used to stain the plant vacuole tonoplast (R.H.B., unpublished results), with the caveat that its excitation and emission maxima are 451/497 nm, and overlapping with emission of GFP.

Correlative transmission electron microscopy (TEM) can be a tremendous aid in interpreting the fluorescent signal. For example, consider the confocal image of TMV movement protein fused with GFP (Fig. 5). During the infection cycle, MP elicits a change in ER structure, inducing it to form large bodies (Reichel and Beachy, 1998). MP is aggregated in these large bodies that were identified as virus replication complexes (VRCs, Asurmendi *et al.*, 2004). Shown in Fig. 6, thin sections of VRCs in vascular parenchyma cells in TMV-infected tobacco leaf tissue reveal in high resolution the specific arrangement of MP with respect to the ER. MP accumulates in rope-like structures intermingled with ER cisternae, binding (arrows) its cytoplasmic surface. The confocal image fails to resolve this level of detail, instead suggesting there is a continuous surface of ER covering the body.

An obvious method to localize fluorescent FP signal with a cellular compartment is to use the same material for immunogold labeling of the FP in thin sections visualized by TEM. In practice, this often gives a low gold particle count, likely due to low reactivity with the antibody. An ideal situation would be when the FP fusion product is highly concentrated in a compartment, as illustrated in Figs. 7 and 8. These figures show soybean cotyledon cells containing ER-targeted GFP:KDEL; the KDEL sequences retain GFP in protein storage vacuoles (M. Schmidt and E. Herman, USDA-ARS, Danforth Plant Science Center; unpublished results). Large amounts of GFP in the vacuoles cause the seeds to appear green when

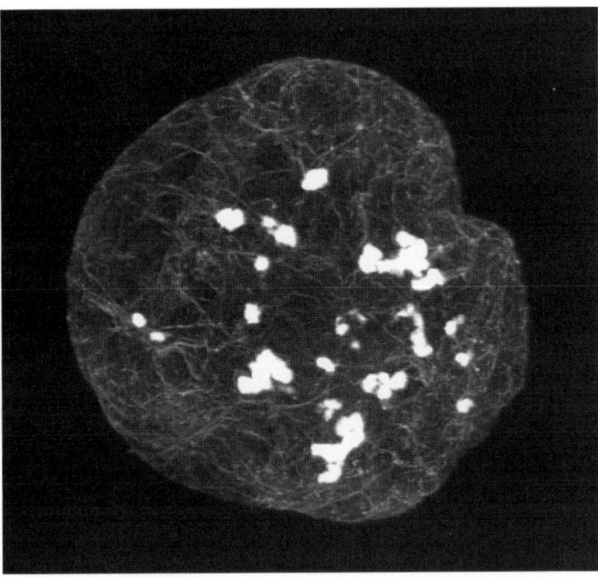

Fig. 5 Confocal image of a tobacco BY-2 cell protoplast, 20 h after infection with tobacco mosaic virus (TMV) expressing movement protein (MP): green fluorescent protein (GFP). MP aggregates in large bodies that contain endoplasmic reticulum (ER) (Reichel and Beachy, 1998) and are viral replication complexes (Asurmendi *et al.*, 2004). Compare with Fig. 6.

viewed in white light; confocal images show a strong fluorescence signal (Fig. 7). High concentrations of GFP in this compartment is confirmed by very high levels of immunogold signal using anti-GFP antibodies, as shown in Fig. 8.

A remarkable instance of the need for TEM in interpreting FP signal is the case of ER bodies (Gunning, 1998). Several studies have shown the formation of spindle-shaped bodies in *Arabidopsis* expressing an ER-targeted FP (e.g., Matsushima *et al.*, 2004). These bodies are highly mobile and until studies of thin sections showed that they were due to accumulation of proteins in a locally dilated region of the ER (Gunning, 1998), there was speculation that a unique organelle had been discovered. Similar spindle-shaped inclusions are common in the Brassicaceae, including in *Arabidopsis*. Stamen filament cells expressing an ER-targeted GFP show these spindle bodies (Fig. 9, time-lapse movies online at www/danforthcenter.org/imf/4danimation.htm), and thin sections of the same cells show that these are ER inclusions, with ribosomes binding the surrounding membrane (Fig. 10).

In plants, proteins smaller than 40–60 kDa diffuse into nuclei (Grebenok *et al.*, 1997). Therefore, FPs not fused to a second protein are observed in the cytoplasm as well as in the nucleus. Chytilova *et al.* (1999) specifically localized GFP to the nucleus by tagging it with a nuclear localization sequence (NLS).

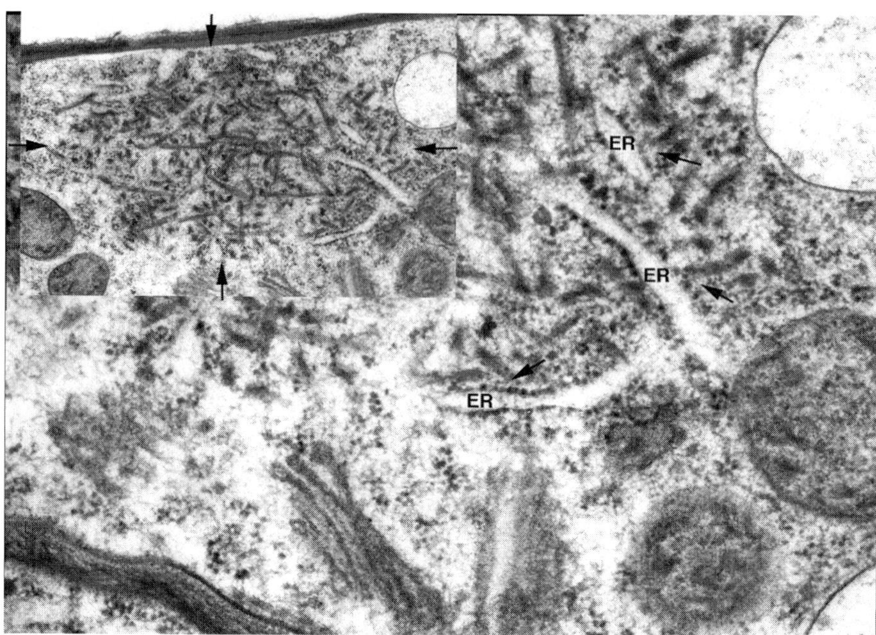

Fig. 6 Thin section electron microscopy of tobacco mosaic virus (TMV)-infected tobacco leaf tissue containing viral replication complexes. One complex is shown in the inset (delineated by the arrows), and is comparable in function to the bodies shown by confocal microscopy in Fig. 5. Electron microscopy is essential for showing the relationship of movement protein (MP), present in the rope-like structures, to the endoplasmic reticulum (ER) in these complexes. Rather than forming a continuous surface, as suggested by confocal images, the ER maintains cisternal structure that occurs throughout the complex. The rope-like structures bind the ER surface (arrows) (freeze-substituted specimen, R.H.B and R.N.B., unpublished results).

III. Imaging

It is helpful to use a fluorescence dissecting microscope when screening tissues that may produce fluorescence. In nongreen tissues a long pass emission filter allows detection of a weak GFP signal: a band-pass filter is used to reject chlorophyll autofluorescence and to verify the specificity of any signal detected with the long-pass filter. If the dissecting microscope has a zoom lens, when operated at high magnification the intensity of emission signal is higher. Putative transformed cells or plants must be rigorously tested by observing them with narrow bandpass filters during cellular imaging. In addition, it is important to compare samples with those from comparable nontransgenic materials for autofluorescence. Particular attention should be paid to autofluorescence that may result from biolistic protocols or leaf infiltration techniques.

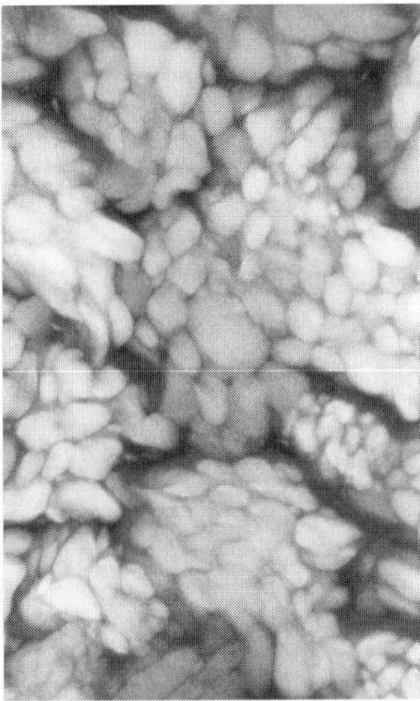

Fig. 7 Confocal micrograph of soybean cotyledon cells expressing an endoplasmic reticulum (ER)-targeted green fluorescent protein (GFP). GFP is stored in high concentration, in protein storage vacuoles. Compare to Fig. 8 (M. Schmidt and E. Herman, USDA-ARS, Danforth Plant Science Center, unpublished results).

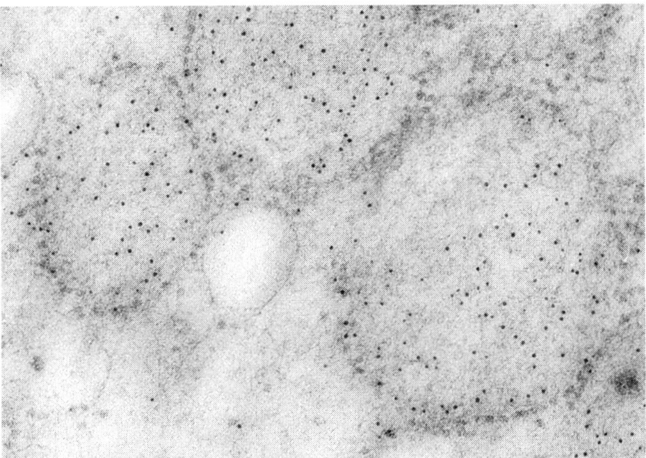

Fig. 8 Thin section electron micrograph of the material from Fig. 7, showing immunogold detection of green fluorescent protein (GFP) antigen in a high-titer tissue (freeze-substituted specimen, R.H.B., M. Schmidt, and E. Herman, USDA-ARS, Danforth Plant Science Center, unpublished results).

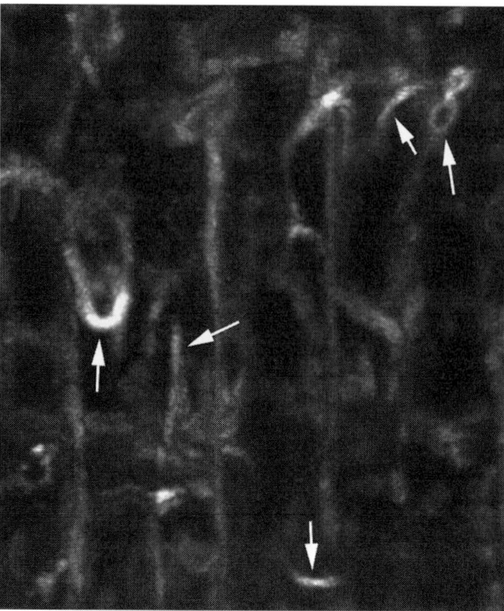

Fig. 9 Endoplasmic reticulum (ER) bodies in *Arabidopsis* stamen filament cells. ER-targeted green fluorescent protein (GFP) in these cells accumulates in elongated, spindle-shaped structures (arrows). Compare to Fig. 10 (R.H.B., unpublished results).

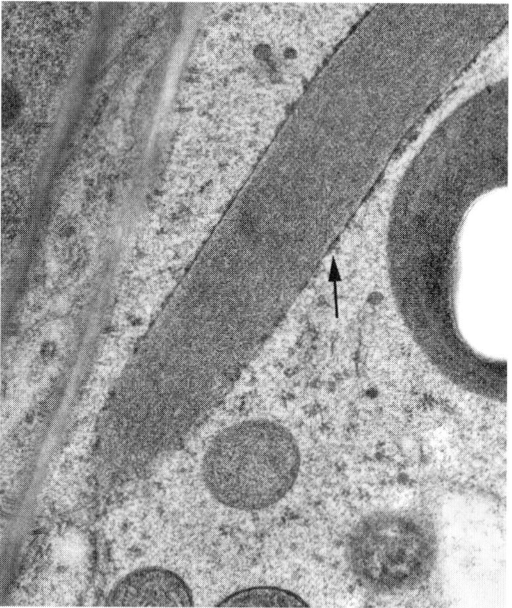

Fig. 10 Thin section electron micrograph of the material from Fig. 9, showing that the elongated bodies are endoplasmic reticulum (ER) (coated with ribosomes, arrow), dilated by an accumulated product (R.H.B., unpublished results).

For cellular imaging one has the choice of widefield or confocal/multiphoton microscopy (Shaw, 2006). Widefield microscopy has the advantage of permitting rapid image acquisition, high camera sensitivity [especially with electron multiplying charge coupled detector (EMCCD) cameras, Pawley, 2006b], and availability of mercury or xenon light sources (more excitation bandwidth available for a variety of fluorophores). The chief disadvantage of widefield imaging is the out-of-focus haze arising from fluorophore excitation outside of the focal plane of the objective; this can reduce image contrast. Computational techniques (deconvolutions) can be used to reduce or eliminate this haze (Shaw, 2006), but adds another step to data analysis, and requires additional software and expertise. Thin tissue specimens such as root hairs, pollen, pollen tubes, and cultured cells do not generate much out-of-focus haze and can be imaged with high sensitivity with a widefield microscope.

Confocal microscopy reduces out-of-focus haze, and produces an optical section that is the standard for 2D and 3D imaging, with appropriate software. It is important to keep in mind that, as in widefield microscopy, fluorophores above and below the focal plane of the objective are excited during confocal imaging; however, this fluorescence is not detected because it is rejected by the confocal pinhole. Thus, acquisition of 3D confocal image stacks causes a tremendous amount of photodamage to the cell (discussed below). In addition, confocal is a beam-scanning microscopy and requires a relatively long acquisition time compared with widefield microscopy.

Multiphoton (typically two photons) microscopy produces optical sections without the need for a pinhole, limiting excitation to only those fluorophores in the focal plane (Denk *et al.*, 2006). Multiphoton imaging uses IR laser excitation and these longer wavelengths penetrate deeper into tissues compared with wavelengths used in confocal microscopy. These features combine to make two photon microscopy the imaging method of choice when there is a need for time-lapse imaging. Multiphoton microscopy is more user-friendly than in the past because the laser is computer controlled, and laser components are packed into a single component that does not require user alignment. Use of multiphoton microscopy in plant cell imaging is described in Tirlapur and Konig (2002) and Feijo and Moreno (2004). Two photon excitation cross sections for FPs are broad (Blab *et al.*, 2001), with maxima for ECFP ~850–925 nm, EGFP ~950–1000 nm, EYFP ~975–1050 nm, and DsRed ~1100–1150 nm.

For live cell imaging, tissue mounting should preserve tissue viability and keep the tissue close to the cover slip, when using close working distance, high resolution objectives. Highest resolution objectives will require immersion medium: water immersion lens should be used since the refractive index of water most closely matches that of live cells. Oil lens are not corrected for imaging in live cells. Given two water immersion objectives with the same 1.2 NA (numerical aperture) but different magnifications, one should use a standard sample to compare signal brightness between them. Our Zeiss 40 × 1.2 NA water immersion lens produces a significantly brighter signal than our Zeiss 63 × 1.2 NA objective, both Plan Apos.

Because they have the same resolution, we use the brighter 40×. Inverted microscopes are convenient for live cell time-lapse imaging because the specimen can be mounted to preserve viability more easily than in upright microscopes. We use glass bottom Petri dishes (www.biosciencetools.com/catalog/WillCo.htm) that allow the tissue to be mounted in a moist environment while exposed to air. The tissue is covered with a moist piece of filter paper held in place with a stainless steel washer that encircles the specimen, thereby avoiding direct pressure on the specimen. This assembly keeps the tissue close to the cover slip without blocking access to air, and leaves open the possibility of adding experimental reagents to the tissue during imaging. Protoplasts are delicate and easily damaged during mounting. An agarose bed is effective in both maintaining their viability and keeping them positioned for long-term imaging (Mas and Beachy, 1998).

Brightness is usually a significant issue in imaging FPs because many proteins are produced at low levels. Whether using a widefield or a confocal microscope, specimens are subjected to intense excitation light, producing free radicals that damage molecules in the vicinity (Bensasson *et al.*, 1993). When fluorophores absorb excitation light the excited electrons have multiple pathways for deexcitation, only one being via emission of a photon (Clegg *et al.*, 2003). The nonradiative decay pathways include processes that generate free radicals, including reactive oxygen species (ROS). The higher the excitation energy to which the sample is subjected, the higher the proportion of fluorophores that are driven into the excited state. In this situation, the proportion of electrons in the longer-lived excited triplet state increases; this increases the rate of formation of ROS and photobleaching of the fluorophore. Dixit and Cyr (2003) show the effect of excitation light on generation of ROS and the effects on plant cell viability. Photodamage is mediated by cell redox buffering capacity and occurs even at excitation intensities that are below those causing photobleaching. The simple conclusion from this discussion is that the very minimum excitation energy possible should be used to image FPs. Excitation intensity can be kept low in widefield microscopes by using neutral density filters, and in confocal/multiphoton microscopes by using very low laser power. In confocal microscopy, opening the pinhole significantly increases signal, at the expense of *Z*-axis resolution. Another way to compensate for low excitation and low emission is to increase the detector gain and use frame averaging, at the cost of temporal resolution in time-lapse studies. In widefield microscopes equipped with charge coupled detector (CCD) cameras, one can compensate by increasing exposure time.

When multiple spectral forms of FPs are imaged spectral bleed-through is a potential problem. Separation of GFP and DsRed by sequential acquisition or multitracking is problematic: no cutoff filter currently available can eliminate GFP bleed-through into the DsRed channel. However, spectral "unmixing" is an effective means for separating these two FPs and the even closer emission spectra of GFP and YFP (discussed in Section IV—Advanced Techniques). CFP and YFP are commonly used for FRET and their signals can be separated by sequential acquisition/multitracking, during which the signals are collected in two sequential scans set to minimize bleed-through. This is done by using control specimens that each contain a single FP and setting gain and laser power so that no bleed-through occurs.

As previously discussed, plant cells and tissues may contain a variety of autofluorescent compounds. Typically these compounds have a broader emission spectrum than FPs (Fig. 1) and therefore can be identified by the presence of their signal in multiple channels. Generally the signals are low with the exception of chlorophyll. Chlorphyll is localized to chloroplasts in above-ground tissues and can be used as a reference for the FP signal. Chlorophyll emits at a relatively long wavelength; therefore its signal can be minimized in such tissues by avoiding red spectral FPs such as DsRed. An alternative is to separate the signals by spectral unmixing (see Section IV—Advanced Techniques).

A. Protocol for 4D Imaging

Live cell imaging provides an important tool for the study of cellular dynamics, by adapting 3D imaging over time (4D imaging). Technical issues that must be considered in 4D imaging include spatial versus temporal resolution, photodamage to the specimen, controlling movement of the specimen, and maintaining tissue viability. Viability and movement can be optimized by using the mounting method described above, securing the Petri dish firmly on the microscope stage, and maintaining constant ambient temperatures. The greatest negative effect on tissue viability is photodamage: to prevent damage it is most important to keep laser power at a minimum and to keep the number of optical sections acquired to a minimum. The same sample volume can be captured with fewer sections by increasing pinhole size: this increases slice thickness and consequently increases the amount of signal reaching the detector, permitting lower laser power levels and reducing photodamage. On the other hand, thicker optical sections reduce spatial resolution in the *Z*-axis, meaning that each 3D frame will be limited to a small viewing angle near the normal. Fewer slices on the *Z*-axis also improve temporal resolution. There is also a spatial–temporal resolution trade-off in the X–Y plane. Optimal 4D imaging requires as few pixels as possible in the image, which in turn leads to a faster scan and less photodamage. This can be accomplished by either having poor pixel resolution in an image that covers a relatively large area, or by cropping the image to a small area containing the region of interest (ROI) with relatively small pixel size.

An illustration of these concepts is presented in two 4D movies posted at www. danforthcenter.org/imf/4danimation.htm, Agg1-tl1.mov and Agg1-tl3.mov. Made using the same region of leaf tissue of *Arabidopsis*, the signal is a YFP fusion with Agg1, the gamma subunit of heterotrimeric G-protein (Q. Zeng, X. Wang, and M.P. Running, Danforth Plant Science Center, Plant Physiology, In Press). Agg1-tl1 was made using relatively high *Z*-axis spatial resolution and low temporal resolution, and Agg1-tl3 vice versa (technical details listed on the web site). The high spatial resolution image, constructed using a smaller pinhole and eight optical sections, compresses time (74×) too much to see details of signal movement and exhibits significant photobleaching over the 40-min acquisition period. The high temporal resolution image, using a larger pinhole and three optical sections, has eight times better temporal resolution (9× compression) that clearly shows the nature of the signal dynamics and no signal fade over the 5-min acquisition period (same excitation dose for both acquisitions). By cropping the field of view, temporal resolution was improved. Pixel size is similar in both images, voxel size is larger in Agg1-tl3.

IV. Advanced Techniques

A. Spectral Imaging

Spectral imaging produces images that contain quantitative spectral information in each pixel that can be used for a variety of purposes: (1) to verify the identity of FP emission; (2) emission curves can be used to identify appropriate filtration parameters for confocal or multiphoton imaging; (3) identification of autofluorescence; (4) quantitative analysis of ratiometric dyes; (5) 3D/4D analysis of dynamic changes in emission of FP pairs for FRET; and (6) spectral unmixing (Zimmerman *et al.*, 2003). Spectral imaging is possible in many laser scanning microscopes (Berg, 2004).

1. Protocol for Spectral Unmixing of GFP and YFP Signals

In spectral unmixing, signals from mixtures of spectrally different coexpressed FPs can be detected by computational deconvolution of spectral data, using reference spectra of the respective FPs. Spectral pairs such as GFP and DsRed, or GFP and YFP can be analyzed without the problems associated with bleed-through. Likewise, autofluorescent compounds can be unmixed from the FP signal. When done carefully, this is a powerful technique. Of critical importance for successful unmixing is the need for accurate spectra, both in the reference spectra and in the experimental data. These spectra are affected by all components of the optical system, meaning the same components should be used for acquiring reference and experimental spectra. These spectra should have a high dynamic range but should not contain saturated pixels. There must be a sufficiently high signal in the specimen that the spectra are above the noise floor, for example, above ~25% of the bit-depth of the image.

Figure 11 illustrates a spectral unmix of GFP and YFP in an *Arabidopsis* root hair (J. Thole and E. Nielsen, Danforth Plant Science Center; unpublished results). The YFP:RabA4b is a fusion with a membrane-associated (possibly trans Golgi network) Rab GTPase that is involved in polar transport in root hair cells (Preuss *et al.*, 2004). The GFP:ABD2 is fusion protein with the actin-binding domain of fimbrin. Three spectral images were made for this unmixing: a spectral image

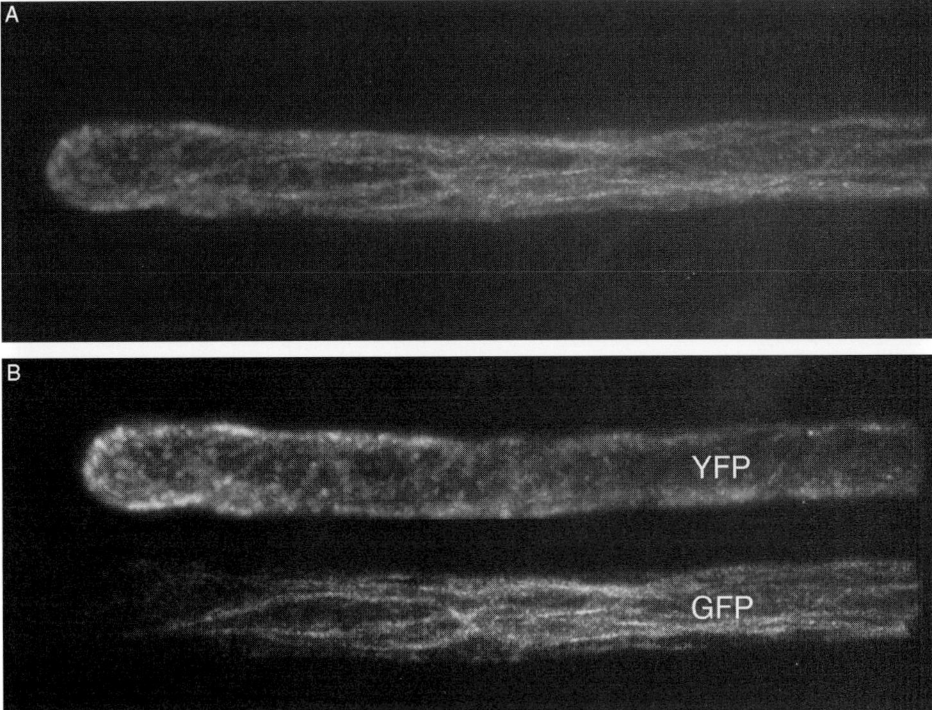

Fig. 11 Separate detection of GFP and yellow fluorescent protein (YFP) signals by spectral unmixing [(a) combined signals and (b) signals displayed separately]. This root hair is from a transgenic *Arabidopsis* plant expressing green fluorescent protein GFP:ABD2 (binds actin filaments) and YFP: RabA4b (membrane-associated, small compartment). The two signals were separated computationally via spectral unmixing (spectra shown in Fig. 12), using a Zeiss LSM 510 META spectral imaging system (see text). These signals cannot be separated by conventional imaging based on interference filters (J. Thole and E. Nielsen, unpublished results). (See Plate no. 20 in the Color Plate Section.)

of each of the parental lines was made to acquire reference spectral data for each FP expressed singly, and then a spectral image of a genetic cross that expresses both FP constructs. In all cases, the signal/noise was maximized. In order to use the same optical components for all images, the 488 nm argon laser line was used to excite both fluorophores, allowing the use of the same primary dichroic. In the system used, a Zeiss LSM 510 META, this primary dichroic is the only filter in the optical path during spectral imaging (Berg, 2004). Detection bandwidth was set to collect emission from 494 to 580 nm, using eight photomultiplier tube (PMT) detectors, each with 10.7 nm bandwidth. This encompasses all of the potential spectral information expected from each image. A reference spectrum was selected from each of the reference images by drawing a ROI over the cleanest signal region in the image, and this was stored in the database. Acquisition of the experimental data made use of the online fingerprinting function in the software, which unmixes continuously by assigning reference spectra to each digital channel to be unmixed, and unmixing each optical section immediately after its acquisition. This avoids having to save spectral image data and saves time that would be used to manually unmix offline. Figure 11 is a 3D projection of a stack of unmixed optical sections: 11(a) is a color composite of the two unmixed signals and 11(b) displays the two separate unmixed signals. Figure 12 shows the reference spectra curves and a spectral curve from the experimental data from a ROI drawn over a region containing both signals. These two signals are impossible to separate using interference filters.

B. Fluorescence Lifetime Imaging

Fluorescence lifetime imaging microscopy (FLIM) produces images in which each pixel is a quantitative measure of the average fluorescence lifetime of the fluorophore (van Munster and Gadella, 2005). FLIM image intensity is a function of lifetime and is independent of fluorophore concentration or light path length, in contrast to steady-state intensity images. Lifetimes are in the range of nanoseconds, requiring ultrafast detection technology. FLIM can be implemented by adding a detector to the appropriate fluorescence microscope. There currently are commercially available detectors for FLIM that make use of the time domain (e.g., Becker-Hickl, www.becker-hickl.com) or frequency domain (e.g., ISS Alba, www.iss.com)

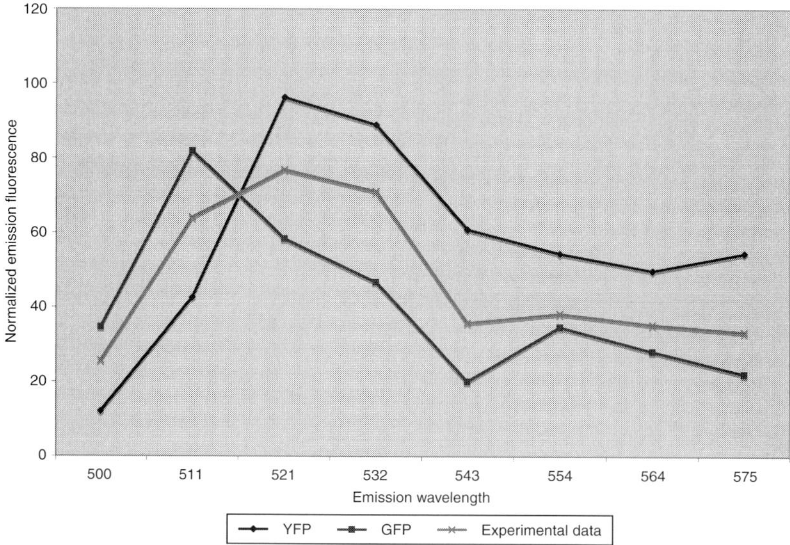

Fig. 12 Reference spectra for green fluorescent protein (GFP) and yellow fluorescent protein (YFP) used in spectral unmixing for Fig. 11. The experimental data spectrum is from a region of interest (ROI) drawn to include both signals. All spectra were made using the same microscope configuration (see text).

methods. FLIM has the potential to quantitatively measure any of the nonradiative processes that affect fluorescence lifetime, that is, to measure the molecular environment around the FP fluorophore (Clegg *et al.*, 2003). With FPs, the most promising use of FLIM is to measure FRET simply by measuring the reduction in donor FP lifetime as it interacts with acceptor FP (Redford and Clegg, 2005). This avoids the problems of intensity-based FRET measurements that depend on unknown FP concentrations and optical path length. Using CFP and YFP pairs, Immink *et al.* (2002) used FRET-FLIM (frequency domain detector) to show formation of heterodimers of a MADS transcription factor in petunia protoplasts, confirming previous yeast two-hybrid data. In addition, FRET-FLIM detected a previously unknown homodimerization of the factor. Bhat *et al.* (2005) investigated calmodulin-mediated activation of the plasma membrane protein mildew resistance locus (MLO) in response to fungal attack in barley leaf epidermal cells. FRET-FLIM (time domain detector) between CFP:calmodulin and YFP:MLO was localized in domains of the plasma membrane and around the nucleus, indicating activation of MLO at these sites. The commercial availability of FLIM detectors likely will bring this technology into use in more labs that are interested in a relatively straight forward means for detecting FRET and other physiological parameters that can be detected using FLIM.

C. Fluorescence Correlation Spectroscopy

In addition to FRET, fluorescence correlation spectroscopy (FCS) is a useful tool for measuring protein–protein interactions on the cellular level (Hink *et al.*, 2002). FCS detectors measure fluorescence fluctuations in femtoliter volumes with sufficient sensitivity to measure single molecule fluctuations (Kohl and Schwille, 2005). Measurement of fluctuations over a period of time permits calculation of an autocorrelation curve from which is derived the diffusion constant of, for example, FP-tagged molecules, and their concentration. Curve-fitting routines can be used to model diffusion kinetics *in vivo*, or to monitor biochemical reactions in solutions in cuvettes mounted in the microscope. Intermolecular interactions can be quantified, even if only one of the molecules is FP tagged. FCS could, theoretically, quantify movement or accumulation of FP in the tip of growing cells such as pollen tubes or root hairs, or accumulation and movement of MP:GFP through plasmodesmata. Goedhart *et al.* (2000) quantified diffusion and binding of BODIPY-labeled Nod factor in root hairs using FCS. Kohler *et al.* (2000) measured the diffusion of free GFP in stromules of plastids using FCS, finding both passive and active components in this movement. FCS detectors are commercially available (Zeiss Confocor™, ISS Alba™), complete with software packages that simplify data analysis.

V. Summary

Plant cell biologists have taken advantage of the wide variety of fluorescent protein (FP) applications that are now available. Nuances in their function, identified by molecular biology and computer-assisted analyses, have given us sophisticated tools to move forward to probe the function of living plant cells. These include a rainbow of spectral mutants encompassing the entire spectral region. Control of their expression is improved by using native or inducible promoters. Biosensors and highlighter proteins are valuable probes of physiological processes and FP dynamics. Electron microscopy is an important tool for interpreting and verifying the location of FP fusion proteins. Microscope technology for analyzing FP function continues to evolve, with the advent of quantitative detectors for FLIM and FCS. This a provocative and exciting time in the study of plant cell biology.

Acknowledgments

We thank the many Danforth Plant Science Center scientists who have contributed images for this work: E. Herman, E. Nielsen, M.P. Running, M. Schmidt, M. Soto-Aguilar, J. Thole, X. Wang, and Q. Zeng. In addition, thanks go to J. Samaj and F. Baluska (Slovak Academy of Sciences) for the ABD construct used for spectral unmixing (Figs. 11 and 12).

References

Asurmendi, S., Berg, R. H., Koo, J. C., and Beachy, R. N. (2004). Coat protein regulates formation of replication complexes during tobacco mosaic virus infection. *Proc. Natl. Acad. Sci. USA* **101,** 1415–1420.

Avila, E. L., Zouhar, J., Agee, A. E., Carter, D. G., Chary, S. N., and Raikhel, N. V. (2003). Tools to study plant organelle biogenesis. Point mutation lines with disrupted vacuoles and high-speed confocal screening of green fluorescent protein-tagged organelles. *Plant Physiol.* **133,** 1673–1676.

Baulcombe, D. (2005). RNA silencing. *Trends Biochem. Sci.* **30,** 290–293.

Bensasson, R. V., Land, E. J., and Truscott, T. G. (1993). "Excited States and Free Radicals in Biology and Medicine." Oxford University Press, Oxford.

Berg, R. H. (2004). Evaluation of spectral imaging for plant cell analysis. *J. Microsc.* **214,** 174–181.

Bhat, R. A., Miklis, M., Schmelzer, E., Schulze-Lefert, P., and Panstruga, R. (2005). Recruitment and interaction dynamics of plant penetration resistance components in a plasma membrane microdomain. *Proc. Natl. Acad. Sci. USA* **102,** 3135–3140.

Blab, G. A., Lommerse, P. H. M., Cognet, L., Harms, G. S., and Schmidt, T. (2001). Two-photon excitation action cross-sections of the autofluorescent proteins. *Chem. Phys. Lett.* **350,** 71–77.

Campbell, R. E., Tour, O., Palmer, A. E., Steinbach, P. A., Baird, G. S., Zacharias, D. A., and Tsien, R. Y. (2002). A monomeric red fluorescent protein. *Proc. Natl. Acad. Sci. USA* **99,** 7877–7882.

Chalfie, M., Tu, Y., Euskirchen, G., Ward, W. W., and Prasher, D. C. (1994). Green fluorescent protein as a marker for gene expression. *Science* **263,** 802–805.

Chapman, S., Oparka, K. J., and Roberts, A. G. (2005). New tools for *in vivo* fluorescence tagging. *Curr. Opin. Plant Biol.* **8,** 565–573.

Chiu, W. L., Niwa, Y., Zeng, W., Hirano, T., Kabayashi, H., and Sheen, J. (1996). Engineered GFP as a vital reporter in plants. *Curr. Biol.* **6,** 325–330.

Chytilova, E., Macas, J., and Galbraith, D. W. (1999). Green fluorescent protein targeted to the nucleus, a transgenic phenotype useful for studies in plant biology. *Ann. Bot.* **83,** 645–654.

Clegg, R. M., Holub, O., and Gohlke, C. (2003). Fluorescence lifetime-resolved imaging: Measuring lifetimes in an image. *In* "Biophotonics, Part A", Methods in Enzymology (G. Marriott and I. Parker, eds.), Vol. 360, pp. 509–542. Academic Press, San Diego.

Denk, W., Piston, D. W., and Webb, W. W. (2006). Multi-photon molecular excitation in laser-scanning microscopy. *In* "Handbook of Biological Confocal Microscopy" (J. B. Pawley, ed.), 3rd edn. pp. 535–549. Springer, New York.

Dixit, R., and Cyr, R. (2003). Cell damage and reactive oxygen species production induced by fluorescence microscopy: Effect on mitosis and guidelines for non-invasive fluorescence microscopy. *Plant J.* **36,** 280–290.

Dixit, R., Cyr, R., and Gilroy, S. (2006). Using intrinsically fluorescent proteins for plant cell imaging. *Plant J.* **45,** 599–615.

Earley, K. W., Haag, J. R., Pontes, O., Opper, K., Juehne, T., Song, K., and Pikaard, C. S. (2006). Gateway-compatible vectors for plant functional genomics and proteomics. *Plant J.* **45,** 616–629.

Ehrhardt, D. (2003). GFP technology for live cell imaging. *Curr. Opin. Plant Biol.* **6,** 622–628.

Escobar, N. M., Haupt, S., Thow, G., Boevink, P., Chapman, S., and Oparka, K. (2003). High-throughput viral expression of cDNA-green fluorescent protein fusions reveals novel subcellular addresses and identifies unique proteins that interact with plasmodesmata. *Plant Cell* **15,** 1507–1523.

Feijo, J. A., and Moreno, N. (2004). Imaging plant cells by two-photon excitation. *Protoplasma* **223,** 1–32.

Fricker, M., Runions, J., and Moore, I. (2006). Quantitative fluorescence microscopy: From art to science. *Annu. Rev. Plant Biol.* **57,** 79–107.

Goedhart, J., Hink, M., Visser, A. J. W. G., Bisseling, T., and Gadella, T. W. J. (2000). *In vivo* fluorescence correlation microscopy (FCM) reveals accumulation and immobilization of Nod factors in root hair cell walls. *Plant J.* **21,** 109–119.

Grebenok, R. J., Pierson, E., Lambert, G. M., Gong, F. C., Afonso, C. L., Haldeman-Cahill, R., Carrington, J. C., and Galbraith, D. W. (1997). Green fluorescent protein fusions for efficient characterization of nuclear targeting. *Plant J.* **11,** 573–586.

Gunning, B. E. S. (1998). The identity of mystery organelles in *Arabidopsis* plants expressing GFP. *Trends Plant Sci.* **3,** 417.

Haseloff, J., and Siemering, K. R. (2006). The uses of green fluorescent protein in plants. *In* "Green Fluorescent Protein: Properties, Applications, and Protocols" (M. Chalfie and S. R. Kain, eds.), pp. 259–284. John Wiley and Sons, Hoboken, NJ.

Haseloff, J., Siemering, K. R., Prasher, D. C., and Hodge, S. (1997). Removal of a cryptic intron and subcellular localisation of green fluorescent protein are required to mark transgenic *Arabidopsis* plants brightly. *Proc. Natl. Acad. Sci. USA* **94,** 2122–2127.

Hashimoto, T. (2002). Molecular genetic analysis of left-right handedness in plants. *Philos. Trans. R. Soc. Lond. B* **357,** 799–808.

Hepler, P., and Gunning, B. E. S. (1998). Confocal fluorescence microscopy of plant cells. *Protoplasma* **201,** 121–157.

Hink, M., Bisseling, T., and Visser, A. J. W. G. (2002). Imaging protein–protein interactions in living cells. *Plant Mol. Biol.* **50,** 871–883.

Immink, R. G. H., Gadella, T. W. J., Ferrario, S., Busscher, M., and Angenent, G. C. (2002). Analysis of MADS box protein–protein interactions in living plant cells. *Proc. Natl. Acad. Sci. USA* **99,** 2416–2421.

Inselman, A. L., Gathman, A. C., and Lilly, W. W. (1999). Two fluorescent markers identify the vacuolar system of *Schizophyllum commune*. *Curr. Microbiol.* **38,** 295–299.

Kohl, T., and Schwille, P. (2005). Fluorescence correlation spectroscopy with autofluorescent proteins. *Adv. Biochem. Eng. Biotechnol.* **95,** 107–142.

Kohler, R. H., Schwille, P., Webb, W. W., and Hanson, M. R. (2000). Active protein transport through plastid tubules: Velocity quantified by fluorescence correlation spectroscopy. *J. Cell Sci.* **113,** 3921–3930.

Koo, J. C., Asurmendi, S., Bick, J., Woodford-Thomas, T., and Beachy, R. N. (2004). Ecdysone agonist-inducible expression of a coat protein gene from tobacco mosaic virus confers viral resistance in transgenic Arabidopsis. *Plant J.* **37,** 439–448.

Lalonde, S., Ehrhardt, D. W., and Frommer, W. B. (2005). Shining light on signaling and metabolic networks by genetically encoded biosensors. *Curr. Opin. Plant Biol.* **8,** 574–581.

Lazarowitz, S. G., and Beachy, R. N. (1999). Viral movement proteins as probes for intracellular and intercellular trafficking in plants. *Plant Cell* **11,** 535–548.

Lorence, A., and Verpoorte, R. (2004). Gene transfer and expression in plants. *Methods Mol. Biol.* **267,** 329–350.

Lukyanov, K. A., Chudakov, D. M., Fradkov, A. F., Labas, Y. A., Matz, M. V., and Lukyanov, S. (2006). Discovery and properties of GFP-like proteins from nonbioluminescent anthozoa. *In* "Green Fluorescent Protein: Properties, Applications, and Protocols" (M. Chalfie and S. R. Kain, eds.), pp. 121–138. John Wiley and Sons, Hoboken, NJ.

Mas, P., and Beachy, R. N. (1998). Distribution of TMV movement protein in single living protoplasts immobilized in agarose. *Plant J.* **15,** 835–842.

Matsushima, R., Fukao, Y., Nishimura, M., and Hara-Nishimura, I. (2004). *NAI1* gene encodes a basic-helix-loop-helix-type putative transcription factor that regulates the formation of an endoplasmic reticulum-derived structure, the ER body. *Plant Cell* **16,** 1536–1549.

Moore, I., Samalova, M., and Kurup, S. (2006). Transactivated and chemically inducible gene expression in plants. *Plant J.* **45,** 651–683.

Moreno, N., Bougourd, S., Haseloff, J., and Feijo, J. A. (2006). Imaging plant cells. *In* "Handbook of Biological Confocal Microscopy" (J. B. Pawley, ed.), 3rd edn., pp. 769–787. Springer, New York.

Padidam, M., Gore, M., Lu, D. L., and Smirnova, O. (2003). Chemical-inducible, ecdysone receptor-based gene expression system for plants. *Transgenic Res.* **12,** 101–109.

Pawley, J. B. (2006a). "Handbook of Biological Confocal Microscopy," 3rd edn., p. 985. Springer, New York.

Pawley, J. B. (2006b). More than you ever really wanted to know about charge-coupled devices. *In* "Handbook of Biological Confocal Microscopy" (J. B. Pawley, ed.), 3rd edn., pp. 918–931. Springer, New York.

Periasamy, A., and Day, R. N. (2005). "Molecular Imaging, FRET Microscopy and Spectroscopy." Oxford University Press, Oxford.

Perroud, P. F., and Quatrano, R. S. (2006). The role of ARPC4 in tip growth and alignment of the polar axis in filaments of *Physcomitrella patens. Cell Motil. Cytoskelet.* **63,** 162–171.

Preuss, M. L., Serna, J., Falbel, T. G., Bednarek, S. Y., and Nielsen, E. (2004). The Arabidopsis Rab GTPase RabA4b localizes to the tips of growing root hair cells. *Plant Cell* **16,** 1589–1603.

Redford, G., and Clegg, R. M. (2005). Real-time fluorescence lifetime imaging and FRET using fast-gated image intensifiers. *In* "Molecular Imaging, FRET Microscopy and Spectroscopy" (A. Periasamy and R. N. Day, eds.), pp. 193–226. Oxford University Press, Oxford.

Reichel, C., and Beachy, R. N. (1998). Tobacco mosaic virus infection induces severe morphological changes of the endoplasmic reticulum. *Proc. Natl. Acad. Sci. USA* **95,** 11169–11174.

Rizzo, M. A., Springer, G. H., Granada, B., and Piston, D. W. (2004). An improved cyan fluorescent protein variant useful for FRET. *Nat. Biotechnol.* **22,** 445–449.

Rost, F. W. D. (1995). "Fluorescence Microscopy, Volume II." Cambridge University Press, Cambridge.

Ruzin, S. E. (1999). "Plant Microtechnique and Microscopy." Oxford University Press, Oxford.

Shaner, N. C., Campbell, R. E., Steinbach, P. A., Giepmans, B. N. G., Palmer, A. E., and Tsien, R. Y. (2004). Improved monomeric red, orange and yellow fluorescent proteins derived from *Discosoma* sp. Red fluorescent protein. *Nat. Biotechnol.* **22,** 1567–1572.

Shaw, P. J. (2006). Comparison of widefield/deconvolution and confocal microscopy for three-dimensional imaging. *In* "Handbook of Biological Confocal Microscopy" (J. B. Pawley, ed.), 3rd edn., pp. 453–467. Springer, New York.

Tamura, K., Shimada, T., Ono, E., Tanaka, Y., Nagatani, A., Higashi, S., Watanabe, M., Nishimura, M., and Hara-Nishimura, I. (2003). Why green fluorescent fusion proteins have not been observed in the vacuoles of higher plants. *Plant J.* **35,** 545–555.

Tian, G. W., Mohanty, A., Chary, N., Li, S., Paap, B., Drakakaki, G., Kopec, C. D., Li, J., Ehrhardt, D., Jackson, D., Rhee, S. Y., Raikhel, N. V., *et al.* (2004). High-throughput fluorescent tagging of full-length Arabidopsis gene products in planta. *Plant Physiol.* **135,** 25–38.

Tirlapur, U. K., and Konig, K. (2002). Two-photon near-infrared femtosecond laser scanning microscopy in plant biology. *In* "Confocal and Two-Photon Microscopy: Foundations, Applications, and Advances" (A. Diaspro, ed.), pp. 449–468. Wiley-Liss, New York.

Tzfira, T., Tian, G. W., Lacroix, B., Vyas, S., Li, J., Leitner-Dagan, Y., Krichevsky, A., Taylor, T., Vainstein, A., and Citovsky, V. (2005). pSAT vectors: A modular series of plasmids for autofluorescent protein tagging and expression of multiple genes in plants. *Plant Mol. Biol.* **57,** 503–516.

van Munster, E. B., and Gadella, T. W. J. (2005). Fluorescence lifetime imaging microscopy (FLIM). *Adv. Eng. Biotechnol.* **95,** 143–175.

Wortman, J. R., Haas, B. J., Hannick, L. I., Smith, R. K., Maiti, R., Ronning, C. M., Chan, A. P., Yu, C., Ayele, M., Whitelaw, C. A., White, O. R., and Town, C. D. (2003). Annotation of the *Arabidopsis* genome. *Plant Physiol.* **132** 461–488.

Zacharias, D. A., and Tsien, R. Y. (2006). Molecular biology and mutation of green fluorescent protein. *In* "Green Fluorescent Protein: Properties, Applications, and Protocols" (M. Chalfie and S. R. Kain, eds.), pp. 83–120. John Wiley and Sons, Hoboken, NJ.

Zimmerman, T., Rietdorf, J., and Pepperkok, R. (2003). Spectral imaging and its applications in live cell microscopy. *FEBS Lett.* **546,** 87–92.

CHAPTER 9

Expression and Imaging of Fluorescent Proteins in the *C. elegans* Gonad and Early Embryo

Rebecca A. Green,* Anjon Audhya,* Andrei Pozniakovsky,† Alexander Dammermann,* Hayley Pemble,* Joost Monen,* Nathan Portier,* Anthony Hyman,† Arshad Desai,* and Karen Oegema*

*Ludwig Institute for Cancer Research
Department of Cellular and Molecular Medicine
University of California
San Diego, La Jolla, California 92093

†Max-Planck Institute of Molecular and Cellular Biology and Genetics
Dresden 01307, Germany

0091-679X/08 $35.00
DOI: 10.1016/S0091-679X(08)85009-1

Abstract

The *Caenorhabditis elegans* gonad and early embryo have recently emerged as an attractive metazoan model system for studying cell and developmental biology. The success of this system is attributable to the stereotypical architecture and reproducible cell divisions of the gonad/early embryo, coupled with penetrant RNAi-mediated protein depletion. These features have facilitated the development of visual assays with high spatiotemporal resolution to monitor specific subcellular processes. Assay development has relied heavily on the emergence of methods to circumvent germline silencing to allow the expression of transgenes encoding fluorescent fusion proteins. In this chapter, we discuss methods for the expression and imaging of fluorescent proteins in the *C. elegans* germline, including the design of transgenes for optimal expression, the generation of transgenic worm lines by ballistic bombardment, the construction of multimarker lines by mating, and methods for live imaging of the gonad and early embryo.

I. Introduction

A. The *Caenorhabditis elegans* Gonad and Early Embryo: A Model System for Cell and Developmental Biology

The *Caenorhabditis elegans* gonad and early embryo have recently emerged as powerful model systems for studying cell and developmental biology (for reviews see Hubbard and Greenstein, 2000; Hyman and Oegema, 2005). Each arm of the gonad in the adult *C. elegans* hermaphrodite is an assembly line of ~800 nuclei progressing through the various stages of meiotic prophase in an ordered fashion (Fig. 1). The distal region of the gonad is a syncytium of partially enclosed nuclei that share a common maternal cytoplasm. In the proximal region of the gonad, individual nuclei become packaged into distinct oocytes. Oocytes are fertilized as they pass through the spermatheca, which sits at the proximal tip of the gonad. Hermaphrodites produce sperm at an earlier stage of larval development, prior to the switch to oogenesis (for details on the spermatogenesis–oogenesis switch, see Kimble and Strome, 2005). As the oocytes are fertilized, they pass into the uterus, where the resulting embryos complete meiotic segregation of the oocyte-derived

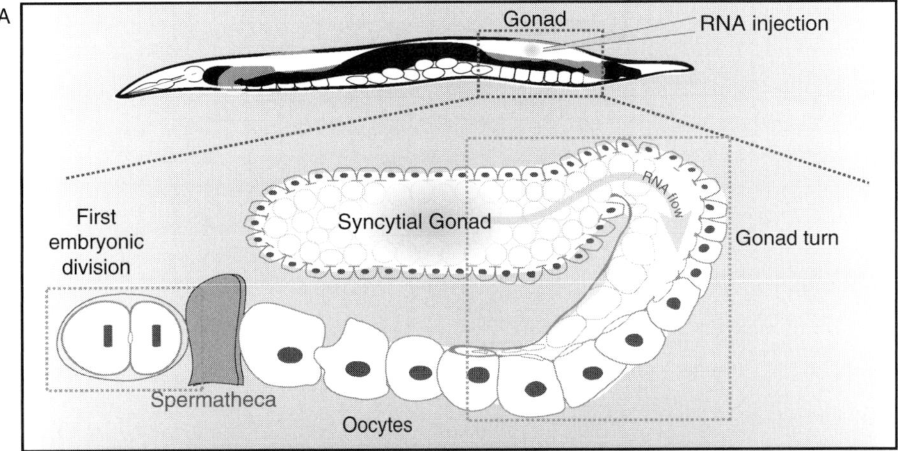

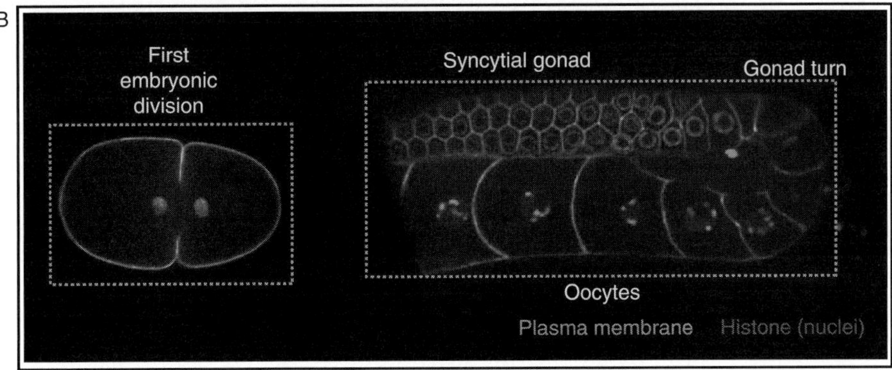

Fig. 1 The *C. elegans* gonad and early embryo. (A) The schematic shows one arm of the gonad in the adult *C. elegans* hermaphrodite to illustrate the assembly-line process that gives rise to oocytes and embryos. As they pass the turn in the gonad arm, individual compartments containing a meiotic nucleus are loaded with cytoplasm from the syncytial gonad and expand to form the developing oocytes. Oocytes are fertilized as they pass through the spermatheca, giving rise to early embryos. Introduction of dsRNA against a specific target rapidly catalyzes the destruction of the corresponding mRNA. The target protein present when the RNA is introduced is depleted by the continual packaging of the maternal cytoplasm into oocytes. (B) Live imaging of the gonad (right) and an early embryo (left) in a *C. elegans* strain expressing mCherry-histone and a GFP fusion with a PH domain (GFP-PH$^{PLC\delta1}$) to mark chromosomes and the plasma membrane, respectively (A. Audhya, Oegema Laboratory, unpublished strain). (See Plate no. 21 in the Color Plate Section.)

chromosomes (generating the oocyte pronucleus and two polar bodies), and subsequently undergo several rounds of mitotic cell division.

One of the primary advantages of the gonad/early embryo as a model system is efficient RNAi-mediated protein depletion, which facilitates quantitative analysis of the loss-of-function phenotypes for essential proteins. The reproducibility of protein depletion is due to the syncytial architecture of the gonad. Introduction of dsRNA rapidly catalyzes the destruction of the corresponding mRNA in many different systems. However, depletion of preexisting protein is generally a slow process that depends on the half-life of the targeted protein. In contrast, in the *C. elegans* gonad, the protein present when the dsRNA is introduced is depleted by the continual packaging of maternal cytoplasm into oocytes. Since depletion depends on the rate of embryo production, the kinetics tend to be similar for different targets. By 36–48 h after introduction of the dsRNA, newly formed oocytes are typically >95% depleted of the target protein (reviewed in Oegema and Hyman, 2005). The reproducibility of RNAi-mediated protein depletion has led to a series of genome-wide screens (Fernandez *et al.*, 2005; Fraser *et al.*, 2000; Gonczy *et al.*, 2000; Kamath *et al.*, 2003; Maeda *et al.*, 2001; Piano *et al.*, 2000; Rual *et al.*, 2004; Simmer *et al.*, 2003; Sonnichsen *et al.*, 2005) that have complemented genetic approaches (for some examples, see Encalada *et al.*, 2000; Golden *et al.*, 2000; Gonczy *et al.*, 1999; Kemphues *et al.*, 1988; O'Connell *et al.*, 1998) to define the sets of genes required for embryo production and viability.

B. Quantitative Imaging-Based Assays Capitalize on the Rapid, Invariant Early Embryonic Cell Divisions

In addition to efficient RNAi-mediated protein depletion, the stereotypical architecture of the gonad, and the rapid and highly reproducible mitotic divisions of the early embryo facilitate the development of imaging-based methods to assess the consequences of molecular perturbations. The invariant nature of the first few mitotic divisions (Fig. 2) allows any parameter of interest (such as the fluorescence intensity of a localized protein, the size or the distance between subcellular structures, the number of microtubules touching the cortex, the extent of chromosome condensation, etc.) to be measured as a function of time with respect to a specific temporal landmark (such as the permeabilization of the nuclear envelope, the onset of chromosome segregation in anaphase, or the initiation of furrow ingression). The average value of the parameter measured in multiple embryos can then be plotted as a function of time to generate kinetic profiles that reveal behaviors not evident from qualitative viewing of individual time-lapse sequences. For example, quantitative analysis has provided detailed information on *pronuclear movement* (O'Connell *et al.*, 2000), *cortical flow* (Cheeks *et al.*, 2004; Hird and White, 1993; Munro *et al.*, 2004), *positioning of the spindle and anaphase spindle elongation* (Cheeseman *et al.*, 2004; Colombo *et al.*, 2003; Labbe *et al.*, 2004; Tsou *et al.*, 2002), *chromosome segregation and kinetochore-microtubule attachment* (Cheeseman *et al.*, 2004; Labbe *et al.*, 2004), and *chromosome condensation* (Maddox *et al.*, 2006).

The development of imaging-based assays has relied on the parallel emergence of methods to circumvent germline silencing, allowing for the expression of transgenes encoding fluorescent fusion proteins in the gonad and early embryo. Critical for this has been the development of vectors containing regulatory

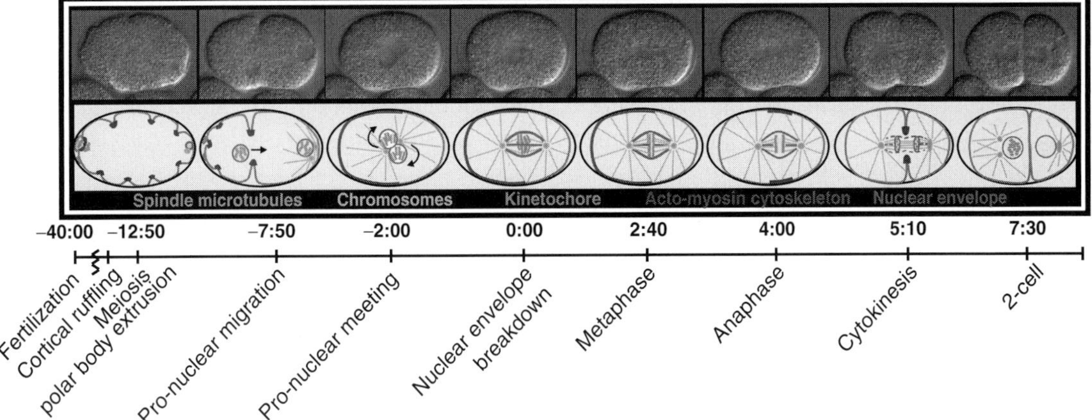

Fig. 2 Timeline of events in the *C. elegans* early embryo. Differential interference contrast images (top) and schematics (bottom) illustrate the timing of major events during the first mitotic division. Approximate times are in minutes:seconds with respect to nuclear envelope breakdown. Images are courtesy of Amy Maddox. Figure adapted from Oegema and Hyman (2005). (See Plate no. 22 in the Color Plate Section.)

sequences that direct germline expression of transgenes (Strome *et al.*, 2001) and the emergence of efficient ballistic bombardment techniques to integrate vectors into the genome at low-copy number (Praitis *et al.*, 2001; Wilm *et al.*, 1999).

II. Fluorescent Proteins in the *C. elegans* Gonad and Early Embryo

A. Fluorescent Proteins Commonly Used in *C. elegans*

To date, the majority of *C. elegans* work with fluorescent proteins has utilized an uncommon enhanced green fluorescent protein (GFP) variant in which serine 65, one of the three residues of the chromophore, is mutated to cysteine (S65C). GFP S65C is reported to have similar excitation and emission spectra to GFP S65T, the most commonly used enhanced variant due to its greater brightness, faster oxidation to final fluorescent product, slower photobleaching, and lack of photoisomerization when compared to wild-type GFP (Cubitt *et al.*, 1995; Heim and Tsien, 1996; Heim *et al.*, 1994). The common use of the unusual S65C variant in *C. elegans* is a historical consequence of its inclusion in the original vectors for *C. elegans* GFP expression, developed by Andrew Fire's group (Fire *et al.*, 1998). The decision to use S65C was based on its greater photostability *in vivo* in worms (A. Fire, personal communication). However, a systematic test comparing the properties of GFP S65C to GFP S65T in the *C. elegans* gonad and early embryo has not been performed.

In addition to GFP (S65C), the common color variants yellow fluorescent protein (YFP) (S65G, V68A, S72A, and T203Y) and cyan fluorescent protein (CFP) (Y66W, N146I, M153T, and V163A) have also been engineered for use in *C. elegans* (Miller *et al.*, 1999), and sequences encoding these fluorescent proteins have been incorporated into germline expression vectors (Franz *et al.*, 2005; Galy *et al.*, 2003).

Variants of the coral red fluorescent protein, DsRed, are emerging as the most popular choice as a dual color marker with GFP for live imaging. Due to its naturally tetrameric state, DsRed is poorly suited for *in vivo* studies (Matz *et al.*, 1999). Extensive mutagenesis generated a monomeric version from which a spectrum of fluorescent proteins, called the mFruits, has been developed (Campbell *et al.*, 2002; Shaner *et al.*, 2004, 2005). Of these, the most widely used is mCherry. Initial attempts to express mCherry in the *C. elegans* germline were unsuccessful; however, codon optimization and the inclusion of introns have overcome these problems (McNally *et al.*, 2006). Below we outline the strategy used to adapt mCherry as a means of providing guidelines for optimization of other fluorescent proteins for *C. elegans* germline expression.

B. Engineering New Fluorescent Proteins for Expression in the *C. elegans* Gonad/Early Embryo: The mCherry Experience

To reengineer mCherry for expression in *C. elegans*, we changed the codon usage and inserted artificial introns. Due to the variation in codon usage between species, optimal expression of a gene derived from one species in another typically requires "codon optimization." However, changing each of the codons for a specific amino acid to the most-favored codon is also not ideal; balance is important. Based on the analysis of numerous highly expressed genes, a favored codon usage value between 60 and 75% is predicted to yield optimal gene expression (Duret and Mouchiroud, 1999). We used this as a guideline in reengineering the mCherry coding sequence. As a second step, we introduced introns, which are known to stimulate expression in *C. elegans* (Fire *et al.*, 1990). We note, however, that although the reengineering of mCherry was successful, the contribution of each of these steps to the final outcome is unclear, since neither step was performed on its own. This point should be kept in mind when reading the mCherry reengineering procedure and using it to reengineer other fluorescent proteins.

1. Optimizing Codon Usage

For codon optimization of mCherry, we first listed all amino acids in the primary sequence noting their frequency and distribution. Then, based on the *C. elegans* codon usage table (see Fig. 3), we reassigned 70% of the codons to the optimal triplet for each amino acid in a random fashion. Next, we reassigned 5–10% of the remaining codons to the most poorly used triplet. The optimal and poorly used codons were well spaced to avoid prolonged stretches of poorly used codons. Finally, we reassigned the residual 20% of codons to other triplets. After completing this process, we used web-based bioinformatic programs to check for five potentially adverse features in the designed sequence:

1. *Guanine-cytosine (GC) content bias*: The GC content of the entire *C. elegans* coding genome is ~43% and we checked to make sure that the GC content of the reencoded sequence approximated this value.

2. *Repetitive sequences*: Such sequences can lead to ribosome slippage and should be eliminated.

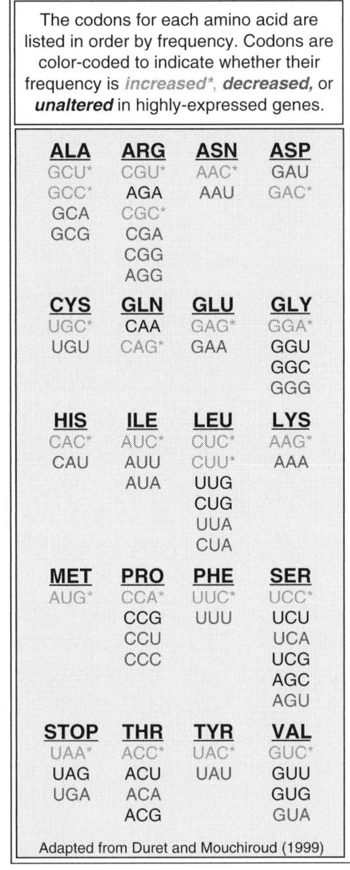

Fig. 3 Codon bias in *C. elegans*. Codons are listed by frequency of usage (in order of most common to least used) for expressed short proteins. Codons listed in green (*) are biased for usage in highly expressed sequences (more frequently used in highly expressed sequences when compared to low to nonexpressed sequences). Codons listed in red are biased against usage in highly expressed sequences. Codons listed in black are used at an equivalent frequency irrespective of expression level. This information was adapted from data presented in Duret and Mouchiroud (1999). (See Plate no. 23 in the Color Plate Section.)

3. *mRNA secondary structure*: Available programs such as the Vienna RNA package (http://www.tbi.univie.ac.at/~ivo/RNA/) are useful in predicting mRNA stability and structure. Incorporation of hairpin loops or other secondary structure elements that increase mRNA stability should be avoided.

4. *Cryptic splice sites*: The Alternative Splice Site Predictor (http://es.embnet.org/~mwang/assp.html) will identify potential cryptic splice sites. Such sites must be eliminated prior to the synthesis of the reencoded gene.

5. *tRNA steric hindrance*: Steric interactions between tRNAs have been shown to impact their ability to simultaneously occupy the two available ribosomal binding sites during translation (Smith and Yarus, 1989). Commercial programs, including DNA 2.0 (http://www.dnatwopointo.com/commerce/misc/opt.jsp), can be used to predict tRNA steric hindrance and eliminate it prior to finalizing the design of the reencoded gene.

2. Insertion of Introns

Incorporation of multiple introns has been shown to improve gene expression in *C. elegans* (Fire *et al.*, 1990). Stimulation by intronic sequences does not necessarily require the introns to be inserted into coding regions; inclusion within the 5′ or 3′ UTR can also increase gene expression by several folds. To prevent intramolecular recombination after transformation, use of identical introns is not recommended. Although a number of potential introns are available from the genome database, care must be exercised in choosing an appropriate sequence. Most importantly, introns with cryptic splice sites should be avoided. Although insertion of introns into a target gene is relatively straightforward, it requires several cloning steps. The use of unique blunt-end restriction sites is most convenient, allowing quick insertion of introns encoded by synthetic oligonucleotides. Alternatively, introns can be added to a target gene prior to its synthesis. The three artificial introns that were inserted into the coding sequence of mCherry are identical to those that were inserted into GFP (S65C) by Fire *et al.* (1998) [these sequences were: (1) gtaagtttaaacatatatatactaactaaccctgattatttaaattttcag, (2) gtaagtttaaacagttcggtactaactaaccatacatatttaaattttcag, (3) gtaagtttaaacatgattttactaactaactaatctgatttaaattttcag].

Recent data indicate that introns in germline-expressed genes exhibit a high level of periodicity in their DNA structure, often showing a strong bias for AA/TT dinucleotides along one face of the DNA helix (Fire *et al.*, 2006). This bias is postulated to affect the association of intronic sequences with nucleosomes. Improper spacing of introns may affect their ability to form an appropriate DNA structure, and thereby compromise germline expression. However, further study is needed to examine this issue.

3. Other Considerations in Designing Transgenic Constructs

A final consideration is the position of the fluorescent tag with respect to the coding sequence. Potential protein modifications must be considered when incorporating the tag. For example, the N-terminus of myristoylated proteins should be left free to allow accessibility to N-myristoyltransferase. Also, although most fluorescent proteins encode flanking sequences both upstream and downstream of the fluorophore, an additional spacer is often useful to prevent protein misfolding.

III. Transgene Expression in the *C. elegans* Germline: Breaking the Silence

The efficiency of germline silencing has posed significant challenges to the expression of transgenes (Seydoux and Strome, 1999) encoding fluorescent fusion proteins. Although no strategy has fully circumvented this difficulty, it can be rendered manageable by the appropriate selection of regulatory sequences (Strome *et al.*, 2001) combined with the use of ballistic bombardment as the means of transgene delivery (Praitis *et al.*, 2001; Wilm *et al.*, 1999). In this section, we describe the most commonly used vectors and highlight important practical considerations when designing constructs for germline expression.

A. Promoter and 3′ UTR Choice

Although there are selected examples of successful germline expression under the control of endogenous regulatory sequences, the most commonly used strategy is to place coding regions under the control of the regulatory sequences for PIE-1, a transcriptional repressor that is highly expressed in the germline. This strategy, developed by Seydoux and coworkers, has proven remarkably effective in minimizing the germline silencing of transgenes (Strome *et al.*, 2001). In addition to the *pie-1* promoter, the *pie-1* 3′ UTR is also important to avoid silencing. In our experience, proteins expressed under control of the *pie-1* regulatory sequences are typically present at 5–50% of the levels of the endogenous protein when analyzed by Western blotting. The fact that this is consistently true, despite wide variation in the normal expression levels of the proteins tagged, likely reflects the importance of coding region sequence structure and potentially codon usage, in controlling protein levels. Functional rescue of mutants has been observed for many *pie-1* controlled transgenes. In some cases, increased levels of the transgenic protein are seen in the mutant background, possibly reflecting selection against silencing when transgene expression becomes essential for survival.

B. Currently Available Vectors for Expression of Fluorescent Proteins in the Germline

The currently used vectors for germline expression are based on pPD95.75, a vector generated in Andrew Fire's laboratory containing a version of GFP modified for expression in *C. elegans* (available from Addgene, www.addgene.org/pgvec1/f=c&identifier=1494&cmd=findpl). Geraldine Seydoux's laboratory modified this vector to place the GFP fusion under the control of the *pie-1* regulatory sequences, resulting in pJH4.52 (Strome *et al.*, 2001). pJH4.52 was designed to be linearized, mixed with a second linearized plasmid containing a dominant or selectable marker and an excess of cut genomic DNA, and injected into the gonad to generate stably transmitted complex extrachromosomal arrays (Kelly *et al.*, 1997), a popular technique that works well for expression in somatic tissues. Since arrays suffer from silencing, due to the presence of repetitive sequences, pJH4.52 was further modified in Judith Austin's laboratory to create pAZ132 (Praitis *et al.*, 1991), a vector designed for direct integration by ballistic bombardment. pAZ132 contains a cassette that drives the expression of UNC-119, the most popular selectable transformation marker for bombardments. In addition, it contains a cassette that directs the expression of protein fusions with GFP at their N-terminus under the control of the *pie-1* regulatory regions. Insertion of a tandem protein purification tag between the GFP and the target gene was performed in Arshad Desai's laboratory (pIC26) to allow biochemical isolation of the GFP fusion and its associated proteins [for detailed protocols, see Cheeseman and Desai (2005), Cheeseman *et al.* (2004)]. A derivative of pAZ132 that fuses GFP to the C-terminus of the target protein, pAZ132-C, has also recently been engineered in Tony Hyman's laboratory. In addition, the Hyman laboratory has generated a series of vectors that eliminate extraneous restriction sites, incorporate a multiple cloning site, reduce the size of these relatively large vectors by introducing a shortened region of *unc-119*, and carry distinct resistance and fluorescence markers. These smaller constructs offer greater convenience for cloning; however, occasionally incomplete rescue of the Unc phenotype makes the identification of moving worms, following bombardment, more challenging. Vectors incorporating other fluorescent proteins, such as mCherry (pAA64 and pAA65), YFP (pPAG4 and pAZ-YFP), and CFP (pUP9 and pAZ-CFP) have also been made in the laboratories of Oegema, Hyman, and Askjaer. A list of the available vectors for germline expression of fusions with fluorescent proteins can be found in Fig. 4.

IV. Constructing Fluorescent Worm Lines

A. Integration of Constructs by Ballistic Bombardment

Ballistic bombardment, in which pressurized helium is used to "bombard" small transgene-coated gold particles into worm tissue at high speeds, is the method of choice for introduction of germline expression constructs into *C. elegans*. Bombardment has proven vastly superior to the more traditional method of injecting

Germline fluorescence expression vector- N-terminal FP

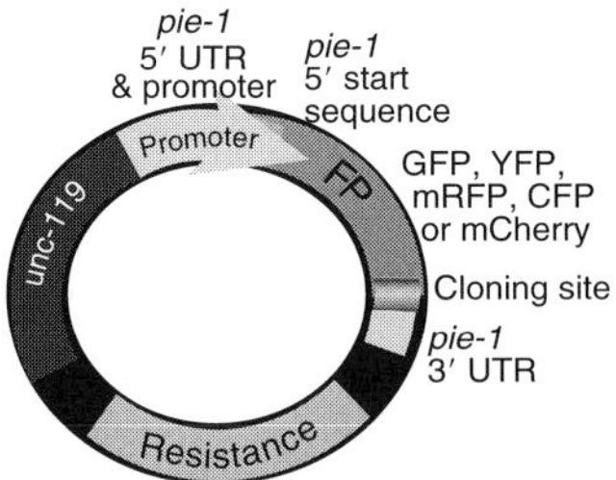

Germline fluorescence expression vector- C-terminal FP

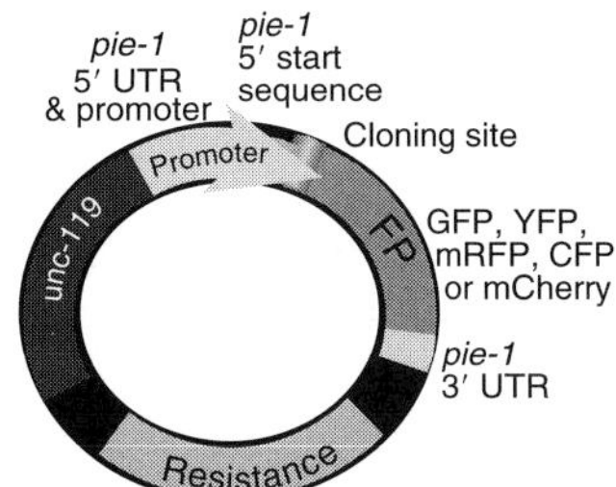

	Plasmid name	Promoter	Fluorescent protein	Location	Resistance	Selection marker	Additional features	Source
Parental vectors	pPD95.75		GFP					Fire lab
	pJH4.52	*pie-1*	GFP	N-terminus	Amp			Seydoux lab
	TH317-NoTag-AMP	*pie-1*	No tag	N-terminus	Amp	*unc-119*	short *unc-119*	A. Pozniakovsky; Hyman lab
	TH318-NoTag-CM	*pie-1*	No tag	N-terminus	Cm	*unc-119*	short *unc-119*	A. Pozniakovsky; Hyman lab
	TH319-NoTag-KAN	*pie-1*	No tag	N-terminus	Kan	*unc-119*	short *unc-119*	A. Pozniakovsky; Hyman lab
GFP vectors	pAZ132 (N-GFP)	*pie-1*	GFP	N-terminus	Amp	*unc-119*		Fire lab
	pIC26 (N-LAP)	*pie-1*	GFP	N-terminus	Amp	*unc-119*	LAP	I Cheeseman; Desai lab
	TH304-GFP (C) AMP	*pie-1*	GFP	C-terminus	Amp	*unc-119*	short *unc-119*, MCS	A. Pozniakovsky; Hyman lab
	TH303-GFP (N) AMP	*pie-1*	GFP	N-terminus	Amp	*unc-119*	short *unc-119*, MCS	A. Pozniakovsky; Hyman lab
	TH305-GFP (C) CM	*pie-1*	GFP	C-terminus	Cm	*unc-119*	short *unc-119*, MCS	A. Pozniakovsky; Hyman lab
	TH314-LAP (N) AMP	*pie-1*	GFP	N-terminus	Amp	*unc-119*	short *unc-119*, MCS, LAP	A. Pozniakovsky; Hyman lab
	TH315-LAP(N)GW-AMP	*pie-1*	GFP	N-terminus	Amp/Cm	*unc-119*	short *unc-119*, MCS, GateWay, LAP	A. Pozniakovsky; Hyman lab
mCherry vectors	pAA65(LAP)	*pie-1*	mCherry	N-terminus	Amp	*unc-119*	LAP, codon optim.	A. Audhya; Oegema lab
	pAA64	*pie-1*	mCherry	N-terminus	Amp	*unc-119*	codon optim.	A. Audhya; Oegema lab
	TH313-Cherry (C) AMP	*pie-1*	mCherry	C-terminus	Amp	*unc-119*	short *unc-119*, MCS codon optim.	A. Pozniakovsky; Hyman lab
	TH312-Cherry (N) AMP	*pie-1*	mCherry	N-terminus	Amp	*unc-119*	short *unc-119*, MCS codon optim.	A. Pozniakovsky; Hyman lab
mRFP vectors	TH309-mRFP (N) AMP	*pie-1*	mRFP	N-terminus	Amp	*unc-119*	short *unc-119*, MCS	A. Pozniakovsky; Hyman lab
	TH310-mRFP (C) AMP	*pie-1*	mRFP	C-terminus	Amp	*unc-119*	short *unc-119*, MCS	A. Pozniakovsky; Hyman lab
	TH311-mRFP (N) CM	*pie-1*	mRFP	N-terminus	Cm	*unc-119*	short *unc-119*, MCS	A. Pozniakovsky; Hyman lab
CFP vector	TH306-CFP(N) AMP	*pie-1*	CFP	N-terminus	Amp	*unc-119*	short *unc-119*, MCS	A. Pozniakovsky; Hyman lab
YFP vectors	TH307-YFP(N) AMP	*pie-1*	YFP	N-terminus	Amp	*unc-119*	short *unc-119*, MCS	A. Pozniakovsky; Hyman lab
	TH308-3xYFP(N) AMP	*pie-1*	Triple YFP	N-terminus	Cm	*unc-119*	short *unc-119*, MCS	A. Pozniakovsky; Hyman lab

Fig. 4 Vectors for the expression of fluorescent fusion constructs in the *C. elegans* gonad and early embryo. Schematics illustrate the common features for expression vectors that fuse a fluorescent protein to the N-terminus (left) or C-terminus (right) of the target protein. Each contains the 5′ UTR, promoter, and 5′ initial coding sequence for PIE-1, followed by fluorescent protein and 3′ *pie-1* UTR. The plasmids also contain a resistance marker for growth in *E. coli* and a selection marker (*unc-119)* for integration into the *C. elegans* genome. The list outlines features of currently available vectors, including promoter, fluorescent protein, location of fluorescent protein (N- or C-terminus), type of resistance marker (Amp, Cm, Kan), *C. elegans* selection marker (*unc-119)*, specific features of each plasmid, and the source of each vector. Specialized features include (1) the addition of multiple cloning sites, (2) the presence of a localization and affinity purification tag, (2) optimized fluorescent proteins for expression in *C. elegans* (intron incorporation and codon bias), (3) GateWay compatible cloning vector, and (4) truncated versions of *unc-119* (to reduce overall plasmid size). Some key parental vectors are also listed. (See Plate no. 24 in the Color Plate Section.)

the construct into the gonad to generate heritable and repetitive extrachromosomal arrays. Although expression from extrachromosomal arrays is relatively straightforward in somatic tissues, germline expression from even the best arrays is silenced within a few generations (Kelly and Fire, 1998; Seydoux and Strome, 1999).

Bombardment techniques were adapted to *C. elegans* to avoid problems associated with extrachromosomal arrays (Jackstadt *et al.*, 1999; Praitis *et al.*, 1991; Wilm *et al.*, 1999). The most widely used method for the generation of low-copy number transformants was developed by Praitis *et al.* (2001). This approach reduces the likelihood of generating large extrachromosomal arrays because a single bombarded bead can carry only a limited amount of DNA into a cell.

Bombardments are typically performed in the DP38 strain harboring the *unc-119(ed3)* mutation. These worms are unable to move and importantly also fail to undergo the starvation-induced developmental transition to the dauer stage. Integration of the *unc-119* genomic region, which is present in the bombarded plasmid, allows transformed worms to produce moving progeny that can survive starvation. Untransformed worms on the same plate remain uncoordinated and die. This strong selection is necessary since large quantities of worms must be bombarded to generate the relatively rare desired integration events.

1. Standard Small-Scale Bombardment Protocol

The protocol discussed below is for bombardments on standard small (60 mm diameter) plates using the Bio-Rad PDS-1000 apparatus (see Fig. 5) that requires a vacuum source (a pump or a good house vacuum line) and a helium tank. For each construct, we recommend seeding 10–12 large (100 mm) plates with DP38 worms to generate sufficient quantities of worms for 10 small bombardment plates. Bombardments can also be performed on large (100 mm) plates using the hepta-adaptor attachment (Bio-Rad, Hercules, CA; see details below).

a. Preparation of the Worms

1. Evenly seed DP38 (*unc-119(ed3)*; Maduro and Pilgrim, 1995) worms from two slightly starved 60-mm plates onto 10–12 large (100 mm diameter) peptone/c600 plates. This can be done by dividing each starved out DP38 plates into six chunks with a scalpel and dabbing each chunk onto a peptone/c600 plate. Peptone plates and the fast growing C600 strain of *Escherichia coli* are used to increase the number of worms per plate. The plates should be maintained at 20 °C until they begin to starve out (about 1 week).

2. Using sterile M9 and a 10-ml pipette, wash the worms off the peptone plates by pipeting up and down. Transfer the washed worms from all 10 to 12 plates into a 50-ml conical tube. After allowing the worms to settle to the bottom of the tube, remove excess supernatant, leaving ∼1–2 ml of liquid/worms in the bottom of each conical tube.

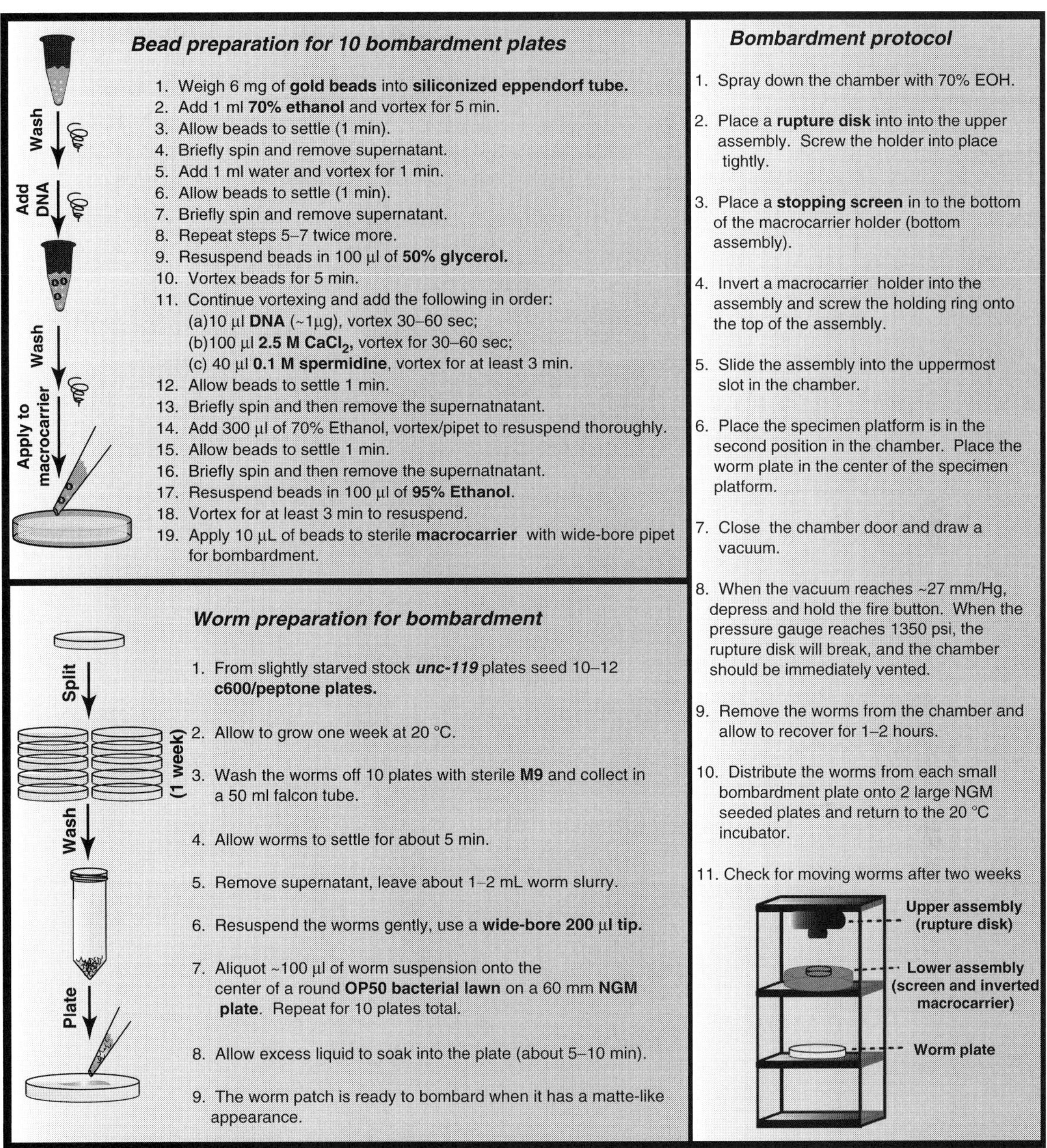

Fig. 5 Standard small-scale bombardment protocol. This figure details the preparation of gold beads and worms for a small-scale bombardment (one or two constructs). A protocol for bombarding the worms is also included (see chapter text for more detailed information). (See Plate no. 25 in the Color Plate Section.)

3. Gently resuspend the worms using wide-bore pipette tips and pipet 100 μl of worms onto the bacterial lawn in the center of an NGM/OP50 60-mm plate. Each conical tube should have enough worms for ~10 plates. Excess worms can be divided between the 10 plates. Plates with a circular lawn roughly the size of a quarter are the best for this since the worms will spread to the edge of the lawn.

4. The worms are ready to bombard when the excess liquid has absorbed into the plate and the worm patch has a matte-like appearance (usually after about 10 min).

b. Preparation of the Beads

The instructions below are for the preparation of beads for 10 bombardments. A vortex mixer with a tube holder (such as the TurboMix attachment from Vortex Genie) is very useful for this procedure.

1. Weigh out 6 mg of gold beads (Bio-Rad 1 μm beads #1652263) into a siliconized microfuge tube (prevents beads from sticking to the tube walls).
2. Add 1 ml of 70% ethanol (EtOH) and vortex the beads continuously for 5 min.
3. Allow the beads to settle for 1 min, then briefly spin and remove the supernatant.
4. Wash the beads three times with 1 ml water; after each wash vortex the beads for 1 min, allow the beads to settle, briefly spin, and carefully remove the supernatant for each wash.
5. Resuspend the beads in 100 μl of 50% glycerol and vortex the beads for 5 min.
6. Continue to vortex, pausing only briefly to add the following in order (vortex for 1 min between additions):

   i. 10 μl of DNA (about 1 μg)
   ii. 100 μl 2.5 M $CaCl_2$
   iii. 40 μl 0.1 M spermidine tissue culture grade, Sigma S-4139, St. Louis, MO, store at −20 °C

7. Vortex the beads for 3 min.
8. Allow the beads to settle to the bottom for 1 min, spin briefly, and remove the supernatant.
9. Perform a final wash with 70% EtOH (300 μl), vortexing and pipeting to thoroughly resuspend the beads. Allow them to settle out, spin briefly, and remove the supernatant.
10. Resuspend the beads in 100 μl 95% EtOH, vortexing and pipeting to thoroughly resuspend the beads. Using a wide-bore pipette tip, pipet 10 μl of bead slurry onto the center of each sterile bombardment macrocarrier (10 in total).

Note: The DNA-coupled beads have a tendency to clump. To avoid this problem, proceed expeditiously through steps 6–10 (without stopping) and load the beads onto the macrocarriers as soon as possible after the final resuspension in 100% EtOH. Macrocarriers are small plastic discs that are mounted in a metal holder for the bombardment (see Bio-Rad manual for more details). It is convenient to mount them in the holders in advance and sterilize them by autoclaving in a foil-covered beaker. Alternatively, the macrocarriers can be sterilized by dipping in 70% EtOH, placing them in the holders, and putting them in the hood with the ultraviolet light on in an open petri dish.

11. Place the macrocarriers in an open petri dish inside a desiccator chamber to facilitate drying. *Use the prepared carriers within 2 h.*

c. Bombarding the Worms

1. Some contamination is inevitable, but this can be minimized by taking some precautions. Before beginning, sterilize the bombardment apparatus by spraying down the inside of the chamber and the upper and lower assemblies with 70% EtOH.

2. Using forceps, dip a rupture disk (Bio-Rad, #1652330) into 70% isopropanol and place it flush into the base of the upper assembly disk holder unit. Screw the disk holder into place and tighten with a torque wrench.

3. Next, place a stopping screen (Bio-Rad, #1652336) at the base of the lower assembly unit, then invert a macrocarrier holder on top of the screen and screw the holding ring into place.

4. Slide the lower unit into the top slot in the bombardment chamber. Place an OP50/NGM plate (seeded packed worms) on the specimen platform below the lower assembly unit.

5. Close the chamber door and draw a vacuum of 28 mm/Hg. Depress and hold the "fire" button until the rupture disk pops (around 1350 psi).

6. Immediately vent the chamber and remove the plate of worms. Repeat this procedure for the remaining macrocarriers/worm plates.

7. Allow the worms to recover for 1 h and then transfer the worms from each 60 mm plate to two or three 10 cm NGM/OP50 plates using an M9 wash.

8. Place all plates at 20°C and wait $\sim$3 generations (about 12 days). The vast majority of the worms will remain uncoordinated and die; however, some worms will be successfully transformed. Since three generations have passed, plates on which a worm was transformed will be clearly identifiable by the presence of multiple moving siblings (progeny of the original transformed worm).

9. Single three young moving worms from each plate with movers onto individual small plates and screen for expression of the fluorescent transgene in their progeny. Usually, the three progeny will yield similar results, since they are most

likely siblings—but it is also possible for more than one worm to have been transformed per plate.

The progeny of these worms can be visually screened under the microscope for the incorporation of the fluorescently tagged gene of interest to the appropriate location (i.e., germline). If after several generations, you notice that one third of the progeny are uncoordinated, it is likely that the bombardment generated an obligate heterozygous transgenic line. This can occur, for example, if the transgene inserts into an essential gene. While these lines can be quite useful, they cannot be readily used to generate double (crossed) fluorescent lines. If all progeny are coordinated for several generations, then you have successfully generated a homozygous transgenic line. These lines can easily be mated with other fluorescent strains to develop multimarker and/or multicolor worm lines (see below).

2. Adapting the Protocol for Large-Scale Bombardments

For large-scale bombardments, larger quantities of DP38 can be obtained by culturing them in liquid. If bombardments are ongoing in the laboratory, a synchronous culture can be initiated and maintained for multiple generations. In the protocol below, we describe a two-step procedure to initiate a synchronous liquid culture. In the *first* step, an asynchronous liquid culture is started from plates and grown until the majority of worms are adults. The adults are isolated and their embryos obtained by treatment with a basic bleach solution. After hatching the embryos overnight without food to synchronize the population as starved L1s, the worms are reseeded into a *second* synchronous round of liquid culture. After a synchronous culture of DP38s is started, it can be maintained indefinitely by harvesting two third of the worms at each generation and obtaining embryos from the remaining one third to seed the subsequent round of culture.

a. Step 1: Starting an Asynchronous Liquid Culture of DP38s

Approximately 6 days prior to starting the culture: Seed 10–15 large (100 mm) OP50 plates with 30 clean worms each; incubate the worms at 20°C for ~6 days (or until worms are just starved).

Two days prior to starting the culture: Seed a 50-ml overnight bacterial culture with OP50–1 in LB plus 50 μg/ml streptomycin.

One day prior to starting the culture: Use the 50 ml culture to seed 2 × 1 liter overnight cultures of OP50–1 in LB plus 50 μg/ml streptomycin. This will be enough bacteria to start 2 × 500 ml worm cultures.

First day of culture:

1. *Prepare the media*: Pellet the bacteria in sterile 1 liter bottles by spinning at 4200 rpm for 15 min, resuspend the bacteria corresponding to each liter of bacterial culture in 500 ml S complete media and transfer to a sterile 2.8-liter flask in preparation for worm growth. Alternatively, pellets can be stored at

4°C for 1–2 weeks until use. To prevent contamination, set up the cultures using sterile technique in a tissue culture hood.

2. *Harvest the worms from the plates and seed the liquid culture*: Rinse the NGM/OP50 worm plates with sterile M9 and collect the worms in a 50-ml conical tube. Spin down the worms at 600 × *g* for 2 min in a clinical centrifuge to pellet. Remove the supernatant and resuspend the worms in 10 ml M9. Divide the worms between the two flasks containing OP50–1 in S complete media (from step 1). Shake the worms at 230 rpm in a cooling incubator maintained at 20°C for 3–4 days. Monitor the progress of the cultures carefully and harvest when the majority of worms are at the adult stage. Check the cultures by removing a 5-ml aliquot from the flask with a sterile pipette, spinning down the worms at 600 × *g* for 2 min, removing the supernatant, and washing the worms 1× with cold M9. You can then resuspend the worm pellet in a little amount of M9 and examine them on a dissecting scope after spotting a 4-μl aliquot onto a slide under a coverslip.

Harvesting the worms and isolating embryos:

1. Transfer each 500 ml culture of worms to a sterile 1 liter centrifuge bottle and spin for 7 min at 700 × *g*.
2. Aspirate off excess media and transfer worm slurry from each bottle to a 50-ml conical tube. Pellet at 600 × *g* for 3 min and aspirate off supernatant.
3. Combine worms into one 50-ml conical tube (not more than 10 ml of worms per conical tube) and wash by filling the conical tube with cold M9 buffer and pelleting again at 600 × *g* for 3 min.
4. Isolate the adults from the other larval stages and debris by performing a "sucrose float." Aspirate off the supernatant and resuspend the worm pellet in 25 ml of cold M9. Add 25 ml of cold 60% w/v sucrose dissolved in water. Mix by inverting and immediately spin at 1500 × *g* for 5 min. Initiate this spin right away or the worms will slowly desiccate in the sucrose.
5. During the spin, the adult worms will float to the top of the conical tube. Transfer the adults to a new conical tube with a transfer pipette and wash them once by filling the tube up with cold M9 and pelleting them at 600 × *g* for 3 min. Remove the supernatant and resuspend the worms in 25 ml of 0.1 M NaCl, mix by pipeting up and down and incubate on ice for exactly 5 min. After 5 min, aspirate off worms that have not settled (nongravid) in preparation for bleaching of gravid worms.
6. Aspirate off the supernatant and resuspend the worm pellet to 30 ml with 0.1 M NaCl. To this, add a mixture of 5 ml of 5 M NaOH and 10 ml of bleach [Fisher SS290–1 (4–6%), store sealed container, wrapped with parafilm at 4°C]. Vortex briefly at maximum speed and allow the mixture to stand at room temperature for 2 min.

7. Repeat vortexing every 2 min for a total of 8–15 min. Monitor the bleaching periodically by pipeting 4 μl samples onto a slide, covering with a coverslip and examining under a dissecting scope to assess degree to which worm debris is dissolved. Do not overbleach.

8. Centrifuge for 1 min at 800 × *g* at 4 °C. Wash the pellet once by filling the conical tube up with ddH_2O (4 °C) and repellet by spinning at 700 × *g* for 2 min.

9. Wash once more by filling up the conical tube with room temperature M9 and pelleting the embryos at 700 × *g* for 2 min.

10. Add 50 ml of M9 to the pellet and transfer the embryo pellet to a sterile 500 ml Erlenmeyer culture flask.

11. Shake in a temperature-controlled 22 °C incubator until the embryos hatch (~12–24 h) yielding a synchronous culture of starved L1s.

b. Step 2: Synchronous Growth of DP38s in Liquid Culture

Two days prior to starting the synchronous culture: Seed 100 ml culture of OP-50–1 in LB plus 50 μg/ml streptomycin.

One day prior to starting the synchronous culture: Use the 100 ml culture to seed 6 × 1 liter overnight cultures of OP50–1 in LB plus 50 μg/ml streptomycin. This will be enough bacteria to start 6 × 500 ml worm cultures.

First day of synchronous culture:

1. *Prepare the media*: Pellet the bacteria from each 1 liter culture in a sterile 1 liter bottle by spinning the first liter at 4200 rpm for 15 min, adding the remaining 500 ml of culture to the bottle, and repeating the spin. Resuspend the bacteria from each bottle in 20 ml of S complete media and transfer to a sterile 2.8-liter flask containing 500 ml of S complete media. After this step, you should have six flasks containing S complete media plus the bacteria from 1 liter of culture.

2. *Seed the synchronous culture*: Pellet the L1s from step 11 above by spinning at 600 × *g* for 3 min. Bring up to 50 ml with sterile M9 and repellet.

3. Resuspend the worms in 12 ml of sterile M9.

4. Seed each flask with the equivalent of 50–100 μl of pure L1 pellet.

5. Shake the worms at 230 rpm in a cooling incubator maintained at 20 °C for 3–4 days. Monitor the progress of the cultures carefully by periodically removing a 5-ml aliquot from the flask with a sterile pipette, spinning down the worms at 600 × *g* for 2 min, removing the supernatant, and washing the worms 1× with cold M9. Examine the resuspended worm pellet after spotting a 4-μl aliquot onto a slide under a coverslip. Harvest four of the flasks for bombardments (see below) when the worms are at the L4 stage (do not use young adult worms). Each flask should yield enough worms for 8–10 heptad bombardments (depending on worm density).

6. If you want to maintain the liquid culture, continue to incubate the remaining two flasks until the worms are adults; harvest embryos by bleaching (as described above) to seed the next round of synchronous culture.

c. Step 3: Modifications for Large-Scale Bombardment Using the Hepta-Adaptor

When bombarding worms cultured in liquid, a *hepta-adaptor (Bio-Rad #165–2225)* can be attached to the bombardment apparatus allowing greater quantities of worms to be bombarded at one time.

1. Pellet the L4 culture (as prepared above) at 700 × *g* for 5 min. Worms should be washed with M9 to remove excess bacteria. After spinning 600 × *g* for 2 min, remove as much supernatant as possible in preparation for distribution of worms onto the bombardment plates.
2. On each bombardment plate (10 cm), mark the approximate position of each macrocarrier in the hepta-adaptor using a pen.
3. Distribute 100 μl clumps of washed L4 worms onto these marked areas.
4. Prepare DNA beads and macrocarriers as described in the small-scale protocol, but use 10 μg of DNA for each heptad bombardment. Also, macrocarrier disks should not be placed in individual metal holders (as described in the small-scale protocol); instead, they are inserted directly into the hepta-adaptor unit.
5. Position each bombardment plate on the shelf directly below the hepta-adaptor (~2 cm). Line up each worm clump with each macrocarrier in the hepta-adaptor apparatus. Verify that the air outlets are also in alignment.
6. Bombard using 1350 psi rupture disks and hepta-adaptor stopping screens (Bio-Rad 1652226). Stopping screens can be used for several bombardments with the same construct. Because of the hepta-adaptor setup, the rupture disks will break closer to 2000 psi.
7. Let worms recover for 1–2 h at 20 °C.
8. Distribute worms from one heptad bombardment plate onto 10 small (60 mm) plates and incubate at 20 °C for 2–3 weeks. Expect approximately five moving lines per heptad bombardment.

3. Current Challenges: Endogenous Gene Replacement

In other systems, such as yeast and mice, gene targeting with homologous recombination can replace endogenous genes with constructed transgenes or knock them out to create null mutations. While homologous recombination occurs with low incidence in *C. elegans* following large-scale bombardment (Berezikov *et al.*, 2004) or injection of DNA into oocyte nuclei (Broverman *et al.*, 1993), its low frequency has prevented this method from being extensively utilized. By scaling up biolistic bombardment methods, Berezikov and colleagues have demonstrated, for two independent loci, that ~300 transformants are required to isolate one homologous recombination event. Future work improving the frequency of homologous transformation events either by technological advances or by identification of mutants that facilitate efficient recombination will be an important advance.

4. Controlling Growth Temperature to Influence Expression

Germline silencing is temperature sensitive and weakest at 25°C, which is effectively the highest temperature at which worms will remain fertile (Strome *et al.*, 2001). For difficult-to-obtain transgenic lines, growth at 24.5°C can be used to maximize the chances of obtaining a line that expresses the transgene. However, in most cases, returning these transgenic lines to lower temperatures results in rapid silencing that can sometimes be permanent. The use of these strains has many limitations. For example, such transgenes cannot be used in temperature-sensitive mutants due to silencing at the permissive temperature. In general, we recommend that transgenic strains initially be isolated at 20°C. In cases where fluorescence is absent, strains should be shifted to 24.5°C and rescreened shortly thereafter.

B. Making Dual/Triple Fluorescent Marker Lines by Mating

It is usually straightforward to generate multimarker lines once single fluorescent transgenic lines have been established. To do this, males generated for one of the fluorescent strains are crossed to hermaphrodites homozygous for the second fluorescent marker (see Fig. 6). The F2 progeny from this mating are singled and their progeny are screened to identify lines homozygous for both fluorescent markers. In our experience, GFP and mCherry provide an optimal combination of fluorescent markers.

1. Generating and Maintaining Males

Male worms, which have a single X chromosome (XO), arise at low frequency in a hermaphrodite (XX) population by chromosome nondisjunction. Below we describe our version of a commonly used heat shock protocol to increase the frequency of males in strains bearing fluorescent transgenes.

1. Set up an incubator (or water bath) at 30°C and another at 31.5°C.
2. Transfer 10 L4 hermaphrodites onto each of two 10-cm plates seeded with OP50 bacteria.
3. Parafilm and place one plate at 30°C and one at 31.5°C for 8 h.
4. Remove excess condensation and return plates to 20°C incubator for 3–4 days.

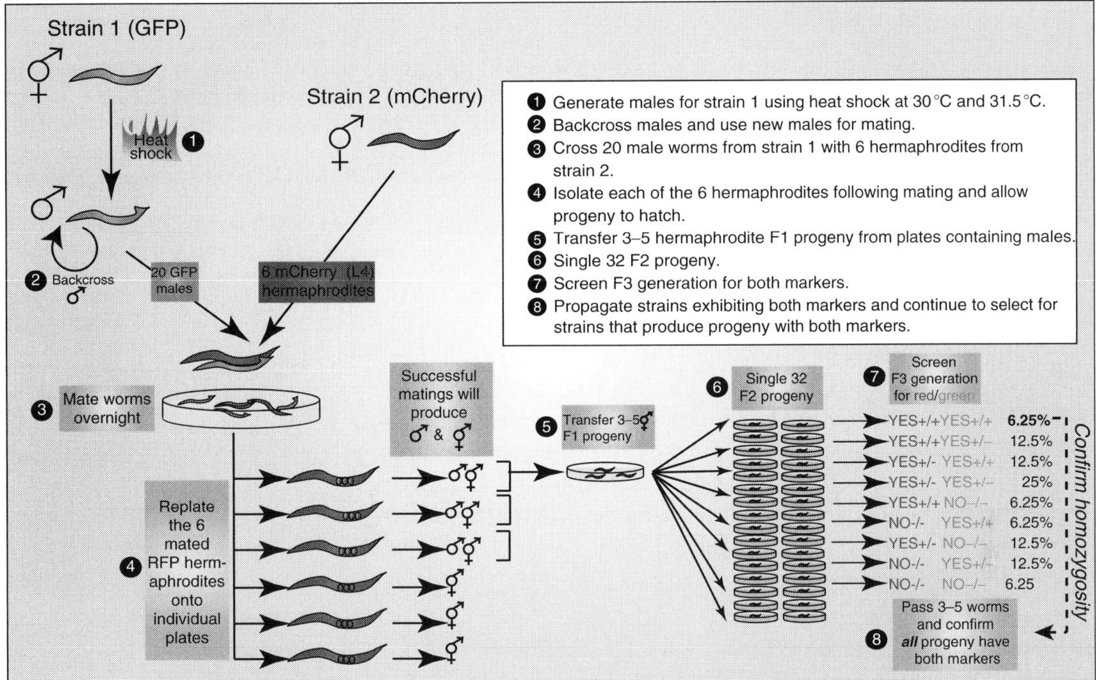

Fig. 6 Standard mating protocol to generate multimarker lines. Begins by generating males from the first strain by subjecting L4 hermaphrodites to heat shock at 30 and 31.5° C. Allow hatching of progeny and isolate males for backcrossing. Backcross males and use newly isolated males for mating. Cross 20 male worms (strain 1) with 6 hermaphrodite worms (strain 2). Single the six hermaphrodites onto individual plates and allow progeny to hatch. Transfer three to five hermaphrodite F1 progeny from plates that have ~50% males (indicative of successful mating). Single out 32 F2 progeny and screen their progeny (the F3 generation) for both markers to determine the genotype of the F2 parent. The expected incidence for each genotype is listed (assuming the markers are on different chromosomes). Propagate strains where all F3 progeny express both markers, indicating that the F2 parent was homozygous for both markers. If the F2 parent was heterozygous for a marker, then ~1/4 of its progeny will lack the marker. Check again at the next generation to confirm that both markers are homozygous. (See Plate no. 26 in the Color Plate Section.)

5. Check plates for males when the progeny are L4 stage. Male worms are easily distinguished by their smaller size and the presence of a copulatory apparatus or "hook" at the end of the tail.

This method generally only produces a few males per plate, and due to heat shock these males may not mate efficiently. Therefore, it is best to backcross the males before using them for mating. Male sperm out compete hermaphrodite sperm (LaMunyon and Ward, 1995; Ward and Carrel, 1979). Since 50% of male sperm carry the X chromosomes and 50% do not, successfully mated hermaphrodites produce ~50% male progeny.

2. Backcrossing Males

1. Place 5–20 males and 6 L4 hermaphrodites onto a mating plate [60 mm dish seeded with small spot (10 μl of saturated OP50 culture)].
2. Incubate at 16 or 20 °C overnight.
3. Move hermaphrodite worms to a standard NGM/OP50 60-mm dish and look for male progeny after several days.
4. Male worms can be maintained by setting up a mating with 20 L4 males and 6 L4 hermaphrodites on a standard 60-mm dish seeded with bacteria at each generation.

If only a few males are generated by the heat shock, you can maximize your chance of "capturing" the males by mating them with an excess (~20) of hermaphrodites for 24 h and then plating the hermaphrodites, three onto a plate, to simplify identification of plates with male progeny. It is often worth trying to make males from both of the strains containing the two transgenes that you are trying to combine, since the frequency of males and their ability to mate efficiently can vary between strains.

3. Crossing Males from One Strain with Hermaphrodites from a Strain Carrying a Second Fluorescent Marker

1. Once you have males from one strain, cross the two strains by placing 10 L4 males from the first strain and 6 L4 hermaphrodites from the second strain on a mating plate.

2. Allow them to mate for 24–36 h at 20 °C. If one strain requires growth at 16 °C, allow them to mate longer at 16 °C.

3. Single the hermaphrodites to fresh plates (six separate plates) and allow progeny to hatch. A plate carrying ~50% males indicates successful mating. Throw away the other plates.

4. The hermaphrodites on the plates with ~50% males are cross progeny and should be heterozygous for both fluorescent markers. Make two small plates with 3–5 heterozygous worms each and let these lay progeny to obtain the F2 generation.

5. The F2 worms will have a variety of genotypes and can be heterozygous, homozygous, or null for each fluorescent marker. If the two markers are on different chromosomes, 1/16 of the F2 progeny will have neither marker, 3/8 will be null for one marker and either heterozygous or homozygous for the second marker, 1/4 will be heterozygous for both markers, 1/4 will be homozygous for one marker and heterozygous for the second marker, and 1/16 will be homozygous for both markers. To identify the desired double homozygotes, we single the F2

worms onto individual plates and use fluorescence microscopy to screen expression in the gonad/embryos of their adult progeny (screening procedure described below). The progeny of the worms homozygous for both markers (~1/16 of the plates) will all have both markers. Depending on how many F2 worms you single, you may not find a plate where this is the case. The next best thing is plates where the parental F2 was homozygous for one marker and heterozygous for the second marker. The progeny of these worms will all have the first marker and ~1/4 of the progeny will lack the second marker. If you do not get a homozygote the first time, you can single worms from one of these plates and screen again in the next generation to homozygose the second marker—this time 1/4 of the worms should be homozygous for both markers. If the two markers are on the same chromosome, obtaining a double homozygote requires a recombination event. Recombinants can be straightforward to obtain if the markers are far apart—or nearly impossible to obtain if they are closely linked.

4. Screening for Strains Homozygous for Two Fluorescent Markers

1. Single ~32 F2 L4 hermaphrodites onto individual plates and incubate until there are adult progeny on the plate.
2. Whole-mount adult hermaphrodites from each plate. To whole mount, spot 2 μl of M9 onto a glass slide, transfer worm into M9 spot, place coverslip gently on top, and lightly press down on the edges of the coverslip to restrain worm movement (careful not to squish worm). Place it under the microscope and screen for the presence of both fluorescent markers in the germline/embryos. If one or both markers are absent from any of the progeny, throw the plate away. If both markers are present in all progeny (say >32 progeny), then the strain is most likely homozygous for both markers. In this case, pass three to five worms onto a new plate and screen again at the next generation to confirm that the line is homozygous for both markers. If one marker is present in all progeny but the second is absent in ~1/4 of the progeny, keep the plate but note that the parent was likely heterozygous for the second marker. Single worms from this plate and check again for double homozygotes in the next generation.

C. Benefits and Challenges with Multimarker Lines

Establishing dual color *C. elegans* lines typically results in beautiful images and valuable data. However, the usefulness of double or triple *single*-color lines is often overlooked. When two fluorescently marked proteins are spatially distinct in distribution and similar enough in signal intensity that a single acquisition setting can be employed to image both; dual single-color strains can be incredibly useful.

A nice example of this is a line that expresses GFP-γ-tubulin and GFP-histone, marking the spindle poles and chromosomes, respectively (see Fig. 7). This line offers the benefit of requiring less light exposure, compared to a dual colored line marking the same cellular components, which allows for longer filming. For laboratories with limited access to elaborate microscope facilities, these types of lines permit sophisticated analysis of multiple proteins of interest at the same time, using a basic fluorescence microscope setup. In addition, these lines can be crossed with other fluorescent lines to allow visualization of three or four cellular components simultaneously.

V. Using Fluorescent Worm Strains

A. Confirming Functionality of Transgenes

Before using fluorescent fusion proteins to obtain quantitative data, it is important to confirm that the fusion is functional and therefore likely to reflect the behavior of the endogenous protein. If there is an appropriate mutant available,

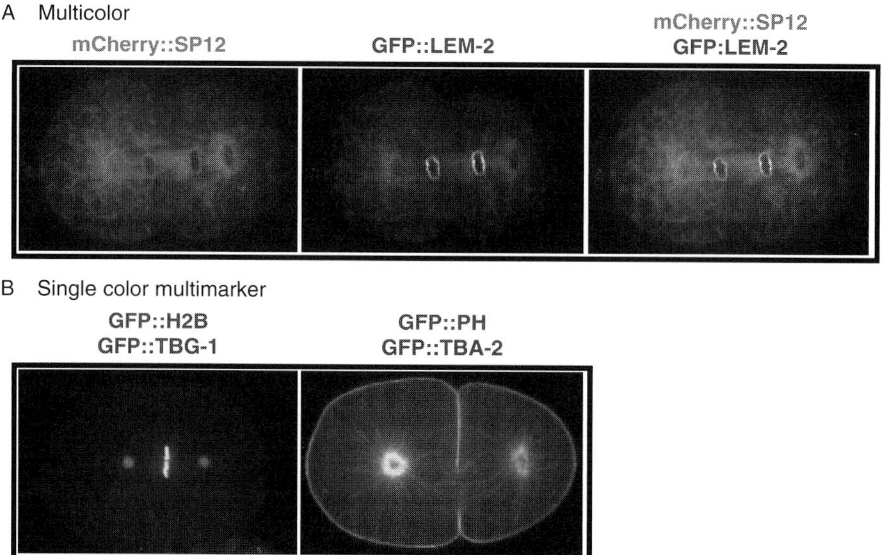

Fig. 7 Multicolor versus single-color multimarker strains. (A) In this strain, two proteins with a partially overlapping distribution are labeled with different fluorophores. mCherry::SP12 (left panel) labels the ER lumen. GFP:LEM-2 also localizes to the ER, but unlike SP12, is specifically enriched in the inner nuclear membrane (A. Audhya, Oegema Laboratory, unpublished strain). (B) Two examples are presented of single-color, multimarker strains to illustrate their usefulness when the spatial distributions of the two marked populations are distinct. The left panel shows GFP marking in both chromosomes (GFP::H2B) and centrosomes (GFP::TBG-1) at metaphase [strain TH32 (Cheeseman *et al.*, 2004)]. The right panel shows the plasma membrane (GFP::PH$^{PLC\delta1}$) and microtubules (GFP::TBA-2) in an embryo undergoing cytokinesis [strain OD73 (Audhya *et al.*, 2005)]. (See Plate no. 27 in the Color Plate Section.)

introduction of the GFP fusion into the mutant background can accomplish this goal. For transgenes expressed using the *pie-1* promoter, maternal effect or conditional mutants are ideal, since worms homozygous for a null mutation in an essential gene will often not survive to the developmental stage when the *pie-1* regulatory sequences drive expression. Below we discuss the use of RNAi-resistant (RR) transgenes, a useful alternative approach when an appropriate mutant is not available.

1. RR Forms of Transgenes

The generation of RR transgenes can be useful to test the localization and function of an introduced fusion protein in the absence of the endogenous protein (Fig. 8). RR transgenes are engineered to contain silent mutations that diverge from the original gene at the nucleotide level but not at the protein level. The first step in generating an RR transgene is to identify a small region of the endogenous gene against which an effective dsRNA can be made. We typically choose a region that is 300–500 bp in size, flanked by convenient restriction sites, and composed of mostly exonic (coding) sequence. Before recoding, it is important to make a dsRNA against this region and ensure that injection of this RNA effectively depletes the endogenous protein. In some cases, use of an RNA against the 3′ UTR of the endogenous gene (not present in transgenes driven by the *pie-1* regulatory sequences) may suffice, obviating the need to make a special transgene. However, in many cases, the 3′ UTRs are too short to be effective. In general, longer dsRNAs improve depletion efficiency (although they make reencoding more expensive).

After an appropriate target region is selected, the nucleotide sequence must be "recoded" without altering the original amino acid sequence or overall codon bias. To achieve this, the simplest method is to shuffle alternative codons encoding the same amino acid within the region. For example, if the first serine in the selected region is encoded by AGU and the second serine in the region is encoded by AGC, swap the two codons. Continue swapping codons until the coding sequence for all the serine residues have been altered. This approach should be repeated for each amino acid within the targeted sequence until there are no long (>10 bp) stretches of homology between original and reencoded sequence at the nucleotide level. The final sequence can then be compared with the endogenous sequence using a web-based program such as Graphical Codon Usage Analyser (http://gcua.schoedl.de/) to verify that (1) the amino-acid sequence remains unchanged, (2) there are no stretches of conserved nucleotide sequence >10 bp, and (3) the overall codon bias remains the same. Any introns present in the targeted region should be replaced entirely with synthetic introns (see Section II.B.2, for example, synthetic intron sequences). The recoded region is then synthesized (usually by a commercial vendor) and introduced into the transgene in place of the original coding sequence via the flanking restriction sites.

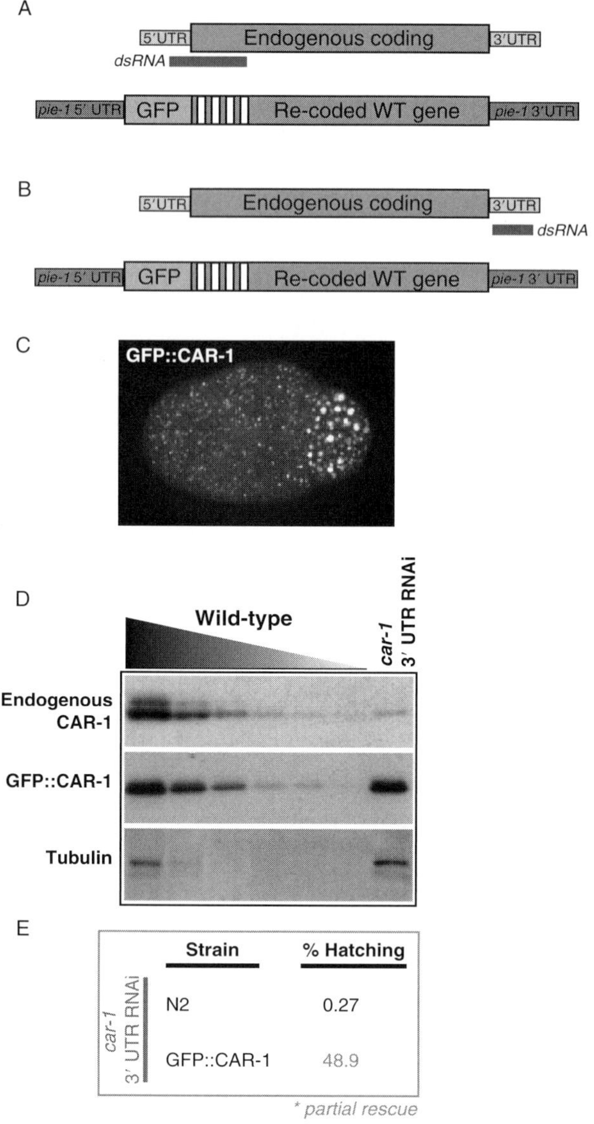

Fig. 8 Two RNAi strategies to selectively deplete the endogenous protein without affecting expressed tagged fusions. (A) The nucleotide sequence of part of the coding region in the introduced transgene is modified so that a dsRNA directed against part of the endogenous coding region can be used to selectively deplete the endogenous protein. (B) A dsRNA directed against the 3′ UTR of the endogenous gene selectively targets the endogenous mRNA because the transgene utilizes the *pie-1* 3′ UTR. (C–E) Using 3′ UTR RNAi to test functionality of GFP fusions. Panels C–E have been adapted from Audhya *et al.* (2005). (C) GFP::CAR-1, like endogenous CAR-1, localizes to cytoplasmic particles and P-granules. (D) Western blots of extracts from GFP::CAR-1 expressing worms injected with 3′ *car-1* UTR dsRNA, probed using anti-CAR-1 and anti-GFP antibodies reveal the selective depletion of the endogenous protein without alteration of the levels of the GFP fusion. Serial twofold dilutions of extracts prepared from untreated worms were loaded to quantify depletion. (E) Expression of GFP:: CAR-1 rescued embryonic viability following selective depletion of endogenous CAR-1. (See Plate no. 28 in the Color Plate Section.)

Once an RR transgene has been generated and a strain produced by integration, it is possible to use the RNA against the recoded region to selectively deplete the endogenous protein without altering levels of the transgenic protein, allowing the functionality of the transgenic protein to be tested. Once it is established that a fluorescent fusion with a coding region rescues function, it is also possible to compare the functionality of the control RR transgene with that of transgenes containing specific mutations of interest. A nice example of how this technique has been employed is presented in Ozlu *et al.* (2005) examining the function of TPXL-1 in targeting of AIR-1 to the mitotic spindle.

B. Available Worm Strains for Imaging in the Gonad and Early Embryo

A number of useful worm strains, carrying fluorescent transgenes, have been published and are available for general use. These strains include markers for studying various cellular components, including centrosomes, microtubules, kinetochores, chromosomes, actin cytoskeleton, microtubule-associated proteins, polarity determinants, and membrane-bound organelles. To view a comprehensive list of currently available strains (see Fig. 9).

C. Practical Techniques for Gonad/Embryo Imaging: Specimen Mounting and Drug Treatments

Mounting techniques for imaging the first embryonic division include *ex vivo* methods, either on agarose pads or in meiosis media, or *in utero* methods, via a worm whole mount. Each technique will be discussed briefly below.

1. Mounting Dissected Embryos for Imaging under Compression

Below we outline the standard method used to image living embryos. In this method, the embryos are dissected out of the mother and are mounted under compression on an agarose pad.

1. Make an agarose pad for mounting embryos by placing a slide between two slides that are each covered with two strips of laboratory tape, so that their tops are slightly higher than the slide in the middle. Put an M&M-sized drop of melted agarose (melted by placing an eppendorf tube containing 2% agarose in a heating block at 95 °C) on the center of the middle slide. After waiting for a second or two for the agarose to slightly cool, place a second slide crosswise over the agarose drop, so that it rests on the tape-covered slides. This should generate a circular pad of agarose on the center of the slide. After the agarose has cooled, pick up the agarose pad-slide sandwich and twist the two slides until you have one slide with a thin pad of agarose on it ready for embryo mounting. *Note:* The agarose breaks down over time at high temperatures reducing its usefulness for making pads, so put a new aliquot into the heating block every hour or so.

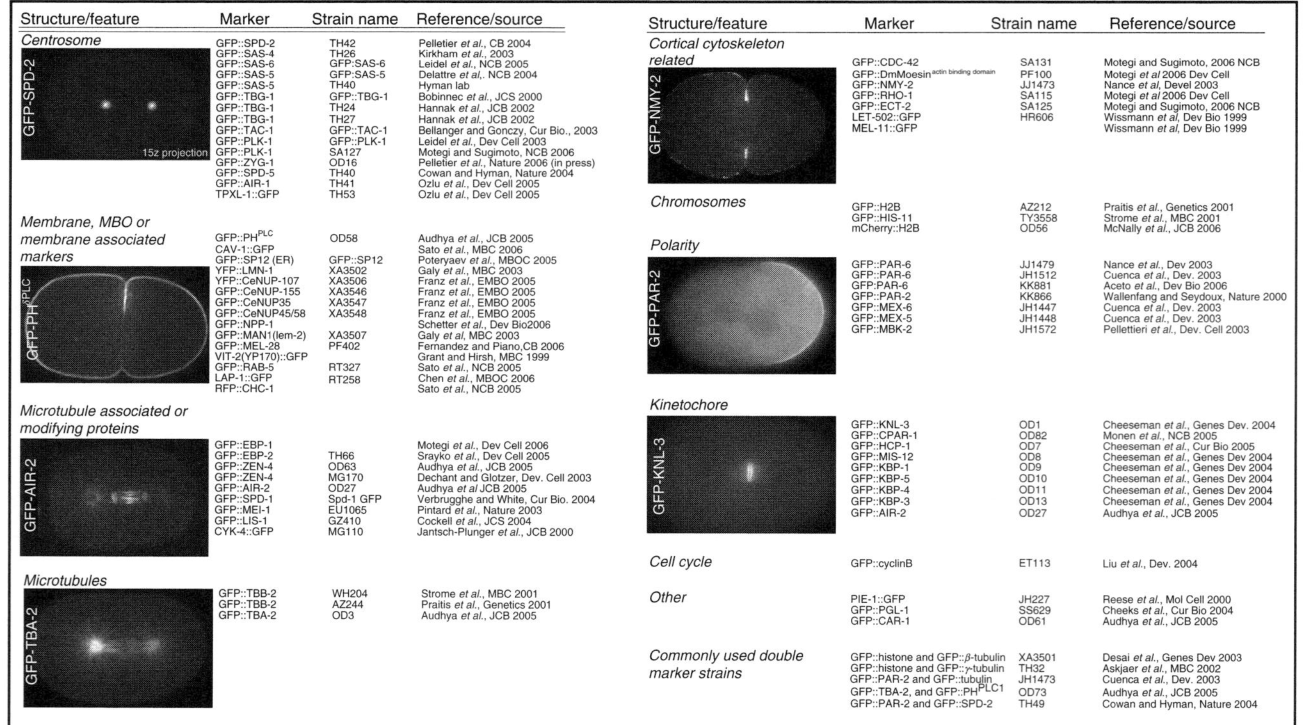

Structure/feature	Marker	Strain name	Reference/source
Centrosome	GFP::SPD-2	TH42	Pelletier *et al.*, CB 2004
	GFP::SAS-4	TH26	Kirkham *et al.*, 2003
	GFP::SAS-6	GFP:SAS-6	Leidel *et al.*, NCB 2005
	GFP::SAS-5	GFP:SAS-5	Delattre *et al*,. NCB 2004
	GFP::SAS-5	TH40	Hyman lab
	GFP::TBG-1	GFP::TBG-1	Bobinnec *et al.*, JCS 2000
	GFP::TBG-1	TH24	Hannak *et al.*, JCB 2002
	GFP::TBG-1	TH27	Hannak *et al.*, JCB 2002
	GFP::TAC-1	GFP::TAC-1	Bellanger and Gonczy, Cur Bio., 2003
	GFP::PLK-1	GFP::PLK-1	Leidel *et al.*, Dev Cell 2003
	GFP::PLK-1	SA127	Motegi and Sugimoto, NCB 2006
	GFP::ZYG-1	OD16	Pelletier *et al.*, Nature 2006 (in press)
	GFP::SPD-5	TH40	Cowan and Hyman, Nature 2004
	GFP::AIR-1	TH41	Ozlu *et al.*, Dev Cell 2005
	TPXL-1::GFP	TH53	Ozlu *et al.*, Dev Cell 2005
Membrane, MBO or membrane associated markers	GFP::PH^{PLC}	OD58	Audhya *et al.*, JCB 2005
	CAV-1::GFP		Sato *et al.*, MBC 2006
	GFP::SP12 (ER)	GFP::SP12	Poteryaev *et al.*, MBOC 2005
	YFP::LMN-1	XA3502	Galy *et al.*, MBC 2003
	YFP::CeNUP-107	XA3506	Franz *et al.*, EMBO 2005
	GFP::CeNUP-155	XA3546	Franz *et al.*, EMBO 2005
	GFP::CeNUP35	XA3547	Franz *et al.*, EMBO 2005
	GFP::CeNUP45/58	XA3548	Franz *et al.*, EMBO 2005
	GFP::NPP-1		Schetter *et al.*, Dev Bio2006
	GFP::MAN1(lem-2)	XA3507	Galy *et al*, MBC 2003
	GFP::MEL-28	PF402	Fernandez and Piano,CB 2006
	VIT-2(YP170)::GFP		Grant and Hirsh, MBC 1999
	GFP::RAB-5	RT327	Sato *et al.*, NCB 2005
	LAP-1::GFP	RT258	Chen *et al.*, MBOC 2006
	RFP::CHC-1		Sato *et al.*, NCB 2005
Microtubule associated or modifying proteins	GFP::EBP-1		Motegi *et al.*, Dev Cell 2006
	GFP::EBP-2	TH66	Srayko *et al.*, Dev Cell 2005
	GFP::ZEN-4	OD63	Audhya *et al.*, JCB 2005
	GFP::ZEN-4	MG170	Dechant and Glotzer, Dev. Cell 2003
	GFP::AIR-2	OD27	Audhya *et al* JCB 2005
	GFP::SPD-1	Spd-1 GFP	Verbrugghe and White, Cur Bio. 2004
	GFP::MEI-1	EU1065	Pintard *et al.*, Nature 2003
	GFP::LIS-1	GZ410	Cockell *et al.*, JCS 2004
	CYK-4::GFP	MG110	Jantsch-Plunger *et al.*, JCB 2000
Microtubules	GFP::TBB-2	WH204	Strome *et al.*, MBC 2001
	GFP::TBB-2	AZ244	Praitis *et al.*, Genetics 2001
	GFP::TBA-2	OD3	Audhya *et al.*, JCB 2005
Cortical cytoskeleton related	GFP::CDC-42	SA131	Motegi and Sugimoto, 2006 NCB
	GFP::DmMoesin$^{actin\ binding\ domain}$	PF100	Motegi *et al* 2006 Dev Cell
	GFP::NMY-2	JJ1473	Nance *et al*, Devel 2003
	GFP::RHO-1	SA115	Motegi *et al* 2006 Dev Cell
	GFP::ECT-2	SA125	Motegi and Sugimoto, 2006 NCB
	LET-502::GFP	HR606	Wissmann *et al*, Dev Bio 1999
	MEL-11::GFP		Wissmann *et al*, Dev Bio 1999
Chromosomes	GFP::H2B	AZ212	Praitis *et al.*, Genetics 2001
	GFP::HIS-11	TY3558	Strome *et al.*, MBC 2001
	mCherry::H2B	OD56	McNally *et al.*, JCB 2006
Polarity	GFP::PAR-6	JJ1479	Nance *et al.*, Dev 2003
	GFP::PAR-6	JH1512	Cuenca *et al.*, Dev. 2003
	GFP:PAR-6	KK881	Aceto *et al.*, Dev Bio 2006
	GFP::PAR-2	KK866	Wallenfang and Seydoux, Nature 2000
	GFP::MEX-6	JH1447	Cuenca *et al.*, Dev. 2003
	GFP::MEX-5	JH1448	Cuenca *et al.*, Dev. 2003
	GFP::MBK-2	JH1572	Pellettieri *et al.*, Dev. Cell 2003
Kinetochore	GFP::KNL-3	OD1	Cheeseman *et al.*, Genes Dev. 2004
	GFP::CPAR-1	OD82	Monen *et al.*, NCB 2005
	GFP::HCP-1	OD7	Cheeseman *et al.*, Cur Bio 2005
	GFP::MIS-12	OD8	Cheeseman *et al.*, Genes Dev 2004
	GFP::KBP-1	OD9	Cheeseman *et al.*, Genes Dev 2004
	GFP::KBP-5	OD10	Cheeseman *et al.*, Genes Dev 2004
	GFP::KBP-4	OD11	Cheeseman *et al.*, Genes Dev 2004
	GFP::KBP-3	OD13	Cheeseman *et al.*, Genes Dev 2004
	GFP::AIR-2	OD27	Audhya *et al.*, JCB 2005
Cell cycle	GFP::cyclinB	ET113	Liu *et al.*, Dev. 2004
Other	PIE-1::GFP	JH227	Reese *et al.*, Mol Cell 2000
	GFP::PGL-1	SS629	Cheeks *et al.*, Cur Bio 2004
	GFP::CAR-1	OD61	Audhya *et al.*, JCB 2005
Commonly used double marker strains	GFP::histone and GFP::β-tubulin	XA3501	Desai *et al.*, Genes Dev 2003
	GFP::histone and GFP::γ-tubulin	TH32	Askjaer *et al.*, MBC 2002
	GFP::PAR-2 and GFP::tubulin	JH1473	Cuenca *et al.*, Dev. 2003
	GFP::TBA-2, and GFP::PH^{PLC1}	OD73	Audhya *et al.*, JCB 2005
	GFP::PAR-2 and GFP::SPD-2	TH49	Cowan and Hyman, Nature 2004

Fig. 9 Published fluorescence strains for imaging in the gonad and early embryo. The fluorescence marker::gene, strain name (when available), and reference/source are listed for all currently *published* germline expression strains. Fluorescent strains are grouped according to the cellular features that they mark; these include centrosomes, membranes/membrane-bound organelles, microtubules, microtubule-associated proteins, cytoskeleton, chromosomes, polarity determinants, kinetochores, cell cycle regulators, and miscellaneous targets.

2. To obtain embryos, transfer an adult worm to a specimen watch glass (Electron Microscopy Sciences #71570–01) containing a small amount (~1–2 ml) of M9. Hold the worm at one end with fine forceps (Dumont-Dumostar #10570) and use a scalpel to cut the worm in half. Grab each half of the worm with the forceps and scrape the embryos out of the worm with the scalpel, like toothpaste from a tube. Using a scalpel with a small blade (Becton-Dickinson #371615) can be very helpful for this.

3. Pick up the embryos with a mouth pipette and transfer them to the agarose pad, trying to minimize the amount of liquid transferred. Mouth pipettes, with a diameter slightly larger than an embryo, are pulled from 25 μl capillaries over a Bunsen burner. The mouth pipette/capillary holders can be found in the capillary packages.

4. After transferring the embryos to the pad, use the mouth pipette to remove some of the excess liquid if necessary and then use an eyelash tool (an eyelash affixed to the end of a toothpick with glue or nail polish) to herd the desired embryos together (making them easier to find under the microscope).

5. Place an 18 mm × 18 mm coverslip over the embryos and transfer the slide to the microscope for imaging.

An alternative method that can be simpler for beginners is to place a worm in a small drop (2–3 μl of M9) on an 18 mm × 18 mm coverslip and use a pair of fine needles to dissect the embryos from the worm on the coverslip. The coverslip can then be inverted onto the agarose pad and the embryos imaged.

2. Imaging the Gonad in Anesthetized Worms

Below we describe our standard protocol for imaging the gonad in anesthetized worms. In some cases, the protocol can also be used to image the division of embryos "*in utero,*" which can be useful if the embryos are osmotically sensitive and lyse when imaged using the standard procedure described above.

To anesthetize the worms for *in utero* imaging:

1. Make up a fresh mixture of 1 mg/ml Tricaine (ethyl 3-aminobenzoate methanesulfonate salt) and 0.1 mg/ml of tetramisole hydrochloride in M9.

2. Place worms into a pool of anesthetic (a depression slide works well for this) for 15–30 min, or until worms stop moving.

3. Transfer the anesthetized worms to an agarose pad (prepared as described above) and carefully place a coverslip on top. The orientation of the anesthetized worm on the agarose pad is somewhat random, obscuring a clear view of the embryos in some cases. Therefore, it is recommended that several worms be anesthetized for each condition.

3. Mounting for Meiosis Imaging

Meiotic embryos do not tolerate compression and are osmotically sensitive, since the eggshell is not yet fully formed. Nevertheless, it can be convenient to image them *ex vivo* to obtain a high-resolution view of meiotic events, such as meiotic chromosome segregation, in the embryo. To do this, we use the alternative mounting technique described below.

1. First, assemble a meiosis filming chamber as follows:
   (a). Use two strips of double-sided tape to secure a 60 mm × 20 mm coverslip onto the top of an aluminum "slide" with a hole in the center. The aluminum slide is a holder that provides structural rigidity to the coverslip, while allowing imaging through the hole.
   (b). Apply a ring of vaseline on top of the coverslip, inside the area defined by the hole in the aluminum slide.
   (c). Pipet 8 μl of meiosis media into the center of the vaseline ring.
2. Place three adult worms into the meiosis media drop and dissect out the embryos.
3. Push worm debris to the edges of the drop, leaving the embryos centered.
4. Place an 18 mm × 18 mm coverslip on top of the drop. The vaseline should prevent the embryos from experiencing compression.
5. Place the slide on the microscope stage and image with the lens closest to the large coverslip (through the hole in the aluminum slide).

4. Drug Treatments

The eggshell of the early embryo serves as a barrier, making introduction of experimental drugs, such as nocodazole, difficult at this stage. While several laboratories have had some success introducing drugs into embryos by cracking the eggshell or incubating dissected embryos (with intact eggshells) in drugs (Encalada *et al.*, 2005; Kurz *et al.*, 2002; Stear and Roth, 2004; Strome and Wood, 1983), the reproducibility of these approaches can be variable. We have had success introducing both nocodazole and latrunculin A into early embryos by dissecting worms in meiosis media, supplemented with drug. Because there is a delay between fertilization and eggshell formation, newly fertilized embryos can still take up the drug until approximately the end of meiosis II. These embryos can be filmed under meiosis imaging conditions or on an agarose pad following drug exposure.

For introduction of drugs at later timepoints, we have had some success with a chitinase treatment protocol that maintains the eggshell in a permeable state. In this protocol, three worms are dissected in 8 μl egg salts (48 mM NaCl, 118 mM KCl) on a 24 mm × 60 mm coverslip on which a drop of 4 μl polylysine (1 mg/ml) has been dried in an oven for 10 min. The polylysine spot helps to immobilize the embryos through subsequent treatments. The coverslip is mounted with a ring of vaseline on an aluminum slide, as described above for meiosis filming. The buffer is removed with a mouth pipette and replaced with 8 μl of 9:1 ddH_2O:bleach by volume. After 2 min, the bleach solution is replaced with 8 μl egg salts buffer, followed sequentially by 8 μl L-15 blastomere culture medium (Edgar, 1995) and 8 μl chitinase (5 U/ml in L-15 blastomere culture medium). After 4 min, the chitinase solution is replaced with 8 μl L-15 blastomere culture medium. Drugs can be introduced by replacing this medium with 8 μl L-15 blastomere culture medium containing the drug.

D. Guidelines for Live Imaging of *C. elegans* Embryos

Once the specimens are prepared, the *C. elegans* gonad and early embryo are imaged in a fashion similar to other living specimens, and are subject to the same considerations. We remind experimenters to choose conditions to favor the health and viability of the embryo (temperature, illumination intensity) and to optimize imaging parameters (optics, filters, on-chip binning). Because general recommendations for live imaging are well described elsewhere, we will limit our discussion to *C. elegans*-specific considerations for live imaging.

The *C. elegans* gonad and early embryos have been imaged with great success using both wide-field and confocal microscopes. Ideally, the choice of imaging system should be tailored for each experiment, depending on the goal of the study. When imaging the cytoskeleton or the cell cortex, we recommend using a spinning disk confocal microscope. The spinning disk maximizes the number of confocal images that can be acquired before bleaching or phototoxicity become problematic (Audhya *et al.*, 2005; Maddox *et al.*, 2003). For chromosome or centrosome imaging, a standard wide-field microscope or a spinning disk confocal can be used to acquire a short *z*-series (5–6 planes, 2 μm intervals) through the middle region of the embryo. For fixed embryos analysis, deconvolution wide-field imaging at full resolution is the method of choice (Agard, 1984; Cheeseman *et al.*, 2004; Maddox *et al.*, 2005; Monen *et al.*, 2005). Other methods of imaging, such as multiphoton analysis, have also been successfully used to image *C. elegans* embryos (Strome *et al.*, 2001).

Filters used for fluorescence imaging should be closely matched to the peak wavelengths of the fluorophores being used to avoid the need for longer exposure conditions, which may be harmful to the specimen. For GFP-only imaging, we use a 488-nm band-pass filter and for dual GFP and mCherry imaging, we use 488/568 nm dual band-pass dichroic with individual band-pass excitation and emission filters mounted in filter wheels. For simultaneous imaging of two fluorescent proteins, we prefer GFP/mCherry to CFP/YFP since the excitation light for CFP is not well tolerated by *C. elegans* embryos.

For high-resolution live imaging of *C. elegans* embryos, 100× oil immersion (1.4 NA), 60× oil immersion (1.4 NA), and 60× water immersion (1.2 NA) objectives are most commonly used. Water immersion lenses can reduce spherical aberration, which becomes noticeable when imaging deep in ~20 μm thick embryos.

For maximal detail, the net magnification must be sufficient to capture full resolution onto the detector, which is typically a charge coupled detector camera. Magnification beyond this optimum reduces intensity without providing any added information. Typically, the theoretical resolution limit (~0.25 μm for green light) is split over three pixels on the camera (i.e., for a camera with 6.45 μm square pixels, 77× magnification is optimal). The 1.5× auxiliary magnification, built into the microscope (often referred to as an optivar), can be used to increase net magnification onto the detector by an additional 50%.

In practice, we rarely image living embryos at full resolution because many GFP transgenes are expressed at low levels and the signal is limiting. Instead, we sacrifice spatial resolution by using 2 × 2 on-chip binning to amplify the signal without increasing noise. Binning converts four adjacent pixels into a single large pixel (for a 6.45-μm square camera, after binning the total pixel number will be reduced fourfold and each pixel will be 12.9 μm^2); effectively, binning results in a fourfold increase in signal-to-noise and a twofold decrease in spatial resolution. Binning is essential for many of our imaging procedures because it allows the use of illumination and exposure conditions that permit long-term imaging of embryos without significant photodamage.

VI. Summary

The *C. elegans* germline has recently emerged as a powerful model system for studying meiosis and mitosis. The advantages of this system include reproducible RNAi-mediated protein depletion, stereotypical gonad morphology, and the rapid and reproducible progression of the first embryonic cell divisions. The advent of fluorescent proteins engineered for *C. elegans* expression and the means to avert the germline silencing of transgenes has made it possible to image fluorescent proteins in this system. In this chapter, we have outlined methods currently used to engineer fluorescent transgenes for germline expression, bombardment methods for the generation of stable lines, approaches to verify transgene function and expression, methods to generate multimarker lines, techniques for mounting worms and embryos, and basic imaging guidelines. In addition, we provide a list of currently available vectors and strains for live imaging analysis in the gonad and the early embryo.

Acknowledgments

We are grateful to Andy Fire for providing intron sequences and useful discussions, and members of the Oegema and Desai laboratories for sharing techniques and expertise on all subjects relating to this chapter. In addition, we are thankful to A. Maddox, I. Cheeseman, L. Lewellyn, S. Kline, and F. Motegi for images provided in figures and help with protocols described in this chapter. R.G. is an American Cancer Society postdoctoral fellow (PF-06–254–01-CCG). A.A. is a Helen Hay Whitney postdoctoral fellow. J.M. is supported by the University of California, San Diego Genetics Training Grant.

Work in the Oegema and Desai laboratories is supported by grants from the NIH (R01-GM074215 to A.D. and R01-GM074207–01 to K.O.). A.D. and K.O. also receive salary and other support from the Ludwig Institute for Cancer Research. K.O. is a Pew Scholar in the Biomedical Sciences. A. D. is a Damon Runyon Scholar supported by the Damon Runyon Cancer Research Foundation (DRS 38–04).

Appendix

Media and Supplements

1. *LB (2×)*

   100 g of LB
   1800 ml ddH_2O
   *Autoclave 35 min

2. *LB plates (1 liter)*

   500 ml ddH_2O
   15 g agar
   Add 450 ml 2× LB to the bottle
   *Autoclave 35 min

3. *Meiosis media*

   Stocks needed to make media:
   500 mg/100 ml inulin in culture safe H_2O (autoclaved)
   0.25 M HEPES pH 7.4 (5.95 g/100 ml)
   Leibowitz L-15 media
   Heat-inactivated FBS

   For 10 ml of meiosis media:
   6 ml of L-15 media
   1 ml of HEPES (0.25 M pH 7.4)
   1 ml inulin solution
   2 ml FBS

4. *M9 (2 liter)*

   10 g NaCl
   12 g Na_2HPO_4
   6 g KH_2PO_4
   0.5 g $MgSO_4 \cdot 7H_2O$
   Add water to 2 liter
   *Autoclave 35 min

5. *NGM agarose plates (1 liter)*

   3 g NaCl
   25 g agarose
   2.5 g peptone
   1.0 ml cholesterol (5 mg/ml in EtOH)
   975 ml ddH_2O
   *Autoclave 35 min and place in 55°C water bath
   *When cooled sterilely add:
   1.0 ml 1 M $CaCl_2$
   1.0 ml 1 M $MgSO_4$
   25 ml 1 M KH_2PO_4 (pH 6.0)

6. *Nystatin solution (500 ml)*

   5 g nystatin
   250 ml EtOH
   250 ml 7.5 M NH_4Ac
   *Filter sterilize and store at −20°C

7. *Peptone plates (1 liter)*

   1.2 g NaCl
   20 g bactopeptone
   25 g agar
   *Autoclave then sterilely add:
   1 ml 5 mg/ml cholesterol
   1 ml 1 M $MgSO_4$
   25 ml 1 M KH_2PO_4 pH 6.0

8. *Potassium citrate (1 M), pH 6.0*

   268.8 g tripotassium citrate
   26.3 g citric acid monohydrate
   Add water to 900 ml
   *Adjust pH to 6.0 using 10N KOH and bring up to 1 liter
   *Autoclave and store at room temperature

9. *S basal (per liter)*

   5.9 g NaCl
   50 ml of 1 M KH_2PO_4 pH 6.0
   1 ml of cholesterol (5 mg/ml in EtOH)
   *Autoclave and store at room temperature

10. *S basal (complete media):*

    To S basal 500 ml bottle add:
    5 ml 1 M potassium citrate pH 6.0
    5 ml trace metals solution
    1.5 ml 1 M $MgSO_4$
    1.5 ml 1 M $CaCl_2$

11. *Trace metals solution:*

Disodium EDTA	1.86 g	(5 mM)
$FeSO_4 \cdot 7H_2O$	0.69 g	(2.5 mM)
$MnCl_2 \cdot 4H_2O$	0.20 g	(1 mM)
$ZnSO_4 \cdot 7H_2O$	0.29 g	(1 mM)
$CuSO_4 \cdot 5H_2O$	0.025 g	(0.1 mM)

*Dissolve in 1 liter water; aliquot into 50 ml conical tubes and store in dark

Bombardment reagents

1. Bio-Rad 1.0 μm gold beads (Cat# 1652263)
2. Stopping screens Cat#1652336
3. Macrocarriers Cat# 9202964 120502
4. Rupture discs Cat# 1652330
5. 50% glycerol (filter sterilized)
6. pAZ-based plasmid for transforming (uncut) ~1 mg/ml (1 μl per bombardment)
7. 2.5 M $CaCl_2$ (sterile filtered)
8. 1.0 M spermidine (Sigma S-0266 in water, stored at−20°C)
9. 70% EtOH
10. 70% isopropanol
11. 100% EtOH
12. Sterile water

References

Aceto, D., Beers, M., and Kemphues, K. J. (2006). Interaction of PAR-6 with CDC-42 is required for maintenance but not establishment of PAR asymmetry in *C. elegans*. *Dev. Biol.* **299**(2), 386–397.

Agard, D. A. (1984). Optical sectioning microscopy: Cellular architecture in three dimensions. *Annu. Rev. Biophys. Bioeng.* **13,** 191–219.

Audhya, A., Hyndman, F., McLeod, I. X., Maddox, A. S., Yates, J. R., 3rd, Desai, A., and Oegema, K. (2005). A complex containing the Sm protein CAR-1 and the RNA helicase CGH-1 is required for embryonic cytokinesis in *Caenorhabditis elegans*. *J. Cell Biol.* **171,** 267–279.

Berezikov, E., Bargmann, C. I., and Plasterk, R. H. (2004). Homologous gene targeting in *Caenorhabditis elegans* by biolistic transformation. *Nucleic Acids Res.* **32**(4), e40.

Bobinnec, Y., Fukuda, M., and Nishida, E. (2000). Identification and characterization of *Caenorhabditis elegans* gamma-tubulin in dividing cells and differentiated tissues. *J. Cell Sci.* **113**(Pt 21), 3747–3759.

Broverman, S., MacMorris, M., and Blumenthal, T. (1993). Alteration of *Caenorhabditis elegans* gene expression by targeted transformation. *Proc. Natl. Acad. Sci. USA* **90**(10), 4359–4363.

Campbell, R. E., Tour, O., Palmer, A. E., Steinbach, P. A., Baird, G. S., Zacharias, D. A., and Tsien, R. Y. (2002). A monomeric red fluorescent protein. *Proc. Natl. Acad. Sci. USA* **99**(12), 7877–7882.

Cheeks, R. J., Canman, J. C., Gabriel, W. N., Meyer, N., Strome, S., and Goldstein, B. (2004). *C. elegans* PAR proteins function by mobilizing and stabilizing asymmetrically localized protein complexes. *Curr. Biol.* **14**(10), 851–862.

Cheeseman, I. M., and Desai, A. (2005). A combined approach for the localization and tandem affinity purification of protein complexes from metazoans. *Sci. STKE* **2005**(266), pl1.

Cheeseman, I. M., Niessen, S., Anderson, S., Hyndman, F., Yates, J. R., III, Oegema, K., and Desai, A. (2004). A conserved protein network controls assembly of the outer kinetochore and its ability to sustain tension. *Genes Dev.* **18**(18), 2255–2268.

Cockell, M. M., Baumer, K., and Gönczy, P. (2004). lis-1 is required for dynein-dependent cell division processes in *C. elegans* embryos. *J. Cell Sci.* **117**(Pt 19), 4571–4582.

Colombo, K., Grill, S. W., Kimple, R. J., Willard, F. S., Siderovski, D. P., and Gönczy, P. (2003). Translation of polarity cues into asymmetric spindle positioning in *Caenorhabditis elegans* embryos. *Science* **300**(5627), 1957–1961.

Cowan, C. R., and Hyman, A. A. (2004). Centrosomes direct cell polarity independently of microtubule assembly in *C. elegans* embryos. *Nature* **431**(7004), 92–96.

Cubitt, A. B., Heim, R., Adams, S. R., Boyd, A. E., Gross, L. A., and Tsien, R. Y. (1995). Understanding, improving and using green fluorescent proteins. *Trends Biochem. Sci.* **20**(11), 448–455.

Cuenca, A. A., Schetter, A., Aceto, D., Kemphues, K., and Seydoux, G. (2003). Polarization of the *C. elegans* zygote proceeds via distinct establishment and maintenance phases. *Development* **130**(7), 1255–1265.

Dechant, R., and Glotzer, M. (2003). Centrosome separation and central spindle assembly act in redundant pathways that regulate microtubule density and trigger cleavage furrow formation. *Dev. Cell* **4**(3), 333–344.

Duret, L., and Mouchiroud, D. (1999). Expression pattern and, surprisingly, gene length shape codon usage in *Caenorhabditis, Drosophila,* and *Arabidopsis. Proc. Natl. Acad. Sci. USA* **96**(8), 4482–4487.

Edgar, L. G. (1995). Blastomere culture and analysis. *Methods Cell Biol.* **48,** 303–321.

Encalada, S. E., Martin, P. R., Phillips, J. B., Lyczak, R., Hamill, D. R., Swan, K. A., and Bowerman, B. (2000). DNA replication defects delay cell division and disrupt cell polarity in early *Caenorhabditis elegans* embryos. *Dev. Biol.* **228**(2), 225–238.

Encalada, S. E., Willis, J., Lyczak, R., and Bowerman, B. (2005). A spindle checkpoint functions during mitosis in the early *Caenorhabditis elegans* embryo. *Mol. Biol. Cell* **16**(3), 1056–1070.

Fernandez, A. G., Gunsalus, K. C., Huang, J., Chuang, L. S., Ying, N., Liang, H. L., Tang, C., Schetter, A. J., Zegar, C., Rual, J. F., Hill, D. E., Reinke, V., *et al.* (2005). New genes with roles in the *C. elegans* embryo revealed using RNAi of ovary-enriched ORFeome clones. *Genome Res.* **15**(2), 250–259.

Fernandez, A. G., and Piano, F. (2006). MEL-28 is downstream of the Ran cycle and is required for nuclear-envelope function and chromatin maintenance. *Curr. Biol.* **16**(17), 1757–1763.

Fire, A., Ahnn, J., Kelly, W. G., Harfe, B., Kostas, S. A., Hsieh, J., Hsu, M., and Xu, S. (1998). GFP applications in *C. elegans. In* "GFP Strategies and Applications" (M. Chalfie and S. Kain, eds.), pp. 153–168. John Wiley and Sons, New York.

Fire, A., Alcazar, R., and Tan, F. (2006). Unusual DNA structures associated with germline genetic activity in *Caenorhabditis elegans. Genetics* **173**(3), 1259–1273.

Fire, A., Harrison, S. W., and Dixon, D. (1990). A modular set of lacZ fusion vectors for studying gene expression in *Caenorhabditis elegans*. *Gene* **93**(2), 189–198.

Franz, C., Askjaer, P., Antonin, W., Iglesias, C. L., Haselmann, U., Schelder, M., de Marco, A., Wilm, M., Antony, C., and Mattaj, I. W. (2005). Nup155 regulates nuclear envelope and nuclear pore complex formation in nematodes and vertebrates. *EMBO J.* **24**(20), 3519–3531.

Fraser, A. G., Kamath, R. S., Zipperlen, P., Martinez-Campos, M., Sohrmann, M., and Ahringer, J. (2000). Functional genomic analysis of *C. elegans* chromosome I by systematic RNA interference. *Nature* **408**(6810), 325–330.

Galy, V., Mattaj, I. W., and Askjaer, P. (2003). *Caenorhabditis elegans* nucleoporins Nup93 and Nup205 determine the limit of nuclear pore complex size exclusion in vivo. *Mol. Biol. Cell* **14**(12), 5104–5115.

Golden, A., Sadler, P. L., Wallenfang, M. R., Schumacher, J. M., Hamill, D. R., Bates, G., Bowerman, B., Seydoux, G., and Shakes, D. C. (2000). Metaphase to anaphase (mat) transition-defective mutants in *Caenorhabditis elegans*. *J. Cell Biol.* **151**(7), 1469–1482.

Gonczy, P., Echeverri, C., Oegema, K., Coulson, A., Jones, S. J., Copley, R. R., Duperon, J., Oegema, J., Brehm, M., Cassin, E., Hannak, E., Kirkham, M., *et al.* (2000). Functional genomic analysis of cell division in *C. elegans* using RNAi of genes on chromosome III. *Nature* **408**(6810), 331–336.

Gonczy, P., Pichler, S., Kirkham, M., and Hyman, A. A. (1999). Cytoplasmic dynein is required for distinct aspects of MTOC positioning, including centrosome separation, in the one cell stage *Caenorhabditis elegans* embryo. *J. Cell Biol.* **147**(1), 135–150.

Grant, B., and Hirsh, D. (1999). Receptor-mediated endocytosis in the *Caenorhabditis elegans* oocyte. *Mol. Biol. Cell* **10**(12), 4311–4326.

Hannak, E., Oegema, K., Kirkham, M., Gönczy, P., Habermann, B., and Hyman, A. A. (2002). The kinetically dominant assembly pathway for centrosomal asters in *Caenorhabditis elegans* is gamma-tubulin dependent. *J. Cell Biol.* **157**(4), 591–602.

Heim, R., Prasher, D. C., and Tsien, R. Y. (1994). Wavelength mutations and posttranslational autoxidation of green fluorescent protein. *Proc. Natl. Acad. Sci. USA* **91**(26), 12501–12504.

Heim, R., and Tsien, R. Y. (1996). Engineering green fluorescent protein for improved brightness, longer wavelengths and fluorescence resonance energy transfer. *Curr. Biol.* **6**(2), 178–182.

Hird, S. N., and White, J. G. (1993). Cortical and cytoplasmic flow polarity in early embryonic cells of *Caenorhabditis elegans*. *J. Cell Biol.* **121**(6), 1343–1355.

Hubbard, E. J., and Greenstein, D. (2000). The *Caenorhabditis elegans* gonad: A test tube for cell and developmental biology. *Dev. Dyn.* **218**(1), 2–22.

Hyman, A. A., and Oegema, K. (2005). Cell division. *In* "WormBook." The *C. elegans* Research Community, WormBook.http://www.wormbook.org.

Jackstadt, P., Wilm, T. P., Zahner, H., and Hobom, G. (1999). Transformation of nematodes via ballistic DNA transfer. *Mol. Biochem. Parasitol.* **103**(2), 261–266.

Jantsch-Plunger, V., Gonczy, P., Romano, A.,, Schnabel, H., Hamill, D., Schnabel, R., Hyman, A. A., and Glotzer, M. (2000). CYK-4: A Rho family gtpase activating protein (GAP) required fo central spindle formation and cytokinesis. *J. Cell Biol.* **149**(7), 1391–1404.

Kamath, R. S., Fraser, A. G., Dong, Y., Poulin, G., Durbin, R., Gotta, M., Kanapin, A., Le Bot, N., Moreno, S., Sohrmann, M., Welchman, D. P., Zipperlen, P., *et al.* (2003). Systematic functional analysis of the *Caenorhabditis elegans* genome using RNAi. *Nature* **421**(6920), 231–237.

Kelly, W. G., and Fire, A. (1998). Chromatin silencing and the maintenance of a functional germline in *Caenorhabditis elegans*. *Development* **125**(13), 2451–2456.

Kelly, W. G., Xu, S., Montgomery, M. K., and Fire, A. (1997). Distinct requirements for somatic and germline expression of a generally expressed *Caernorhabditis elegans* gene. *Genetics* **146**(1), 227–238.

Kemphues, K. J., Priess, J. R., Morton, D. G., and Cheng, N. S. (1988). Identification of genes required for cytoplasmic localization in early *C. elegans* embryos. *Cell* **52**(3), 311–320.

Kimble, J., and Strome, S. (2005). The germline. *In* "WormBook." The *C. elegans* Research Community, WormBook.http://www.wormbook.org.

Kirkham, M., Muller-Reichert, T., Oegema, K., Grill, S., and Hyman, A. A. (2003). SAS-4 is a *C. elegans* centriolar protein that controls centrosome size. *Cell* **112**(4), 575–587.

Kurz, T., Pintard, L., Willis, J. H., Hamill, D. R., Gönczy, P., Peter, M., and Bowerman, B. (2002). Cytoskeletal regulation by the Nedd8 ubiquitin-like protein modification pathway. *Science* **295** (5558), 1294–1298.

Labbe, J. C., McCarthy, E. K., and Goldstein, B. (2004). The forces that position a mitotic spindle asymmetrically are tethered until after the time of spindle assembly. *J. Cell Biol.* **167**(2), 245–256.

LaMunyon, C. W., and Ward, S. (1995). Sperm precedence in a hermaphroditic nematode (*Caenorhabditis elegans*) is due to competitive superiority of male sperm. *Experientia* **51**(8), 817–823.

Leidel, S., Delattre, M., Cerutti, L., Baumer, K., and Gönczy, P. (2005). SAS-6 defines a protein family required for centrosome duplication in *C. elegans* and in human cells. *Nat. Cell Biol.* **7**(2), 115–125.

Liu, J., Vasudevan, S., and Kipreos, E. T. (2004). CUL-2 and ZYG-11 promote meiotic anaphase II and the proper placement of the anterior-posterior axis in *C. elegans*. *Development* **131**(15), 3513–3525.

Maddox, A. S., Habermann, B., Desai, A., and Oegema, K. (2005). Distinct roles for two *C. elegans* anillins in the gonad and early embryo. *Development* **132**(12), 2837–2848.

Maddox, P. S., Moree, B., Canman, J. C., and Salmon, E. D. (2003). Spinning disk confocal microscope system for rapid high-resolution, multimode, fluorescence speckle microscopy and green fluorescent protein imaging in living cells. *Methods Enzymol.* **360,** 597–617.

Maddox, P. S., Portier, N., Desai, A., and Oegema, K. (2006). Molecular analysis of mitotic chromosome condensation using a quantitative time-resolved fluorescence microscopy assay. *Proc. Natl. Acad. Sci. USA* **103**(41), 15097–15102.

Maduro, M., and Pilgrim, D. (1995). Identification and cloning of unc-119, a gene expressed in the *Caenorhabditis elegans* nervous system. *Genetics* **141**(3), 977–988.

Maeda, I., Kohara, Y., Yamamoto, M., and Sugimoto, A. (2001). Large-scale analysis of gene function in *Caenorhabditis elegans* by high-throughput RNAi. *Curr. Biol.* **11**(3), 171–176.

Matz, M. V., Fradkov, A. F., Labas, Y. A., Savitsky, A. P., Zaraisky, A. G., Markelov, M. L., and Lukyanov, S. A. (1999). Fluorescent proteins from nonbioluminescent *Anthozoa* species. *Nat. Biotechnol.* **17**(10), 969–973.

McNally, K., Audhya, A., Oegema, K., and McNally, F. J. (2006). Katanin controls mitotic and meiotic spindle length. *J. Cell Biol.* **175**(6), 881–891.

Miller, D. M., III, Desai, N. S., Hardin, D. C., Piston, D. W., Patterson, G. H., Fleenor, J., Xu, S., and Fire, A. (1999). Two-color GFP expression system for *C. elegans*. *Biotechniques* **26**(5), 914–918, 920–921.

Monen, J., Maddox, P. S., Hyndman, F., Oegema, K., and Desai, A. (2005). Differential role of CENP-A in the segregation of holocentric *C. elegans* chromosomes during meiosis and mitosis. *Nat. Cell Biol.* **7**(12), 1248–1255.

Motegi, F., and Sugimoto, A. (2006). Sequential functioning of the ECT-2 RhoGEF, RHO-1 and CDC-42 establishes cell polarity in *Caenorhabditis elegans* embryos. *Nat. Cell Biol.* **8**(9), 978–985.

Munro, E., Nance, J., and Priess, J. R. (2004). Cortical flows powered by asymmetrical contraction transport PAR proteins to establish and maintain anterior-posterior polarity in the early *C. elegans* embryo. *Dev. Cell* **7**(3), 413–424.

Nance, J., Munro, E. M., and Priess, J. R. (2003). *C. elegans* PAR-3 and PAR-6 are required for apicobasal asymmetries associated with cell adhesion and gastrulation. *Development* **130**(22), 5339–5350.

O'Connell, K. F., Leys, C. M., and White, J. G. (1998). A genetic screen for temperature-sensitive cell-division mutants of *Caenorhabditis elegans*. *Genetics* **149**(3), 1303–1321.

O'Connell, K. F., Maxwell, K. N., and White, J. G. (2000). The spd-2 gene is required for polarization of the anteroposterior axis and formation of the sperm asters in the *Caenorhabditis elegans* zygote. *Dev. Biol.* **222**(1), 55–70.

Oegema, K., and Hyman, A. A. (2005). Cell Division. *In* "WormBook." The *C.elegans* Research Community, WormBook. http://www.wormbook.org.

Ozlu, N., Srayko, M., Kinoshita, K., Habermann, B., O'toole, E. T., Müller-Reichert, T., Schmalz, N., Desai, A., and Hyman, A. A. (2005). An essential function of the *C. elegans* ortholog of TPX2 is to localize activated aurora A kinase to mitotic spindles. *Dev. Cell* **9**(2), 237–248.

Pelletier, L., Ozlu, N., Hannak, E., Cowan, C., Habermann, B., Ruer, M., Müller-Reichert, T., and Hyman, A. A. (2004). The *Caenorhabditis elegans* centrosomal protein SPD-2 is required for both pericentriolar material recruitment and centriole duplication. *Curr. Biol.* **14**(10), 863–873.

Pellettieri, J., Reinke, V., Kim, S. K., and Seydoux, G. (2003). Coordinate activation of maternal protein degradation during the egg-to-embryo transition in *C. elegans*. *Dev. Cell* **5**(3), 451–462.

Piano, F., Schetter, A. J., Mangone, M., Stein, L. D., and Kemphues, K. J. (2000). RNAi analysis of genes expressed in the ovary of *Caenorhabditis elegans*. *Curr. Biol.* **10**(24), 1619–1622.

Pintard, L., Willis, J. H., Willems, A., Johnson, J. L., Srayko, M., Kurz, T., Glaser, S., Mains, P. E., Tyers, M., Bowerman, B., and Peter, M. (2003). The, BTB protein MEL-26 is a substrate-specific adaptor of the CUL-3 ubiquitin-ligase. *Nature* **425**(6955), 311–316.

Poteryaev, D., Squirrell, J. M., Campbell, J. M., White, J. G., and Spang, A. (2005). Involvement of the actin cytoskeleton and homotypic membrane fusion in ER dynamics in *Caenorhabditis elegans*. *Mol. Biol. Cell* **16**(5), 2139–2153.

Praitis, V., Casey, E., Collar, D., and Austin, J. A. (2001). Creation of low-copy integrated transgenic lines in *Caenorhabditis elegans*. *Genetics* **157**(3), 1217–1226.

Praitis, V., Katz, W. S., and Solomon, F. (1991). A codon change in beta-tubulin which drastically affects microtubule structure in *Drosophila melanogaster* fails to produce a significant phenotype in *Saccharomyces cerevisiae*. *Mol. Cell Biol.* **11**(9), 4726–4731.

Rual, J. F., Ceron, J., Koreth, J., Hao, T., Nicot, A. S., Hirozane-Kishikawa, T., Vandenhaute, J., Orkin, S. H., Hill, D. E., van den Heuvel, S., and Vidal, M. (2004). Toward improving *Caenorhabditis elegans* phenome mapping with an ORFeome-based RNAi library. *Genome Res.* **14**(10B), 2162–2168.

Sato, K., Sato, M., Audhya, A., Oegema, K., Schweinsberg, P., and Grant, B. D. (2006). Dynamic regulation of caveolin-1 trafficking in the germline and embryo of *Caenorhabditis elegans*. *Mol. Biol. Cell* **17**(7), 3085–3094.

Sato, M., Sato, K., Fonarev, P., Huang, C.-J., Liou, W., and Grant, B. D. (2005). *Caenorhabditis elegans* RME-6 is a novel regulator of RAB-5 at the clathrin-coated pit. *Nat. Cell Biol.* **7**(6), 559–569.

Schetter, A., Askjaer, P., Piano, F., Mattaj, I., and Kemphues, K. (2006). Nucleoporins NPP-1, NPP-3, NPP-4, NPP-11 and NPP-13 are required for proper spindle orientation in *C. elegans*. *Dev. Biol.* **289**(2), 360–371.

Seydoux, G., and Strome, S. (1999). Launching the germline in *Caenorhabditis elegans*: Regulation of gene expression in early germ cells. *Development* **126**(15), 3275–3283.

Shaner, N. C., Campbell, R. E., Steinbach, P. A., Giepmans, B. N. G., Palmer, A. E., and Tsien, R. Y. (2004). Improved monomeric red, orange and yellow fluorescent proteins derived from *Discosoma sp.* red fluorescent protein. *Nat. Biotechnol.* **22**(12), 1567–1572.

Shaner, N. C., Steinbach, P. A., and Tsien, R. Y. (2005). A guide to choosing fluorescent proteins. *Nat. Methods* **2**(12), 905–909.

Simmer, F., Moorman, C., van der Linden, A. M., Kuijk, E., van den Berghe, P. V., Kamath, R. S., Fraser, A. G., Ahringer, J., and Plasterk, R. H. (2003). Genome-wide RNAi of *C. elegans* using the hypersensitive rrf-3 strain reveals novel gene functions. *PLoS Biol.* **1**(1), E12.

Smith, D., and Yarus, M. (1989). Transfer RNA structure and coding specificity. I. Evidence that a D-arm mutation reduces tRNA dissociation from the ribosome. *J. Mol. Biol.* **206**(3), 489–501.

Sonnichsen, B., Koski, L. B., Walsh, A., Marschall, P., Neumann, B., Brehm, M., Alleaume, A. M., Artelt, J., Bettencourt, P., Cassin, E., Hewitson, M., Holz, C., *et al.* (2005). Full-genome RNAi profiling of early embryogenesis in *Caenorhabditis elegans*. *Nature* **434**(7032), 462–469.

Stear, J. H., and Roth, M. B. (2004). The *Caenorhabditis elegans* kinetochore reorganizes at prometaphase and in response to checkpoint stimuli. *Mol. Biol. Cell* **15**(11), 5187–5196.

Strome, S., Powers, J., Dunn, M., Reese, K., Malone, C. J., Seydoux, G., Saxton, W., and White, J. G. (2001). Spindle dynamics and the role of gamma-tubulin in early *Caenorhabditis elegans* embryos. *Mol. Biol. Cell* **12**(6), 1751–1764.

Strome, S., and Wood, W. B. (1983). Generation of asymmetry and segregation of germ-line granules in early *C. elegans* embryos. *Cell* **35**(1), 15–25.

Tsou, M. F., Hayashi, A., DeBella, L. R., McGrath, G., and Rose, L. S. (2002). LET-99 determines spindle position and is asymmetrically enriched in response to PAR polarity cues in *C. elegans* embryos. *Development* **129**(19), 4469–4481.

Verbrugghe, K. J., and White, J. G. (2004). SPD-1 is required for the formation of the spindle midzone but is not essential for the completion of cytokinesis in *C. elegans* embryos. *Curr. Biol.* **14**(19), 1755–1760.

Wallenfang, M. R., and Seydoux, G. (2000). Polarization of the anterior-posterior axis of *C. elegans* is a microtubule-directed process. *Nature* **408**(6808), 89–92.

Ward, S., and Carrel, J. S. (1979). Fertilization and sperm competition in the nematode *Caenorhabditis elegans*. *Dev. Biol.* **73**(2), 304–321.

Wilm, T., Demel, P., Koop, H. U., Schnabel, H., and Schnabel, R. (1999). Ballistic transformation of *Caenorhabditis elegans*. *Gene* **229**(1–2), 31–35.

Wissmann, A., Ingles, J., and Mains, P. E. (1999). The *Caenorhabditis elegans* mel-11 myosin phosphatase regulatory subunit affects tissue contraction in the somatic gonad and the embryonic epidermis and genetically interacts with the Rac signaling pathway. *Dev. Biol.* **209**(1), 111–127.

CHAPTER 10

Fluorescent Proteins in Zebrafish Cell and Developmental Biology[1]

H. William Detrich, III

Department of Biology
Northeastern University
Boston, Massachusetts 02115

[1] Throughout this chapter genes and proteins are designated according to the nomenclature system of the relevant organism: zebrafish, human, or mouse. Fluorescent protein genes/cDNAs and their encoded proteins are indicated in capital letters.

0091-679X/08 $35.00
DOI: 10.1016/S0091-679X(08)85010-8

Abstract

The zebrafish is a compelling vertebrate model for understanding cellular processes in the context of the developing embryo and for analysis of cellular defects that lead to diseases such as cancer. Major advances in fluorescent protein technology have been, and will continue to be, combined with novel experimental strategies to explore these biological phenomena. Furthermore, fluorescent proteins can be used in the design of forward genetic and chemical modifier screens of ever increasing sophistication. Here I review three noteworthy applications of fluorescent proteins in zebrafish: (1) analysis of kinesin motor function in the cleaving zebrafish embryo, (2) determination of the roles of semaphorins in axonal guidance, and (3) creation of transgenic models of leukemia and other cancers.

I. Introduction

The introduction of fluorescent protein technology to the zebrafish model system has led to remarkable advances in our understanding of diverse cellular and developmental phenomena. The functions of members of the kinesin superfamily of microtubule motors (Lawrence *et al.*, 2004) in early developmental processes are being examined by injection of zebrafish embryos with expression constructs that encode the green fluorescent protein (GFP)-tagged motors (Chen and Detrich, 1999; Chen *et al.*, 2002). GFP has been used as a reporter transgene for dissection of lineage-specific *cis*-acting gene regulatory elements in an organismal context (Meng *et al.*, 1999a,b). The differentiation and movement of hematopoietic lineages, including erythroid, thrombocytic, myeloid, and lymphoid cells (Berman *et al.*, 2003; Hsu *et al.*, 2004; Traver *et al.*, 2003; Trede *et al.*, 2004), have been tracked using GFP. The zebrafish is well known as a system for study of neural development (Chapouton and Bally-Cuif, 2004), and fluorescent proteins have been used to monitor retinal neurogenesis (Avanesov and Malicki, 2004), retinotectal axon guidance (Bak and Fraser, 2003), and cerebellar regeneration (Koster and Fraser, 2006) by time-lapse microscopy (Koster and Fraser, 2004). Transgenic zebrafish expressing GFP in discrete organs or organ systems are contributing to our understanding of cardiac (Bartman *et al.*, 2004), liver (Field *et al.*, 2003b), and pancreatic development (Field *et al.*, 2003a). Similarly, the role of the neural crest in craniofacial development is being pursued using GFP

(Wada *et al.*, 2005). Laser ablation of neurons that express GFP under the control of the immediate early *c-fos* promoter is under development as a global approach to mapping neural activity to specific behaviors (O'Malley *et al.*, 2004). The efficiency of transgenesis in zebrafish mediated by the *Sleeping Beauty* transposon (Davidson *et al.*, 2003; Hermanson *et al.*, 2004), meganuclease I-SceI (Grabher *et al.*, 2004; Thermes *et al.*, 2002), *Tol2* transposable element (Kawakami, 2004; Urasaki *et al.*, 2006), and *Cre* recombinase (Dong and Stuart, 2004) has been evaluated using GFP and its color variants as reporters. As this admittedly partial summary indicates, fluorescent proteins have almost unlimited potential to illuminate biological processes in the zebrafish, a compelling vertebrate genetic and developmental system whose advantages complement those of the mouse (Orkin and Zon, 1997).

The goal of this chapter is to review several novel and noteworthy applications of fluorescent proteins in zebrafish and to emphasize the synergy between these molecular tags and the zebrafish model. The features of the zebrafish most obviously valuable to cell biologists include the ease of generating large clutches of embryos, the optical transparency of the early embryo, and the formation of most tissue and organ anlage within 24–36 hours postfertilization (hpf) (Detrich *et al.*, 1999a). Many cellular processes studied using fluorescent proteins in cell culture can now be examined in the developing vertebrate embryo. Furthermore, the availability of numerous forward and reverse genetic technologies and the thousands of mutants derived from general and targeted genetic screens provide the cell biologist with a treasure trove of cellular research opportunities (Detrich *et al.*, 1999b,c, 2004a,b). I encourage the cell biological community to embrace the zebrafish as a metazoan model that complements the fly, the worm, and the mouse.

II. Zebrafish Kinesin Genes in Early Development: A Cytokinetic Role for zMklp1

The cytoplasmic microtubules of metazoan embryos and their associated motors play critical roles in cleavage (Chen *et al.*, 2002; Raich *et al.*, 1998), in the localization of the developmental determinants that specify the embryonic axes (Brendza *et al.*, 2000; Januschke *et al.*, 2002), and in segregation of the germ plasm (Quaas and Wylie, 2002; van Eeden *et al.*, 2001). Recent work demonstrates that assembly of the central spindle and regulation of cytokinesis entail cell cycle control (Mishima *et al.*, 2004) of several kinesin proteins (Neef *et al.*, 2006) via intermediary kinases (Gruneberg *et al.*, 2004; Guse *et al.*, 2005). Relatively little of this work has been performed in developing embryos, in which the activities of individual cells must be integrated within their tissue and organ environments. Furthermore, very few studies (Chen *et al.*, 2002; Minestrini *et al.*, 2003) have exploited the advantage of fluorescent protein/kinesin fusions to image motor function *in vivo* and to localize kinesins in postfixation specimens. Here I outline the strategy that my laboratory has employed to study the role of Mklp1, a member of the Kinesin-6 family [for the newly standardized kinesin nomenclature, see Lawrence *et al.* (2004)], in cleaving zebrafish embryos.

A. Cloning and Sequence Analysis of Zebrafish Kinesin cDNAs

The motor domains of kinesins contain two highly conserved peptides, IFAYGQT and DLAGSE (Goldstein, 1993). Applying degenerate primers [one forward, four reverse (Fig. 1)] deduced from these sequences (Aizawa *et al.*, 1992), we isolated zebrafish kinesin motor-domain gene fragments of ~450 bp by reverse transcription polymerase chain reaction (PCR) using total RNA from 50 shield-stage (6-h) gastrulae. Standard precautions for RNA handling and PCR amplification (gloves, DEPC-treated H_2O, prevention of aerosols by use of stuffed pipette tips, separation of pre- and post-PCR work areas) are essential. We also found that magnesium ion concentration had to be optimized for each primer pair: ZKF paired with ZKR1, 2, or 3 required 1 mM $MgCl_2$, whereas 2.25 mM was necessary for ZKF and ZKR4. The *Taq* polymerase-amplified products were cloned into pUC19 using the T-vector system of Marchuk *et al.* (1991); the pGEM-T Easy Vector kit (Promega, WI) or equivalent can be substituted. The short cDNA inserts are readily sequenced by automated methods, now typically available through sequencing centers. Using the NCBI BLAST sequence algorithms, we

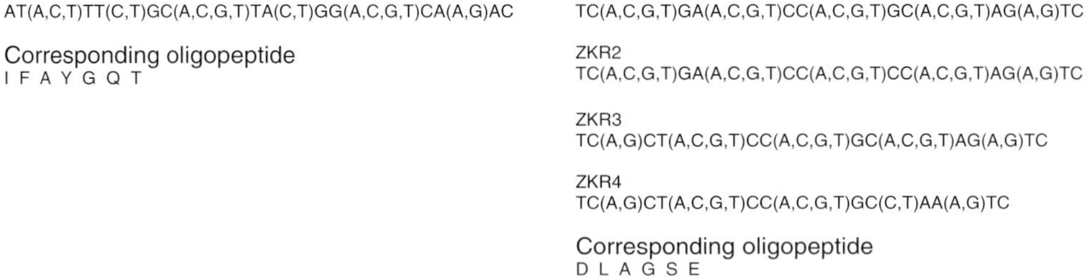

Fig. 1 Polymerase chain reaction (PCR) amplification strategy for zebrafish kinesin gene fragments. Beneath the schematic structure of an N-type kinesin (motor domain at the N-terminus) are shown degenerate sense (forward) and antisense (reverse) primers that correspond to two highly conserved motor-domain oligopeptides. Reverse transcription PCR of total gastrula RNA should yield motor-domain fragments of ~450 bp. Reprinted from Chen and Detrich (1999) with permission. Copyright (1999) Elsevier.

determined that we obtained cDNAs for KHC (Kinesin-1 family), KRP85/95 and KIF3 (Kinesin-2 members), Unc104/KIF1A (Kinesin-3), Eg5/BimC (Kinesin-5), Mklp1 (Kinesin-6), and Ncd (Kinesin-14).

With kinesin probes in hand, full-length cDNAs can be cloned from an appropriate cDNA library. The oligo-dT-primed zebrafish kidney cDNA library in Lambda ZAP Express described by Liao *et al.* (1998) yields a high percentage of full-length cDNAs. The remainder of Section II will be developed using our work with zebrafish Mklp1 (zMklp1) as an example (Chen *et al.*, 2002).

B. Engineering of Expression Constructs That Encode GFP-Tagged Wild-Type and Mutant zMklp1s

A major consideration in the design of a GFP-tagged kinesin is the availability of its N- and/or C-termini on the surface of the motor. Based on structural analyses, most kinesins with the motor located at the N-terminus (N-type motors, kinesins 1–8, 11) (Lawrence *et al.*, 2004; Miki *et al.*, 2005) appear to have accessible N-termini (Marx *et al.*, 2005) and should be amenable to tagging. However, the availability of the N- and/or C-termini of kinesins with motor domains located internally or at the C-terminus [I- (or M-) and C-type motors; kinesins 13 and 14, respectively] is unclear, and their "tagability" would have to be determined experimentally.

To examine zMklp1 function prior to the onset of zygotic transcription at the midblastula transition, we developed expression construct in which the *GFP* coding sequence was fused in frame to the 5′ end of the zMklp1 cDNA. *pGFP-Mklp1* encodes GFP fused to the N-terminus of zMklp1 via an 18-residue spacer. Dominant-negative variants were generated by deletion of portions of the Mklp1 coding sequence [e.g., *pGFP-Mklp1(ΔN1–275)*, which deletes most of the motor domain] or by point mutation of the ATP-binding site [*pGFP-Mklp1(T119N)*; wild-type residue/position/mutant residue]. Complete details may be found at http://www.biology.neu.edu/documentation_mklp1.pdf.

To determine whether the GFP tag altered the behavior of zMklp1, we transfected *pGFP-Mklp1* into COS-7 cells for 12–24 h and processed the cells for indirect immunofluorescence microscopy with mouse anti-chicken α-tubulin IgG DM1D (Sigma, MO) and rabbit anti-GFP IgG (Clontech, CA) primary antibodies. Secondary antibodies were FITC-conjugated goat anti-mouse IgG and Cy3-conjugated goat anti-rabbit IgG (Jackson ImmunoResearch Laboratories, PA). Our results showed that the GFP-tagged motor was readily detected in the nuclei of cells in interphase, redistributed to spindle microtubules in metaphase, migrated to the spindle interzone in anaphase, and became concentrated at the midbody in telophase and cytokinesis (Chen and Detrich, 1999; Chen *et al.*, 2002). Therefore, the tagged motor behaves during the cell cycle exactly as does the untagged (Kuriyama *et al.*, 1994).

The amino acid sequence of zMklp1 revealed that potential nuclear localization signals were present, one in the N-terminus and two in the C-terminus of the motor. Transfection of COS-7 cells with *pGFP-Mklp1(ΔN1–275)* and with constructs that deleted the C-terminal signals singly or together demonstrated conclusively that the two C-terminal sites were responsible for nuclear localization in interphase (Chen *et al.*, 2002). Furthermore, deletion of both sites was necessary to eliminate nuclear accumulation completely, which indicates that the two elements constitute a bipartite nuclear localization motif.

C. *In Vitro* Synthesis of Capped GFP-Mklp1 mRNAs and Embryo Microinjection

pGFP-Mklp1, pGFP-Mklp1(ΔN1–275), and *pGFP-Mklp1(T119N)* were digested with *Kpn*I, the linear templates were transcribed using T3 RNA polymerase (Promega), and the mRNAs were capped using T3 Cap-Scribe nucleotide solution (Boehringer-Mannheim Biochemicals, IN). mRNA preparations were electrophoresed on denaturing gels, and those that gave a single RNA band of appropriate size were used for microinjection.

Embryos were dechorionated manually and injected with synthetic mRNAs (0.5–4.5 ng in ~5 nl volumes) or sterile water (equivalent volume) at the one-cell stage. Our laboratory uses a PLI-100 Picoinjector (Medical Systems, MA) and a Narashige micromanipulator for injections. To control for injection-mediated damage, we analyzed only those clutches of embryos in which water-injected controls developed through the shield stage. Embryos were examined alive by confocal microscopy to detect the GFP signal.

D. Results

Live zebrafish embryos injected with *GFP-Mklp1* mRNA or the dominant-negative mRNA variants (0.5 ng in all cases) were imaged by confocal epifluorescence microscopy using a Bio-Rad model MRC-600 microscope (Chen *et al.*, 2002). Figure 2 shows representative images for the GFP–Mklp1 (panels A and B) and the GFP–Mklp1(Δ1–275) (panels C and D) fusion proteins. Zebrafish blastulae expressing wild-type GFP-Mklp1 contained GFP-positive nuclei and midbody remnants (panel A). The wild-type protein appeared to form an annulus as the cleavage furrow contracted (panel B and data not shown). By contrast, the dominant-negative mutant proteins GFP-Mklp1(Δ1–275) (panel C) and GFP-Mklp1(T119N) (not shown) stained nuclei but rarely appeared at the midbody (panel D). Occasional binucleated cells were observed, consistent with partial inhibition of cytokinesis at the low dominant-negative mRNA dose. The GFP signals detected in a small proportion of cleavage furrows are interpreted to represent the rare movement to the midbody of translocation-deficient heterodimers composed of one endogenous Mklp1 subunit and one GFP-tagged dominant-negative subunit.

When injected at a higher dose (3.0 ng), ~30% of embryos (11 of 37) failed to develop beyond the four- to eight-cell stage (not shown). These embryos repeatedly

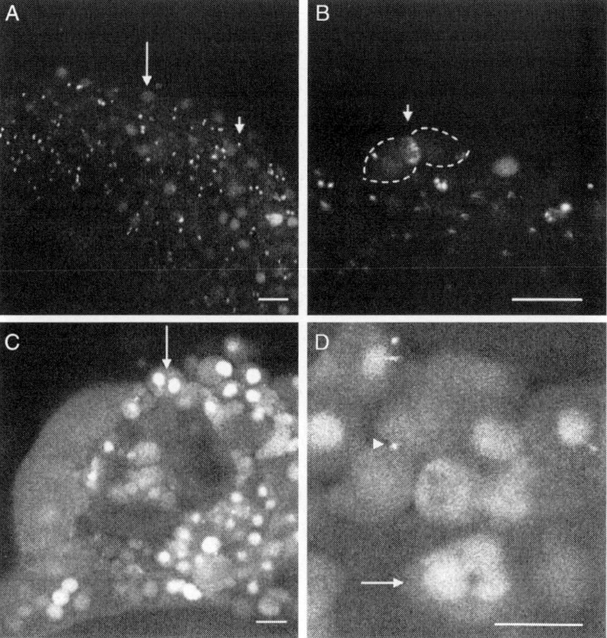

Fig. 2 Distribution of wild-type and dominant-negative GFP-Mklp1 in living zebrafish embryos. One- or two-cell embryos were injected with mRNA (0.5 ng) encoding either wild-type GFP-Mklp1 or GFP-Mklp1(ΔN1–275), and the GFP signal of the expressed fusion protein was detected by confocal microscopy of late blastulae. (A, B) GFP-Mklp1. Numerous labeled nuclei and midbody remnants (examples indicated by large and small arrows, respectively) are visible (A). In dividing blastomeres (B), the fusion protein forms a ring (small arrow) equidistant between the nascent daughter cells (boundaries indicated by dashed lines). (C, D) GFP-Mklp1(ΔN1–275). Many labeled nuclei are present (C), but midbody staining is rare (D). The arrows in C and D show multinucleated blastomeres, and the arrowhead in D indicates a rare midbody containing GFP-Mklp1(ΔN1–275). Bars = 50 μm. Reprinted from Chen *et al.* (2002) with permission. Copyright (2002) The American Physiological Society.

initiated cytokinesis, but their cleavage furrows retracted prior to achieving cell partition. Ten percent of the embryos (4 of 37) completed cleavage but failed to initiate epiboly [the spreading of the blastodisc and the yolk syncytial layer over the yolk cell, which begins during the blastula period and continues through gastrulation (Kimmel *et al.*, 1995)]. The remaining 60% appeared to develop normally to the bud stage but had larger, multinucleated blastomeres during cleavage and slower epibolic movement than did embryos that received wild-type *GFP-Mklp1* mRNA. This last group developed through the somatic stages but showed poor tissue organization and failed to move in response to mechanical stimulation at 24 h. Embryos that received 3.0 ng of *GFP-Mklp1* or *GFP* mRNAs showed few developmental abnormalities, which supports the specificity of the dominant-negative phenotypes (Chen *et al.*, 2002).

E. Future Applications and Improvements

Fluorescent proteins, whether derived from GFP, DsRedII, or others, now cover a wide color palette (Heim and Tsien, 1996; Shaner *et al.*, 2004) and have been adopted widely by cell biologists as molecular tags, but their use in developmental biology has been limited largely to lineage markers under the control of lineage-specific promoters in transgenic organisms. The application developed in this section, GFP tagging and analysis of Mklp1 in developing zebrafish embryos, illustrates the considerable potential for exploring cytoskeletal processes during vertebrate development. Using combinations of fluorescent tags fused to different kinesins and tubulin, for example, it should be feasible to monitor motor movements on microtubules of the blastomeres and the yolk cell and to examine changes in the dynamics of cellular microtubules and motors during major developmental transitions, such the convergent extension movements of gastrulation and early segmentation (Solnica-Krezel and Cooper, 2002). Such studies would be greatly facilitated by the development of transgenic zebrafish lines that express a microtubule motor and tubulin, each tagged with a different fluorescent protein variant (e.g., cyan and yellow fluorescent proteins), under the control of ubiquitous or cell-specific promoters. More broadly, one can imagine the development of fluorescence correlation spectroscopy and fluorescence resonance energy transfer methods to explore the diffusion, tracking, interaction, and conformational changes of proteins fused to appropriate fluorescent proteins (Giepmans *et al.*, 2006; Shaner *et al.*, 2005) in embryonic cells, particularly in the context of the nearly pigmentless *nacre* zebrafish (Lister *et al.*, 1999). The future is clearly bright for mechanistic analysis of protein function in the cells of zebrafish embryos by means of fluorescent protein technology.

III. Cell-Specific, Laser-Induced Transgene Expression in the Zebrafish Embryo: The *Sema3a1* Gene in Axonal Guidance

The ability to control the spatial and temporal expression of protein-encoding transgenes in the developing embryo would provide a powerful analytical tool that complements the forward and reverse genetic methods already available. Wataru Shoji (Tohoku University), John Kuwada (University of Michigan), and their colleagues have engineered transgenic zebrafish in which EGFP or EGFP-tagged proteins can be expressed under the control of the zebrafish *hsp70-4* promoter (Halloran *et al.*, 2000). Global expression in F2 progeny can be induced by heat shock, and cell-specific expression is achieved by focusing an attenuated laser beam on individual cells. I will review their technologies in the context of their elegant analysis of the role of the semaphorin *sema3a1* gene in guidance of spinal motor neurons in the developing myotome.

The connection of neuronal axons to their appropriate targets depends upon several families of guidance factors that attract or repel the motile growth cone at the tip of the elongating axon. The semaphorin family is large, diverse (eight classes), and highly conserved among metazoans. Some of these proteins are secreted, others reside in the plasma membrane, and all share a conserved Sema domain (Kruger *et al.*, 2005; Yazdani and Terman, 2006). Originally recognized for their importance in the developing nervous system (Fiore and Puschel, 2003), they have now been found to be important in the formation and functioning of virtually all organ systems (Yazdani and Terman, 2006).

The zebrafish spinal motor system is a simple, ideal model for analysis of axonal guidance. Figure 3 shows that each somatic hemisegment is innervated by three primary motoneurons, designated caudal primary (CaP, red), middle primary (MiP, green), and rostral primary (RoP, blue), that originate in the spinal cord. Led by the CaP growth cone, the CaP, MiP, and RoP axons exit the spinal cord and extend ventrally along the "common pathway" until they reach the horizontal myoseptum or "choice point." Here the growth cones pause in contact with the muscle pioneer cells; subsequently, they take different pathways to their final destinations (Eisen *et al.*, 1986).

The zebrafish genome contains two copies of the *sema3a* gene, undoubtedly due to a whole-genome duplication event that occurred after the divergence of ray- and lobe-finned fish but before the teleost radiation (Amores *et al.*, 1998). *sema3a2* is expressed transiently in the posterior half of each somite during early maturation and is followed by *sema3a1* expression in the same domain (Shoji *et al.*, 2003). *sema3a1* expression then changes to the dorsal and ventral regions of each somite but is absent in the intervening horizontal myoseptum as motoneuron growth cones enter the region. Meanwhile, *sema3a2* expression declines. Misexpression

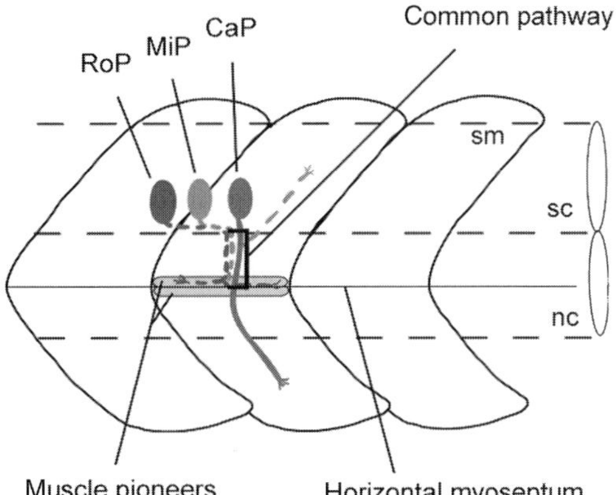

Fig. 3 Schematic representation of the three primary motoneurons (CaP, MiP, and RoP) and their axonal trajectories. CaP axons (red) project first to establish the common pathway. The common pathway ends at the muscle pioneers located at the horizontal myoseptum, which divides the myotome into dorsal and ventral halves. At this choice point, the three axons diverge and follow cell-specific pathways to innervate the ventral, dorsal, and horizontal myoseptal myotomes [CaP (red), MiP (green), and RoP (blue), respectively]. Abbreviations: sc, spinal cord; nc, notochord; sm, somite. Reprinted from Sato-Maeda *et al.* (2006) with permission. Copyright (2006) The Company of Biologists, Ltd. (See Plate no. 29 in the Color Plate Section.)

studies suggest that the Sema3a1 protein repels motor growth cones (Halloran *et al.*, 2000; Shoji *et al.*, 2003).

Consideration of growth cone pathways and of the repulsive role of Sema3a1 reveals a conundrum; the CaP growth cones are able to penetrate into the ventral region in which Sema3a1 concentration is high. Sato-Maeda *et al.* (2006) hypothesized that Sema3a1 initially restricts the CaP growth cone to the common pathway but this influence is lost at the choice point as the growth cones lose responsiveness to the semaphorin. To test this hypothesis, they made use of the transgenic fish line *hsp:EGFP-sema3a1-myc* that they had previously engineered (Shoji *et al.*, 2003).

A. Generation of *sema3a1* Transgenic Zebrafish

The zebrafish *sema3a1* cDNA (Yee *et al.*, 1999) was tagged by insertion of the *EGFP* coding sequence (Clontech) between the signal peptide sequence and the *sema*-domain coding region (between codons 25 and 26), and DNA encoding six Myc (MEQKLISEEDL) epitopes (Roth *et al.*, 1991) was fused in frame to the 3′ end of the cDNA. Subsequently, the 1.5-kb zebrafish *hsp70-4* promoter was attached to the 5′ end of the *EGFP-sema3a1-myc* construct. Zebrafish embryos were injected with the *hsp:EGFP-sema3a1-myc* DNA at the one- to four-cell stage, the embryos were raised to adulthood, and injected adults were pairwise mated to wild-type zebrafish (Shoji *et al.*, 2003). The F1 progeny were screened by PCR for the presence of the transgene to identify founder fish and for the ability to express EGFP following heat treatment (37–40 °C, 1 h). Two lines were generated that expressed EGFP ubiquitously after heat shock.

Statistically, 4 of 96 injected fish passed the transgene to F1 offspring (Shoji *et al.*, 2003). Progeny of F1 crosses from different founders expressed the transgene at 2.4–28% when tested for EGFP fluorescence and the Myc epitope by immunohistochemistry, consistent with mosaicism in the germ lines of the founders. The positive F1s, when crossed to wild type, produced 50% transgenic F2 offspring. Fish homozygous for the *hsp:EGFP-sema3a1-myc* transgene were generated by incrossing of F1 or F2 hemizygotes and were identified by their ability to pass the transgene to 100% of offspring in crosses with wild-type fish. Homozygous transgenic embryos produced EGFP fluorescence throughout the body only when exposed to elevated temperature, and the induction of the transgene was confirmed by detection of the 150-kDa fusion protein on Western blots probed with anti-Myc antibody (data not shown).

B. Laser Induction of Transgene Expression

To control the location and timing of induction of the *hsp:EGFP-sema3a1-myc* transgene, single muscle fibers of dechorionated embryos at the 16–18 somite stages (17–18 hpf) were exposed to pulses from a MicroPoint Coumarin 440 dye laser (Photonic Instruments, Arlington Heights, IL) as described previously (Halloran *et al.*, 2000). The embryos were placed on microscope slides in 1-mm-diameter Teflon rings containing a Ringer's solution, and the chambers were sealed with coverslips. The dye laser beam was focused on single muscle cells by use of a collinear helium/neon laser (Sato-Maeda *et al.*, 2006). Heating of single muscle fibers was produced by a 2-min burst of 4-ns dye laser pulses delivered at 4 Hz.

C. Detection of Axons and EGFP-Sema3a1-Myc by Immunohistochemistry

Primary motor axons in whole zebrafish embryos were labeled with the mouse monoclonal antibody (mAb) Znp1 (Melancon *et al.*, 1997; Trevarrow *et al.*, 1990) at a 1:50 dilution. The Myc and EGFP tags were stained using a 1:10 dilution of the mAb 9E10 (Evan *et al.*, 1985; Roth *et al.*, 1991) and a 1:400 dilution of a polyclonal anti-GFP antibody (Clontech), respectively. Bound primary antibodies were detected using appropriate horse radish peroxidase (HRP)-conjugated secondary antibodies. Briefly, embryos were fixed in 4% paraformaldehyde in phosphate-buffered saline and stained sequentially for Znp1 and for one of the tags. The Znp1 antibody was detected by incubation with anti-mouse IgG-HRP and 0.5-mg/ml diaminobenzidine, which results in a brown peroxidase reaction product. The embryos were then refixed, and the mouse anti-Myc mAb or rabbit anti-GFP polyclonal antibodies were detected by incubation with anti-mouse or anti-rabbit IgG-HRP. $NiCl_2$ (0.08%) and $CoCl_2$ (0.08%) were added to the diaminobenzidine (0.3 mg/ml) to produce a blue-black peroxidase reaction product.

D. Results

In preliminary experiments to test the effects of Sema3a1 on CaP growth cones, fertilized wild-type eggs were injected with either the *hsp:EGFP-sema3a1-myc* construct or with a control *hsp:myc* construct, and embryos were induced to express the corresponding proteins (EGFP-Sema3a1-Myc and Myc_6, respectively) at 15 hpf by exposure to 38 °C for 30 min. The embryos were returned to 28.5 °C and assayed at 28 hpf; as expected, mosaic expression of the proteins in cells throughout the embryo was observed. Figure 4B and C shows CaP axons whose migration along the common pathway were diverted laterally (panel B) or stalled (panel C) upon encountering muscle cells that expressed exogenous Sema3a1. Axons were observed to stall or turn away in 72% of the cases examined ($n = 65$). By contrast, axons continued past the horizontal myoseptum on their normal ventral route in embryos ($n = 20$) expressing the control Myc tag (Fig. 4A). Thus, CaP axons were repulsed by Sema3a1 when the semaphorin was encountered on the common pathway.

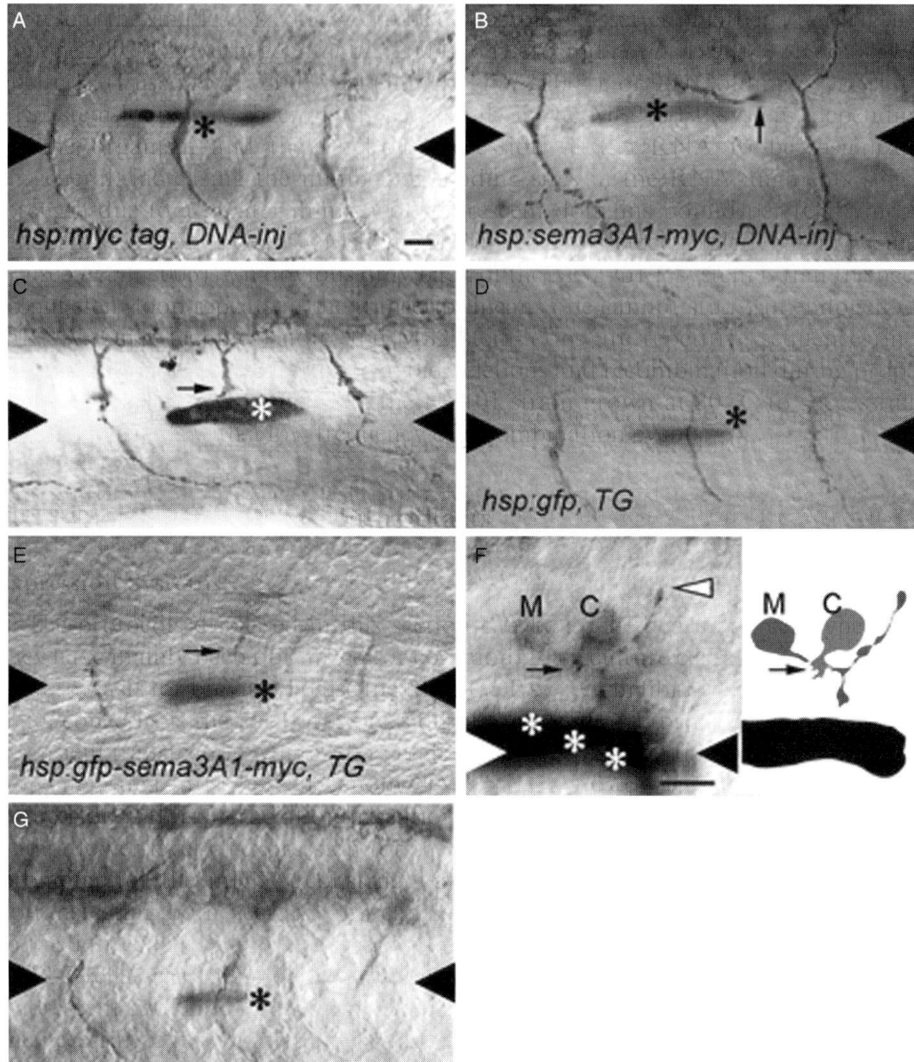

Fig. 4 CaP axons are repulsed by myotome cells that express ectopic Sema3a1 along the common pathway but not by those along the CaP-specific ventral pathway. (A, B) CaP axons are not perturbed by expression of the Myc epitope (asterisk) in myotome cells along the common pathway following heat induction of embryos injected with an *hsp70:myc* construct (A) but are turned away (arrow) by heat-induced ectopic Sema3a1 (asterisk) expressed from an injected *hsp70:sema3a1-myc* construct (B). CaP axons in segments anterior and posterior of the repulsed CaP axon in B follow normal trajectories. In the case of the posterior segment, a lateral myotome cell (out of the plane of focus) also expressed Sema3a1 but, since this cell does not line the common pathway, the CaP axon was unaffected. In these and all other panels, the triangles on each side demarcate the horizontal myoseptum. (C) A CaP axon (arrow) is stalled at a myotome cell (asterisk) that expressed Sema3a1 following heat induction of an embryo injected with *hsp70:sema3a1-myc*. (D) Laser induction of EGFP expression in a muscle pioneer

The spatial and temporal effects of Sema3a1 were tested more thoroughly in the *hsp:EGFP-sema3a1-myc* transgenic animals, in which the transgene could be expressed in any muscle cell by laser induction. Once again, the CaP axons stalled on the common pathway at muscle pioneers that expressed EGFP-Sema3a1-Myc (Fig. 4E) but were unaffected by expression of EGFP alone (Fig. 4D). In cases in which both CaP and MiP axons could be followed, the former stalled near the zone of Sema3a1 expression, whereas the latter completed the common pathway and then moved normally to innervate the dorsal myoseptal myotomes (Fig. 4F).

The fate of CaP axons on their specific ventral pathway was examined by laser induction of transgene expression in muscle fibers that were ventral to the horizontal myoseptal choice point. Figure 4G shows that CaP axons on the ventral pathway migrated normally despite encountering Sema3a1-producing muscle cells. Evidently CaP axons become insensitive or less sensitive to Sema3a1 once they have passed the horizontal myoseptum.

E. Future Applications and Improvements

The potential applications of *hsp70* promoter control and laser induction of the expression of transgenes in single cells at specific developmental stages are almost limitless. Ectopic expression of virtually any gene of choice can be readily engineered by production of transgenic zebrafish using the methods described here. As emphasized previously, cellular processes, such as the analysis of axonal pathfinding described here, can be easily accomplished within the physiological context of the developing embryo.

The production of transgenic fish by injection of DNA alone is an inefficient process. One can anticipate that the efficiency of transgenesis will be approved upon application of transposon-mediated gene insertion by the *Tol2* transposon (Kawakami, 2004), by the *Sleeping Beauty* transposon (Hermanson *et al.*, 2004) or through use of the Cre-loxP (Dong and Stuart, 2004; Langenau *et al.*, 2005) or I-SceI meganuclease (Grabher *et al.*, 2004) systems.

---

cell (asterisk) of a transgenic embryo did not prevent the CaP axon from continuing on its cell-specific ventral route. (E) A CaP axon is stalled near a muscle pioneer cell that was laser induced to express ectopic Sema3a1 in an embryo transgenic for the *hsp70:EGFP-sema3a1-myc* hybrid gene. (F) A CaP axon (arrow) is stalled on the common pathway in the vicinity of three myotome cells (asterisks) that express the *hsp70:EGFP-sema3a1-myc* transgene upon laser induction. By contrast, the adjacent MiP axon (white arrowhead) traversed its normal dorsal route after leaving the common pathway. The camera lucida drawing (right panel) shows the CaP and MiP cells and their axons. C, the CaP cell body; M, the MiP cell body. (G) A CaP axon migrated normally on its cell-specific ventral route upon leaving the common pathway, despite encountering a ventral muscle fiber (asterisk, below the horizontal myoseptum) expressing ectopic Sema3a1 in a laser-induced *hsp70:EGFP-sema3a1-myc* transgenic embryo. Reprinted from Sato-Maeda *et al.* (2006) with permission. Copyright (2006) The Company of Biologists, Ltd. (See Plate no. 30 in the Color Plate Section.)

IV. Transgenic Zebrafish Models of Myc-Induced T-Cell Acute Lymphoblastic Leukemia

T-cell acute lymphoblastic leukemia (T-ALL) is a malignant disease of thymocytes that arises when these cells acquire mutations that arrest differentiation, promote proliferation, and suppress apoptosis (Ferrando *et al.*, 2002). In molecular terms, many T-cell malignancies are caused by chromosomal translocations that place proto-oncogenic T-cell transcription factors under the control of the strong promoter elements of the T-cell receptor (Look, 1997). In human T-ALL, five different multistep molecular pathways that involve overexpression of transcription factors have been identified by Tom Look (Dana-Farber Cancer Institute) and his colleagues (Ferrando *et al.*, 2002): (1) *TAL1/SCL* plus *LMO1* or *LMO2*, (2) *LYL1* plus *LMO2*, (3) *HOX11*, (4) *HOX11L2*, and (5) *MLL-ENL* (Ferrando *et al.*, 2002). Each subgroup has distinct molecular signatures, and they differ in clinical prognosis (Ballerini *et al.*, 2002; Ferrando *et al.*, 2002, 2004; Yeoh *et al.*, 2002). In four of these subgroups, *MYC* or *MYCN* are expressed at high levels (Ferrando *et al.*, 2002), which argues that MYC may play a central role with respect to proliferation or genomic instability in T-ALL.

The zebrafish has emerged as a valuable vertebrate model of cancer, including Myc-induced T-ALL (Langenau *et al.*, 2003). Figure 5B shows that stably transgenic *rag2:EGFP-mMyc* (*rag2* represents the promoter of the zebrafish *rag2* gene; *mMyc* is the mouse *c-Myc* gene) zebrafish develop EGFP-labeled thymic lymphoma, in which malignant cells infiltrate regions adjacent to the thymus.

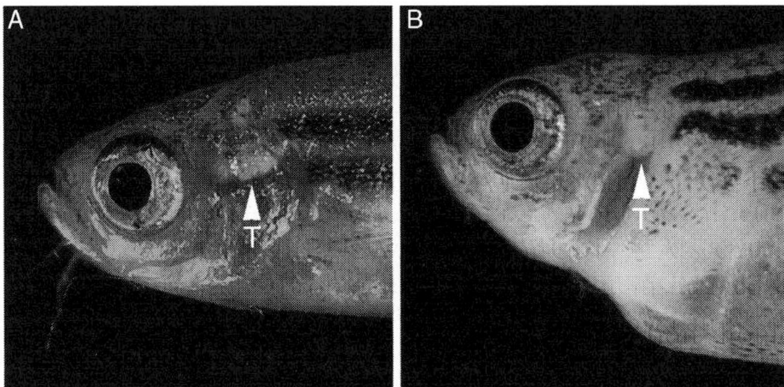

Fig. 5 Stably transgenic *rag2:EGFP-mMyc* zebrafish develop EGFP-labeled thymic lymphoma, which progresses to T-ALL. The thymus of a control, 50-dpf *rag2:GFP* transgenic fish (A) appears normal when imaged by fluorescence microscopy, whereas a 50-dpf *rag2:EGFP-mMyc* transgenic fish (B) shows expansion of EGFP-labeled T cells in the thymus and massive dissemination of EGFP-labeled leukemic lymphoblasts into regions adjacent to the thymus. Arrowheads mark the location of the thymus (T). Reprinted from Langenau *et al.* (2005) with permission. Copyright (2005) National Academy of Sciences. (See Plate no. 31 in the Color Plate Section.)

Subsequently, the transformed T cells spread throughout the body, as occurs in human T-ALL (Langenau *et al.*, 2005). Further characterization of this line demonstrated that the Myc-induced leukemias were transplantable into irradiated adult fish and that the lymphoblasts were arrested at a stage in which *scl* and *lmo2* are coexpressed; *lmo1* expression was not observed. By contrast, *rag2:GFP* control transgenic fish show GFP labeling of a normal thymus (Fig. 5A), and their thymocytes expressed very low levels of *scl* mRNA, although *lmo2* expression was comparable to that observed in leukemic lymphoblasts. Together, these results and others not described indicate that the transgenic *rag2:EGFP-mMyc* zebrafish line is a faithful model of the most common and severe subtype of human T-ALL, in which *TAL1/SCL* and *LMO1* or *LMO2* are coexpressed (Langenau *et al.*, 2003, 2005).

Among the advantages of the zebrafish is the ability to perform chemical (Peterson *et al.*, 2004) or genetic modifier (Driever *et al.*, 1996; Golling *et al.*, 2002; Haffter *et al.*, 1996) screens to identify enhancers or suppressors of diseases such as T-ALL. Unfortunately, *rag2:EGFP-mMyc* transgenic zebrafish (Langenau *et al.*, 2003) often develop severe T-ALL by the time they reach sexual maturity, which makes the line difficult to breed and maintain. Therefore, Langenau, Look, and their colleagues applied a Cre/lox strategy to develop a conditional zebrafish line for T-ALL in which fluorescent proteins are used to monitor the recombination status of the transgene (Langenau *et al.*, 2005). This transgenic line, the creation of which will be described here, should greatly facilitate chemical-modifier and forward-genetic screens, and the methodology should be broadly applicable to the creation of conditional zebrafish models of a variety of cancers and other diseases.

A. Generation of Zebrafish Containing the *rag2:loxP-dsRED2-loxP-EGFP-mMyc* Transgene

Starting with the vector *zRag2:EGFP-mMyc* (Langenau *et al.*, 2003), Langenau *et al.* (2005) introduced a "floxed" (flanked by *loxP* sites) *dsRED2* cDNA upstream of the *EGFP-mMyc* sequence by use of *Bam*HI restriction enzyme sites to generate the plasmid *zRag2:loxP-dsRED2-loxP-EGFP-mMyc* (Fig. 6A). This plasmid was digested with *Xho*I to generate linear DNA, which was then extracted using phenol/chloroform and precipitated using ethanol. The DNA [100 ng/μl in 0.5× TE (1× TE = 10 mM Tris, 1 mM EDTA, pH ≈ 7.0) plus 100 mM KCl] was injected into one-cell embryos (AB strain), which were raised to reproductive maturity. These fish were screened for the ability to produce offspring carrying the transgene by monitoring the developing thymocytes of the latter for dsRED2 fluorescence at 6 dpf. Two lines, G7 and G16, that stably incorporated the *rag2:loxP-dsRED2-loxP-EGFP-mMyc* transgene were identified.

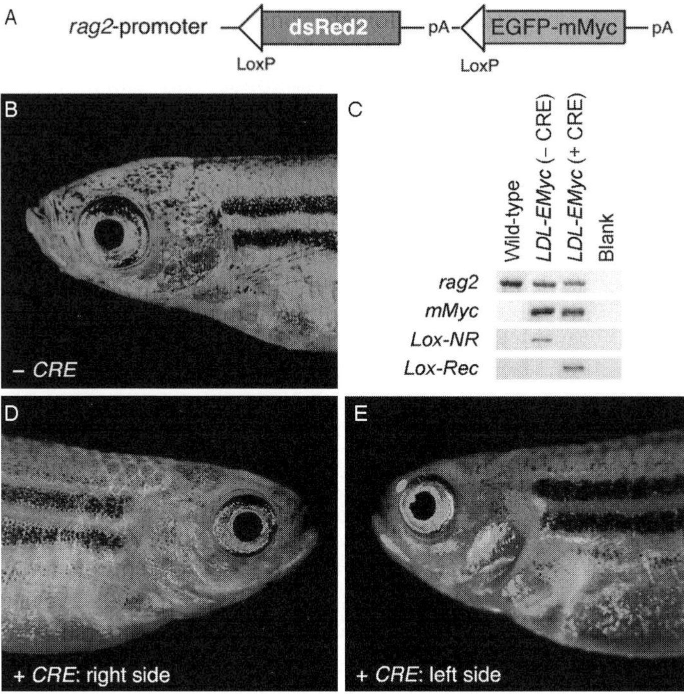

Fig. 6 *Cre* RNA injection into stable transgenic *rag2:loxP-dsRED2-loxP-EGFP-mMyc* zebrafish embryos (one-cell stage) leads to transgene recombination and rapid onset of mMyc-induced T-ALL. (A) Diagram of *rag2:loxP-dsRED2-loxP-EGFP-mMyc* construct. (B) Thymocytes from a 73-day-old *rag2:loxP-dsRED2-loxP-EGFP-mMyc* transgenic zebrafish express dsRED2 (red fluorescence) in the absence of *Cre* expression. (C) Transgene recombination in blood cells of wild-type or transgenic *rag2:loxP-dsRED2-loxP-EGFP-mMyc* fish (*LDL-EMyc*) analyzed by PCR. The *rag2* primers amplify the wild-type *rag2* gene present in the genomes of all zebrafish strains examined in this study. *mMyc* primers amplify the *mMyc* transgene. The Lox primers amplify either a 1.7-kb nonrecombined fragment (*Lox-NR*) or a 0.4-kb fragment (*Lox-Rec*) that results from *Cre*-mediated recombination. (D, E) One-cell *rag2:loxP-dsRED2-loxP-EGFP-mMyc* embryos were injected with *Cre* recombinase RNA (25 ng/μl) and raised to 51 days of development, at which time they had leukemias labeled with EGFP and dsRED2 (D) or with EGFP alone (E). Bilateral mosaicism of the labels was observed in some individuals, such as the fish shown, whose right-side leukemia expresses both labels (D) whereas the left-side leukemia expresses EGFP alone (E). Images are composites of dsRED2 and EGFP fluorescence and brightfield images. Reprinted from Langenau *et al.* (2005) with permission. Copyright (2005) National Academy of Sciences. (See Plate no. 32 in the Color Plate Section.)

B. Activation of the Conditional *mMyc* Transgene by Injection of *Cre* Recombinase RNA and Analysis of Leukemic Cells

One-cell embryos produced by F1 or F2 crosses of G7 or G16 transgenic fish were injected with RNA (25 ng/μl) that encoded the Cre recombinase. Leukemic cells from the *Cre* RNA-injected fish were collected for (1) transplantation into irradiated adult fish, (2) characterization of lymphoblast morphology by May-Grunwald/Giemsa staining of cytospin preparations, (3) FACS analysis of EGFP and dsRED2 fluorescence in lymphoblasts, and (4) genomic PCR analysis for Cre-mediated recombination of the transgene. For PCR, the forward primer (specific to the *rag2* promoter region) was 5′–ATGCTAATTTGAAGCACTAGCA–3′, and the reverse primer (specific to the *EGFP* coding sequence) was 5′–GTGCA-GATGAACTTCAGGGT–3′.

C. Results

In preliminary experiments with transiently transfected one-cell wild-type embryos, Langenau *et al.* (2005) found that coinjection of the plasmid construct *CMV:loxP-dsRED2-loxP-EGFP* with *Cre* RNA caused excision of the floxed *dsRED2* locus and expression of EGFP, whereas embryos injected with the construct alone fluoresced red due to expression of dsRED2. Encouraged by these preliminary results, they created the stably transgenic G7 and G16 lines that contain the *rag2:loxP-dsRED2-loxP-EGFP-mMyc* transgene. In the absence of Cre recombinase, these fish lines produced thymocytes that expressed high levels of dsRED2, but expression of the EGFP–mMyc fusion protein and development of lymphoma or leukemia were absent (Fig. 6B). By contrast, injection of *Cre* RNA into embryos of the two lines led to excision of the *dsRED2* gene (Fig. 6C), but only G7 fish developed T-ALL. Furthermore, leukemias in the G7 line usually expressed both dsRED2 and EGFP–mMyc (Fig. 6D), which suggests that recombination of the transgene was incomplete. However, some G7 leukemias expressed only the *EGFP-mMyc* locus, indicated by the appearance of green fluorescence but not red (Fig. 6E).

Microscopic analysis of G7 fish with the recombined transgene demonstrated that the leukemias were of T-cell origin. These leukemias contained lymphoblasts of typical morphology and were transplantable into irradiated zebrafish hosts. Thus, the leukemias observed in this conditional *rag2:loxP-dsRED2-loxP-EGFP-mMyc* G7 line are similar to those developed by the original *rag2:EGFP-mMyc* transgenic line (Langenau *et al.*, 2003).

D. Future Applications and Improvements

Langenau *et al.* (2003, 2005) have generated constitutive and conditional transgenic zebrafish models of T-ALL that are very similar to the human malignancy. The mMyc-induced leukemias develop after a latency period and involve clonal T-cell receptor α gene rearrangements (data not shown), which is consistent with a requirement for additional mutations to transform the T cell. Furthermore, the zebrafish T-ALLs mimic the most common and severe subclass of human T-ALL that coexpresses *TAL1/SCL* plus *LMO2*. Therefore, the stage has been set for future chemical and genetic screens to identify suppressors and enhancers of T-ALL in this valuable vertebrate model system.

Although the conditional T-ALL zebrafish model eases the maintenance of the fish stock, only one of two independent lines, G7, gave offspring that develop T-ALL. Failure of G16 offspring to progress to the disease after Cre-mediated transgene recombination strongly suggests that transgene function may be affected by position or orientation effects at its site of integration (Langenau *et al.*, 2005). Multiple transgenic lines are likely to be needed when creating conditional models of cancer and other diseases. Hence, one may anticipate that future work will make use of more efficient transgenic technologies (see Section III.E).

The G7 conditional strain also gave low frequencies of progeny that progressed to disease (13%) when injected with *Cre* RNA. Hence, it is likely that recombination to yield the *EGFP-mMyc* transgene was suboptimal under the conditions used, a conclusion supported by the observation that most leukemias that arose from *Cre* RNA-injected embryos expressed both dsRED2 and EGFP. This problem should be solved by the development of transgenic zebrafish lines that express Cre recombinase in T cells. Mating of the *rag2:loxP-dsRED2-loxP-EGFP-mMyc* G7 line to such a stably transgenic, *Cre*-expressing line should yield ~100% expression of the *EGFP-mMyc* transgene in doubly transgenic offspring.

Langenau *et al.* (2005) have established the Cre/lox strategy as a viable mechanism for production of conditional zebrafish models of cancer and other diseases. They note that these tools can also be applied to assess the plasticity of stem cells and the commitment of cells to distinct lineages and to generate conditional gene knockouts in developing embryos. To achieve the latter goal, methods for isolation and culture of stem cells and for targeted homologous recombination must be developed. Rapid progress in these technical arenas has already been made (Alvarez *et al.*, 2007; Fan *et al.*, 2004a,b, 2006; Ma *et al.*, 2001; Traver, 2004; Wu *et al.*, 2006).

V. Summary

Since Chalfie's original demonstration of GFP as a fluorescent reporter of great utility in *Caenorhabditis elegans* (Chalfie *et al.*, 1994), fluorescent protein technology has become ubiquitous in cell and developmental biology. Peak excitation and emission wavelengths of GFP variants and other fluorescent proteins now cover almost the entire visible spectrum, from far-red through cyan (Giepmans *et al.*, 2006; Shaner *et al.*, 2004, 2005). Investigators may also select fluorescent proteins based on brightness and stability criteria. Furthermore, many biophysical techniques, including fluorescence correlation spectroscopy, fluorescence resonance energy transfer, and fluorescence recovery after photobleaching, can now be employed to study protein localization, interaction, and activity in the cells of living embryos (e.g., zebrafish) or organisms (e.g., *C. elegans*) (Giepmans *et al.*, 2006). The potential of fluorescent protein technology is so great that biologists of all disciplines are likely to feel as if they are the proverbial "kids in a candy store."

Acknowledgments

This work was supported by NSF grants OPP-0089451 and OPP-0336932 (H.W.D.).

References

Aizawa, H., Sekine, Y., Takemura, R., Zhang, Z., Nangaku, M., and Hirokawa, N. (1992). Kinesin family in murine central nervous system. *J. Cell Biol.* **119,** 1287–1296.

Alvarez, M. C., Bejar, J., Chen, S., and Hong, Y. (2007). Fish ES cells and applications to biotechnology. *Mar. Biotechnol. (NY)* **9**(2), 117–127.

Amores, A., Force, A., Yan, Y. L., Joly, L., Amemiya, C., Fritz, A., Ho, R. K., Langeland, J., Prince, V., Wang, Y. L., Westerfield, M., Ekker, M., *et al.* (1998). Zebrafish hox clusters and vertebrate genome evolution. *Science* **282,** 1711–1714.

Avanesov, A., and Malicki, J. (2004). Approaches to study neurogenesis in the zebrafish retina. *In* "The Zebrafish, 2nd edn.: Cellular and Developmental Biology" (H. W. Detrich, III, M. Westerfield, and L. I. Zon, eds.), Vol. 76, pp. 333–384. Elsevier Academic Press, San Diego.

Bak, M., and Fraser, S. E. (2003). Axon fasciculation and differences in midline kinetics between pioneer and follower axons within commissural fascicles. *Development* **130,** 4999–5008.

Ballerini, P., Blaise, A., Busson-Le Coniat, M., Su, X. Y., Zucman-Rossi, J., Adam, M., van den Akker, J., Perot, C., Pellegrino, B., Landman-Parker, J., Douay, L., Berger, R., *et al.* (2002). HOX11L2 expression defines a clinical subtype of pediatric T-ALL associated with poor prognosis. *Blood* **100,** 991–997.

Bartman, T., Walsh, E. C., Wen, K. K., McKane, M., Ren, J., Alexander, J., Rubenstein, P. A., and Stainier, D. Y. (2004). Early myocardial function affects endocardial cushion development in zebrafish. *PLoS Biol.* **2,** E129.

Berman, J., Hsu, K., and Look, A. T. (2003). Zebrafish as a model organism for blood diseases. *Br. J. Haematol.* **123,** 568–576.

Brendza, R. P., Serbus, L. R., Duffy, J. B., and Saxton, W. M. (2000). A function for kinesin I in the posterior transport of oskar mRNA and Staufen protein. *Science* **289,** 2120–2122.

Chalfie, M., Euskirchen, G., Ward, W., and Prasher, D. C. (1994). Green fluorescent protein as a marker for gene expression. *Science* **263,** 802–805.

Chapouton, P., and Bally-Cuif, L. (2004). Neurogenesis. *In* "The Zebrafish, 2nd edn.: Cellular and Developmental Biology" (H. W. Detrich, III, M. Westerfield, and L. I. Zon, eds.), Vol. 76, pp. 163–206. Elsevier Academic Press, San Diego.

Chen, M. C., and Detrich, H. W., III (1999). Kinesin-like microtubule motors in early development. *In* "Methods in Cell Biology, The Zebrafish: Biology" (H. W. Detrich, III, M. Westerfield, and L. I. Zon, eds.), Vol. 59, pp. 227–250. Academic Press, San Diego, CA.

Chen, M. C., Zhou, Y., and Detrich, H. W., III (2002). Zebrafish mitotic kinesin-like protein 1 (Mklp1) functions in embryonic cytokinesis. *Physiol. Genomics* **8,** 51–66.

Davidson, A. E., Balciunas, D., Mohn, D., Shaffer, J., Hermanson, S., Sivasubbu, S., Cliff, M. P., Hackett, P. B., and Ekker, S. C. (2003). Efficient gene delivery and gene expression in zebrafish using the Sleeping Beauty transposon. *Dev. Biol.* **263,** 191–202.

Detrich, H. W., III, Westerfield, M., and Zon, L. I. (1999a). Overview of the zebrafish system. *Methods Cell Biol.* **59,** 3–10.

Detrich, H. W., III, Westerfield, M., and Zon, L. I. (1999b). "The Zebrafish: Biology." Academic Press, San Diego.

Detrich, H. W., III, Westerfield, M., and Zon, L. I. (1999c). "The Zebrafish: Genetics and Genomics." Academic Press, San Diego.

Detrich, H. W., III, Westerfield, M., and Zon, L. I. (2004a). "The Zebrafish, 2nd edn.: Cellular and Developmental Biology." Elsevier Academic Press, San Diego.

Detrich, H. W., III, Westerfield, M., and Zon, L. I. (2004b). "The Zebrafish, 2nd edn.: Genetics, Genomics, and Informatics." Elsevier Academic Press, San Diego.

Dong, J., and Stuart, G. W. (2004). Transgene manipulation in zebrafish by using recombinases. *In* "Methods in Cell Biology, The Zebrafish, 2nd edn.: Genetics, Genomics, and Informatics" (H. W. Detrich, III, M. Westerfield, and L. I. Zon, eds.), Vol. 77, pp. 363–379. Elsevier Academic Press, San Diego.

Driever, W., Solnica-Krezel, L., Schier, A. F., Neuhauss, S. C. F., Malicki, J., Stemple, D. L., Stainier, D. Y. R., Zwartkruis, F., Abdelilah, S., Rangini, Z., Belak, J., and Boggs, C. (1996). A genetic screen for mutations affecting embryogenesis in zebrafish. *Development* **123** 37–46.

Eisen, J. S., Myers, P. Z., and Westerfield, M. (1986). Pathway selection by growth cones of identified motoneurones in live zebra fish embryos. *Nature* **320,** 269–271.

Evan, G. I., Lewis, G. K., Ramsay, G., and Bishop, J. M. (1985). Isolation of monoclonal antibodies specific for human c-myc proto-oncogene product. *Mol. Cell. Biol.* **5,** 3610–3616.

Fan, L., Crodian, J., and Collodi, P. (2004a). Culture of embryonic stem cell lines from zebrafish. *In* "The Zebrafish, 2nd edn.: Cellular and Developmental Biology" (H. W. Detrich, III, M. Westerfield, and L. I. Zon, eds.), Vol. 76, pp. 151–160. Elsevier Academic Press, San Diego.

Fan, L., Crodian, J., and Collodi, P. (2004b). Production of zebrafish germline chimeras by using cultured embryonic stem (ES) cells. *In* "The Zebrafish, 2nd edn.: Genetics, Genomics, and Informatics" (H. W. Detrich, III, M. Westerfield, and L. I. Zon, eds.), Vol. 77, pp. 113–119. Elsevier Academic Press, San Diego.

Fan, L., Moon, J., Crodian, J., and Collodi, P. (2006). Homologous recombination in zebrafish ES cells. *Transgenic Res.* **15,** 21–30.

Ferrando, A. A., Neuberg, D. S., Dodge, R. K., Paietta, E., Larson, R. A., Wiernik, P. H., Rowe, J. M., Caligiuri, M. A., Bloomfield, C. D., and Look, A. T. (2004). Prognostic importance of TLX1 (HOX11) oncogene expression in adults with T-cell acute lymphoblastic leukaemia. *Lancet* **363,** 535–536.

Ferrando, A. A., Neuberg, D. S., Staunton, J., Loh, M. L., Huard, C., Raimondi, S. C., Behm, F. G., Pui, C. H., Downing, J. R., Gilliland, D. G., Lander, E. S., Golub, T. R., *et al.* (2002). Gene expression signatures define novel oncogenic pathways in T cell acute lymphoblastic leukemia. *Cancer Cell* **1,** 75–87.

Field, H. A., Dong, P. D., Beis, D., and Stainier, D. Y. (2003a). Formation of the digestive system in zebrafish. II. Pancreas morphogenesis. *Dev. Biol.* **261,** 197–208.

Field, H. A., Ober, E. A., Roeser, T., and Stainier, D. Y. (2003b). Formation of the digestive system in zebrafish. I. Liver morphogenesis. *Dev. Biol.* **253,** 279–290.

Fiore, R., and Puschel, A. W. (2003). The function of semaphorins during nervous system development. *Front Biosci.* **8,** s484–s499.

Giepmans, B. N., Adams, S. R., Ellisman, M. H., and Tsien, R. Y. (2006). The fluorescent toolbox for assessing protein location and function. *Science* **312,** 217–224.

Goldstein, L. S. B. (1993). With apologies to Sheharazde: Tails of 1001 kinesin motors. *Annu. Rev. Genet.* **27,** 319–351.

Golling, G., Amsterdam, A., Sun, Z., Antonelli, M., Maldonado, E., Chen, W., Burgess, S., Haldi, M., Artzt, K., Farrington, S., Lin, S. Y., Nissen, R. M., *et al.* (2002). Insertional mutagenesis in zebrafish rapidly identifies genes essential for early vertebrate development. *Nat. Genet.* **31,** 135–140.

Grabher, C., Joly, J. S., and Wittbrodt, J. (2004). Highly efficient zebrafish transgenesis mediated by the meganuclease I-SceI. *In* "Methods in Cell Biology, The Zebrafish, 2nd edn.: Genetics, Genomics, and Informatics" (H. W. Detrich, III, M. Westerfield, and L. I. Zon, eds.), Vol. 77, pp. 381–401. Elsevier Academic Press, San Diego.

Gruneberg, U., Neef, R., Honda, R., Nigg, E. A., and Barr, F. A. (2004). Relocation of Aurora B from centromeres to the central spindle at the metaphase to anaphase transition requires MKlp2. *J. Cell Biol.* **166,** 167–172.

Guse, A., Mishima, M., and Glotzer, M. (2005). Phosphorylation of ZEN-4/MKLP1 by Aurora B regulates completion of cytokinesis. *Curr. Biol.* **15,** 778–786.

Haffter, P., Granato, M., Brand, M., Mullins, M. C., Hammerschmidt, M., Kane, D. C., Odenthal, J., van Eeden, F. J. M., Jiang, Y.-J., Heisenberg, C.-P., Kelsh, R. N., Furutani-Seiki, M., *et al.* (1996).

The identification of genes with unique and essential functions in the development of the zebrafish, *Danio rerio*. *Development* **123,** 1–36.

Halloran, M. C., Sato-Maeda, M., Warren, J. T., Su, F., Lele, Z., Krone, P. H., Kuwada, J. Y., and Shoji, W. (2000). Laser-induced gene expression in specific cells of transgenic zebrafish. *Development* **127,** 1953–1960.

Heim, R., and Tsien, R. Y. (1996). Engineering green fluorescent protein for improved brightness, longer wavelengths and fluorescence resonance energy transfer. *Curr. Biol.* **6,** 178–182.

Hermanson, S., Davidson, A. E., Sivasubbu, S., Balciunas, D., and Ekker, S. C. (2004). Sleeping Beauty transposon for efficient gene delivery. *In* "Methods in Cell Biology, The Zebrafish, 2nd edn.: Genetics, Genomics, and Informatics" (H. W. Detrich, III, M. Westerfield, and L. I. Zon, eds.), Vol. 77, pp. 349–362. Elsevier Academic Press, San Diego.

Hsu, K., Look, A. T., and Kanki, J. P. (2004). Lessons from transgenic zebrafish expressing the green fluorescent protein (GFP) in the myeloid lineage. *In* "Methods in Cell Biology, The Zebrafish, 2nd edn.: Genetics, Genomics, and Informatics" (H. W. Detrich, III, M. Westerfield, and L. I. Zon, eds.), Vol. 77, pp. 333–347. Elsevier Academic Press, San Diego.

Januschke, J., Gervais, L., Dass, S., Kaltschmidt, J. A., Lopez-Schier, H., St Johnston, D., Brand, A. H., Roth, S., and Guichet, A. (2002). Polar transport in the Drosophila oocyte requires Dynein and Kinesin I cooperation. *Curr. Biol.* **12,** 1971–1981.

Kawakami, K. (2004). Transgenesis and gene trap methods in zebrafish by using the Tol2 transposable element. *In* "Methods in Cell Biology, The Zebrafish, 2nd edn.: Genetics, Genomics, and Informatics" (H. W. Detrich, III, M. Westerfield, and L. I. Zon, eds.), Vol. 77, pp. 201–222. Elsevier Academic Press, San Diego.

Kimmel, C. B., Ballard, W. W., Kimmel, S. R., Ullmann, B., and Schilling, T. F. (1995). Stages of embryonic development of the zebrafish. *Dev. Dyn.* **203,** 253–310.

Koster, R. W., and Fraser, S. E. (2004). Time-lapse microscopy of brain development. *In* "The Zebrafish, 2nd edn.: Cellular and Developmental Biology" (H. W. Detrich, III, M. Westerfield, and L. I. Zon, eds.), Vol. 76, pp. 207–235. Elsevier Academic Press, San Diego.

Koster, R. W., and Fraser, S. E. (2006). FGF signaling mediates regeneration of the differentiating cerebellum through repatterning of the anterior hindbrain and reinitiation of neuronal migration. *J. Neurosci.* **26,** 7293–7304.

Kruger, R. P., Aurandt, J., and Guan, K. L. (2005). Semaphorins command cells to move. *Nat. Rev. Mol. Cell Biol.* **6,** 789–800.

Kuriyama, R., Dragas-Granoic, S., Maekawa, T., Vassilev, A., Khodjakov, A., and Kobayashi, H. (1994). Heterogeneity and microtubule interaction of the CHO1 antigen, a mitosis-specific kinesin-like protein. *J. Cell Sci.* **107,** 3485–3499.

Langenau, D. M., Feng, H., Berghmans, S., Kanki, J. P., Kutok, J. L., and Look, A. T. (2005). Cre/lox-regulated transgenic zebrafish model with conditional myc-induced T cell acute lymphoblastic leukemia. *Proc. Natl. Acad. Sci. USA* **102,** 6068–6073.

Langenau, D. M., Traver, D., Ferrando, A. A., Kutok, J. L., Aster, J. C., Kanki, J. P., Lin, S., Prochownik, E., Trede, N. S., Zon, L. I., and Look, A. T. (2003). Myc-induced T cell leukemia in transgenic zebrafish. *Science* **299,** 887–890.

Lawrence, C. J., Dawe, R. K., Christie, K. R., Cleveland, D. W., Dawson, S. C., Endow, S. A., Goldstein, L. S., Goodson, H. V., Hirokawa, N., Howard, J., Malmberg, R. L., McIntosh, J. R., *et al.* (2004). A standardized kinesin nomenclature. *J. Cell Biol.* **167,** 19–22.

Liao, E. C., Paw, B. H., Oates, A. C., Pratt, S. J., Postlethwait, J. H., and Zon, L. I. (1998). SCL/Tal-1 transcription factor acts downstream of cloche to specify hematopoietic and vascular progenitors in zebrafish. *Genes Dev.* **12,** 621–626.

Lister, J. A., Robertson, C. P., Lepage, T., Johnson, S. L., and Raible, D. W. (1999). nacre encodes a zebrafish microphthalmia-related protein that regulates neural-crest-derived pigment cell fate. *Development* **126,** 3757–3767.

Look, A. T. (1997). Oncogenic transcription factors in the human acute leukemias. *Science* **278,** 1059–1064.

Ma, C., Fan, L., Ganassin, R., Bols, N., and Collodi, P. (2001). Production of zebrafish germ-line chimeras from embryo cell cultures. *Proc. Natl. Acad. Sci. USA* **98,** 2461–2466.

Marchuk, D., Drumm, M., Saulino, A., and Collins, F. S. (1991). Construction of T-vectors, a rapid and general system for direct cloning of unmodified PCR products. *Nucleic Acids Res.* **19,** 1154.

Marx, A., Muller, J., and Mandelkow, E. (2005). The structure of microtubule motor proteins. *Adv. Protein Chem.* **71,** 299–344.

Melancon, E., Liu, D. W., Westerfield, M., and Eisen, J. S. (1997). Pathfinding by identified zebrafish motoneurons in the absence of muscle pioneers. *J. Neurosci.* **17,** 7796–7804.

Meng, A., Jessen, J. R., and Lin, S. (1999a). Transgenesis. *In* "Methods in Cell Biology, The Zebrafish: Genetics and Genomics" (H. W. Detrich, III, M. Westerfield, and L. I. Zon, eds.), Vol. 60, pp. 133–148. Academic Press, San Diego.

Meng, A., Tang, H., Yuan, B., Ong, B. A., Long, Q., and Lin, S. (1999b). Positive and negative cis-acting elements are required for hematopoietic expression of zebrafish GATA-1. *Blood* **93,** 500–508.

Miki, H., Okada, Y., and Hirokawa, N. (2005). Analysis of the kinesin superfamily: Insights into structure and function. *Trends Cell Biol.* **15,** 467–476.

Minestrini, G., Harley, A. S., and Glover, D. M. (2003). Localization of Pavarotti-KLP in living Drosophila embryos suggests roles in reorganizing the cortical cytoskeleton during the mitotic cycle. *Mol. Biol. Cell* **14,** 4028–4038.

Mishima, M., Pavicic, V., Gruneberg, U., Nigg, E. A., and Glotzer, M. (2004). Cell cycle regulation of central spindle assembly. *Nature* **430,** 908–913.

Neef, R., Klein, U. R., Kopajtich, R., and Barr, F. A. (2006). Cooperation between mitotic kinesins controls the late stages of cytokinesis. *Curr. Biol.* **16,** 301–307.

O'Malley, D. M., Sankrithi, N. S., Borla, M. A., Parker, S., Banden, S., Gahtan, E., and Detrich, H. W., III (2004). Optical physiology and locomotor behaviors of wild-type and nacre zebrafish. *In* "The Zebrafish, 2nd edn.: Cellular and Developmental Biology" (H. W. Detrich, III, M. Westerfield, and L. I. Zon, eds.), Vol. 76, pp. 261–284. Elsevier Academic Press, San Diego.

Orkin, S. H., and Zon, L. I. (1997). Genetics of erythropoiesis: Induced mutations in mice and zebrafish. *Annu. Rev. Genet.* **31,** 33–60.

Peterson, R. T., Shaw, S. Y., Peterson, T. A., Milan, D. J., Zhong, T. P., Schreiber, S. L., Macrae, C. A., and Fishman, M. C. (2004). Chemical suppression of a genetic mutation in a zebrafish model of aortic coarctation. *Nat. Biotechnol.* **22,** 595–599.

Quaas, J., and Wylie, C. (2002). Surface contraction waves (SCWs) in the Xenopus egg are required for the localization of the germ plasm and are dependent upon maternal stores of the kinesin-like protein Xklp1. *Dev. Biol.* **243,** 272–280.

Raich, W. B., Moran, A. N., Rothman, J. H., and Hardin, J. (1998). Cytokinesis and midzone microtubule organization in *Caenorhabditis elegans* require the kinesin-like protein ZEN-4. *Mol. Biol. Cell* **9,** 2037–2049.

Roth, M. B., Zahler, A. M., and Stolk, J. A. (1991). A conserved family of nuclear phosphoproteins localized to sites of polymerase II transcription. *J. Cell Biol.* **115,** 587–596.

Sato-Maeda, M., Tawarayama, H., Obinata, M., Kuwada, J. Y., and Shoji, W. (2006). Sema3a1 guides spinal motor axons in a cell- and stage-specific manner in zebrafish. *Development* **133,** 937–947.

Shaner, N. C., Campbell, R. E., Steinbach, P. A., Giepmans, B. N., Palmer, A. E., and Tsien, R. Y. (2004). Improved monomeric red, orange and yellow fluorescent proteins derived from Discosoma sp. red fluorescent protein. *Nat. Biotechnol.* **22,** 1567–1572.

Shaner, N. C., Steinbach, P. A., and Tsien, R. Y. (2005). A guide to choosing fluorescent proteins. *Nat. Methods* **2,** 905–909.

Shoji, W., Isogai, S., Sato-Maeda, M., Obinata, M., and Kuwada, J. Y. (2003). Semaphorin3a1 regulates angioblast migration and vascular development in zebrafish embryos. *Development* **130,** 3227–3236.

Solnica-Krezel, L., and Cooper, M. S. (2002). Cellular and genetic mechanisms of convergence and extension. *In* "Pattern Formation in Zebrafish" (L. Solnica-Krezel, ed.), pp. 136–165. Springer, Berlin.

Thermes, V., Grabher, C., Ristoratore, F., Bourrat, F., Choulika, A., Wittbrodt, J., and Joly, J. S. (2002). I-SceI meganuclease mediates highly efficient transgenesis in fish. *Mech. Dev.* **118,** 91–98.

Traver, D. (2004). Cellular dissection of zebrafish hematopoiesis. *In* "The Zebrafish, 2nd edn.: Cellular and Developmental Biology" (H. W. Detrich, III, M. Westerfield, and L. I. Zon, eds.), Vol. 76, pp. 127–149. Elsevier Academic Press, San Diego.

Traver, D., Herbomel, P., Patton, E. E., Murphey, R. D., Yoder, J. A., Litman, G. W., Catic, A., Amemiya, C. T., Zon, L. I., and Trede, N. S. (2003). The zebrafish as a model organism to study development of the immune system. *Adv. Immunol.* **81,** 253–330.

Trede, N. S., Langenau, D. M., Traver, D., Look, A. T., and Zon, L. I. (2004). The use of zebrafish to understand immunity. *Immunity* **20,** 367–379.

Trevarrow, B., Marks, D. L., and Kimmel, C. B. (1990). Organization of hindbrain segments in the zebrafish embryo. *Neuron* **4,** 669–679.

Urasaki, A., Morvan, G., and Kawakami, K. (2006). Functional dissection of the *Tol2* transposable element identified the minimal *cis*-sequence and a highly repetitive sequence in the subterminal region essential for transposition. *Genetics* **174,** 639–649.

van Eeden, F. J., Palacios, I. M., Petronczki, M., Weston, M. J., and St Johnston, D. (2001). Barentsz is essential for the posterior localization of oskar mRNA and colocalizes with it to the posterior pole. *J. Cell Biol.* **154,** 511–523.

Wada, N., Javidan, Y., Nelson, S., Carney, T. J., Kelsh, R. N., and Schilling, T. F. (2005). Hedgehog signaling is required for cranial neural crest morphogenesis and chondrogenesis at the midline in the zebrafish skull. *Development* **132,** 3977–3988.

Wu, Y., Zhang, G., Xiong, Q., Luo, F., Cui, C., Hu, W., Yu, Y., Su, J., Xu, A., and Zhu, Z. (2006). Integration of double-fluorescence expression vectors into zebrafish genome for the selection of site-directed knockout/knockin. *Mar. Biotechnol. (NY)* **8,** 304–311.

Yazdani, U., and Terman, J. R. (2006). The semaphorins. *Genome Biol.* **7,** 211.1–14.

Yee, C. S., Chandrasekhar, A., Halloran, M. C., Shoji, W., Warren, J. T., and Kuwada, J. Y. (1999). Molecular cloning, expression, and activity of zebrafish semaphorin Z1a. *Brain Res. Bull.* **48,** 581–593.

Yeoh, E. J., Ross, M. E., Shurtleff, S. A., Williams, W. K., Patel, D., Mahfouz, R., Behm, F. G., Raimondi, S. C., Relling, M. V., Patel, A., Cheng, C., Campana, D., *et al.* (2002). Classification, subtype discovery, and prediction of outcome in pediatric acute lymphoblastic leukemia by gene expression profiling. *Cancer Cell* **1,** 133–143.

CHAPTER 11

Identifying and Quantitating Neural Stem and Progenitor Cells in the Adult Brain

Juan Manuel Encinas and Grigori Enikolopov

Cold Spring Harbor Laboratory
Cold Spring Harbor
New York 11724

0091-679X/08 $35.00
DOI: 10.1016/S0091-679X(08)85011-X

Abstract

Adult brain contains neural stem and progenitor cells that are capable of generating new neurons. Active continuous neurogenesis is limited to the subventricular zone of the lateral ventricles and the subgranular zone of the hippocampal dentate gyrus. Newborn neurons gradually become fully functional and integrated into the existing circuitry of the olfactory bulb and the hippocampus. Transition from stem cells to fully differentiation neurons, the neuronal differentiation cascade, occurs through defined steps, and different classes of neuronal precursors can be distinguished by their morphology, expressed markers, and mitotic activity. Cells in these classes can be identified by immunophenotyping, labeling with thymidine analogues, and infection with retro- and lentiviral vectors. We here describe a transgenic approach that allows identification, *in vivo* visualization, isolation, and accurate enumeration of various classes of stem and progenitor cells in the adult brain. We generated a series of reporter mouse lines in which neural stem and progenitor cells express various fluorescent proteins (GFP, CFPnuc, H2B-GFP, DsRedTimer, and mCherry) under the control of the regulatory elements of the nestin gene. Using these lines, we were able to dissect the neuronal differentiation cascade into several discrete steps and to evaluate the changes induced by various neurogenic and antineurogenic stimuli. In particular, nuclear localization of the fluorescent signal in nestin-CFPnuc mice greatly simplifies the distribution pattern of neural stem and progenitor cells and allows accurate quantitation of changes induced by neurogenic agents in distinct classes of neuronal precursors. We present protocols for applying confocal microscopy, stereology, and electron microscopy to evaluate changes in the neurogenic compartments of the adult brain.

I. Introduction

A. Neural Stem and Progenitor Cells and Neuronal Differentiation Cascade in the Adult Brain

New neurons are continuously generated from neural stem and progenitor cells in the brain of adult rodents and primates (Abrous *et al.*, 2005; Alvarez-Buylla *et al.*, 2001; Gage, 2000; Kempermann, 2006; Kempermann *et al.*, 2004; Lie *et al.*, 2004; Lledo *et al.*, 2006; Ming and Song, 2005; Song *et al.*, 2005; Taupin and Gage, 2002).

Neural stem cells are defined as self-renewing, multipotent cells, usually with long life span, that generate neurons, astrocytes, and oligodendrocytes. Progenitor cells have a limited life span, less self-renewal ability, and may be multipotential or unipotential (e.g., only generating neurons). Persistent production of new neurons is limited to two areas of the adult brain: the olfactory bulb (OB) and the dentate gyrus (DG) of the hippocampus. Neurons of the OB originate in the anterior part of the subventricular zone (SVZ) of the lateral ventricles (LVs), whereas neurons of the DG are generated in the subgranular zone (SGZ) of the DG (Fig. 1). Neural precursors in the SVZ migrate through a network of tangential pathways, and converge onto the rostral migratory stream (RMS) before arriving to the OB and differentiating into granule and periglomerular neurons (Alvarez-Buylla and Garcia-Verdugo, 2002; Lois and Alvarez-Buylla, 1994). Neural precursors of the adult DG are born in the SGZ and then migrate locally to the granule cell layer (GCL) and differentiate into granule neurons.

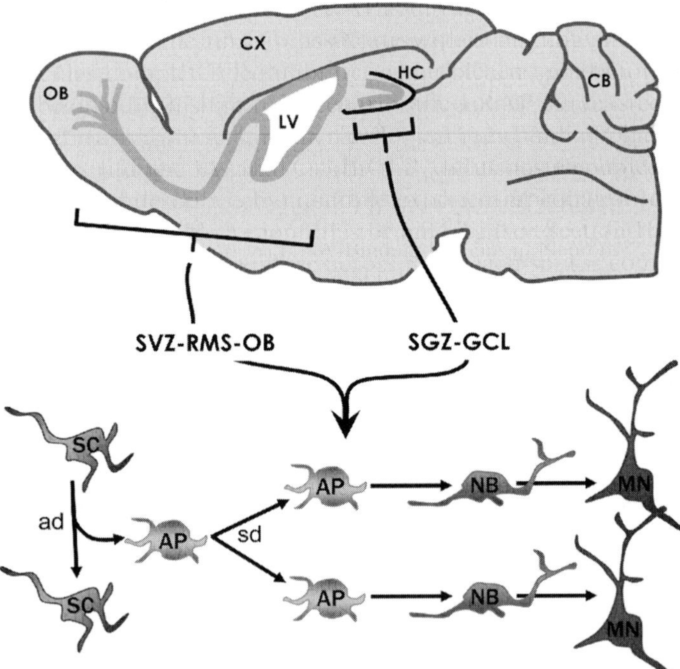

Fig. 1 Adult neurogenesis persists in two areas of the adult brain. The subventricular zone (SVZ) of the lateral ventricle (LV) generates neuronal precursors that will reach the olfactory bulb (OB) through the rostral migratory stream (RMS). The subgranular zone (SGZ) of the dentate gyrus generates neuronal precursors that will integrate locally in the granule cell layer (GCL). In both cases, a similar chain of events is followed to generate new neurons: Neural progenitors with stem-cell capabilities (SC) divide asymmetrically (ad) giving rise to amplifying precursors (AP) that divide symmetrically (sd). Then, they exit the cell cycle and differentiate into neuroblasts (NB) that will finally evolve to mature neurons (MN). CX, cortex; HC, hippocampus; and CB, cerebellum. (See Plate no. 33 in the Color Plate Section.)

Newly generated neurons are fully functional: for instance, new DG neurons start extending axons several days after the last mitosis and, when fully differentiated, receive synaptic input and project functional connections into the CA3 region of the hippocampus (Ge *et al.*, 2006; Song *et al.*, 2002; Tashiro *et al.*, 2006; van Praag *et al.*, 2002). A full transformation of a neural stem cell into a functional neuron takes 25–30 days. During this time, cells undergo asymmetric and symmetric divisions, exit the cell cycle, express a wide range of markers, change their morphology, and establish connections with other cells. This transition from stem cells to fully integrated neurons, the neuronal differentiation cascade, proceeds through defined steps that can be distinguished through a combination of markers, morphological features, and mitotic activity, and will be discussed in more detail below.

B. Adult Neurogenesis Is a Dynamic Process

Adult neurogenesis is modulated by a very wide range of intrinsic and extrinsic factors (Abrous *et al.*, 2005; Lledo *et al.*, 2006; Ming and Song, 2005). It is regulated by growth factors (e.g., epidermal-, fibroblast-, brain-derived, and insulin-like growth factors), neurotransmitters (e.g., serotonin, dopamine, glutamate, acetylcholine, norepinephrine, and nitric oxide), hormones (e.g., estrogen, prolactin, and corticosteroids), and drugs (e.g., antidepressants, opiates, and lithium). Furthermore, it is influenced by aging, pregnancy, stress, disease, physical activity, enriched environment, dietary restrictions, and learning. To give one set of examples, hippocampal neurogenesis is inhibited by chronic stress, depression, and posttraumatic stress disorder; conversely, it is augmented by antidepressant drugs directed at monoamine neurotransmitters (e.g., the selective serotonine reuptake inhibitor fluoxetine), brain-derived growth factor and insulin-like growth factors (both of which show efficacy in animal models of depression), and electroconvulsive shock (Dranovsky and Hen, 2006; Malberg and Blendy, 2005; Warner-Schmidt and Duman, 2006). Moreover, recent evidence indicates that neurogenesis may be an obligatory step in the behavioral action of antidepressants (Santarelli *et al.*, 2003).

Factors that lead to a net increase or decrease in the number of new neurons may, in principle, affect any step in the differentiation cascade that converts neural stem cells into fully differentiated neurons, for example, symmetric and asymmetric divisions of different types of precursors, their survival, or their differentiation; both the cell populations that are targeted by neurogenic stimuli and the molecular mechanisms that mediate the action of these stimuli are only beginning to be understood.

C. Identification and Quantification of Newborn Cells in the Adult Brain

A detailed analysis of neurogenesis requires approaches to describe distinct steps in the differentiation cascade that converts stem cells into differentiation neurons, to identify cell types within the cascade, to label these cells types and follow their lineage, and to quantify the changes induced by neurogenic stimuli. Traditionally, the main approaches used to study adult neurogenesis were to label proliferating cells with thymidine analogues (mainly with 8-bromodeoxyuridine, BrdU) and to categorize by phenotype the subclasses of precursors and differentiated cells using immunocytochemistry. More recently, retroviral and lentiviral labeling, double labeling with halogenated thymidine analogues, and generation of transgenic animals with fluorescently labeled subclasses of precursors have been added to the list of experimental approaches.

Labeling of dividing cells with BrdU has become a standard method for studying adult neurogenesis. Its main advantage is that the cells that have undergone DNA synthesis can be detected in days, weeks, and months after the labeling, thus allowing lineage analysis of the newly generated cells. Moreover, in combination with cell-type specific markers, it allows for double and triple labeling and thus, more precise phenotyping of labeled cells. Importantly, the BrdU signal is restricted to the nucleus, facilitating the scoring of the signal and accurate enumeration of labeled cells. Note, however, that BrdU may be incorporated into damaged cells (Kuan *et al.*, 2004); that high doses of BrdU may be toxic to cells; that changes in the length of the cell cycle or S phase and the details of the labeling schedule may have profound effects on the fraction of labeled cells and thus, the interpretation of the results when different treatments are compared (Hayes and Nowakowski, 2002); and that identification of labeled cells can only be achieved with fixed tissue (precluding, for instance, electrophysiological studies of new neurons and their precursors).

Nucleotide labeling of new neurons can be further elaborated by using two halogenated thymidine analogues, 8-chlorodeoxyuridine (CldU) and 8-iododeoxyuridine (IdU) (Burns and Kuan, 2005; Vega and Peterson, 2005). This allows a much more precise temporal discrimination of the cell cycle progression and elimination of new neurons. When combined with immunophenotyping, this method may become a powerful tool for high-resolution analysis of cell proliferation and lineage determination in the adult brain.

Some of the limitations of BrdU labeling can be overcome by using retroviruses carrying reporter transgenes [such as genes for green, cyan, or yellow fluorescent proteins (GFP, CFP, or RFP)] to label dividing cells and their progeny. Infected cells, after undergoing mitosis, become permanently labeled and, most importantly, can be accessed for the morphological and electrophysiological studies (Ge *et al.*, 2006; van Praag *et al.*, 2002). Note, however, that cells have to undergo division to be labeled with the retroviral vectors [this limitation may be overcome through the use of lentiviral vectors (Geraerts *et al.*, 2006)]; that labeling efficiency is low and variable and, therefore, not amenable to quantitative analysis; and that labeling requires an invasive manipulation (stereotaxic injection) which can elicit an inflammatory reaction.

Selected classes of neural stem and progenitor cells can be identified through immunophenotyping. The most primitive precursor cells both in the SVZ and in the DG express glial fibrillary acidic protein (GFAP), vimentin, brain lipid-binding protein (BLBP), Sox2, and nestin; their progeny start to lose these markers and to express Olig2, Tbr2, neurogenin 1, doublecortin (Dcx), Prox-1, and a host of other markers whose combination defines specific steps of the differentiation cascade. When combined with BrdU labeling, immunophenotyping can provide a detailed view of the birth, maturation, and differentiation of newborn neurons. An obvious limitation of the approach is the requirement to fix the tissue which precludes *in vivo* analyses of these cells, for example, isolation for RNA or protein profiling, imaging, lineage tracing, transplantation, or electrophysiological studies.

D. Nestin Marks Neural Stem and Progenitor Cells

Nestin (for *ne*uroepithelial *st*em; Lendahl *et al.*, 1990) is an intermediate filament protein selectively expressed in neural stem and early progenitor cells of the developing and adult nervous system. It was originally identified in neuroepithelial cells as the protein reacting with the monoclonal antibody Rat401 by McKay and coworkers and, so far, has been the best marker that correlates with neural stem cell potential.

Nestin mRNA and protein are abundantly expressed in the developing and adult nervous system. Nestin is also strongly expressed in the myotomes of the embryo. In addition, nestin expression has been reported in several other tissues and cell types, for example, pancreatic islets, the developing testis, tongue, tooth, and heart (Kachinsky *et al.*, 1994; Sejersen and Lendahl, 1993; Terling *et al.*, 1995; Zulewski *et al.*, 2001). This relatively wide spectrum of nestin expression reflects the presence of various regulatory elements in the nestin gene: for instance, expression of nestin in the embryonic neuroepithelium is dependent on the presence of transcriptional enhancer which resides in the second intron of the gene (Josephson *et al.*, 1998; Yaworsky and Kappen, 1999; Zimmerman *et al.*, 1994), whereas regulatory elements in the first and third introns direct nestin expression to myotomes (Yaworsky and Kappen, 1999; Zimmerman *et al.*, 1994). The neural enhancer in the second intron of the nestin gene is strong and dominant, being sufficient to direct the expression of an exogenous transgene to the developing neuroepithelium in the transient transgenic assay even when combined with a heterologous promoter (e.g., promoter of the herpes virus thymidine kinase gene; Zimmerman *et al.*, 1994). Importantly, the use of this enhancer element seems to "rectify" the expression pattern of the transgene: for instance, although endogenous nestin is expressed both in the nervous system and in the myotomes of the embryo, expression of the transgene in the myotomes is abrogated if only the second intron is used in the transgenic construct. Thus, regulatory elements residing in the second intron of the nestin gene are both necessary and sufficient to direct the expression of a transgene into neural stem and progenitor cells, and these elements have been used to generate a number of transgenic mouse lines (Encinas *et al.*, 2006; Imayoshi *et al.*, 2006; Kawaguchi *et al.*, 2001; Mignone *et al.*, 2004; Tronche *et al.*, 1999; Yamaguchi *et al.*, 2000).

E. Transgenic Reporter Lines for Visualizing Neural Stem and Progenitor Cells

We used neurospecific regulatory elements of the nestin gene to generate several transgenic mouse lines that allow direct visualization of neural stem and progenitor cells in the developing and adult brain (Encinas *et al.*, 2006; Mignone *et al.*, 2004). In these animals, fragments of the nestin gene (5.8 kb of the promoter region and 1.8 kb of the second intron), combined with a polyadenylation signal from SV40, drive expression of GFP (Fig. 2A), fusion of GFP with H2B histone (H2B-GFP), RFPs DsRedTimer and mCherry, or CFP with nuclear localization signal (CFPnuc, Fig. 2B). Several independent lines were obtained for each transgene and all demonstrated a highly similar pattern of transgene expression in the developing and adult brain [note that similar patterns of expression were obtained with other transgenes containing the second intron of nestin (Kawaguchi *et al.*, 2001; Yamaguchi *et al.*, 2000), even though these transgenes employed heterologous promoters].

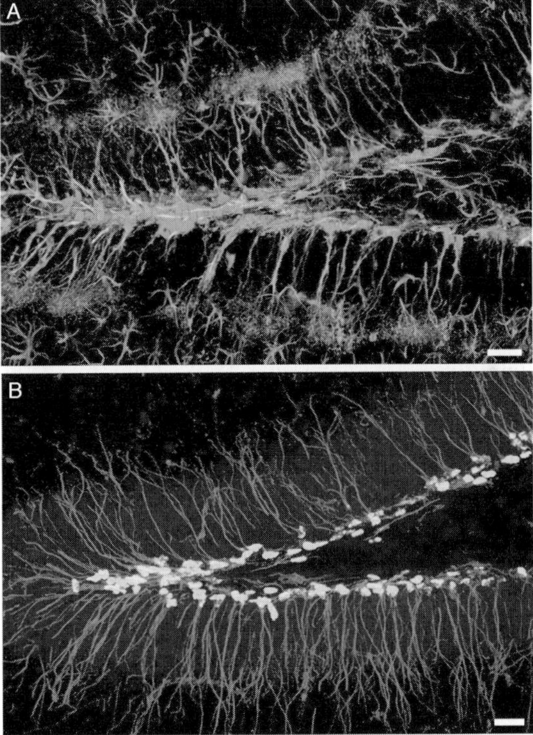

Fig. 2 Transgenic mice to visualize neuroprogenitor cells in the adult brain. (A) Neural stem cells and progenitors (green) in the dentate gyrus of a nestin-GFP mouse. Red is immunostaining for glial fibrillary acidic protein (GFAP). (B) Neural stem and progenitor cells (green) in the dentate gyrus of a nestin-CFPnuc mouse. Red is immunostaining for nestin and blue is 4′,6-diamidino-2-phenylindole (DAPI). Scale bars are 25 μm in (A) and (B). (For Panel A. The Journal of Comparative Neurology, Vol. 469, No. 3, 2004, pp. 311–324. © 2004 Wiley-Liss, Inc. Reprinted with permission of Wiley-Liss, Inc., a subsidiary of John Wiley & Sons, Inc.) (See Plate no. 34 in the Color Plate Section.)

No obvious defects were apparent during development and adulthood of the transgenic mice. The transgene was normally transmitted and, in the case of nestin-GFP mice, the expression pattern in the developing and adult CNS remained invariant over the course of at least 20 generations.

Several lines of evidence demonstrate that expression of GFP or CFPnuc marks neural stem and early progenitor cells in our nestin-GFP and nestin-CFPnuc transgenic mice:

a. The transgene is expressed in those areas of the developing embryo that correspond to the neuroepithelial cells of the developing nervous system.
b. The transgene is expressed in those areas of the adult brain (SVZ, RMS, OB, and DG) that are marked by persistent production of new neurons.
c. GFP and CFPnuc expression is absent in those cells that have already undergone differentiation and in those areas of the brain that only contain fully differentiated cells.
d. The sites of the transgene expression in the developing and adult nervous system overlap with the sites of expression of nestin that has served as a reliable marker of neural stem cells.
e. GFP- and CFPnuc-positive cells are capable of forming neurospheres and producing a variety of types of progeny *in vitro*.
f. GFP-expressing cells are strongly (~1400 fold) enriched in neurosphere-forming cells, and, conversely, most of the neurosphere-forming cells of the adult brain are present within the fraction of GFP-expressing cells.

Together, these results indicate that GFP- and CFPnuc-positive cells in the nestin-GFP and nestin-CFPnuc transgenic animals accurately represent neural stem and early progenitor cells in the developing and adult nervous system.

F. Using Transgenic Reporter Lines to Dissect Neuronal Differentiation Cascade in the DG

We have used the nestin-GFP and the nestin-CFPnuc reporter lines to define discrete steps in the neuronal differentiation cascade in the DG (leading from stem/progenitor cells to differentiated granule neurons), based on the morphology of the cells, the marker proteins that they express, and their mitotic activity (measured by BrdU incorporation) (Encinas *et al.*, 2006). We identify six classes of cells in the neuronal lineage in the DG of nestin-CFPnuc mice; these classes encompass and partially overlap with the categories of neuronal precursors defined by other approaches (Fukuda *et al.*, 2003; Kempermann *et al.*, 2004; Kronenberg *et al.*, 2003; Mignone *et al.*, 2004; Seri *et al.*, 2004).

The first class is represented by GFAP/nestin/vimentin/BLBP/Sox2-positive nestin-GFP and nestin-CFPnuc cells. The triangular soma and the nuclei of these cells reside in the SGZ; they extend a single or double apical process radially across the GCL, terminating as an elaborated arbor of very fine leaf-like processes in the molecular layer. Cells of this class correspond to the most primitive, stem-like population in the DG; note, however, that not all of the criteria of stem cells, for example ability to self renew, have been demonstrated for these cells (Seaberg and van der Kooy, 2003). Only a small fraction of these cells (less than 2%) are labeled by BrdU after a short (2 h) pulse, indicating their low rate of division and consistent with the quiescent state of these cells; we therefore designate these cells as quiescent neural progenitors (QNPs).

The second class is represented by small (somatic diameter ~10 μm) round or oval cells located in the SGZ. These cells also express nestin-GFP or nestin-CFPnuc but they do not stain for GFAP or vimentin and stain very weakly for nestin (this may indicate that reporter fluorescent proteins persist in these cells longer than nestin, or that the nestin is unequally distributed during cell division); they also do not stain for Dcx, for PSA-NCAM, or for markers of differentiated neurons. These cells are labeled with BrdU at high frequency (20–25%, 2 h after a single injection of BrdU) indicating that most of them are involved in mitotic activity; we designate these cells as amplifying neural progenitors (ANP). They are often seen in clusters extending along the SGZ; when the plane of division of cells in these clusters is visible, it is most often perpendicular to the SGZ such that the daughter cells remain in the SGZ. Importantly, a fraction of these cells are seen separating from QNPs after mitosis; in each case, the division plane is parallel or slightly oblique to the SGZ such that the daughter cell is deposited beneath the QNP cell (the plane of division may explain why these cells do not inherit GFAP, vimentin, or nestin which are predominantly localized to the apically positioned processes of the QNPs but not to their soma). Together, our results suggest that QNP cells undergo asymmetric divisions and give rise to ANP cells, which then propagate in the SGZ through a series of symmetric divisions.

The next class of precursor cells, still located in the SGZ, ceases to express nestin or nestin-driven reporters and starts to express Dcx and PSA-NCAM. A small subclass (~1% of cells in this class) morphologically resembles ANPs, carries short (1–5 μm) horizontal processes, and is the final population in the differentiation cascade that is labeled by BrdU. Most of the cells in this class are represented by larger (10–15 μm somatic diameter) cells which extend longer (10–30 μm) horizontal processes in the plane of the SGZ and do not incorporate BrdU. These cells stain for Dcx, PSA-NCAM, and Prox-1. Thus, the bulk of this class is represented by postmitotic neuronal precursors; we designate them as type 1 neuroblasts (NB1).

Cells of the next class, type 2 neuroblasts or NB2, are larger than NB1 cells (somatic diameter ~15 μm) and remain confined to the SGZ. They extend longer (20–40 μm) processes horizontally and obliquely to the plane of the SGZ. They do not express nestin, nestin-GFP, or CFPnuc, and express Dcx, PSA-NCAM, Prox-1, and NeuN.

The next class of cells corresponds to immature neurons (IN). They are larger than the cells of the previous classes (somatic diameter 15–20 μm), and their morphology resembles that of mature granule cells of the DG. Their soma is round or oval and can be found both in the SGZ and, mainly, in the GCL. These cells carry a single apical process that branches in its distal part located in the molecular layer. They express Dcx, PSA-NCAM, Prox-1, and NeuN.

The next class represents differentiated granule neurons, with developed apical dendrites and axons forming the mossy fiber. They cease to express PSA-NCAM and Dcx, but express NeuN and Prox-1.

The differentiation cascade in the DG of nestin-CFPnuc mice can thus be divided into discrete steps based on the expression of markers, morphology, and mitotic activity (Fig. 3).

G. Using Reporter Lines to Quantify Neural Stem and Progenitor Cells

Accurate enumeration of neural precursors using immunocytochemistry is often problematic: high cell density, complex cell morphology, and uncertainties in defining distinct boundaries between subclasses of cells present a real challenge when precise counts are required (for instance, when evaluating the action of a neurogenic stimulus). This reduces the precision of evaluating changes in particular subclasses of neuronal precursors (e.g., in contrast to BrdU or thymidine labeling of cell nuclei, where great precision can be achieved); this problem is particularly acute in the young brain, where the number of neural stem and progenitor cells is particularly high or when the changes evoked by a stimulus are low (note that most of the known inducers of neurogenesis increase the number of newly generated cells only by 30–50%).

For experiments which require both morphological and quantitative analysis of neurogenesis, use of two reporter lines, nestin-GFP and nestin-CFPnuc, is particularly helpful. In nestin-GFP mice, the fluorescent signal highlights all of the soma and the processes of stem and early progenitor cells (Fig. 2A) and these mice are very well suited for the studies of the morphology of neuronal precursors in the developing and adult brain. In contrast, in nestin-CFPnuc mice, the signal is localized in the cell nucleus and the distribution of the stem and progenitor cells is visualized as a punctuate pattern; this nuclear representation of stem and progenitor cells greatly reduces the complexity of their distribution pattern and permits their unambiguous enumeration (thus capturing the power of BrdU- or thymidine-based enumeration of nuclei) (Fig. 2B). Thus, these two reporter lines complement each other and allow visualization and counting of neural stem and progenitor cells. In our pilot experiments, we carefully compared the structures of the SVZ and DG as revealed by immunochemistry for nestin and by expression of nestin-CFPnuc or nestin-GFP. Whereas we were unable to generate accurate counts of nestin- or nestin-GFP-positive cells (particularly in the young brain), we were able to unambiguously count all of the labeled nuclei in the SVZ and DG of the nestin-CFPnuc mice. Importantly, crosses between these two lines allow simultaneous visualization of the soma and the nuclei of stem and progenitor cells, thus we were able to follow the morphological changes in these cells while enumerating them (Encinas, Chiang, and Enikolopov, unpublished data).

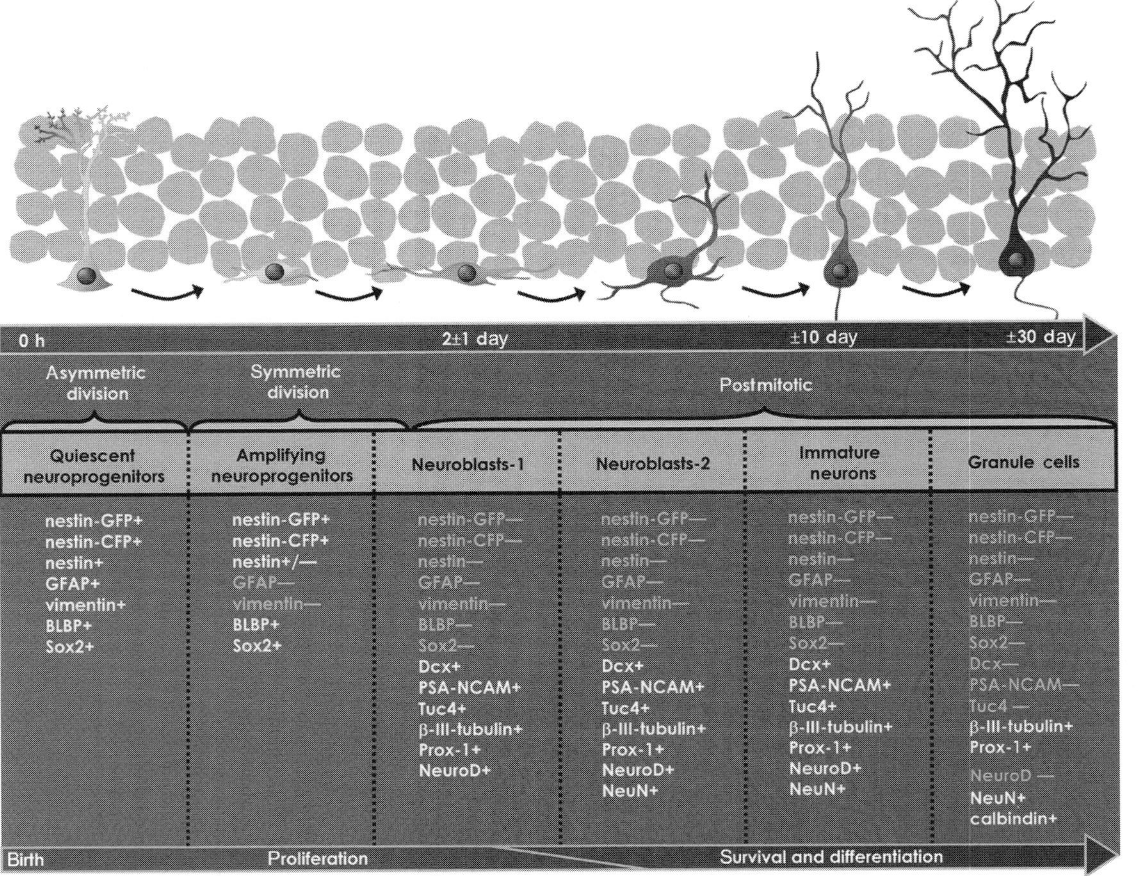

Fig. 3 Schematic representation of the consecutive steps taken to generate a new neuron in the adult dentate gyrus. Each step represents a cellular differentiation stage and is characterized by different cell markers as well as different mitotic capability. Modified from Encinas *et al.* (2006). (See Plate no. 35 in the Color Plate Section.)

In summary, our approach with transgenic reporter lines that circumvents several obstacles is assessing changes in cell number during neurogenesis, for example, high cell density which hinders precise counts or uncertainty in attributing precursor cells to a particular class. It reduces the complex distribution pattern of precursor cells to a readily quantifiable punctuate pattern of labeled nuclei. It allows unambiguous enumeration of cells in a particular precursor class and can be used to analyze changes induced by a wide range of stimuli in the developing or adult brain.

II. Protocol I: Immunofluorescence Microscopy of Nestin-GFP and Nestin-CFPnuc Cells

Confocal microscopy is a crucial tool for the analysis of transgenic reporter mice, allowing quantification of neuroprogenitors and visualization of their anatomical and morphological features. In the nestin-GFP mice, GFP fluorescence reveals the entire stem or progenitor cell, helps to distinguish between the subtypes of neuroprogenitors, and, when combined with immunodetection of other cell-specific markers, makes it possible to investigate changes in protein expression patterns and anatomical relations with other cell types (Fig 4A).

In the case of the nestin-CFPnuc mice, because the fluorescent signal in both QNPs and ANPs is restricted to the nucleus (Fig. 4B), it is difficult to distinguish

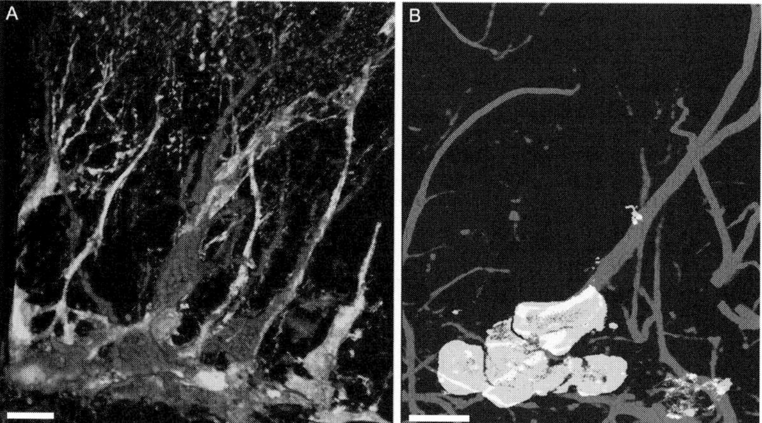

Fig. 4 Confocal images of nestin-GFP and nestin-CFPnuc mice after immunostaining. (A) Neuroblasts (red is PSA-NCAM immunofluorescence) use quiescent neural progenitors (QNPs) (green is nestin-GFP expression) as scaffolds to integrate into the granule cell layer. (B) A cluster of amplifying neural progenitors (ANPs) locates below a QNP in the subgranular zone (SGZ) [red is glial fibrillary acidic protein (GFAP) immunofluorescence; green is nestin-CFPnuc expression detected with an anti-GFP antibody]. Scale bar is 10 μm in (A) and 5 μm in (B). (See Plate no. 36 in the Color Plate Section.)

between the two populations of neuroprogenitors in the hippocampus. This problem can be overcome by using antibodies against BLBP, GFAP, nestin, or vimentin, which highlight the apical processes characteristic of QNPs, thus affording a distinction between QNPs and ANPs (Fig. 4B).

A. Perfusion

1. Anesthetize the animals with an overdose of 15% chloral hydrate (10 μl/g of bodyweight).
2. Perform transcardial perfusion by inserting a cannula into the left ventricle and introducing 30 ml of phosphate-buffered saline (PBS) at fast flow (10 ml/min), followed by 30 ml of 4% (w/v) paraformaldehyde (PF) in PBS (pH 7.4), with the first 15 ml delivered at fast flow (10 ml/min) and the next 15 ml at slower flow (5 ml/min).

B. Postfixation

1. Remove the brain and place in a vial with 4% PF in PBS for 4 h at room temperature.
2. Transfer the brain to PBS with 0.4% PF and keep at 4°C until sectioning.

Comments

- Larger volumes of PBS (~30 ml) improve washing of the brain tissue and result in less background after immunostaining without affecting antigenicity.
- A useful indicator of how well the perfusion is proceeding is the liver becoming white and clean of blood.
- Four hours of postfixation is adequate for the brain tissue to get fixed. Overfixation increases the likelihood of losing antigenicity and can unevenly affect different antibodies.
- Adding a small amount of PF to the PBS (0.01%) for tissue storage prevents growth of bacteria or fungi without affecting the antigenicity.
- It is best to use the samples as soon as possible, as prolonged storage increases the chance of losing antigenicity.

C. Sectioning

1. Slice the brain sagittally into halves, select one hemisphere for sectioning, and store the other.
2. Prepare 50-μm-thick sections using a Vibratome, orienting the medial surface on the platform or stage such that the slices are collected in the lateral to medial direction.
3. Collect the slices into PBS (with 0.01% PF if they are going to be stored).

Comments

- 50-μm-thick sections can be handled and processed readily and most of the antibodies easily penetrate the entire thickness of the section. This thickness is optimal for confocal microscopy; however, thinner sections are preferable for brightfield or epifluorescence microscopy.
- If the tissue is well fixed, the speed of the vibrating blade can be set high to save time. The amplitude or frequency can be lowered to avoid a "ruffle" effect on the surface of the slices. However, for softer tissues, lower speed and higher amplitude or frequency should be used.
- If the brain tissue is too soft (as in samples from perinatal animals), it can be embedded in 2% agarose to provide consistency and support. The agarose does not penetrate the tissue or affect antigenicity.

D. Fluorescence Immunostaining

This is the standard protocol that we routinely use, but it may require modifications specific for different antibodies.

1. Blocking and permeabilization:
   - Incubate the slices for 30 min (up to 2 h), at room temperature with PBS containing 0.2% Triton and 3% bovine serum albumin (BSA).
2. Primary antibody incubation:
   - Incubate the slices with the primary antibodies in blocking and permeabilization solution overnight (12–16 h) at 4 °C or for 3–4 h at room temperature.
   - Rinse with PBS (or PBS with 0.2% Triton X) three times, 5 min per rinse.
3. Secondary antibody incubation:
   - Incubate the slices with the appropriate secondary (fluorochrome-conjugated) antibodies combined together, in blocking and permeabilization solution, for 1 h at room temperature in darkness.
   - Rinse with PBS three times, 5 min per rinse.
4. Mounting:
   - Mount the slices on the slide and let dry in darkness until they are stuck to the glass.
   - Add mounting medium and cover the slices with a coverslip without pressing down. Seal the edges with nail polish.
   - Let dry for 10 min at room temperature in darkness.
   - Store at 4 °C in darkness.

Comments

- Each antibody may require slightly different conditions, for instance, using serum instead of BSA (in this case, use serum from the animal in which the secondary antibody was generated), omitting serum or BSA (BSA still has to be used at the blocking and permeabilization stage), or using different incubation times. In these cases, the primary antibodies can be incubated with the slices consecutively instead of at the same time. Similar considerations (i.e., using them sequentially rather than in combination) apply to dealing with the secondary antibodies.
- Leaving the slices on the slide for too long after mounting or pressing down on the coverslip may flatten the slices and distort the tissue.

E. BrdU Labeling

To make BrdU accessible to antibodies for immunostaining, the samples must be treated with acid prior to incubating them with the primary antibodies:

1. Incubate the brain slices with 2 M HCl, for 30 min at 37 °C.
2. Rinse the slices with PBS two times, 5 min per rinse.
3. Incubate the slices with 0.1 M sodium tetraborate for 10 min at room temperature.
4. Rinse the slices with PBS two times, 5 min per rinse.
5. Continue with blocking and permeabilization.

The acidic treatment affects the ability of the tissue to be stained by some antibodies and thus may cause problems for double labeling with BrdU. In these cases an alternative method may be used:

1. Incubate the slices with 50% formamide in sodium citrate buffer (SSC) two times at 65°C for 10 min.
2. Rinse the slices in SSC two times for 5 min.
3. Incubate with 2 M HCl at 37°C for 30 min.
4. Rinse with 0.1 M boric acid (pH 8.5) for 10 min.
5. Rinse the slides with PBS two times, 5 min per rinse.

F. Triple Labeling

When triple immunolabeling is required, it can be carried out as described above. Some recommendations apply.

- Typically, triple immunolabeling is carried out with green, red, and far red fluorochromes. The spectrum of emission of the far-red fluorochrome to which the secondary antibody is conjugated must be checked, avoiding those fluorochromes which have a noticeable peak in the red wavelength range. While the separation of the green and red fluorochromes should never be problematic, the far-red dyes must be chosen carefully. Many of them have an emission spectrum with two peaks: a major peak around the 633 nm wavelength (the real far-red peak) and another peak around 600 nm. This second, smaller, peak will be detected in the red channel, thus interfering with the true red signal.

• The samples must be analyzed as soon as possible because the far-red fluorochromes decay much faster than the green and red ones. This is especially true for the far-red peak of the emission spectrum, and only the red peak will remain after a few days.

• When preparing images of triple immunolabeling, the dark blue color usually attributed by software default to the far-red channel should be avoided. This dark blue does not contrast enough with the black background and the viewer's eye will miss information. The details are much easier to observe if a lighter blue or light gray or even white is attributed to the signal from the far-red channel (Fig. 5).

III. Protocol II: The Use of Confocal Stereology to Quantify Changes in Defined Classes of Neuronal Precursors

Quantitative analysis of the changes in different classes of neuronal precursors in response to neurogenic stimuli is crucial for identifying the classes that respond to the stimulus, that is, the steps within the differentiation cascade targeted by the stimulus. Accurate cell enumeration can be achieved through the use of design-based stereology (Gundersen *et al.*, 1999; Howell *et al.*, 2002; Peterson, 1999;

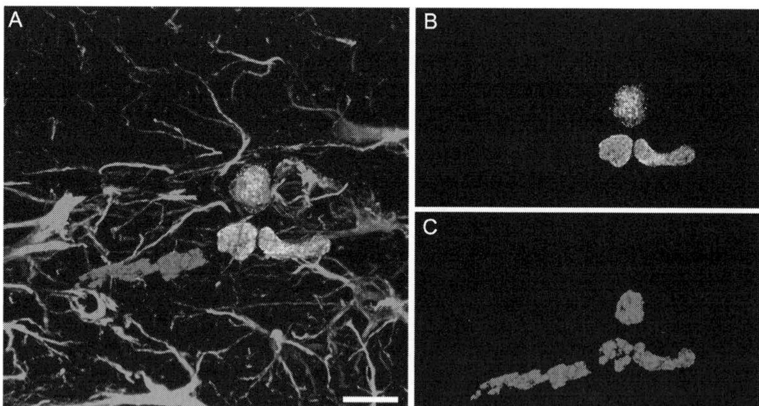

Fig. 5 Confocal image of the dentate gyrus of a nestin-CFPnuc mouse after triple immunolabeling. (A) Blue is glial fibrillary acidic protein (GFAP) immunostaining; (B) green is nestin-CFPnuc expression visualized with green fluorescent protein (GFP) immunostaining; and (C) red is 8-bromodeoxyuridine (BrdU) immunostaining. The animal was administered BrdU and sacrificed 48 h later. Scale bar is 10 μm. (See Plate no. 37 in the Color Plate Section.)

Schmitz and Hof, 2005); this approach is particularly powerful when combined with confocal microscopy which provides high optical resolution, the ability to automatically collect a series of focal planes, and the potential to follow several signals simultaneously. We routinely use confocal stereology to analyze the changes in adult neurogenesis, applying the Fractionator method together, as needed, with the Optical Disector method and the Cavalieri principle. To help explain how an experimental design based on confocal stereology is executed, we will illustrate the method using as an example our study of how antidepressant fluoxetine (Prozac) affects adult neurogenesis (Encinas *et al.*, 2006). Prolonged treatment with fluoxetine increases neurogenesis in the DG (Malberg *et al.*, 2000); moreover, this increase may be necessary for the behavioral action of the drug (Santarelli *et al.*, 2003). In our study, we used nestin-CFPnuc and nestin-GFP reporter mice and confocal stereology to determine the classes of neuronal precursors targeted by fluoxetine and we will refer to this study in the protocols below.

A. The Fractionator Method

1. Sectioning

1. Slice the brain sagittally into halves, select randomly one hemisphere for sectioning, and store the other.
2. Prepare 50-μm-thick sections using a Vibratome, orienting the medial surface on the platform or stage such that the slices can be collected in the lateral to medial direction.
3. Collect the slices in PBS (with 0.01% PF if they are going to be stored) in a multiwell plate.

2. Section Sampling

1. Discard the most lateral sections and start collecting just when the DG appears [coordinates 3.12 mm, lateral (Franklin and Paxinos, 1997)].
2. Collect the slices, following a fractionator scheme (Fig. 6), in 6 parallel sets, each set consisting of 11–12 slices:

The 1st slice goes into the 1st well of the plate. The 2nd slice goes into the 2nd well, the 3rd slice goes into the 3rd well, and so on until the 6th slice. The 7th slice goes into the 1st well, the 8th slice goes into the 2nd well, the 9th slice goes into the 3rd well, and so on until the 12th slice, at which point the 13th slice goes into the 1st well and this process is repeated until the sample is completely sliced. Thus, each slice is 300 μm apart from the next slice within the set in any given well. Typically, in an adult mouse, the DG spans ca. 70 slices, each 50-μm-thick, so that by collecting 6 sets in 6 wells, one can expect having at least 11 slices in each set.

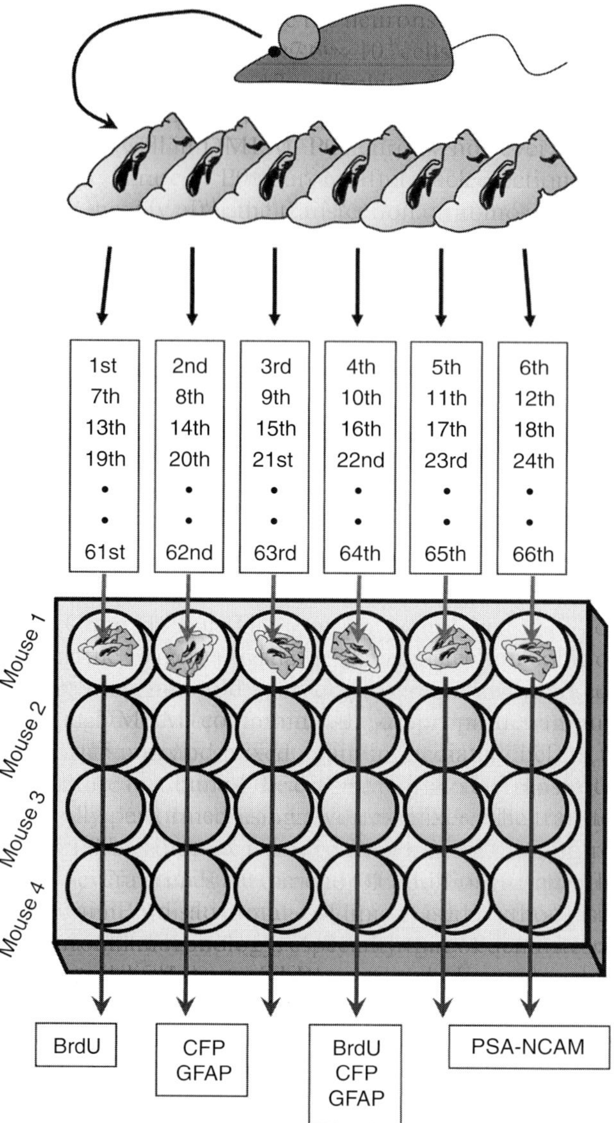

Fig. 6 Collection of samples following the Fractionator principle. Collecting the samples in this unbiased manner allows performing different experiments in equally representative sets of samples. The four rows of six wells each represent the slices collected from four different mice.

Thus, each well contains a set of slices that is a fraction (one sixth in this case) of the reference space (the DG). Each one of these fractions is a representative sample of the total DG, and any one of them can be used for quantification purposes (Fig. 6).

3. Cell Counting for the Analysis of Cell Proliferation

1. For cell counting, select one set of slices and carry out BrdU immunostaining. On each slice, all BrdU$^+$ cells excluding those in the uppermost focal plane should be counted using an epifluorescence microscope or a confocal microscope and a 63× (or similar) objective. In each field, focus on the most superficial cells (those in the uppermost focal plane), but do not count them. Then, move the focal plane deeper into the tissue and count all cells that now appear. The precaution of excluding cells in the first focal plane is crucial: because the cells in the surface layer can be cut, one part of them may be in one slice and the other part in the next slice; thus, if all cell profiles are counted, those cells cut into two will be counted twice, resulting in an error of overestimation.
2. Once all cells in one set of slices are counted, that count must be multiplied by the number of sets (six in our case) to obtain the total number of cells of interest in the examined volume of tissue. Because only one half of the brain was used for analysis, multiplying the result by 2 yields the total number of cells of interest in the entire brain.
3. Other sets of slices can be analyzed for different parameters, for example, the number of nestin-CFPnuc$^+$ cells with or without GFAP labeling, reflecting the number of QNP and ANP cells, respectively. Another set of slices can be used to quantify PSA-NCAM$^+$ cells and, applying the morphological criteria explained above, to distinguish between NB1, NB2, and IN cells.

In our experiments with fluoxetine we found that fluoxetine increases cell proliferation (the number of BrdU$^+$ cells) and the number of ANPs in the adult DG; the number of NB1 cells was also increased as a consequence of the increase in the number of ANPs.

The increase in the number of ANPs could be explained by an increase in either QNP or ANP divisions. To distinguish between these possibilities, we carried out triple immunolabeling (BrdU/CFP/GFAP) on another set of slices and examined all BrdU$^+$ cells in each slice (in each case excluding the uppermost focal plane as described) for colocalization with CFP and GFAP, thus analyzing the number of BrdU$^+$ QNPs and BrdU$^+$ ANPs. The number of BrdU-labeled QNPs did not change whereas the number of BrdU-labeled ANPs increased, suggesting that fluoxetine increases symmetric division of ANPs.

4. Cell Counting for the Analysis of Survival of Newborn Cells

The majority of newborn cells do not survive; furthermore, not all of the surviving cells might differentiate into neurons. Thus, it is important to evaluate the number of labeled cells and their phenotypes at the time when all of the newborn cells have differentiated (3–4 weeks). In our example, one set of slices was chosen randomly from each animal sacrificed 30 days after fluoxetine treatment and BrdU injection. The set was double-immunostained for BrdU and NeuN. Using a confocal microscope, we quantified the number of BrdU$^+$ cells as detailed above. At the same time, we analyzed each $BrdU^+$ cell for colocalization with NeuN to determine how many of the newborn cells became neurons and whether that proportion was changed by treatment with fluoxetine. The number of QNPs, ANPs, NBs-1, NBs-2, and INs was quantified in other sets of slices.

In our experiments, we found that fluoxetine increased the number of newborn differentiated neurons in the DG; that it did not affect the fate of newborn cells (i.e., the proportion of neurons among the newborn cells was not changed); that it did not change the survival of newborn cells (the proportion of labeled cells one month after the fluoxetine treatment and labeling did not differ between the control and treated groups); and that it did not have a long lasting effect on neural progenitor proliferation and neurogenesis (the number of QNPs, ANPs, NBs-1, NBs-2, and INs returned to the baseline levels one month after the treatment).

A similar experimental design (i.e., using reporter lines to quantitate the number, proliferation, and survival of stem and progenitor cells) can be applied to examine the action of any other neurogenic or antineurogenic agent.

B. The Optical Disector

Sometimes the number of cells of interest is too high and counting all of them (as outlined earlier) is very time consuming. In our case, this issue arose when we wanted to quantify the number of $BrdU^+$ and nestin-$CFPnuc^+$ cells in the SVZ (Fig. 7A). In such a case, the preferable option is to sample part of the entire population and then estimate the total cell number using an easy-to-calculate stereological parameter, for example, the volume, as a reference.

1. Sectioning

We made use of the same slices used in the cell counting experiments because they also cover the entire span of the LV. Once we had one set of slices from each animal immunostained for BrdU and CFP, we proceeded with sampling and counting as follows.

2. Cell Counting

Each slice was visualized in the confocal microscope at low magnification so that the entire LV with the SVZ could be observed on the screen. An acetate sheet with a square grid was positioned (with a random orientation) on the screen. Then, we used a high-magnification objective to acquire a Z-stack at those points where gridline intersections fall within the aspect of the SVZ (Fig. 7). The size of the grid squares was adjusted so that there were about 8–10 points of intersection in each slice. In each of these points, the accuracy is the relevant parameter so we used high magnification and adjusted the volume of the Z-stack to contain an average of no more than five cells ($20 \times 20 \times 20$ μm). Each one of these Z-stacks

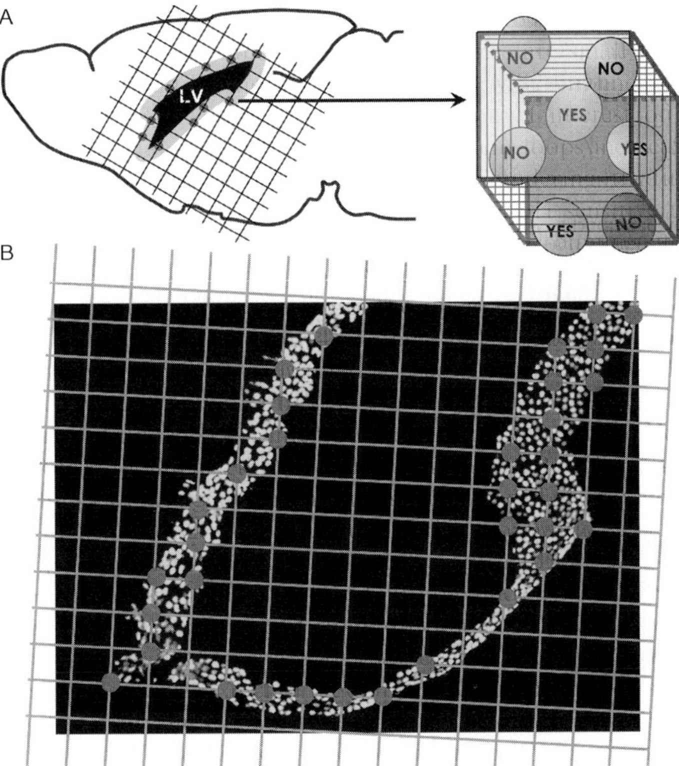

Fig. 7 The Optical Disector and the Cavalieri principle. When there are too many cells of interest to be counted in each slice, the Optical Disector provides an unbiased efficient way to obtain a reliable and accurate estimate. (A) Cell quantification is carried out in "boxes" or counting frames, called optical, disectors distributed in a systematic random manner in the area of interest. The optical disector has guards above and below, and only those cells which are actually inside the "box" are counted. It also has inclusion planes (green) and exclusion planes (red). Any cell located between the top and bottom planes that touches an inclusion plane is counted, even if most of its body is laterally out of the disector. Any cell located between the top and bottom planes that touches an exclusion plane is not counted, even if most of its body is inside the "box." (B) The same method to locate the optical disectors can be used to obtain a reliable unbiased estimation of the total volume of our area of interest. In the figure, the grid is superimposed onto the actual image of fluorescent nuclei in the subventricular zone (SVZ) of nestin-CFPnuc mice. (See Plate no. 38 in the Color Plae Section.)

is an optical box called an optical disector. The following considerations must be taken into account:

1. The Z-stack is acquired from the center of the tissue, avoiding the surfaces of the section, where the artifacts caused by the pass of the blade can lead to miscalculations.

2. In the optical disector, two sides are *inclusion planes*, and any cell touching those sides should be counted, even if the greater part of them is outside the box. The other two sides are *exclusion planes*, and cells touching those sides should not be counted, even if the greater part of them is within the box (Fig. 7A).

Using this method it is possible to estimate the density (number of cell per mm^3) of cells of interest in our control and treated animals. The density, however, must be avoided as a parameter for comparison. Density is the ratio of two parameters: the number of objects and the volume. A change in density does not tell us which parameter is changing. In other words, using density alone we cannot claim that the number of cells of a certain type is altered due to treatment, unless we also quantify the volume of the structure of the reference space, in our case, the SVZ.

In our experiments, we found that in the SVZ, treatment with fluoxetine did not change cell proliferation ($BrdU^+$ cells), the number of neural progenitors, their density, or the volume of the SVZ.

3. Volume Estimation Using the Cavalieri Principle

The Cavalieri principle is based on the notion that the volume of a structure can be estimated without bias from the product of the sum of areas of sections obtained in a systematic random manner and the thickness of the sections. The Fractionator method can be used to obtain a fraction or sample of sections in a systematic random manner that is representative of the total structure. The thickness of the sections can be easily measured with the confocal microscope (measuring the section in the *z*-axis). To measure the area, first one must properly define the structure of interest. In our case this is the SVZ, defined here as the area occupied by nestin-CFP cells surrounding the LV (Fig. 7B). We obtained a low-magnification image of the SVZ in each slice. Then, a square grid was set up on the image (physically, using an acetate sheet on the monitor screen, or virtually, using PowerPoint program). The size of the grid was adjusted so that about 40–80 gridline intersections were contained within the aspect of the SVZ (Fig. 7B). Then the number of these intersection points was counted and multiplied by the area of a single grid square, thus providing an estimate of the area of the SVZ in a given slice (it is critical to include a scale bar so the size of the image can be calculated). Multiplying the area of the SVZ by the thickness gives us the volume of the SVZ in that given slice. Adding up the volumes of all analyzed slices and multiplying by the fraction factor, which is the number of sets (six in our case because we collected six sets of slices), gives the total volume of our structure of interest.

One can now use the cell density obtained by the optical disector and the total volume of the SVZ, calculated by the Cavalieri principle, to estimate the total number of cells of interest (e.g., nestin-$CFPnuc^+$ cells) in the SVZ. In our fluoxetine experiments, we did not find changes in the number of neural stem and progenitor cells between our control and fluoxetine-treated animals.

Comments

- When the number of cells to be quantified is low, for instance when counting nestin-CFP^+ cells or $BrdU^+$ cells in the DG, there is no need to use the Optical Disector method, and all the cells (excluding those in the uppermost focal plane) should be counted on each systematic-random sampled slice.
- In the rest of the cases the Optical Disector method should be used. Systematic random sampling within the slices is required and can be achieved in two ways:

1. In the traditional approach, a grid is placed on the screen with a low-magnification image of the structure of interest. In those places where the grid intersects with the structure of interest, a high magnification is used to acquire a Z-stack of images. As a rule of thumb, the size of the grid has to be adjusted so that 10–20 optical disectors (10–20 points of intersection) are analyzed in each slice to minimize the sampling error (Gundersen *et al.*, 1999; Howell *et al.*, 2002; Peterson, 1999; Schmitz and Hof, 2005), thus leaving biological variance between individuals as the main source of error.
2. Another approach, used in an automated or a semiautomated version, also relies on random-systematic sampling but makes use of software specifically developed for stereology and a modified X-Y stage which uses software to automatically move the section from one counting frame to the next. X-Y-Z stages are also available and these also perform automated *z*-axis movement. The efficiency of this approach is significantly higher than that of the more traditional manual technique as every step is automated. Such programs as Stereo Investigator, Stereologer, and Histometrix include software suitable for these studies.

IV. Protocol III: Electron Microscopy of Nestin-GFP/CFPnuc Cells

The following protocol is developed for the *optic-electronic microscopy transfer* technique (Fig. 8), whose main feature is immunostaining of the samples at the preembedding stage. This is our method of choice for ultrastructural analysis of adult neural stem and progenitor cells because preembedding immunostaining allows visualization and tracking of cells of interests throughout the entire processing of the tissue, and because of the strength of the signal, two features that are absent in postembedding techniques.

A. Perfusion

1. Anesthetize the animals with a 15% chloral hydrate overdose (10 μl/g of bodyweight).

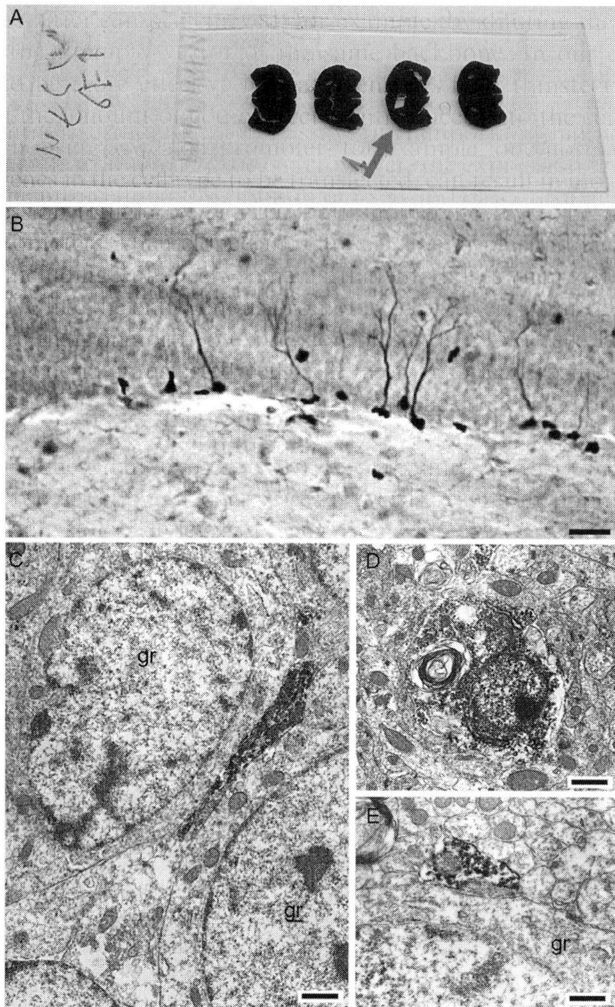

Fig. 8 Electron microscopy of neural stem and progenitor cells in nestin-GFP mice. (A) Following this protocol, 3,3′-diaminobenzidine (DAB)-immunolabeled slices can be observed and imaged under a normal brightfield microscope. (B) An area of interest can be photographed and dissected out (arrow in A). (C, D, E) Ultrathin sections can be obtained from the area of interest, and therefore the immunolabeled neural stem and progenitor cells can be observed and imaged under an electron microscope. (C) It shows the apical process of a quiescent neural progenitor (QNP) passing between two granule cells (gr). (D) It shows an amplifying neural progenitor (ANP). (E) It shows a tiny cytoplasmic expansion from the arborization of a QNP apposed to cytoplasm of a granule cell. Scale bar is 25 μm in (B), 1 μm in (D), and 0.5 μm in (E). (See Plate no. 39 in the Color Plate Section.)

2. Perform transcardial perfusion by inserting a cannula into the left ventricle and introducing 30 ml of PBS at fast flow (10 ml/min), followed by 30 ml of 4% (w/v) PF in PBS (pH 7.4), with the first 15 ml delivered at fast flow (10 ml/min) and the next 15 ml at slower flow (5 ml/min).

B. Postfixation

1. Remove the brain and place in a vial with 4% PF in PBS for 4 h at room temperature.
2. Transfer the brain to 30% sucrose in PBS and keep at 4 °C until samples sink to the bottom of the vial.

C. Sectioning

1. Cut the brain coronally in two blocks and immerse them in liquid nitrogen for 5 s, then immediately put them into PBS at room temperature.
2. Prepare 40-μm-thick coronal sections using a Vibratome, attaching the brain blocks by their caudal surface to the platform or stage.
3. Collect the slices (as many as necessary) that contain the area of interest in PBS (with a 0.01% of PF if they are going to be stored).

Comments

- The immersion in liquid nitrogen will create microfissures that will allow the antibody to penetrate the tissue, as it will freeze and thaw very quickly. To avoid formation of ice crystals which would cause structural damage of the tissue, the sample is soaked in sucrose.
- It is best to use the samples as soon as possible, since the preservation of antigenicity is crucial for obtaining a good immunostaining in electron microscopy (EM).

D. 3,3′-Diaminobenzidine Immunostaining

1. Inhibition of endogenous peroxidase activity:
   - Incubate the slices with 0.3% H_2O_2 in PBS for 20 min at room temperature.
   - Rinse with PBS four times, 5 min per rinse or until the slices sink to the bottom and no bubbles can be observed.
2. Blocking and permeabilization:
   - Incubate the slices for at least 30 min (up to 2 h) at room temperature with PBS containing 3% BSA (important: do not add Triton X-100).

3. Primary antibody incubation:
   - Incubate the slices with the primary antibody (GFP/CFP) in blocking and permeabilization solution (no Triton X-100), overnight (12–16 h) at 4 °C or for 3–4 h at room temperature.
   - Rinse with PBS three times, 5 min per rinse.
4. Secondary antibody incubation:
   - Incubate the slices with the secondary (biotin-conjugated) antibody in blocking and permeabilization solution (no Triton X-100), for 1 h at room temperature.
   - Rinse with PBS three times, 5 min per rinse.
5. ABC complex incubation:
   - Incubate the slices with ABC complex (A and B compounds should be mixed together 20 min prior to incubation), for 90 min at room temperature.
   - Rinse with PBS three times, for 5 min each time.
6. Staining reaction development:
   - Incubate the slices with a known volume of 6 mg/ml 3,3′-diaminobenzidine (DAB) in PBS solution for 10 min.
   - Add an equal volume of 3% H_2O_2 and incubate for 3–5 min, gently shaking the slices occasionally.
   - Stop the reaction by rinsing with PBS four times, 5 min per rinse.

   Comments

   - The use of Triton X-100 should be avoided when preparing the samples for the EM because it will distort the membranous structures of the cell. The liquid nitrogen shock of cryoprotected (with sucrose) slices improves the permeabilization of the tissue and penetration of the antibodies. The penetration is, however, weaker than when Triton X-100 is used. If this technique is not enough to get an acceptable staining, one can use Triton X-100 but at tenfold lower concentration than usual.

E. Contrasting and Dehydration

1. Further fix the slices with 5% glutaraldehyde in PBS for 30 min, at room temperature.
2. Transfer the slices to a glass Petri dish.
3. Rinse with PBS three times, for 5 min each time.
4. Incubate the slices with 0.1% osmium tetraoxide, for 1 h, at room temperature.
5. Rinse with PBS three times, for 5 min per rinse.
6. Incubate the slices with:
   - 50° ethanol two times for 20 min
   - 70° ethanol for 20 min

- 1% uranyl acetate, in 70° ethanol for 30 min
- 100° ethanol two times for 20 min each time
- propylene oxide, for 10 min
- propylene oxide in Durcupan resin (Fluka, MO), or similar, 1:1 volume for 10 min

7. Add Durcupan to the Petri dish (enough to generously cover the slices) and place in a vacuum chamber. Slowly apply a light vacuum and leave overnight.

F. Mounting and Embedding

1. Spread a thin layer of Durcupan on uncoated slides.
2. Mount the slices on the Durcupan-coated slides.
3. Cover the slides with coverslips cut out from acetate transparencies.
4. Incubate the slides at 50 °C for 48–72 h.
5. Remove the acetate coverslips by gently pulling up on them.
6. Observe under the microscope and dissect the area of interest using a scalpel.
7. Place the area of interest, together with a drop of Durcupan, on top of an already solidified Durcupan capsule (1 cm in length and 0.5 cm in diameter).
8. Place at 50 °C for 72 h.
9. Sculpt the capsule and prepare ultrathin sections with an ultramicrotome.

Comments

- The better a tissue is fixed, the easier it is to get good ultrathin sections. Typically, this is achieved by adding 0.1% glutaraldehyde in the perfusion fixative. However, this can interfere with the immunostaining, and impede the use of that tissue for immunofluorescence. We prefer to fix the tissue using the standard method, in order to avoid problems with immunostaining and to be able to use the same sample for different immunolabeling techniques. We then postfix the slices when immunostaining is completed.
- Several resins are available for preparing samples for the EM. We routinely use Durcupan as it offers better results at the slice-embedding steps; note, however, that Durcupan is not as hard as other resins (e.g., Araldite).
- The vacuum must be generated slowly in order to prevent the formation of large bubbles which can damage the fragile slices.
- Osmium tetraoxide fixes the tissue and contrasts lipids, proteins, and the reduced DAB, whereas uranyl acetate contrasts DNA. Therefore, no further contrasting is required to visualize the cell ultrastructure and the immunostaining pattern (Fig. 8). However, lead citrate counterstaining of the ultrathin section can be used to improve contrast if necessary.
- Use ethanol of the purest quality possible for dehydration of the samples.

- Once the Durcupan on the slides with mounted slices has solidified, the samples will last for years. The immunostaining can be visualized and photographed in a standard manner, and additional ultrathin section can be prepared.

References

Abrous, D. N., Koehl, M., and Le Moal, M. (2005). Adult neurogenesis: From precursors to network and physiology. *Physiol. Rev.* **85,** 523–569.

Alvarez-Buylla, A., and Garcia-Verdugo, J. M. (2002). Neurogenesis in adult subventricular zone. *J. Neurosci.* **22,** 629–634.

Alvarez-Buylla, A., Garcia-Verdugo, J. M., and Tramontin, A. D. (2001). A unified hypothesis on the lineage of neural stem cells. *Nat. Rev. Neurosci.* **2,** 287–293.

Burns, K. A., and Kuan, C. Y. (2005). Low doses of bromo- and iododeoxyuridine produce near-saturation labeling of adult proliferative populations in the dentate gyrus. *Eur. J. Neurosci.* **21,** 803–807.

Dranovsky, A., and Hen, R. (2006). Hippocampal neurogenesis: Regulation by stress and antidepressants. *Biol. Psychiatry* **59,** 1136–1143.

Encinas, J. M., Vaahtokari, A., and Enikolopov, G. (2006). Fluoxetine targets early progenitor cells in the adult brain. *Proc. Natl. Acad. Sci. USA* **103,** 8233–8238.

Franklin, K. B. J., and Paxinos, G. (1997). "The Mouse Brain in Stereotaxic Coordinates." Academic Press, San Diego.

Fukuda, S., Kato, F., Tozuka, Y., Yamaguchi, M., Miyamoto, Y., and Hisatsune, T. (2003). Two distinct subpopulations of nestin-positive cells in adult mouse dentate gyrus. *J. Neurosci.* **23,** 9357–9366.

Gage, F. H. (2000). Mammalian neural stem cells. *Science* **287,** 1433–1438.

Ge, S., Goh, E. L., Sailor, K. A., Kitabatake, Y., Ming, G. L., and Song, H. (2006). GABA regulates synaptic integration of newly generated neurons in the adult brain. *Nature* **439,** 589–593.

Geraerts, M., Eggermont, K., Hernandez-Acosta, P., Garcia-Verdugo, J. M., Baekelandt, V., and Debyser, Z. (2006). Lentiviral vectors mediate efficient and stable gene transfer in adult neural stem cells *in vivo*. *Hum. Gene Ther.* **17,** 635–650.

Gundersen, H. J., Jensen, E. B., Kieu, K., and Nielsen, J. (1999). The efficiency of systematic sampling in stereology—reconsidered. *J. Microsc.* **193,** 199–211.

Hayes, N. L., and Nowakowski, R. S. (2002). Dynamics of cell proliferation in the adult dentate gyrus of two inbred strains of mice. *Brain Res. Dev. Brain Res.* **134,** 77–85.

Howell, K. K., Hopkins, N. N., and McLoughlin, P. P. (2002). Combined confocal microscopy and stereology: A highly efficient and unbiased approach to quantitative structural measurement in tissues. *Exp. Phys.* **87,** 747–756.

Imayoshi, I., Ohtsuka, T., Metzger, D., Chambon, P., and Kageyama, R. (2006). Temporal regulation of Cre recombinase activity in neural stem cells. *Genesis* **44,** 233–238.

Josephson, R., Muller, T., Pickel, J., Okabe, S., Reynolds, K., Turner, P. A., Zimmer, A., and McKay, R. D. (1998). POU transcription factors control expression of CNS stem cell-specific genes. *Development* **125,** 3087–3100.

Kachinsky, A. M., Dominov, J. A., and Miller, J. B. (1994). Myogenesis and the intermediate filament protein, nestin. *Dev. Biol.* **165,** 216–228.

Kawaguchi, A., Miyata, T., Sawamoto, K., Takashita, N., Murayama, A., Akamatsu, W., Ogawa, M., Okabe, M., Tano, Y., Goldman, S. A., and Okano, H. (2001). Nestin-EGFP transgenic mice: Visualization of the self-renewal and multipotency of CNS stem cells. *Mol. Cell. Neurosci.* **17,** 259–273.

Kempermann, G. (2006). "Adult Neurogenesis: Stem Cells and Neuronal Development in the Adult Brain." Oxford University Press, New York.

Kempermann, G., Jessberger, S., Steiner, B., and Kronenberg, G. (2004). Milestones of neuronal development in the adult hippocampus. *Trends Neurosci.* **27,** 447–452.

Kronenberg, G., Reuter, K., Steiner, B., Brandt, M. D., Jessberger, S., Yamaguchi, M., and Kempermann, G. (2003). Subpopulations of proliferating cells of the adult hippocampus respond differently to physiologic neurogenic stimuli. *J. Comp. Neurol.* **467,** 455–463.

Kuan, C. Y., Schloemer, A. J., Lu, A., Burns, K. A., Weng, W. L., Williams, M. T., Strauss, K. I., Vorhees, C. V., Flavell, R. A., Davis, R. J., Sharp, F. R., and Rakic, P. (2004). Hypoxia-ischemia induces DNA synthesis without cell proliferation in dying neurons in adult rodent brain. *J. Neurosci.* **24,** 10763–10772.

Lendahl, U., Zimmerman, L. B., and McKay, R. D. (1990). CNS stem cells express a new class of intermediate filament protein. *Cell* **60,** 585–595.

Lie, D. C., Song, H., Colamarino, S. A., Ming, G. L., and Gage, F. H. (2004). Neurogenesis in the adult brain: New strategies for central nervous system diseases. *Annu. Rev. Pharmacol. Toxicol.* **44,** 399–421.

Lledo, P. M., Alonso, M., and Grubb, M. S. (2006). Adult neurogenesis and functional plasticity in neuronal circuits. *Nat. Rev. Neurosci.* **7,** 179–193.

Lois, C., and Alvarez-Buylla, A. (1994). Long-distance neuronal migration in the adult mammalian brain. *Science* **264,** 1145–1148.

Malberg, J. E., and Blendy, J. A. (2005). Antidepressant action: To the nucleus and beyond. *Trends Pharmacol. Sci.* **26,** 631–638.

Malberg, J. E., Eisch, A. J., Nestler, E. J., and Duman, R. S. (2000). Chronic antidepressant treatment increases neurogenesis in adult rat hippocampus. *J. Neurosci.* **20,** 9104–9110.

Mignone, J. L., Kukekov, V., Chiang, A. S., Steindler, D., and Enikolopov, G. (2004). Neural stem and progenitor cells in nestin-GFP transgenic mice. *J. Comp. Neurol.* **469,** 311–324.

Ming, G. L., and Song, H. (2005). Adult neurogenesis in the mammalian central nervous system. *Annu. Rev. Neurosci.* **28,** 223–250.

Peterson, D. D. A. (1999). Quantitative histology using confocal microscopy: Implementation of unbiased stereology procedures. *Methods* **18,** 493–507.

Santarelli, L., Saxe, M., Gross, C., Surget, A., Battaglia, F., Dulawa, S., Weisstaub, N., Lee, J., Duman, R., Arancio, O., Belzung, C., and Hen, R. (2003). Requirement of hippocampal neurogenesis for the behavioral effects of antidepressants. *Science* **301,** 805–809.

Schmitz, C. C., and Hof, P. P. R. (2005). Design-based stereology in neuroscience. *Neuroscience* **130,** 813–831.

Seaberg, R. M., and van der Kooy, D. (2003). Stem and progenitor cells: The premature desertion of rigorous definitions. *Trends Neurosci.* **26,** 125–131.

Sejersen, T., and Lendahl, U. (1993). Transient expression of the intermediate filament nestin during skeletal muscle development. *J. Cell Sci.* **106**(Pt. 4), 1291–1300.

Seri, B., Garcia-Verdugo, J. M., Collado-Morente, L., McEwen, B. S., and Alvarez-Buylla, A. (2004). Cell types, lineage, and architecture of the germinal zone in the adult dentate gyrus. *J. Comp. Neurol.* **478,** 359–378.

Song, H., Kempermann, G., Wadiche, L. O., Zhao, C., Schinder, A. F., and Bischofberger, J. (2005). New neurons in the adult mammalian brain: Synaptogenesis and functional integration. *J. Neurosci.* **25,** 10366–10368.

Song, H. J., Stevens, C. F., and Gage, F. H. (2002). Neural stem cells from adult hippocampus develop essential properties of functional CNS neurons. *Nat. Neurosci.* **5,** 438–445.

Tashiro, A., Sandler, V. M., Toni, N., Zhao, C., and Gage, F. H. (2006). NMDA-receptor-mediated, cell-specific integration of new neurons in adult dentate gyrus. *Nature* **442,** 929–933.

Taupin, P., and Gage, F. H. (2002). Adult neurogenesis and neural stem cells of the central nervous system in mammals. *J. Neurosci. Res.* **69,** 745–749.

Terling, C., Rass, A., Mitsiadis, T. A., Fried, K., Lendahl, U., and Wroblewski, J. (1995). Expression of the intermediate filament nestin during rodent tooth development. *Int. J. Dev. Biol.* **39,** 947–956.

Tronche, F., Kellendonk, C., Kretz, O., Gass, P., Anlag, K., Orban, P. C., Bock, R., Klein, R., and Schutz, G. (1999). Disruption of the glucocorticoid receptor gene in the nervous system results in reduced anxiety. *Nat. Genet.* **23,** 99–103.

van Praag, H., Schinder, A. F., Christie, B. R., Toni, N., Palmer, T. D., and Gage, F. H. (2002). Functional neurogenesis in the adult hippocampus. *Nature* **415,** 1030–1034.

Vega, C. J., and Peterson, D. A. (2005). Stem cell proliferative history in tissue revealed by temporal halogenated thymidine analog discrimination. *Nat. Methods* **2,** 167–169.

Warner-Schmidt, J. L., and Duman, R. S. (2006). Hippocampal neurogenesis: Opposing effects of stress and antidepressant treatment. *Hippocampus* **16,** 239–249.

Yamaguchi, M., Saito, H., Suzuki, M., and Mori, K. (2000). Visualization of neurogenesis in the central nervous system using nestin promoter-GFP transgenic mice. *Neuroreport* **11,** 1991–1996.

Yaworsky, P. J., and Kappen, C. (1999). Heterogeneity of neural progenitor cells revealed by enhancers in the nestin gene. *Dev. Biol.* **205,** 309–321.

Zimmerman, L., Parr, B., Lendahl, U., Cunningham, M., McKay, R., Gavin, B., Mann, J., Vassileva, G., and McMahon, A. (1994). Independent regulatory elements in the nestin gene direct transgene expression to neural stem cells or muscle precursors. *Neuron* **12,** 11–24.

Zulewski, H., Abraham, E. J., Gerlach, M. J., Daniel, P. B., Moritz, W., Muller, B., Vallejo, M., Thomas, M. K., and Habener, J. F. (2001). Multipotential nestin-positive stem cells isolated from adult pancreatic islets differentiate ex vivo into pancreatic endocrine, exocrine, and hepatic phenotypes. *Diabetes* **50,** 521–533.

CHAPTER 12

Using Fluorescent Proteins to Study mRNA Trafficking in Living Cells

Emmanuelle Querido and Pascal Chartrand

Département de Biochimie
Université de Montréal
2900 Edouard-Montpetit
Montréal, Québec H3C 3J7, Canada

Abstract

This chapter presents the MS2-GFP system, a method to study the trafficking of RNA molecules in living cells. This system is based on two components: a fusion of the MS2 coat protein to a fluorescent protein and a reporter mRNA containing multimers of the RNA stem-loop recognized by the MS2 coat protein. The MS2-GFP protein bound to the RNA stem-loops acts as a beacon that allows the detection of this mRNA within a cell by epifluorescence or confocal microscopy. This chapter focuses on the use of this system in mammalian fibroblast cells and in

0091-679X/08 $35.00
DOI: 10.1016/S0091-679X(08)85012-1

yeast cells, and discusses several technical considerations of the MS2-GFP system. Detailed protocols for validating the MS2-GFP signal in fixed cells by fluorescent *in situ* hybridization of the target RNA using fluorophore-labeled oligonucleotide probes are also provided.

I. Introduction

The "RNA World" theory of evolution posits that life originated from RNA, a molecule that can be both a gene and an enzyme. The past few years have certainly seen tremendous excitement in the RNA world of research, with the discovery of a new class of small noncoding RNAs that play a regulatory role in mRNA translation and transcription. The study of RNA trafficking, the movement and intracellular localization of RNAs, has revealed another level of regulation of gene expression. Splicing and mRNA nuclear export are tightly coupled and regulated processes (reviewed in Cullen, 2003). Once in the cytoplasm, mRNA transport and local translation have now been documented in vertebrates, invertebrates, and unicellular organisms (reviewed in Kindler *et al.*, 2005). Polarized cells in particular localize RNAs to locally translate specific proteins, which restricts their distribution to a particular subregion of the cytoplasm. This mechanism of protein sorting is involved in major biological processes such as asymmetric cell division, oogenesis, cellular motility, and synapse formation. Complex machineries are also involved in the transport of noncoding RNAs, like snRNAs, snoRNAs, and rRNAs (reviewed in Cullen, 2003). To properly study RNA trafficking, new methods have been developed to visualize RNA in living cells, and this chapter presents one such method, the MS2-GFP system. The first part of this chapter introduces general considerations and the use of this system in mammalian cells, while the second part focuses on its utilization in the budding yeast.

II. The MS2-GFP System

The coat protein from the MS2 bacteriophage binds with high specificity to an RNA stem-loop structure of 19 nucleotides containing the initiation codon of the phage replicase gene (Bernardi and Spahr, 1972). By binding to a unique site in the RNA genome of the phage, the coat protein represses translation of the RNA replicase gene and also guides packaging into phage particles. Over the years, the MS2 coat protein has been engineered so that it can be fused to any protein and tethered to any RNA containing the MS2 stem-loop motif. This approach is notably at the core of the yeast three-hybrid system, which has been developed to study RNA–protein interactions (Jaeger *et al.*, 2003).

The MS2-GFP system involves the expression of two constructs: a fusion of the MS2 coat protein to the green fluorescent protein (GFP) or any fluorescent protein, and a multimer of the MS2 stem-loop sequence in the mRNA to be tracked. When the MS2-GFP protein binds to the RNA stem-loops, it acts as a beacon that allows the detection of this mRNA within a cell by microscopy. The MS2-GFP protein is highly specific for RNAs containing MS2 stem-loops in bacteria, plant, and higher eukaryote cells (Bertrand *et al.*, 1998; Fusco *et al.*, 2003; Golding *et al.*, 2005; Rook *et al.*, 2000; Zhang and Simon, 2003). The advantage of using this system over the expression of a GFP-tagged endogenous RNA-binding protein is that MS2-GFP is specific to the RNA containing MS2 stem-loops, while the endogenous RNA-binding protein may bind several mRNAs and reflect the behavior of all of them. Therefore, this system offers two benefits: the specific detection of MS2-tagged mRNAs by standard epifluorescence microscopy and the study of RNA dynamics in living cells.

A. Designing the MS2 Fluorescent Protein Fusion

The MS2 phage coat protein has been studied extensively over the past 30 years and its structure, binding specificity, and affinity are well known. Once bound to the phage RNA, the native MS2 coat protein multimerizes to form the phage capsid. Several mutations to the wild-type MS2 protein sequence have been created in order to decrease multimer formation and enhance specificity and avidity of binding to the MS2 RNA stem-loop. Mutant versions that inhibit multimer formation should preferentially be used in order to avoid the aggregation of the MS2-GFP proteins in the cell. Among these mutations, the dlFG mutation, described in Fig. 1, is a deletion in the 15-amino acid FG loop that prevents the MS2 coat protein dimers from assembling into a phage capsid (Peabody and Ely, 1992). The FG loop is involved in the interdimer interactions that permit capsid assembly. Another mutation in the FG loop, W82R, can be used, which also inhibits capsid assembly (Peabody and Ely, 1992; Rook *et al.*, 2000). The replacement of valine at position 29 with isoleucine (V29I) results in tighter binding of the MS2 dimer to the RNA stem-loop (Peabody and Ely, 1992). The V29I-dlFG double mutant illustrated in Fig. 1 decreases the K_d of the RNA–protein interaction to 40 nM *in vitro* (Lim and Peabody, 1994). The formation of an MS2-MS2 protein dimer is essential for binding to the RNA stem-loop (Valegard *et al.*, 1994), and mutations that disrupt dimer formation should be avoided.

Expression of very high levels of MS2-GFP can cause many problems when using this system, by making it nearly impossible to discriminate between signals coming from the unbound versus the RNA-bound MS2-GFP protein. To eliminate this problem, a nuclear localization signal (NLS) can be added to the fusion protein which will result in the nuclear accumulation of the MS2-GFP protein. To accumulate in the cytoplasm, the MS2-GFP protein must be exported from the nucleus via its interaction with an mRNA containing MS2 stem-loops (Fusco *et al.*, 2003). In the absence of a target transcript, the MS2-GFP protein stays in the nucleus. This approach has been very useful to visualize single RNA molecules in the cytoplasm, as demonstrated in the study by Fusco *et al.* (2003).

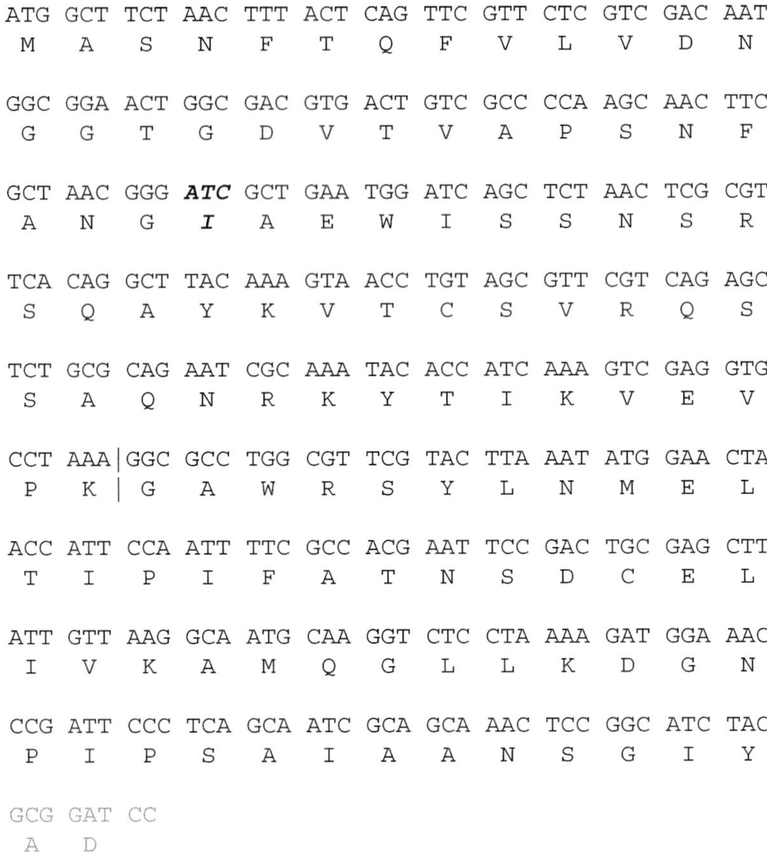

Fig. 1 Sequence of the V29I dlFG mutant MS2 protein used for this study. The V29I mutated amino acid is highlighted in bold, and the site of the FG loop deletion is indicated by a vertical line. A linker sequence added at the C-terminus of the protein that includes a BamHI site is shown in gray.

For the visualization of transcripts in the nucleus, it is of course preferable to not include an NLS (Shav-Tal *et al.*, 2004). The MS2-GFP protein will be distributed throughout the cell without accumulating in any compartment (see Fig. 3B). Different approaches can be used to carefully time and dose the expression of MS2-GFP so that the nuclear MS2-tagged mRNA can be visualized successfully, and some of these approaches are described in Section III.A on mammalian expression systems.

An ingenious new method to eliminate background from the unbound MS2-GFP fusion protein was recently described in the laboratory of Chris Brown, who used split Venus yellow fluorescent protein (YFP) fusions (Rackham and Brown, 2004). In their system, each half of the split Venus fluorescent protein was fused to a different RNA-binding protein (MS2 and either IMP1 or FMRP), and coexpressed with a reporter mRNA containing an MS2 stem-loop and the IMP1- or FMRP-binding motif in proximity. The recruitment of both MS2-Venus and IMP1- or FMRP-Venus proteins on the reporter mRNA brings the two portions of the split Venus in proximity and result in a functional fluorescent protein. This system has a very low background of fluorescence in the absence of the reporter mRNA and is very specific since it requires the simultaneous binding of both fusion proteins in order to produce fluorescence.

B. Designing the Reporter mRNA

The MS2 RNA hairpin is composed of 19 nucleotides; a seven-base-paired stem containing a bulged adenine and a four-nucleotide loop. When the uracil at position −5 of the loop is replaced by a cytosine (illustrated in bold in Fig. 2), it results in a 15-fold increase in binding affinity of the MS2 coat protein (Talbot *et al.*, 1990; Valegard *et al.*, 1994). This C-variant mutant was thus used for the MS2-GFP system. Figure 2 shows the sequence of two MS2 stem-loops separated by a linker containing common restriction enzyme sites designed to express a convenient MS2 stem-loop tandem array (from the plasmid pSL-MS2–6; Bertrand *et al.*, 1998). Multimerization of the tandem MS2 stem-loops can be achieved using the compatible BamHI/BglII sites.

The location of the MS2 stem-loop repeats in the RNA must be considered. If the stem-loop cassette is placed in the 5′untranslated region (UTR) of the transcript, it gives the advantage that the MS2-GFP protein may locate the site of transcription of the reporter transcript on the genomic DNA, as shown in Fig. 4. However, the possibility of interference with translation needs to be considered if the stable MS2 stem-loop repeats are placed in the 5′UTR or coding sequence of the message. If the MS2 RNA repeats are placed in the 3′UTR of the transcript, as is commonly done, the GFP signal will correlate with the mature free mRNA.

The need to place multiple stem-loops in the transcript may limit which kind of RNA one can study with the MS2-GFP system. Visualization of snoRNAs with this system was recently achieved in yeast (Verheggen *et al.*, 2001), and this may

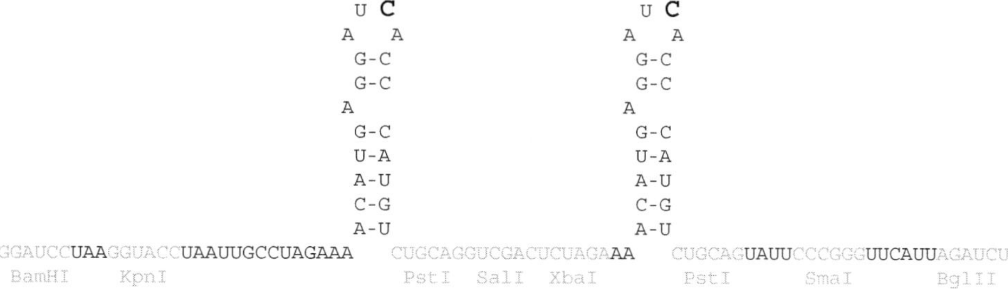

Fig. 2 Sequence of the C-variant MS2 stem-loop tandem array used for this study. The mutated cytidine residue is shown in bold.

represent the smallest sized RNA to be detectable. Cassettes of either 1 (Zhang and Simon, 2003), 2 (Verheggen *et al.*, 2001), 6 (Bertrand *et al.*, 1998), 8 (Rook *et al.*, 2000), 12 (Fusco *et al.*, 2003), 24 (Janicki *et al.*, 2004), or 96 (Golding *et al.*, 2005) MS2 stem-loops have been used successfully in the MS2-GFP system, depending on the organism and the level of expression of the RNA to be visualized. In general, increasing the number of binding sites in the RNA augments signal strength due to a greater number of fluorescent domains bound per RNA molecule. However, one needs 24 MS2 repeats in order to detect single RNA molecules (Fusco *et al.*, 2003). An obvious drawback of the MS2-GFP system is that the need to put stem-loop repeats in the transcript means one cannot study an endogenous mRNA. On a technical note, the MS2 stem-loops are repetitive sequences in a plasmid and care must be taken to avoid deletions and recombination during propagation in bacteria. We have used the HB101 strain grown at 30 °C to successfully amplify plasmids with a low frequency of recombination events.

III. RNA Trafficking in Fibroblasts

A. Mammalian Expression Systems

The protocols discussed in this section can be applied to many cell types, but they have mainly been optimized for common tissue culture cells, hence referred to as fibroblasts for brevity. To use the MS2-GFP system in fibroblast cells, one of the main concerns is the level of expression of the MS2-GFP protein one will obtain. Transient transfection will result in high levels of expression from plasmids that are not integrated into the cell genome. Adapting the promoter and the timing of expression is thus a crucial optimization step. One approach is to use a promoter with a lower level of expression for the MS2-GFP protein and a high expressing promoter for the mRNA construct. One can then do an optimization step by cotransfecting both vectors, and can select vector amount ratios that will result in the desired expression level. Rackham and Brown (2004) transiently transfected COS-7 cells with four times the amount of RNA reporter compared to the MS2 fluorescent protein expression plasmids. Fusco *et al.* (2003) also cotransfected all the necessary expression plasmids into COS-7 cells for their experiments.

When experimental goals include the detection of nuclear RNA, a low level of expression of MS2 fluorescent protein such as that seen in Fig. 3B and C is even more critical. Janicki *et al.* (2004) electroporated an MS2-YFP expressing plasmid into U2OS cells and were able to detect appropriate low levels of fluorescent proteins 2.5 h later. Additionally, for nuclear RNA tracking, the stable integration of the MS2-tagged RNA expression construct into chromosomal DNA is advisable (Janicki *et al.*, 2004; Shav-Tal *et al.*, 2004). Retroviral vectors offer a quick and convenient method for stable integration into mammalian cells that can be maintained with selectable markers.

The experiment presented here was designed to visualize the nuclear accumulation of mRNA containing the mutant 3′UTR of dystrophia myotonica protein

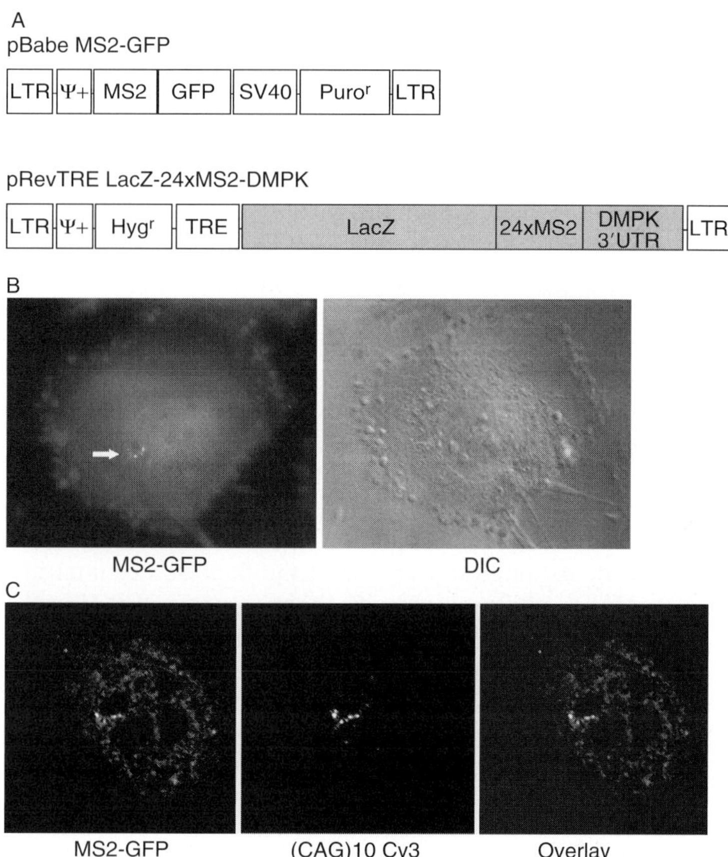

Fig. 3 Application of the MS2-GFP system in myoblast cells to study myotonic dystrophy. (A) Schematic representation of the two retroviral vectors used. (B) GFP-tagged LacZ-MS2-DMPK mRNA foci in living cells. Arrow points to nuclear foci. (C) Colocalization of the MS2-GFP signal with FISH on the LacZ-MS2-DMPK mRNA (Cy3 CAG probe).

kinase (DMPK) with expanded CUG repeats as a model system to study type 1 myotonic dystrophy. While the DMPK 3′UTR normally contains 5–37 CUG repeats, patients with an expanded repeat allele may express several thousand CUG repeats (reviewed in Machuca-Tzili *et al.*, 2005). In Fig. 3A, a schematic representation of the two retroviral vectors used for this experiment is shown. C2C12 myoblast clones stably expressing the desired level of LacZ-24× MS2-DMPK 3′UTR mRNA were derived by hygromycin selection. In the pRevTRE retroviral vector (BD Biosciences Clontech, Palo Alto, CA), the MS2-tagged mRNA is expressed from the strongly inducible tetracycline-responsive element promoter. The MS2-GFP cDNA was cloned into pBabe puro, with the low expressing long terminal repeat promoter. Myoblasts were infected only once with diluted retroviral supernatant, to promote single integration events. The cells were visualized as soon as 24 h postinfection with pBabe MS2-GFP retrovirus, in a Focht Live-Cell Chamber System 2 (FCS2) (Bioptechs, Inc., Butler, PA). A single time-point image is shown in Fig. 3B, both in the differential interference contrast and in the GFP channel, showing a cluster of nuclear RNA foci that were clearly visible in the focal plane shown.

B. Fluorescent *In Situ* Hybridization on Fibroblasts

Because there is the possibility of unbound MS2-GFP protein being visualized in this system, it is important to confirm that the GFP signal one is tracking actually colocalizes with the MS2-tagged reporter mRNA. To do this, one should perform fluorescent *in situ* hybridization (FISH) as part of the experimental setup. In Fig. 3C, C2C12 myoblasts expressing the mRNA reporter with expanded CUG repeats, as in Fig. 3B, were fixed, and FISH was performed with a Cy3-labeled DNA probe consisting of 10 CAG repeats. This allowed us to verify the colocalization of the MS2-GFP foci that we observed with CUG repeat nuclear mRNA.

The following protocol is modified from Chartrand *et al.* (2000) and allows the simultaneous detection of a specific mRNA and the GFP signal. Paraformaldehyde fixation does not affect the fluorescence intensity or the distribution of the MS2-GFP fusion protein. When designing FISH probes to detect a nonrepetitive sequence in an mRNA, one may need to combine four to seven probes specific for different regions of the mRNA, in order to achieve sufficient fluorescence intensity. Probes consist of oligonucleotides of 50 nucleotides in length containing four to five amino-modified C6 thymidine residues. Labeling the probes with fluorophores, like Cy3, occurs on the amino-modified C6-dT residues. To image LacZ transcripts (as shown in Fig. 4C), we use a mix of five oligonucleotides spread across the coding sequence. The sequences of the LacZ probes can be found in the study by Long *et al.* (1995). For the detection of the transcript containing MS2 stem-loops by FISH, a probe that targets the MS2 stem-loops present in the reporter mRNA can be used (Fusco *et al.*, 2003). It has the advantage that it can be used to detect any MS2-tagged RNA.

MS2 probe: 5′ AT*GTCGACCTGCAGACAT*GGGTGATCCTCAT*GTTT TCTAGGCAATT*A 3′

1. Labeling Amino-Modified Oligonucleotides Probes with Cy3

Important: All solutions for fixation and *in situ* hybridization should be diethyl pyrocarbonate (DEPC) treated or prepared with DEPC-treated distilled water, and gloves should be worn at all times to avoid RNase contamination.

1. Dissolve 10 μg of oligonucleotide in 35 μl of freshly made sodium carbonate buffer pH 8.8.
2. Dissolve the contents of one foil pack of Cy3 fluorophore (Amersham Biosciences, United Kingdom) in 30 μl of ddH_2O. See package insert for additional details (kit no. PA23001). Ambient light in the room should be dimmed during manipulation of Cy3 dye to minimize bleaching.

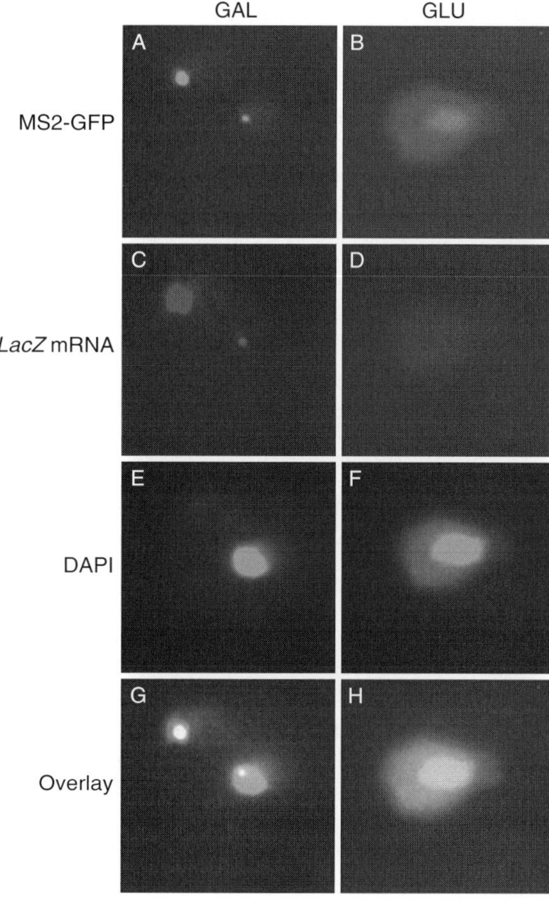

Fig. 4 Visualization of a specific mRNA in the yeast nucleus using the MS2-GFP system. Yeast cells expressing an MS2-GFP protein without nuclear localization signal (NLS) were transformed with a plasmid containing the *LacZ* gene with 24 copies of the MS2 RNA motif in its 5′UTR. The gene also contained an *ASH1* bud-localization zipcode. GAL: galactose induction of the single *LacZ-MS2* gene (panels A, C, E, and G). GLU: glucose repression of the *LacZ-MS2* gene (panels B, D, F, and H). MS2-GFP: Signal from the MS2-GFP protein bound to the LacZ-MS2 mRNA (panel A) or free (panel B). LacZ mRNA: FISH signal from LacZ-specific probes. DAPI: Chromosomal DNA marker. Overlay: Overlap between MS2-GFP, LacZ probes, and DAPI signals. These images are of formaldehyde-fixed yeast cells. (See Plate no. 40 in the Color Plate Section.)

3. Add 15 μl of monoreactive dye to the oligo (we use one foil pack to label two probes). Incubate for 24–36 h in the dark at room temperature, mixing vigorously at regular intervals.
4. Remove unincorporated dye by passing the dye/oligo on a commercial (RNase free) Sephadex G-25 or G-50 column.

5. Determine the labeling efficiency (dye/oligo ratio):
   a. Dilute the labeled oligonucleotide and measure OD_{260} and OD_{552}.
   b. Calculate the total molar extinction coefficient (MEC) of the oligonucleotide. The individual base values are adenosine 15,400 Mol^{-1} cm^{-1}, cytidine 7400 Mol^{-1} cm^{-1}, guanine 11,500 Mol^{-1} cm^{-1}, and uracil/thymidine 8700 Mol^{-1} cm^{-1}; or one can consider that the average MEC is 10,000 Mol^{-1} cm^{-1} per base for DNA in aqueous solution. The MEC of the Cy3 dye at 552 nm is 150,000 Mol^{-1} cm^{-1}.
   c. Use the following equations to measure labeling efficiency:

$$[\text{Cy3 dye}] = \frac{A_{552}}{150,000}$$

$$[\text{Oligo}] = \frac{(A_{260} - 0.08A_{552})}{\text{oligo MEC}}$$

$$\text{Dye/oligo ratio} = \frac{[\text{Cy3 dye}]}{[\text{Oligo}]}$$

While 100% labeling efficiency is ideal, we have obtained good results with calculated labeling efficiencies of around 70% of amino-reactive residues per probe.

Sodium carbonate buffer
106 mg Na_2CO_3
84 mg $NaHCO_3$
Up to 10 ml ddH_2O
Adjust pH to 8.8 with HCl

2. Gelatin-Coated Coverslips for FISH

1. Set up a beaker with 0.1 N HCl on a hot plate in a fume hood. Boil 22 × 22 mm glass coverslips for 30 min then let cool at room temperature.
2. Rinse with ddH_2O 10 times.
3. Prepare 0.5% gelatin in ddH_2O in a large beaker. The volume of liquid must not be higher than halfway up the beaker.
4. Autoclave the coverslips in the beaker, then store at 4 °C.

3. FISH Protocol to Colocalize RNA with GFP Signal

a. Fixation

1. Plate fibroblasts on 22 × 22 mm glass coverslips coated with gelatin.
2. When the cells are ready, remove the growth media and replace with 4% formaldehyde in 1× PBS. Allow fixation to proceed for 15–30 min at room temperature.
3. After fixation, wash the coverslips twice with 1× PBS.
4. Dehydrate in 70% EtOH DEPC at 4 °C for more than 2 h.
5. Coverslips stored in 70% EtOH should be protected from light and can remain usable for FISH for several weeks.

Notes: The quality of the formaldehyde is crucial for the preservation and detection of small details by FISH. We always use electron microscopy grade ultrapure, single-usage sealed ampoules of formaldehyde from Electron Microscope Sciences (Fort Washington, PA). A concentration of 10% acetic acid in the fixation buffer may be optimal for some FISH experiments; however, acetic acid is omitted here because it negatively affects GFP fluorescence. Additional details of the FISH procedure can be found in the study of Chartrand *et al.* (2000).

b. Hybridization

1. Prepare FISH probes by diluting them to 1 ng/μl in ddH_2O. If a combination of probes is used, prepare a mix so that each individual probe has a final concentration of 1 ng/μl. Ambient light in the room should be dimmed during the FISH procedure to minimize bleaching of the Cy3 probes.
2. In a tube, combine 10 μl of probe mix and 20 μg of tRNA.
3. Lyophilize completely on medium heat. Cover from light in Speed-Vac and at all times.
4. Add 10 μl of 2× solution A and denature for 3 min at 95 °C.
5. Let the tube cool down to room temperature and add 10 μl of 2× solution B. Mix carefully, the solution will be very viscous and bubbles should be avoided.

Notes: The final concentration indicated here for the probe solution and the washes is 40% formamide. This is a good starting point, but concentrations ranging from 25% to 50% formamide may need to be tested.

2× Solution A

4× SSC

80% Formamide

2× Solution B

20% Dextran sulphate (weigh powder directly in 50-ml tube, add ddH_2O DEPC gradually, dissolve by agitation at 37 °C, and adjust final volume then keep at 4 °C)

0.4% BSA (commercial 2% BSA solution certified RNase free)

4 mM Vanadyl ribonucleotide complex

6. Rehydrate coverslips in 2× SSC 40% formamide for 5 min at room temperature.
7. Incubate coverslips with 20 μl of probe overnight at 37 °C.

Notes: We use Parafilm® to create an RNase-free-humidified chamber that will protect the probe from drying out in these conditions. Cover a 16 × 20 cm glass plate with a layer of Parafilm by placing its paper side up and rubbing firmly to make it adhere to the glass. Ultrafine tip tweezers should be used to manipulate the coverslips. Flame tweezer tips prior to RNase-free use. Place a 20 μl drop of probe mix on the Parafilm and lay the coverslip on the drop, making sure no bubbles are trapped. The surface of the coverslip containing the cells should face the drop. Cut another length of Parafilm, remove paper, and place the protected side down onto the coverslips. Do not move the coverslips after they have been lain down. Seal all four sides of the chamber by rubbing firmly with a blunt object. Wrap the glass plate in aluminum foil and place in a 37 °C incubator.

8. Wash coverslips twice in 8 ml of 2× SSC 40% formamide for 30 min at 37 °C.

Notes: To remove the inverted coverslips from the Parafilm without damaging the cells, first pinch and raise up the Parafilm near the coverslip to break the seal, then carefully lift the coverslip and place in a Coplin jar containing 8 ml of preheated wash solution. Wrap the Coplin jar in foil during washes.

9. Wash once with 1× PBS DEPC [containing 4′,6-diamidino-2-phenylindole (DAPI) if desired] at room temperature.
10. Mount with fresh antifade mounting media and visualize as soon as possible. Fluorescence microscopy followed by deconvolution is the preferred method to visualize both Cy3 and GFP/YFP signal. Some FISH Cy3 signal can be intense enough to be successfully imaged by laser confocal microscopy, but this is often not the case.

C. Visualization of RNA Movement in Living Cells

The experimental details described here are for the FCS2 closed perfusion system (Bioptechs, Inc.), but may apply to the use of other systems as well. A closed chamber system enables the acquisition of data from cells in optimal viability conditions, and the perfusion of fresh media will help minimize photodamage. For short-term observation of living cells, a closed perfusion chamber may not be needed. One can simply invert a 22 × 22 mm coverslip onto a slide and image the cells immediately. Phenol red-free media or PBS can be used as the imaging media. However, to film the movement of mRNP particles, a more elaborate setup is required. Movement of mRNPs can be temperature dependent (Shav-Tal *et al.*, 2004) and fibroblast cells should be in optimal conditions for accurate experiments.

1. Plate Cells on FCS2 Coverslips

1. Place 40-mm-round Bioptechs coverslips in 60 mm dishes for cell culture. For typical fibroblasts no special coating of the glass is needed.
2. Cell placement can be restricted to the center of the coverslip by using glass culture cylinders (Bioptechs, PA). Before adding media, position one or two cylinders on the coverslip.
3. Dilute fibroblasts to the desired concentration and drop them into the culture cylinders, then apply media to the rest of the 60 mm dish. This will restrict cell growth to only one side of the coverslip, and allow a proper seal between the edges of the coverslip and the chamber gaskets. A different population of cells can also be plated in each of two cylinders on the same coverslip.
4. Allow at least 12 h for the fibroblasts to grow and settle on the glass.

2. Prepare Equipment in the Microscope Room

1. Turn on the 37 °C warmer oven in the microscope room so that it has time to reach the set temperature (Boekel Desk Top Warmer, Bioptechs, PA). This small no-humidity oven can be used to keep perfusion media, assembled chambers, objectives, extra cells, etc. in proximity to a microscope.
2. Calibrate the chamber system controller and turn it off—this controller must be off when the assembled chamber is connected to it.
3. Mount the objective heater and turn it on so that the temperature stabilizes at 37 °C.

3. Prepare Perfusion Media

1. Prepare phenol red-free media by adding tissue culture grade 1 M HEPES to a final concentration of 25 mM and preheating the media to 37 °C. Alternatively, the pH of the media can be regulated with a CO_2-bubbling apparatus.
2. To avoid contamination and evaporation, place the perfusion media in a small Erlenmeyer caped with a two-hole rubber stopper.
3. Thread 1/16 in. ID Tygon tubing through one hole so that the intake is kept permanently submerged in media. The second hole acts as a vent.

4. Place the perfusion media container in the Boekel warmer keeping the door slightly open for the tubing.

4. Set up the Perfusion System

1. Install the single-channel tube in the microperfusion pump (Instech peristaltic pump, Bioptechs, PA).
2. Position the pump at the same height and close to the microscope stage.
3. Connect tubing from the perfusion media to the pump intake port.
4. Attach tubing to the outflow port of the pump. Keep all tubing lengths as short as possible.
5. Calibrate the pump to a low-flow rate, such as 2 ml/h. Fill up all the inflow lines with perfusion media to remove air bubbles from the tubing.

5. Assemble the FCS2 Chamber

1. Remove the glass culture cylinder(s) and rinse the coverslip three times in phenol red-free media.
2. Mount the coverslip into the chamber according to manufacturer's instructions.
3. Immediately attach inflow and outflow tubing and start infusing the chamber with fresh media. Check the assembly for leaks at this time.
4. Securely mount the chamber onto the stage adapter.
5. Plug the chamber into the FCS2 controller and turn it on.
6. Tape the inflow and outflow tubing to the stage to avoid displacement when moving in *xy*-direction. The outflow tubing should extend only as far as necessary to reach the edge of the microscope stage.
7. Tape a cotton strip to the end of the outflow tube to allow the exiting media to reach a collection container by smoothly flowing through the hydrophilic material. This should prevent the formation of drops that create a vacuum that could cause flexing of the coverslip. This is especially important for prolonged imaging sessions where maintaining precise focus over time is critical.
8. The chamber must not be left unattended when the microperfusion pump is running. For obvious reasons, extreme care must be taken to avoid leaks.

6. Observe Living Cells

1. Focus on the cells using brightfield illumination.
2. Switch to the appropriate filter set and find a cell with the desired characteristics, taking care not to bleach the cell during focusing.

3. Determine the optimal duration and frequency of data acquisition for time-lapse imaging by trial and error. Shav-Tal *et al.* (2005) have recently published an outline of procedures to track and quantify single mRNP particles with the MS2-GFP system.

IV. Following RNA Trafficking in Living Yeasts

The MS2-GFP system to observe mRNA trafficking was first developed in yeast cells (Bertrand *et al.*, 1998; Chartrand *et al.*, 1999). Since then, several studies have used this system to follow the transport and/or localization of specific transcripts in living yeast cells. Other systems to track mRNA trafficking in living yeasts have been developed subsequently, using GFP fusion with either the U1A RNA-binding protein (Brodsky and Silver, 2000; Takizawa and Vale, 2000) or the MS2 coat protein (Beach *et al.*, 1999), but they all are based on the same principle.

A. Expression in the Budding Yeast

1. The MS2-GFP Expression Vectors

For the expression of the MS2-GFP protein in yeast, the original expression vector (pG14-MS2-GFP) developed by the Singer laboratory was based on the pG14 vector (2 μ, *LEU2)* containing the MS2-GFP fusion under the control of the *GPD1* promoter (Bertrand *et al.*, 1998). An SV40 NLS and an HA tag were inserted upstream of the MS2-GFP open reading frame (ORF). This plasmid allows constitutive expression of a nuclear-localized MS2-GFP protein. Only the MS2 fluorescent protein tethered to an mRNA containing MS2 stem-loops can be exported out of the nucleus and accumulate in the cytoplasm. New versions of this expression plasmid are now available, with lower expression levels (a centromere, *LEU2* plasmid YCP111-MS2-GFP) or without NLS(YCP111-MS2-GFP-ΔNLS) for detection of transcripts in the nucleus (see Fig. 4). Another version with the RedStar fluorescent protein has been recently published (Schmid *et al.*, 2006).

2. Tagging an mRNA with MS2 Stem-Loops

For the expression of the mRNA with MS2 stem-loops, two approaches can be used. In both cases, expression of the reporter transcript with an inducible promoter, like the galactose-inducible *GAL1* promoter, is strongly suggested. The use of an inducible promoter makes it possible to follow the trafficking of the reporter mRNA immediately after induction rather than at steady state. One approach is to use a fusion reporter mRNA containing a heterologous ORF with MS2 stem-loops in its 3′ UTR. For instance, the vector expressing a *lacZ*-6 × MS2 reporter (YEP195-lacZ-6 × MS2) can be used to study the role of specific RNA zipcodes or localization elements involved in RNA trafficking. Besides the *lacZ* ORF and six MS2 stem-loops, it contains an *ADH2* 3′ UTR with a multiple cloning site for insertion of heterologous DNA. The other approach is to directly insert the MS2 stem-loops in the 3′UTR of the ORF in the mRNA of interest. In this case, one can use the six MS2 stem-loops from the pSL-MS2–6 plasmid, which contain several restriction sites for subcloning. The MS2 stem-loops can also be inserted in the 5′ UTR of the gene of interest, as is shown in Fig. 4. The *LacZ* gene was expressed with 24 copies of the MS2 stem-loop in its 5′UTR as well as the *ASH1* bud-localization zipcode in its 3′UTR (Chartrand *et al.*, 1999). The site of transcription in the nucleus is visible in Fig. 4 A, as well as the mature MS2-tagged mRNA localized to the bud tip.

3. Induction and Visualization in Yeast

Yeast cells are transformed with the plasmids encoding the MS2-GFP fusion protein and the mRNA with MS2 stem-loops using standard yeast transformation procedures (Schiestl and Gietz, 1989). Yeasts are grown overnight in culture tubes at 30 °C in 5 ml of the appropriate selection media with either 2% raffinose/0.02% glucose or 2% lactic acid/3% glycerol/0.05% glucose. These carbon sources neither repress nor activate the *GAL1* promoter, so the reporter mRNA is not transcribed. The following day, the cultures are diluted to $OD_{600} = 0.1$ in the same medium but without glucose (glucose, even in small amount, represses the *GAL1* promoter) and the yeasts are grown to mid-log phase to an OD_{600} between 0.2 and 0.4. Galactose is added to a final concentration of 3% to activate the transcription of the MS2-tagged RNA (as seen in Fig. 4). The yeast cells can then be directly put under the microscope (for kinetic analysis) or maintained in culture for the desired period before imaging.

An important aspect of the growth medium is the yeast autofluorescence. Several yeast strains are *ade2*$^-$ and a red pigment (an oxidized metabolite of phosphoribosylaminoimidazole) accumulates in the vacuole. This pigment is fluorescent in the green emission channel and may interfere with the GFP signal. Yeast cells grown in yeast peptone dextrose medium, which is poor in adenine, show more autofluorescence. Use selection medium with extra adenine (~20 μg/ml) to avoid this problem. Also, use yeast cultures at OD_{600} below 0.5, since yeasts grown to high concentration (~1 OD_{600}) show more autofluorescence.

For acquisition of videos or time-lapse images with the microscope, a 10 μl sample of the yeast culture can be spotted on a microscope slide, covered with a coverslip (do not seal with nail polish) and put directly under the microscope. However, the yeasts float in the medium and they can rapidly move out of the focal plane, making acquisition a difficult task. To acquire images from the same yeast cell over a long period, it is better to immobilize the yeasts on a solid medium. To do so, prepare a 2% agarose solution containing the appropriate yeast selection medium with 3% galactose. Deposit 200 μl of the melted agarose solution on a microscope slide and drop a second slide over the agarose solution in order to create a sandwich of two slides separated by the thin agarose layer. Once solidified, the sandwich can be open, leaving a slide coated with a thin layer of agarose gel. On this layer of gel, the sample of yeast culture can be spotted and covered with a coverslip (do not seal). The yeasts will be immobilized on the agarose gel, making it easier to perform time-lapse imaging. Moreover, the gel is transparent enough not to interfere with image acquisition. Finally, the presence of selection medium in the agarose gel maintains the yeasts actively dividing over several hours.

B. *In Situ* Hybridization on Yeast Cells

Since the MS2-GFP system allows the visualization of the MS2-tagged RNA, this method provides an indirect tracking of the GFP-bound transcripts. To confirm that the GFP signal is associated with the MS2-tagged mRNA, a FISH can be performed to detect the transcript. In Fig. 4C and D, the cells were hybridized with a Cy3-labeled mix of five probes directed against LacZ (Long *et al.*, 1995). In Fig. 4G, the overlay shows that the MS2-GFP and the Cy3 signal colocalize in the bud and at the mRNA transcription site in the nucleus.

The following protocol is modified from Chartrand *et al.* (2000) and allows the simultaneous detection of a specific mRNA and the GFP signal. Paraformaldehyde fixation has no effect on the fluorescence intensity of the MS2-GFP fusion protein or on its distribution. The probes used are the same as described in Section III.B.

Important: All solutions for fixation, spheroplasting, and *in situ* hybridization should be DEPC-treated or prepared with DEPC-treated distilled water. Individuals must also wear gloves to avoid RNase contamination.

1. Preparation of the Coverslips

To maintain the yeast cells at the surface of the coverslip during the hybridization and the washing steps, the coverslips must be coated with poly-L-lysine.

1. Boil type 1 coverslips (22 × 22 mm) in 250 ml of 0.1 N HCl for 30 min. Let the coverslips cool down at room temperature.
2. Rinse 10 times with distilled water in a beaker.
3. Autoclave in distilled water. Coverslips can be stored at 4 °C for several months.
4. Put one coverslip in each well of a six-well tissue culture plate and drop 200 μl of poly-L-lysine 0.01% on each coverslip.
5. Incubate 2 min at room temperature, aspirate the excess, and let dry at room temperature (about 2–3 h).
6. When dry, wash each well three times with distilled water for 10 min.
7. Rest each coverslip on the wall of the wells, the face treated with poly-L-lysine on the top, and let dry (do not let the coverslips air-dry on the bottom of the wells, they will stick to the plastic).

2. Fixation and Spheroplasting of the Yeast Cells

1. Yeasts are grown in 50 ml cultures in the appropriate media until they reach early log phase (OD_{600} between 0.2 and 0.4, about 10^8 cells).
2. Fix cells for 45 min at room temperature by directly adding to the medium 6.25 ml of 32% formaldehyde.
3. Spin down cells for 5 min at 3500 rpm at 4 °C.
4. Wash cells three times with 10 ml of ice-cold 1× buffer B (1.2 M sorbitol, 0.1 M potassium phosphate, pH 7.5) and centrifugation (5 min at 3500 rpm at 4 °C).
5. Resuspend cells (do not vortex) in 1 ml of buffer B containing 20 mM vanadyl ribonucleoside complex, 28 mM β-mercaptoethanol, 0.06 mg/ml phenylmethylsulfonyl fluoride, 5 μg/ml of pepstatin, 5 μg/ml of leupeptin, 5 μg/ml of aprotinin, and 120 U/ml of RNase Inhibitor.
6. Transfer cells to a tube containing 60 μg of dried Zymolyase 100T (Seikagaku, Japan).
7. Incubate the cells for 20 min at 30 °C.
8. Centrifuge for 4 min at 3500 rpm at 4 °C.
9. Wash with 1 ml of ice-cold buffer B and spin.
10. Resuspend spheroplasts in 750 μl of buffer B.
11. Add 100 μl of spheroplasts on each poly-L-lysine-coated coverslips in the six well tissue culture plates.
12. Cells are left to adhere to the coverslips by incubating for 30 min at 4 °C.
13. Wash spheroplasts carefully with 3 ml of buffer B and remove by aspiration.
14. Add 5 ml of 70% ethanol in each well. Incubate at least 15 min at −20 °C before performing the *in situ* hybridization. At this stage, the coverslips can be stored for few weeks at −20 °C.

Notes: The Zymolyase 100T should be resuspended in 1× buffer B, aliquoted at 60 μg per tube, and lyophilized. These aliquots can be stored at −20 °C in a dessicator.

3. *In Situ* Hybridization

a. Preparation of the Probes

For each coverslip used in the hybridization, prepare one tube of probes (we suggest the use of two coverslips per experiment in order to have a duplicate if one is broken during manipulations).

1. Mix 10 μl of a 1 ng/μl probes solution with 4 μl of a 5 mg/ml solution of 1:1 sonicated salmon sperm DNA:*Escherichia coli* tRNA.
2. Lyophylize and resuspend in 12 μl of 80% formamide, 10-mM sodium phosphate pH 7.0.
3. Heat the probe solution at 95 °C for 3 min. Keep covered at room temperature.

b. Hybridization

1. Rehydrate the cell-coated coverslips in a Coplin jar with two washes in 8 ml of 2× SSC for 5 min at room temperature.
2. Incubate coverslips in 8 ml of 40% formamide, 2× SSC for 5 min at room temperature.
3. Add to the probes 12 μl of 4× SSC, 20-mM vanadyl ribonucleoside complex, 4 μg/μl of RNase-free BSA, 50 U of RNase inhibitor.
4. Incubate the coverslips with 24 μl of the probe solution overnight at 37 °C in a humidified chamber, as described in Section III.B.

c. Washing

1. After the incubation, remove the coverslips from the Parafilm sheet and put them back in the Coplin jar.
2. Wash the coverslips twice with 8 ml of 40% formamide, 2× SSC (preheated at 37 °C) for 15 min at 37 °C.
3. Wash with 8 ml of 2× SSC, 0.1% Triton X-100 for 15 min at room temperature.
4. Wash twice with 8 ml of 1× SSC for 15 min at room temperature.
5. Add 8 ml of 1× PBS containing 1 ng/ml of DAPI.
6. Mount coverslips on a glass slides (1-mm thick). Drop 5 μl of mounting medium on the slide, lay down the coverslip on the drop (the surface of the coverslip containing the cells should face the drop), and remove the excess of medium with Kimwipes®.
7. Seal the sides of the coverslips with nail polish. Image cells as soon as possible.

Acknowledgments

We thank Edouard Bertrand for the sequence of the MS2 probe. Research in the author's laboratory was supported by grants from the Canadian Institutes of Health Research (CIHR). P.C. is supported by a fellowship from the Fond de Recherche en Santé du Québec (FRSQ).

References

Beach, D. L., Salmon, E. D., and Bloom, K. (1999). Localization and anchoring of mRNA in budding yeast. *Curr. Biol.* **9,** 569–579.

Bernardi, A., and Spahr, P. F. (1972). Nucleotide sequence at the binding site for coat protein on RNA of bacteriophage R17. *Proc. Natl. Acad. Sci. USA* **69,** 3033–3037.

Bertrand, E., Chartrand, P., Schaefer, M., Shenoy, S. M., Singer, R. H., and Long, R. M. (1998). Localization of ASH1 mRNA particles in living yeast. *Mol. Cell* **2,** 437–445.

Brodsky, A. S., and Silver, P. A. (2000). Pre-mRNA processing factors are required for nuclear export. *RNA* **6,** 1737–1749.

Chartrand, P., Meng, X.-H., Singer, R. H., and Long, R. M. (1999). Structural elements required for the localization of ASH1 mRNA and of a green fluorescent protein reporter particle *in vivo*. *Curr. Biol.* **9,** 333–336.

Chartrand, P., Singer, R. H., and Long, R. M. (2000). Sensitive and high-resolution detection of RNA in situ. *Methods Enzymol.* **318,** 493–506.

Cullen, B. R. (2003). Nuclear RNA export. *J. Cell Sci.* **116,** 587–597.

Fusco, D., Accornero, N., Lavoie, B., Shenoy, S. M., Blanchard, J.-M., Singer, R. H., and Bertrand, E. (2003). Single mRNA molecules demonstrate probabilistic movement in living mammalian cells. *Curr. Biol.* **13,** 161–167.

Golding, I., Paulsson, J., Zawilski, S. M., and Cox, E. C. (2005). Real-time kinetics of gene activity in individual bacteria. *Cell* **123,** 1025–1036.

Jaeger, S., Eriani, G., and Martin, F. (2003). Results and prospects of the yeast three-hybrid system. *FEBS Lett.* **556,** 7–12.

Janicki, S. M., Tsukamoto, T., Salghetti, S. E., Tansey, W. P., Sachidanandam, R., Prasanth, K. V., Ried, T., Shav-Tal, Y., Bertrand, E., Singer, R. H., and Spector, D. L. (2004). From silencing to gene expression: Real-time analysis in single cells. *Cell* **116,** 683–698.

Kindler, S., Wang, H., Richter, D., and Tiedge, H. (2005). RNA transport and local control of translation. *Annu. Rev. Cell Dev. Biol.* **21,** 223–245.

Lim, F., and Peabody, D. S. (1994). Mutations that increase the affinity of a translational repressor for RNA. *Nucleic Acids Res.* **22,** 3748–3752.

Long, R. M., Elliott, D. J., Stutz, F., Rosbash, M., and Singer, R. H. (1995). Spatial consequences of defective processing of specific yeast mRNAs revealed by fluorescent in situ hybridization. *RNA* **1,** 1071–1078.

Machuca-Tzili, L., Brook, D., and Hilton-Jones, D. (2005). Clinical and molecular aspects of the myotonic dystrophies: A review. *Muscle Nerve* **32,** 1–18.

Peabody, D. S., and Ely, K. R. (1992). Control of translational repression by protein–protein interactions. *Nucleic Acids Res.* **20,** 1649–1655.

Rackham, O., and Brown, C. M. (2004). Visualization of RNA–protein interactions in living cells: FMRP and IMP1 interact on mRNAs. *EMBO J.* **23,** 3346–3355.

Rook, M. S., Lu, M., and Kosik, K. S. (2000). CaMKIIα 3$'$ untranslated region-directed mRNA translocation in living neurons: Visualization by GFP linkage. *J. Neurosci.* **20,** 6385–6393.

Schiestl, R. H., and Gietz, R. D. (1989). High efficiency transformation of intact yeast cells using single stranded nucleic acids as a carrier. *Curr. Genet.* **16,** 339–346.

Schmid, M., Jaedicke, A., Du, T.-G., and Jansen, R.-P. (2006). Coordination of endoplasmic reticulum and mRNA localization to the yeast bud. *Curr. Biol.* **16,** 1538–1543.

Shav-Tal, Y., Darzacq, X., Shenoy, S. M., Fusco, D., Janicki, S. M., Spector, D. L., and Singer, R. H. (2004). Dynamics of single mRNPs in nuclei of living cells. *Science* **304,** 1797–1800.

Shav-Tal, Y., Shenoy, S. M., and Singer, R. H. (2005). Visualization and quantification of single RNA molecules in living cells. *In* "Live Cell Imaging: A Laboratory Manual" (R. D. Goldman and D. L. Spector, eds.), pp. 603–615. Cold Spring Harbor Laboratory Press, Cold Spring Harbor, NY.

Takizawa, P. A., and Vale, R. D. (2000). The myosin motor, Myo4p, binds Ash1 mRNA via the adapter protein, She3p. *Proc. Natl. Acad. Sci. USA* **97,** 5273–5278.

Talbot, S. J., Goodman, S., Bates, S. R. E., Fishwick, C. W. G., and Stockley, P. G. (1990). Use of synthetic oligoribonucleotides to probe RNA-protein interactions in the MS2 translational operator complex. *Nucleic Acids Res.* **18,** 3521–3528.

Valegard, K., Murray, J. B., Stockley, P. G., Stonehouse, N. J., and Liljas, L. (1994). Crystal structure of an RNA bacteriophage coat protein-operator complex. *Nature* **371,** 623–626.

Verheggen, C., Mouaikel, J., Thiry, M., Blanchard, J.-M., Tollervey, D., Bordonné, R., Lafontaine, D. L. J., and Bertrand, E. (2001). Box C/D small nucleolar RNA trafficking involves small nucleolar RNP proteins, nucleolar factors and a novel nuclear domain. *EMBO J.* **20,** 5480–5490.

Zhang, F., and Simon, A. E. (2003). A novel procedure for the localization of viral RNAs in protoplasts and whole plants. *Plant J.* **35,** 665–673.

CHAPTER 13

Visualizing mRNA Localization and Local Protein Translation in Neurons

Ralf Dahm, Manuel Zeitelhofer, Bernhard Götze,[1] Michael A. Kiebler, and Paolo Macchi[2]

Center for Brain Research
Division of Neuronal Cell Biology
Medical University of Vienna
Spitalgasse 4, A-1090 Vienna, Austria

[1] Present address: Carl Zeiss MicroImaging GmbH, Carl Zeiss Promenade 10, D-07745 Jena, Germany.

[2] Present address: Center for Integrative Biology, Laboratory of Molecular and Cellular Neurobiology, University of Trento, via delle Regole 101, 38060 Mattarello, Trento, Italy.

0091-679X/08 $35.00
DOI: 10.1016/S0091-679X(08)85013-3

Abstract

Fluorescent proteins (FPs) have been successfully used to study the localization and interactions of proteins in living cells. They have also been instrumental in analyzing the proteins involved in the localization of RNAs in different cell types, including neurons. With the development of methods that also tag RNAs via fluorescent proteins, researchers now have a powerful tool to covisualize RNAs and associated proteins in living neurons. Here, we review the current status of the use of FPs in the study of transport and localization of ribonucleoprotein particles (RNPs) in neurons and provide key protocols used to introduce transgenes into cultured neurons, including calcium-phosphate-based transfection and nucleofection. These methods allow the fast and efficient expression of fluorescently tagged fusion proteins in neurons at different stages of differentiation and form the basis for fluorescent protein-based live cell imaging in neuronal cultures. Additional protocols are given that allow the simultaneous visualization of RNP proteins and cargo RNAs in living neurons and aspects of the visualization of fluorescently tagged proteins in neurons, such as colocalization studies, are discussed. Finally, we review approaches to visualize the local synthesis of proteins in distal dendrites and axons.

I. Introduction

RNA localization is a widely used mechanism to restrict gene products to specific regions of a cell or organism (Dahm and Kiebler, 2007; Palacios and St Johnston, 2001; St Johnston, 2005). In the nervous system, RNA localization plays a crucial role both during development and in the adult organism. During development, it is essential for the differential sorting of cell fate determinants in the asymmetric divisions of neuroblasts (Betschinger and Knoblich, 2004) as well as for the local synthesis of proteins in the growth cones of exploratory neurites (Piper and Holt, 2004). In the adult nervous system, the localization of RNAs is important for the formation of new synapses as well as their subsequent modulation (Ashraf *et al.*, 2006; Dahm *et al.*, 2007; Klann and Dever, 2004; Schuman *et al.*, 2006). Mature neurons, for example, localize certain mRNAs into their dendrites. The local translation of these mRNAs upon specific stimuli is believed to be particularly important for the fine tuning of synaptic connections in the CNS and thus to form the basis for the acquisition of new memories. It has been demonstrated that a dysregulation of this process can result in severe impairment of mental functions, such as the fragile-X mental retardation syndrome (Antar and Bassell, 2003; Bagni and Greenough, 2005; Dahm and Macchi, 2007; Miller *et al.*, 2002). In addition to neurons, also glia cells have been demonstrated to localize specific RNAs. Oligodendrocytes, for instance, localize the mRNA for myelin basic protein (MBP), a major component of the myelin sheaths formed by these cells, to the cells' periphery (Carson *et al.*, 2001). RNA localization is thus crucial both for the formation and for the maintenance of a functioning nervous system.

The best-studied paradigm of RNA localization in the nervous system is the transport of RNAs into neuronal processes. This occurs both during nerve cell differentiation and in mature neurons. During differentiation, RNAs localize to the leading edges of the exploratory growth cones of outgrowing neurites. In mature neurons, RNAs are transported to postsynaptic sites (Martin and Zukin, 2006) as well as (in rare examples) presynaptic sites (Piper and Holt, 2004). The localization of RNAs to leading edges is a basic process also observed in other migratory cell types and has been extensively studied, for example, in fibroblasts (Mingle *et al.*, 2005). The localization of mRNAs into dendrites and axons, however, is an adaptation of the nervous system to allow for a rapid regulation of local protein content at distant sites (Kiebler and DesGroseillers, 2000; Steward and Schuman, 2003). This is particularly important during the formation of new memories.

While short-term memory is mediated via transient modifications, for instance phosphorylation and dephosphorylation of *existing* molecules, long-term memory relies to a large extent on the *de novo* synthesis of RNAs and proteins (Schuman *et al.*, 2006; Steward and Schuman, 2001). The proteins required for the remodeling of synapses are synthesized not only in the perinuclear cytosol but also locally at those synapses that are remodeled during learning. The local translation of mRNAs at activated synapses is viewed as a mechanism to rapidly produce postsynaptic proteins in an activity-dependent and input-specific manner. While ever more dendritically localized RNAs are being discovered—estimates range up to 400 different transcripts localized in dendrites (Job and Eberwine, 2001)—it is generally believed that the vast majority of mRNAs in a neuron is not transported into distal dendrites or axons, but are restricted to the neuron's cell body.

The finding that neurons synthesize proteins in distal dendritic regions—at a distance of up to several cell body diameters from the nucleus—raises several questions: Which mRNAs are localized at distal sites? How are these localized RNAs recognized and targeted to their distal locations? What is the molecular nature of the transport machinery that binds these mRNAs and ferries them into the neuronal processes? How are mRNAs captured, anchored, and ultimately selected for local protein synthesis by, for example, activated synapses? And how is the translation of mRNAs repressed during transport and activated on demand?

The process of dendritic RNA localization can be subdivided into three broad phases. First, the localized mRNAs have to be packaged into transport-competent particles. Part of this packaging is thought to occur already in the nucleus following or in parallel to transcription and nuclear processing of the RNAs (Kiebler *et al.*, 2005). Additional factors are then added in the cytoplasm subsequent to nuclear export (Lopez de Heredia and Jansen, 2004). The assembly of ribonucleoprotein particles (RNPs) relies both on the recognition of motifs in the RNA (*cis*-acting elements) by RNA-binding proteins (*trans*-acting factors) and on protein–protein interactions between *trans*-acting factors. Second, when RNP assembly is complete, the transport-competent particles are transported to their destinations, that is, sites of local protein synthesis. This transport occurs along the cytoskeleton with the help of molecular motors. While at least some of the cytoskeletal components and molecular motors involved in the directed translocation of transport RNPs have been identified (Hirokawa, 2006; Hirokawa and Takemura, 2005; St Johnston, 2005), it is currently unclear which adapter proteins couple transport RNPs to these motors. Moreover, little is known about the molecules and mechanisms that anchor RNPs at their final destinations. Third, specific stimuli have to be integrated to induce the translation of the localized mRNAs (Govindarajan *et al.*, 2006). During transport localized mRNAs have to be kept translationally silent to avoid any inappropriate production of the corresponding proteins. The signaling events and molecular mechanisms mediating this switch from translational silencing to translational activation are only recently being unraveled (Dahm and Kiebler, 2005; Huttelmaier *et al.*, 2005).

Fluorescent proteins (FPs) have been instrumental in elucidating many of the processes underlying RNA localization in neurons and other cell types. They make *trans*-acting factors visible and allow the experimenter to follow their movements in living cells. Moreover, the use of mutant versions of *trans*-acting factors fused to FPs can yield valuable insights into the roles these domains play. The recent development of an FP-based method to label RNAs now also allows a covisualization of *trans*-acting factors with their cargo RNAs. In the current chapter, we provide detailed protocols for introducing transgenes into cultured neurons. These methods allow a fast and efficient expression of fluorescently tagged fusion proteins in neurons at different stages of differentiation and form the basis for FP-based live cell imaging. Further protocols are provided that allow the simultaneous visualization of RNP proteins and cargo RNAs in living neurons as well as the local translation of mRNAs in distal dendrites and axons. The methods described below are generally optimized for use with primary cultured neurons. They can, however, (possibly with slight adaptations) also be used for neuronal cell lines.

II. Visualization of RNA Transport via RNA-Binding Proteins in Neurons

A. Methods for Transient Transfection of Cells

The transient transfection of cells with expression constructs encoding FP-tagged proteins is a rapid and efficient way of introducing labeled proteins into living cells. Several methods for a transient transfection of neurons have been established. These include lipofection, electroporation, microinjection, biolistics, adeno-, and retrovirus-based methods as well as the transfection with a DNA/calcium-phosphate coprecipitate (DNA/CaP_i) [reviewed in Goetze *et al.* (2004) and Washbourne and McAllister (2002)]. However, few of these methods are suitable for delivering transgenes into primary neurons. In the following sections, we will focus on two transfection methods, (i) the DNA/CaP_i method and (ii) nucleofection, an advanced electroporation technique.

The transfection with a DNA/CaP_i coprecipitate was first employed over 30 years ago (Graham and van der Eb, 1973) and has since been modified for use with neurons. It has several advantages over other methods. First, it is an easy to use protocol that consistently yields transfection rates sufficient for, for example, microscopy analyses. Second, the chemicals required are easily available and significantly less expensive than commercial transfection agents. No specialized equipment is needed. Third, the DNA/CaP_i method can be used to transfect both established cell lines and primary cell cultures. Also, nonadherent cells can be transfected using DNA/CaP_i coprecipitates. Importantly, it allows the transfection of postmitotic cells and also the creation of stably transfected cell lines is possible. Finally, in contrast to many liposome-based methods, the neuronal morphology is preserved after DNA/CaP_i transfection and the detachment of neuronal processes is not an issue.

The basic principle underlying the transfection with DNA/CaP_i is the following: The DNA construct to be transfected, generally an expression vector encoding the desired transgene, and Ca^{2+} are mixed in a phosphate-containing buffer to form crystals composed of DNA, Ca^{2+}, and PO_4^{2-} ions. The formation and growth of these crystals crucially depends on the pH of the solution they form in. The transfection solution containing the DNA/CaP_i crystals is subsequently added to the medium overlying the cells to be transfected. The crystals continue to grow in the medium and with time sink onto the cells, which endocytose the DNA/CaP_i coprecipitate. Inside the cells, the DNA leaves the endocytic compartment and is believed to enter the nucleus as the nuclear envelope breaks down during cell division. Since also postmitotic cells, such as neurons, can be transfected with this method, it is likely that a second as yet unidentified pathway exists that allows the DNA to enter the nucleus.

The following protocol is based on a method originally published in 1982 (Ishiura *et al.*, 1982) and has been optimized since for the transfection of cultured primary neurons. This method allows not only the transfection of freshly isolated, immature neurons but also the cultured neurons of all stages of differentiation, including of mature, fully polarized neurons.

B. Protocol for the Transfection with a DNA/CaP_i Coprecipitate

The volumes of buffers and solutions used in this procedure are optimized for 2 ml of transfection medium and a total amount of 3 μg of plasmid DNA. They can be linearly scaled up or down according to the volume of media required.

To produce the DNA/CaP_i coprecipitate solution (termed *transfection solution*), mix $CaCl_2$ solution, ddH_2O, DNA plasmid solution (equivalent of 3 μg DNA), and 2× BBS buffer in a 1.5 ml Eppendorf tube according to the protocol detailed in Table I. This results in a total of 120 μl of transfection solution. The transfection solution is then added to the 2 ml of transfection medium (Table II) in the culture dish containing the cells to be transfected. Take care to disperse the transfection solution as evenly as possible and mix by gently swirling the dish to ensure optimal mixing of the transfection solution with the transfection medium.

Following the addition of the transfection solution to the transfection medium, the cells are incubated in a humidified incubator (without CO_2 supply) at 36.5 °C and inspected for the presence of the DNA/CaP_i coprecipitate on the cells at

Table I
Pipetting Scheme for the Transfection of Neurons with the Calcium-Phosphate Method

Solution	Volume	Notes
2.5 M $CaCl_2$	6 μl	
ddH_2O	× μl (adjust to yield total volume of 60 μl after addition of plasmid DNA solution)	Mix well by pipetting the solution up and down ~10 times.
Plasmid DNA	× μl (equivalent of 3 μg DNA)	Briefly centrifuge the plasmid solution to bring down any evaporated liquid and small particles, which can act as nuclei for the crystallization process and thus influence the size and speed of formation of the DNA/CaP_i coprecipitate in an unpredictable manner. Add the plasmid DNA solution very slowly while stirring with the pipette to thoroughly mix the components. For cotransfections of more than one plasmid, the total amount of DNA used should be 3 μg, for example, use 1.5 μg DNA of plasmid 1 plus 1.5 μg DNA of plasmid 2. Note: It is crucial that the plasmid DNA is added after the ddH_2O and $CaCl_2$ have been mixed.
Total volume:	60 μl	
2× BBS	60 μl	Add BBS dropwise to the DNA/CaP_i solution. Flick the tube with your finger after each addition of a few drops to mix. After all BBS has been added, pipet up and down ~10 times and/or bubble air through the tube to mix the solution well. Do not vortex! For a note on the pH of the 2× BBS used, see below.
Final volume:	120 μl	

Table II
Solutions Required for the Transfection of Neurons with the Calcium-Phosphate Method

Solution	Composition
2× BBS buffer	50 mM BES [*N*,*N*-Bis(2-hydroxyethyl)-2-aminoethanesulfonic acid]
	1.5 mM Na_2HPO_4
	280 mM $NaCl_2$ in ddH_2O
Transfection medium	1× MEM (modified Eagle's medium) from a 10× MEM stock
	15 mM HEPES [4-(2-hydroxyethyl)-1-piperazineethanesulfonic acid]
	1 mM sodium pyruvate
	2 mM L-glutamine, stable
	33 mM D-glucose
	1× B-27 supplement in ddH_2O; adjust pH to 7.45
HBSS (Hank's balanced saline solution) washing buffer	135 mM $NaCl_2$
	20 mM HEPES [4-(2-hydroxyethyl)-1-piperazineethanesulfonic acid]
	1 mM Na_2HPO_4
	4 mM KCl
	2 mM $CaCl_2$
	1 mM $MgCl_2$
	10 mM D-glucose in ddH_2O; adjust pH to 7.3

regular intervals (approximately every 20–30 min). Inspections are done with an inverted microscope and a 10× or a 20× air objective. Typically, a precipitate can first be detected after 45 min. Depending on the conditions of the transfection solution (plasmid, pH of 2× BBS, time of incubation before addition to transfection medium; for explanations, see notes below), it can, however, take up to 3–4 h for the precipitate to form and settle down onto the neurons.

In case the precipitate is too fine and takes long to sink onto the cells, a centrifugation step may be inserted to bring down the DNA/CaP_i crystals. This should, however, only be done for fine to intermediate-sized crystals. Big crystals damage neurons when brought down by the centrifugation. To centrifuge, tape the culture dishes with the cells facing up onto swing-out tables and centrifuge. We typically use the following setup for centrifugation: Eppendorf centrifuge 5810, rotor A-4–62 (18 cm), swing-out tables (fitting standard 96-well plates/ 200-ml tissue culture flasks). Tape the culture dishes to the center of the table and centrifuge at 1000 rpm for 1 (intermediate-sized precipitate) to 2 min (fine precipitate). After the centrifugation, continue the incubation in transfection medium for a maximum of 30 min. Longer incubations lead to damage to the cells.

Ultimately, the cells should be covered evenly with a fine precipitate barely visible with a 10× objective. The overall time of exposure of the cells to the precipitate depends on the size and density of the DNA/CaP_i crystals. The larger the crystals and the higher their density, the shorter the exposure time should be as large crystals or a high density are detrimental to the cells. Generally, as long as the cells appear healthy, the precipitate can be left on. As soon as cells show the first signs of stress, however, the precipitate should be removed by washing it off with prewarmed (36.5 °C) Hank's balanced saline solution (HBSS) washing buffer (see Table II). Inspect the cells after 3–5 min under a microscope to ensure that the precipitate has dissolved completely (do not wash for more than 5 min) and transfer the cells back into the culture dish containing the glia and conditioned neuronal tissue culture medium (NMEM; Table III).

In addition to neurons, the method described can also be used to transfect glia cells. Generally, these cells require larger precipitates than neurons. Therefore, increasing the pH and/or the incubation time of the transfection solution before adding it to the transfection medium is recommended. Similarly, a longer washing period to remove the precipitate following the transfection may be needed.

1. Note on the Formation of the DNA/CaP$_i$ Coprecipitate

The size of the DNA/CaP$_i$ crystals is critical to the success and efficiency of the transfection (an example of a precipitate with optimal size and density is shown in Fig. 1). Its initial formation in the transfection solution and subsequently in the transfection medium crucially depends on the pH. In the transfection solution, increasing the pH of the 2× BBS will result in a faster formation and fewer but larger DNA/CaP$_i$ crystals. Conversely, if the pH of the 2× BBS is lowered, the DNA/CaP$_i$ crystals will form more slowly and be smaller. Generally, the optimal

Table III
Solutions Required for Culturing and Live Imaging of Neurons

Solution	Composition
NMEM-B27 medium (neuronal culture medium)	1× MEM (modified Eagle's medium) from a 10× MEM stock
	26 mM $NaHCO_3$
	1 mM sodium pyruvate
	200 mM L-glutamine, stable
	33 mM D-glucose
	1× B-27 supplement in ddH_2O; pH adjusts automatically via the CO_2 concentration (5%) in the incubator
Modified HBSS (Hank's balanced saline solution)	137 mM $NaCl_2$
	7 mM HEPES [4-(2-hydroxyethyl)-1-piperazineethanesulfonic acid]
	4.2 mM $NaHCO_3$
	0.35 mM Na_2HPO_4
	0.45 mM KH_2PO_4
	5.4 mM KCl
	1.3 mM $CaCl_2$
	0.8 mM $MgSO_4$
	5.5 mM D(+)-glucose
	100 U/ml penicillin
	100 μg/ml streptomycin in ddH_2O; adjust pH to 7.3

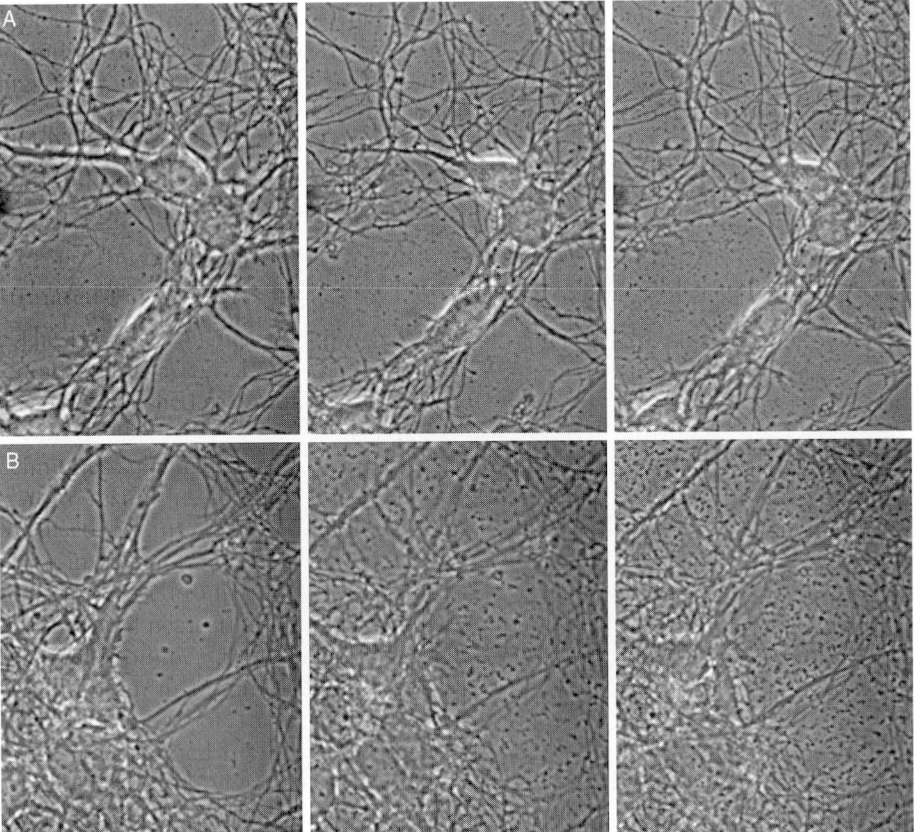

Fig. 1 Formation of a DNA/CaP_i coprecipitate on primary cultures of hippocampal neurons. (A) The formation and deposition of the DNA/CaP_i coprecipitate on cultured neurons was monitored over a period of 4 h. Assessment of the size and density of the precipitate is best done in areas of the coverslip that are free of neuronal somata and processes. Immediately after the addition of the transfection solution to the medium, no precipitate is visible. After 2 h of incubation several crystals have settled onto the coverslip and cells. By 4 h a uniform and fine precipitate at optimal density for high-transfection rates has formed. (B) The formation and deposition of the DNA/CaP_i coprecipitate on cultured neurons was monitored over a period of 8.5 h. No crystals are visible immediately following the addition of the transfection solution. After 4 h numerous large crystals have settled onto the coverslip and neurons. The size of these crystals exceeds the optimal size of crystals for the transfection of neurons. By 8.5 h a dense carpet of large crystals covers the neurons. Such crystal sizes and densities likely result in low survival rates of transfected neurons.

pH for the 2× BBS lies between pH 7.0 and 7.2. Moreover, the time that elapses between the addition of the 2× BBS to the transfection solution and the addition of the transfection solution to the transfection medium in the culture dish influences the formation of the DNA/CaP_i crystals. The longer this time, the larger the crystals will grow. The experimenter thus has two parameters to play with in order to obtain the optimal DNA/CaP_i coprecipitate for their purposes.

The formation of the DNA/Ca^{2+} crystals continues in the transfection medium. Therefore, in addition to depending on the pH of the 2× BBS, it also depends on the pH of the transfection medium. In order to maintain a stable pH during the incubation of the cells, the transfection medium is HEPES buffered. Unlike standard carbonate-buffered media, this renders it independent of CO_2 and allows for a controlled and linear formation of the coprecipitate increasing both transfection efficacy as well as cell survival during the transfection. The pH value of the transfection medium is adjusted to 7.45 and thus corresponds to the physiological pH of the medium used to culture the neurons before and after the transfection [for detailed protocols on the culture of primary mammalian neurons, see Goetze *et al.* (2003) and Goslin *et al.* (1998)]. In order to obtain high transfection rates, it is important to minimize the stress the neurons experience during the entire procedure. Therefore, maintaining a constant pH across the different solutions is advisable.

The formation of the DNA/CaP_i coprecipitate also tends to vary with the plasmid used and even between two preparations of the same plasmid. When working with a new plasmid or a novel combination of plasmids (see below), it is advisable to test variations of these parameters in order to obtain optimal transfection conditions. Typically, a series of 3 different pH values of 2× BBS (e.g., 6.90, 7.00, and 7.10) as well as 3 different incubation times should result in the identification of near optimal conditions for any given plasmid. Please also note that there may be significant differences between different batches of 2× BBS and that new batches will have to be tested for their properties with well-characterized plasmids [for details, see Goetze and Kiebler (2006)].

2. Note on the Preparation of Plasmid DNA for Transfections

For the transfection of neurons, the purity of the DNA is essential. We recommend that endotoxin-removing DNA isolation kits are used to remove any neurotoxic substances, such as bacterial endotoxins. Alternatively (but less conveniently), plasmid DNA for transfections can be purified using a CsCl density gradient centrifugation. Dried plasmid pellets can be redissolved either in ddH_2O or in TE buffer.

3. Note on Addition of 2× BBS to Transfection Solution

Before the addition of 2× BBS, no DNA/CaP_i crystals will form and the mixture can be left sitting at this stage. This can be helpful when, for example, preparing a large number of transfection solutions. Once the 2× BBS has been added, however, the transfection solution must rapidly be added to the transfection medium in order to avoid the crystals growing too large.

4. Note on the Preparation and Storage of 2× BBS and $CaCl_2$

The 2× BBS (see Table II for chemical composition) and the 2.5 M $CaCl_2$ solutions are sterilized by filtering (0.22 μm pore size). Freezing and thawing of these solutions affect the efficiency of the transfections. Storage of 2× BBS in plastic containers (e.g., Falcon tubes) at room temperature results in a deterioration of the 2× BBS. We therefore suggest storing all solutions (and media) in glass bottles at 4 °C.

It is advisable to produce 2× BBS solutions with a range of pH values (e.g., from pH 6.95 to 7.20 in steps of 0.05) for use with different plasmids and plasmid combinations. This is important as the pH of the transfection solution crucially influences the formation of the DNA/Ca^{2+} crystals and different plasmids tend to precipitate better at different pH values (see above). Use a 2× BBS aliquot that has been used in a successful transfection as a pH-reference to adjust the pH of a new batch of 2× BBS buffers. Test a new batch of 2× BBS by performing control experiments with well-characterized plasmids. When testing a pH series, the size of the DNA/Ca^{2+} crystals should increase linearly with increasing pH of the 2× BBS. Should this not be the case, discard the entire 2× BBS batch. It is much more efficient to produce 2× BBS of very good quality than trying to adjust for the effects of poor quality 2× BBS in transfections.

5. Note on the Preparation and Storage of Transfection Medium

In contrast to normal neuronal culture medium (NMEM; see Table III), the pH of the medium used for calcium-phosphate-based transfection is independent of CO_2. As a consequence, the pH of the medium has to be adjusted to 7.45 before use. A reference medium that is not used for transfections can be used for comparison to adjust the pH accurately and reproducibly by comparing the color of the phenol red. Subsequently, the transfection medium is sterilized by filtering (0.22 μm pore size) and can be stored in sealed 50-ml glass bottles for up to 2 months.

C. Cotransfection Using More than One Plasmid

When two or more plasmids are to be cotransfected, the total amount of plasmid DNA in the transfection solution should be 3 μg (as is the case when transfecting a single plasmid). Mix volumes corresponding to 1.5 μg of each plasmid DNA in a separate tube before addition to transfection solution. Usually the efficacy of cotransfections of two plasmids mixed at a ratio of 1:1 is >95% (Goetze *et al.*, 2004). By contrast, the use of three or more plasmids at a time has proven to be much less efficient and less reproducible (Goetze *et al.*, unpublished data). It therefore needs careful controls at the single cell level with appropriate markers. If the efficiency of cotransfection for two plasmids is not good, try mixing the two plasmid solutions (in an equimolar ratio), reprecipitate, and redissolve them. Add the equivalent of 3 μg DNA to the transfection solution.

D. High Efficiency Transfection of Neurons by Nucleofection

Generally, the DNA/CaP_i transfection method described above results in transfection efficiencies that are sufficient for experiments, which analyze individual cells, for example, microscopy analyses. Similarly, the transfection efficiencies of lipid-based methods tend to be sufficient for such applications. For approaches requiring very high transfection rates, for example, to control the knockdown efficiency in RNAi experiments, other techniques are recommended. Virus-based methods are more efficient, but require additional safety precautions and time-consuming viral particle preparation. The nucleofection technique overcomes these limitations while ensuring consistently high-transfection rates (Gartner *et al.*, 2006; Gresch *et al.*, 2004; Leclere *et al.*, 2005; Zeitelhofer *et al.*, 2007). This technology allows biochemical analyses of transfected primary neurons to be performed, for example, Western blot analyses of protein levels after RNAi knockdown (Goetze *et al.*, 2006; Li *et al.*, 2005). As a consequence, in many cases nucleofection is the method of choice to achieve very high levels of transfection in cultured primary neurons (Fig. 2).

Nucleofection is based on electroporation. A transfection by electroporation relies on the application of voltage pulses that temporarily alter the physical properties of the cells' plasma membranes and thus allow extracellular material (including plasmids in the surrounding medium) to enter the cells. In contrast to conventional electroporation where a single voltage pulse is applied, the nucleofection technique employs a series of complex voltage patterns. These electrical parameters are combined with cell type-specific reagents. On the one hand, this ensures a good viability of cells. On the other, it facilitates the transfer of transfected constructs directly into the nucleus. This is of particular importance when postmitotic cells, such as primary neurons, are to be made accessible to efficient gene transfer (Gresch *et al.*, 2004; Leclere *et al.*, 2005; Zeitelhofer *et al.*, 2007).

A drawback of current nucleofection protocols is that, in contrast to lipofection or the DNA/CaP_i transfection, nucleofection is more expensive, both in terms of dedicated equipment and solutions required and in terms of the number of cells that have to be used per experiment. The latter is partially overcome by the recent development of a second generation of nucleofection devices (Zeitelhofer *et al.*, 2007). The biggest limitation of nucleofection when working with primary neurons, however, is that it has to be done in suspension, that is, immediately after isolation of the neurons from the animal. This can be a problem if mature neurons are to be studied. In culture, neurons require at least 10 days to differentiate, that is, to develop axons, a dendritic tree and functional synapses. As a consequence, plasmids with inducible promoters may have to be used in cases where prolonged overexpression of a transgene may interfere with the differentiation and/or function of neurons or when only late effects of transgene expression are to be studied. Furthermore, in some instances electroporation might delay the development of hippocampal neurons with even mock-treated neurons displaying delays (Goetze and Kiebler, unpublished data).

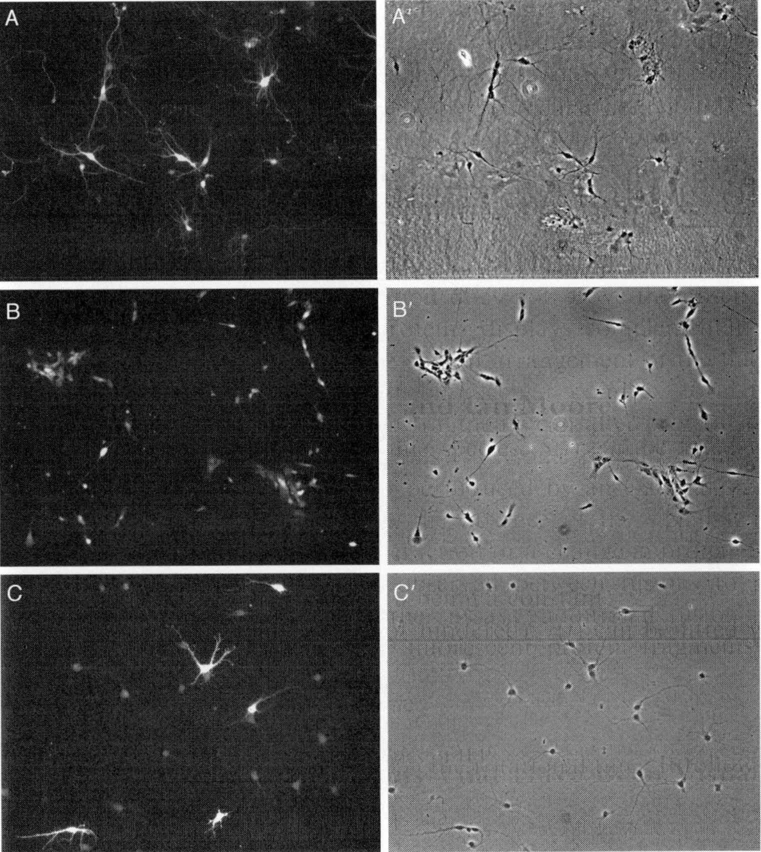

Fig. 2 Hippocampal neurons nucleofected with expression plasmids encoding green fluorescent protein (GFP). Freshly isolated rat hippocampal neurons (P0) were nucleofected with pmaxGFP (A) or pSyn-GFP (B) and cultured in 96-well plates. After 1 day *in vitro* (DIV; A) or 7 DIV (B) neurons were fixed and analyzed by fluorescence (A, B) and phase contrast microscopy (A′, B′). The transfection efficiency with optimized conditions for the Nucleofector 96-well Shuttle System ranged between 30% and 50%. The morphology of transfected neurons (assessed after 7 DIV) was unaltered compared with untransfected neurons. Panel C shows a representative field of view of rat hippocampal neurons transfected with a plasmid encoding ZBP1-EGFP at 11 DIV with the DNA/CaP$_i$ method and imaged 12 h after the transfection. The comparison of C with panels A and B shows the lower-transfection efficiency achieved with this method. (See Plate no. 41 in the Color Plate Section.)

E. Protocol for Nucleofection of Neurons

1. Note on the Quality of Neurons Used for Nucleofection

The quality of neurons is an important factor for the success of any transfection. All transfection procedures are stressful for neurons. Therefore, the neuronal preparation should be as good as possible to assure maximal survival rate after transfection. This is particularly important for nucleofection procedures. It is therefore important to ensure optimal conditions during the isolation of the neurons as well as before and after the transfection. In particular, the time required for the isolation of the neurons and mechanical stress (e.g., shear forces during pipetting) should be minimized, and shifts in temperature, pH, and osmolarity should be avoided in cases where media are changed.

For a nucleofection, the following materials are needed: a nucleofector device that allows the delivery of defined sets of electrical pulses as well as commercially available nucleofector kits that contain optimized and cell type-specific nucleofection solutions (Leclere *et al.*, 2005). Since the nucleofector technology is currently only commercially available, there is no information in the public domain as to the exact voltage parameters applied and chemical composition of the solutions used in the nucleofection procedure.

Generally, primary neurons are nucleofected immediately after isolation from embryonic brains. For example, rat hippocampi isolated from embryonic (E17) or from neonatal rats (P0 to P2) can be used. After dissecting the hippocampi, the neurons are dissociated with a trypsin-EDTA solution for 10 min at 37 °C (Table IV). The trypsin solution is removed and the neurons are washed twice with HBSS buffer. DMEM-HS [10% (v/v) horse serum in DMEM] is then added, and the neurons are dissociated by pipetting 10× with a fire-polished glass Pasteur pipette with normal opening diameter and 1× for 1 min with a fire-polished Pasteur pipette having half the normal opening diameter. The dissociated neurons are subsequently pelleted by centrifugation (80 *g* for 5 min/ml at room temperature) and resuspended in fresh DMEM-HS.

F. Protocol for Nucleofection with First Generation Device

Each nucleofection (i.e., one cuvette) requires between 5×10^5 and 2×10^6 cells (use separate tubes). The aliquoted cells are pelleted by centrifugation (5 min at 80 *g*) and resuspended in 100 μl of prewarmed (room temperature) nucleofection solution.

Table IV
Solutions Required for the Nucleofection of Neurons

Solution	Composition
Trypsin-EDTA solution	50 mg/l trypsin 0.2% (w/v) EDTA (ethylenediaminetetraacetic acid)
	0.1% HEPES [4-(2-hydroxyethyl)-1-piperazineethanesulfonic acid]
	100 U/ml penicillin
	10 μg/ml streptomycin in PBS
PBS (phosphate buffered saline)	137 mM NaCl
	6.5 mM Na_2HPO_4
	2.7 mM KCl
	1.5 mM KH_2PO_4 in ddH_2O; adjust pH to 7.1

The cells are then transferred into Eppendorf tubes containing the plasmid DNA (3 μg in total) in ddH_2O or TE buffer. It is advisable that the cells are resuspended in transfection solution before the plasmid is added to the transfection solution. If the plasmid is mixed with the transfection solution before resuspending the cells, the cells may form aggregates, which can reduce cell viability.

The 100 μl cell/DNA mixture is subsequently transferred into a dedicated nucleofection cuvette (take care to avoid air bubbles as they would interfere with the flow of the electrical currents during nucleofection), which is placed into the nucleofector. A nucleofection program suited to the cell type used is selected and applied to the cells. After the nucleofection, 500 μl DMEM-HS are added into the cuvette to minimize shearing forces when pipetting the cells out of the cuvette. The cells are subsequently seeded onto coverslips or into dishes and incubated under standard conditions with B27-supplemented NMEM (Goetze *et al.*, 2003, 2004). Both coverslips and dishes should be coated to facilitate adhesion of the neurons. Plastic dishes should be coated with poly-L-lysine; for coverslips, it is advisable to add an additional laminin coat on top of the lysine-coat. After coating both plastic dishes and coverslips are incubated with DMEM-HS. Two to three hours before use, the DMEM-HS is replaced by B27-supplemented NMEM. For details on the coating procedure, please refer to Goetze *et al.* (2003).

Recently, a new nucleofection system using 96-well plates has been developed (Zeitelhofer *et al.*, 2007). This system has several advantages over the older nucleofection system. First, it allows high throughput transfections of plasmid DNA (or RNAi duplexes) in a wide range of cell types, including primary neurons. Second, it requires an order of magnitude fewer cells than the conventional nucleofector. This is particularly valuable when working with cell types that are time consuming to isolate, such as hippocampal neurons. This system allows the simultaneous testing of up to 96 plasmids or conditions, thus eliminating variations in culture conditions, which can mar comparisons between experiments performed on separate neuronal preparations. Moreover, the 96-well plate format is suitable for large-scale analyses, such as ELISA screening or target identification and validation with RNAi libraries.

G. Protocol for Nucleofection with the 96-Well Shuttle System

Each nucleofection (i.e., 1 cuvette or well of a 96-well plate) requires between 5×10^4 and 5×10^5 cells. The cells are pelleted by centrifugation (5 min at 80 *g*) and resuspended in 20 μl of prewarmed (room temperature) Amaxa nucleofector solution. The cells are then transferred into Eppendorf tubes containing the plasmid DNA (0.5 μg in total) in ddH_2O or TE buffer and mixed gently. The 20 μl of cell/DNA mixture is subsequently transferred into each well of a dedicated 96-well plate (take care to avoid air bubbles) and the 96-well plate is placed into the nucleofector.

A nucleofection program suited to the cell type and plasmid used is selected and applied to the cells. After the nucleofection 80 μl of prewarmed (37 °C) DMEM medium is immediately added to retrieve the neurons from the well(s). Neurons are subsequently seeded at a density of 4 to 7.5×10^4 cells per well (96-well plate) or at a density of 10^5 per coverslip in a 12-well plate. Four hours after plating, the medium is replaced with fresh medium. For E17 neurons, NMEM should be used as they tend not to grow well in DMEM. P0 neurons, however, should be cultured in DMEM as the survival rate of P0 neurons after nucleofection tends to be lower in NMEM medium. One day after the transfection, arabinose (5 μM final concentration) is added to the DMEM used to culture P0 neurons. This inhibits the proliferation of glial cells, which are more abundant in preparations from postnatal brains.

In our experience, culturing transfected neurons in 96-well plates does not affect their differentiation as judged by their morphology. Nucleofected neurons display axons and extended dendritic trees with boutons that later develop into dendritic spines and establish contacts with other neurons (Zeitelhofer *et al.*, 2007).

H. Transfection of Neurons with Lipid-Based Methods

Lipid-based methods (e.g., LipofectAMINE2000) can be used as an alternative to the DNA/CaP_i and nucleofection methods [see Kaech *et al.* (1996) and Ma *et al.* (2002); reviewed in Craig (1998)]. The protocols employed vary according to the companies providing the reagents. Usually, the transfection solution is incubated at room temperature for 5 min and subsequently mixed with an equal volume of DNA solution (e.g., DMEM containing the appropriate amount of plasmid DNA), incubated for 15 min and mixed again immediately before the addition of the mixture to the culture medium of the cells that are to be transfected. The mixing of solutions is generally performed using a vortex mixer. The transfection solution containing the lipid–DNA complex is added dropwise to the culture medium, and the plate is swirled several times at the end of addition to mix the transfection solution and the medium. A disadvantage of lipid-based methods is that liposomes often alter of the neuronal morphology, especially that of dendrites and can lead to their detachment from the substrate.

III. Visualization of RNP Transport

Analyzing RNA localization by *in situ* hybridization (ISH) can only yield static images of the distribution of RNAs at a given time point. It does not allow observations of how this distribution emerged and how it might change. A given pattern observed for an RNA might, however, be caused by different mechanisms, for example, directed transport, anchoring, and spatially differential stabilization of the RNA as well as changes in the accessibility of the RNA to the ISH probe, which cannot be gleaned from static information. As a consequence, elucidating the mechanisms that lead to a certain distribution of an RNA can only be achieved by live imaging of the dynamic behavior of this RNA.

The labeling of RNAs in living cells is technically more demanding than that of proteins. Several methods have been developed to follow RNA transport in living nerve cells, including the microinjection of RNA that has been transcribed and fluorescently labeled *in vitro* (Schratt *et al.*, 2006; Shan *et al.*, 2003) and the use of molecular beacons to label endogenous RNAs (Bratu, 2006; Bratu *et al.*, 2003). The analysis of RNA transport using a microinjection approach, however, requires an injection setup. Also, care has to be taken to minimize damage to the cells during the injection procedure. Moreover, RNAs whose localization depends on nuclear processing, such as pre-mRNA splicing, have to be injected into the nucleus to ensure correct assembly of all *trans*-acting factors on the RNA.

In contrast to the injection of *in vitro* transcribed RNAs to detect nucleic acids in living cells, the use of molecular beacons allows the detection of endogenous RNAs. Molecular beacons are oligonucleotides, which are complementary to a sequence in the RNA that is to be imaged. They carry both a fluorophore and a quencher. In molecular beacons that are not bound to a complementary sequence the fluorophore and the quencher are in close contact thus precluding fluorescence of the unbound probe (Bratu, 2006; Bratu *et al.*, 2003). When the molecular beacons hybridize to their target sequence, the quencher is separated from the fluorophore and a fluorescence signal is emitted. Nuclease-resistant molecular beacons are designed to efficiently hybridize to accessible regions within RNAs and can be detected via fluorescence microscopy (Bratu, 2006). Microinjected molecular beacons have been used to trace, for instance, the movement of RNAs in *Drosophila* oocytes (Bratu *et al.*, 2003), but no application in neurons has been published to date.

An alternative to these methods is the use of FPs to label RNAs. Theoretically, if a *trans*-acting factor for an RNA has been identified, FP-tagged versions of this factor can be used to track the movement of its cargo RNA inside a cell. In practice, however, this approach is limited by two factors. First, relatively few *trans*-acting factors recognizing complex localization elements in specific RNAs have been identified, particularly in neurons. Second, at least some *trans*-acting factors can bind more than one species of RNA. This promiscuity in their RNA-binding properties makes it impossible to unambiguously assign a signal emanating from an FP-tagged *trans*-acting factor to a specific RNA. These limitations can be overcome by introducing binding sites for a heterologous RNA-binding protein into the RNA that is to be imaged. Cotransfection of constructs encoding this RNA and the corresponding FP-tagged RNA-binding protein, respectively, will result in a specific labeling of the target RNA with the FP. This approach has been realized with the development of the MS2 system.

A. The MS2 System to Visualize RNAs in Living Cells

The MS2-based system provides a means to indirectly label specific RNAs with a fluorescent protein. It thus allows live cell imaging of RNA dynamics and gives the possibility to examine the transport of a specific RNA in real time. In 1998,

Bertrand and colleagues used the system for the first time to visualize the localization of the *ASH1* mRNA in living yeast cells (Bertrand *et al.*, 1998). The system was subsequently utilized in a large variety of model organisms, such as bacteria, plants, and *Drosophila*, as well as cell types, including invertebrate and mammalian neurons (Ashraf *et al.*, 2006; Rook *et al.*, 2000).

For an MS2 system-based visualization of RNA, two constructs have to be generated and coexpressed by the cell. The first encodes a fusion of a fluorescent protein and a mutated MS2 protein as well as a nuclear localization signal (NLS) (Fig. 3). The MS2 protein is a phage RNA-binding protein, which normally forms the capsid of the MS2 bacteriophage. Mutated versions of the MS2 protein are available that do not multimerize but bind MS2 stem loops as dimers. The second

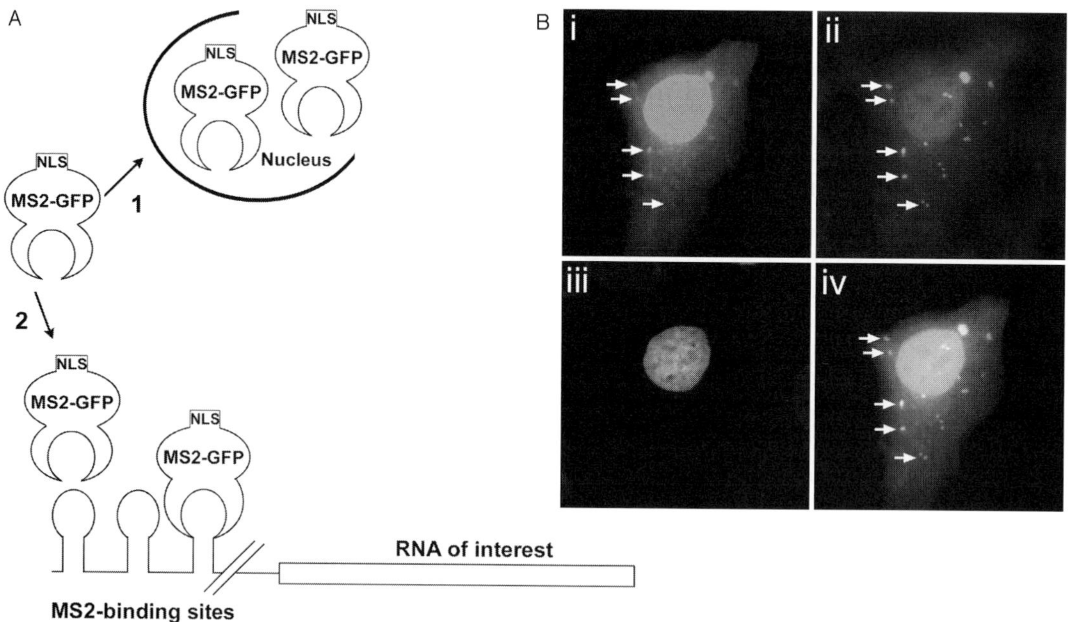

Fig. 3 The MS2 system. (A) Schematic representation of the MS2 system used to label RNAs. The MS2-GFP fusion protein and the RNA to be visualized containing MS2-binding sites are coexpressed in cells. The NLS in the MS2-GFP fusion protein ensures that MS2-GFP not bound to RNA is sequestered by the nucleus (1). When the MS2 protein binds to an MS2 site in the target RNA (2), it follows this RNA to its location(s) inside the cell. The fluorescence of the green fluorescent protein (GFP) is used to locate the RNA molecules. (B) Osteosarcoma cell cotransfected with a plasmid encoding the MS2-GFP fusion protein and a plasmid coding for the Luciferase-ZIP RNA flanked by 6 MS2-binding sites. (i) The strong GFP fluorescence in the nucleus reflects sequestration of unbound MS2-GFP proteins to this compartment via the nuclear localization signal (NLS). Bright GFP-positive foci in the cytoplasm (arrows) indicate the location of the RNA of interest. (ii) Staining with an antibody against the stress granule marker protein TIAR. (iii) Labeling with the DNA stain DAPI to show the location of the nucleus. (iv) Merging of images (i) and (ii) reveals that MS2-GFP-positive cytoplasmic foci colocalize with stress granules (arrows). Images (i) to (iv) kindly provided by Stefan Hüttelmaier, Halle. (See Plate no. 42 in the Color Plate Section.)

construct produces a hybrid RNA containing the RNA to be visualized in addition to several MS2-binding sites. These MS2-binding sites are small RNA hairpins that are specifically bound by the MS2 protein. The MS2-binding motifs can be cloned either upstream or downstream of the region of interest of a localized transcript. However, it should be considered that stem loops in the 5′-UTR are prone to downregulate or prevent translation by hindering the initiation of translation.

For live cell imaging applications, between 6 and 48 MS2-binding sites per RNA construct are used. The number of MS2-binding sites inserted into an RNA construct depends on the intensity of the signal required. Live cell imaging, for example, usually works well with 24 sites. For single molecule applications 24 or 48 MS2-binding sites may be warranted (Fusco *et al.*, 2003, 2004). When these constructs are coexpressed, the XFP-MS2-NLS fusion proteins bind to the RNA hairpins thus labeling the RNA of interest. The larger the number of MS2-binding motifs in the RNA, the higher the number of bound XFP molecules will be and therefore the fluorescent signal per target RNA molecule will increase. It was estimated that on average 32–34 MS2 proteins [each fused with a green fluorescent protein (GFP)] bind per 24 MS2 repeats (Fusco *et al.*, 2003, 2004). Unbound FP-MS2 fusion proteins will be sequestered into the nucleus via the NLS ensuring that cytoplasmic FP signals can be attributed to the presence of the target RNA. This method allows the assessment of parameters like the path and dynamics of mRNA movement as well as anchorage at specific subcellular sites, such as the leading edges of growth cones, dendrites and dendritic spines, and the axon hillock. Also the association of specific RNAs with organelles or macromolecular complexes, such as stress granules, processing bodies (P-bodis) or transport RNPs can be determined.

B. Experimental Procedure

1. Note on the Amplification of Plasmids Containing MS2-Binding Sites

The presence of several MS2-binding sites in the RNA construct increases the risk of recombination events taking place during amplification of the plasmid in bacteria. It is therefore recommended that a specific bacterial strain that minimizes recombination is used, for example, the Stbl strain (Invitrogen).

The transfection procedure of the MS2 system constructs is identical to the methods described for the transfection of FP-tagged transgenes. The 1.5 μg each of the target RNA–MS2-binding site construct and the XFP-MS2 protein construct are mixed and cultures of neurons are transfected as described above. To avoid overexpression and a subsequent overload of the nucleus with unbound MS2 protein, which can be detrimental to the cells, neurons should be imaged between 8 and 9 h after transfection as described below. Alternatively, the RNA–MS2-binding site construct and the XFP-MS2 protein construct can be mixed in a ratio of 5:1 to reduce the amount of XFP-MS2 protein expressed by the cells.

Also, the use of weaker promoters driving the expression of the XFP-MS2 mRNA may be considered to lower the cellular levels of XFP-MS2 protein.

Please note that the overexpression of RNAs, like that of proteins, can result in artifacts, for example, by titrating out proteins and thus limiting their availability to the endogenous transcripts.

C. Imaging of Transfected Cells

The period between transfection and the time point when imaging can be started depends on several factors. First, the time required for a particular protein to be expressed by transfected cells varies. Second, the FP tag needs time to fold into a functional fluorophore. Together, these processes typically take ~3–4 h for a range of FP-tagged proteins in primary hippocampal neurons (Goetze *et al.*, 2004). Moreover, if the protein has to undergo posttranslational modifications or be localized to a specific subcellular compartment, time has to be given for these processes to occur. Finally, the transfection method can influence the time required before the expressed protein can first be detected. For instance, nucleofection tends to result in a more rapid expression of transgenes than the DNA/CaP_i method. The optimal time point for imaging should be chosen to be when the protein is expressed at sufficiently high levels for imaging and has localized to its endogenous locations in the cell. The efficacy of a transfection should also be determined only when the expression of the transgene can clearly be detected in the transfected cells.

If a plasmid is expressed for over 24 h in culture, overexpression artifacts are frequent, especially if strong promoters such as the *Cytomegalovirus* promoter are used to drive expression. These can be direct consequences of the increased amounts of overexpressed molecule or result indirectly from it. Direct consequences can, for instance, be that unphysiologically high amounts of a normally tightly regulated protein unbalance the cellular processes this protein is involved in. Moreover, overexpressed proteins can accumulate and clog cellular processes, such as directed transport and protein degradation. Indirect consequences could be that the increased amounts of protein, RNA, or indeed the expression plasmid itself bind and thus due to their overabundance titrate out and thus limit the availability of essential factors, such as *trans*-acting factors or motor proteins generally involved in transport mediated by the cytoskeleton; general RNA-binding proteins, such as components of the exon junction complex , elongation initiation factors, poly-A-binding protein, and cytoplasmic polyadenylation element-binding protein; or factors of the RNA transcription machinery, respectively.

A frequently observed overexpression artifact of many RNA-binding proteins is their aggregation into large foci. In neurons, these foci tend to be very pronounced at bifurcations in dendrites (Fig. 4C). Often overexpressed proteins also aggregate into stress granules [e.g., Vessey *et al.* (2006) and references therein]. This can affect the normal pattern of these proteins in a cell. It is therefore important to optimize the transfection conditions until the pattern of the transgene reflects that of the endogenous protein (Fig. 4A,B).

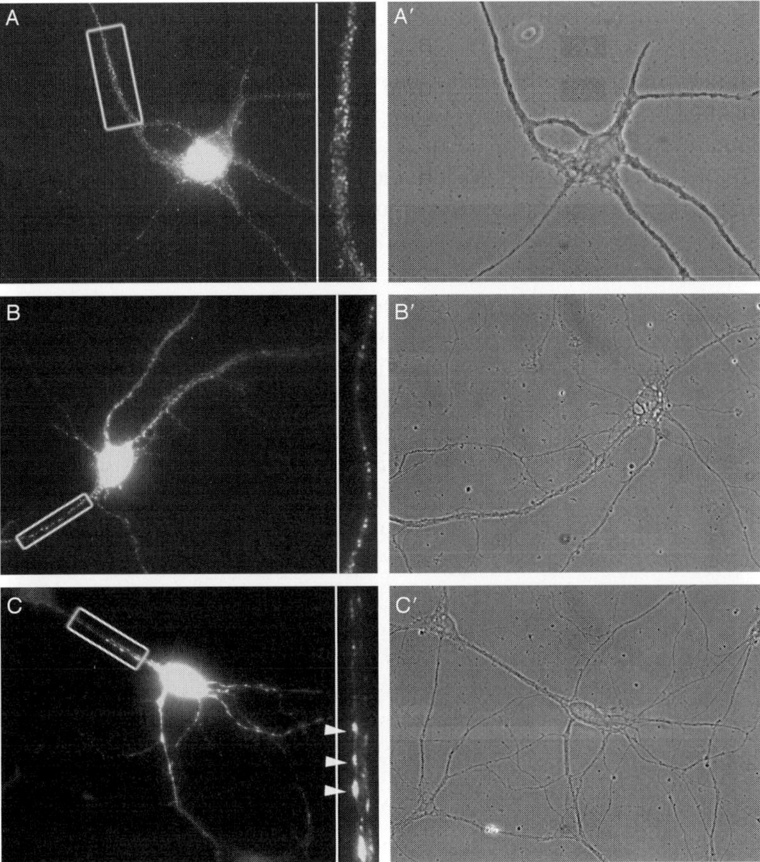

Fig. 4 Hippocampal neurons transfected with the Ca^{2+}-phosphate method. (A) Staining of cultured rat hippocampal neurons with antibodies against the conserved transport ribonucleoprotein particle (RNP) protein Staufen2 (Stau2). (B, C) Cultured rat hippocampal neurons transfected with a plasmid encoding Stau2-EGFP at 13 DIV were imaged 12 (B) and 18 (C) hours after the transfection, respectively. Note that while 12 h of overexpression of the plasmid (B) results in a pattern that closely resembles the size, shape and localization of particles containing the endogenous protein (A), a prolonged overexpression (C) leads to the aggregation of the tagged protein in large foci (arrowheads) that are generally not observed in cells stained with antibodies. A′–C′: phase contrast images corresponding to A–C, respectively. (See Plate no. 43 in the Color Plate Section.)

Such artifacts can be avoided by determining the time window in which the overexpressed protein localizes like its endogenous counterpart, that is, it can already be detected in its endogenous locations but is still expressed at near physiological levels and thus does not disturb the normal functions inside the cells. To increase this window, the amount of plasmid transfected per cell can be lowered. This can be achieved either by decreasing the time the cells are exposed to the DNA/CaP_i coprecipitate or by lowering the amount of expression vector in the precipitate. The latter can be achieved, for example, by diluting the actual expression vector with an empty vector of the same backbone. In our experience, the latter proves to be more effective in that it ensures high transfection efficiencies while reducing the amount of coding vector introduced into the cells.

Moreover, choosing a weaker promoter, for example, one normally driving the expression of genes in the cell type to be transfected, can result in more physiological levels of expression (Kugler *et al.*, 2003). It should, however, be noted that also endogenous promoters when cloned into an expression plasmid lack their normal environment, including enhancer elements that modulate their activity, and thus might not behave like the endogenous promoter. Moreover, a transfection generally introduces more than one or two expression plasmids into a cell. The higher copy number of the gene in transfected cells will likely also lead to increased levels of protein. Finally, when working with neuronal cell lines instead of primary neurons, creating stable lines expressing the desired transgene can avoid overexpression artifacts since clones that express excessive amounts of the transgene tend to die. This negative selection of overexpressing cells results in populations of cells displaying near physiological levels of expression.

D. Protocol for Live Imaging of Transfected Neurons

Neurons are cultured on poly-L-lysine-coated coverslips. For imaging, they are transferred into a live cell imaging chamber mounted onto the table of an inverted fluorescence microscope. Several types of imaging chambers are commercially available. It is generally advisable to choose large chambers, for example, encasing the entire microscope. The larger volume of air in these chambers greatly reduces disturbances of the culture medium due to air flow. Ideally, the chamber is equipped to maintain a controlled temperature and CO_2 level (CO_2-dependent media only). If no such chamber is available, also simpler versions in which the neurons are exposed to ambient temperatures and air can be used. In this case, care should be taken, however, when measuring reactions that are temperature dependent. If the neurons are to be exposed to different environments during an experiment, perfusion chambers that allow an exchange of the medium have to be used. If only a certain region of a coverslip is to be exposed to a specific substance or medium, local perfusion pipettes can be placed into the chamber. This is, for example, useful when exposing part of a neuron's dendritic tree to a chemical while the remaining (unexposed) dendrites serve as an internal control (Veselovsky *et al.*, 1996). Before starting the actual experiment, it is advisable to perform several trial runs with perfusion solutions containing a water-soluble dye as a marker (e.g., hydrazine-conjugated fluorescent dyes) to determine the optimum flow rate and to avoid undesired diffusion of the perfusion solution. Such dyes should also be included in the perfusion solution during experiments to be able to monitor the flow rate and assess the diffusion of the perfusion solution.

For imaging, it is generally advisable to replace the normal neuronal culture medium (Table III) with a medium lacking pH indicators as these substances display autofluorescence. This autofluorescence can overlay the actual fluorescence signal and reduce the signal-to-noise ratio, especially when nonconfocal microscope setups are used for image acquisition. We recommend a modified HBSS (see Table III). This solution can be sterile filtered through a 0.22 μm pore size filter and stored at 4 °C for up to 1 year. Neurons can be kept in prewarmed HBSS buffer for several hours without any obvious adverse effects. When the imaging is performed over longer periods (i.e., above 1 h at room temperature), however, care should be taken that evaporation of water from the HBSS does not result in hyperosmolarity that may affect the physiology of the neurons. In these cases, it is recommended either to use a closed humidified chamber or to regularly exchange the imaging buffer, for example, by global perfusion. Because of its lack of essential nutrients, HBBS is, however, not suitable for longer periods (>4 h). In these cases, a slow perfusion with transfection medium or culture medium (possibly lacking phenol red when not using a confocal microscope) is recommended. As the culture medium is CO_2-dependent, care has to be taken to maintain physiological pH values during the experiment, for example, by using preequilibrated medium and encasing the cells in an imaging chamber that allows an adjustment of atmospheric CO_2 levels.

As with all fluorescence microscopes, a highly sensitive detection setup is recommended to optimize the signal-to-noise ratio and to reduce the phototoxicity. Especially, appropriate filter sets are important to ensure maximum excitation and detection of emission from the FPs. This becomes particularly important when changes in FP fluorescence intensities are to be imaged, for example, during fluorescence resonance energy transfer (FRET) and fluorescence recovery after photobleaching (FRAP) applications. Moreover, the use of sensitive cameras reduces the exposure time needed to excite the FPs. This is important to minimize photobleaching of the FPs, but even more so in experiments involving living cells, which can be harmed by prolonged exposure to high intensities of light. Phototoxic damage is a particular concern in time-lapse experiments that require repeated exposure over longer periods.

Living cells, including neurons, generally show substantial levels of autofluorescence, especially of the perinuclear ER and Golgi. This autofluorescence can be discriminated from genuine fluorescence emanating from a fluorophore by the fact that it shows a broader emission spectrum, that is, it can be detected in different channels. Importantly, dead neurons show strong, sometimes punctuate autofluorescence that can easily be confused with genuine fluorescence patterns of living transfected cells by inexperienced experimenters.

When trying to assess the pattern of overexpressed proteins in cells, care also has to be taken during the processing of images following their acquisition. Excessive adjustment of, for example, the brightness, contrast or the gamma value (scaling) can easily result in artifacts that can be mistaken for genuine patterns. For instance, when neurites of varying thickness are imaged, a homogeneously distributed protein (e.g., GFP) may appear to form discrete foci after scaling. This effect may be, however, due to the different amounts of cytoplasm (and hence fluorescent protein) in different parts of the neurite, as removing the lower intensity pixels from the image can result in continuous regions of varying signal intensity (in this case due to varying amounts of cytoplasm) can break up into discrete signals. As a consequence, any final processed image should be compared with the actual appearance of a cell's fluorescence pattern in the microscope.

IV. Visualization of RNP Assembly and Composition

A. Imaging the Colocalization of *trans*-Acting Factors

The classical way to assess whether two proteins colocalize in a cell is to use specific antibodies directed against these proteins. While being a very powerful approach in that *endogenous* proteins in their *native subcellular locations* are detected, this method has several drawbacks. For example, antibodies against a specific protein are not always available and are time consuming and costly to generate. Also, it is difficult to costain with antibodies raised in the same species. Most importantly, antibodies can generally be applied only to fixed cells and hence preclude live cell imaging approaches. This is particularly important when the process of RNA localization is being investigated as a colocalization of two proteins does not necessarily mean that these proteins are also present in the same RNP and cotransported. Instead a colocalization might reflect the presence of different RNPs anchored at common sites. This question is best addressed by visualizing moving RNPs in cells transfected with fluorescently tagged versions of the *trans*-acting factors under investigation.

Generating expression constructs in which the gene of interest is tagged with an FP also overcomes many of the other shortcomings of antibody staining. Expression constructs are comparatively fast and inexpensive to generate. The availability of a wide range of FPs with different spectral properties (Giepmans *et al.*, 2006; Shaner *et al.*, 2004), that is, excitation and emission spectra, allows for more color combinations and the parallel visualization of larger numbers of proteins than is generally possible with antibody staining in which a wide range of secondary antibodies linked to various fluorophores and directed against (primary) antibodies from a large number of species would be required. This is, however, only an advantage if excitation/emission filter combinations allowing for the spectral separation of the different FPs or software permitting linear unmixing are available. Importantly, fluorescently tagged proteins can be imaged in living cells allowing the visualization of dynamic processes, such as following the assembly or disassembly of protein complexes, such as transport RNPs, under certain conditions or the velocities of protein movements.

The expression of fluorescently tagged proteins has, however, also several drawbacks. When interpreting results obtained with transfected proteins, one has to keep in mind that *exogenous* proteins are visualized that might behave differently than their endogenous counterparts. For one, the attachment of the fluorescent protein to the protein to be studied may alter or even obliterate the function(s) of the latter. The fluorescent protein may, for instance, sterically hinder the normal interactions of the tagged protein with other molecules. It may also interfere with posttranslational modifications that modulate the function of the endogenous protein. These problems can generally be overcome by generating different fusion proteins in which the FP is attached to the target protein at different positions, for example, N- and C-terminal. In addition to these effects, the overexpression of the exogenous protein could lead to a number of phenotypes due to the overabundance of the protein (see above).

Generally, analyses aimed at determining colocalization of proteins require confocal image acquisition. In the case of mature neurons, the axon and distal dendrites are thin enough to be imaged without a confocal microscope. However, even when imaging with a confocal microscope, care has to be taken when trying to determine if two coincident fluorescent signals arise from a single complex or two complexes that are in close proximity. Moreover, the technical properties of the imaging setup and postacquisition image processing have to be considered when interpreting putative colocalization events. When imaging faint signals with a confocal microscope, for instance, opening up the pinhole through which emitted light has to pass before reaching the detector increases the signal. However, it also increases the thickness of the optical section up to a point where confocality is lost and the image is no different from one acquired with a conventional microscope. This in turn means that two signals, which appear in the same image plane (*X–Y*-plane), may in fact be located at different positions along the *Z*-axis.

Care should also be taken during image acquisition that signals never exceed the maximum range of the detection system as overexposed signals generally appear larger than they really are. Two such signals when adjacent may subsequently be interpreted as coincident or overlapping when the adequately exposed signals would appear discrete. Similar phenomena can occur during the adjustment of brightness and contrast during the postacquisition image processing. Further important factors when trying to assess the colocalization of two proteins in a common complex are, for example, also the relative size and shape of the two signals. If two signals overlap but have a significantly different shape, it is unlikely that the two proteins are part of a single complex. Importantly, only particles with sizes and shapes also seen in antibody stainings should be considered for analysis. Overexpressed proteins tend to nonspecifically cluster into unphysiologically large aggregates. In neurons, such aggregates are often located at bifurcations. A colocalization at these "known aggregation hot spots" should generally be viewed with caution if not ignored entirely.

As a consequence of these constraints of colocalization experiments involving FP-tagged overexpressed proteins, any putative colocalization of proteins in a common complex has to be verified by alternative methods. Such methods can be FRET or bimolecular fluorescence complementation (BiFC; see below) that show direct interactions between proteins. These methods fail, however, when the proteins are in the same complex but separated by other proteins. In this case, biochemical methods (e.g., coimmunoprecipitation, pull-down experiments) have to be applied to confirm the presence of two proteins in the same complex.

The different refraction properties of different wavelengths of light can lead to the phenomenon that colocalizing signals detected are slightly offset when images of two (or more) fluorophores are superimposed. This pixel shift can easily be recognized by the fact that the colors are always shifted in exactly the same direction and to the same extent. Such pixel shifts can generally be corrected in image analysis programs subsequent to the acquisition of the data. It should also be noted that a colocalization of proteins by no means has to be 100% even if two proteins are known to be present in the same complex. One protein may be part of different complexes. Furthermore, many protein complexes are dynamic and can change composition, for example, as a consequence of different physiological states of a cell. When making statements about the colocalization of proteins, it is therefore essential to always include the degree of colocalization observed.

B. Visualization of Direct Interactions Between *trans*-Acting Factors

There are several methods to visualize direct interactions between two proteins. FRET is a well-established method and has been used widely to demonstrate close physical contact between two proteins (see Chapter 16 by Shimozono and Miyawaki and Chapter 18 by Piston and Rizzo, this volume). This technique is, however, technically not trivial. Recently, a powerful alternative method has been developed (Hu *et al.*, 2002; Hu and Kerppola, 2003): BiFC (see also chapter 19 by Kerppola, this volume). In this method, an FP is split into two halves, each of which is subsequently fused to one of a pair of proteins whose interaction is to be probed. On their own, these two halves show no fluorescence. A physical interaction of the two proteins, however, brings the two halves of the FP into close proximity and thus allows the FP to reconstitute and hence fluoresce. The detection of fluorescence is thus a sign for a direct interaction of the two proteins. This method can be used to visualize both interactions between *trans*-acting factors within an RNP and interactions between a *trans*-acting factor and its cargo RNA. In the latter case, one FP half is not attached directly to the RNA, but instead to a protein that in turn specifically binds the RNA. Since this approach requires three molecules to interact, it has been termed trimolecular fluorescence complementation (see below).

V. Visualization of Interactions Between RNAs and *trans*-Acting Factors

When studying the localization of RNAs, one of the key questions is which proteins are associated with a specific RNA at different stages of the localization process. In large cells, such as oocytes, fluorescently labeled RNAs can be injected and their association with endogenous or fluorescently tagged overexpressed proteins tested. Unfortunately, microinjection of small cells like primary neurons is not straightforward and while it has been achieved (Schratt *et al.*, 2006), does not lend itself to routine experiments. As a consequence, two alternative avenues have been taken to covisualize RNAs and associated proteins in neurons. The first combines the overexpression of one (or more) fluorescently tagged proteins and a fluorescent *in situ* hybridization (FISH; Munchow *et al.*, 1999). The second relies entirely on the expression of transgenes to label both the RNA and the proteins to be visualized. While the first approach can only be carried out on fixed cells, the latter is amenable to live cell imaging and hence allows the visualization of dynamic processes in living cells.

A. Colocalization of Fluorescent Proteins with RNAs via ISH Staining

In order to covisualize RNAs and proteins in neurons with this method, neurons are first transfected with the plasmids encoding a tagged version of the respective protein (see above). Cells at this stage are processed further for fluorescent *in situ* detection of endogenous RNAs as described elsewhere (Munchow *et al.*, 1999). Importantly, FPs tend not to retain their fluorescence during the ISH procedure. As a consequence, FPs have to be detected with specific antibodies following the ISH. The most important drawback of this approach is that it only allows the visualization of RNA in fixed cells. Moreover, the ISH procedure generally results in a deterioration of the cellular morphology. From a practical point of view, the procedure is not easy to establish as RNase-free working materials (work space, tools, solutions, etc.) and procedures have to be ensured to avoid degradation of the RNA to be detected. Finally, the procedure required is prone to both false positive and false negative results. Each experiment involving a new target RNA therefore has to be carefully controlled.

B. Colocalization of Fluorescent Proteins with RNAs Using the MS2 System

The MS2 system (see above) has the major advantage of allowing the visualization of RNAs in living cells. It thus allows the imaging of dynamic processes, such as the movement of RNAs. Moreover, no RNase-free working conditions have to be established. A disadvantage of this method over a FISH-based detection of RNA, however, is that—instead of the endogenous transcript—only overexpressed RNA can be visualized. Compared to microinjected, directly labeled RNAs, the method also has the advantage that the RNA is transcribed by the cells themselves and will thus be processed in a similar fashion. In order to ensure that all proteins normally associated with a eukaryotic mRNA are assembled on an RNA transcribed from an exogenously supplied template, it might be advisable to include at least one intron that is spliced out during the RNA processing in the nucleus.

Labeling of an RNA with the MS2 system requires the expression of two plasmids: one encoding the XFP-tagged MS2 protein and other encoding the target RNA flanked by MS2-binding sites (see above). If an RNA and a *trans*-acting factor are to be covisualized, this would necessitate a transfection with a third plasmid encoding the tagged *trans*-acting factor. The efficiency of triple transfections, that is, the number of cells being transfected with all three plasmids, is typically very low and difficult to control for (see also above). As a consequence, it is recommended that two of the transcripts are expressed from a common plasmid, for example, via an internal ribosomal entry site. This reduces the number of plasmids to be transfected to two.

If in addition to a mere colocalization, also direct interactions between an RNA and a *trans*-acting factor are to be assessed, a modified version of the BiFC protocol can be employed [trimolecular fluorescence complementation; see Rackham and Brown (2004)]. In this protocol, the RNA is labeled via the MS2 system (see above), but instead of the MS2 protein being attached to a functional XFP, it carries a nonfluorescent half of an FP. The *trans*-acting factor under investigation carries the complementary second half of the FP. If the MS2-binding sites are located adjacent to a binding site for this *trans*-acting factor, the two halves of the FP can interact and reconstitute the fluorescent protein. This method allows the identification of binding sites for *trans*-acting factors in a target RNA.

VI. Visualization of Local mRNA Translation

A. Protein Synthesis in Dendrites

The fact that certain mRNAs are localized to specific subcellular regions does not entail that all of the corresponding protein at these sites is synthesized from localized transcripts. Instead, at least in some cases, some of the local protein may have been produced in the perinuclear cytoplasm and subsequently have diffused or been transported to these sites. This mixing of proteins synthesized in the soma and those produced locally makes it difficult to discriminate between general and local protein synthesis.

Biochemical preparations such as synaptosomes (isolated presynaptic nerve endings) and synaptoneurosomes (pre- and postsynaptic nerve endings that are still associated) have been used to monitor the synthesis of specific proteins at the synapse. These approaches, however, have several important drawbacks. First, synaptosome and synaptoneurosome preparations are not of sufficient purity, since they are often contaminated by cell types other than neurons and they are very heterogeneous in their composition. Therefore, results obtained from these preparations are difficult to interpret. Second, the postsynaptic nerve endings need to rapidly reseal after breaking off the dendritic shaft during the isolation procedure in order to ensure the presence of all components of the protein synthesis machinery. Third, this biochemical approach precludes *in vivo* analyses of protein synthesis.

An alternative approach to demonstrate protein synthesis in dendrites is to sever the dendrites from the cell body and assess their ability to, for example, translate localized reporter constructs (Aakalu *et al.*, 2001; Crino *et al.*, 1998). This approach excludes the possibility of proteins produced in the cell body from moving into dendrites, but precludes studies aimed at analyzing local protein synthesis in the context of functional neuronal networks.

Fluorescent protein-based reporter assays can overcome these limitations. Flanking the coding sequence for an FP with *cis*-acting elements that mediate the localization of an RNA ensures that the reporter transcript is localized like the endogenous RNA. Several strategies can be employed to ensure that only protein synthesized locally will be present at distal sites. For one, the FP can contain an NLS, which directs all protein produced in the perinuclear cytoplasm to the nucleus. Alternatively, experiments detecting FRAP can yield information on the sites and dynamics of protein synthesis. Finally, sequences that anchor the FP locally can be introduced. Such anchors can, for example, be myristylation sites that tether the FP to the plasma membrane (Aakalu *et al.*, 2001) and PDZ or talin motifs that anchor the FP to the postsynaptic density and actin cytoskeleton (Kost *et al.*, 1998), respectively. This restriction of newly produced protein allows to better discriminate, which proteins are synthesized locally.

Biarsenical dyes have been developed, which do not show fluorescence until they bind to a tetracysteine motif, at which point they become strongly green (FlAsH-EDT2) or red (ReAsH-EDT2) fluorescent (Zhang *et al.*, 2002). Using these dyes, it was demonstrated that AMPA receptor subunits are synthesized in dendrites and that inhibition of synaptic activity enhances dendritic translation of GluR1 subunits (Ju *et al.*, 2004). In this study, neurons were transfected with constructs expressing tetracysteine-tagged GluR1. Pulse-chase experiments were then performed, applying first ReAsH-EDT2 and then FlAsH-EDT2 to cells. In this way ReAsH-EDT2 labeled all preexisting Cys4-tagged proteins, while the FlAsH-EDT2 labeled only those synthesized during the chase period. The combination of a small genetically encoded peptide tag with a small molecule detection reagent makes this particularly suitable for the investigation of biochemical changes in living cells that are difficult to approach with FPs as molecular tags. A detailed protocol of this technique to image protein dynamics in living cells has been published by Machleidt and colleagues (Machleidt *et al.*, 2006).

FRAP measures the recovery of fluorescence after photobleaching of FP in a defined region over time. FRAP can be used to measure the dynamics of molecular movements during, for example, diffusion, transport, or any other kind of movement of fluorescence-labeled molecules in living cells. Analyzing the spatiotemporal dynamics of fluorescence recovery in the bleached region allows the discrimination between the movement of unbleached FPs from the surrounding area into the bleached region and the synthesis of FPs in the bleached region itself. Since translation of reporter mRNAs can still occur after photobleaching and therefore be visualized in a cell, this technique has been utilized to study the translation of localized reporter mRNAs in growth cones and dendrites of neurons (Aakalu *et al.*, 2001; Gong *et al.*, 2006) and even protein degradation in dendritic spines (Bingol and Schuman, 2006).

In order to limit the detection of fluorescent signal in the bleached area to locally synthesized FP, reporter constructs encoding FPs with tethering sequences can be used. For example, a myristylated version of GFP has recently been utilized to dissect the molecular mechanism controlling mRNA translation in dendrites (Aakalu *et al.*, 2001). For this approach, the coding sequence for a destabilized (1 h turnover time) and myristylated GFP was flanked by the 5′- and 3′-UTRs of dendritically localized mRNAs (e.g., CaMKII alpha). This construct was subsequently used to visualize translation in dendrites of living hippocampal neurons. To limit GFP diffusion from the cell body into the dendrites, a myristylation consensus sequence (MGTVLSLSPS) was introduced at the N-terminus of the GFP molecule. This results in the covalent attachment of a myristyl group to the GFP protein, which anchors the protein to the dendritic plasma membrane immediately subsequent to local synthesis. This system was used to show that after photobleaching of the dendrites, the dynamics of fluorescence recovery in the dendrites was indicative of dendritic protein synthesis (Aakalu *et al.*, 2001; Gong *et al.*, 2006).

B. The IRE-Based System to Detect Local Protein Synthesis

In order to detect local protein synthesis in living neurons, a GFP-based reporter system that allows for the uncoupling of dendritic RNA transport from its subsequent translation has been designed (Goetze *et al.*, 2003; Macchi *et al.*, 2003). The iron responsive element (IRE) discovered by Hentze and colleagues (Hentze *et al.*, 1987) was adapted in such a way that translation of a GFP-based reporter can be induced in a specific subcellular compartment upon the local addition of iron to the extracellular medium. The ferritin promoter containing the IRE drives the transcription of the following two constructs: the first construct codes for a GFP containing an NLS; the second construct encodes GFP-NLS plus the full-length 3′-UTR of the *CaMKII alpha* mRNA. This sequence contains a targeting element that localizes the resulting reporter mRNA into dendrites. Cells that are transfected with the IRE-containing construct 2 express GFP in the presence of iron, whereas no translation of the reporter is detected in the absence of iron.

Taking advantage of this new reporter system, it was shown that translation of GFP in dendrites of hippocampal neuron depends—in addition to the presence of iron—on synaptic activity (Macchi *et al.*, 2003). The presence of the dendritic targeting element within the *CaMKII alpha* 3′-UTR increased the translation of the reporter construct on chemical stimulation. If this element is missing in the reporter construct, translation is only dependent on iron in the medium, but not affected by synaptic activity. These results provided further evidence for a model postulating that the activity status of a synapse regulates protein synthesis. In future experiments, this GFP-based fluorescent system might allow the detection of local protein synthesis at individual synapses that have been locally perfused with chemicals affecting synaptic activity. This could represent an important assay to further dissect the regulation of protein synthesis at individual synapses.

C. Protocol for the Detection of Local Protein Synthesis via the IRE System

Hippocampal neurons are transiently transfected as described above and incubated in either iron-containing medium or mock treated. The concentration of iron in the medium can be reduced by adding an iron chelator (desferroxyamine mesylate). GFP expression then is almost exclusively detected in the presence of an iron source [ferric ammonium citrate or holotransferrin; for details on the procedure see Goetze *et al.* (2003)]. As an additional control, a mutated IRE should be used, for example, one harboring a deletion of an essential nucleotide (ΔC165) in its stem-loop structure (Goossen *et al.*, 1990; Rouault *et al.*, 1990). Neurons transfected with this construct express GFP independently of the level of iron. Moreover, experiments performed in the presence of the transcriptional inhibitor actinomycin D (applied shortly before the iron pulse) should be included to demonstrate that the rise in GFP levels is due to the translational activation of preexisting reporter mRNA instead of a rise in transcription.

To target the reporter mRNA into a specific compartment of the cell, for example, the dendrites of mature hippocampal neurons, and to detect local protein synthesis at individual synapses, an appropriate localization element should be included in the transcript. FISH or an MS2-system-based assay should be employed on neurons cultured in the absence of iron to demonstrate that the localization element is sufficient and necessary for dendritic localization of the reporter mRNA.

The induction of local protein synthesis in a restricted compartment of living neurons, in this case dendrites, is achieved via local perfusion of part of the dendritic tree with ion and a stimulator of synaptic activity. Since previous work indicated that local protein synthesis is dependent on the synaptic activity (Aakalu *et al.*, 2001; Casadio *et al.*, 1999; Feig and Lipton, 1993; Kang and Schuman, 1996; Martin *et al.*, 1997; Steward and Schuman, 2001), various chemical stimulation protocols can be employed to investigate whether any of these treatments yield a reproducible stimulation of GFP expression in transfected hippocampal neurons.

VII. Outlook

The field of light microscopy, especially the area of *in vivo* imaging, is currently undergoing rapid change. Novel tools and techniques are allowing progressively less invasive imaging of a range of biological processes at ever greater resolution. While in the past, the study of transport RNPs was often confined to individual molecules, the wide range of FPs now allows the simultaneous visualization of several *trans*-acting factors and RNA species in parallel. This will allow the elucidation of complex and dynamic processes, such as RNP assembly, transport, anchoring, and translational control during transport and at sites of local protein synthesis. Increasingly, these questions are also being addressed not only in neurons grown in culture but also directly in the intact brains of living animals (Svoboda and Yasuda, 2006).

Acknowledgments

The authors are grateful to Stefan Hüttelmaier, Yaron Shav-Tal, and Sabine Thomas for fruitful discussions pertaining to this topic and for critically reading the manuscript. The financial support of the Schram-Stiftung, the HFSP, and the Medical University of Vienna is gratefully acknowledged.

References

Aakalu, G., Smith, W. B., Nguyen, N., Jiang, C., and Schuman, E. M. (2001). Dynamic visualization of local protein synthesis in hippocampal neurons. *Neuron* **30,** 489–502.

Antar, L. N., and Bassell, G. J. (2003). Sunrise at the synapse: The FMRP mRNP shaping the synaptic interface. *Neuron* **37,** 555–558.

Ashraf, S. I., McLoon, A. L., Sclarsic, S. M., and Kunes, S. (2006). Synaptic protein synthesis associated with memory is regulated by the RISC pathway in *Drosophila. Cell* **124,** 191–205.

Bagni, C., and Greenough, W. T. (2005). From mRNP trafficking to spine dysmorphogenesis: The roots of fragile X syndrome. *Nat. Rev. Neurosci.* **6,** 376–387.

Bertrand, E., Chartrand, P., Schaefer, M., Shenoy, S. M., Singer, R. H., and Long, R. M. (1998). Localization of ASH1 mRNA particles in living yeast. *Mol. Cell* **2,** 437–445.

Betschinger, J., and Knoblich, J. A. (2004). Dare to be different: Asymmetric cell division in *Drosophila, C. elegans* and vertebrates. *Curr. Biol.* **14,** R674–R685.

Bingol, B., and Schuman, E. M. (2006). Activity-dependent dynamics and sequestration of proteasomes in dendritic spines. *Nature* **441,** 1144–1148.

Bratu, D. P. (2006). Molecular beacons: Fluorescent probes for detection of endogenous mRNAs in living cells. *Methods Mol. Biol.* **319,** 1–14.

Bratu, D. P., Cha, B. J., Mhlanga, M. M., Kramer, F. R., and Tyagi, S. (2003). Visualizing the distribution and transport of mRNAs in living cells. *Proc. Natl. Acad. Sci. USA* **100,** 13308–13313.

Carson, J. H., Cui, H., Krueger, W., Schlepchenko, B., Brumwell, C., and Barbarese, E. (2001). RNA trafficking in oligodendrocytes. *Results Probl. Cell Differ.* **34,** 69–81.

Casadio, A., Martin, K. C., Giustetto, M., Zhu, H., Chen, M., Bartsch, D., Bailey, C. H., and Kandel, E. R. (1999). A transient, neuron-wide form of CREB-mediated long-term facilitation can be stabilized at specific synapses by local protein synthesis. *Cell* **99,** 221–237.

Craig, A. M. (1998). Transfecting cultured neurons. *In* "Culturing Nerve Cells" (G. Banker and K. Goslin, eds.), pp. 79–111. MIT Press, Cambridge, MA.

Crino, P., Khodakhah, K., Becker, K., Ginsberg, S., Hemby, S., and Eberwine, J. (1998). Presence and phosphorylation of transcription factors in developing dendrites. *Proc. Natl. Acad. Sci. USA* **95,** 2313–2318.

Dahm, R., and Kiebler, M. (2005). Cell biology: Silenced RNA on the move. *Nature* **438,** 432–435.

Dahm, R., and Kiebler, M. A. (2007). RNA localisation: New roles for an evolutionarily ancient mechanism. *Semin. Cell Dev. Biol.* **18**(2), 161–162.

Dahm, R., Kiebler, M., and Macchi, P. (2007). RNA localisation in the nervous system. *Semin. Cell Dev. Biol.* **18**(2), 216–223.

Feig, S., and Lipton, P. (1993). Pairing the cholinergic agonist carbachol with patterned Schaffer collateral stimulation initiates protein synthesis in hippocampal CA1 pyramidal cell dendrites via a muscarinic, NMDA-dependent mechanism. *J. Neurosci.* **13,** 1010–1021.

Fusco, D., Accornero, N., Lavoie, B., Shenoy, S. M., Blanchard, J. M., Singer, R. H., and Bertrand, E. (2003). Single mRNA molecules demonstrate probabilistic movement in living mammalian cells. *Curr. Biol.* **13,** 161–167.

Fusco, D., Bertrand, E., and Singer, R. H. (2004). Imaging of single mRNAs in the cytoplasm of living cells. *Prog. Mol. Subcell. Biol.* **35,** 135–150.

Gartner, A., Collin, L., and Lalli, G. (2006). Nucleofection of primary neurons. *Methods Enzymol.* **406,** 374–388.

Giepmans, B. N., Adams, S. R., Ellisman, M. H., and Tsien, R. Y. (2006). The fluorescent toolbox for assessing protein location and function. *Science* **312,** 217–224.

Goetze, B., Grunewald, B., Baldassa, S., and Kiebler, M. (2004). Chemically controlled formation of a DNA/calcium phosphate coprecipitate: Application for transfection of mature hippocampal neurons. *J. Neurobiol.* **60,** 517–525.

Goetze, B., Grunewald, B., Kiebler, M. A., and Macchi, P. (2003). Coupling the iron-responsive element to GFP—an inducible system to study translation in a single living cell. *Sci. STKE* **2003,** PL12.

Goetze, B., and Kiebler, M. (2006). Transfection of hippocampal neurons using DNA/calcium phosphate co-precipitation. *In* "DNA Delivery/Gene Transfer Manual." Cold Spring Harbor Laboratory Press, Cold Spring Harbor.

Goetze, B., Tuebing, F., Xie, Y., Dorostkar, M. M., Thomas, S., Pehl, U., Boehm, S., Macchi, P., and Kiebler, M. A. (2006). The brain-specific double-stranded RNA-binding protein Staufen2 is required for dendritic spine morphogenesis. *J. Cell Biol.* **172,** 221–231.

Gong, R., Park, C. S., Abbassi, N. R., and Tang, S. J. (2006). Roles of glutamate receptors and the mammalian target of rapamycin (mTOR) signaling pathway in activity-dependent dendritic protein synthesis in hippocampal neurons. *J. Biol. Chem.* **281,** 18802–18815.

Goossen, B., Caughman, S. W., Harford, J. B., Klausner, R. D., and Hentze, M. W. (1990). Translational repression by a complex between the iron-responsive element of ferritin mRNA and its specific cytoplasmic binding protein is position-dependent *in vivo*. *EMBO J.* **9,** 4127–4133.

Goslin, K., Asmussen, H., and Banker, G. (1998). Rat hippocampal neurons in low-density culture. *In* "*Culturing Nerve Cells*" (G. Banker and K. Goslin, eds.), pp. 339–370. MIT Press, Cambridge, MA.

Govindarajan, A., Kelleher, R. J., and Tonegawa, S. (2006). A clustered plasticity model of long-term memory engrams. *Nat. Rev. Neurosci.* **7,** 575–583.

Graham, F. L., and van der Eb, A. J. (1973). A new technique for the assay of infectivity of human adenovirus 5 DNA. *Virology* **52,** 456–467.

Hentze, M. W., Caughman, S. W., Rouault, T. A., Barriocanal, J. G., Dancis, A., Harford, J. B., and Klausner, R. D. (1987). Identification of the iron-responsive element for the translational regulation of human ferritin mRNA. *Science* **238,** 1570–1573.

Hirokawa, N. (2006). mRNA transport in dendrites: RNA granules, motors, and tracks. *J. Neurosci.* **26,** 7139–7142.

Hirokawa, N., and Takemura, R. (2005). Molecular motors and mechanisms of directional transport in neurons. *Nat. Rev. Neurosci.* **6,** 201–214.

Hu, C. D., Chinenov, Y., and Kerppola, T. K. (2002). Visualization of interactions among bZIP and Rel family proteins in living cells using bimolecular fluorescence complementation. *Mol. Cell* **9,** 789–798.

Hu, C. D., and Kerppola, T. K. (2003). Simultaneous visualization of multiple protein interactions in living cells using multicolor fluorescence complementation analysis. *Nat. Biotechnol.* **21,** 539–545.

Huttelmaier, S., Zenklusen, D., Lederer, M., Dictenberg, J., Lorenz, M., Meng, X., Bassell, G. J., Condeelis, J., and Singer, R. H. (2005). Spatial regulation of beta-actin translation by Src-dependent phosphorylation of ZBP1. *Nature* **438,** 512–515.

Ishiura, M., Hirose, S., Uchida, T., Hamada, Y., Suzuki, Y., and Okada, Y. (1982). Phage particle-mediated gene transfer to cultured mammalian cells. *Mol. Cell. Biol.* **2,** 607–616.

Job, C., and Eberwine, J. (2001). Localization and translation of mRNA in dendrites and axons. *Nat. Rev. Neurosci.* **2,** 889–898.

Ju, W., Morishita, W., Tsui, J., Gaietta, G., Deerinck, T. J., Adams, S. R., Garner, C. C., Tsien, R. Y., Ellisman, M. H., and Malenka, R. C. (2004). Activity-dependent regulation of dendritic synthesis and trafficking of AMPA receptors. *Nat. Neurosci.* **7,** 244–253.

Kaech, S., Kim, J. B., Cariola, M., and Ralston, E. (1996). Improved lipid-mediated gene transfer into primary cultures of hippocampal neurons. *Brain Res. Mol. Brain Res.* **35,** 344–348.

Kang, H., and Schuman, E. M. (1996). A requirement for local protein synthesis in neurotrophin-induced hippocampal synaptic plasticity. *Science* **273,** 1402–1406.

Kiebler, M. A., and DesGroseillers, L. (2000). Molecular insights into mRNA transport and local translation in the mammalian nervous system. *Neuron* **25,** 19–28.

Kiebler, M. A., Jansen, R. P., Dahm, R., and Macchi, P. (2005). A putative nuclear function for mammalian Staufen. *Trends Biochem. Sci.* **30,** 228–231.

Klann, E., and Dever, T. E. (2004). Biochemical mechanisms for translational regulation in synaptic plasticity. *Nat. Rev. Neurosci.* **5,** 931–942.

Kost, B., Spielhofer, P., and Chua, N. H. (1998). A GFP-mouse talin fusion protein labels plant actin filaments *in vivo* and visualizes the actin cytoskeleton in growing pollen tubes. *Plant J.* **16,** 393–401.

Kugler, S., Lingor, P., Scholl, U., Zolotukhin, S., and Bahr, M. (2003). Differential transgene expression in brain cells *in vivo* and *in vitro* from AAV-2 vectors with small transcriptional control units. *Virology* **311,** 89–95.

Leclere, P. G., Panjwani, A., Docherty, R., Berry, M., Pizzey, J., and Tonge, D. A. (2005). Effective gene delivery to adult neurons by a modified form of electroporation. *J. Neurosci. Methods* **142,** 137–143.

Li, Y., Jia, Y. C., Cui, K., Li, N., Zheng, Z. Y., Wang, Y. Z., and Yuan, X. B. (2005). Essential role of TRPC channels in the guidance of nerve growth cones by brain-derived neurotrophic factor. *Nature* **434,** 894–898.

Lopez de Heredia, M., and Jansen, R. P. (2004). mRNA localization and the cytoskeleton. *Curr. Opin. Cell Biol.* **16,** 80–85.

Ma, H., Zhu, J., Maronski, M., Kotzbauer, P. T., Lee, V. M., Dichter, M. A., and Diamond, S. L. (2002). Non-classical nuclear localization signal peptides for high efficiency lipofection of primary neurons and neuronal cell lines. *Neuroscience* **112,** 1–5.

Macchi, P., Hemraj, I., Goetze, B., Grunewald, B., Mallardo, M., and Kiebler, M. A. (2003). A GFP-based system to uncouple mRNA transport from translation in a single living neuron. *Mol. Biol. Cell* **14,** 1570–1582.

Machleidt, T., Robers, M., and Hanson, G. T. (2006). Protein labeling with FlAsH and ReAsH. *Methods Mol. Biol.* **356,** 209–220.

Martin, K. C., Casadio, A., Zhu, H., Yaping, E., Rose, J. C., Chen, M., Bailey, C. H., and Kandel, E. R. (1997). Synapse-specific, long-term facilitation of aplysia sensory to motor synapses: A function for local protein synthesis in memory storage. *Cell* **91,** 927–938.

Martin, K. C., and Zukin, R. S. (2006). RNA trafficking and local protein synthesis in dendrites: An overview. *J. Neurosci.* **26,** 7131–7134.

Miller, S., Yasuda, M., Coats, J. K., Jones, Y., Martone, M. E., and Mayford, M. (2002). Disruption of dendritic translation of CaMKIIalpha impairs stabilization of synaptic plasticity and memory consolidation. *Neuron* **36,** 507–519.

Mingle, L. A., Okuhama, N. N., Shi, J., Singer, R. H., Condeelis, J., and Liu, G. (2005). Localization of all seven messenger RNAs for the actin-polymerization nucleator Arp2/3 complex in the protrusions of fibroblasts. *J. Cell Sci.* **118,** 2425–2433.

Munchow, S., Sauter, C., and Jansen, R. P. (1999). Association of the class V myosin Myo4p with a localised messenger RNA in budding yeast depends on She proteins. *J. Cell Sci.* **112**(Pt. 10), 1511–1518.

Palacios, I. M., and St Johnston, D. (2001). Getting the message across: The intracellular localization of mRNAs in higher eukaryotes. *Annu. Rev. Cell Dev. Biol.* **17,** 569–614.

Piper, M., and Holt, C. (2004). RNA translation in axons. *Annu. Rev. Cell Dev. Biol.* **20,** 505–523.

Rackham, O., and Brown, C. M. (2004). Visualization of RNA–protein interactions in living cells: FMRP and IMP1 interact on mRNAs. *EMBO J.* **23,** 3346–3355.

Rook, M. S., Lu, M., and Kosik, K. S. (2000). CaMKIIalpha 3′ untranslated region-directed mRNA translocation in living neurons: Visualization by GFP linkage. *J. Neurosci.* **20,** 6385–6393.

Rouault, T. A., Tang, C. K., Kaptain, S., Burgess, W. H., Haile, D. J., Samaniego, F., McBride, O. W., Harford, J. B., and Klausner, R. D. (1990). Cloning of the cDNA encoding an RNA regulatory protein—the human iron-responsive element-binding protein. *Proc. Natl. Acad. Sci. USA* **87,** 7958–7962.

Schratt, G. M., Tuebing, F., Nigh, E. A., Kane, C. G., Sabatini, M. E., Kiebler, M., and Greenberg, M. E. (2006). A brain-specific microRNA regulates dendritic spine development. *Nature* **439,** 283–289.

Schuman, E. M., Dynes, J. L., and Steward, O. (2006). Synaptic regulation of translation of dendritic mRNAs. *J. Neurosci.* **26,** 7143–7146.

Shan, J., Munro, T. P., Barbarese, E., Carson, J. H., and Smith, R. (2003). A molecular mechanism for mRNA trafficking in neuronal dendrites. *J. Neurosci.* **23,** 8859–8866.

Shaner, N. C., Campbell, R. E., Steinbach, P. A., Giepmans, B. N., Palmer, A. E., and Tsien, R. Y. (2004). Improved monomeric red, orange and yellow fluorescent proteins derived from *Discosoma* sp. red fluorescent protein. *Nat. Biotechnol.* **22,** 1567–1572.

Steward, O., and Schuman, E. M. (2001). Protein synthesis at synaptic sites on dendrites. *Annu. Rev. Neurosci.* **24,** 299–325.

Steward, O., and Schuman, E. M. (2003). Compartmentalized synthesis and degradation of proteins in neurons. *Neuron* **40,** 347–359.

St Johnston, D. (2005). Moving messages: The intracellular localization of mRNAs. *Nat. Rev. Mol. Cell Biol.* **6,** 363–375.

Svoboda, K., and Yasuda, R. (2006). Principles of two-photon excitation microscopy and its applications to neuroscience. *Neuron* **50,** 823–839.

Veselovsky, N. S., Engert, F., and Lux, H. D. (1996). Fast local superfusion technique. *Pflugers Arch.* **432,** 351–354.

Vessey, J. P., Vaccani, A., Xie, Y., Dahm, R., Karra, D., Kiebler, M. A., and Macchi, P. (2006). Dendritic localization of the translational repressor Pumilio 2 and its contribution to dendritic stress granules. *J. Neurosci.* **26,** 6496–6508.

Washbourne, P., and McAllister, A. K. (2002). Techniques for gene transfer into neurons. *Curr. Opin. Neurobiol.* **12,** 566–573.

Zeitelhofer, M., Vessey, J. P., Xie, Y., Tübing, F., Thomas, S., Kiebler, M., and Dahm, R. (2007). High efficiency transfection of mammalian neurons via nucleofection. *Nat. Protoc.* **2**(7), 1692–1704.

Zhang, J., Campbell, R. E., Ting, A. Y., and Tsien, R. Y. (2002). Creating new fluorescent probes for cell biology. *Nat. Rev. Mol. Cell Biol.* **3,** 906–918.

CHAPTER 14

Quantitative FRAP in Analysis of Molecular Binding Dynamics *In Vivo*

James G. McNally
Laboratory of Receptor Biology and Gene Expression
National Cancer Institute
Bethesda, Maryland 20892

Abstract

Fluorescence recovery after photobleaching (FRAP) reveals the dynamics of fluorescently tagged molecules within live cells. These molecular dynamics are governed by diffusion of the molecule and its *in vivo* binding interactions. As a result, quantitative estimates of the association and dissociation rates of binding can be extracted from the FRAP. This chapter describes a systematic procedure to acquire the FRAP data, and then fit it with appropriate mathematical models to estimate *in vivo* association and dissociation rates of binding. Also discussed are the applicability and limitations of the models, the utility of the estimated parameters, and the prospects for increased accuracy and confidence in the estimates.

0091-679X/08 $35.00
DOI: 10.1016/S0091-679X(08)85014-5

I. Introduction

Fluorescence recovery after photobleaching (FRAP) is now widely used to study protein mobility in living cells. FRAP is performed by photobleaching fluorescent molecules at a specified location in a cell, and then monitoring the rate at which the bleached molecules are replaced by unbleached ones. The rate of recovery reflects the rate of movement of the fluorescently tagged molecules at that location within the cell.

Molecular mobilities as obtained from FRAP are informative for several reasons. First, they can provide information about the rates of cellular diffusion in different subcellular compartments. Indeed, most of the early FRAP studies focused on the diffusion of proteins and lipids within membranes (Edidin, 1994; Liebman and Entine, 1974; Poo and Cone, 1974).

Second, molecular mobilities often reveal that apparently static structures within cells are actually constructed from highly dynamic protein constituents. FRAP uncovers this hidden flux by selectively marking a subset of molecules, and so discloses the exchange of proteins or other molecules that occur within and between cellular compartments even when the system is at equilibrium (Misteli, 2001; Webb *et al.*, 2003).

Third, molecular mobilities can also be used for the measurement of cellular binding interactions. FRAP of different green fluorescent protein (GFP)-fusion proteins will sometimes reveal that their mobilities are considerably slower than expected for a purely diffusing molecule of that size, or even for a large molecular complex. This often indicates retardation of the protein's mobility by cellular binding interactions. Since stronger interactions will retard mobilities more than weaker interactions, the FRAP curve can be used to estimate the strengths of *in vivo* molecular binding interactions (Sprague and McNally, 2005).

This chapter focuses on how to extract quantitative information about molecular binding interactions from the FRAP data.

II. Rationale

Most estimates of protein binding affinity have been performed *in vitro*. This typically involves isolating the protein and its binding target, incubating the binding partners under appropriate conditions, and then measuring binding affinities by one of several established techniques, such as surface plasmon resonance, calorimetry, capillary electrophoresis, or filter binding assays (He *et al.*, 2004; Leavitt and Freire, 2001; Riggs *et al.*, 1970; Schuck, 1997).

How close are these *in vitro* affinity measurements to *in vivo* affinities? Some differences should be expected. Binding interactions *in vivo* occur within a totally different environment, where both the protein of interest and its binding target may be parts of more complex structures, which might, among other things, alter the accessibility of binding sites. In addition, the *in vivo* milieu contains numerous molecules that could directly modulate binding affinities.

A case in point is the binding of a transcription factor, the glucocorticoid receptor, to its promoter target. *In vitro* measurements with naked DNA and purified protein yielded an estimated residence time of the transcription factor on the promoter of ~90 min (Perlmann *et al.*, 1990). *In vivo*, however, the promoter DNA is packaged as chromatin, and additional factors such as chaperones and proteasomes are thought to influence binding (Stavreva *et al.*, 2004). Perhaps as a result of all of these differences, the *in vivo* estimates of glucocorticoid receptor residence times on its promoter are two to three orders of magnitude shorter than the *in vitro* estimates (Perlmann *et al.*, 1990; Sprague *et al.*, 2006). This is not likely to be an isolated case, and so provides a compelling argument to measure binding affinities *in vivo*.

In vivo measurements of binding will provide an understanding of the chemistry of reactions within a living cell. Such measurements are the raw material for the systems level models of biological networks that are now being constructed (Janes and Lauffenburger, 2006; Schlitt and Brazma, 2006). At present, these models typically contain either *in vitro* estimates of binding affinities or just educated guesses for these parameters (Bhalla *et al.*, 2002; von Dassow *et al.*, 2000). However, more accurate, and therefore more informative models, will require accurate *in vivo* measurements.

FRAP is one of several techniques with the potential to provide *in vivo* binding estimates. A complementary approach is fluorescence correlation spectroscopy (FCS; see Chapter 20 by Langowski, this volume). One advantage of FRAP over FCS is that it can be performed on most confocal microscopes, while FCS requires special instrumentation. A second advantage of FRAP is that the basic procedure is straightforward to troubleshoot. At a simple level at least, FRAP curves are easy to understand and therefore many artifacts are easy to recognize. This is not the case with FCS where the raw data is simply the intensity fluctuations over time at a point in the image. Despite these disadvantages, FCS has the advantage that no perturbation of the system is necessary, unlike the intentional photobleach that is required by FRAP. Both of these approaches are likely to be used in parallel in the future as one way to cross-validate *in vivo* measurements of binding (Schmiedeberg *et al.*, 2004).

III. Methods

There are two principal steps in using FRAP to extract *in vivo* estimates of binding parameters: data acquisition and data analysis.

A. Data Acquisition

1. Cell Preparation

Standard procedures for live-cell imaging can be used to prepare specimens for FRAP experiments. Typically, this involves a cell line with either a stably expressing or transiently transfected GFP-fusion protein of interest. As a baseline for comparison, a cell line expressing unconjugated GFP is also required.

Cells are grown on or attached in some way to a coverslip surface, and enclosed in a live-cell chamber that includes the appropriate media and optimally temperature and CO_2 controls. Many such chambers are commercially available, and the selection of the appropriate one will depend on the particular cells under study. Since many FRAP recoveries occur in a minute or less, it may be possible in some cases to compromise on optimal incubation conditions, and instead frequently replace specimens with fresh ones from a nearby incubator.

2. Choice of Microscope

In principle, the FRAP experiment can be performed on any microscope equipped with a laser to produce the photobleach and with a detector operating in the linear regime of intensity measurements (Cho and Lockett, 2006; Coscoy *et al.*, 2002). These requirements can be met by wide-field, confocal, or two-photon microscopes. However, the analysis methods to extract binding information described here have been developed uniformly for bleaching patterns that extend throughout the depth of the specimen, a requirement that can be reasonably met on wide-field and confocal microscopes, but not two photon microscopes. In the remaining part of this chapter, we presume that the FRAP experiment is performed with a confocal microscope, which is currently the most widely used system for such experiments.

3. FRAP Acquisition Parameters

Nine parameters must be determined and preset. Although listed separately here, the parameters, as noted below, are in some cases interdependent. Thus, several competing factors must often be balanced in order to select the appropriate parameter settings.

1. Number of pre-bleach images

Pre-bleach images are acquired before the intentional photobleach. Twenty pre-bleach images are recommended in order to obtain an accurate estimate of the specimen's intensity prior to the photobleach. The first 10 images often show a rapid, unintentional photobleaching that soon dissipates to be replaced by a slower and steadier unintentional bleaching rate that continues for the remainder of the FRAP experiment. Thus, for GFP-fusion proteins, the pre-bleach intensity is measured from the average of only the last 10 pre-bleach images.

2. Shape of the bleached region

The shape of the bleached region will usually affect the FRAP recovery. Relatively simple equations are available to analyze recoveries for various simple shapes, such as rectangles and circles (Carrero *et al.*, 2004; Hinow *et al.*, 2006; Lele *et al.*, 2004; Sprague *et al.*, 2004). More complex shapes require a more time-consuming computational analysis in which the underlying differential equations must be solved numerically. The equations outlined below are the solutions obtained for a circular bleach spot.

3. Size of the bleached region

The size of the bleached region must be selected. If the recovery is diffusion coupled (see below), then bleach-spot size will have an impact. Ideally, the bleach spot should be large enough to contain enough pixels such that the average intensity of the spot is well determined. However, if the spot is too large, then a substantial amount of cellular fluorescence will be photobleached, and this will significantly reduce the amplitude of the fluorescence recovery. A good rule of thumb is to begin whenever possible with a spot size that is not more than 10% of the size of the cellular compartment under study.

4. Laser power for photobleaching

This should be high enough such that the photobleach reduces the intensity within the bleach spot by at least 50%. Such a large bleached fraction provides sufficient recovery data for accurate fitting analysis. Note that the size of the bleached fraction is also influenced by the duration of the bleach and the laser power (Parameters 5 and 7).

5. Duration of the bleach

This should be as short as possible, otherwise significant amounts of recovery can take place while the bleach is ongoing. Most models for extracting binding information from FRAP presume that the bleach is instantaneous.

6. Time lag between the photobleach and first image

To permit analysis of rapid kinetics, this time lag should be as small as possible. Some confocal microscopes are available in which bleaching is performed with one laser and scanner, and imaging with a second laser and scanner. On these systems, a very short time lag is possible. However, most confocal microscopes use the same laser to perform both the photobleach and, at reduced intensity, the subsequent imaging. Consequently, there is an unavoidable time lag as the instrument converts from photobleaching to imaging mode. This time lag may be reduced by increasing the imaging scan rate (Parameter 9).

7. Laser power for imaging

Here, the compromise is to find a laser power to give sufficient signal from the bleached spot as fluorescence recovers there, without introducing too much additional bleaching as the recovery phase is monitored. This unintentional bleaching due to imaging confounds the quantitative FRAP analysis, and so should be minimized. This can be accomplished by reducing as much as possible the laser power used for imaging.

Note that for confocal microscopes that use the same laser for photobleaching and imaging, an additional compromise must be struck. On these systems, the imaging is performed at reduced intensity levels by controlling the laser throughput with an acousto-optical tunable filter (AOTF). The minimum setting of the AOTF can be used for imaging, while the maximum setting for photobleaching. The laser power can be increased to the point at which a sufficiently deep photobleach (Parameter 5) can be achieved with the AOTF at its maximum, but with minimal bleaching during imaging with the AOTF at its minimum.

8. Duration of post-bleach imaging

This should be long enough such that the recovery curve has reached a plateau. This ensures that there is sufficient recovery data to capture the complete kinetics of the binding process. Some recoveries will plateau well below 100%, indicating that there is likely to be a fraction of molecules that remain tightly bound. This fraction can also be analyzed in principle by further extending the time of measurement to allow a second plateau to be reached.

The equations below presume that the recovery plateaus at 100%. If this is not the case, the FRAP curve can be renormalized a second time such that the final recovery level is 100%. This second renormalization follows the first renormalization in which the pre-bleach intensities were set to one, but is applied to only the recovery phase, expanding it such that it plateaus at one. The estimates of binding obtained from the fit to this doubly renormalized curve will be applicable only to the subset of binding sites responsible for that part of the curve. The binding responsible for the plateau portion of the curve would have to be investigated by measuring the FRAP over a much longer timescale thereby revealing the gradual recovery within this part of the curve.

9. Time interval between post-bleach images

To obtain reasonable sampling of the FRAP curve, at least 20 time points should be collected over the duration of the recovery. The time points can be equally spaced, but if the FRAP curve exhibits fast and slow phases then the image sampling rate can be adjusted accordingly. To analyze very rapid kinetics, these time intervals may be quite short. This can be accomplished using fast scan speeds, and also by shrinking the size of the image that must be scanned (Fig. 1A).

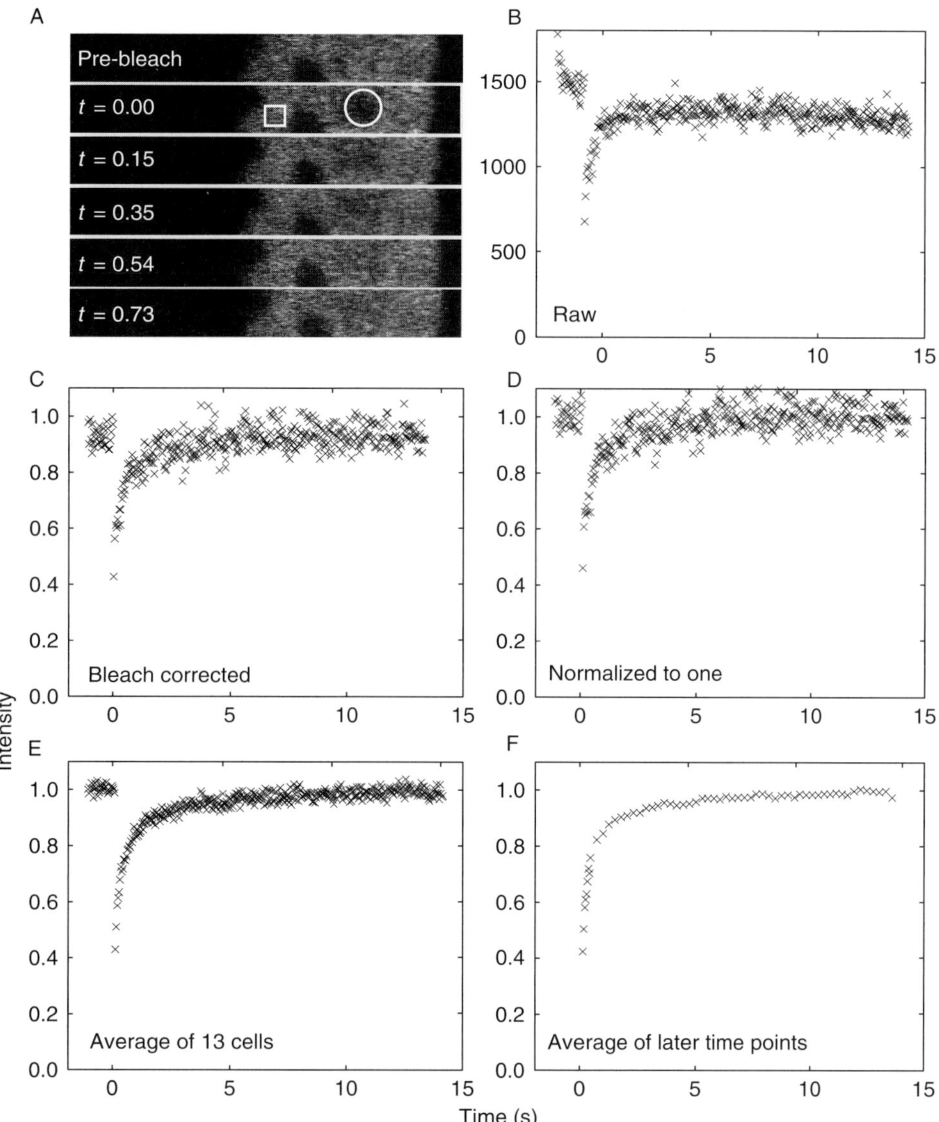

Fig. 1 Generating FRAP data. (A) Sample images from a FRAP sequence are shown. A strip-shaped image is used for rapid acquisition on a scanning confocal microscope. The bleach spot is indicated by the white circle, and the region used for correction of unintentional bleaching by the white square. (B) Average intensity within the bleach spot is plotted as a function of time, yielding the raw FRAP curve. Note the rapid decay of fluorescence in the first pre-bleach images as well as the steady decline evident for $t > 10$ s. (C) This unintentional photobleaching is corrected by dividing the raw data by the average intensity within the square outlined in (A) at each time point. (D) The bleach-corrected data is normalized to one by dividing each time point by the average intensity of the last 10 pre-bleach images. (E) Noise in single FRAP curves is reduced by averaging normalized data from 13 cells. (F) Noise is further reduced by averaging of all time points beyond the ninth time point, as described in the text.

4. Generating FRAP Recovery Curves from Image Data

Once imaging parameters have been established and images have been collected for a FRAP experiment, the image data must be quantified. The first step in this analysis is to measure the intensity within the bleach spot as a function of time. Typically, spatial information is discarded at this point, and the average of all pixels within the bleach spot is calculated. Future analysis models, however, might well be based on average intensities as a function of distance from the center point of the bleach (Seiffert and Oppermann, 2005), which could in principle supply more information for more accurate parameter estimation.

Next, the average background intensity is subtracted from the average intensity within the bleach spot. Background is measured as the average intensity in a region of the image that contains no cell.

These background-subtracted average bleach-spot intensities can be plotted as a function of time to generate the raw FRAP curve (Fig. 1B). These raw data typically exhibit at least two clear defects. First, the FRAP curve suffers from unintentional bleaching due to imaging. This is revealed by the decline in the intensity curve for either the pre-bleach images or the post-bleach images at long times after the intentional photobleach (Fig. 1B). Second, the intentional photobleach permanently removes a fraction of the cellular fluorescence such that the fluorescence cannot fully recover to pre-bleach levels.

Various procedures can be used to correct for this loss of fluorescence due to the intentional and unintentional photobleaching (Coscoy *et al.*, 2002; Phair *et al.*, 2004). The fitting models described below presume that the most widely used bleach-correction protocol has been performed (Phair *et al.*, 2004). This involves measuring the average fluorescent intensity as a function of time at a spot as far as possible from where the intentional photobleach was performed, and then dividing the averaged FRAP data by this decaying intensity measured elsewhere in the cell. This procedure yields a FRAP curve that no longer decays at long times (Fig. 1C). Future mathematical models for FRAP are likely to offer refinements in these bleaching-correction procedures.

Another defect of the raw FRAP data (Fig. 1B) is that they are noisy. This is a consequence of the need to obtain images rapidly and to curtail unintentional photobleaching by operating at low laser illumination. To reduce noise, FRAP curves from a number of cells are typically averaged to achieve a sufficiently smooth recovery curve.

To permit averaging, recovery curves are normalized such that the pre-bleach intensity is set to one. This is done by first calculating the average intensity within the spot to be bleached for 10 images just preceding the photobleach. Dividing all of the intensity data after the photobleach by this average pre-bleach intensity then yields a normalized FRAP curve for each cell (Fig. 1D).

Averaging over some number of cells generates a smoother curve for subsequent quantitative analysis (Fig. 1E). The cells of comparable brightness should be selected for this averaging process, as the fractions of free and bound molecules may change substantially in cells of markedly different brightness, and this will change the FRAP curve from cell to cell.

Finally, averaging can also be performed at later time points where the FRAP curve changes slowly. In this flatter region of the curve, anywhere from three to ten adjacent time points can be averaged. This further smooths the FRAP curve (Fig. 1F). In addition, it provides a better balance between the number of time points in the fast and the slow phases of the FRAP curve, such that they are more equally weighted during the fitting procedure.

B. Data Analysis

1. Applicability of the Models

Once FRAP data have been properly prepared, then a series of mathematical models can be systematically applied to obtain estimates of binding parameters (Sprague *et al.*, 2004). Different experimental scenarios demand different mathematical models; however, the general strategy in applying these models to experimental data remains the same (Fig. 2).

The specific models for FRAP outlined below are appropriate for a fluorescent molecule that is at the outset uniformly distributed throughout a cellular compartment, and then bleached with a circular spot whose diameter is small relative to the size of the compartment. The molecule is presumed to diffuse and bind at sites throughout the compartment. Binding sites are fixed in place, at least on a timescale relative to the rate of FRAP recovery. The models also address the case where more than one type of binding site is present, that is, two or more independent binding states.

Practical examples to which the equations can be applied are the following: (1) A GFP-tagged nuclear protein that may bind to DNA throughout the nucleus. The DNA is relatively immobile, so reflects a fixed binding state. If the nuclear protein also binds to the nuclear matrix, then a second binding state would also be present. (2) A GFP-tagged actin-binding protein that may bind to microfilaments throughout the cytoplasm. If actin rearrangements occur slowly on the timescale of the FRAP recovery, then these microfilaments serve as a fixed binding state. If the actin-binding protein also interacts with microtubules, then a second binding state would be present.

Practical examples to which the equations do not apply are small cellular compartments where the bleach-spot size is large relative to the compartment size. Here, the finite size of the cellular compartment becomes important, requiring modified forms of the equations presented here [see Eq. (53) in Sprague *et al.* (2004)]. Also not accounted for by the models discussed here are all heterogeneous distributions of fluorescent intensity such as a localized binding site cluster

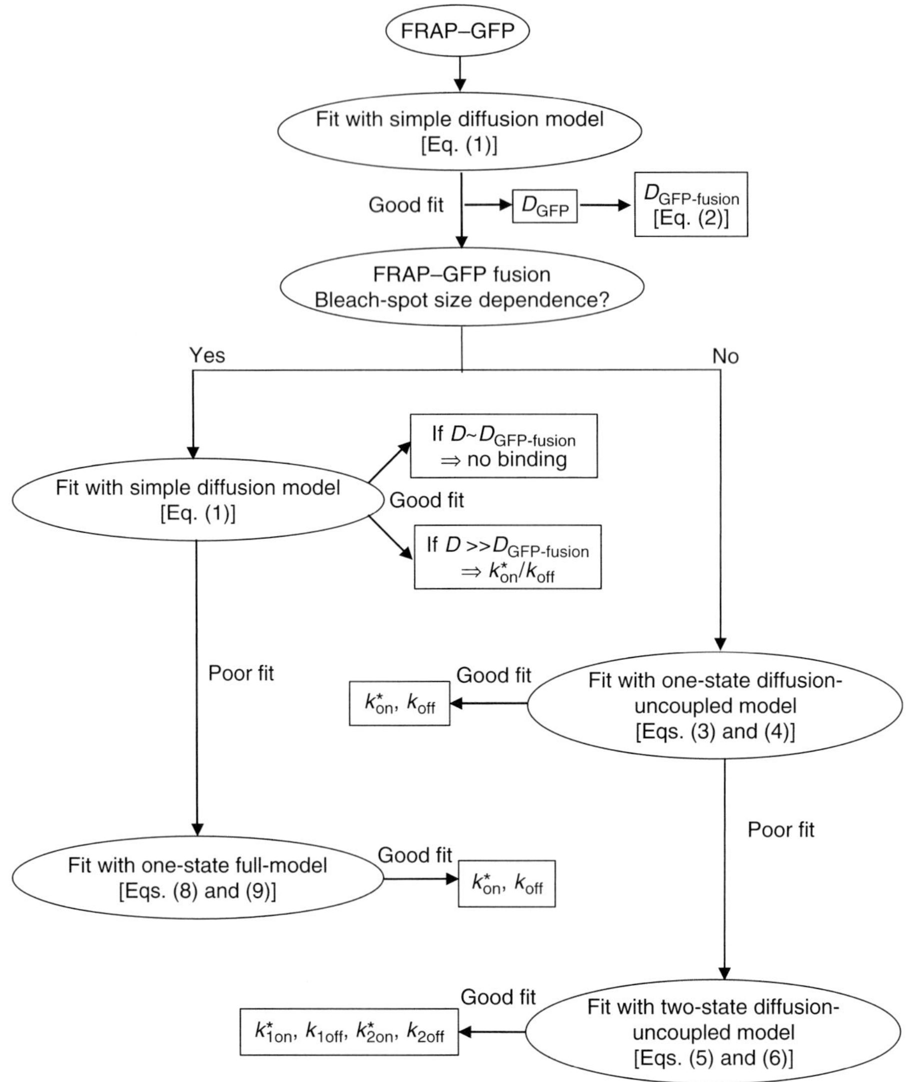

Fig. 2 Flowchart summarizing the steps to be taken in fitting FRAP data.

(e.g., a centriole or telomere) or a more complex structure like the Golgi or endoplasmic reticulum. Many of these situations however can also be handled by more complex models (Beaudouin *et al.*, 2006; Sprague *et al.*, 2004). As noted above, when applying any of these more complicated models, the flowchart strategy outlined in Fig. 2 remains the same, only the specific modeling equations change.

2. Experimental Parameters

The presumption that the GFP-fusion protein undergoes diffusion and binding leads to reaction diffusion equations describing the dynamics of the GFP-fusion protein during the FRAP. The reaction diffusion equations contain up to five parameters: the bleach-spot radius (w), the bleached fraction (ϕ), the cellular diffusion constant of the GFP-tagged protein ($D_{\text{GFP-fusion}}$), and the association and dissociation rates of the binding reactions (k^*_{on}, k_{off}).

Three of these parameters can be predetermined. The bleach-spot size can be measured by photobleaching fixed cells and estimating the radius of the bleach spot by direct measurement.

The bleached fraction can be estimated by extrapolating the FRAP curve to $t = 0$, and determining where the curve intersects the y-axis. Subtracting this y-intercept value from 1 yields the bleached fraction. Thus, a FRAP curve that begins at 0.3 for $t = 0$ has a bleached fraction of 0.7 indicating that 70% of the molecules within the circular spot have been bleached.

The diffusion constant of the GFP-tagged protein can be estimated by determining the diffusion constant of GFP and then extrapolating from this value based on the size of the GFP-fusion protein [see Eq. (2)].

This leaves the two parameters of interest, namely the association and dissociation rates of the binding reaction. These are free parameters that can be varied by the fitting routine to find the combination that will yield the best fit of the model equation to the FRAP data.

In principle, a single model equation (see Section III.B.6.c) is available to predict all FRAP recoveries in the presence of diffusion and binding, subject of course to the assumptions of the model as detailed in Section III.B.1. In practice, however, there are two limiting cases of the full-model equations that give rise to simpler equations, namely effective-diffusion FRAPs and diffusion-uncoupled FRAPs [the latter are also known as reaction-dominant (Sprague *et al.*, 2004), diffusion-limited (Beaudouin *et al.*, 2006; Kaufman and Jain, 1990), or well-mixed (Phair *et al.*, 2004) FRAPs]. It is advisable to first test if either of these simpler situations holds before applying the full model because these simplified scenarios also provide for more simplified fitting procedures. With this in mind, a step-by-step experimental and fitting protocol is outlined in Sections III.B.3–6, and summarized in Fig. 2.

3. FRAP of Unconjugated GFP

First, a control FRAP using unconjugated GFP expressed in the cell line of interest should be performed in both fixed and live cells. The fixed cell FRAP is used to assess whether there is any reversible photobleaching, which may occur under some conditions when bleached GFP molecules spontaneously refluoresce (Sinnecker *et al.*, 2005). If the fixed cell FRAP yields no recovery, then the photobleach is irreversible. If however there is some recovery, then either the cells have not been fixed well enough or there is some reversible photobleaching, which would have to be subtracted from FRAP recoveries occurring in live cells.

The live-cell FRAP of GFP provides both an estimate of the cellular diffusion constant and a test of whether cellular diffusion can even be described by the conventional diffusion equation. The premise of this experiment is that unconjugated GFP should not bind to cellular structures, and should simply diffuse. Because of the extremely rapid kinetics of pure diffusion, FRAP of unconjugated GFP in live cells typically requires (1) a large spot size, (2) a short bleach duration, (3) a short time lag between the bleach and first image, and (4) rapid image acquisition. All of these settings are necessary to obtain a good estimate of the cellular diffusion constant.

The GFP–FRAP curve [frap(t)] should be fit with the equation for simple diffusion without binding interactions (Soumpasis, 1983):

$$\mathrm{frap}(t) = 1 - \phi + (\phi \mathrm{e}^{-\alpha})[I_0(\alpha) + I_1(\alpha)] \tag{1}$$

Here ϕ is the bleached fraction, I_0 and I_1 are modified Bessel functions, and $\alpha = w^2/2Dt$ with w the bleach-spot radius and D the diffusion constant. The bleached fraction and bleach-spot radius can be estimated as described above, so the only free parameter in Eq. (1) is the diffusion constant. Fitting this equation to the FRAP data of unconjugated GFP therefore yields an estimate for the diffusion constant of GFP. A simple diffusion model has yielded good fits to GFP in different systems (Braga *et al.*, 2004; Sprague *et al.*, 2004), so if this fit fails, then it indicates that there are either problems in the experimental setup or that the process of cellular diffusion is more complicated than described by the simple diffusion equation. In the latter case, more complicated models accounting for the complexities of anomalous diffusion (Saxton, 2001; Weiss *et al.*, 2004) will be necessary before proceeding to the estimation of binding parameters. The models for binding presented here presume that the diffusion process is not anomalous.

If the fit to the GFP–FRAP data is successful, then it should yield a cellular diffusion constant somewhere in the range from 10 to 80 $\mu m^2/s$ (Arrio-Dupont *et al.*, 2000; Braga *et al.*, 2004; Sprague *et al.*, 2004; Swaminathan *et al.*, 1997). Values well beyond this range are suspect, and likely indicate serious problems in the experimental setup or curve fitting procedure. Note that some of this spread in estimated GFP diffusion values is probably due to cell to cell differences, but some is probably also due to errors in estimating D arising from the rapid kinetics of GFP recovery (Braga *et al.*, 2004; Weiss, 2004).

Once a reasonable estimate of the diffusion constant for unconjugated GFP, D_{GFP}, is obtained, it can be used to derive an estimate of the diffusion constant of the GFP-fusion protein:

$$D_{\mathrm{GFP-fusion}} = D_{\mathrm{GFP}}\left(\frac{M_{\mathrm{GFP}}}{M_{\mathrm{GFP-fusion}}}\right)^{1/3} \tag{2}$$

where M_{GFP} and $M_{GFP\text{-}fusion}$ are the molecular weights of GFP and the GFP-fusion protein. This calculation presumes that both GFP and the fusion protein are roughly spherical molecules, such that the diffusion constants are proportional to their radii. This calculation also presumes that the freely diffusing GFP-fusion protein is not part of a larger complex, which would therefore diffuse more slowly.

4. Testing for Diffusion Dependence in the GFP-Fusion Protein FRAP

If both the fit and the estimate of the GFP diffusion constant are good, then the next step is to perform FRAP of the GFP-fusion protein. The first experiment here should test whether this FRAP depends on diffusion. This is an important determination because it indicates whether simpler equations that ignore diffusion can be used.

Of course diffusion occurs in every FRAP, but the question is whether it occurs so rapidly compared to the time to associate with a binding site that the freely diffusing molecules completely equilibrate in the bleach spot before the binding reaction even begins. In this limiting scenario, the FRAP curve is composed of a very fast initial diffusive phase, and then followed by a binding phase which lasts considerably longer (Fig. 3A and C). In this diffusion-uncoupled FRAP, the binding phase can be described by a simple exponential equation depending on only the association and dissociation rates of binding (see Section III.B.5).

In contrast, when the time to diffuse across the bleach spot is comparable to or longer than the time to begin binding, then the free molecules do not immediately reequilibrate after the bleach. Instead, all molecules repeatedly bind, unbind, and diffuse throughout the FRAP. In this diffusion-coupled FRAP, there is no rapid early diffusive phase, but rather a steady increase in fluorescent intensity immediately after the photobleach (Fig. 3B and D). Depending on just how fast the association rate of the binding reaction is compared to the time to diffuse across the spot, diffusion-coupled FRAPs can be described either by an effective-diffusion equation (Section III.B.6.b) or by a more complicated full-model equation (Section III.B.6.c), both of which can provide information about the binding reaction.

To determine if diffusion can be ignored, FRAPs of the GFP-fusion protein should be performed with different bleach-spot sizes (Lele *et al.*, 2004; Sprague *et al.*, 2004). Diffusion-uncoupled FRAPs will exhibit no difference in recoveries with different bleach-spot sizes while diffusion-coupled FRAPs will do so (Fig. 4A–F). To detect a difference between FRAPs at different bleach-spot sizes, it is important to average enough individual curves at each spot size to obtain sufficiently smooth recovery data. If there is no difference in the FRAP recoveries, then diffusion can be safely ignored in fitting the FRAP data, and the formula for a diffusion-uncoupled FRAP can be used (Section III.B.5). If there is a difference in the FRAP recoveries, then one of the two formulas for a diffusion-coupled FRAP must be used (Section III.B.6).

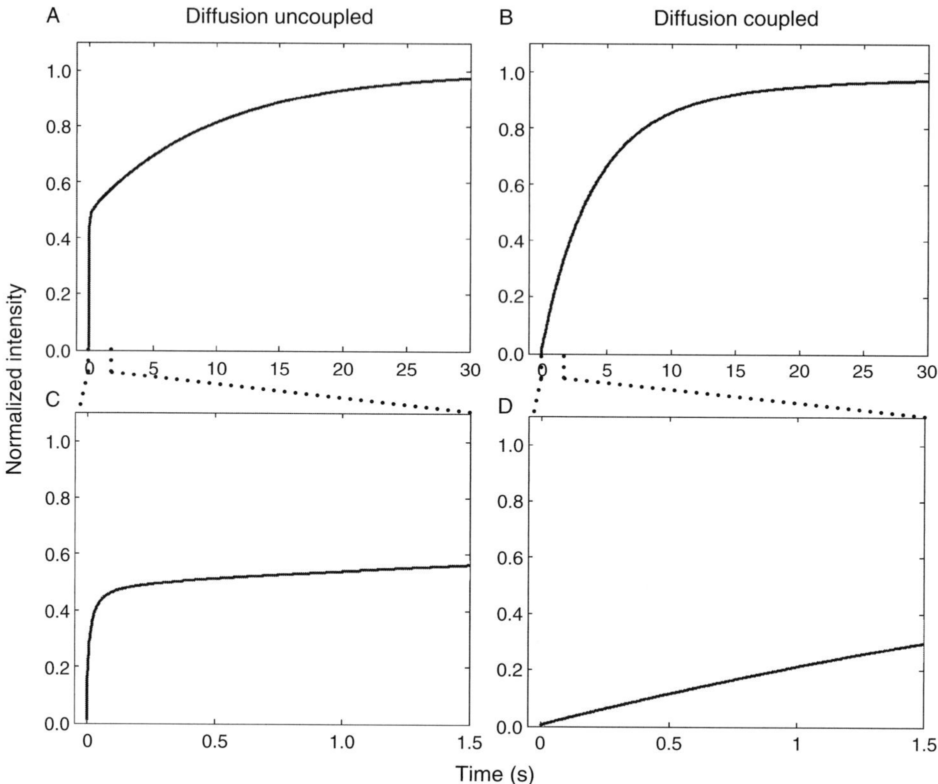

Fig. 3 Comparison of diffusion-coupled and diffusion-uncoupled FRAP curves. In the diffusion-uncoupled case, there is a transition between a very rapid initial recovery due to free diffusion followed by a slower recovery due to binding (A and C). There is no such transition in the diffusion-coupled case at either long or short timescales (B and D). These two curves were generated with the same bleached fraction (100%), the same diffusion constant (30 μm^2/s), and the same bleach spot radius (1 μm), but with different association and dissociation rates ($k^*_{on} = 0.1, k_{off} = 0.1$ for A and C; and $k^*_{on} = 31.6, k_{off} = 0.35$ for B and D).

5. Fitting a Diffusion-Uncoupled FRAP

If there is no bleach-spot size dependence, then the FRAP curve can be fit with

$$\mathrm{frap}(t) = 1 - \phi_{\mathrm{app}} e^{-k_{\mathrm{off}} t} \tag{3}$$

This yields an estimate of both the dissociation rate k_{off} and the apparent bleached fraction ϕ_{app}. Note that ϕ_{app} is not the true bleached fraction, ϕ_{true}, since the diffusive phase of the FRAP is typically not detected when measuring the slower binding phase of the recovery (Fig. 3A). However, an estimate of ϕ_{true} is required

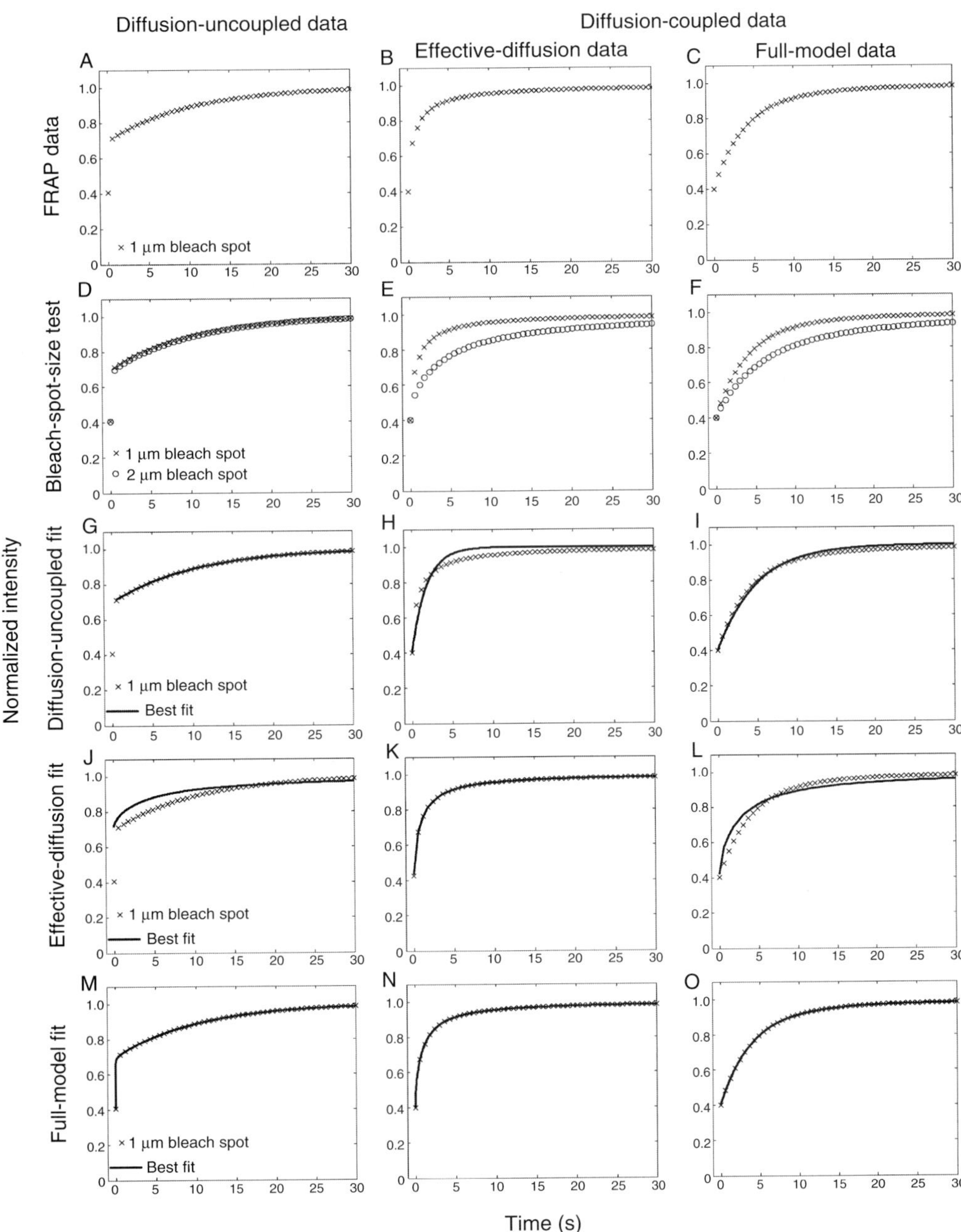

Fig. 4 Fits to FRAP data. Shown in (A–C) are FRAP data generated with a bleached fraction of 60%, a diffusion constant of 30 $\mu m^2/s$ and a bleach spot radius of 1 μm, but with different association and dissociation rates yielding different types of FRAP behavior: a diffusion-uncoupled FRAP [(A), $k^*_{on} = 0.1, k_{off} = 0.1$], an effective-diffusion FRAP [(B), $k^*_{on}/k_{off} = 100$] and a full-model FRAP

to determine the association rate, which for the diffusion-uncoupled case is given by

$$k^*_{on} = \frac{k_{off}\phi_{app}}{\phi_{true} - \phi_{app}} \tag{4}$$

ϕ_{true} can be estimated by collecting an image within a few milliseconds of the photobleach. If this is not possible with the microscope in use, an immobile GFP-fusion protein (such as the histone H2B for nuclear proteins) can be used instead to estimate the true bleach depth under the conditions used for the FRAP. An alternate approach to estimating k^*_{on} is given by Lele *et al.* (2006). In sum, the diffusion-uncoupled fit to the FRAP directly determines the dissociation rate, and this fit combined with an additional measurement provides an estimate of the association rate.

If the preceding fit fails, then it is possible that there is more than one binding state, namely the GFP-fusion protein may be bound to two different types of molecules, each with their own association and dissociation rates. In this case, the FRAP curve can be fit with

$$\text{frap}(t) = 1 - Ae^{-k_{1off}t} - Be^{-k_{2off}t} \tag{5}$$

Then, the estimated constants A and B can be used with an estimate of the true bleach depth ϕ_{true} (see preceding paragraph) to obtain estimates of the two association rates:

$$k^*_{1on} = \frac{k_{off}A}{\phi_{true} - \phi_{app}} \quad k^*_{2on} = \frac{k_{off}B}{\phi_{true} - \phi_{app}} \tag{6}$$

where ϕ_{app} is the apparent bleach depth given by $\phi_{app} = A + B$. Thus, fitting a FRAP curve with Eq. (5) yields an estimate for the dissociation rates of each of the two

---

[(C), $k^*_{on} = 31.6, k_{off} = 0.35$]. Changing the bleach-spot radius to 2 μm leads to detectably different FRAPs for the two diffusion-coupled cases (E and F), but not for the diffusion-uncoupled case (D). Using Eq. (3) for a diffusion-uncoupled fit yields as expected a good fit to the diffusion-uncoupled data (G), but not such good fits for the diffusion-coupled data (H and I). Similarly, using Eq. (1) for an effective diffusion fit yields as expected a good fit to the effective diffusion data (K), but not such good fits for the full-model data (L) or the diffusion-uncoupled data (J). Using Eq. (9) for a full-model fit yields as expected a good fit for the full-model data (O), and also good fits to the diffusion-uncoupled (M) and effective-diffusion (N) data, since these two are simply limiting cases of the full model. Although the full model is capable of fitting all three types of data, the limiting cases should be tested for the following reasons: (1) the formulas for the limiting cases are simpler and therefore easier to fit; (2) for effective diffusion, a full-model fit will not reveal that only the ratio k^*_{on}/k_{off} can be determined; and (3) for diffusion-uncoupled FRAPs, the rapid diffusive phase may not be recorded and so the full-model fit may fail because it expects to see both the diffusive and the binding phases.

binding states. Then an experimental estimate of the true bleach depth, ϕ_{true}, enables an estimate of the association rates for each of the two binding states using Eq. (6).

Caution is required in the application of the double exponential fit in Eq. (5), as it will fit virtually any FRAP curve regardless of whether the curve is diffusion coupled or uncoupled. Such an inappropriate fit can be ruled out if the FRAP depends on bleach-spot size, since in this case a diffusion-coupled fit must be applied. A second indication that a double-exponential fit may be incorrect is if the measured true bleach depth is very close to the apparent bleach depth. In this case, it is more likely that there is a smooth recovery immediately after the photobleach characteristic of a diffusion-coupled FRAP (Fig. 3B) rather than the sharp rise in fluorescence predicted for the diffusion-uncoupled FRAP (Fig. 3A).

6. Fitting a Diffusion-Coupled FRAP

If FRAP of the GFP-fusion protein reveals a dependence on bleach-spot size, then the next step is to fit the FRAP data with the equation for diffusion (Sections III.B.6.a and b), namely the same equation that was used to fit unconjugated GFP [Eq. (1)]. This equation may or may not yield a good fit to the fusion protein FRAP. If not, then the data should be fit with the full-model equation (Section III.B.6.c).

a. Pure Diffusion (No Binding Interactions of the GFP-Fusion Protein)

If the fit of the GFP-fusion FRAP to the diffusion model is good, then an estimate for the diffusion constant for the fusion protein will be obtained. This estimate can be compared to the estimated value for the diffusion constant obtained independently by extrapolation from the diffusion constant for GFP [Eq. (2)]. If these two numbers are within experimental error, then it indicates two things: first, the fusion protein is not part of a larger complex; and second, the fusion protein is freely diffusing throughout the cellular compartment of interest without undergoing any significant binding interactions.

It is also possible that the diffusion model will yield a good fit, but the estimated diffusion constant obtained for the fusion protein may be significantly smaller than that obtained independently by extrapolation from the diffusion constant for GFP [Eq. (2)]. This could mean that the fusion protein is part of a larger complex that is freely diffusing without engaging in any binding interactions. The mass of this putative, freely diffusing complex can be determined from Eq. (2) using the estimated value for D from the fit of the fusion protein's FRAP, the estimated value for D from unconjugated GFP's FRAP, and the known mass of GFP (27 kDa). If this predicted mass for the fusion protein complex yields a size that is plausible for a molecular complex, then this is a reasonable explanation for the fusion protein's FRAP behavior. However, in many cases, the predicted mass is implausibly large indicating that binding interactions as described below are responsible for the slowed FRAP recovery.

b. Effective Diffusion (Binding Interactions That Mimic Slow Diffusion)

FRAP of the fusion protein may be well fit by the diffusion model [Eq. (1)], even though the fusion protein is both diffusing and binding. Such "effective diffusion" behavior is predicted mathematically when the time to associate with a binding site is very fast compared to the time to diffuse across the bleach spot (Beaudouin *et al.*, 2006; Kaufman and Jain, 1990; Lele *et al.*, 2004; Sprague *et al.*, 2004). In this case, the fit no longer yields an estimate of the pure diffusion constant but rather an effective diffusion constant D_{eff} that can be used to determine the ratio of the association to dissociation rates of the binding reaction:

$$\frac{k_{\text{on}}^*}{k_{\text{off}}} = \left(\frac{D_{\text{GFP-fusion}}}{D_{\text{eff}}}\right) - 1 \tag{7}$$

where $D_{\text{GFP-fusion}}$ is the estimate of the fusion protein's diffusion constant extrapolated from unconjugated GFP.

There are several precautions that should be considered in interpreting effective diffusion. First, effective diffusion can also arise in the presence of multiple binding states, as long as each state has fast association kinetics relative to free diffusion. In this case, the value obtained for $k_{\text{on}}^*/k_{\text{off}}$ is actually the average of this ratio for the two or more states. Second, if D_{eff} is close to $D_{\text{GFP-fusion}}$ and predicts a reasonable size for a molecular complex (Section III.B.6.a), then the FRAP data by themselves cannot distinguish between two equally plausible possibilities: (1) pure diffusion of a complex containing the fusion protein, or (2) effective diffusion, namely weak binding behavior, of a fusion protein not in a molecular complex.

c. Full Model (Binding Interactions That Do Not Mimic Slow Diffusion)

FRAP of the GFP-fusion protein may be diffusion coupled but may not be well fit with a diffusion model. This is strong evidence that binding interactions are likely to be present and influencing the FRAP.

A full-model FRAP arises when the time to associate with a binding site is roughly comparable with the time it would take to freely diffuse across the bleach spot. This balance between diffusion and binding means that neither process can be ignored or simplified, and so unlike the previous cases there is no analytical expression for the FRAP. However, an expression for the Laplace transform of the solution is available, and so this transform can then be numerically inverted to yield the predicted FRAP recovery. The process of numerical inversion can be done in less than a second on a computer. In practice then, the full-model fit to FRAP data is performed by supplying the computer with the Laplace transform solution, frap(p), and then calling a program that inverts this transform, an operation symbolized by L^{-1}.

The full-model Laplace transform solution is

$$\overline{\mathrm{frap}(p)} = \left(\frac{1 - C_{\mathrm{eq}}}{p}\right)\left(1 + \frac{k^*_{\mathrm{on}}}{p + k_{\mathrm{off}}}\right)(2K_1(qw)I_1(qw)) \tag{8}$$

where p is the Laplace variable which on inversion is converted into time, w is the bleach-spot radius, $C_{\mathrm{eq}} = k^*_{\mathrm{on}}/(k^*_{\mathrm{on}} + k_{\mathrm{off}})$, and $q^2 = [\mathrm{p}/(D_{\mathrm{GPP-fusion}})][1 + (k^*_{\mathrm{on}}/(p + k_{\mathrm{off}}))]$. Note that Eq. (8) contains only two unknown parameters, the binding association and disassociation rates, $k^*_{\mathrm{on}}$and k_{off}. Estimates for these can be obtained by fitting the actual FRAP data [frap(t)] with the following equation which also accounts for the experimental bleached fraction:

$$\mathrm{frap}(t) = 1 - \phi + \phi\left[L^{-1}\left(\overline{\mathrm{frap}(p)}\right)\right] \tag{9}$$

Fitting experimental FRAP data with Eq. (9) usually requires a good starting guess, otherwise many fitting programs will remain trapped at a value close to the initial guess, which is often far from the best solution. A simple method to identify a good starting guess is to perform a search over a range of $k^*_{\mathrm{on}}$and k_{off} values. Typically, this can be accomplished by varying both k^*_{on} and k_{off} in tenfold increments from 10^{-5} to 10^5. For each $(k^*_{\mathrm{on}}, k_{\mathrm{off}})$ pair tested, the sum of squared residuals is calculated between the predicted FRAP curve and the experimental data. A contour plot of these residuals as a function of $(k^*_{\mathrm{on}}, k_{\mathrm{off}})$ should exhibit a valley in the region where the best estimate will be found. Several $(k^*_{\mathrm{on}}, k_{\mathrm{off}})$ values within this valley can then be selected as initial guesses for the fitting routine. Generally, all of these guesses will converge to nearly identical $(k^*_{\mathrm{on}}, k_{\mathrm{off}})$ values, thereby yielding the best estimate for these two binding parameters.

If this full-model fit fails, then it is possible that there are two binding states present. An equation to describe a full-model FRAP curve is also available in this case [Eq. (55) in Sprague *et al.* (2004)], however a grid search over four variables $(k^*_{1\mathrm{on}}, k_{1\mathrm{off}}, k^*_{2\mathrm{on}}, k_{2\mathrm{off}})$ becomes very time consuming, and in most cases is not practical without specialized computer hardware.

A consistency check that can be performed on full-model FRAP data is to repeat the FRAP using a different bleach-spot size, and then fit this new curve using the new value for w to determine if similar estimates are obtained for the association and dissociation rates. This strategy is also recommended for the case where the diffusion-coupled data exhibit effective diffusion, in which case only the ratio of association to dissociation rate can be estimated. Here, a smaller bleach-spot size may sometimes convert the FRAP curve into a full-model curve (Sprague *et al.*, 2004), in which case independent estimates of association and dissociation rates are possible. The ratio of these predicted rates obtained from the full-model fit should then be equal to that found for the effective diffusion fit at the larger spot size.

IV. Materials

In addition to both a confocal microscope and cells expressing GFP and the GFP-fusion protein, software is also required for the quantitative analysis of FRAP data. Image-analysis software is needed for averaging image intensities within the bleach spot. Often these routines are available with the confocal microscope software. If not, the same procedures can be performed in virtually any image-analysis package, such as ImageJ. The resultant quantitative data can be imported into a spreadsheet program, such as Excel, in order to perform the normalization and averaging procedures outlined in Fig. 1. Finally, these data must be fitted to the various model equations as outlined in Fig. 2. Here, a software program such as MATLAB is required that can implement the equations. Specifically, this software must include routines to calculate exponential functions, Bessel functions, and perform inverse Laplace transforms, and also include a fitting algorithm (such as nlinfit in MATLAB) that can determine the combination of model parameters that produces the best match to the experimental FRAP curve.

V. Discussion

A. Utility of Parameters Estimated by the Models

Correct fitting of experimental FRAP data provides estimates for the association and dissociation rates of binding, k^*_{on} and k_{off}. These numbers can be compared to *in vitro* estimates of binding as a way to judge if *in vitro* biochemistry has accurately captured what is transpiring *in vivo*. However, in comparing *in vivo* and *in vitro* estimates, it is important to realize that the association rate, k^*_{on}, measured by FRAP is actually the product of the molecular on rate times the concentration of free binding sites in the cell, and so direct determination of the on rate by FRAP also requires an estimate of the *in vivo* concentration of free binding sites. With this in mind, if large differences are found between the *in vivo* and *in vitro* estimates for either k^*_{on} or k_{off}, it suggests that key molecular or structural components are absent from the *in vitro* system. This in turn highlights a need to identify the missing components and reconstitute an *in vitro* system that accurately mimics the *in vivo* situation.

As an example, this has already been accomplished to some degree in the case of transient glucocorticoid receptor binding to a promoter, where previous *in vitro* measurements using only naked DNA yielded more than two orders of magnitude longer residence times than the *in vivo* estimates. Now, with a reconstituted system including chromatin and chromatin remodeling factors, an order of magnitude decrease in the residence time has been found (Nagaich *et al.*, 2004).

The estimates of the *in vivo* association and dissociation rates permit calculation of several other biologically relevant parameters. The inverse of the association rate ($1/k^*_{on}$) is the average time it takes a molecule to find a binding site, and the inverse of the dissociation rate ($1/k^*_{\text{off}}$) is the average time a molecule remains at a binding site, namely the residence time. The association and dissociation rates can also be used to estimate the fraction of molecules that are free $[F_{eq} = k_{\text{off}}/(k^*_{\text{on}} + k_{\text{off}})]$ and the fraction of molecules that are bound $[C_{\text{eq}} = k^*_{\text{on}}/(k^*_{\text{on}} + k_{\text{off}})]$. Knowledge of these fractions, as well as the times to bind and unbind, offers a substantive picture of the molecular dynamics of the binding process.

B. Accuracy of Estimated Parameters

The prospect of measuring *in vivo* binding parameters is the motivating force behind quantitative FRAP. This is a field still in its infancy. Consequently, there are no established techniques or benchmark live-cell systems to validate the estimates of *in vivo* binding parameters obtained by either FRAP or complementary techniques such as FCS.

As a result, it remains uncertain whether the *in vivo* estimates of binding produced by any of these methods are correct. The reason for this uncertainty is that in FRAP (and also FCS) assumptions must be made in developing a model that will be used to fit experimental data. These assumptions in FRAP have in the past included neglecting a role for diffusion, ignoring the effects of boundaries or a finite cell volume, presuming that diffusion is simple and nonanomalous, and presuming an instantaneous and cylindrical photobleach. As this field develops, more and more of these assumptions are being addressed, thus leading to increasingly refined FRAP models.

These better FRAP models will yield improved estimates of binding parameters, and in addition more information on which current assumptions are reasonably accurate and which are seriously inaccurate. Although, this will increase our confidence in these estimates, real certainty will only come as complementary approaches such as FCS are also applied to the same live-cell systems and agreement in the estimates is achieved by multiple groups. This will finally provide some benchmark systems and well-accepted *in vivo* binding parameters that will firmly establish these procedures as reliable methods for assaying *in vivo* binding.

VI. Summary

Described herein is a method for acquiring FRAP data, followed by a step-by-step procedure for fitting the data to a series of increasingly complex mathematical models. Successful application of this procedure will yield estimates of the *in vivo* association and dissociation binding rates for the GFP-tagged protein under study. The protein is presumed to bind to an immobilized substrate that is uniformly distributed throughout a cellular compartment, and diffusion within the compartment is presumed to be simple and not anomalous, although more complex scenarios can also be treated by analogous approaches. As these methods become increasingly refined, increasingly accurate and reliable estimates of cellular binding parameters will be possible.

Acknowledgments

I thank Waltraud Müller and Tim Stasevich for comments on the manuscript, and Brian Sprague for suggesting the original format for Fig. 2. I also thank Florian Müller for help with the derivations of the diffusion-uncoupled FRAPs and for comments on the manuscript.

References

Arrio-Dupont, M., Foucault, G., Vacher, M., Devaux, P. F., and Cribier, S. (2000). Translational diffusion of globular proteins in the cytoplasm of cultured muscle cells. *Biophys. J.* **78,** 901–907.

Beaudouin, J., Mora-Bermudez, F., Klee, T., Daigle, N., and Ellenberg, J. (2006). Dissecting the contribution of diffusion and interactions to the mobility of nuclear proteins. *Biophys. J.* **90,** 1878–1894.

Bhalla, U. S., Ram, P. T., and Iyengar, R. (2002). MAP kinase phosphatase as a locus of flexibility in a mitogen-activated protein kinase signaling network. *Science* **297,** 1018–1023.

Braga, J., Desterro, J. M., and Carmo-Fonseca, M. (2004). Intracellular macromolecular mobility measured by fluorescence recovery after photobleaching with confocal laser scanning microscopes. *Mol. Biol. Cell* **15,** 4749–4760.

Carrero, G., Crawford, E., Hendzel, M. J., and de Vries, G. (2004). Characterizing fluorescence recovery curves for nuclear proteins undergoing binding events. *Bull. Math. Biol.* **66,** 1515–1545.

Cho, E. H., and Lockett, S. J. (2006). Calibration and standardization of the emission light path of confocal microscopes. *J. Microsc.* **223,** 15–25.

Coscoy, S., Waharte, F., Gautreau, A., Martin, M., Louvard, D., Mangeat, P., Arpin, M., and Amblard, F. (2002). Molecular analysis of microscopic ezrin dynamics by two-photon FRAP. *Proc. Natl. Acad. Sci. USA* **99,** 12813–12818.

Edidin, M. (1994). Fluorescence photobleaching and recovery, FPR, in the analysis of membrane structure and dynamics. *In* "Mobility and Proximity in Biological Membranes" (S. Damjanovich, M. Edidin, J. Szollosi, and L. Tron, eds.), pp. 109–135. CRC Press, Boca Ration, FL.

He, X., Ding, Y., Li, D., and Lin, B. (2004). Recent advances in the study of biomolecular interactions by capillary electrophoresis. *Electrophoresis* **25,** 697–711.

Hinow, P., Rogers, C. E., Barbieri, C. E., Pietenpol, J. A., Kenworthy, A. K., and DiBenedetto, E. (2006). The DNA binding activity of p53 displays reaction-diffusion kinetics. *Biophys. J.* **91,** 330–342.

Janes, K. A., and Lauffenburger, D. A. (2006). A biological approach to computational models of proteomic networks. *Curr. Opin. Chem. Biol.* **10,** 73–80.

Kaufman, E. N., and Jain, R. K. (1990). Quantification of transport and binding parameters using fluorescence recovery after photobleaching. Potential for *in vivo* applications. *Biophys. J.* **58,** 873–885.

Leavitt, S., and Freire, E. (2001). Direct measurement of protein binding energetics by isothermal titration calorimetry. *Curr. Opin. Struct. Biol.* **11,** 560–566.

Lele, T., Oh, P., Nickerson, J. A., and Ingber, D. E. (2004). An improved mathematical approach for determination of molecular kinetics in living cells with FRAP. *Mech. Chem. Biosyst.* **1,** 181–190.

Lele, T., Wagner, S. R., Nickerson, J. A., and Ingber, D. E. (2006). Methods for measuring rates of protein binding to insoluble scaffolds in living cells: Histone H1-chromatin interactions. *J. Cell. Biochem.* **99,** 1334–1342.

Liebman, P. A., and Entine, G. (1974). Lateral diffusion of visual pigment in photoreceptor disk membranes. *Science* **185,** 457–459.

Misteli, T. (2001). Protein dynamics: Implications for nuclear architecture and gene expression. *Science* **291,** 843–847.

Nagaich, A. K., Walker, D. A., Wolford, R., and Hager, G. L. (2004). Rapid periodic binding and displacement of the glucocorticoid receptor during chromatin remodeling. *Mol. Cell* **14,** 163–174.

Perlmann, T., Eriksson, P., and Wrange, O. (1990). Quantitative analysis of the glucocorticoid receptor-DNA interaction at the mouse mammary tumor virus glucocorticoid response element. *J. Biol. Chem.* **265,** 17222–17229.

Phair, R. D., Gorski, S. A., and Misteli, T. (2004). Measurement of dynamic protein binding to chromatin *in vivo*, using photobleaching microscopy. *Methods Enzymol.* **375,** 393–414.

Poo, M., and Cone, R. A. (1974). Lateral diffusion of rhodopsin in the photoreceptor membrane. *Nature* **247,** 438–441.

Riggs, A. D., Suzuki, H., and Bourgeois, S. (1970). Lac repressor-operator interaction. I. Equilibrium studies. *J. Mol. Biol.* **48,** 67–83.

Saxton, M. J. (2001). Anomalous subdiffusion in fluorescence photobleaching recovery: A Monte Carlo study. *Biophys. J.* **81,** 2226–2240.

Schlitt, T., and Brazma, A. (2006). Modelling in molecular biology: Describing transcription regulatory networks at different scales. *Philos. Trans. R. Soc. Lond. B Biol. Sci.* **361,** 483–494.

Schmiedeberg, L., Weisshart, K., Diekmann, S., Meyer Zu, H. G., and Hemmerich, P. (2004). High- and low-mobility populations of HP1 in heterochromatin of mammalian cells. *Mol. Biol. Cell* **15,** 2819–2833.

Schuck, P. (1997). Use of surface plasmon resonance to probe the equilibrium and dynamic aspects of interactions between biological macromolecules. *Annu. Rev. Biophys. Biomol. Struct.* **26,** 541–566.

Seiffert, S., and Oppermann, W. (2005). Systematic evaluation of FRAP experiments performed in a confocal laser scanning microscope. *J. Microsc.* **220,** 20–30.

Sinnecker, D., Voigt, P., Hellwig, N., and Schaefer, M. (2005). Reversible photobleaching of enhanced green fluorescent proteins. *Biochemistry* **44,** 7085–7094.

Soumpasis, D. M. (1983). Theoretical analysis of fluorescence photobleaching recovery experiments. *Biophys. J.* **41,** 95–97.

Sprague, B. L., and McNally, J. G. (2005). FRAP analysis of binding: Proper and fitting. *Trends Cell Biol.* **15,** 84–91.

Sprague, B. L., Müller, F., Pego, R. L., Bungay, P. M., Stavreva, D. A., and McNally, J. G. (2006). Analysis of binding at a single spatially localized cluster of binding sites by fluorescence recovery after photobleaching. *Biophys. J.* **91,** 1169–1191.

Sprague, B. L., Pego, R. L., Stavreva, D. A., and McNally, J. G. (2004). Analysis of binding reactions by fluorescence recovery after photobleaching. *Biophys. J.* **86,** 3473–3495.

Stavreva, D. A., Müller, W. G., Hager, G. L., Smith, C. L., and McNally, J. G. (2004). Rapid glucocorticoid receptor exchange at a promoter is coupled to transcription and regulated by chaperones and proteasomes. *Mol. Cell. Biol.* **24,** 2682–2697.

Swaminathan, R., Hoang, C. P., and Verkman, A. S. (1997). Photobleaching recovery and anisotropy decay of green fluorescent protein GFP-S65T in solution and cells: Cytoplasmic viscosity probed by green fluorescent protein translational and rotational diffusion. *Biophys. J.* **72,** 1900–1907.

von Dassow, G., Meir, E., Munro, E. M., and Odell, G. M. (2000). The segment polarity network is a robust developmental module. *Nature* **406,** 188–192.

Webb, D. J., Brown, C. M., and Horwitz, A. F. (2003). Illuminating adhesion complexes in migrating cells: Moving toward a bright future. *Curr. Opin. Cell. Biol.* **15,** 614–620.

Weiss, M. (2004). Challenges and artifacts in quantitative photobleaching experiments. *Traffic* **5,** 662–671.

Weiss, M., Elsner, M., Kartberg, F., and Nilsson, T. (2004). Anomalous subdiffusion is a measure for cytoplasmic crowding in living cells. *Biophys. J.* **87,** 3518–3524.

CHAPTER 15

Quantitative and Qualitative Analysis of Plant Membrane Traffic Using Fluorescent Proteins

Marketa Samalova, Mark Fricker, and Ian Moore

Department of Plant Sciences
University of Oxford
Oxford OX1 3RB
United Kingdom

0091-679X/08 $35.00
DOI: 10.1016/S0091-679X(08)85015-7

Abstract

Fluorescent proteins have had a great impact on the way in which plant membrane traffic is studied. Here we review the uses to which these molecules have been put in this field of research and discuss the advantages and pitfalls of particular fluorescent protein derivatives in various applications and plant species. We discuss in detail the need for quantitative estimates of expression level and the potential of fluorescent proteins for quantitative assays of biosynthetic membrane traffic. Detailed descriptions and protocols are provided for the use of the newly developed 2A-based ratiometric polyprotein probes of membrane traffic in conjunction with semiautomated image analysis software packages for quantitative analyses. The ratiometric probes and software are available from the authors.

I. Introduction

In recent years, studies of membrane traffic and endomembrane organization in plant cells have made increasing use of intrinsically fluorescent proteins (IFPs) to visualize several endomembrane organelles (Brandizzi *et al.*, 2002, 2004; Zheng *et al.*, 2005) and to isolate mutants with altered endoplasmic reticulum (ER) morphology (Matsushima *et al.*, 2003; Tamura *et al.*, 2005). Green fluorescent protein (GFP) has also been used as a marker of biosynthetic traffic to the apoplast or the vacuole in plants (Batoko *et al.*, 2000; Boevink *et al.*, 1998; DaSilva *et al.*, 2004, 2005; Flückiger *et al.*, 2003; Geelen *et al.*, 2002; Kotzer *et al.*, 2004; Lee *et al.*, 2004; Sohn *et al.*, 2003; Takeuchi *et al.*, 2000; Zheng *et al.*, 2004, 2005). As GFP fails to accumulate in a fluorescent form in either destination (Batoko *et al.*, 2000; Boevink *et al.*, 1999; Tamura *et al.*, 2003), perturbation of anterograde traffic is readily visualized by the accumulation of fluorescence in upstream compartments such as ER, the Golgi apparatus, or prevacuolar compartments (PVC). This strategy is effective in transformed mutant *Arabidopsis* seedlings (Tamura *et al.*, 2005; Zheng *et al.*, 2004) but has been used most frequently in transient expression studies to investigate the effect of genetically dominant derivatives of putative membrane trafficking proteins or of inhibitors. Thus in principle IFP-based assays can provide qualitative morphological information on the steps of the pathway that are disrupted and quantitative information on the extent to which trafficking is disrupted. Genuinely quantitative studies are rare however.

The utility of fluorescent proteins in each of these applications is critically dependent on several of their photochemical and biological properties. Table I summarizes important properties of several fluorescent proteins that are commonly used for membrane trafficking research. Commercially available proteins now span almost the entire visible range and some noncommercial alternatives or derivatives offer significant advantages. These are discussed further below.

A. Photochemical and Biological Properties of IFPs

The native GFP coding sequence was prone to misexpression in *Arabidopsis* owing principally to the presence of a cryptic intron. This was eliminated by the site-directed mutagenesis in a series of GFP derivatives (mGFP4-mGFP6) generated by Jim Haseloff and colleagues (http://www.plantsci.cam.ac.uk/Haseloff/imaging/GFP.htm). However, the codon modifications introduced into the commercial IFPs for optimization of expression in animal cells also eliminate missplicing, and these can be expressed efficiently in *Arabidopsis* and other plant cells. The commercial IFPs, with the exception of DsRed, can also be used effectively in the plant endomembrane system. DsRed exists as a tetramer, and this property often causes fusion proteins to aggregate in a concentration-dependent manner. This can cause significant disruption to endomembrane organelles. Although DsRed can be used with caution (Saint-Jore *et al.*, 2002), the monomeric derivative mRFP1 (Campbell *et al.*, 2002) is preferable by far and has been used in several studies to mark endomembrane organelles (Lee *et al.*, 2002; Samalova *et al.*, 2006; Zheng *et al.*, 2005).

In common with other applications of IFPs, an important consideration is the relative brightness of each protein. Yellow fluorescent protein (YFP) is 50% brighter than GFP and at least three times brighter than cyan fluorescent protein (CFP) or mRFP1. This is important in localization studies as the intrinsic brightness of a protein establishes a minimal abundance for detection with any given imaging system, and this may or may not be similar to the native abundance of the tagged molecule. Thus it will be possible to image a fluorescent fusion to YFP at threefold lower expression levels than the equivalent red fluorescent protein (RFP) or CFP fusion, with correspondingly reduced risks of overexpression artifact. Another advantage to the use of YFP is the existence of diverse GFP-based fusions that can be used for colocalization studies with imaging systems that allow these two fluors to be discriminated effectively. Indeed, one of the major applications of IFPs is in colocalization studies involving multifluorescent imaging of two or more proteins.

Satisfactory discrimination of IFPs using conventional mirrors and filters is usually dependent on selective excitation as well as selective detection. Two non-commercial IFP derivatives offer significant advantages in this regard. The first of these is mGFP5 (Table I) that exhibits an increase in the 480 nm excitation peak but without either the redshift or loss of the 400 nm peak that characterize EGFP. This provides more efficient excitation with the 458 and 405 nm lasers on confocal systems, allowing for increased discrimination from YFP or mRFP1. For these reasons, we routinely use mGFP5 for our GFP work. The second protein with useful spectral properties is mRFP1, which is redshifted for emission and excitation relative to DsRed (Table I). This provides for more efficient discrimination from YFP and GFP which each have long emission tails and excitation spectra that overlap the short-wavelength shoulder of the DsRed and mRFP1 spectra. A derivative, mCherry, offers similar spectral characteristics but with improved brightness and photostability and is now likely to be the protein of choice (Table I).

For most combinations of IFP, sequential excitation is required to discriminate the two fluors. When imaging organelles in live plant cells, this requires the use of a

Table I
Photochemical and Biological Properties of Fluorescent Proteins

FPs	Ex_{max} (nm)	Em_{max} (nm)	Relative brightness	pK_a	Ex laser (nm)	Note	Ref.
ECFP (CFP)	430	475	3	<5.0	405,458	–	Clontech
mGFP5	405/477	508	?	<5.0	405,458,488,(477)	Maybe dimmer than EGFP	[a]
EGFP	488	508	8	5.5	458,488,(477)	–	Clontech [d]
EYFP (YFP)	514	527	12	7.0	488,514	Cl^- sensitive	Clontech [d]
Venus YFP	515	528	13	6.1	488,514	Less Cl^- sensitive but also less photostable	[b,d]
DsRed	550	580	15	<5.0	–	aggregates	Clontech
mRFP1	580	610	3	<5.0	514,543(488)	Protease insensitive	[c]
mCherry	587	610	4	<4.5	543	–	[d]
mPlum	590	649	1	<4.5	–	–	[d]

[a]Haseloff, J. <http://www.plantsci.cam.ac.uk/Haseloff/imaging/GFP.htm>
[b]Nagai *et al.* (2002).
[c]Campbell *et al.* (2002).
[d]Shaner *et al.* (2005).

confocal laser scanning microscope with line-sequential scanning capability. This is because streaming in the cytoplasm of vacuolate plant cells is too fast (ca. 5 μm/s) for frame sequential imaging, even with electronic switching between detection channels. In the absence of a line-sequential scanning system, unless cells are fixed, only a limited range of IFP combinations can be used for simultaneous detection. These include CFP with either YFP or mRFP1, and mGFP5 with mRFP1. However, in our experience the long emission tail of mGFP5 results in significant bleed-through when its abundance is high relative to that of mRFP1, and we usually use a line-sequential scanning routine with 458 nm (mGFP5) and 543 nm (mRFP1) excitation lines when imaging with this pair. The redshifted, normalized excitation spectrum of EGFP relative to mRFP1 predicts that bleed-through may be greater with this fluor (Fig. 1).

With sequential excitation at 458 and 514 nm, it is possible to discriminate mGFP5 and YFP satisfactorily in most applications (Samalova *et al.*, 2006). The configuration we routinely use for this on a Zeiss LSM510 confocal is outlined in Fig. 2. When mGFP5 signals are weak relative to YFP, bleed-through into the GFP channel can nevertheless be an issue as YFP is excited to about 3% of maximum by the 458 nm line and a correction may be required. Alternatively, if a 405 nm excitation line is available, this can in principle be used to selectively excite mGFP5, but we have found that some plant cells such as tobacco leaf epidermis exhibit significant autofluorescence in the GFP range when illuminated by 405 nm light. Sequential excitation with 458 and 514 nm lines can also be used to image EGFP and YFP within a narrow range of relative abundance at least

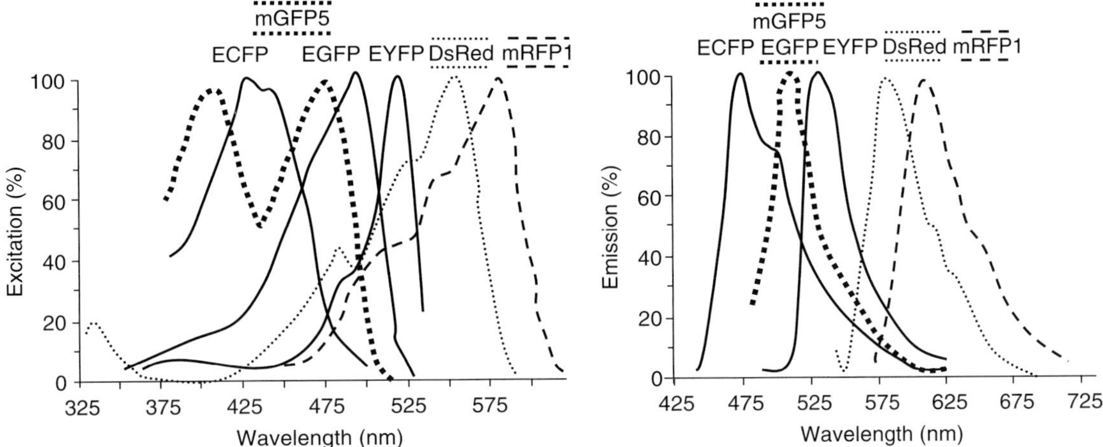

Fig. 1 Excitation and emission spectra of fluorescent proteins. Normalized excitation and emission spectra for proteins discussed in the text. Data extracted from the sources listed in Table I.

(L. Camacho and I. Moore, University of Oxford, Oxford, UK, unpublished). With this combination of fluors, bleed-through from GFP to YFP is also a concern owing to the greater excitation of EGFP at 514 nm, so the EGFP signal in the specimen needs to be relatively weak. The ability to discriminate these fluors at all is perhaps surprising given the normalized spectra shown in Fig. 1. The relative brightness of EGFP and mGFP5 is unknown, however, so it is possible that the weaker excitation of EGFP than mGFP5 by 458 nm light is compensated by an increased brightness. Nevertheless, rigorous bleed-through controls are essential when using EGFP and YFP, so mGFP5 or ECFP are preferable partners for YFP.

An important consideration when working with IFPs in the endomembrane system is the pH sensitivity of their fluorescence. Table I shows that YFP has the highest pK_a, which means it will exhibit minimal fluorescence even in weakly acidic compartments. Consequently, the location of a fusion protein that carries YFP on the luminal face of an endomembrane compartment may be misrepresented by YFP fluorescence. This has been illustrated well by the coexpression of equivalent secreted mGFP5 and YFP fusions in tobacco epidermal cells that revealed both proteins in the ER but only mGFP5 was detectable in post-Golgi compartments and cell wall (Zheng *et al.*, 2005). A variant of YFP, Venus, has a 10-fold lower pK_a and is also effectively chloride insensitive (Table I), so it is preferable to EYFP when these parameters are critical, but note that it is far less photostable than EYFP. Also note that the plant cell wall, vacuole, and PVC can have the pH values well below 6 (Grignon and Sentenac, 1991), so YFP can be used with confidence only in the ER and the Golgi (Zheng *et al.*, 2005) or on the cytoplasmic face of transmembrane proteins. Some membrane proteins carrying GFP derivatives on their cytoplasmic face have been observed to form large inclusions in both plant and animal cells however (Irons *et al.*, 2003; Runions *et al.*, 2006; Snapp *et al.*, 2003).

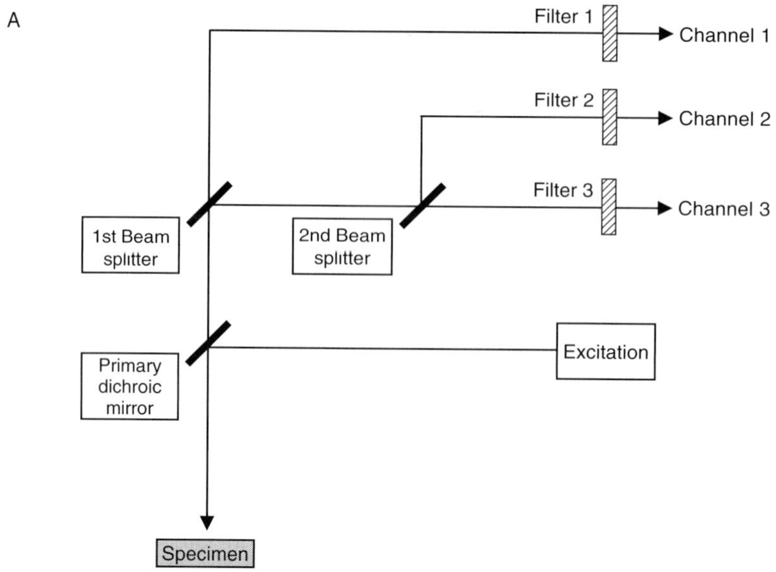

B

FPs	Ex (nm)	Primary dichroic mirror	1st Beam splitter	2nd Beam splitter	Filters	Channels
mGFP5 **YFP** chlorophyll	458 514	HFT 458/514	NFT 635vis	NFT 515	1. LP 650 2. BP 475-525 3. BP 535-590IR	1. chlorophyll (track 1) 2. mGFP5 (track 1) 3. YFP (track 2)
mGFP5 **mRFP1** chlorophyll	458 543	HFT 458/543	NFT 635Vis	NFT 545	1. LP 650 2. BP 475-525 3. BP 585-615	1. chlorophyll (track 1) 2. mGFP5 (track 1) 3. mRFP1 (track 2)
mRFP1 **YFP** **mGFP5**	543 514 405	HFT 458/543	NFT 635Vis	NFT 515	1. BP 592-635 2. BP 475-525 3. BP 535-590IR	1. mRFP1 (track 1) 2. mGFP5 (track 3) 3. YFP (track 2)
mRFP1 **YFP** **mGFP5**	543 514 458	NT 80/20	NFT 545	NFT 515	1. BP 581-635 2. BP 500/20IR 3. BP 535-590IR	1. mRFP1 (track 1) 2. mGFP5 (track 3) 3. YFP (track 2)

Fig. 2 Suggested confocal configurations for dual and triple fluorescent labeling. (A) Simple schematic diagram of confocal optical configuration showing the elements referred to in B. (B) Table listing the multitrack (sequential imaging) configurations used for imaging pairs of fluorescent proteins. Individual laser lines and detection channels are switched on and off in sequence as each y-line in the image is scanned in each track, configurations are taken from Samalova *et al.* (2006).

The pH sensitivity of YFP has been exploited to probe the topology of a plasma membrane (PM) protein as fluorescence is observed only when the fluor is present in an internal loop (Swarup *et al.*, 2004). Immmunodetection can be used to confirm the PM localization of nonfluorescent proteins. A similar strategy may be applied to membrane proteins of other acidic organelles. As discussed below, the pH sensitivity of IFPs can also be exploited to develop visual and quantitative assays of biosynthetic membrane traffic. EGFP is less pH sensitive than EYFP (Table I), but CFP, mGFP5, and particularly mRFP1 are the proteins of choice in this respect.

In addition to the intrinsic pH sensitivity of the fluorophore, GFP derivatives suffer pH- and light-dependent proteolysis in the plant vacuole (Tamura *et al.*, 2003). Similar observations have been made with respect to the cell wall in tobacco, *Arabidopsis*, and onion epidermis (Batoko *et al.*, 2000; Kamiya *et al.*, 2006; Scott *et al.*, 1999; Zheng *et al.*, 2005). In each case, incubating material in the dark or in higher pH media restored fluorescence and protein accumulation in the cell wall or vacuole. The membrane-permeant protease inhibitor E64-d can be used to stabilize GFP in the vacuole of *Arabidopsis* and tobacco leaves (Tamura *et al.*, 2003; Zheng *et al.*, 2005). In the cell wall, a relatively stable degradation product appears to lack ~1 kDa from the C-terminus and the ratio of full-length-to-processed-GFP can therefore act as an indicator of trafficking efficiency to the cell wall (Zheng *et al.*, 2004). In contrast to the sensitivity of GFP derivatives, mRFP1 is apparently resistant to both the pH and the proteolytic environment of the cell wall and vacuole. In our experience, derivatives of mRFP1 targeted to either location have readily been detected, accumulating to high level and remaining stable for several days after transient expression (Samalova *et al.*, 2006; Zheng *et al.*, 2005).

B. Vacuolar Sorting Signals in IFPs

Fluorescent proteins are often fused to plant proteins or protein sorting signals with a view to determining their localization in the endomembrane system. In addition to the usual caveats about the effects of tagging on native sorting signals or on protein folding and ER export, an assumption implicit in these experiments is that the IFP has no active sorting determinants. In the case of plant secretory and vacuolar traffic, this assumption can be unsafe. Zheng *et al.* (2005) have shown that a proportion of a secreted mGFP5 molecule, secGFP, is sorted to the vacuole of tobacco leaf epidermal cells where it can be visualized if the leaves are incubated in the dark with the protease inhibitor E64-d. The secGFP molecule studied by Zheng *et al.* (2005) has a c-myc epitope tag at its C-terminus that could carry the vacuolar sorting signal. However, our unpublished analysis of various GFP and mRFP1 derivatives locates the sorting signal to the GFP moiety (J. Legen and I. Moore, University of Oxford, Oxford, UK, unpublished). Similarly in yeasts, GFP is sorted efficiently to the vacuole by a Vps10-mediated pathway (Kunze *et al.*, 1999). As GFP stability and fluorescence are weak in both the cell wall and the vacuole, the partial targeting of secGFP molecules to the plant vacuole does not compromise its use as a reporter of biosynthetic traffic between the ER and the *trans*-Golgi. By contrast, its use to report inhibition of post-Golgi secretory events may well be compromised by the ability of the protein to escape to the vacuole where it is unstable. Indeed, Zheng *et al.* (2005) showed that a dominant-negative mutant in a Rab-E GTPase, which was proposed to act between the Golgi and the PM, caused a relatively small increase in the secGFP accumulation compared with mutants that inhibited the secretory pathway upstream of the Golgi. At the same time, there was an increase in the amount of secGFP trafficked to the vacuole, so the weaker accumulation of secGFP appears to be attributable in part to its continued traffic to the vacuole when the secretory route is impaired (Zheng *et al.*, 2005).

In contrast to the secGFP marker discussed above, a secreted mRFP1 derivative used in the same study was found exclusively in the cell wall of tobacco leaf epidermis (Zheng *et al.*, 2005) (Fig. 3G). mRFP1 is readily visualized in the vacuole of tobacco epidermal cells when provided with a vacuolar sorting signal (Samalova *et al.*, 2006) (Fig. 3H), so it appears clear that mRFP1 lacks the serendipitous sorting signal of mGFP5. The same observations have been made in transgenic *Arabidopsis* leaves and roots (Samalova *et al.*, 2006; O. Teh and I. Moore, University of Oxford, Oxford, UK, unpublished observations). This, together with the high stability of mRFP1 in both the cell wall and the vacuole, makes mRFP1 the fluorescent protein of choice to study targeting to the vacuole or cell wall or in reporters of vacuolar sorting. It does not however provide the same quantitative change in fluorescence that is associated with the secretion of GFP, so it is a less convenient marker for studies of secretion. Nevertheless pH-sensitive derivatives of mRFP1 are available (Shaner *et al.*, 2005), and these may provide ideal constructs for the assay of biosynthetic traffic to the cell wall.

The situation is however more complex as the vacuolar sorting of IFPs may be dependent on the experimental system used. In transgenic *Arabidopsis* plants, Zheng *et al.* (2004) found no evidence for the vacuolar accumulation of the same secGFP marker that is accumulated in the vacuoles of tobacco leaf epidermal cells. A small proportion of an ER-resident mGFP5 derivative, GFP-HDEL, could however be detected in the vacuole, consistent with reports that escaped ER residents are usually delivered to the vacuole (Tamura *et al.*, 2004). The differing behavior of secGFP in these studies may reflect species differences or the different expression systems (stable transgenic plants vs transient expression). In both these species, secreted mRFP1 molecules were excluded from vacuoles. However, Yang and colleagues report that secreted mRFP1 is transported with high efficiency to the vacuoles of transgenic tobacco BY-2 suspension cultured cells (Yang *et al.*, 2005). The reason for this difference is unclear, but these findings highlight the need for caution when applying IFPs to study trafficking in any new system.

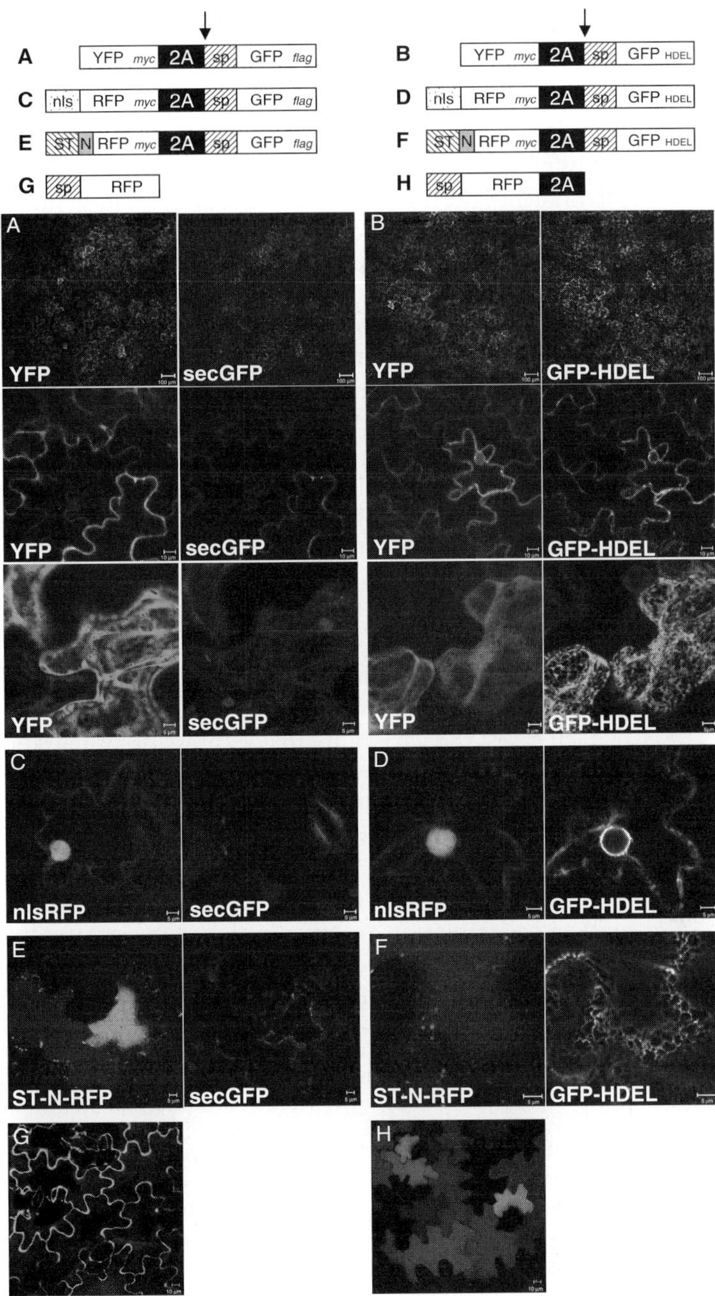

Fig. 3 Selected examples of FMDV-2A-constructs for stoichiometric expression of two proteins. Schematic representation and confocal images of tobacco leaf epidermal cells transiently expressing (A) Y_m-2A-secG_f, (B) Y_m-2A-GH, (C) nlsR_m-2A-secG_f, (D) nlsR_m-2A-GH, (E) STN-R_m-2A-secG_f,

II. Rationale

A. Need to Control Expression Level

In both quantitative and qualitative assays, it is important that the marker does not itself perturb the trafficking process. However, the expression level of a marker can influence the intracellular distribution of itself and other proteins. For example during the early stages of transient expression, secreted, vacuolar, and the Golgi-localized markers are present in the ER to varying degrees, depending on the expression efficiency in individual cells (Batoko *et al.*, 2000; Di Sansebastiano *et al.*, 1998; Flückiger *et al.*, 2003; Zheng *et al.*, 2005). Furthermore, the retrieval of proteins to the ER and sorting to the vacuole is dependent on protein–protein interactions that can become saturated (DaSilva *et al.*, 2005; Frigerio *et al.*, 1998). In transient expression studies, these considerations demand that transfection rates are kept low. Given the stochastic nature of the transfection process, this inevitably leads to wide cell-to-cell variation in marker expression. Consequently, images of individual cells represent essentially anecdotal evidence and the effect of a transgene on trafficking can be inferred only from analyses of large cell populations. In the absence of a simple quantitative metric, performing and presenting such analyses are not straightforward. When transfection efficiencies are low or when high-magnification is required to image changes in marker distribution or organelle morphology, analyses often rely on the subjective scoring of individual cells in the population (Kotzer *et al.*, 2004; Lee *et al.*, 2004; Sohn *et al.*, 2003; Zheng *et al.*, 2005). In such cases, the variability in coexpression of marker and test constructs significantly complicates the analysis. Even when the test construct can be tagged with a visible marker, the expression level of the trafficked marker cannot be known. With higher-efficiency transfection procedures such as *Agrobacterium*-mediated expression in tobacco leaf epidermis, low-magnification images encompassing multiple-transfected cells may be sufficient to illustrate major changes in marker accumulation or distribution between treatments (Batoko *et al.*, 2000;

(F) STN-R_m-2A-GH, (G) secRFP, and (H) secRFP-2A. (A–H) The components of each construct are named and illustrated according to the following scheme: G or GFP = green fluorescent protein (mGFP5); Y or YFP = yellow fluorescent protein; R or RFP = red fluorescent protein (mRFP1); sp = signal peptide; sec = secreted form of fluorescent protein; H or HDEL = presence of endoplasmic reticulum (ER) retrieval signal in addition to signal peptide; nls = nuclear localization signal; ST = Golgi-targeting signal from rat sialyl transferase; N = engineered *N*-glycosylation site (Batoko *et al.*, 2000); $_m$ or myc = c-myc epitope tag; $_f$ or flag = FLAG epitope tag; 2A = foot and mouth disease virus (FMDV) 2A peptide, sequence taken from the constructs described by Halpin *et al.* (1999); the arrow indicates the site at which 2A activity disrupts the polypeptide backbone. The YFP labels cytoplasm (A and B, left) while GFP is secreted out of the cells (A, C, and E, right) or appears in the ER network (B, D, and F, right). The RFP is successfully targeted into nucleus (C and D, left) or Golgi apparatus but unexpectedly accumulates in the central vacuole of cells expressing high levels of the STN-R_m-2A constructs (E and F, left). secRFP accumulates exclusively in the appoplast (G) while secRFP-2A is almost exclusively vacuolar. Shown are single sections (A and B first row; C and D) and projections (A and B, second and third row; E–H). Scale bar 100 μm (A and B, first row), 10 μm (A and B, second row, G and H), and 5 μm (A and B, third row; C–F).

Kotzer *et al.*, 2004; Zheng *et al.*, 2005). We have also used such images to quantify the intracellular accumulation of secreted GFP (Zheng *et al.*, 2005). However, stochastic variation in transient expression efficiency and variable sampling of three-dimensional (3D) space during imaging limit the ability of this approach to resolve differences in marker accumulation between treatments. Furthermore, the approach cannot be applied to individual cells or protoplasts, nor can it be used to quantify changes in the intracellular distribution of markers. Sampling problems associated with the stochasticity of transient expression are exacerbated when individual cells are analyzed at higher magnification and are compounded in differentiated vacuolated cells the by 3D organization of the cytoplasm into a thin-curved cortical layer connected by dynamic transvacuolar strands.

B. Ratiometric Approaches to Quantify Marker Expression and Accumulation

A potential solution to these problems is to provide a stoichiometric baseline reference for the expression efficiency of the trafficked marker. Ideally this should be measurable under the same conditions as the trafficked marker in either transfected protoplasts, single cells, or whole tissues over a broad range of magnifications. This would correct for the variability in marker expression and imaging efficiency while providing a means to normalize between experiments. The availability of spectrally distinct fluorescent proteins and the instrumentation to distinguish their signals facilitate this approach using ratiometric imaging techniques. When cell populations are analyzed, simply cotransfecting the effector or assay construct with a visible marker on the same or separate plasmids can help to normalize for experiment-to-experiment variation in either protoplast or leaf systems (Lee *et al.*, 2002; Samalova *et al.*, 2006). Experiments involving the cotransfection of two visible markers show, however, that this can still result in significant cell-to-cell variation in the relative expression level of each marker (Samalova *et al.*, 2006). This precludes the use of such strategies to determine the expression level of a cotransfected effector or assay construct in individual cells. Here we discuss in detail the utility of a recently developed approach that uses polyproteins based on the foot and mouth disease virus (FMDV) "self-cleaving" 2A peptide (Halpin *et al.*, 1999; Ryan *et al.*, 1999). These polyproteins express the fluorescent marker and a spectrally distinct fluorescent reference marker from a single open-reading frame, which is translated to generate two separate polypeptides in stoichiometric amounts (Fig. 3). The FMDV-2A-based constructs offer greatly improved sensitivity, objectivity, and statistical robustness in the quantitative assays of biosynthetic membrane traffic in plant cells (Samalova *et al.*, 2006). They allow the expression level of a trafficked marker to be inferred from the accumulation of a spectrally distinct fluorescent marker in another cellular compartment so that cells expressing similar and appropriate amounts can be compared. This should help to avoid problems associated with markers becoming missorted or accumulated in upstream compartments simply through overexpression. We describe imaging protocols and analysis tools to quantify marker expression and to facilitate ratiometric membrane trafficking assays with these FMDV-2A-based constructs in cell populations or individual cells of either stable transgenic plants or transfected cell populations.

C. FMDV-2A-Based Ratiometric Assays of Marker Expression and Accumulation

The FMDV 2A peptide is a 20 amino-acid peptide that promotes the separation of the 2A and 2B viral translation products from a polyprotein. It disrupts the polypeptide backbone between the terminal glycine and the proline residues of a highly conserved Pro-Gly-Pro motif at the C-terminus of the 2A sequence. The mechanism is apparently protease independent and occurs early relative to the emergence of the polypeptide from the ribosome exit channel (Ryan *et al.*, 1999). Current models suggest that the FMDV 2A peptide (hereafter referred to simply as 2A) acts as an esterase to hydrolyze the link between the nascent polypeptide and the t-RNA in the ribosome P-site before the formation of the terminal Gly-Pro bond of 2A (Ryan *et al.*, 1999). As the translation of the remainder of the ORF can proceed after 2A-mediated hydrolysis, sequences upstream and downstream of 2A are translated as distinct polypeptides from the same ORF in a fixed stoichiometry. The precise stoichiometry can vary between constructs in a sequence-dependent manner, depending on the frequency of ribosome dissociation at the FMDV 2A sequence (Ryan *et al.*, 1999).

As the two translation products emerge independently from the ribosome, it is expected that targeting information in each one may be processed independently by the cell (de Felipe and Ryan, 2004; de Felipe *et al.*, 2003; El Amrani *et al.*, 2004; Samalova *et al.*, 2006). In accordance with this, it has proven possible to generate 2A-based polyprotein fusions in which a GFP molecule is targeted to the secretory pathway or ER while a coexpressed reference marker is targeted to the cytoplasm or nucleus (Fig. 3). It has also been possible to target two proteins to different compartments of the endomembrane system (Fig. 3), though in this case it was observed that the upstream moiety of the polyprotein was sorted efficiently to the vacuole owing to serendipitous vacuolar sorting determinant within the 19 residues of 2A sequence that remain attached to the C-terminus of the amino-terminal cleavage product (Samalova *et al.*, 2006) (Fig. 3G and H). The 2A-cleavage product was observed in the same PVC as vacuolar-targeted GFP, and its sorting to the vacuole was sensitive to inhibition by a dominant-negative mutant of a Rab-F2 GTPase, suggesting that it followed the conventional route from the Golgi.

D. Quantitative Imaging of Secreted GFP Accumulation Using FMDV-2A-Based Polyproteins

The pH sensitivity of GFP derivatives in the cell wall has allowed secreted GFP molecules to be used to report on biosynthetic membrane traffic in tobacco and *Arabidopsis* tissues (Batoko *et al.*, 2000; Zheng *et al.*, 2004). Reduced traffic to the cell wall is revealed as increased intracellular accumulation with a concomitant increase in the total fluorescence that accumulates in the tissue. While this provides an obvious qualitative visual assay, its quantification is less straightforward, as discussed above. We have developed two approaches to using the FMDV-2A-based constructs to quantify secGFP accumulation. The first uses a cytoplasmic YFP as a reference construct and is most suitable for the analysis of cell populations imaged at relatively low magnification when cytoplasm and ER are not well resolved. The second is applicable to individual cells and uses a nuclear-targeted RFP in conjunction with a specific image analysis package. The imaging protocols below are developed to analyze transfected tobacco leaf epidermal cells transiently expressing the fluorescent markers, but they can be adapted to other types of sample such as protoplasts or transgenic plants. The technical specifications relate to the Zeiss LSM 510 confocal laser scanning microscope.

III. Material

I. Plant material

Nicotiana tabacum SR1 (*cv* Petit Havana) plants grown in potting compost at 20–22 °C constant light for 6–8 weeks.

II. Reagents

A. Bacterial LB medium: 1% (w/v) bacto tryptone, 0.5% (w/v) bacto yeast extract, and 1% (w/v) sodium chloride; pH 7.0; autoclaved.
B. Infiltration medium: 50 mM MES pH 5.6, 0.5% (w/v) glucose, 2 mM Na_3PO_4, and 100 μM acetosyringone (Aldrich) prepared from 1 M stock in DMSO.
C. 5% pleuronic (Molecular Probes)
D. Insulating tape and 3 mm micropore surgical tape

III. CLSM and software

Zeiss LSM 510 laser-scanning microscope or equivalent
Zeiss AIM software version 3.0 or 3.2
Image analysis software available from the authors

IV. Methods

A. Method I: Ratiometric Analysis of GFP Secretion in Populations of Cells Using YFP-2A-secG and YFP-2A-GH

1. Background

YFP-2A-secG and YFP-2A-GH express cytoplasmic YFP in stoichiometric quantities with either secreted or ER-resident GFP, respectively (Fig. 3). When YFP-2A-GH was expressed in tobacco leaf epidermal cells, a close correlation was observed between the YFP and GFP signals in individual cells over a wide range of intensities (Samalova *et al.*, 2006). Using the protocol outlined below, the GFP and the YFP fluorescence intensities can be measured in low-magnification confocal images and used to express the GFP accumulation either as an absolute value or as a ratio normalized to the YFP value for each image. The high correlation between the GFP and YFP accumulation was manifested in a marked improvement in the quality of ratiometric over absolute measurements of GFP accumulation, with the mean coefficient-of-variance being more than twofold lower when the ratio was calculated (Samalova *et al.*, 2006). Furthermore, when Y_m-2A-GH was transfected using a 30-fold range of *Agrobacterium* titres, the absolute GFP accumulation varied ~10-fold but the ratio of GFP:YFP varied only 1.4-fold. The ratio was unchanged above OD_{600} 0.05 and was linear up to this point, suggesting that an OD_{600} close to 0.05 is optimal for the use of constructs like Y_m-2A-GH. The coefficient of variance for the ratio data was threefold lower than that for the absolute fluorescence data at this OD_{600} or less. Thus the 2A-constructs offer an effective strategy for the ratiometric analysis of GFP expression levels in individual cells and in cell populations, with YFP accumulation reliably predicting GFP expression efficiency in each cell.

Y_m-2A-secG_f, which expresses secreted GFP rather than ER-resident GFP, provides a ratiometric assay for biosynthetic traffic. When this construct is expressed in tobacco leaf epidermis, GFP fluorescence is almost undetectable under imaging conditions that detect the GFP signal from the ER of cells expressing Y_m-2A-GH. In cells expressing Y_m-2A-secG_f, inhibition of secGFP traffic is expected to increased GFP fluorescence owing to the accumulation of fluorescent GFP in intracellular compartments (Samalova *et al.*, 2006). For example, traffic between the ER and Golgi can be inhibited by the expression of the dominant-negative N121I mutant of the *Arabidopsis* Rab GTPase *At*Rab-D2[a] (ARA5; AtRab1b; At1g02130). When absolute and ratiometric measurements were used to quantify the accumulation of GFP in leaf areas transiently expressing Y_m-2A-secG_f or Y_m-2A-GH either alone or with *At*Rab-D2[a] [N121I], both measures revealed an increase in secGFP fluorescence but they differed in two notable ways. First, the ratiometric approach revealed that secreted GFP accumulated to levels comparable to that of the ER-resident GFP-HDEL (expressed from Y_m-2A-GH), whereas the absolute expression data in this and previous studies suggested a figure of only 50–75% of GFP-HDEL. This can be explained most simply by a previously undetected reduction in transient expression efficiency in leaf areas coinfiltrated with the *At*Rab-D2[a] [N121I] mutant strain. Second, the Rab mutant also caused a small but significant ($P = 0.05$) increase in the accumulation of GFP-HDEL owing most probably to the escape of some GFP-HDEL molecules from the ER in cells expressing Y_m-2A-GH alone. As in other cases, ratiometric analysis increased the statistical significance of the data as illustrated by the lower coefficients of variance between means of different experiments.

2. Protocol

A. *Agrobacterium tumefaciens*-mediated transient expression

1. Using standard cloning techniques, prepare and transform the desired fluorescent constructs into *Agrobacterium tumefaciens.*
2. Grow the *Agrobacterium* in LB medium supplemented with appropriate antibiotics at 28 °C for 12–24 h.
3. Spin down 1 ml of the culture, wash the pellet in infiltration medium, firstly without and secondly with acetosyringone, and resuspend to an OD_{600} of 0.05–0.06 for 2A-based markers and empirically determined OD_{600} for effector constructs (the lowest OD_{600} that exerts a statistically significant measurable effect is optimal).
4. Infiltrate the bacterial suspension into the lower (abaxial) leaf epidermis of tobacco leaf using a 1 ml of plastic syringe by applying light pressure with a gloved finger at the opposite side of the leaf. Use sections of leaf separated by primary veins for each sample and mark infiltrated (darkened) areas with a permanent pen.
5. Incubate tobacco plants for further 36–60 h at 20–22 °C before examining by CLSM.

B. Sampling and confocal imaging

1. Prepare a slide by placing two 5–10 mm strips of insulating tape ~40 mm apart and place a drop of 0.5% pleuronic in the center.
2. Excise a piece of leaf from within each marked area and mount with the abaxial side upward into the pleuronic.
3. Place a 50 mm cover slip onto the slide ensuring that the edges do not extend beyond the insulating tape.
4. Cut two 60–70 mm pieces of 3 MM micropore tape approximately and attach to the lower side of the microscope slide beneath the edges of the cover slip. Fold the tape over the top of the cover slip to fix it in place. This ensures that the slide and cover slip will remain flat and level on the stage.
5. Gently tap the cover slip to eliminate trapped air bubbles but avoid any saturation of leaf tissue by the mounting fluid.
6. Adjust the quantity of 0.5% pleuronic under the cover slip so that the leaf pieces are surrounded by a thin film of liquid maximizing the air space between them.
7. Set up the microscope configuration for simultaneous imaging of mGFP5 and YFP as summarized in Fig. 2. Adding a chlorophyll channel can also be useful to aid focusing.

8. Use a 10×/0.3 NA objective lens, 0.7–1× zoom and pinhole aperture to give an optical section of 30 μm.
9. Set the detector gains to avoid saturation in the brightest samples and amplifier offset to minimize pixels with a value of 0 in the vacuoles of the dimmest samples in the experiment.
10. Collect at least nine images (12-bit) of each sample and also uninfiltrated areas of the leaf to estimate the background fluorescence. Focusing on the chloroplasts helps to ensure that the plane of focus is at a similar point in the tissue to that of the other samples.

C. Extracting data and calculations

1. Record the average GFP and YFP pixel intensity in each image using the histogram function of the Zeiss AIM software version 3.0 or 3.2.
2. Calculate the average background fluorescence for the GFP and YFP channels from the uninfiltrated sample and subtract this value from the GFP and YFP values measured.
3. Calculate the GFP:YFP ratio for each image by dividing the background subtracted values.
4. Calculate the average GFP:YFP ratio and standard deviation for each sample.

B. Method II: Ratiometric Analysis of Biosynthetic Traffic in Single Cells Using nlsRFP-2A-secG and nlsRFP-2A-GH

1. Background

This method utilizes nlsRFP$_{m}$-2A-secG$_{f}$ of Samalova *et al.* (2006), which expresses a nuclear-targeted mRFP1 in stoichiometreic quantities with a secreted GFP (Fig. 3). An image analysis tool was developed to quantify the intracellular accumulation of GFP and relate it to the expression level in individual cells imaged at high magnification. The ER-resident GFP signal from the nlsRFP$_{m}$-2A-GH construct acted as a positive control for the signal intensity that could be expected when anterograde traffic was inhibited (Fig. 3). The example described below is from transfected tobacco leaf epidermal cells, but this method could be applied equally well to transfected protoplasts or cells of transgenic plants. Similarly, the same method could be adopted with YFP-2A-secG and YFP-2A-GH.

Figure 4 presents a flow chart of the procedures that are implemented by the software, depending on the parameters and options chosen by the user in various check boxes on the interface. The step-by-step instructions are given below, but the procedure is presented in outline here. To measure signals unambiguously from specific cells, analysis is performed on the nucleus and perinuclear region of a series

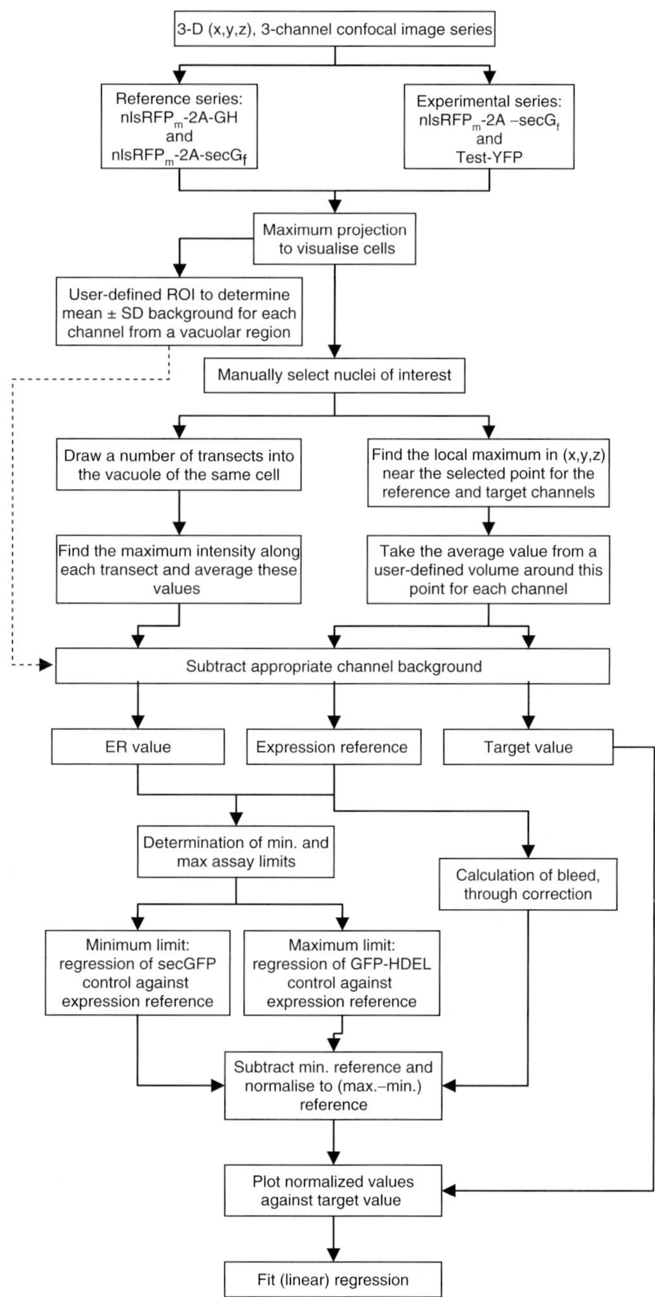

Fig. 4 Flow diagram of the semiautomated ratiometric image-analysis procedure. Flow diagram of the semiautomated ratiometric image-analysis procedure implemented by the software package illustrated in Figs. 5 and 6.

of confocal sections on the *z*-axis. The software imports the *z*-stacks and presents a maximum projection for the interface with the user (Fig. 5). Starting from a position in the nucleus, the user draws a series of transects across the cytoplasm into the surrounding vacuole of the same cell (see Fig. 5). The nuclear RFP intensity is automatically extracted from the 3D data set from a user-defined seed. The corresponding GFP signal is then sampled along the transects. The GFP values can be plotted against the RFP values to give a slope that can be considered a retention index (RI) for any particular set of imaging conditions. For example, in the data shown in Fig. 6, the RI for nlsRFP$_m$-2A-GH was 2.6 and for nlsR$_m$-2A-secG$_f$ it was 0.022. Thus GFP-HDEL accumulated in the ER 100-fold more efficiently than secGFP in these samples. Note that the dynamic range of this analytical method can be orders of magnitude greater than that of Method I which is typically 5–10 fold (Samalova *et al.*, 2006; Zheng *et al.*, 2005). If the output control of the laser on the confocal system is linear, the dynamic range in this assay can also be extended by increasing the laser intensity used for the excitation of GFP and reducing the intensity used for the excitation of RFP when imaging nlsR$_m$-2A-secG$_f$. This allows more sensitive detection of secGFP without saturation of the nlsRFP values. The GFP and RFP values can simply be divided or multiplied, respectively, by the appropriate factors to render the data equivalent to that obtained for nlsR$_m$-2A-GH. No other imaging parameter should be altered however. The AOTF control on the Zeiss LSM 510 is sufficiently linear for this correction to be made. Our ratiometric image analysis package allows laser intensity correction factors to be entered (Fig. 6).

The effect of a coexpressed protein on secreted GFP accumulation can be most easily determined if this protein can be directly visualized by tagging with YFP. The image analysis tool will automatically extract the YFP value for each cell. The YFP intensity can then be plotted against the GFP-to-RFP ratio for individual cells to generate a slope that gives a measure of the inhibition of biosynthetic traffic. This measure is however an arbitrary statistic that is trivially dependent on the imaging parameters chosen and gives no indication of the biological significance of the inhibition measured. To provide a standard against which to compare the relative data generated by this analysis, the secGFP signal in each cell can usefully be expressed as a percentage of the GFP signal that would be expected for a cell expressing nlsR$_m$-2A-GH at the same level. This can be calculated from the known RFP signal for each cell and a plot of GFP against RFP for nlsR$_m$-2A-GH in the same experiment. These percentages can then be plotted against the YFP values for each cell. The software package will perform this calculation by directly using the appropriate control and the reference data within each experiment. Various metrics can be extracted from such a plot to describe the relative effects of different test constructs. This method can detect secGFP accumulation to just 10% of the GFP-HDEL value (Samalova *et al.*, 2006), which would not be measurable using the low-magnification approach owing to the error in the data even with ratiometric analysis. If the YFP-tagged test construct exhibits only

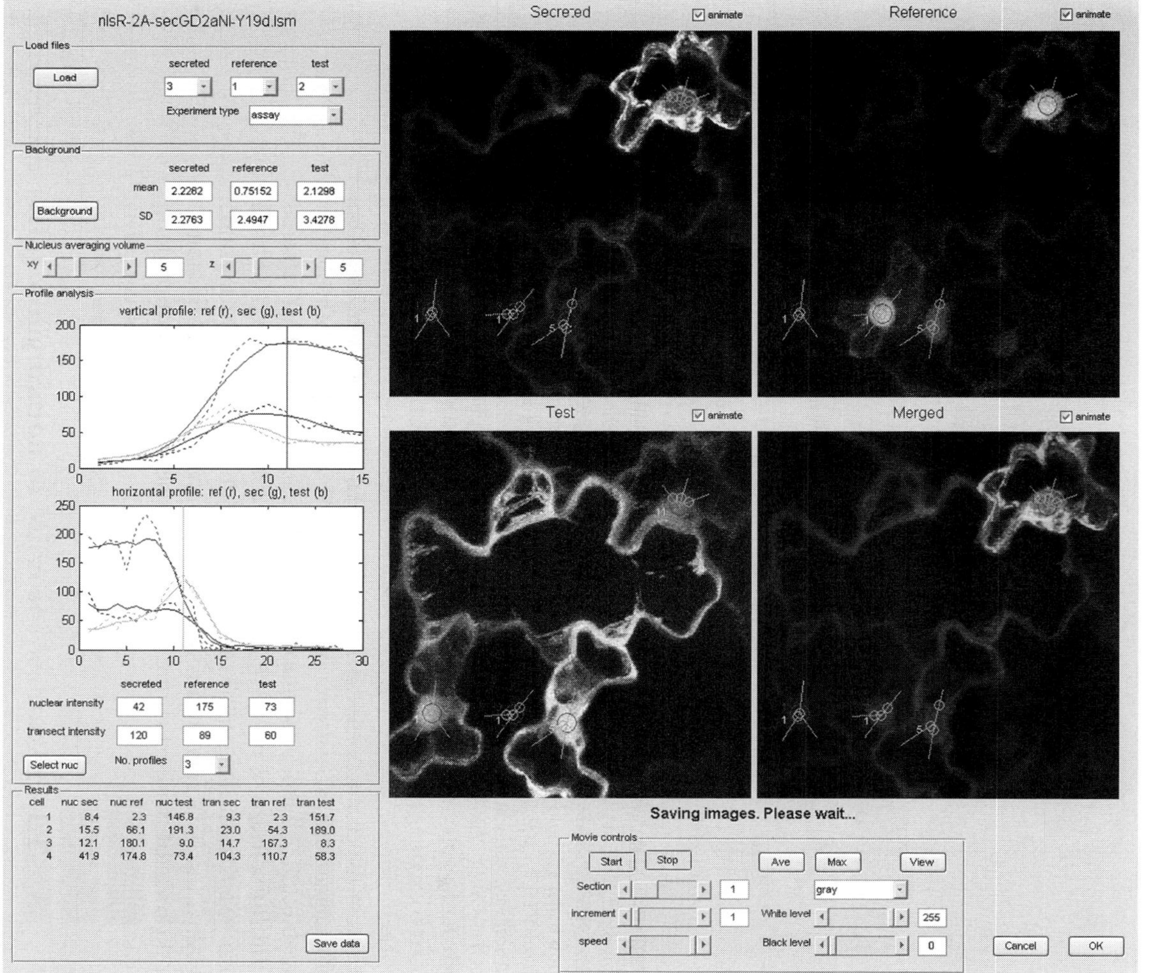

Fig. 5 Screenshot of the data-extraction interface of the image analysis software. Screenshot (reproduced in black and white) of the data-extraction interface of the image analysis software, illustrated with an image from Fig. 8E–H of Samalova *et al.* (2006) as an example. The three user-defined transects from each selected nucleus are indicated in each image. Secreted refers to the secGFP channel, reference to the nlsRFP channel, and test to the YFP-tagged test or effector construct whose influence on secGFP accumulation is under investigation. Various parameters can be set by the user.

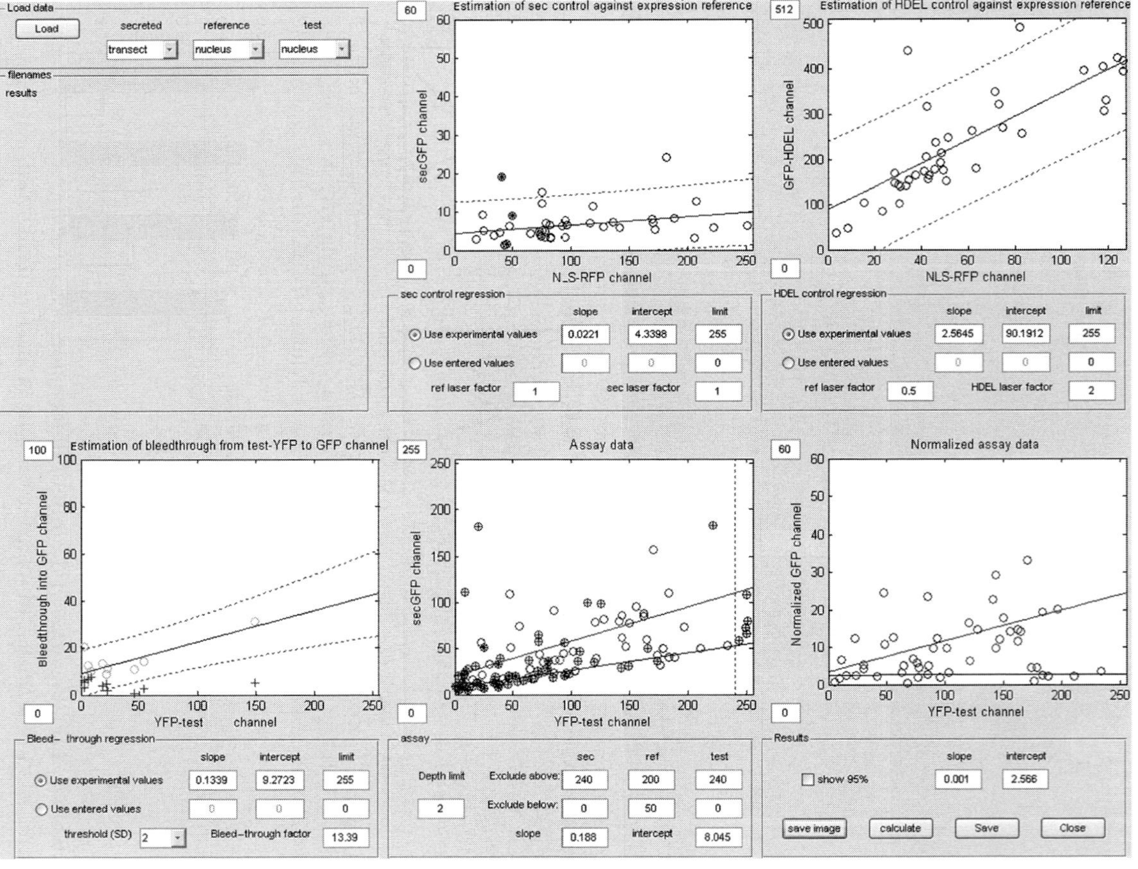

Fig. 6 Screenshot of the data-analysis interface of the ratiometric analysis software. Screenshot of the data-analysis interface (reproduced in black and white) of the ratiometric analysis software in which the relationships between the GFP, RFP, and YFP values can be determined. The top two graphs show the relationship between the nuclear RFP and the secreted or retained GFP to determined factoring in differences in laser intensities used during image acquisition. Bleed-through from YFP into the GFP channel can also be determined (bottom left). The accumulation of secGFP, normalized for nlsRFP data, is plotted in the lower central graph using the bleed-through correction and user-defined limits on the nlsRFP values (i.e., secGFP expression levels) that are considered worth evaluating. The graph at the lower right expresses the secGFP accumulation data relative to the upper (100%) and lower (0%) limits determined by the analysis of cells that express the ratiometric secGFP or GFP-HDEL markers only.

a weak effect on secGFP accumulation, meaningful data will be obtained only from relatively highly expressing cells. Under these circumstances, it is likely that a bleed-through correction will be necessary to eliminate the YFP contribution to the apparent secGFP accumulation values (Samalova *et al.*, 2006). The software allows this to be calculated and factored into the final analysis (Fig. 6). Bleed-through can be calculated either from a sample infected by the YFP-tagged construct only or by instructing the software to extract data from the cotransfected cell population in which the nlsRFP values are below a user-defined threshold indicating that the cell expresses minimal secGFP.

If the protein under investigation cannot directly be visualized, its effect on secGFP accumulation will be more difficult to measure with this method than with the low-magnification method. At sufficiently high coexpression levels, the average RI for a population of cells will give an indication of the strength of the inhibition, but the data is likely to be noisy unless coexpression approaches 100%. At lower coexpression levels, alternative statistical approaches based on the frequency distribution of GFP:RFP values could be used to identify and quantify a shift in the population as a result of marker expression.

2. Protocol

A. *Agrobacterium tumefaciens*-mediated transient expression

See above (Method I).

B. Sampling and confocal imaging

1. Prepare the sample for imaging following steps B1–B6 of Method I above.
2. Set up the microscope configuration for simultaneous imaging of mGFP5, mRFP1, and YFP, as summarized in Fig. 2, using whichever of the two triple-track line-sequential imaging configurations gives the best balance of sensitivities for each fluorescent protein at the requisite expression levels.
3. Use 40× objective lens (C-Apochromat 40×/1.2 NA W corr), 1–1.5 zoom, and 1 Airey unit pinhole aperture (ca. 1 μm optical sections).
4. Set the detector gains to avoid saturation in the brightest samples and amplifier offset to minimize pixels with a value of 0 in the vacuoles of the dimmest samples in the experiment.
5. Collect 3D image stacks (12-bit or 8-bit if the image size is too big) with pixel spacing of 0.3 × 0.3 × 2 μm or 0.22 × 0.22 × 1 μm in *x, y*, and *z*, respectively, of each sample and also uninfiltrated areas of the leaf to estimate the background fluorescence. The number required will depend on the infection rate and the statistical significance required.

C. Extracting data and calculations

1. Import the 3D image stacks into the Matlab environment (The MathWorks, Natick, MA).
2. Spatially, average the images with a user-defined kernel, typically $3 \times 3 \times 3$ or $3 \times 3 \times 5$, in x, y, and z, respectively.
3. Visualize data as a maximum brightness projection.
4. Measure the average background signal from each channel using a user-defined region-of-interest chosen to avoid any morphological features, typically from the vacuole in a single plane of the z-stack.
5. Manually select nuclei in cells of interest from the maximum brightness projection.
6. Extract the location of the brightest pixel in z for the selected (x and y) pixel. In the nuclear region, this value typically corresponds to the midplane of the nucleus.
7. Measure the average nlsRFP and YFP intensity from the nucleus of selected cells at the selected (x, y, and z) pixel coordinates. Although the YFP signal is distributed throughout the cytoplasm, the nucleus is chosen to quantify the YFP intensity, as intensity values here are more homogeneous. Nevertheless, the software does offer a choice between nuclear kernel and transect for extraction of intensity values in each channel.
8. Subtract the appropriate average background signal to give the estimated RFP and YFP signals:
9. Draw a number of user-defined transects into the adjacent vacuole from the seed pixel at the selected z-plane. Each transect thus spans the nuclear envelope and a thin layer of cytoplasm adjacent to the nucleus. The level of ER-localized GFP is estimated from the average of the brightest features along each transect, following the subtraction of the appropriate background. The corresponding YFP signal is also recorded to calculate the bleed-through component (see later).
10. Filter the data to exclude cells whose nuclei exhibit nuclear nlsRFP fluorescence below an arbitrary threshold, typically 50–90, to ensure sufficient marker expression for quantification of secGFP accumulation, or above a maximum limit, typically between 200 and 240 to avoid saturation. Similarly, cells exhibiting the YFP and GFP values near saturation are not considered in the analysis, as it is not possible to quantify these intensities accurately. These limits can be entered at the user interface.
11. Determine the bleed-through correction factor to compensate for the YFP emission in the GFP detection channel using cells expressing only the YFP-tagged test fusion. The absence of the ratiometric nlsR$_m$-2A-secG$_f$ fusion is determined from cells with nuclear nlsRFP pixel intensity within $2 \times$ SD of background.

12. Subtract the estimated bleed-through component from the YFP signal in the GFP channel using the YFP transect estimate multiplied by the correction factor.
13. Determine the minimum and maximum expected limits for the overall assay from cells transfected with nlsR_m-2A-secGFP and nlsR_m-2A-GH, respectively. Note, to bring the signals from nlsR_m-2A-GH into the same range as the test constructs, it may be necessary to reduce the relative intensity of the 458 nm laser for the GFP-HDEL signal by up to twofold, and increase the 543 nm excitation intensity for the nlsRFP signal by a similar amount. As fluorescence brightness scales linearly with changes in laser intensity, the corresponding intensity values are rescaled postcapture by introducing the appropriate factors in the interface boxes for the secGFP and GFP-HDEL data.
14. Fit (linear) regressions to the secGFP and rescaled GH data to determine the relationship between the expected minimum and maximum limits for the assay with the overall expression level of the reporter construct, estimated from the nlsRFP signal. As the biological system can saturate with very high ER-accumulation in the HDEL calibration, it may be necessary to only fit a subset of the GH data. The exclusion limits can be set by the user based on the nlsRFP values for each cell.
15. Normalize the corrected GFP signal to the assay limits determined from the regression equations for the corresponding nlsRFP signal and express as a percentage.
16. Plot the normalized secGFP signal against the level of test construct (the YFP signal) and estimate the relationship using (linear) regression.

All the operations described above have been implemented in the stand-alone MatLab software package available from the authors.

V. Discussion

A. Quantitative Ratiometric Analysis of secGFP Accumulation

While nonratiometric approchaes to secGFP accumulation (Zheng *et al.*, 2005) are suitable for quantifying large changes in secreted GFP accumulation in transfected tobacco epidermis, the two approaches described by Samalova *et al.* (2006) improve these assays in several ways. First, when used in conjunction with the low-magnification imaging approach described by Zheng *et al.* (2005), ratiometric Method I reduces the error associated with the variable transfection rates and the variable sampling of 3D space within and between experiments. It also normalizes for systematic variations in transfection rates between treatments. These allow the influence of a test construct to be established with greater accuracy and statistical support. Second, Method II allows a more powerful quantitative analysis of marker accumulation in individual cells. This procedure has a far higher dynamic range, with the secreted and ER-resident GFP markers exhibiting up to a 1000-fold difference in the slope of a plot of accumulation versus expression, in contrast to the approximately fivefold difference they exhibit in absolute fluorescence signals that are measured by the low-magnification assay described by Zheng *et al.* (2005). The low-dynamic range of the low-magnification approach may result from an underestimate of the background fluorescence in the control samples and this may be an aspect of the method that can be improved. Nevertheless, the ability to distinguish between treatments with very similar means is limited principally by the sampling error in the assay, which demands that multiple independent assays are conducted.

In contrast, the single-cell assay allows numerous data points to be extracted from a single transfection, while the low but accurately measured background and high dynamic range allows much improved estimates of small variations in secreted GFP accumulation. It also allows saturation phenomena to be identified and data to be collected only from cells with appropriate expression levels (Samalova *et al.*, 2006). Furthermore, it is more versatile than the low-magnification approach being insensitive to transfection efficiency and equally applicable to transfected tobacco epidermal cells, transfected protoplasts, or cells of whole transgenic plants. In the latter case, it has the potential to correct for variations in marker expression or imaging efficiency in various cell types and for epigenetic variation in marker expression within and between individuals or between wild-type and mutant individuals. Indeed, epigenetic cell-to-cell variation within the cells of a transgenic plant or between siblings in a transgenic line could be beneficially utilized to plot accumulation against expression level.

The ratiometric analysis of individual cells requires software which is available on request (mark.fricker@plants.ox.ac.uk) and an imaging system that can discriminate mGFP5 (K. Siemering, S. Hodge, J. Haseloff, MRC Laboratory of Molecular Biology, Cambridge, UK), YFP, and mRFP1. mGFP5 has the advantage over EGFP that it retains the 405 nm excitation peak, allowing it to be more easily discriminated from YFP if a suitable excitation source is available. Alternatively, as YFP can be used in place of GFP in assays of biosynthetic membrane traffic (Geelen *et al.*, 2002), nlsRFP$_m$-2A-secYFP or CFP$_m$-2A-secYFP constructs could be assembled and used in conjunction with test proteins tagged with CFP or RFP, respectively. The image analysis software currently measures secreted GFP accumulation in the nuclear envelope and adjacent cytoplasm, as this represents an easily identified and standardized region of the cell. This clearly allows quantification of defects in early stages of biosynthetic traffic, but it may also be applicable to proteins that act at later stages, as it has been shown in two such cases that secGFP accumulates in the ER as well as the Golgi in transfected epidermal cells (Geelen *et al.*, 2002; Zheng *et al.*, 2005). We envisage that the image analysis tool could be modified to measure accumulation in punctate compartments and to measure other aspects of trafficking, such as vacuolar versus either cytoplasmic or apoplastic accumulation of secN-R$_m$ relative to GFP-HDEL using the secN-R$_m$-2A-GH construct. Interestingly, the steady-state levels of the secN-R_m moiety of secN-RFP_m-2A-secG_f in the Golgi relative to ER or PVC were substantially lower than those of the secG_f moiety. It is possible, therefore, that the secN-R_m moiety reaches the PVC and the vacuole without traveling through the Golgi. However, its traffic to the vacuole is inhibited by dominant mutants of the Rab-F2 subclass that also act on the bacuolar trafficking of secGFP and other markers such as aleu-GFP that traffic via the Golgi (Kotzer *et al.*, 2004; Samalova *et al.*, 2006; Sohn *et al.*, 2003). Alternatively, therefore we suspect that secN-R_m is removed from the Golgi more efficiently than secG_f, perhaps as a result of the active vacuolar sorting machinery.

Importantly, nlsR_m-2A and Y_m-2A fusions behave similarly in the two commonly used model organisms, tobacco and *Arabidopsis*. We also envisage that this assay may allow the minimally invasive *Agrobacterium*-mediated transient expression system to be applied in *Arabidopsis* where transfection rates are too low for the low-magnification approaches that have been applied in tobacco. This would obviate the need to use highly perturbed protoplast systems for transfection. Our current attempts suggest that transfection rates are sufficient for ratiometric assays of membrane traffic to be assayed in leaves of mutant or transgenic plants but efficiency will need to be improved if ratiometric markers are to be effectively coexpressed with dominant-interfering proteins.

B. Future Developments of 2A-Mediated Ratiometry of Membrane Traffic in Single Cells

Simple use of the single cell approach to determine the effect of a coexpressed protein requires either that all cells indeed express saturating levels of the protein of interest or that the protein can be visualized and its expression level determined directly. If the protein cannot be tagged, the low-magnification approach is most easily applied. The principal reason to leave a protein of interest untagged will be that the tag impairs or alters activity, as with some Rab GTPases for example (Kotzer *et al.*, 2004; Samalova *et al.*, 2006). In an attempt to circumvent this problem, Samalova and colleagues (2006) asked whether the 2A peptide could be used to indirectly monitor the expression level of a Rab GTPase that is sensitive to tagging at its amino-terminus. They used 2A to link a nuclear-targeted RFP to the amino-terminus of the Rab GTPase. 2A-mediated cleavage was expected to leave a single proline at the amino-terminus of the protein. When coexpressed with YFP_m-2A-secG_f, cells that expressed the highest levels of nuclear RFP also accumulated the highest amounts of secreted GFP relative to YFP. However, the nlsR_m-2A-Rab fusion exhibited lower levels of activity than the untagged fusion and it was shown that 2A-mediated cleavage was inefficient. This low efficiency of separation was surprising given the efficient separation of YFP and RFP from GFP in xFP_m-2A-secG_f and xFP_m-2A-GH. Subsequent mutational analysis indicated that 2A-mediated cleavage at the C-terminus of mRFP1 and especially YFP is indeed inefficient in plant cells and that the efficient separation of xFP_m-2A-secG_f and xFP_m-2A-GH was dependent on the signal peptide of the secGFP and GFP-HDEL moieties. These observations are consistent with previous reports that 2A activity in plant cells is influenced markedly by the upstream moiety of the fusion protein (Ma and Mitra, 2002). One potential solution to the problem therefore is to place a polypeptide that promotes efficient 2A-mediated cleavage between the fluorescent protein and the downstream 2A peptide.

VI. Summary

Fluorescent proteins have had a substantial impact on the way in which plant membrane traffic is investigated. The key advantages and disadvantages of various improved fluorescent proteins have become clear and have been exploited to assay function as well as morphology and location of trafficking components. The potential of fluorescent proteins to reveal quantitative information about membrane traffic is only recently being realized. The use of ratiometric trafficking assays facilitated by software packages discussed here will greatly increase the ease and quality of quantitative fluorescence assays of membrane traffic.

References

Batoko, H., Zheng, H. Q., Hawes, C., and Moore, I. (2000). A Rab1 GTPase is required for transport between the endoplasmic reticulum and Golgi apparatus for normal Golgi movement in plants. *Plant Cell* **12,** 2201–2217.

Boevink, P., Martin, B., Oparka, K., Cruz, S. S., and Hawes, C. (1999). Transport of virally expressed green fluorescent protein through the secretory pathway in tobacco leaves is inhibited by cold shock and brefeldin A. *Planta* **208,** 392–400.

Boevink, P., Oparka, K., Cruz, S. S., Martin, B., Betteridge, A., and Hawes, C. (1998). Stacks on tracks: The plant Golgi apparatus traffics on an actin/ER network. *Plant J.* **15,** 441–447.

Brandizzi, F., Fricker, M., and Hawes, C. (2002). A greener world: The revolution in plant bioimaging. *Nat. Rev. Mol. Cell Biol.* **3,** 520–530.

Brandizzi, F., Irons, S. L., Johansen, J., Kotzer, A., and Neumann, U. (2004). GFP is the way to glow: Bioimaging of the plant endomembrane system. *J. Microsc.* **214,** 138–158.

Campbell, R. E., Tour, O., Palmer, A. E., Steinbach, P. A., Baird, G. S., Zacharias, D. A., and Tsien, R. Y. (2002). A monomeric red fluorescent protein. *Proc. Natl. Acad. Sci. USA* **99,** 7877–7882.

DaSilva, L. L. P., Snapp, E. L., Denecke, J., Lippincott-Schwartz, J., Hawes, C., and Brandizzi, F. (2004). Endoplasmic reticulum export sites and golgi bodies behave as single mobile secretory units in plant cells. *Plant Cell* **16,** 1753–1771.

DaSilva, L. L. P., Taylor, J. P., Hadlington, J. L., Hanton, S. L., Snowden, C. J., Fox, S. J., Foresti, O., Brandizzi, F., and Denecke, J. (2005). Receptor salvage from the prevacuolar compartment is essential for efficient vacuolar protein targeting. *Plant Cell* **17,** 132–148.

De Felipe, P., Hughes, L. E., Ryan, M. D., and Brown, J. D. (2003). Co-translational, intraribosomal cleavage of polypeptides by the foot-and-mouth disease virus 2A peptide. *J. Biol. Chem.* **278,** 11441–11448.

De Felipe, P., and Ryan, M. D. (2004). Targeting of proteins derived from self-processing polyproteins containing multiple signal sequences. *Traffic* **5,** 616–626.

Di Sansebastiano, G. P., Paris, N., Marc-Martin, S., and Neuhaus, J.-M. (1998). Specific accumulation of GFP in a non-acidic vacuolar compartment via a C-terminal propeptide-mediated sorting pathway. *Plant J.* **15,** 449–457.

El Amrani, A., Barakate, A., Askari, B. M., Li, X., Roberts, A. C., Ryan, M. D., and Halpin, C. (2004). Coordinate expression and independent subcellular targeting of multiple proteins from a single transgene. *Plant Physiol.* **135,** 16–24.

Flückiger, R., De Caroli, M., Piro, G., Dalessandro, G., Neuhaus, J.-M., and Di Sansebastiano, G.-P. (2003). Vacuolar system distribution in *Arabidopsis* tissues, visualized using GFP fusion proteins. *J. Exp. Bot.* **54,** 1577–1584.

Frigerio, L., de Virgilio, M., Prada, A., Faoro, F., and Vitale, A. (1998). Sorting of phaseolin to the vacuole is saturable and requires a short C-terminal peptide. *Plant Cell* **101,** 1031–1042.

Geelen, D., Leyman, B., Batoko, H., Di Sansabastiano, G.-P., Moore, I., and Blatt, M. R. (2002). The abscisic acid-related SNARE homolog NtSyr1 contributes to secretion and growth: Evidence from competition with its cytosolic domain. *Plant Cell* **14,** 387–406.

Grignon, C., and Sentenac, H. (1991). pH and ionic conditions in the apoplast. *Annu. Rev. Plant Physiol. Plant Mol. Biol.* **42,** 103–128.

Halpin, C., Cooke, S. E., Barakate, A., El Amrani, A., and Ryan, M. D. (1999). Self-processing 2A-polyproteins—a system for co-ordinate expression of multiple proteins in transgenic plants. *Plant J.* **17,** 453–459.

Haseloff, J. <http://www.plantsci.cam.ac.uk/Haseloff/imaging/GFP.htm>

Irons, S., Evans, D., and Brandizzi, F. (2003). The first 238 amino acids of the human lamin B receptor are targeted to the nuclear envelop in plants. *J. Exp. Bot.* **54,** 943–950.

Kamiya, T., Akahori, T., Ashikari, M., and Maeshima, M. (2006). Expression of the vacuolar Ca^{2+}/H^{+} exchanger, OsCAX1a, in rice: Cell and age specificity of expression, and enhancement by Ca^{2+}. *Plant Cell Physiol.* **47,** 96–106.

Kotzer, A. M., Brandizzi, F., Neumann, U., Paris, N., Moore, I., and Hawes, C. (2004). AtRabF2b (Ara7) acts on the vacuolar trafficking pathway in tobacco leaf epidermal cells. *J. Cell Sci.* **117,** 6377–6389.

Kunze, I., Hensel, G., Adler, K., Bernard, J., Neubohn, B., Nilsson, C., Stoltenburg, R., Kohlwein, S., and Kunze, G. (1999). The green fluorescent protein targets secretory proteins to the yeast vacuole. *Biochim. Biophys. Acta* **1410,** 287–298.

Lee, G. J., Sohn, E. J., Lee, M. H., and Hwang, I. (2004). The *Arabidopsis* Rab5 homologs Rha1 and Ara7 localize to the prevacuolar compartment. *Plant Cell Physiol.* **45,** 1211–1220.

Lee, M. H., Min, M. K., Lee, Y. J., Jin, J. B., Shin, D. H., Kim, D. H., Lee, K.-H., and Hwang, I. (2002). ADP-Ribosylation Factor 1 of *Arabidopsis* plays a critical role in intracellular trafficking and maintenance of endoplasmic reticulum morphology in *Arabidopsis. Plant Physiol.* **129,** 1507–1520.

Ma, C. L., and Mitra, A. (2002). Expressing multiple genes in a single open reading frame with the 2A region of foot-and-mouth disease virus as a linker. *Mol. Breed.* **9,** 191–199.

Matsushima, R., Kondo, M., Nishimura, M., and Hara-Nishimura, I. (2003). A novel ER-derived compartment, the ER body, selectively accumulates a β-glucosidase with an ER retention signal in. *Arabidopsis. Plant J.* **33,** 493–502.

Nagai, T., Ibata, K., Park, E. S., Kubota, M., Mikoshiba, K., and Miyawaki, A. (2002). A variant of yellow fluorescent protein with fast and efficient maturation for cell-biological applications. *Nat. Biotechnol.* **20,** 87–90.

Runions, J., Brach, T., Kuhner, S., and Hawes, C. (2006). Photoactivation of GFP reveals protein dynamics within the endoplasmic reticulum membrane. *J. Exp. Bot.* **57,** 43–50.

Ryan, M. D., Donnelly, M., Lewis, A., Mehrotra, A. P., Wilkie, J., and Gani, D. (1999). A model for nonstoichiometric, cotranslational protein scission in eukaryotic ribosomes. *Bioorg Chem.* **27,** 55–79.

Saint-Jore, C. M., Evins, J., Batoko, H., Brandizzi, F., Moore, I., and Hawes, C. (2002). Redistribution of membrane proteins between the Golgi apparatus and endoplasmic reticulum in plants is reversible and not dependent on cytoskeletal networks. *Plant J.* **29,** 661–678.

Samalova, M., Fricker, M., and Moore, I. (2006). Ratiometric fluorescence-imaging assays of plant membrane traffic using polyproteins. *Traffic* **7,** 1701–1723.

Scott, A., Wyatt, S. E., Tsou, P.-L., Robertson, D., and Allen, N. S. (1999). A model system for plant cell biology: GFP imaging in living onion epidermal cells. *BioTechniques* **26,** 1127–1132.

Shaner, N. C., Steinbach, P. A., and Tsien, R. Y. (2005). A guide to choosing fluorescent proteins. *Nat. Methods* **2,** 905–909.

Snapp, E. L., Hegde, R. S., Francolini, M., Lombardo, F., Colombo, S., Pedrazzini, E., Borgese, N., and Lippincott-Schwartz, J. (2003). Formation of stacked ER cisternae by low affinity protein interactions. *J. Cell Biol.* **163,** 257–269.

Sohn, E. J., Kim, E. S., Zhao, M., Kim, S. J., Kim, H., Kim, Y. W., Lee, Y. J., Hillmer, S., Sohn, U., Jiang, L. W., and Hwang, I. W. (2003). Rha1, an *Arabidopsis* Rab5 homolog, plays a critical role in the vacuolar trafficking of soluble cargo proteins. *Plant Cell* **15,** 1057–1070.

Swarup, R., Kargul, J., Marchant, A., Zadik, D., Rahman, A., Mills, R., Yemm, A., May, S., Williams, L., Millner, P., Tsurumi, S., Moore, I., *et al.* (2004). Structure-function analysis of the presumptive *Arabidopsis* auxin permease AUX1. *Plant Cell* **16,** 3069–3083.

Takeuchi, M., Ueda, T., Sato, K., Abe, H., Nagata, T., and Nakano, A. (2000). A dominant negative mutant of Sar1 GTPase inhibits protein transport from the endoplasmic reticulum to the Golgi apparatus in tobacco and *Arabidopsis* cultured cells. *Plant J.* **23,** 517–525.

Tamura, K., Shimada, T., Kondo, M., Nishimura, M., and Hara-Nishimura, I. (2005). KATAMARI1/MURUS3 is a novel Golgi membrane protein that is required for endomembrane organization in *Arabidopsis*. *Plant Cell* **17,** 1764–1776.

Tamura, K., Shimada, T., Ono, E., Tanaka, Y., Nagatani, A., Higashi, S., Watanabe, M., Nishimura, M., and Hara-Nishimura, I. (2003). Why green fluorescence fusion proteins have not been observed in the vacuoles of higher plants. *Plant J.* **35,** 545–555.

Tamura, K., Yamada, K., Shimada, T., and Hara-Nishimura, I. (2004). Endoplasmic reticulum-resident proteins are constitutively transported to vacuoles for degradation. *Plant J.* **39,** 393–402.

Yang, Y. D., Elamawi, R., Bubeck, J., Pepperkok, R., Ritzenthaler, C., and Robinson, D. G. (2005). Dynamics of COPII vesicles and the Golgi apparatus in cultured Nicotiana tabacum BY-2 cells provides evidence for transient association of Golgi stacks with endoplasmic reticulum exit sites. *Plant Cell* **17,** 1513–1531.

Zheng, H. Q., Camacho, L., Wee, E., Henri, B. A., Legend, J., Leaver, C. J., Malho, R., Hussey, P. J., and Moore, I. (2005). A Rab-E GTPase mutant acts downstream of the Rab-D subclass in biosynthetic membrane traffic to the plasma membrane in tobacco leaf epidermis. *Plant Cell* **17,** 2020–2036.

Zheng, H. Q., Kunst, L., Hawes, C., and Moore, I. (2004). A GFP-based assay reveals a role for RHD3 in transport between the endoplasmic reticulum and Golgi apparatus. *Plant J.* **37,** 398–414.

CHAPTER 16

Engineering FRET Constructs Using CFP and YFP

Satoshi Shimozono and Atsushi Miyawaki

Laboratory for Cell Function Dynamics, Brain Science Institute
The Institute of Physical and Chemical Research (RIKEN), 2-1 Hirosawa, Wako-city, Saitama 351-0198, Japan

Abstract

Fluorescence resonance energy transfer (FRET) technology has been used to develop genetically encoded fluorescent indicators for various cellular functions. Here we discuss how to engineer constructs for FRET between the cyan- and yellow-emitting variants of green fluorescent protein (GFP) from *Aequorea victoria* (CFP and YFP, respectively). Throughout this chapter, we stress the fact that FRET is highly sensitive to the relative orientation and distance between the donor and the acceptor. The chapter consists of two parts. First, we discuss FRET-based indicators encoded by single genes, which were developed in our laboratory. In this approach, a number of different constructs can be made for a comparative assessment of their FRET efficiencies. For example, the length and sequence of the linker

0091-679X/08 $35.00
DOI: 10.1016/S0091-679X(08)85016-9

between the fluorescent protein and the host protein should be optimized for each specific application. In the second part, we describe the use of long and flexible linkers for engineering FRET constructs, including an introduction to a general and efficient tool for making successful fusion proteins with long and flexible linkers. When CFP and YFP are fused through floppy linkers to two protein domains that interact with each other, the two fluorescent proteins will associate due to the weak dimerization propensity of *Aequorea* GFP, which results in moderate FRET. This approach has become even more powerful due to the construction of a new pair of fluorescent proteins for FRET: CyPet and YPet.

I. Introduction

Although green fluorescent protein (GFP) was identified as a fluorescent protein more than 40 years ago, it has only been used as a tool for biological experiments during the past dozen years. In much the same way, fluorescence resonance energy transfer (FRET), which was first theorized in 1948 (Förster, 1948), had been primarily used by a small proportion of biologists as a spectroscopic ruler (Stryer, 1978). FRET is the radiationless transfer of excited-state energy from an initially excited donor to an acceptor (Lakowicz, 1999). This transfer depends on the spectral overlap of the donor and the acceptor, the distance between them, and the relative orientation of the transition dipoles of the chromophores. To allow FRET technology to be used by the larger biological community, a technological advance was needed. With the emergence of spectral variants of GFP, an increasing number of researchers have looked for tools that enable the direct visualization of biological functions. In the past decade, the combination of the increasing variety of fluorescent proteins and FRET has made a significant impact on many molecular and cellular investigations. In the beginning, the blue-emitting variant of GFP was used as the FRET donor in combination with a GFP acceptor. Now most FRET constructs, however, use CFP and YFP as the FRET donor and acceptor, respectively. In this chapter, we focus on the ways in which the fluorescent proteins are fused to engineer FRET constructs.

II. Rationale

A. Optimization of the Relative Positions of CFP and YFP in a Protein or Complex

To develop a unimolecular fluorescent probe encoded by a single gene, one approach is to sandwich a protein composed of multiple domains between CFP and YFP. A variety of unimolecular fluorescent indicators have been developed. Among them, yellow cameleons (YCs) are composed of a linear combination of CFP, calmodulin (CaM), a glycylglycine linker, the CaM-binding peptide of myosin light chain kinase (M13), and YFP (Miyawaki *et al.*, 1997). Binding of Ca^{2+} to the CaM moiety of the YC initiates an intramolecular interaction between the CaM and M13 domains, causing the chimeric protein to shift from an extended conformation to a more compact one. Because FRET is highly sensitive to the relative orientation and distance between the two fluorophores (Miyawaki, 2003a), this shift increases the efficiency of FRET from CFP to YFP. This technique is also amenable to emission ratioing, which is more quantitative than single-wavelength monitoring, and is also an ideal readout for fast imaging using laser-scanning confocal microscopy. A large dynamic response range is an important factor for detecting subtle but significant signals. Expansion of the Ca^{2+} responses of YCs has been achieved by combining FRET and circular-permutation techniques. To achieve a larger Ca^{2+}-dependent change in the relative orientation and distance between the fluorophores of CFP and YFP, recent studies have employed circularly permuted fluorescent protein variants in which the amino and carboxyl portions of a protein are interchanged and reconnected by a short spacer between the original termini (Baird *et al.*, 1999). Circular permutation was conducted on YFP, and new termini were introduced into surface-exposed loop regions of the β-barrel. Compared to conventional YCs, YC3.60, in which YFP has been replaced by a circularly permuted YFP, produces equally bright signals while showing a five- to sixfold larger dynamic range (Nagai *et al.*, 2004).

B. FRET Constructs Containing Long and Flexible Linkers

Fluorescent proteins can be incorporated into many other proteins using recombinant DNA techniques. There are three important characteristics to consider when fusing a fluorescent protein to a protein of interest.

1. Fluorescence: the fluorescent protein should fold correctly and fluoresce.
2. Function: the host protein should fold correctly and retain its function.
3. Integrity: the fluorescent protein and the host protein should be resistant to separation by proteolysis.

In most cases, unless the linker joining the two constituents of the fusion protein is sufficiently long and flexible, steric hindrance or folding interference between the fluorescent protein and the protein of interest may occur. Generally, the linker should be flexible, soluble, and resistant to proteolysis and should not have a secondary structure or aggregate (Miyawaki *et al.*, 2003b). The amino acid that confers the greatest flexibility to a peptide chain is Gly, which has the smallest side chain. Thus, fusion proteins are often designed to include a number of Gly residues between the two components. Single-chain Fvs are recombinant antibody fragments consisting of only the variable light chain (V_L) and variable heavy chain (V_H) domains, which are covalently connected by a polypeptide linker of about 20 amino acids. Longer linkers that span ~4 nm have been extensively used to engineer single-chain Fvs (Huston *et al.*, 1988). The most widely used linker structures for Fvs have sequences consisting primarily of stretches of Gly and

Ser residues; the Ser residues are interspersed along the linker to improve the solubility of the poly-Gly peptides.

Although sequences coding for these linkers are usually appended to PCR primers (Shimozono *et al.*, 2004), it is laborious to prepare a long primer that includes Gly- and Ser-encoding stretches for each fusion protein. Thus, we have constructed a series of DNA plasmids that facilitate the fusion of coding sequences for any two protein domains with a DNA linker that encodes a triple repeat of the amino acid linker Gly-Gly-Gly-Gly-Ser [$(GGGGS)_3$]. The sequence encoding the linker was inserted into the *Eco*RV site in the center of the multicloning site (MCS) of pBlueScript (pBS) (Stratagene) (Fig. 1A). To account for all three potential reading frames and both orientations, six DNA plasmids were constructed and named pBS Couplers 1–6. Two sequential ligation reactions with a pBS Coupler using appropriate restriction enzymes allow the coding sequence of a fluorescent protein variant to be fused to the long linker and the protein of interest. The large number of restriction enzyme sites in the MCS of pBS provides a number of possibilities for the ligation reactions.

Due to the weak dimerization propensity of *Aequorea* GFP (Zacharias *et al.*, 2002), the CFP and YFP moieties, which are linked by a long and flexible linker, can associate to yield moderate FRET. If the linker contains a sequence that is recognized by a protease, the chimeric protein can serve as an indicator for the activity of the protease (Takemoto *et al.*, 2003). Moreover, attachment of CFP and YFP through such linkers to two protein domains that interact with each other may create bimolecular FRET-based indicators. This approach using long and flexible linkers may obviate the extensive work required for optimization of the FRET reaction discussed in Section I.

Furthermore, variants of CFP and YFP have recently been identified that improve the efficiency of FRET between these two fluorescent proteins. Mutagenesis and screening of both CFP and YFP were performed using a loosely concatenated construct to identify a new CFP–YFP pair. Nguyen and Daugherty (2005) applied a quantitative evolutionary strategy using FACS (fluorescence activated cell sorter) to obtain the new pair, CyPet and YPet, which exhibit a more efficient level of FRET.

III. Methods

A. Optimization of the Tight Concatenation of CFP and YFP for Highly Efficient FRET

Recently, we tightly concatenated CFP and YFP and attempted to optimize the concatenation to maximize the FRET efficiency (Shimozono *et al.*, 2006). The C-terminus of CFP and the N-terminus of YFP were truncated to various degrees before the variants were fused using a dipeptide (Leu-Glu: LE) encoded by an *Xho*I restriction site (Fig. 2A). The chimeric proteins are referred to as $C\Delta X_1$-LE-ΔX_2Y; X_1 and X_2 are the numbers of amino acids that were deleted from CFP and YFP, respectively. The bacterially expressed proteins were purified and spectrally characterized by measuring the emission spectra resulting from excitation at 440 nm.

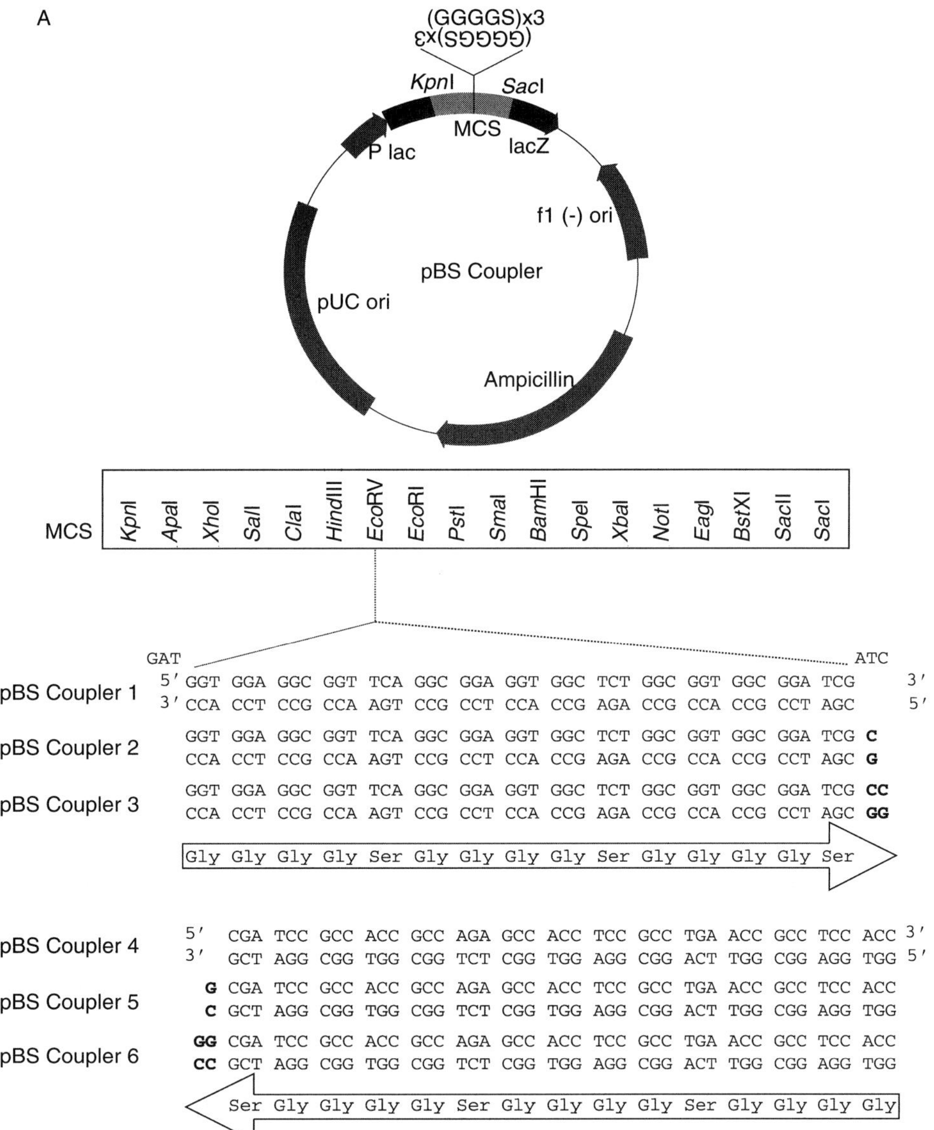

Fig. 1 (*continues*)

Numerous constructs with different truncations were examined. The emission ratio of YFP (527 nm) to CFP (476 nm) generally increased as larger deletions were made (Fig. 2B). CFP and/or YFP, however, were misfolded in some of the chimeric proteins. For example, N-terminal truncations of more than 10 amino acids appeared to prevent proper folding of YFP. Among the chimeric proteins examined, CΔ11-LE-Δ5Y (Fig. 2B) produced a bright signal and showed the highest emission ratio (527 nm/476 nm), which reached 20. The emission signal at around 480 nm, which was emitted from the CFP of CΔ11-LE-Δ5Y, was almost negligible (Fig. 2C); thus, the emission spectra of CFP and CΔ11-LE-Δ5Y can be separated when the proteins are excited at 440 nm. Next, the dependence of the

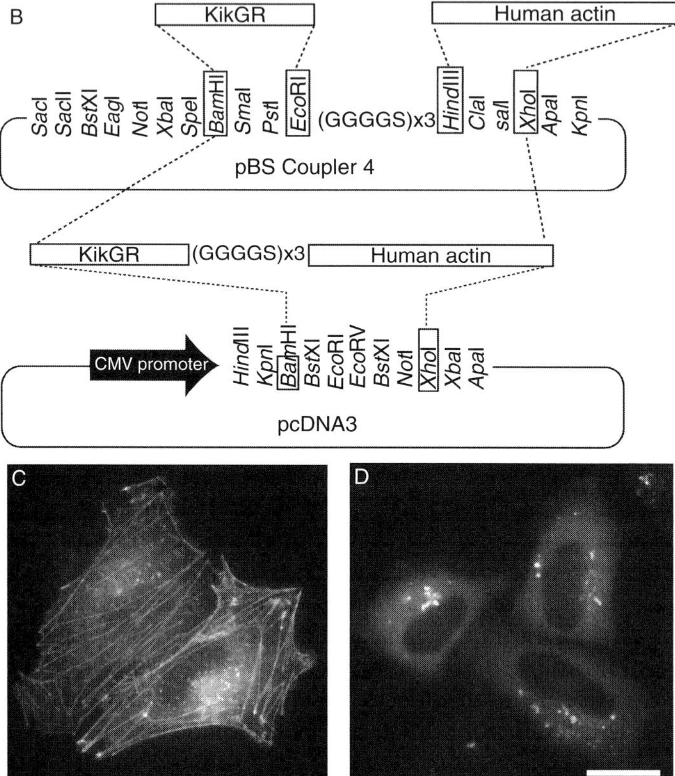

Fig. 1 Sequence of the multicloning site in the pBS Couplers. (A) Linker sequence was inserted into the *Eco*RV site of pBlueScript (pBS). pBS Couplers 1–3 are *Kpn*I–*Sac*I oriented, whereas pBS Couplers 4–6 are *Sac*I–*Kpn*I oriented. Three vectors were constructed for each direction to account for all of the potential reading frames for the C-terminal protein. A schematic drawing of the vectors is shown at the lower right. (B) The scheme for the construction of the expression plasmid for KikGR-tagged β-actin using pBS Coupler 4 (KikGR:L:β-actin:) is shown. DNA fragments coding for KikGR and β-actin were subcloned into the *Bam*HI/*Eco*RI site and the *Hin*dIII/*Xho*I site, respectively, so that the reading frames were in frame with the linker. The resulting fragment coding for KikGR:L:β-actin was extracted from pBS Coupler 4 and subcloned into pcDNA3 using the *Bam*HI/*Xho*I site. (C) Expression pattern of KikGR:L:β-actin in HeLa cells. The fusion protein was expressed in HeLa cells and the expression pattern was evaluated using a wide-field microscope. Stress fibers were clearly observed. (D) Expression pattern of KikGR:β-actin lacking a linker. Few stress fibers were visible and a diffuse expression pattern with fluorescent puncta was observed. Scale bar = 20 μm.

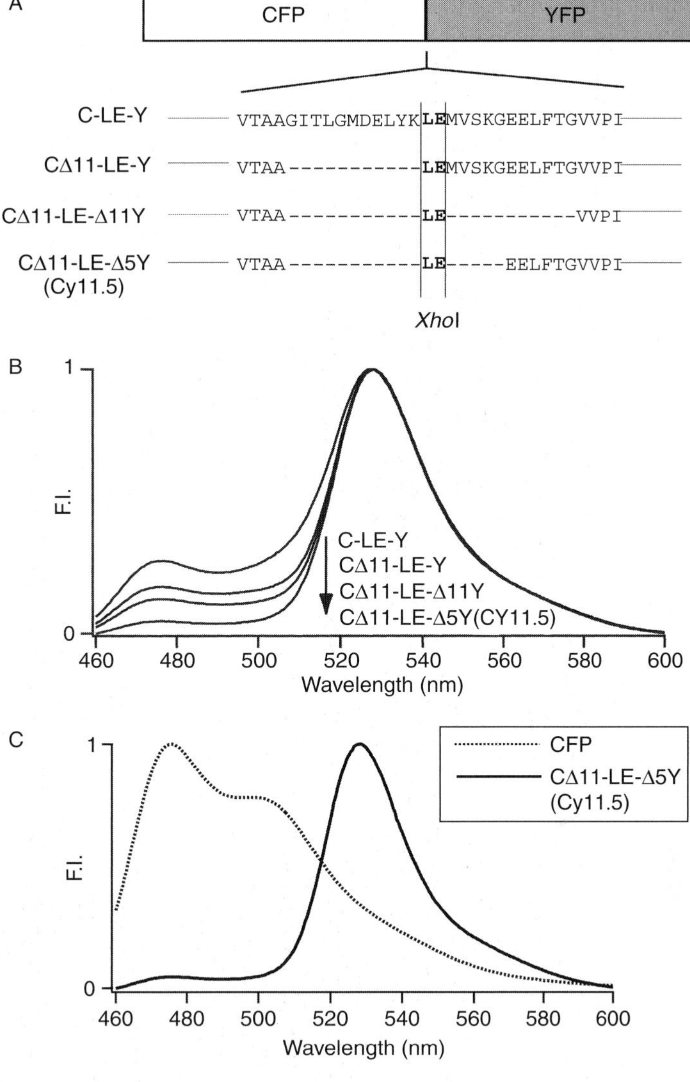

Fig. 2 (*continues*)

emission spectrum of CΔ11-LE-Δ5Y on its concentration was examined. Within a range from 50 to 1.6 nM (a significantly low concentration range considering the dissociation constant for CFP and YFP is 100–300 μM), the shape of the emission spectra was constant (Fig. 2D), indicating that FRET from CFP to YFP primarily takes place within a single CΔ11-LE-Δ5Y molecule. The protein was named "Cy11.5".

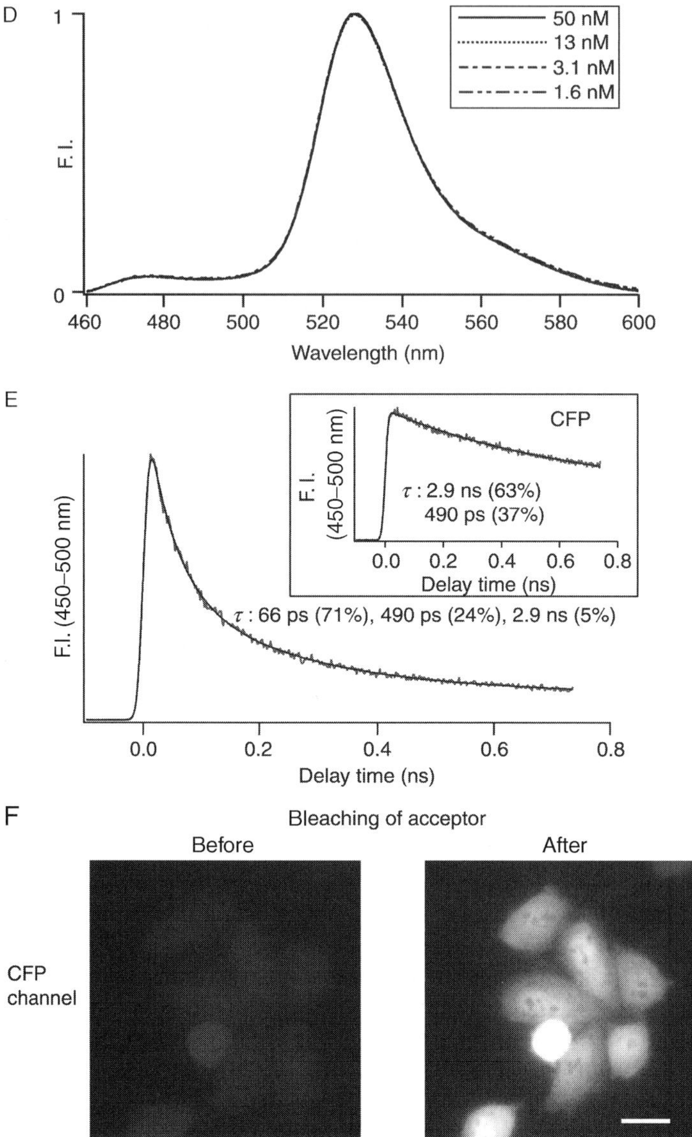

Fig. 2 CFP–YFP fusion proteins including the Cy11.5 variant. (A) A diagram of the CFP–YFP fusion proteins. The amino acid sequences of the linking regions of four representative chimeric proteins are shown. (B) Normalized emission spectra of the four CFP–YFP fusion proteins shown in (A). (C) The emission spectra of Cy11.5 and CFP after excitation at 440 nm. (D) The emission spectrum does not change for Cy11.5 at concentrations of 50 nM (solid line), 13 nM (dotted line), 3.1 nM (dashed line), and 1.6 nM (two-dot dashed line). (E) A trace representing the time-resolved emission spectrum between 450 and 500 nm of Cy11.5. (Inset) A trace representing the time-resolved emission spectrum between 450 and 500 nm of free CFP. (F) Images taken through the CFP channel (excitation at 440 nm and emission at 480 nm) before (left) and after (right) photobleaching. Scale bar = 20 μm. [This figure was reproduced with permission from Shimozono *et al.* (2006). Copyright 2006, American Chemical Society.]

For a CFP–YFP tandem construct in a cuvette, the increase in the fluorescence intensity of CFP after proteolysis by trypsin allows us to obtain the efficiency of FRET from CFP to YFP (Miyawaki and Tsien, 2000). This approach, however, cannot be used for Cy11.5 because the chimeric protein is resistant to proteases. Thus, we measured the excited-state lifetime of the CFP in Cy11.5 using a streak camera. Cy11.5 was excited with a 400-nm pulse generated from the second harmonic of a Ti:Sa laser. The emission was monitored in a short (450–500 nm) and a long (550–600 nm) wavelength region. The emission between 450 and 500 nm showed an initial fast-decay component with a lifetime of 66 ps (71% of the emission), which was followed by long-lived emission with two lifetimes of 490 ps (24%) and 2.9 ns (5%) (Fig. 2E). The two long-lifetime components were characteristic of unquenched CFP; when free CFP was used under the same conditions, the emission decayed with average lifetimes of 490 ps (37%) and 2.9 ns (63%) (Fig. 2E, inset). The 2.9-ns lifetime reflects the radiative process to the ground state; the reported fluorescence lifetime of CFP is around 3 ns (Rizzo *et al.*, 2004). In contrast, the 490-ps lifetime may be related to CFP being converted to a distinct conformation that has been proposed for this protein (Hyun Bae *et al.*, 2003). Because Cy11.5 and CFP showed similar fractions of the emissions in the 490-ps decay components (24% vs 37%), it is assumed that the deactivation process was not involved in the FRET. It is also likely that the decay of Cy11.5 characterized by the 2.9-ns lifetime resulted from CFP molecules that were linked to a misfolded YFP. Despite the contamination of the samples with these partially mature products, it is possible to quantify the FRET efficiency (E) in fully mature Cy11.5 using the following equation:

$$E = 1 - \frac{(\tau_{DA})}{\tau_D} \tag{1}$$

where τ_{DA} and τ_D are the excited-state lifetimes of the donor in the presence and absence of the acceptor, respectively. In this situation, τ_{DA} and τ_D were 66 ps and 2.9 ns, respectively. Thus, the FRET efficiency was calculated to be 98%.

There was a fast rise in the emission between 550 and 600 nm (data not shown), which occurred concomitantly with the fast decay of the emission between 450 and 500 nm (Fig. 2E). Both the rise and the decay shared the same lifetime of 66 ps, indicating that the lifetime indeed represented the FRET. The emission between 550 and 600 nm then decayed slowly with a lifetime of 3.9 ns, which is characteristic of YFP (data not shown).

It should be noted that Cy11.5 folds efficiently in bacterial cells as evidenced by the fact that a large percentage of the prepared molecules were fully mature. The efficient folding of Cy11.5 was also confirmed in mammalian cells. Cy11.5 expressed in HeLa cells was distributed uniformly in the cytosol and the nuclei. Spectral imaging revealed that the color of the emitted signals (or the fluorescence intensity ratio of YFP to CFP) was identical in the cytosol and the nucleus, and the signals were stable as long as the cells were alive, suggesting again that Cy11.5 was stable. After the cells were illuminated with intense green light from a xenon lamp (4.4 W/cm^2 at the specimen), no fluorescence signal was detected from YFP. On the contrary, the signal from CFP increased approximately tenfold (Fig. 2F). The FRET efficiency was calculed to be about 90% using the following equation:

$$E = 1 - \frac{(F_{\mathrm{DA}})}{F_{\mathrm{D}}} \tag{2}$$

where F_{DA} and F_{D} are the fluorescence intensities of the donor in the presence and absence of the acceptor, respectively.

B. Engineering FRET Constructs Using Long and Flexible Linkers

1. Making a Successful Fusion Protein with a Long Flexible Linker

To demonstrate the utility of the vectors shown in Fig. 1A, we generated a mammalian expression plasmid that encodes a chimeric protein comprising the photoconvertible fluorescent protein KikGR (Tsutsui *et al.*, 2005), the $(GGGGS)_3$ linker, and human β-actin.

Steps:

1. The cDNAs for KikGR and human β-actin were inserted into pBS Coupler 4 at the *Bam*HI/*Eco*RI and *Hin*dIII/*Xho*I sites, respectively (Fig. 1B). The encoded chimeric protein was named KikGR:L:β-actin.
2. The DNA fragment coding for KikGR:L:β-actin was transferred into the *Bam*HI/*Xho*I site of the mammalian expression vector pcDNA3 (Invitrogen, Carlsbad, CA).
3. As control, KikGR and β-actin were directly concatenated via a Lys-Leu dipeptide (encoded by the *Hin*dIII sequence AAGCTT) to produce KikGR:β-actin.
4. HeLa cells were transfected with these DNA plasmids using Lipofectamine reagent (Invitrogen, Carlsbad, CA).
5. After 20 h, the cells were imaged using a wide-field epifluorescence microscope (IX-70, Olympus, Tokyo, Japan) equipped with a cooled CCD camera (CoolSNAP HQ, Photometrics, Tucson, AZ).

In our experiments, stress fibers were clearly visible in the cells expressing KikGR:L:β-actin (Fig. 1C), whereas a diffused cytosolic distribution with fluorescent puncta was observed in the cells expressing KikGR:β-actin (Fig. 1D). The six DNA plasmids in the pBS Coupler series are expected to be useful for the concatenation of any two proteins through the flexible $(GGGGS)_3$ linker, thereby circumventing the need to produce long PCR primers.

2. Development of an Indicator for a Protein–Protein Interaction Using CyPet and YPet

Caspase-activated DNase (CAD) is the primary nuclease that causes DNA fragmentation during apoptosis. In proliferating cells, CAD is expressed as an inactive form due to an association with the inhibitor of CAD (ICAD). During apoptosis, caspase-3 and caspase-7 cleave ICAD, resulting in the release of active CAD. To visualize the spatiotemporal pattern of CAD activation, we used the $(GGGGS)_3$ linker to fuse CyPet and YPet with CAD and the functional ICAD isoform ICAD-L, respectively (Fig. 3A).

Steps:

1. HeLa cells were attached to a coverslip in a Petri dish. The cells were transfected with 1 μg per dish of cDNAs coding for CAD:L:CyPet and ICAD-L:L:Ypet using Lipofectamine and Plus reagent according to the manufacturer's instructions.
2. Two days after cDNA transfection, HeLa cells were imaged on an inverted microscope (IX-70) with a cooled CCD camera (Micromax, Photometrics, Tucson, AZ). Cells were kept at 37 °C in Hank's balanced salt solution (Invitrogen, Carlsbad, CA) containing 15-mM HEPES (pH 7.4). Image acquisition and processing were controlled using a personal computer and the MetaFluor program (Universal Imaging, West Chester, PA). The excitation filter wheel in front of the xenon lamp and the emission filter wheel (Lambda 10–2, Sutter Instruments, San Rafael, CA) immediately below the CCD camera were also under computer control. Excitation light from a 75-W xenon lamp passed through a 440 DF20 (440 ± 10-nm) excitation filter. The light was reflected onto the sample using a 455-nm long-pass (455DRLP) dichroic mirror. The emitted light was collected with a 40× (numerical aperture: 1.35) objective lens and passed through a 480 ± 15-nm or 480 ± 25-nm band-pass filter (480DF30 or 480DF50, donor channel) for CyPet, and a 535 ± 12.5-nm band-pass filter (535DF25, FRET channel) for YPet. Interference filters were obtained from Omega Optical (Brattleboro, VT) or Chroma Technologies (Rockingham, VT).
3. Several factors for image acquisition were defined, including (1) the excitation power, which depends on the type of light source and the neutral density filter; (2) the numerical aperture of the objective; (3) the time period of exposure to the light; (4) the image acquisition interval; and (5) binning. The last three factors were considered in terms of whether temporal or spatial resolution was required.
4. We chose moderately bright cells and cells that expressed equal amounts of CAD:L:CyPet and ICAD-L:L:YPet.
5. The cells were treated with 500 ng/ml anti-Fas antibodies (CH-11; MBL, Nagoya, Japan) and 10 μg/ml cycloheximide at 37 °C, and time-lapse imaging was initiated.

In our experiments, the fluorescence signals of both CAD:L:CyPet and ICAD-L:L:YPet were localized in the nuclei. In the beginning, the ratio of YPet to CyPet signals was high, indicating that CAD and ICAD-L associated, thereby allowing

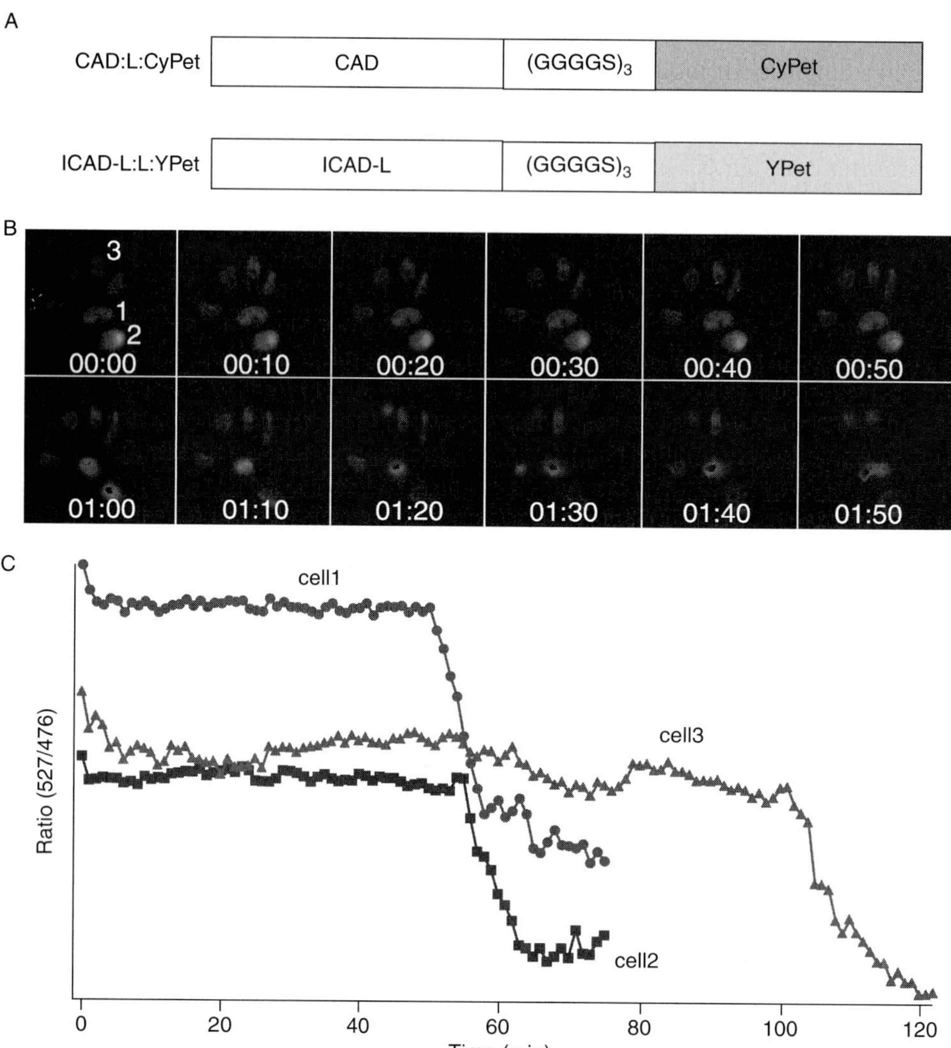

Fig. 3 Spatiotemporal pattern of caspase-activated DNase (CAD) activity using CyPet and YPet as a fluorescence resonance energy transfer (FRET) pair. (A) Schematic drawings of the constructs. (B) A sequence of images showing the ratio of the signal from YPet to that from CyPet is shown. CAD:L:CyPet and inhibitor of CAD (ICAD)-L:L:YPet were localized in the nuclei. In the beginning, the ratio of YPet to CyPet was high, indicating CAD was associated with ICAD-L. As apoptosis proceeded, the ratio dropped, indicating CAD was released from ICAD-L inhibition. (C) Time courses of the YPet to CyPet ratios in the three cells indicated in (B) are plotted. (See Plate no. 44 in the Color Plate Section.)

FRET to occur. As apoptosis proceeded, this ratio dropped, indicating ICAD-L was cleaved and CAD was released from inhibition (Fig. 3B, time is shown in hour: minute). The time courses of the ratios from three cells are shown in Fig. 3C.

Acknowledgments

This work was partly supported by grants from the Special Postdoctoral Researcher Program of RIKEN to S.S.; the CREST project of the Japan Science and Technology Agency; the Special Coordination Fund for the promotion of the Ministry of Education, Culture, Sports, Science and Technology; the Japanese Government; the New Energy and Industrial Technology Development Organization; and the Human Frontier Science Program.

References

Baird, G. S., Zacharias, D. A., and Tsien, R. Y. (1999). Circular permutation and receptor insertion within green fluorescent proteins. *Proc. Natl. Acad. Sci. USA* **96,** 11241–11246.

Förster, Th. (1948). Intermolecular energy migration and fluorescence. *Ann. Phys.* **2,** 55–75.

Huston, J. S., Levinson, D., Mudgett-Hunter, M., Tai, M. S., Novotny, J., Margolies, M. N., Ridge, R. J., Bruccoler, R. E., Haber, E., Crea, R., and Oppermann, H. (1988). Protein engineering of antibody binding sites: Recovery of specific activity in an anti-digoxin single-chain Fv analogue produced in *Escherichia coli. Proc. Natl. Acad. Sci. USA* **85,** 5879–5883.

Hyun Bae, J., Rubini, M., Jung, G., Wiegand, G., Seifert, M. H. J., Azim, M. K., Kim, J. S., Zumbusch, A., Holak, T. A., Moroder, L., Huber, R., and Budisa, N. (2003). Expansion of the genetic code enables design of a novel "gold" class of green fluorescent proteins. *J. Mol. Biol.* **328,** 1071–1081.

Lakowicz, J. R. (1999). Energy transfer. *In* "Principles of Fluorescence Spectroscopy" (J. R. Lakowicz, ed.), 2nd edn., pp. 367–394. Kluwer Academic/Plenum Publishers, New York.

Miyawaki, A. (2003a). Visualization of the spatial and temporal dynamics of intracellular signaling. *Dev. Cell.* **4,** 295–305.

Miyawaki, A., Llopis, J., Heim, R., McCaffery, J. M., Adams, J. A., Ikura, M., and Tsien, R. Y. (1997). Fluorescent indicators for Ca^{2+} based on green fluorescent proteins and calmodulin. *Nature* **388,** 882–887.

Miyawaki, A., Sawano, A., and Kogure, T. (2003b). Lighting up cells: Labeling proteins with fluorophores. *Nat. Cell Biol.* Suppl. S1–S7.

Miyawaki, A., and Tsien, R. Y. (2000). Monitoring protein conformations and interactions by fluorescence resonance energy transfer between mutants of green fluorescent protein. *Methods Enzymol.* **327,** 472–500.

Nagai, T., Yamada, S., Tominaga, T., Ichikawa, M., and Miyawaki, A. (2004). Expanded dynamic range of fluorescent indicators for Ca^{2+} by circularly permutated yellow fluorescent proteins. *Proc. Natl. Acad. Sci. USA* **101,** 10554–10559.

Nguyen, A. W., and Daugherty, P. S. (2005). Evolutionary optimization of fluorescent proteins for intracellular FRET. *Nat. Biotechnol.* **23,** 355–360.

Rizzo, M. A., Springer, G. H., Granada, B., and Piston, D. W. (2004). An improved cyan fluorescent protein variant useful for FRET. *Nat. Biotechnol.* **22,** 445–449.

Shimozono, S., Fukano, T., Kimura, K. D., Mori, I., Kirino, Y., and Miyawaki, A. (2004). Slow Ca^{2+} dynamics in pharyngeal muscles in *Caenorhabditis elegans* during fast pumping. *EMBO Rep.* **5,** 521–526.

Shimozono, S., Hosoi, H., Mizuno, H., Fukano, T., Tahara, T., and Miyawaki, A. (2006). Concatenation of cyan and yellow fluorescent proteins for efficient resonance energy transfer. *Biochemistry* **45,** 6267–6271.

Stryer, L. (1978). Fluorescence energy transfer as a spectroscopic ruler. *Annu. Rev. Biochem.* **47,** 819–846.

Takemoto, K., Nagai, T., Miyawaki, A., and Miura, M. (2003). Spatio-temporal activation of caspase revealed by indicator that is insensitive to environmental effects. *J. Cell Biol.* **160,** 235–243.

Tsutsui, H., Karasawa, S., Shimizu, H., Nukina, N., and Miyawaki, A. (2005). Semi-rational engineering of a coral fluorescent protein into an efficient highlighter. *EMBO Rep.* **6,** 233–238.

Zacharias, D. A., Violin, J. D., Newton, A. C., and Tsien, R. Y. (2002). Partitioning of lipid-modified monomeric GFPs into microdomains of live cells. *Science* **296,** 855–857.

CHAPTER 17

Fluorescence Anisotropy Imaging Microscopy for Homo-FRET in Living Cells

Marc Tramier and Maïté Coppey-Moisan

Institut Jacques Monod
UMR 7592 CNRS
University Paris 6/University Paris 7
2 Place Jussieu, 75251
Paris Cedex 05, France

Abstract

In this chapter, we present the basic physical principles of the fluorescence anisotropy imaging microscopy (FAIM) and its application to study FP-tagged protein dynamics and interaction in live cells. The Förster mechanism of electronic

0091-679X/08 $35.00
DOI: 10.1016/S0091-679X(08)85017-0

energy transfer can occur between like chromophores (homo-fluorescence resonance energy transfer, homo-FRET) inducing fluorescence depolarization and can be monitored by fluorescence anisotropy. The energy transfer rate is fast compared to the rotational time of proteins, and therefore its detection as a fast depolarization process in the fluorescence anisotropy can be easily discriminated from rotational motion. Quantitative analysis of fluorescence anisotropy decays provides information on structural parameters: distance between the two interacting chromophores and spatial orientation between the chromophores within dimeric proteins. Fluorescence anisotropy decay is not easy to measure in living cells under the microscope and the instrumentations are necessarily sophisticated. In contrast, any type of microscope can be used to measure the steady-state anisotropy. Interestingly, two-photon excitation steady-state FAIM is a powerful tool for qualitative analysis of macromolecule interactions in living cells and can be used easily for time-lapse homo-FRET.

I. Introduction

The interaction of light with matter at the molecular level can reveal a great deal of information about molecules by investigating the physical properties of light. For example, the polarization of the emission can provide structural and dynamical information about the macromolecules and their associations. The polarization of light is defined as the orientation in time and space of the electric field vector. This orientation is spatially determined by decomposition of the electric field vector into three orthogonal components oriented according to given coordinated axes. The state of polarization is determined by the relative amplitudes of these components.

The polarization state can be modified when light interacts with matter, through absorption (dichroism) or transmission (birefringence) for example. In the case of fluorescence, electronic transitions involved in absorption and emission are characterized by electric dipoles of absorption and emission transition moments. The radiation from the emission dipole is linearly polarized, the electric vector of the fluorescence light being in the same plane as the dipole.

II. Photoselection Process

Linearly polarized light preferentially excites fluorophore molecules that have a dipole of absorption transition moments parallel to the electric vector of the excitation beam, the probability of excitation depending on the angle between the two vectors. This process of photoselection of isotropic distribution of an ensemble of chromophores leads to anisotropic spatial distribution of the excited species and thus to anisotropy of the emitted fluorescence. The fluorescence anisotropy r is defined by the following expression:

$$r = \frac{I_{||} - I_{\perp}}{I_{||} + 2I_{\perp}} \tag{1}$$

where $I_{||}$ and $I_{\perp}$ are the linearly polarized components of emission parallel and perpendicular to the exciting vector. $I_{||} + 2I_{\perp}$ is the total polarization-independent emission.

In the absence of rotation during the excited state and when there is no energy transfer, the emission anisotropy of a rigid isotropic solution of fluorophores excited by linearly polarized light is determined by photoselection and is given by the following expression (Gryczynski *et al.*, 2005):

$$r_0(\theta, \beta) = \left[\frac{3}{2}\langle\cos^2\theta\rangle - \frac{1}{2}\right]\left[\frac{3}{2}\cos^2\beta - \frac{1}{2}\right] \tag{2}$$

where θ is the angle of the absorption transition moment from the z-axis, and β is the angle between the absorption and the emission transition moments. The average value of $\cos^2\theta$ depends on the type of photoselection and is given by the following equation:

$$\langle\cos^2\theta\rangle = \frac{\int_0^{\pi/2}\cos^2\theta f_i(\theta)\mathrm{d}\theta}{\int_0^{\pi/2} f_i(\theta)\mathrm{d}\theta} \tag{3}$$

where $f_i(\theta)$ is the directional distribution of the excited state.

In the case of one-photon excitation and isotropic distribution, $f_i(\theta) = \cos^2\theta \sin\theta$ and Eq. (2) becomes

$$r_{01} = \frac{2}{5}\left(\frac{3}{2}\cos^2\beta - \frac{1}{2}\right) \tag{4}$$

In the case of two-photon excitation and isotropic distribution, $f_i(\theta) = \cos^4\theta \sin\theta$ and Eq. (2) becomes

$$r_{02} = \frac{4}{7}\left(\frac{3}{2}\cos^2\beta - \frac{1}{2}\right) \tag{5}$$

$\beta = 0$ if the absorption and emission dipoles are parallel, as in a single transition moment. In this latter case, the maximal anisotropies, r_{01} and r_{02}, are 0.4 and 0.571 for one-photon and two-photon excitation, respectively (Fig. 1).

In the case of anisotropic distribution of fluorophores (as in membranes), the expression of $f_i(\theta)$ becomes more complex (Axelrod, 1989; Baumann and Fayer, 1986; Chen and Van Der Meer, 1993).

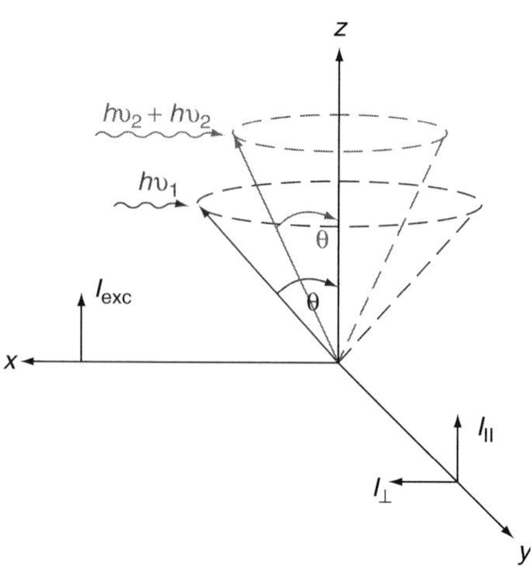

Fig. 1 Photoselection process by one- or two-photon excitation. Linearly polarized excitation light (I_{exc}) preferentially excites absorption dipole transition moment located in the blue cone for one-photon excitation and in the red cone for two-photon excitation. This property corresponds to a directional distribution of excited states proportional to $\cos^2\theta$ for a single absorbed photon and to $\cos^4\theta$ for two simultaneous absorbed photons. (See Plate no. 45 in the Color Plate Section.)

III. Rotational Depolarization

The fluorescence consists of uncorrelated individual events, which means that there is no dependence between photons emitted by the different fluorophores. The polarization of fluorescence emission, therefore, depends on the structural and dynamical features of the molecular system and geometric orientation of the excitation and observation. Rotational motion of the chromophore during the excited state randomizes the orientation of the emission dipoles with respect to the initial photoselected dipole and thus reduces the fluorescence anisotropy (Fig. 2). Time-resolved fluorescence anisotropy decay following the pulse excitation is defined by

$$r(t) = \frac{I_{||}(t) - I_{\perp}(t)}{I_{\mathrm{Total}}(t)} \tag{6}$$

Assuming a freely rotating spherical (or oblate) fluorophore, and in the absence of energy transfer, the decay of fluorescence anisotropy can be analyzed by the following expression:

$$r(t) = r_0 \mathrm{e}^{-t/\phi} \tag{7}$$

where r_0 is the anisotropy at time 0 and ϕ denotes the rotational correlation time of the fluorescent molecule. ϕ is correlated to the hydrodynamic volume of the

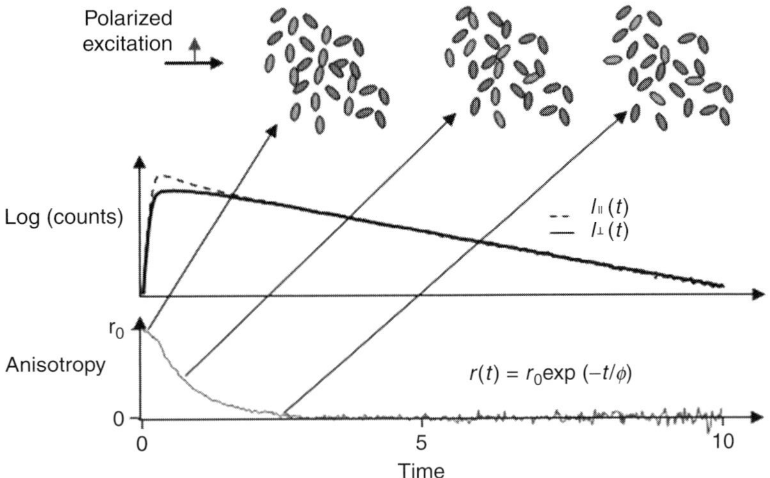

Fig. 2 Fluorescence anisotropy decay of Rhodamine 6G (Rd6G) in butanol. Pulsed polarized light preferentially excites parallel oriented absorption transition moments at time zero. During the excited state, the molecules loose their initial orientation by rotational diffusion (Top). Parallel fluorescence intensity at time zero is higher than the perpendicular one and the difference decreases during the fluorescence decay concomitantly to the rotation of the molecules (Middle). The corresponding fluorescence anisotropy decay exhibits exponential decay from r_0 to 0 with a relaxation time directly dependent on the rotational diffusion [Eqs. (7) and (8)] (Bottom). (See Plate no. 46 in the Color Plate Section.)

molecule, V, by the Stokes-Einstein equation:

$$\phi = \frac{\eta V}{kT} \tag{8}$$

where η is the viscosity of the medium, k the Boltzmann's constant, and T the temperature.

IV. Experimental Measurement of Fluorescence Anisotropy Decay in Confocal Microscopy

Anisotropy decays can be performed in either the time or the frequency domain (Gratton and Limkeman, 1983; Spencer and Weber, 1969) and rotational correlation times determined in living cells (Clayton *et al.*, 2002; Gautier *et al.*, 2001; Swaminathan *et al.*, 1997; Tramier *et al.*, 2000). In the time domain, a pulsed laser is focused through a high numerical aperture objective and the fluorescence collected by the objective is detected by time-correlated single photon counting. The optical design of the microscope results in four geometric components of the fluorescence polarization, where i_{vh} and i_{hv} pertain to the parallel direction and i_{vv} and i_{hh} to the perpendicular direction, relative to the direction of the laser excitation, with the first index standing for excitation and the second for emission. For anisotropy measurements, parallel ($i_{vh}(t)$) and perpendicular ($i_{hh}(t)$) decays can be acquired sequentially or simultaneously from the same sample spot (Gautier *et al.*, 2001; Swaminathan *et al.*, 1997; Tramier *et al.*, 2000) or image (Buehler *et al.*, 2000; Clayton *et al.*, 2002; Siegel *et al.*, 2003). A correction has to be made to take into account the different transmission efficiencies of each geometric component of the excited and emitted polarized light, as well as any depolarizing (or polarizing) effects linked to the microscope optics (Tramier *et al.*, 2000). The fluorescence anisotropy decay then becomes

$$r(t) = \frac{i_{vh}(t) - Gi_{hh}(t)}{i_{vh}(t) - 2Gi_{hh}(t)} \quad (9)$$

where G is determined from the steady-state intensity measurements of the four geometric components of the fluorescence polarization (Fig. 3).

Sv and Sh are the transmission efficiencies for, respectively, vertical and horizontal polarization of the excitation light, and S'v and S'h are the transmission efficiencies for, respectively, the vertical and horizontal polarization of the emission.

In the case of sequential acquisitions of parallel and perpendicular polarization, an additional correction has to be made to take into account possible photobleaching and laser fluctuations (Tramier *et al.*, 2000).

The anisotropy function is defined by

$$r(t) = \frac{D(t)}{S(t)} \quad (10)$$

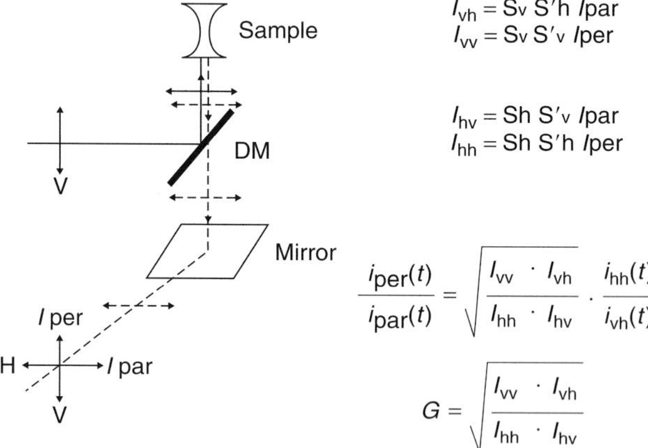

Fig. 3 *G* factor in confocal microscopy (see text for details).

where

$$D(t) = i_{||}(t) - i_{\perp}(t) \tag{11}$$

is the difference between the parallel and the perpendicular decay, and

$$S(t) = i_{||}(t) + 2i_{\perp}(t) \tag{12}$$

is the decay of the total intensity. By combining Eqs. (10–12) we obtain the following expressions:

$$3i_{||}(t) = S(t)[1 + 2r(t)] \tag{13}$$

and

$$3i_{\perp}(t) = S(t)[1 - r(t)] \tag{14}$$

In the case of single fluorescence lifetime species, $S(t)$ can be fitted with the theoretical model

$$S(t) = \frac{a}{3}\mathrm{e}^{-t/\tau} \tag{15}$$

with τ being the fluorescence lifetime and a being a constant.

For a freely rotating molecule (i.e., not hindered), $r(t)$ can be fitted with the theoretical model

$$r(t) = r_0\mathrm{e}^{-t/\phi} \tag{16}$$

with r_0 being the anisotropy at time 0 and ϕ the rotational time. It follows that

$$i_{||}(t) = a\mathrm{e}^{-t/\tau}(1 + 2r_0\mathrm{e}^{-t/\phi}) \tag{17}$$

and

$$i_{\perp}(t) = a\mathrm{e}^{-t/\tau}(1 - r_0\mathrm{e}^{-t/\phi}) \tag{18}$$

The experimental decays, $i_{\mathrm{vh}}(t)$ and $i_{\mathrm{hh}}(t)$, corresponding to the parallel and perpendicular polarized decays, respectively, for the microscope setup are distorted by the measurement apparatus and related to the real time behavior, $i_{||}(t)$ and $i_{\perp}(t)$ by the convolution product of the instrument response function IRF(t),

$$i_{\mathrm{vh}}(t) = G \times \mathrm{IRF}(t) \times i_{||}(t) \tag{19}$$

and

$$i_{\mathrm{hh}}(t) = \mathrm{IRF}(t) \times i_{\perp}(t) \tag{20}$$

where G is the transmission correction factor. The rotational time ϕ is determined from the fit of the following two experimental decays:

$$i_{\mathrm{vh}}(t) = G \times \mathrm{IRF}(t) \times [a\mathrm{e}^{-t/\tau}(1 + 2r_0\mathrm{e}^{-t/\phi})] \quad (21)$$

and

$$i_{\mathrm{hh}}(t) = \mathrm{IRF}(t) \times [a\mathrm{e}^{-t/\tau}(1 - r_0\mathrm{e}^{-t/\phi})] \quad (22)$$

The use of a standard molecule (known rotational time and fluorescence lifetime) is important for calibration of the system, and especially for determining the depolarization due to the objective. Indeed, the high numerical aperture of the objective generates a decrease in the anisotropy value at time 0 (Fig. 4) [described theoretically by Axelrod (1979)] without, however, modifying the kinetic parameter of the fluorescence anisotropy decay (Tramier *et al.*, 2000).

For a small molecule, such as Rhodamine 6G in butanol solution (usually used for calibration), the rotational time of the molecule is fast (820 ps) compared to the fluorescence lifetime (3.4 ns) (Tramier *et al.*, 2000). The molecules excited by photoselection have relaxed during the lifetime and the fluorescence anisotropy decay is zero at long time. The fluorescence lifetime determines the timescale of the motion that can be observed by fluorescence anisotropy decay measurements.

V. Fluorescence Anisotropy Decay of GFP-Tagged Proteins

For bigger molecules, such as fluorescent proteins, the rotational time of the whole protein is much longer than the excited state lifetime of the chromophore tag (Gautier *et al.*, 2001; Swaminathan *et al.*, 1997; Volkmer *et al.*, 2000). The reorientation of the emission dipole of the photoselected molecules is incomplete during the time window of the fluorescence emission. Analysis of the fluorescence

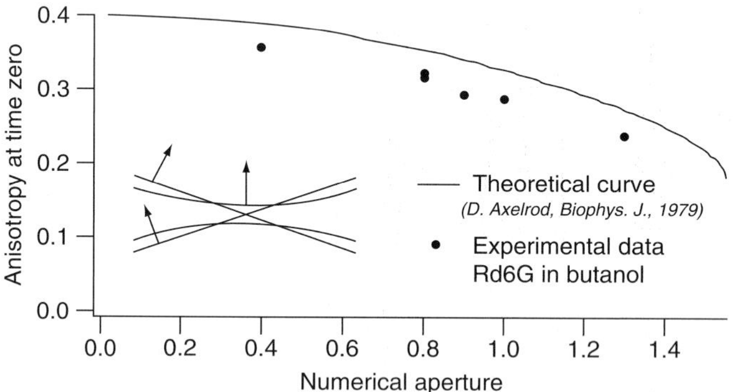

Fig. 4 Anisotropy value at time zero in function of the numerical aperture of the objective. High numerical aperture induces geometrical depolarization of the polarized focusing beam (sketch in the figure) theoretically calculated by Axelrod (1979). The predicted decrease of the anisotropy at time zero was characterized experimentally. The difference between theory and experimental data would correspond to additional depolarization into microscope optics.

anisotropy decay of green fluorescent protein (GFP) in cells gives a rotational time of 24 ns, which corresponds to the apparent volume of a sphere of 50 Å in diameter when assuming a subcellular viscosity of 1.5 cP (Fig. 5). The theoretical value for the apparent diameter of GFP assimilated to a sphere based on a 27-kDa molecular mass and a hydrated volume of a protein of ~1 cm^3/g, would be 44 Å.

The amplitude of the relaxation of the fluorescence anisotropy decay of GFP-tagged proteins during the time window of the fluorescence measurement decreases when the molecular mass increases. For GFP-tagged proteins with a molecular mass of up to 70 kDa, the fit of the decay can be determined with a monoexponential decay model; the rotational time of the chimeric protein can be obtained and its proportion in relation to the molecular mass measured (Fig. 6). The imprecision of the rotational time value, however, increases with the molecular mass, which limits the use of this technique to determine the apparent volume (molecular mass) of a relatively small protein tagged with fluorophore (Gautier et al., 2001; Tramier *et al.*, 2003).

VI. Fluorescence Depolarization by Homo-FRET

The Förster mechanism of electronic energy transfer (Förster, 1948) can occur between like chromophores if they are close enough (R <~$1.6R_0$, the Förster radius). Because the photophysical properties of the two donor molecules are the same, the excitation energy is reversibly transferred between the fluorescent tags. This transfer neither changes the fluorescence steady-state intensity nor the fluorescence lifetime. This homotransfer can only be monitored by fluorescence anisotropy (Weber, 1954). Indeed, if the dipoles of the donor and acceptor are not parallel, the emitted fluorescence from the acceptor due to fluorescence resonance energy transfer (FRET) from the photoselected donor is depolarized compared to

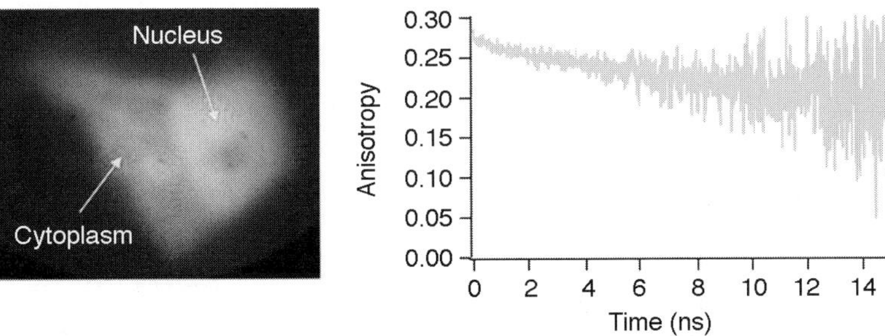

Fig. 5 Fluorescence anisotropy decay of GFP in the live cell. Using polarized confocal beam (480 nm) located in the cytoplasm or in the nucleus, the fluorescence anisotropy decays are similar with a relaxation correlation time of 24 ns in accordance with the Stokes-Einstein formula for rotational correlation time. As the fluorescence lifetime of the GFP is 2.6 ns, the anisotropy decay is incomplete during the accessible time window.

Protein	Number of experiments	τ (ns)	r_0	ϕ (ns)	Molecular weight (kDa)
$TK_{27}GFP$	6	2.62 ± 0.09	0.27 ± 0.02	35.6 ± 8.6	32
$TK_{210}GFP$	5	2.57 ± 0.05	0.25 ± 0.02	54.6 ± 4.2	52
$TK_{366}GFP$	3	2.64 ± 0.15	0.26 ± 0.01	81.0 ± 15.3	70

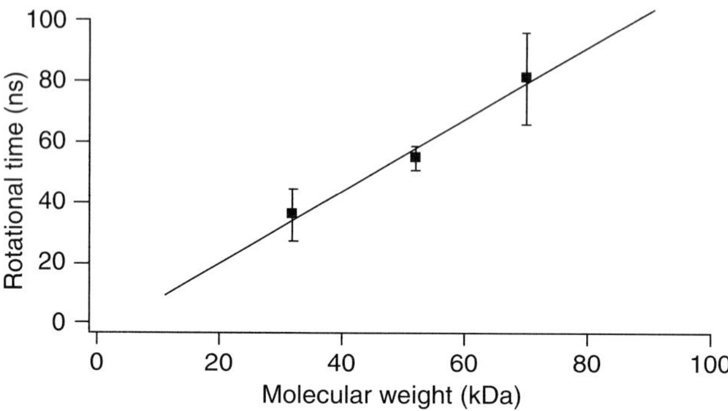

Fig. 6 Rotational time of different GFP-fusion proteins.

the excitation (Fig. 7). Interestingly, the energy transfer rate is fast compared to the rotational time of proteins, and therefore can be detected as a fast depolarization correlation time in the fluorescence anisotropy decay.

Homo-FRET between identical GFP chromophores fused to monomers of Herpes simplex virus (HSV-1) thymidine kinase (TK) was unveiled directly in live cells by fluorescence anisotropy decay measurements (Gautier *et al.*, 2001). The chimeric protein $TK_{366}GFP$ can exist in cells in the form of monomeric or dimeric protein as shown from anisotropy decay data. The rotational time of the monomer was estimated to be 81.0 ± 15.3 ns (Fig. 8A), which gives a value of $\sim$162 ns for the rotation of the dimer. This rotational time cannot be revealed from the anisotropy decay because the de-excitation process of GFP is much faster (2.6 ns). The fast depolarization relaxation time of 2.4 ± 0.3 ns revealed from the fit of the fluorescence anisotropy decay arises from homotransfer between two

Fig. 7 Depolarization of fluorescence light by homo-FRET. After polarized excitation, transfer of energy between like fluorophores oriented differently induces a depolarization of the emitted fluorescence. (See Plate no. 47 in the Color Plate Section.)

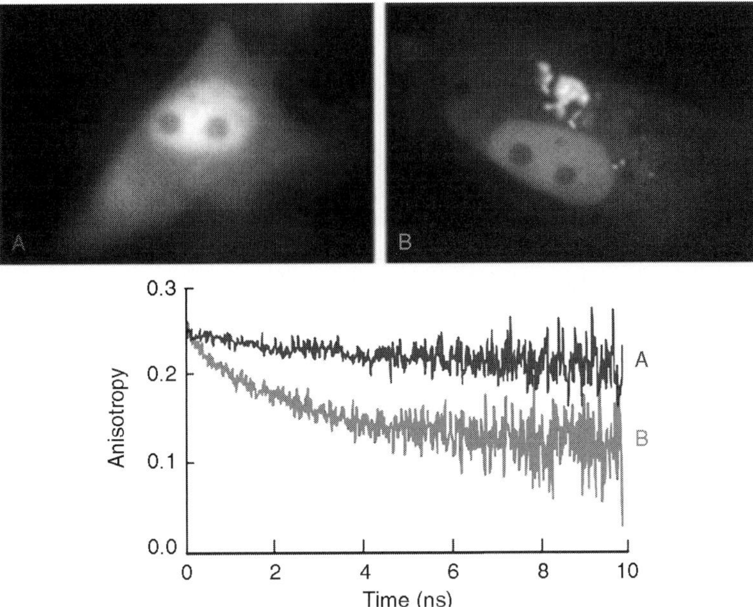

Fig. 8 Subcellular fluorescence anisotropy decays of $TK_{366}GFP$ proteins. (Top) Steady-state fluorescence images of Vero cells expressing $TK_{366}GFP$ presenting only a diffuse cytoplasmic and nuclear fluorescence pattern (A) and containing fluorescent aggregates (B). (Bottom) Time-resolved fluorescence depolarization from a cytoplasmic area of diffuse fluorescence (A) and from an area inside an aggregate (B). For cells containing aggregates, anisotropy decays from nuclear or cytoplasmic area of diffuse fluorescence were similar to that obtained from aggregates (B). The rapid decrease of the anisotropy decay in (B) is a signature of homo-FRET between GFPs. (See Plate no. 48 in the Color Plate Section.)

GFP chromophores of interacting $TK_{366}GFP$ monomers (Fig. 8B). Indeed, since the GFP chromophore is rigidly fixed inside the proteic barrel because rapid motion is not evidenced either *in vitro* (Volkmer *et al.*, 2000) or in the living cells (Gautier *et al.*, 2001), the fast fluorescence depolarization relaxation cannot be due to a wobbling movement of the GFP chromophore. Moreover, the GFP moieties themselves do not move significantly because no such rotation could be evidenced in the fluorescence anisotropy decay of the $TK_{366}GFP$ monomer.

In the case of a symmetric dimer, the time-dependent anisotropy can be modeled by (Tanaka and Mataga, 1979):

$$r(t) = \frac{r_0}{4}[(3 - 3\cos^2\theta)e^{-2\omega t} + 3\cos^2\theta + 1] \tag{23}$$

where θ is the mutual orientation between the two GFP chromophores and ω is the transfer rate. In the static limits (no reorientation of the transition moments during the fluorescence lifetime), the transfer rate is linked to the distance R between the two interacting chromophores by

$$\omega = \frac{3}{2}\langle\kappa^2\rangle\left(\frac{R_0}{R}\right)^6\tau^{-1} \tag{24}$$

where $\langle\kappa^2\rangle$ is the average angular dependence of dipole–dipole coupling and R_0 is the dynamical average Förster radius. With this type of symmetric dimer model, it becomes possible to estimate the structural parameters of $TK_{366}GFP$ in living cells (Fig. 9): the mutual orientation of the transition dipoles of the two GFP chromophores was calculated from the residual anisotropy to be $44.6^\circ \pm 1.6^\circ$, and the upper intermolecular limit between the two fluorescent tags, $R = 70$ Å, was calculated from the energy transfer rate (Gautier *et al.*, 2001; Tramier *et al.*, 2003).

VII. Steady-State Fluorescence Anisotropy Imaging

Fluorescence anisotropy decay is not easy to measure in living cells under the microscope. Whether in the time domain or in the frequency domain, the instrumentations are necessarily sophisticated. In contrast, any type of microscope can be used to measure steady-state anisotropy measurements (Lidke *et al.*, 2003). The steady-state anisotropy r is given by the average of $r(t)$ weighted by $i(t)$:

$$r = \frac{\int r(t)i(t)\mathrm{d}t}{\int i(t)\mathrm{d}t} \tag{25}$$

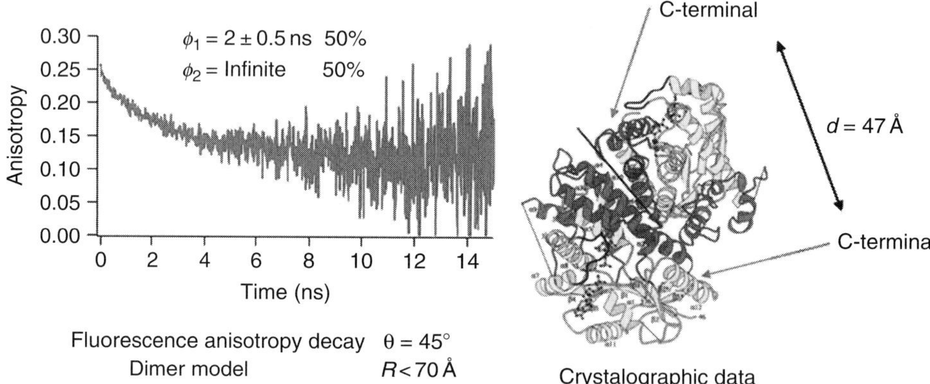

Fig. 9 Fluorescence anisotropy decay analyzed with a model of a $TK_{366}GFP$ symmetric dimer. The relaxation correlation time and the anisotropy at infinite time give informations about the transfer rate (then upper distance) and the angle between the two GFP inside the dimer. The upper distance is in good agreement with crystallographic data (Wild *et al.*, 1997) and the distance between the two C-terminals in the dimer. (See Plate no. 49 in the Color Plate Section.)

In the simplest case corresponding to the free rotation of a single lifetime fluorophore,

$$i(t) = a\mathrm{e}^{-t/\tau} \quad (26)$$

and

$$r(t) = r_0\mathrm{e}^{-t/\phi} \quad (27)$$

we obtain the equation called the Perrin equation, so called because it was originally described by F. Perrin in 1932 (Perrin, 1932):

$$r = \frac{r_0}{1 + (\tau/\phi)} \quad (28)$$

For small fluorescent molecules (such as Rhodamine 6G), the rotational time is fast in comparison to the lifetime. Measurements of r can thus afford insight into the viscosity changes through variation of ϕ. In the case of fluorescent proteins, the rotational time is longer than the fluorescence lifetime. Since the reorientation of the emission dipole by rotation is quasi-null during the fluorescence lifetime for high molecular mass proteins (or macromolecular complexes), $r \simeq r_0$ in the absence of homo-FRET and wobbling movements of the GFP moieties.

We took advantage of the dual polarization possibility of the multifocal two-photon excitation of the TriMScope (LaVision biotech, Germany) to perform fluorescence anisotropy imaging microscopy (FAIM) measurements in living cells (Fig. 10). In this pseudo-wide-field two-photon microscope, a single beam (average power of ~1.5 W) from a mode-locked Ti:Sa laser (~100-fs pulses at a repetition rate of 80 MHz, Tsunami pumped by a Millenia 10 W, Spectra-Physics, France) is split up into two sets of 32 beams by a 50/50 beam splitter and mirrors, one set of which has its polarization rotated by 90° with a half-wave plate creating h- and v-polarization. The two sets of beams are recombined through a polarizer cube before passing through a 2000-Hz scanner. A rotating shutter is used to select h- or v-polarized beams before illuminating the back aperture of a 60× NA 1.3 infrared water immersion objective (Olympus). A line of foci is then created at the focal plane, which can be scanned across the sample, making a pseudo-wide-field illumination with two-photon excitation. A filter wheel placed in front of a CCD camera selects the analyzer for polarization measurements.

In two-photon excitation the theoretical r_0 value is higher than in one-photon excitation ($r_{02} = 0.571$ and $r_{01} = 0.4$, see earlier). The dynamical range of r variations is thus higher in two-photon than in one-photon excitation, even if some depolarization arises from the use of high numerical aperture objectives.

The steady-state anisotropy of Rhodamine 6G is weak because the rotational time of the molecule is much smaller than the lifetime. The dynamic range of the r measurements is, however, sufficient to discriminate between the values of r in butanol and in water solutions (Fig. 11, Top). Similarly, the r values of GFP

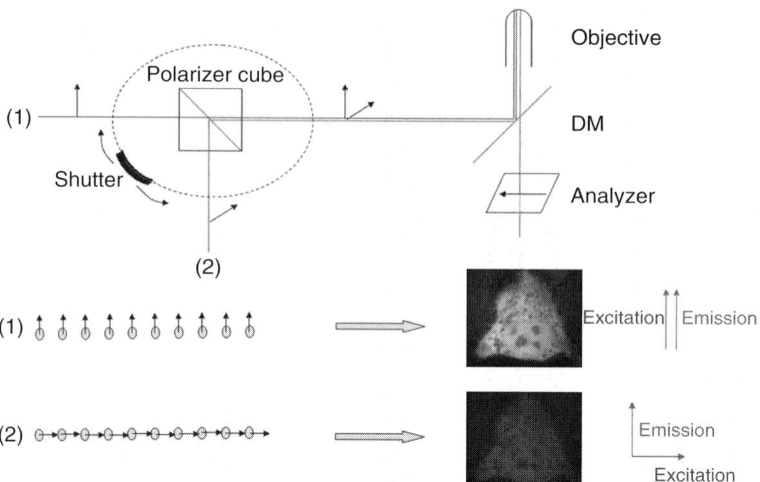

Fig. 10 Schematic of steady-state anisotropy measurement with the TriMScope. Originally, the 64 foci of the TriMScope are obtained by combining into a polarizer cube two lines of 32 beams with cross linear polarization. Taking advantage of it, a shutter is used to selected position (1) or position (2) and to excite the sample with polarized light. A fixed analyzer at the emission is then used to acquire sequentially parallel and perpendicular intensity by moving the shutter.

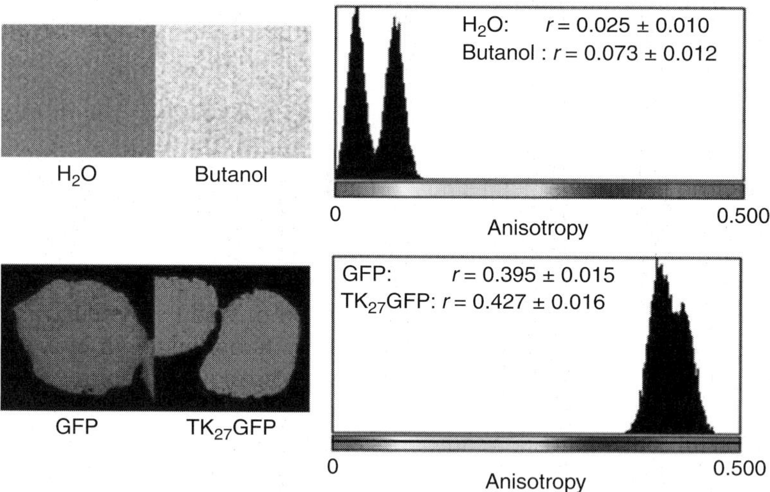

Fig. 11 Steady-state fluorescence anisotropy image of rotating fluorophores. Fluorescence anisotropy images microscopy was carried out with the TriMScope at 850 nm excitation. (Top) Anisotropy images of Rhodamine 6G (Rd6G) in solution of different viscosity, water and butanol. (Bottom) Anisotropy images of cells expressing fluorescent proteins of different size, GFP and $TK_{27}GFP$. (See Plate no. 50 in the Color Plate Section.)

(28 kDa) and of $TK_{27}GFP$ (32 kDa) can be discriminated due to the variation in the molecular mass of these monomeric proteins (Fig. 11, Bottom). By using the previously determined lifetime and rotational time of these proteins (Gautier *et al.*, 2001), the experimental initial anisotropy value is found to be $r_0 = 0.444$ and 0.454 from GFP and $TK_{27}GFP$ measurements, respectively. These experimental values should be compared to the theoretical value $r_0 = 0.571$ for two-photon excitation. The decrease of r_0 arises from the depolarization effect of the high numerical aperture objective.

VIII. Imaging Homo-FRET by Two-Photon FAIM

In the presence of homo-FRET, the expression of the fluorescence anisotropy decay, $r(t)$, is different from that of Eq. (27) and depends on the energy migration between the chromophores in the oligomer. For example, in the case of a symmetric dimer (as described earlier) by combining the Eqs. (26) and (23) in the Eq. (25), the steady-state anisotropy becomes:

$$r = \frac{r_0}{4}\left[\frac{3 - 3\cos^2\theta}{2\omega\tau + 1} + 3\cos^2\theta + 1\right] \tag{29}$$

The value of the steady-state anisotropy depends on τ, the fluorescence lifetime; ω, the energy transfer rate; and θ, the mutual orientation between the transition moments of the two chromophores in the dimer. The fluorescence lifetime is a known fixed parameter, and the two other unknown parameters are independent from each other. It appears, thus, that θ plays a critical role in the value of anisotropy. For example, if $\theta = \pi/2$ and $\omega\tau = 1$ (50% efficiency of energy transfer), $r = r_0/2$. If $\theta = 0$ (the two transition moments are parallel or antiparallel), however, $r = r_0$ whatever the transfer rate: there is no fluorescence depolarization due to homo-FRET.

FAIM performed with the TriMScope on the tandem, GFP-L-GFP (where L is a peptide linker), gives an r value of 0.311, which corresponds to the existence of homotransfer between the two GFP chromophores. Indeed, the anisotropy value of the GFP fluorescence of the mCherry-L-GFP tandem is $r = 0.411$, which in comparison to the experimental r_0 value (i.e., 0.45) corresponds to a weak fluorescence depolarization due to rotation of the whole protein (a molecular mass of 55 kDa) (Fig. 12).

In the case of higher order oligomerization, the steady-state anisotropy will decrease due to homo-FRET (if the orientations between the dipoles of the donors and of the acceptors are not parallel or antiparallel); a quantitative analysis is, however, more difficult to perform without a structural model. In particular cases, where very efficient transfer takes place between randomly oriented chromophores,

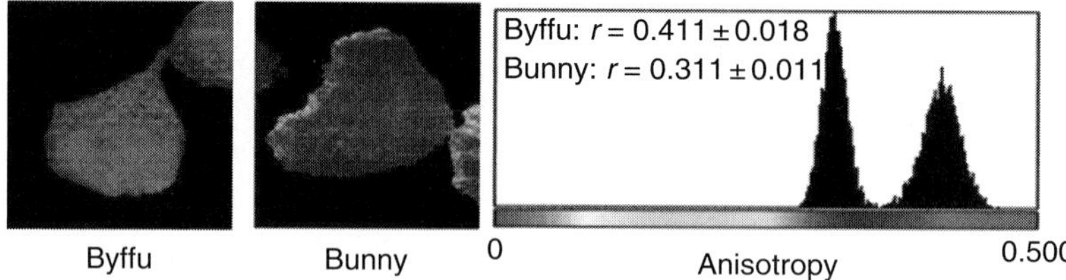

Fig. 12 Homo-FRET by FAIM in GFP tandem protein. FAIM of cells expressing Byffu (mCherry-L-GFP tandem) and Bunny (GFP-L-GFP tandem with the same linker as Byffu) and their corresponding histogram. Acquisitions were done using the TriMScope with 850 nm excitation wavelength and a 515–560 nm emission filter. (See Plate no. 51 in the Color Plate Section.)

the r value of the oligomer constituted of n monomers will be n^{-1} that of the monomer (Runnels and Scarlata, 1995).

IX. Biological Applications with GFP-Tagged Proteins

FKBP (FK506 binding protein) is a cytoplasmic human protein that serves as the initial intracellular target for the natural product immunosuppressive drugs FK506 and rapamycin. The drug AP20187 developed by ARIAD Pharmaceuticals Inc. (Cambridge, MA) cross-links fusion proteins containing FKBP domain, which induces homodimerization of the fusion protein.

By using the ARGENT™-regulated homodimerization kit, GFP was fused to FKBP domain, then expressed in living cells and AP20187-induced homodimerization was monitored in real time by FAIM with the TriMScope (Fig. 13). The decrease in the anisotropy of GFP fluorescence from 0.43 to 0.37 accounts for homo-FRET between two GFP chromophores brought together by the AP20187-triggered homodimerization of FKBP.

Arrestins are important proteins that regulate the function of serpentine heptahelical G-protein–coupled receptors (GPCRs) and contribute to multiple signaling pathways downstream of receptors. We have shown that β-arrestins can exist as homo- and heterooligomers in living cells using different approaches, including coimmunoprecipitation of epitope-tagged β-arrestins and Bioluminescence resonance energy transfer (BRET) and hetero-FRET (Storez *et al.*, 2005). Homo-FRET, however, cannot be revealed in the fusion protein β-arrestin2-GFP, probably because the orientation of the GFP dipoles in the oligomer is parallel or antiparallel as shown in Fig. 14.

Previous studies on visual arrestin *in vitro* have proposed a model in which oligomerization prevents the inappropriate activation of arrestin and subsequent deleterious cellular effects are avoided. To verify whether this hypothesis might be valid for β-arrestins, we studied the effect of forced and controlled β-arrestin oligomerization on β-arrestin activity. We designed a construct in which an FKBP binding motif was fused downstream from the β-arrestin2 C-terminal tail

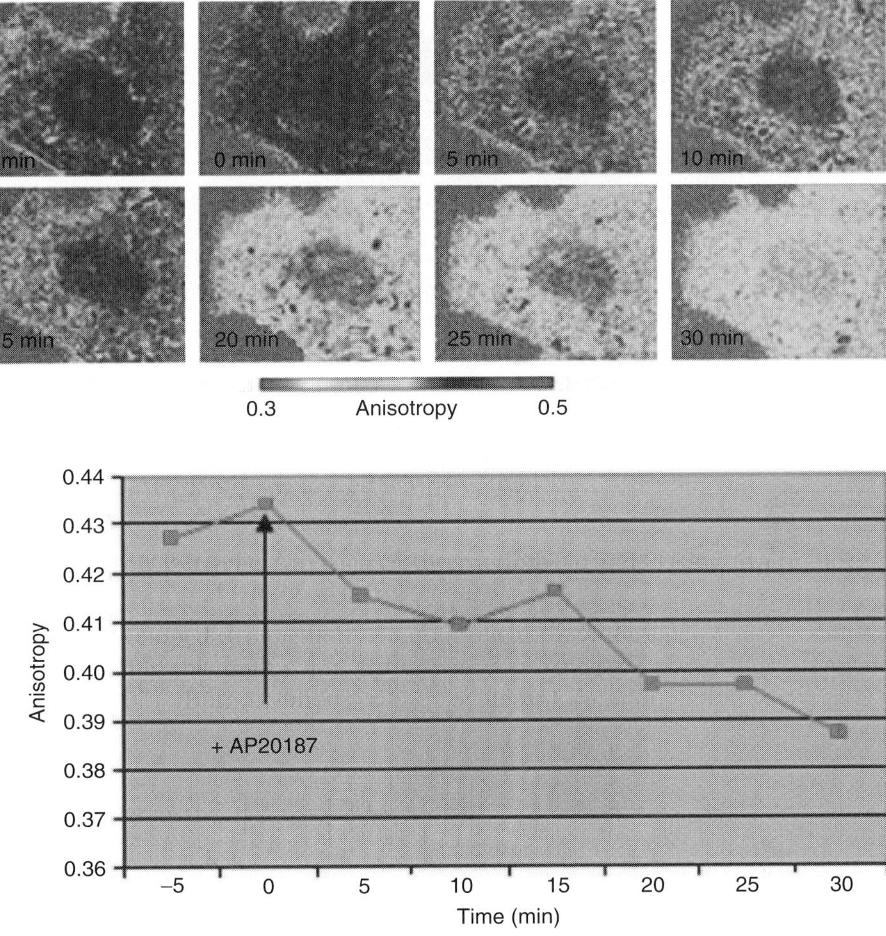

Fig. 13 Time lapse of FKBP-GFP expressing cell after addition of AP20187. FAIM were carried out each 5 min using the TriMScope with 850 nm excitation wavelength. AP20187 was added at time zero to induce dimerization of FKBP-GFP protein. The pseudo-colored FAIM images of FKBP-GFP expressing cell were presented (Top). The corresponding mean value of anisotropy is presented as a curve (Bottom). (See Plate no. 52 in the Color Plate Section.)

and upstream from the GFP (βarr2-FKBP-GFP). Using FAIM we verified that on addition of the FKBP-dimerizing small molecule AP20187, β-arrestin2 fusion proteins were all clustered through FKBP dimerization (Fig. 14). On angiotensin stimulation, pre-oligomerized βarr2-FKBP-GFP translocated to the activated receptors and accumulated in clathrin-coated pits and endosomes. These results indicate that the oligomeric organization of βarr2 molecules per se does not prevent their reactivity to receptor activation (Storez *et al.*, 2005).

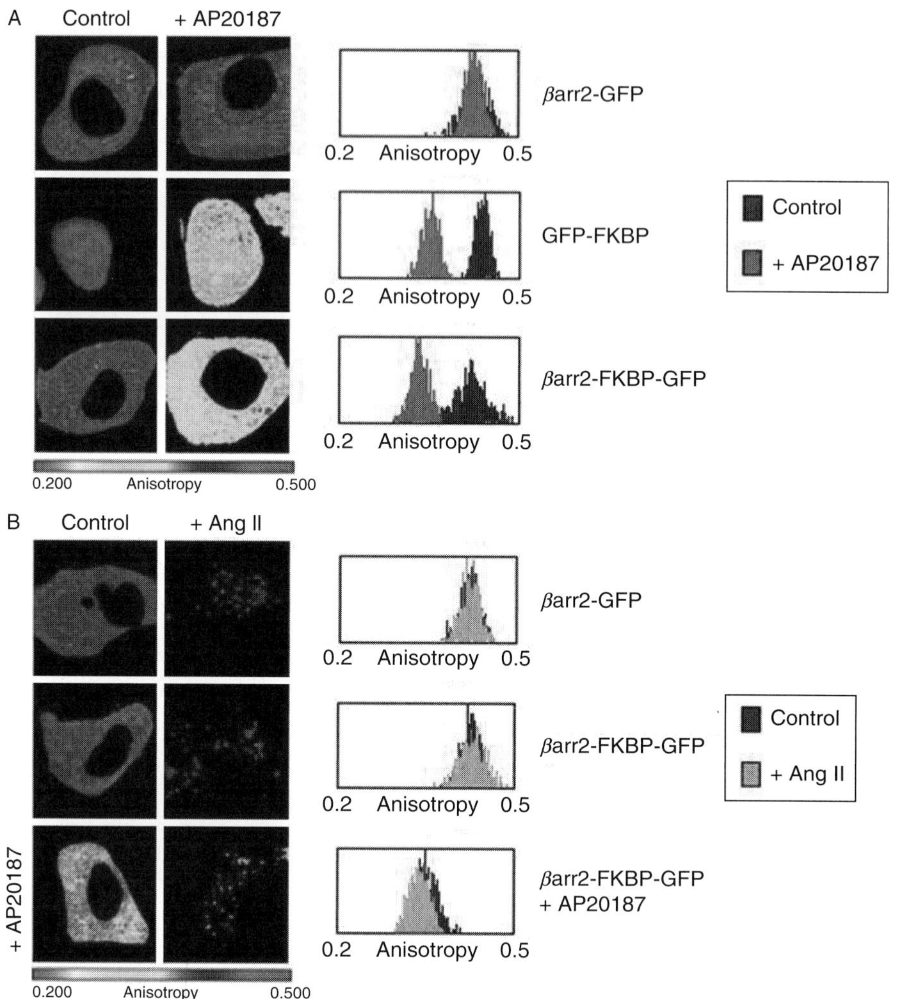

Fig. 14 Two-photon steady-state fluorescence anisotropy images of GFP-tagged proteins in living cells. COS-7 cells expressing control βarr2-GFP, GFP-FKBP, or βarr2-FKBP-GFP, plated on glass coverslips, were incubated or not for 1 h with 50 nM of the homodimerizing compound AP20187. Fluorescence anisotropy images were presented in pseudo-color. The pixel value distribution is shown on the histogram. (A) Anisotropy in cells expressing βarr2-GFP was similar to that measured in cells expressing FKBP-GFP in the absence of the dimerizing ligand AP20187 (first and second blue histograms from the top). For both GFP-FKBP and βarr2-FKBP-GFP, AP20187 induced a decrease of the anisotropy (histograms of second and third row), which corresponds to a dimerization through the FKBP domains. As expected, control βarr2-GFP did not show any decrease of the anisotropy parameter after AP20187 addition (Panel A, first row). (B) Cells coexpressing the AT1 angiotensin receptor (AT1AR) and either βarr2-GFP or βarr2-FKBP-GFP were incubated or not with AP20187 and stimulated or not with angiotensin II (Ang II) for 10 min before being analyzed as in (A). After AT1AR stimulation, both βarr2-GFP and βarr2-FKBP-GFP were accumulated in endosomes (pictures) without any change of their anisotropy parameter, compared to the unstimulated condition. βarr2-FKBP-GFP molecules, which were all predimerized by AP20187 localized to endosomes as the nondimerized control (compare pictures of second and third rows). (See Plate no. 53 in the Color Plate Section.)

X. Conclusion

In this chapter, we have presented the basic physical principals of the fluorescence anisotropy. We have shown that quantitative analysis of fluorescence anisotropy decays provides informations on structural parameters: distance between the two interacting chromophores and spatial orientation between the chromophores within dimeric proteins. Interestingly, two-photon excitation steady-state FAIM is a powerful tool for qualitative analysis of macromolecule interactions in living cells and can be used easily for time-lapse homo-FRET. Hetero-FRET imaging enables measurement of spatiotemporal dynamics of protein–protein interactions and signaling activity in live cells and is becoming a widely used approach in cellular biology (Wallrabe and Periasamy, 2005; Yasuda, 2006). Hetero-FRET requires, however, the use of two spectrally different chromophores. Although an impressive set of fluorescent proteins, from UV-excitable to far-red, provides different donor/acceptor pairs that can be used for hetero-FRET measurements, there are limitations in the use of each hetero-FRET pair and in the different methods to measure hetero-FRET (Shaner *et al.*, 2005; Yasuda, 2006). In contrast, homo-FRET requires a single type of chromophore. EGFP provides the best chormophore for homo-FRET because its fluorescence decay is mainly monoexponential and its fluorescence is less sensitive to light excitation (no photoconversion) than the other fluorescent proteins (Tramier *et al.*, 2006; Valentin *et al.*, 2005). FAIM has limitations, however, to detect homo-FRET in molecules with nonrandom orientation. Indeed, parallel or antiparallel orientations of the dipoles of the donor and the acceptor do not induce fluorescence depolarization whatever the transfer rate of homo-FRET.

Acknowledgments

We are grateful to Dr. Stephano Marullo for providing the FKBP-GFP construct and to Antonia Kropfinger for correction of the English language. Experiments were performed at the Imaging facilities center of the IJM and was supported by the Fondation pour la Recherche Médicale, the Region Ile de France (SESAME), the GEFLUC, the CNRS (ACI BCSM), the Association pour la Recherche sur le Cancer (ARC), and the Agence Nationale de la recherche.

References

Axelrod, D. (1979). Carbocyanine dye orientation in red cell membrane studied by microscopic fluorescence polarization. *Biophys. J.* **26,** 557–574.

Axelrod, D. (1989). Fluorescence polarization microscopy. *Methods Cell Biol.* **30,** 333–352.

Baumann, J., and Fayer, M. D. (1986). Excitation transfer in disordered two-dimensional and anisotropic three-dimensional systems: Effects of spatial geometry on time-resolved observables. *J. Chem. Phys.* **85**(7), 4087–4107.

Buehler, C., Dong, C. Y., So, P. T. C., French, T., and Gratton, E. (2000). Time-resolved polarization imaging by pump-probe (stimulated emission) fluorescence microscopy. *Biophys. J.* **79,** 536–549.

Chen, S. Y., and Van Der Meer, B. W. (1993). Theory of two-photon induced fluorescence anisotropy decay in membranes. *Biophys. J.* **64,** 1567–1575.

Clayton, A. H. A., Hanley, Q. S., Arndt-Jovin, D. J., Subramaniam, V., and Jovin, T. M. (2002). Dynamic fluorescence anisotropy imaging microscopy in the frequency domain (rFLIM). *Biophys. J.* **83,** 1631–1649.

Förster, T. (1948). Zwischenmolekulare energiewanderung und fluoreszenz. *Ann. Phys.* **2,** 55–75.

Gautier, I., Tramier, M., Durieux, C., Coppey, J., Pansu, R. B., Nicolas, J. C., Kemnitz, K., and Coppey-Moisan, M. (2001). Homo-FRET microscopy in living cells to measure monomer-dimer transition of GFP-tagged proteins. *Biophys. J.* **80,** 3000–3008.

Gratton, E., and Limkeman, M. (1983). A continuously variable frequency cross-correlation phase fluorometer with picosecond resolution. *Biophys. J.* **44,** 315–324.

Gryczynski, Z., Gryczynski, I., and Lakowicz, J. R. (2005). Basics of fluorescence and FRET. *In* "Molecular Imaging: FRET Microscopy and Spectroscopy" (R. N. Day, and A Periasamy, eds.), pp. 21–56. Oxford University Press, New York.

Lidke, D. S., Nagy, P., Barisa, B. G., Heintzmann, R., Post, J. N., Lidke, K. A., Clayton, A. H. A., Arndt-Jovin, D. J., and Jovin, T. M. (2003). Imaging molecular interactions in cells by dynamic and static fluorescence anisotropy (rFLIM and emFRET). *Biochem. Soc. Trans.* **31**(5), 1020–1027.

Perrin, F. (1932). Théorie quantique des transferts d'activation entre molécules de même espèce. Cas des solutions fluorescentes. *Ann. Phys. (Paris)* **17,** 283–314.

Runnels, L. W., and Scarlata, S. F. (1995). Theory and application of fluorescence homotransfer to melittin oligomerization. *Biophys. J.* **69,** 1569–1583.

Shaner, N. C., Steinbach, P. A., and Tsien, R. Y. (2005). A guide to choosing fluorescent proteins. *Nat. Methods* **2,** 905–909.

Siegel, J., Suhling, K., Lévêque-Fort, S., Webb, S. E. D., Davis, D. M., Phillips, D., Sabharwal, Y., and French, P. M. W. (2003). Wide-field time-resolved fluorescence anisotropy imaging (TR-FAIM): Imaging the rotational mobility of a fluorophore. *Rev. Sci. Instrum.* **74**(1), 182–192.

Spencer, R. D., and Weber, G. (1969). Measurements of subnanosecond fluorescence lifetimes with a cross-correlation phase fluorometer. *Ann. NY Acad. Sci.* **158,** 361–376.

Storez, H., Scott, M. G., Issafras, H., Burtey, A., Benmerah, A., Muntaner, O., Piolot, T., Tramier, M., Coppey-Moisan, M., Bouvier, M., Labbe-Jullie, C., and Marullo, S. (2005). Homo- and hetero-oligomerization of beta-arrestins in living cells. *J. Biol. Chem.* **280**(48), 40210–40215.

Swaminathan, R., Hoang, C. P., and Verkman, A. S. (1997). Photobleaching recovery and anisotropy decay of green fluorescent protein GFP-S65T in solution and cells: Cytoplasmic viscosity probed by green fluorescent protein translational and rotational diffusion. *Biophys. J.* **72,** 1900–1907.

Tanaka, F., and Mataga, N. (1979). Theory of time-dependent photo-selection in interacting fixed systems. *Photochem. Photobiol.* **29,** 1091–1097.

Tramier, M., Kemnitz, K., Durieux, C., Coppey, J., Denjean, P., Pansu, B., and Coppey-Moisan, M. (2000). Restrained torsional dynamics of nuclear DNA in living proliferative mammalian cells. *Biophys. J.* **78,** 2614–2627.

Tramier, M., Piolot, T., Gautier, I., Mignotte, V., Coppey, J., Kemnitz, K., Durieux, C., and Coppey-Moisan, M. (2003). Homo-FRET versus hetero-FRET to probe homodimers in living cells. *Methods Enzymol.* **360,** 580–597.

Tramier, M., Zahid, M., Mevel, J. C., Masse, M. J., and Coppey-Moisan, M. (2006). Sensitivity of CFP/YFP and GFP/mCherry pairs to donor photobleaching on FRET determination by FLIM in living cells. *Microsc. Res. Tech.* **69**(11), 933–999.

Valentin, G., Verheggen, C., Piolot, T., Neel, H., Coppey-Moisan, M., and Bertrand, E. (2005). Photoconversion of YFP into a CFP-like species during acceptor photobleaching FRET experiments. *Nat. Methods* **2,** 801.

Volkmer, A., Subramaniam, V., Birch, D. J. S., and Jovin, T. M. (2000). One- and two-photon excited fluorescence lifetimes and anisotropy decays of green fluorescent proteins. *Biophys. J.* **78,** 1589–1598.

Wallrabe, H., and Periasamy, A. (2005). Imaging protein molecules using FRET and FLIM microscopy. *Curr. Opin. Biotechnol.* **16,** 19–27.

Weber, G. (1954). Dependence of polarization of the fluorescence on the concentration. *Trans. Faraday Soc.* **50,** 552–555.

Wild, K., Bohner, T., Folkers, G., and Schulz, G. E. (1997). The structure of thymidine kinase from Herpes simplex virus type 1 in complex with substrate analogue. *Protein Sci.* **6,** 2097–2106.

Yasuda, R. (2006). Imaging spatiotemporal dynamics of neuronal signaling using fluorescence resonance energy transfer and fluorescence lifetime imaging microscopy. *Curr. Opin. Neurobiol.* **16,** 551–561.

CHAPTER 18

FRET by Fluorescence Polarization Microscopy

David W. Piston* and Mark A. Rizzo†

*Department of Molecular Physiology & Biophysics
Vanderbilt University Medical Center
Nashville, Tennessee 37232

†Department of Physiology
University of Maryland School of Medicine
Baltimore, Maryland 21201

Abstract

The widespread success in using genetically encoded fluorescent proteins (FPs) to track protein motion in living cells has led to extensive interest in measuring Förster resonance energy transfer (FRET) between two FPs of different colors. FRET occurs over distances less than 10-nm and can thus be used to detect

0091-679X/08 $35.00
DOI: 10.1016/S0091-679X(08)85018-2

protein–protein interactions and changes in protein conformation. However, FP-FRET measurements are complicated by the spectral properties of FPs. Consequently, extensive correction or photo-destructive approaches have been used to detect the presence of FRET. Since these methods limit the temporal and spatial resolution of FRET measurements, they are not well suited for many live-cell imaging applications. Here, we describe an alternative approach to detect FP-FRET by measuring fluorescence anisotropies (AFRET). Since FPs are large in size, excitation of FPs with polarized light results in highly polarized emission. In this case, FRET to a second FP that lies outside the photoselection plane will depolarize the fluorescence. This method provides high contrast and unambiguous indication of FRET using a simple image collection strategy that can be easily adapted to any modality including widefield and laser scanning approaches. In this chapter, we will discuss the theory behind AFRET imaging, calculation of FP anisotropies using fluorescent microscopes, and configuration of microscopes for AFRET experiments.

I. Introduction

Fluorescent protein (FP)–Förster resonance energy transfer (FRET) experiments generally incorporate a blue-shifted donor FP such as a cyan FP (CFP), and a red-shifted acceptor FP such as a yellow FP (YFP) (Miyawaki *et al.*, 1997). Provided that the donor and acceptor molecules are in close proximity (~5 nm apart), energy can transfer from the excited state of the donor FP to the acceptor FP. This leads to quenching of donor fluorescence and enhanced fluorescence emission from the acceptor fluorophore. Importantly, the mechanism of FRET involves interaction of the two fluorescence dipole moments. Transfer between these two electric fields must occur between vibrational states that have equivalent, or resonant energy levels. As a consequence of the physical nature of this interaction, FRET can only occur over very short distances and is very sensitive to the relative orientations of the FRET pair. If the FRET pair consists of two FPs, the separation between the two dipole moments must be on the order of 5 nm, or about the length of a single FP. The range of FRET makes it potentially very useful for measuring the dynamics of protein biology, being either conformational changes or protein–protein interactions.

While an understanding of the physical basis of FRET is essential for engineering FP biosensors, measurement of FRET in fluorescence microscopes requires much more practical considerations. Success of a given FP-FRET experiment depends on sufficient contrast between the dynamic states being measured (e.g., FRET vs no FRET). In this respect, FRET imaging is no different than imaging any other fluorescent sensor. Higher contrast enables faster sampling and better resolution, whereas lower contrast diminishes the overall usefulness of the assay. Optimization of FP-FRET contrast poses a number of technical challenges that are related to the spectral and physical properties of FPs. First, there is a disparity in brightness between the possible donor and acceptor FPs. Typically, cyan or blue FPs are paired with green or yellow FPs. The brightness of the most common donor, ECFP is approximately fivefold less than that of the YFPs, and even the brightest CFP (Cerulean) is still twofold less brighter than the latest generation YFP counterparts (Citrine and Venus) (Shaner *et al.*, 2005). This enhances excitation crosstalk of the acceptor protein relative to the amount of the excited donor excitation, even at wavelengths that are selective for donor excitation. Second, the emission spectra of the donor FPs are very broad, and produce crosstalk emission into the acceptor collection channel. Thus, FRET emission of the acceptor FP is convolved with emission from non-FRET excitation. Contrast of intensity-based measurements is reduced by both of these crosstalk mechanisms and the associated corrective measurements required to separate FRET from the crosstalk emission.

Several approaches to minimize and correct for FP crosstalk in FRET measurements have been developed. An optimized excitation wavelength can be used to minimize acceptor excitation (Van Rheenen *et al.*, 2004) and spectral imaging techniques, such as linear unmixing, can be used to separate out donor and acceptor fluorescence (Thaler *et al.*, 2005; Zimmermann *et al.*, 2002). This approach works especially well for intramolecular FRET using a single probe containing both donor and acceptor FPs (Rizzo *et al.*, 2006). Intermolecular FRET experiments are much more problematic, since the ratio of donor–acceptor molecules is harder to control on a single cell basis. In this case, corrective images can be collected to mathematically extract FRET emission from fluorescence crosstalk (Jares-Erijman and Jovin, 2003). At least a dozen corrective algorithms have been derived, but these methods are not easily applied to live-cell imaging for two reasons. First, many of these algorithms require image collection protocols that are time consuming. Second, accurate correction requires images with a good signal-to-noise ratio. For widefield systems, this is generally not a problem, but optical sectioning techniques such as laser scanning confocal microscopy collect far fewer photons and generally produce low signal-to-noise images that may be unsuitable for corrective approaches. Alternatively, acceptor photobleaching and fluorescence lifetime microscopy (FLIM) have been used. The contrast of these methods tends to be low because it is directly related to the FRET efficiency. For FP-FRET, FRET efficiencies are typically below 20%, because the size of the FPs (Ormö *et al.*, 1996) occupies much of the useful FRET distance (Patterson *et al.*, 2000; Rizzo *et al.*, 2006). Photobleaching methods have additional drawbacks, including sample destruction, and tend to be error prone (Rizzo *et al.*, 2006). Adaptation of FLIM to live-cell imaging has made substantial progress over recent years, but development of this technology is ongoing and more progress is required before its application to live-cell imaging becomes widespread.

One solution for high contrast intermolecular FRET imaging is the use of polarized light to detect the presence of FRET on the basis of FP anisotropy (AFRET). Fluorescence from directly excited FPs is highly polarized, and is easily distinguished from FRET fluorescence, which is highly depolarized. Thus, AFRET can simply and unambiguously resolve FRET with 10-fold greater contrast than FLIM (Rizzo and Piston, 2005). Furthermore, accurate fluorescence anisotropy images can be collected rapidly using a minimal set of images, and this method can be easily adapted to any imaging modality.

II. Measuring FRET by Polarization Microscopy

Detection of FRET by measuring fluorescence anisotropies is based on two principles. First, the excitation of FPs with polarized light produces highly polarized fluorescence, and second, fluorescence from acceptor molecules excited by FRET is depolarized. To understand these principles, we must first consider that excitation of fluorophores in solution with polarized light will selectively excite molecules with a favorable orientation. This basis for photoselection arises from the physical requirement for alignment between the transition dipole moment of the fluorophore and the electric field of the incoming photon. Similarly, relaxation of the excited state emits light that is polarized parallel to the orientation of the dipole moment. Thus, an immobile fluorophore will emit light with a well-defined polarization distribution around the polarization of the excitation light. If the fluorophore rotates during the lifetime of the excited state, the emission polarization becomes randomized or *depolarized* with respect to the excitation polarization. Such a solution emission is said to be *isotropic*, and the extent of polarization remaining in the fluorescence is called the *anisotropy* of the solution. Solutions of FPs have a very high anisotropy (~0.3 out of a maximum of 0.4). This is a consequence of the very large size of FPs (~30 kDa) which slows rotational diffusion (rotational correlation time of 20 nsec) compared to the lifetime of the fluorescence state (2.9 nsec) (Swaminathan *et al.,* 1997).

FRET has the ability to excite molecules outside the photoselection plane (Fig. 1). Excitation of the acceptor through FRET is dependent on the relative positions of the donor and acceptor dipole moments. For molecules in solution, there is a very high probability that the orientation of the acceptor dipole moment *will be different than* the precise orientation of the photoselected donor. Thus, vertically polarized excitation of donor results in nonvertically polarized, or depolarized, fluorescence from the acceptor. Since anisotropy is measured with respect to excitation polarization, FRET fluorescence can have a much lower anisotropy.

Measurement of fluorescence anisotropies is performed by using a vertically polarized excitation source and collection of vertically (parallel to excitation) and horizontally (perpendicular) polarized emission. Figure 2 shows agarose beads labeled with mCerulean and mVenus FPs and imaged using polarization microscopy. Beads labeled with a single FP have highly polarized fluorescence (Fig. 2A, compare vertical image with horizontal image). FRET beads containing a mixture of mCerulean and mVenus have highly polarized cyan fluorescence but depolarized yellow fluorescence. The fluorescence spectra of each bead is shown in Fig. 2B, and even a small amount of FRET (ROI 4) has strongly depolarized fluorescence. Donor anisotropy increases during FRET (Fig. 2C, black bars, $r_{2p} = 0.4$ for

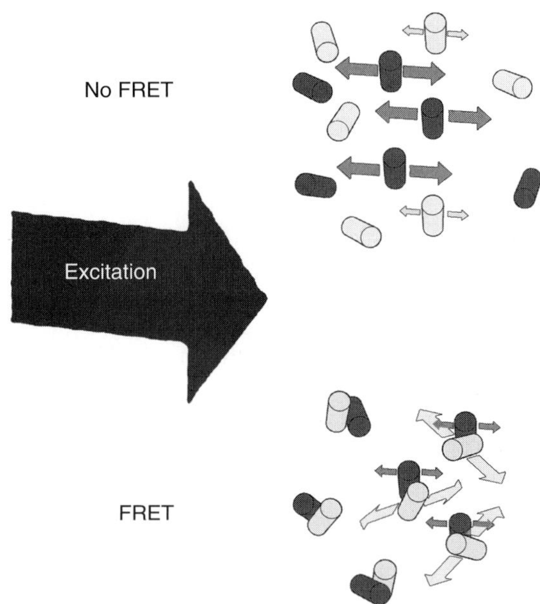

Fig. 1 Depolarization of fluorescence by Förster resonance energy transfer (FRET). Polarized light selectively excites fluorescent molecules with a specific orientation. The large size of FP restricts movement during the fluorescence lifetime. Therefore, fluorescence from directly excited FPs is emitted with a polarization similar to that of the excitation (top). During FRET, excited donor FPs (darkened cylinders) transfer energy to an acceptor FP (lightened cylinders). Since FRET excitation is based on proximity of the donor and the acceptor, the orientation of the excited acceptor can be outside the original photoselection plane. Photons emitted from the FRET acceptor are therefore depolarized in comparison to directly excited FPs.

mCerulean) because of a reduction in the fluorescence lifetime of the donor and the acceptor anisotropy from FRET approaches 0. The magnitude of decreased anisotropy is proportional to the amount of FRET collected in the acceptor channel compared to the excitation and emission crosstalk. For this reason, higher contrast is obtained using wavelengths optimized for exclusive excitation of the donor. For the mCerulean–mVenus FRET pairing, 430 nm is the optimal single photon excitation wavelength, and 820 nm is optimal for two-photon excitation (Rizzo *et al.*, 2006).

III. Configuration of Microscopes for AFRET

Modification of a fluorescence microscope for AFRET requires control over the polarization of excitation and emission light. In general, it is valuable to first inspect the microscope and remove optics that provide transmitted light contrast enhancement. These include phase contrast, Hoffman illumination contrast, and

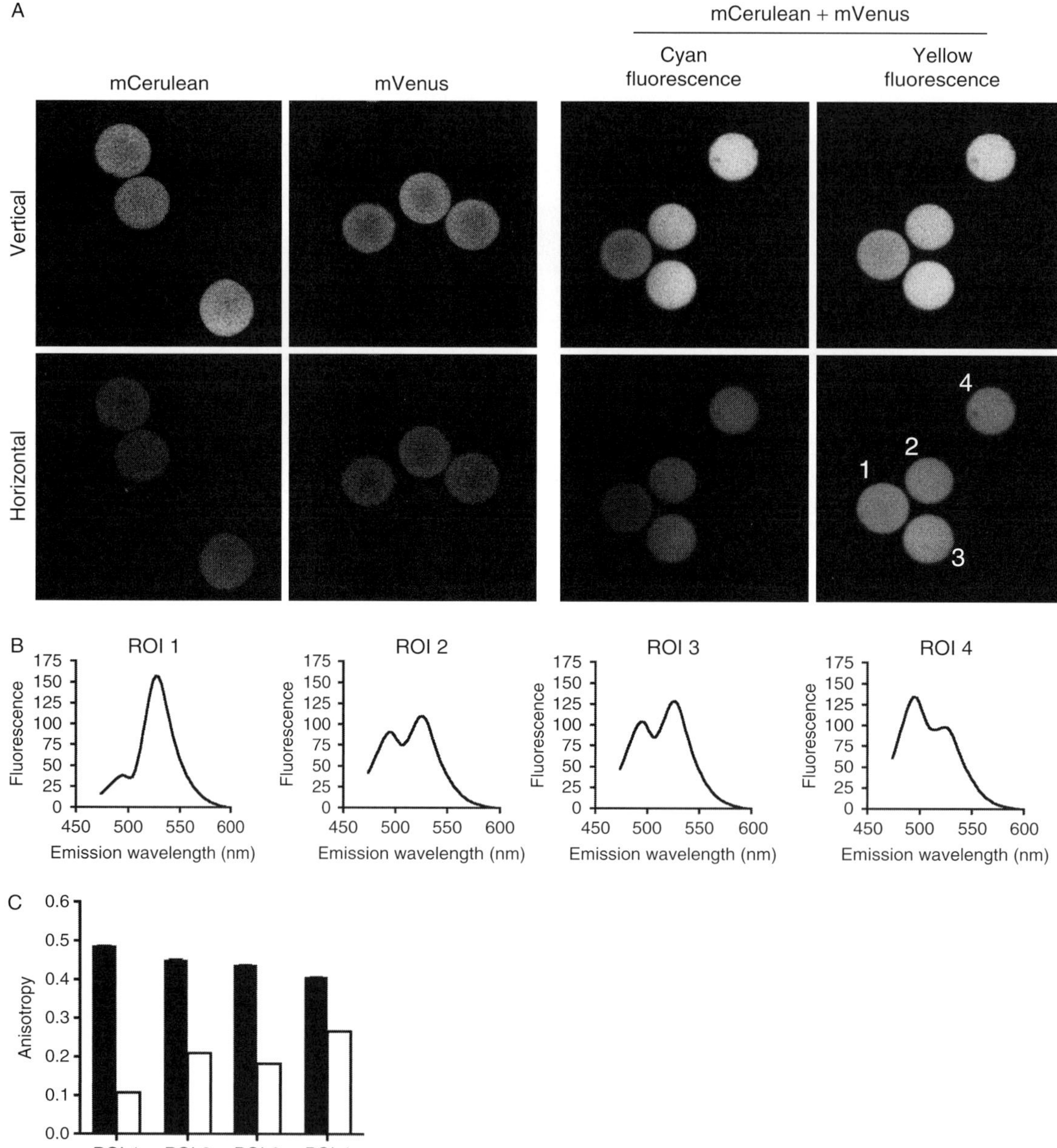

Fig. 2 Detection of Förster resonance energy transfer (FRET) using AFRET microscopy. (A) Agarose beads were labeled with recombinant mCerulean, mVenus, or a mixture of mCerulean, mVenus, and linked mCerulean–mVenus tandem dimers to generate a range of FRET efficiencies. Beads were imaged using the two-photon AFRET configuration shown in Fig. 3B. Vertical and horizontal polarization images are shown for collection of cyan fluorescence (460–500 nm) from mCerulean-labeled beads (column 1) and FRET pair-labeled beads (column 3). Collection of yellow fluorescence (500–550 nm filter) of mVenus beads is shown in column 2, and FRET pair-labeled beads (column 4).

Nomarski differential image contrast. Transmitted light contrast techniques generally apply polarization and interference to enhance the contrast. As a consequence, these polarizers and objective lenses cause distortions in polarized light collection that will interfere with AFRET measurements.

A. Configuring a Widefield System for Polarization Microscopy

Widefield systems are commonly set up using an arc lamp as an excitation source. A linear plate polarizer can be used to select vertically polarized light (Fig. 3A). Polarizers differ in efficiency with the wavelength. Excitation of CFP falls at the blue range for polarizers designed for visible light excitation, and the loss of selectivity is quantifiable. For mCerulean, we observe $r = 0.30$ for 470/40 excitation, but only $r = 0.25$ for a 436/20 filter. However, exclusive donor excitation is a greater priority. We use the ET436/20× excitation filter from Chroma, along with a 460-nm long pass dichroic mirror to deflect light to the source. For collection, two different strategies can be employed. First, donor (460–500 nm for cyan) and acceptor anisotropies (520–550 nm for yellow) can be collected separately. A dual filter wheel strategy works well for this method, since separate filtration of vertically and horizontally polarized fluorescence is required for anisotropy measurements. A more simplified collection strategy can also be employed using a single fluorescence filter, along with an Optical Insights Dual-view for splitting the polarization into two channels (Fig. 4). The advantage of this approach is that only a single image is required.

An example of this approach is shown in Fig. 4. Cells expressing a protease sensor containing CFP and YFP separated by a caspase-3 cleavage site. In healthy cells, the construct is intact, and FRET between the FPs is observed. Initiation of apoptosis results in activation of caspase-3, cleavage of the probe, and loss of FRET. A single image was collected at each time point, and used for calculation of fluorescence anisotropies. As shown in Fig. 4B, the fluorescence anisotropy is initially very low, and rises as the sensor is cleaved.

B. Configuring a Laser Scanning System for Polarization Microscopy

A key difference between configuring widefield and laser scanning systems is control over the polarization of the excitation source. Unlike arc lamps, which provide depolarized light that can be filtered using thin film polarizers, laser

---

The FRET pair-labeled beads contain a wide variety of FRET efficiencies, and the spectral profiles of beads 1–4 are shown in (B). Even small amounts of FRET show a large difference in the fluorescence anisotropies calculated for cyan (black bars) and yellow (white bars) (C). Increased FRET results in a decrease in the observed anisotropy in the yellow channel. However, the magnitude of the depolarization is derived from the relative contributions of donor and acceptor fluorescence to the fluorescence captured in the acceptor channel. Thus, straight acceptor/donor fluorescence ratio outperforms AFRET for intramolecular FRET-based sensors (Rizzo *et al.*, 2006).

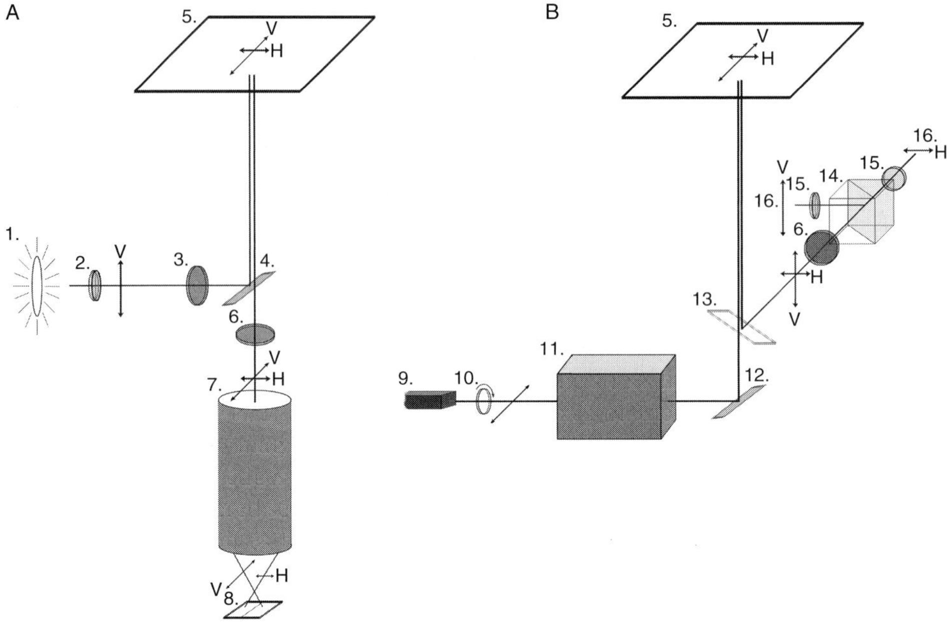

Fig. 3 Single pass collection strategies for AFRET microscopy. (A) Widefield microcopy elements. 1. Arc lamp illumination source passes through a 2. vertically oriented polarizer and a 3. CFP excitation filter (436/20x). 4. A 460-nm-long pass dichroic mirror deflect polarized light to the sample 5. fluorescence emission is filtered by a 6. 500–550 nm emission filter. Vertically and horizontally polarized fluorescence are then separated by the 7. Optical Insights Dual-view and focused on separate halves of a 8. CCD camera chip. (B) Two-photon excitation microscopy configuration. Illumination is provided by a 9. Ti:Saph laser source (820 nm). Polarization is controlled by a 10. half-wave plate, and laser scanning is controlled by a 11. confocal scan head. 12. A mirror in the microscope reflects the illumination toward the 5. sample. Fluorescence emission is reflected toward 16. two non-descanned detectors using a 13. short pass 680 nm dichroic mirror. Emission is filter using a 6. IR blocking GFP filter set (500–550 nm), and polarization are separated using a 14. polarizing beam splitting cube. 15. Thin plate polarizers can be added to enhance separation of vertically and horizontally polarized light.

sources are usually highly polarized. Rotation of the excitation polarization can be performed by insertion of a half-wave retardation plate in the beam path. Insertion of this optic is simple for direct-coupled lasers, however, most commercial confocal systems utilize fiber optics to deliver laser light to the microscope. This makes the polarization of the beam difficult to manipulate. An additional barrier to configuring high-end confocal microscopes is the inaccessibility of the emission filter sets. Unfortunately, modification of existing high-end confocals for polarization microscopy is impractical at best. In contrast, the lower-cost, personal confocals, such as the Olympus Fluoview FV300 are much more accessible to modification. The main drawback for this system are fiber-coupled lasers and the lack of a reliable 430-nm excitation source.

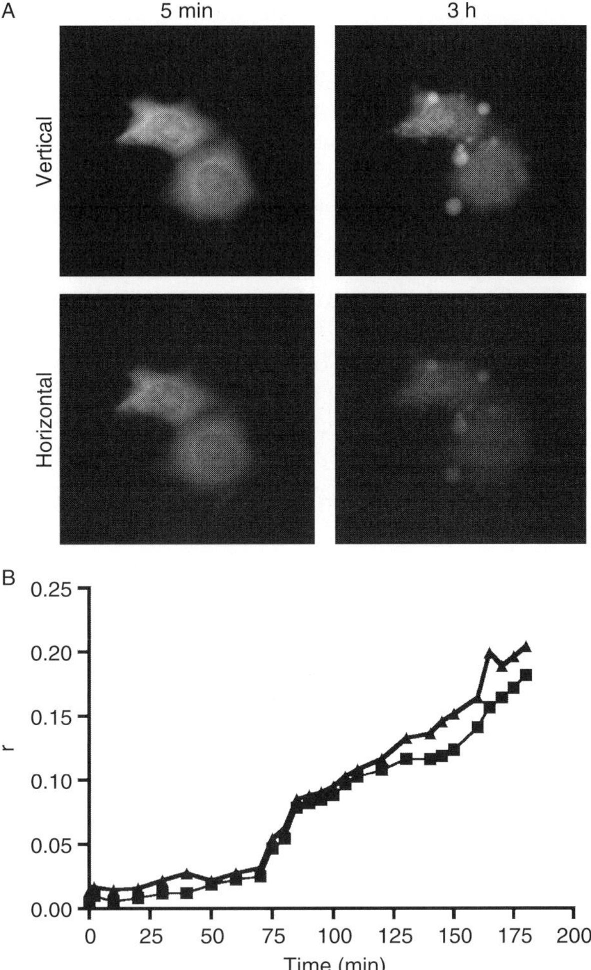

Fig. 4 Detection of caspase-3 activation using widefield AFRET microscopy. Pancreatic insulinoma cells (βTC3) were transfected with the SCAT3.1 protease sensor, consisting of cyan and yellow FPs separated by a caspase-3 cleavage site (Nagai and Miyawaki, 2004), and imaged using the widefield microscopy configuration described in Fig. 3A. Cells were treated with H_2O_2, and images were collected at 5-min intervals. Panel (A) shows vertical and horizontal polarization images 5 min and 3 h after H_2O_2 treatment. (B) Fluorescence anisotropies were calculated for regions of interest for both cells and plotted against time. The FPs in the intact sensor undergo Förster resonance energy transfer (FRET), and results in highly depolarized fluorescence emission. Induction of apoptosis and caspase-3 activation causes cleavage of the sensor, and FRET is lost. This results in a gradual increase in polarized fluorescence as SCAT3.1 biosensor is cleaved.

A more practical solution is modification of a two-photon excitation microscopy system, since the configuration is more accessible to placement of polarizing optics (Fig. 3B). Laser sources are generally guided into the scan head using a series of mirrors, and this allows adequate space for placement of a half-wave plate in the beam path. On the detection side, two non-descanned detectors can be configured around a 1-in. polarizing cube beam splitter. Alternatively, a rotatable polarizer or filter wheel can be placed in front of a single detector. An additional benefit is greater photoselection that can result from the simultaneous absorption of two photons. Thus, the two-photon anisotropy of FPs is even greater (for mCerulean $r_{2p} = 0.40$) than single photon absorption (for mCerulean $r_{1p} = 0.30$).

IV. Calculation of Fluorescence Anisotropy

Fluorescence anisotropy (r) is calculated from the fluorescence intensities collected parallel ($I_{\|}$) and perpendicular ($I_{\perp}$) to the excitation plane.

$$r = \frac{I_{\|} - I_{\perp}}{I_{\|} + 2I_{\perp}} \tag{1}$$

In a fluorometer, collection of these intensities is very simple, since detectors and polarizers can be easily positioned in the proper configuration. Collection of $I_{\|}$ and $I_{\perp}$ in a fluorescence microscope is more complicated, but in the simplest sense, $I_{\|}$ is approximated by vertically polarized excitation and collection of vertically polarized emission (*VV*), and $I_{\perp}$ is approximated by vertically polarized excitation and collection of horizontally polarized emission (*VH*). In a general sense, the ratio of *VV*/*VH* images can very informative, especially if the dynamics of FRET over time is the most important parameter. However, for accurate calculation of the fluorescence anisotropy, the effects of the microscope optics on polarization must also be taken into account.

A. Correction for Polarization Bias in the Microscope Configuration

The first consideration is for the fact that the vertical and horizontal collection channels are not equally sensitive. This is most obvious if separate detectors are used to collect *VV* and *VH*, but in general, microscope optics are not equally transmissive to both polarizations. The extent of the instrumental bias is highly dependent on the configuration of the system. On our microscopes, we have found that some are more transmissive to horizontal polarization, while others are more transmissive to vertically polarized light. The polarization bias in the microscope configuration can be easily determined using any one of a variety of methods to calculate a corrective value, known as the *g*-factor. The *g*-factor is then used to normalize values for *VH* and *VV*.

The simplest method for calculation of the g-factor is to use image of an isotropic solution of fluorescein. Thus

$$g = \frac{VV_{\text{isotropic}}}{VH_{\text{isotropic}}} \tag{2}$$

We use a 100 nM solution for widefield experiments, and a higher concentration for laser scanning.

Alternatively, the excitation polarization can be rotated 90° to the horizontal polarization and horizontal (HH) and vertical (HV) images can be collected. This swaps the polarization of the detectors, where

$$\frac{VV}{g \times VH} = \frac{g \times HH}{HV} \tag{3}$$

or, more simply,

$$g = \sqrt{\frac{VV \times HV}{VH \times HH}} \tag{4}$$

The advantage of this method is that g can be calculated using a single sample preparation. However, in some systems there may be a significant drop in illumination intensity or the polarizer may not be very accessible. For laser excitation, a half-wave plate can be used to rotate the polarization of the laser. Since the change in polarization is equal to twice the rotation of the half-wave plate, the half-wave plate must be rotated 45°.

For a two-detector system, a fluorescein solution can also be used [Eq. (2)], however, the image-based g-factor correction must be modified to take into account differences in sensitivity of the detectors. Perhaps the simplest method is to rotate the *excitation* polarization to a corrective angle (C) that will send an equal number of photons to each detector. The corrective angle can be determined by imaging a solution of 5 μM mCerulean in Tris–HCl (pH 8.0) at various angles. VV and VH are measured for the mCerulean solution, and r is calculated for VV and VH using g factors calculated from multiple excitation angles where

$$g = \frac{CV}{CH} \tag{5}$$

Thus, C is empirically determined as the polarization angle that yields the same correction as an isotropic fluorescence solution. The value for C should be ~45°. For our two-photon system $C = 43.5°$.

B. Correction for High NA Objective Lenses

In addition to correcting for polarization bias in the instrument, the effects of the objective lens on collecting and focusing polarized light should be taken into account. When light is collected by a high numerical aperture (NA) lens, the steep angle of the angular aperture will mix the polarizations as function of the NA of the objective lens and the refractive index of the sample (n). Therefore, *VV and VH are not equivalent to* $I_{\parallel}$ and $I_{\perp}$. In actuality, a three-dimensional treatment of fluorescence is required, thus $I_{\parallel}$ becomes I_z with respect to the coordinate plane of the sample. $I_{\perp}$ requires two components, I_x and I_y, since there are two possible perpendicular geometries in three-dimensional space. As defined by Axelrod (1979):

$$VV = K_c I_z + K_b I_y + K_a I_x \tag{6}$$

$$g \times VH = K_b I_z + K_c I_y + K_a I_x \tag{7}$$

Where

$$K_a = \frac{1}{3}(2 - 3\cos\sigma_0 + \cos^3\sigma_0) \tag{8}$$

$$K_b = \frac{1}{12}(1 - 3\cos\sigma_0 + 3\cos^2\sigma_0 - \cos^3\sigma_0) \tag{9}$$

$$K_c = \frac{1}{4}(5 - 3\cos\sigma_0 - \cos^2\sigma_0 - \cos^3\sigma_0) \tag{10}$$

Angle σ_0 is the angular aperture of the lens, as defined by NA $= n \sin \sigma_0$. Assuming the fluorophore is randomly oriented, $I_x = I_y$. Solving for I_z and I_y then allows calculation of the anisotropy using

$$r = \frac{I_z - I_y}{I_z + 2I_y} \tag{11}$$

This correction is most accurate when applied NA $\sim$ 1.0. For very high NA lenses (NA $\sim$ 1.3 and above), additional effects of the steep focusing angle on polarized light transmission become more apparent (Axelrod, 1989), and will reduce the measured anisotropy.

V. Sample Preparation

A. Choice of Fluorescent Proteins

A wide variety of FPs is available for FRET experiments (Shaner *et al.*, 2005). Selection of an appropriate pairing depends on several criteria, including color, monomeric character, and availability. We have found the best success using the monomeric variants (A206K) (Zacharias *et al.*, 2002) of Cerulean CFP (Rizzo *et al.*, 2004) as a donor and Venus YFP (Nagai *et al.*, 2002) for the acceptor. Among the Aequorea FPs, we find this pairing has enough spectral separation to permit an effective imaging strategy and that it is also the most efficient FRET pair (Rizzo *et al.*, 2006). The monomeric A206K mutation is essential to prevent nonspecific interaction, especially in membrane systems (Zacharias *et al.*, 2002). Another interesting pairing is the monomeric green FP together with a monomeric red FP. However, we find that the current generation of monomeric red FPs (e.g., mCherry) (Shaner *et al.*, 2004) still have issues with aggregation and photostability that limit their usefulness for AFRET experiments. In addition, the excitation spectra of the monomeric red FPs have a long blue-shifted tail, and this leads to considerable acceptor excitation crosstalk. Currently, the mCerulean–mVenus pairing is the most suitable for AFRET experiments.

B. Preparing Fluorescent Protein Standards

Fluorescence standards are useful for calibrating and aligning the microscopy system. We use recombinant FPs as standards in solution and to label beads.

1. Generation of Recombinant mCerulean

Recombinant mCerulean, mVenus, and a tandem mVenus–mCerulean pair (SGLRSPPVAT linker) were expressed in bacteria using the T5 expression system (Qiagen). Sequences encoding the FPs were cloned into the pQE9 vector, and transformed into competent M15(pREP4) cells. Twenty milliliters of an overnight starter culture was used in inoculate 1 l of prewarmed Luria Bertani broth containing 50 μg/ml carbenicillin and 30 μg/ml kanamycin. Cultures were shaken (250 RPM) at 37 °C. Protein production was induced by addition of 1 mM isopropyl-beta-D-thiogalactopyranoside (IPTG) when the culture density reached $OD_{600\ nm}$ of 0.4. Cultures were shaken for 5 h, and bacteria were harvested by centrifugation. Pellets were then frozen in a dry ice/ethanol bath and stored at −80 °C.

Pellets were thawed on ice (15 min) resuspended in lysis buffer [5 ml/g of 1X Bugbuster protein extraction reagent, 25 units/ml Benzonase (Novagen), 1 kUnit/ml rLyzozyme (Novagen), and 1X Protease Inhibitor Cocktail for use with Bacterial cells (Sigma)]. The suspension was placed on a rotating platform for 20 min, and insoluble material was removed by centrifugation (20 min, 12,000 × *g*). The cleared lysate was adjusted to 10 mM imidizole, and bound to 1-ml Ni-NTA agarose resin (Qiagen) per gram of cell paste by shaking at 300 RPM on an orbital shaker for 1 h (4 °C). The resin was then loaded on a 5 ml disposable column (Qiagen) and washed with 3 ml of wash buffer (20 mM imidizole, 300 mM NaCl, 50 mM NaH_2PO_4, pH 8.0), and elution in buffer containing 250 mM imidizole, 300 mM NaCl, 50 mM NaH_2PO_4 (pH 8.0).

For preparation of slides, 1–5 μM of recombinant protein was prepared in 50 mM Tris–HCl (pH 8.0). Hundred microliters was pipetted in a hanging drop slide (Cat 12–560A, Fisher Scientific), covered with a glass coverslip, and sealed with nail polish. Isotropic fluorescein solutions (100 nM for widefield, 10 μM for two photon) were also prepared using the same methods. Once sealed, these preparations are good for at least 1 week when kept at 4 °C.

2. Preparation of Fluorescent Protein-Labeled Beads

Agarose beads were also labeled using recombinant FPs (Fig. 2) using a protocol slightly modified from Youvan *et al.* (1997) in order to reduce homotransfer. A 1 ml HiTRAP chelating HP column (Amersham Biosciences) was washed with 5 ml of sterile distilled water prior to charging with 0.5 ml of a 0.1 M $NiSO_4$ solution. The column was then washed with 5 ml of distilled water, and 5 ml of 50 mM Tris–HCl. The column was then cut open and the beads removed, and stored in 50% slurry of 50 mM Tris–HCl, pH 8.0, and 0.01% sodium azide. Ten microliter of beads were labeled with 100 μl of 1 μM FP solution in 50 mM Tris–HCl, pH 8.0 by mixing on an orbital shaker for 1 h. Beads were isolated by centrifugation (30 sec, 2200 × *g*), and washed twice with 100 μl 30 mM Tris–HCl, pH 8.0, 0.5 M NaCl, and 5 mM imidizole, and once with 50 mM Tris–HCl, pH 8.0. Washed beads were resuspended in 10 μl 50 mM Tris–HCl, pH 8.0. To prepare slides, 1 μl of the bead slurry was spread on a coverslip and dried for 1 h at room temperature. Coverslips were then mounted in Prolong Gold (Invitrogen) and cured for 24 h prior to sealing with nail polish. These slides have been kept for at least 6 months at −20 °C.

C. Controls and Other Considerations

Typically, we prepare three types of samples for AFRET measurements: transfection of the donor alone, the acceptor alone, and a cotransfection with both donor and acceptor. Since energy transfer can occur between two FPs of the same color (homotransfer), the donor and the acceptor controls can be used to quantify the extent of energy transfer from multimerization of donors or acceptors. In practice, using the Cerulean: Venus FRET pair, we have not found that homotransfer prohibits AFRET measurements. The efficiency of homotransfer between two mCerulean FPs is far less than heterotransfer (Rizzo *et al.*, 2006). Clustering of His-tagged mCerulean on agarose beads can reduce the overall anisotropy by threefold due to homotransfer. However, FRET to mVenus reduces the polarization by an additional 10-fold. Ultimately, donor homotransfer can be easily corrected for by measuring anisotropy in both donor and acceptor channels, and subtracting the difference, for a Δr value.

$$\Delta r = r_{\text{Donor}} - r_{\text{Acceptor}} \tag{12}$$

A nonzero Δr indicates FRET from heterotransfer. Homotransfer between dimerized acceptors can be excluded from the AFRET signal by using excitation wavelengths that are optimized for donor excitation (430 nm for single photon, 820 nm for two-photon excitation). During these conditions, we have found that direct acceptor excitation is negligible, and thus, potential acceptor homotransfer does not significantly contribute to the AFRET signal.

VI. Conclusions

AFRET is a simple, high contrast, and unambiguous method for detecting the presence of intermolecular FRET between two FPs. Since a single image acquisition can be used for live-cell imaging, this method favorably compares to other FRET imaging methodologies. Furthermore, AFRET can be easily adapted to any imaging modality, including widefield and laser scanning microscopy. The main limitation of AFRET is that changes in fluorescence anisotropies are not directly related to the FRET efficiency, and therefore other methodologies are better suited for some FRET applications.

Acknowledgments

Funding for this work was provided by the US National Institutes of Health grants DK067415 (M.A.R.) and DK53434, CA86283, GM072048 (to D.W.P.), the US National Science Foundation grant BBI-9871063 (to D.W.P.), the US Department of Defense Medical Free-Electron Laser program grant F49620-01-1-0429 (to D.W.P.), and the University of Maryland School of Medicine (M.A.R.).

References

Axelrod, D. (1979). Carbocyanine dye orientation in red cell membrane studied by microscopic fluorescence polarization. *Biophys. J.* **26,** 557–573.

Axelrod, D. (1989). Fluorescence polarization microscopy. *Methods Cell Biol.* **30,** 333–352.

Jares-Erijman, E. A., and Jovin, T. M. (2003). FRET imaging. *Nat. Biotechnol.* **21,** 1387–1395.

Miyawaki, A., Llopis, J., Heim, R., McCaffery, J. M., Adams, J. A., Ikura, M., and Tsien, R. Y. (1997). Fluorescent indicators for Ca^{2+} based on green fluorescent proteins and calmodulin. *Nature* **388,** 882–887.

Nagai, T., Ibata, K., Park, E. S., Kubota, M., Mikoshiba, K., and Miyawaki, A. (2002). A variant of yellow fluorescent protein with fast and efficient maturation for cell-biological applications. *Nat. Biotechnol.* **20,** 87–90.

Nagai, T., and Miyawaki, A. (2004). A high-throughput method for development of FRET-based indicators for proteolysis. *Biochem. Biophys. Res. Commun.* **319,** 72–77.

Ormö, M., Cubitt, A. B., Kallio, K., Gross, L. A., Tsien, R. Y., and Remington, S. J. (1996). Crystal structure of Aequorea victoria green fluorescent protein. *Science* **273,** 1392–1395.

Patterson, G. H., Piston, D. W., and Barisas, B. G. (2000). Förster distances between green fluorescent protein pairs. *Anal. Biochem.* **284,** 438–440.

Rizzo, M. A., and Piston, D. W. (2005). High-contrast imaging of fluorescent protein FRET by fluorescence polarization microscopy. *Biophys. J.* **88,** L14–L16.

Rizzo, M. A., Springer, G., Segawa, K., Zipfel, W., and Piston, D. W. (2006). Optimization of pairings and detection conditions for measurement of FRET between cyan and yellow fluorescent proteins. *Microsc. Microanal.* **12,** 238–254.

Rizzo, M. A., Springer, G. H., Granada, B., and Piston, D. W. (2004). An improved cyan fluorescent protein variant useful for FRET. *Nat. Biotechnol.* **22,** 445–449.

Shaner, N. C., Campbell, R. E., Steinbach, P. A., Giepmans, B. N., Palmer, A. E., and Tsien, R. Y. (2004). Improved monomeric red, orange and yellow fluorescent proteins derived from *Discosoma* sp. red fluorescent protein. *Nat. Biotechnol.* **22,** 1567–1572.

Shaner, N. C., Steinbach, P. A., and Tsien, R. Y. (2005). A guide to choosing fluorescent proteins. *Nat. Methods* **2,** 905–909.

Swaminathan, R., Hoang, C. P., and Verkman, A. S. (1997). Photobleaching recovery and anisotropy decay of green fluorescent protein GFP-S65T in solution and cells: Cytoplasmic viscosity probed by green fluorescent protein translational and rotational diffusion. *Biophys. J.* **72,** 1900–1907.

Thaler, C., Koushik, S. V., Blank, P. S., and Vogel, S. S. (2005). Quantitative multiphoton spectral imaging and its use for measuring resonance energy transfer. *Biophys. J.* **89,** 2736–2749.

Van Rheenen, J., Langeslag, M., and Jalink, K. (2004). Correcting confocal acquisition to optimize imaging of fluorescence resonance energy transfer by sensitized emission. *Biophys. J.* **86,** 2517–2529.

Youvan, D. C., Silva, C. M., Bylina, E. J., Coleman, W. J., Dilworth, M. R., and Yang, M. M. (1997). Calibration of fluorescence resonance energy transfer in microscopy using genetically engineered GFP derivatives on nickel chelating beads. *Biotechnology et alia* **3,** 1–18.

Zacharias, D. A., Violin, J. D., Newton, A. C., and Tsien, R. Y. (2002). Partitioning of lipid-modified monomeric GFPs into membrane microdomains of live cells. *Science* **296,** 913–916.

Zimmermann, T., Rietdorf, J., Girod, A., Georget, V., and Pepperkok, R. (2002). Spectral imaging and linear un-mixing enables improved FRET efficiency with a novel GFP2-YFP FRET pair. *FEBS Lett.* **531,** 245–249.

CHAPTER 19

Bimolecular Fluorescence Complementation: Visualization of Molecular Interactions in Living Cells

Tom K. Kerppola

Department of Biological Chemistry
Howard Hughes Medical Institute
University of Michigan Medical School
Ann Arbor, Michigan 48109

METHODS IN CELL BIOLOGY, VOL. 85

0091-679X/08 $35.00
DOI: 10.1016/S0091-679X(08)85019-4

Abstract

A variety of experimental methods have been developed for the analysis of protein interactions. The majority of these methods either require disruption of the cells to detect molecular interactions or rely on indirect detection of the protein interaction. The bimolecular fluorescence complementation (BiFC) assay provides a direct approach for the visualization of molecular interactions in living cells and organisms. The BiFC approach is based on the facilitated association between two fragments of a fluorescent protein when the fragments are brought together by an interaction between proteins fused to the fragments. The BiFC approach has been used for visualization of interactions among a variety of structurally diverse interaction partners in many different cell types. It enables detection of transient complexes as well as complexes formed by a subpopulation of the interaction partners. It is essential to include negative controls in each experiment in which the interface between the interaction partners has been mutated or deleted. The BiFC assay has been adapted for simultaneous visualization of multiple protein complexes in the same cell and the competition for shared interaction partners. A ubiquitin-mediated fluorescence complementation assay has also been developed for visualization of the covalent modification of proteins by ubiquitin family peptides. These fluorescence complementation assays have a great potential to illuminate a variety of biological interactions in the future.

I. Introduction

Most cellular functions are regulated and executed by protein complexes. Extensive networks of protein interactions have been identified through both biochemical and genetic approaches. Visualization of protein complexes in living cells enables determination whether putative interactions occur in the normal cellular environment and identification of the subcellular locations of these complexes. The bimolecular fluorescence complementation (BiFC) approach enables visualization of protein complexes in living cells and organisms. This approach was made possible by the discoveries that the association between two nonfluorescent fragments of a fluorescent protein can produce a fluorescent complex and that this association can be facilitated by an interaction between proteins fused to the fragments. The BiFC assay has been used to visualize interactions between proteins from many different structural classes in a variety of cell types and species. The BiFC approach provides several unique advantages for the investigation of molecular complexes in living cells. Modified versions of the BiFC assay have been used to visualize the competition between alternative interaction partners and the covalent modification of proteins by ubiquitin family peptides. The BiFC approach is applicable for the visualization of a wide range of molecular interactions.

The following sections will briefly discuss the significance of protein interactions and the methods that are available for their study. The historical development of complementation strategies for the analysis of protein interactions is outlined briefly and the direct visualization of interactions using BiFC analysis is compared with a subset of other imaging approaches. The fundamental principles of BiFC analysis and the requirements for the application of this approach are reviewed. Instrumentation that can be used to conduct BiFC assays is described in general terms. A selection of interactions that have been visualized using the BiFC assay and the cell types and organisms in which this assay has been used are listed. Some of the limitations and pre-conditions of the BiFC approach are discussed. The extensions of BiFC analysis for the simultaneous visualization of multiple protein interactions and the visualization of ubiquitin conjugates are described. Finally, some future opportunities and challenges are presented.

A. Roles of Protein Interactions in Regulatory Complexity

The mammalian genomes that have been sequenced contain only slightly more annotated genes than the genomes of plants, insects, and nematodes. Thus, the arguably more complex developmental programs and adaptive responses of mammals are not specified by a dramatically larger number of genes. The principal mechanism whereby mammals achieve a greater number of biological functions than the number of proteins encoded in their genomes is through combinatorial interactions among these proteins. Indeed, most regulatory mechanisms in mammalian cells rely on the activities of protein complexes rather than the actions of individual proteins. Through interactions with different partners under different conditions, each protein can perform different functions in different cell types and in response to different extracellular stimuli (see Box 1 on the complexity generated by combinatorial interactions). For instance, a transcription factor can bind to thousands of loci in a mammalian genome. However, in a particular cell type at any one time, the protein may regulate only a few genes. This combination of great versatility and context-dependent specificity is made possible by interactions that specify the functions of a protein in each cell.

Box 1 Combinatorial Interactions Can Generate Complexity

The connectivity of a protein interaction network can be characterized by the average number of interaction partners for each protein. The distribution of the number of interaction partners about this mean varies for different types of networks (random, scale-free, etc.). The number of binary complexes that can be formed among proteins encoded by a genome is a product of the number of proteins and the number of interaction partners for each protein. Thus, for interaction networks with the same connectivity, the number of complexes is proportional to the square of the number of proteins. The total number of protein complexes is also dramatically affected by the average number of proteins in a complex and the distribution about that average. An increase in the average number of components in each complex results in a dramatic increase in the total number of complexes. The combination of higher connectivity and a larger number of components in each complex can therefore produce a large increase in the number of protein complexes without a change in the size of the proteome.

II. Approaches for the Investigation of Protein Interactions

Protein interactions have been investigated using many approaches under different experimental conditions. Most of the experimental approaches that enable direct detection of interactions such as biochemical copurification require removal of the proteins from their normal environment. Conversely, most studies of protein interactions in their normal environment such as genetic analysis of the combined effects of mutations rely on indirect consequences of the interactions. The combined use of genetic and biochemical approaches has identified thousands of potential protein interactions. The largely nonoverlapping interaction networks identified by high-throughput interaction screens at the proteomic level (Gavin *et al.*, 2002; Giot *et al.*, 2003; Ito *et al.*, 2001; Li *et al.*, 2004; Stelzl *et al.*, 2005) suggest that the number of potential protein complexes is orders of magnitude larger than the number of proteins encoded by the genome. The cell specificity and the subcellular localization of the vast majority of these complexes remain unknown. It is also possible that some of the interactions identified under nonnative conditions do not occur in cells.

A. Studies of Protein Interactions Using Complementation Assays

Complementation approaches enable investigation of protein interactions in their normal environment and can allow direct detection of the interaction. Complementation between protein fragments was first observed between proteolytic fragments of ribonuclease (Richards, 1958) and has been subsequently described for many proteins. Importantly, complementation between some protein fragments is conditional and can be facilitated by tethering the fragments in proximity to each other (Table I; Kerppola, 2006). This enables the detection of protein interactions and other molecular processes in living cells (see Box 2 on the detection of protein interactions using complementation approaches). Complementation between fragments of a variety of different proteins enables detection of protein interactions in living cells. However, visualization of the subcellular localization of protein complexes requires that the function produced by complementation can be detected with high spatial resolution. This is possible by using fluorescent ligands that bind to the complementation complex or by using BiFC, which produces an intrinsically fluorescent complex.

B. Visualization of Protein Interactions in Living Cells

Direct visualization of protein interactions in living cells enables validation of complex formation in the normal environment and determination of their subcellular localization. Two principal methods have been used to visualize the localization of protein interactions in living cells. Fluorescence resonance energy transfer (FRET) analysis is based on changes in the fluorescence intensities and lifetimes of two fluorophores that are brought sufficiently close together. BiFC analysis is based on the formation of a fluorescent complex by fragments of fluorescent proteins whose association is facilitated by an interaction between proteins fused to the fragments. Other methods such as fluorescence correlation spectroscopy (FCS; Brock and Jovin, 1998), image correlation spectroscopy (Petersen *et al.*, 1993), and complementation approaches using fragments of other proteins

Table I
Comparison of Complementation Methods Using Fragments of Different Proteins

Proteins	Detection	Spatial resolution[a]	Time resolution[a]	Experimental systems[a]	References
Ubiquitin	Ub-protease coupled reporters	Cell population	Day	Yeast	(Rossi *et al.*, 1997)
β-Galactosidase	FDG hydrolysis	Cellular	Hours	Cultured cells, *Drosophila melanogaster*	(Johnsson and Varshavsky, 1994)
Dihydrofolate reductase	Fl-MTX binding	Subcellular	Minutes	Cultured cells, plants	(Pelletier *et al.*, 1998)
GFP variants	Intrinsic fluorescence	Subcellular	Minutes–Hours	Cultured cells, plants, fungi	(Hu *et al.*, 2002; Ghosh *et al.*, 2000)
*Synechocystis*d-naE intein	Reporter ligation	Cell population	Hours	Cultured, implanted cells	(Ozawa *et al.*, 2001)
β-Lactamase	CCF2/AM hydrolysis	Cellular	Minutes	Cultured cells, primary neurons	(Galarneau *et al.*, 2002; Spotts *et al.*, 2002; Wehrman *et al.*, 2002)
Firefly luciferase	Luciferin luminescence	Cell population	Hours	Cultured, implanted cells	(Paulmurugan *et al.*, 2002)
Renilla luciferase	Coelenterazine luminescence	Cell population	Minutes–Hours	Cultured, implanted cells	(Paulmurugan and Gambhir, 2003)
Gaussia luciferase	Coelenterazine luminescence	Cell population	Minutes	Cultured cells	(Remy and Michnick, 2006)
TEV protease	Coupled reporters	Cellular	Minutes	Cultured cells	(Wehr *et al.*, 2006)

[a]The spatial and time resolution as well as the experimental systems used reflect those reported in publications using these approaches, and are not intended to represent the limits of performance of these methods.

Box 2 Complementation Approaches for the Detection of Protein Interactions

Many proteins can be divided into fragments that can associate to produce a functional complex. This phenomenon is classically known as complementation by analogy with the ability of different mutations to complement each other to produce an organism with a normal phenotype. Complementation between protein fragments is due to the large favorable free energy of folding of most proteins and the fact that many proteins fold in multiple stages that involve initial interactions among neighboring amino acid residues and subsequent interactions between partially folded secondary structure elements. Nevertheless, the great majority of protein fragments are unable to produce a functional complex because of misfolding or the inability of partially folded fragments to associate with each other. Only a small subset of protein fragments has the potential to associate to form a functional complex. Experimental applications of protein fragment complementation are therefore critically dependent on the identification of fragments that can associate with each other under relevant conditions.

The correct folding of a protein occurs in competition with misfolding, which is often irreversible. The ability of protein fragments to associate with each other can depend on the frequency of their collisions, which depends on the effective local concentrations of the fragments. Thus, the association of some protein fragments can be facilitated by tethering them in proximity to each other. The association of such fragments can be conditional on their molecular proximity even if the fragments can associate independent of tethering when they are present at sufficiently high concentrations. The conditional association of protein fragments is a powerful reporter of molecular proximity and can be used to investigate many biological processes that involve changes in molecular proximity, including protein interactions, nucleoprotein complex formation, and covalent protein modifications (Fang and Kerppola, 2004; Hu et al., 2002; Rackham and Brown, 2004; Stains et al., 2005).

Fragments of many proteins have been identified that can complement each other under specific experimental conditions (see Table I for a small subset). The nature of the fragments that can undergo complementation varies between different proteins. In many cases the fragments are predicted to exist in a partially unfolded state due to the absence of interactions that are necessary for formation of some secondary structure elements within the individual fragments. Unfortunately, the structures of protein fragments that can undergo complementation as well as intermediates in their folding pathway have not been characterized. Thus, the characteristics of protein fragments required for complementation are unknown. The identification of protein fragments that can support complementation therefore remains a largely empirical process.

It is not clear why some protein fragments exhibit conditional complementation whereas others apparently associate with each other independent of mechanisms that would tether them in proximity to each other. It is possible that differences in the folding pathways of the fragments alter their susceptibility to misfolding and affect the probability of productive association. Nevertheless, it is likely that all fragments that exhibit conditional complementation also have the potential to associate with each other independent of tethering under some conditions such as high protein concentrations.

(Table I) have also been used. Bioluminescence resonance energy transfer (BRET) analysis enables detection of protein interactions in live cells, but current methods do not enable visualization of the subcellular locations of the protein complexes (Pfleger and Eidne, 2006). Studies of protein interactions using the FRET assay have been described in several recent reviews (Jares-Erijman and Jovin, 2003; Miyawaki, 2003; Schmid and Neumeier, 2005; Zal and Gascoigne, 2004; Zhang *et al.*, 2002). This review focuses on critical principles and assumptions underlying the BiFC assay and on selected applications of this approach.

C. Challenges for the Visualization of Protein Interactions

One major challenge for the direct detection of protein interactions is that most proteins have multiple partners in the cell, and only a small subpopulation interacts with a particular partner. In most experimental approaches, interactions between the proteins under study with other cellular proteins interfere with detection of the complex being investigated. This problem is often addressed by overexpression of the proteins under study to outcompete endogenous interaction partners and to produce a larger amount of complexes. This strategy carries the risk that protein overexpression produces nonnative complexes or alters the characteristics of the complexes that are formed. One advantage of complementation approaches is that complexes that are formed with other partners are not detected, enabling selective observation of the complex under investigation. A fundamental difference between the FRET and BiFC approaches is that FRET analysis is based on measurement of the difference in fluorescence intensity or lifetime of one fluorophore in the presence and absence of a second fluorophore. In contrast, BiFC analysis is based on the formation of a fluorescent complex from nonfluorescent constituents. This makes BiFC analysis potentially more sensitive and avoids interference from changes in fluorescence intensity or lifetime caused by cellular conditions unrelated to protein interactions. Conversely, FRET provides the potential for real-time observation of complex formation and dissociation, whereas BiFC analysis does not enable real-time detection of complex dynamics.

III. Bimolecular Fluorescence Complementation Analysis

BiFC analysis is based on the formation of a fluorescent complex when two fragments of a fluorescent protein are brought together by an interaction between proteins fused to the fragments (see Fig. 1). In *Escherichia coli*, fluorescence complementation was first detected using fragments of a green fluorescent protein (GFP) variant fused to artificial peptide sequences designed to form an antiparallel coiled coil (Ghosh *et al.*, 2000). In mammalian cells, expression of yellow fluorescent protein (YFP) fragments fused to calmodulin and its target peptide M13 was shown to produce fluorescence that was modulated by the Ca^{2+} concentration (Nagai *et al.*, 2001). However, fluorescence complementation by these fragments was not shown to require calmodulin binding to M13 and the fragments have not been used for studies of other interactions. Conditional fluorescence complementation was demonstrated between fragments of the enhanced YFP fused to transcription factors containing the basic region—leucine zipper (bZIP) domain or the Rel domain (Hu *et al.*, 2002). Mutations that prevented the interactions reduced the fluorescence, demonstrating that the native protein interactions facilitated fluorescence complementation (Hu *et al.*, 2002). The same fragments were shown to support complementation when fused to other bZIP or Rel family proteins (Hu *et al.*, 2002). Subsequent studies have shown that these fragments can be used to study interactions among a variety of structurally unrelated proteins, validating the BiFC approach as a general strategy for the visualization of molecular interactions in living cells (Kerppola, 2006 a,b).

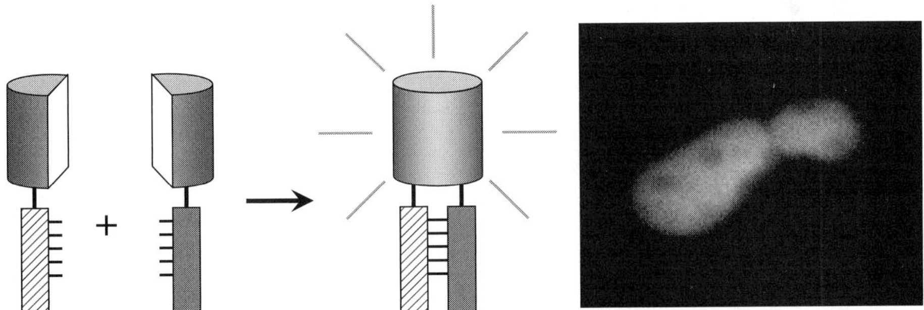

Fig. 1 Principle of bimolecular fluorescence complementation (BiFC). BiFC analysis is based on the facilitated association between two fluorescent protein fragments when they are brought together by an interaction between proteins fused to the fragments. The individual fragments are nonfluorescent. Please see the text for factors that can influence bimolecular fluorescent complex formation. The image on the right shows BiFC analysis of complexes formed between Fos and Jun transcription factors. Image acquired by Changdeng Hu (Hu *et al.*, 2002) reprinted with permission from Elsevier © 2002. (See Plate no. 54 in the Color Plate Section.)

A. Requirements for BiFC Analysis

BiFC analysis does not require information about the structures of the interaction partners or of their interaction interface. The association of the fluorescent protein fragments fused to the interaction partners does not require that the interaction partners position the fragments in a specific orientation or within a fixed distance from each other. Nevertheless, steric constrains can prevent association of the fragments within a complex. The fragments must have sufficient freedom of motion in the complex to collide with each other and to undergo the mutually induced folding required to form the β-barrel structure. Flexible linkers between the interaction partners and the fluorescent protein fragments can uncouple the motions of the fluorescent protein fragments from those of the interaction partners in the complex and may facilitate bimolecular fluorescent complex formation. The interaction partners do not need to form a complex with a long half-life since transient interactions can be trapped by the association of the fluorescent protein fragments (see Box 3 on the dynamics of BiFC complexes). Additionally, it is not necessary for a large fraction of the interaction partners to associate with each other in order to detect fluorescence complementation because cells have low background fluorescence and unassociated fragments do not interfere with detection of an interaction using the BiFC assay.

Box 3 Dynamics of Bimolecular Fluorescent Complexes

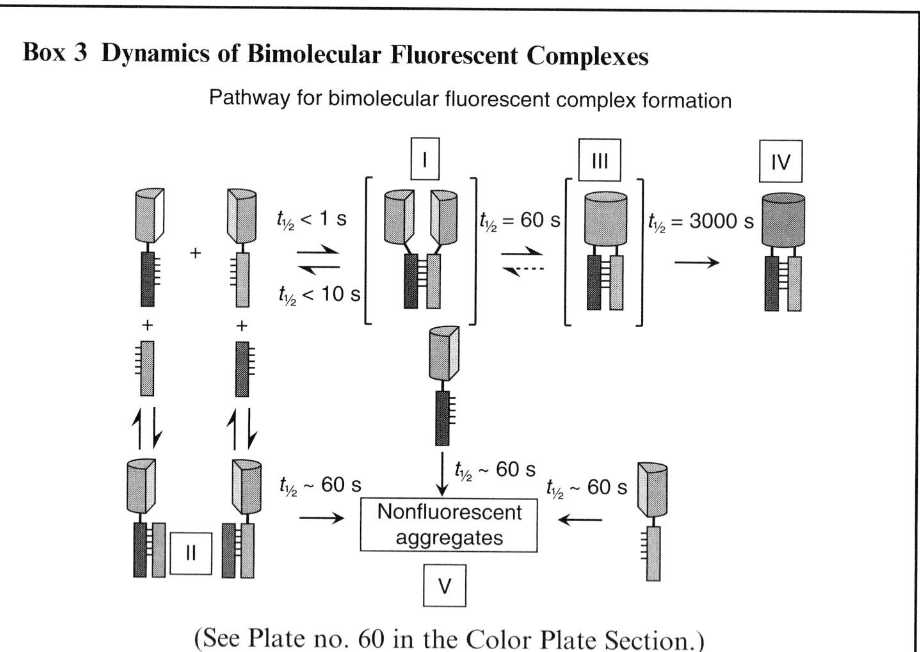

(See Plate no. 60 in the Color Plate Section.)

The dynamics of BiFC have been investigated in order to elucidate the pathway for fluorescent complex formation (Hu *et al.*, 2002). *In vitro* studies using purified proteins indicate that the initial association between the fusion proteins (complex I) is mediated by the interaction partners. This interaction occurs in competition with mutually exclusive interactions with alternative interaction partners (complexes II). The association between the fluorescent protein fragments is slower and produces an intermediate (complex III) that undergoes the slow chemical reactions (maturation) required to produce the peptide fluorophore (complex IV). This causes a delay in detection of BiFC following formation of the protein complex (Hu *et al.*, 2002). The length of this delay depends on the intensity of the signal and the sensitivity and background of the detector. The latter two steps are irreversible under some conditions (Hu *et al.*, 2002; Magliery *et al.*, 2005). Some fluorescent protein fragments purified from *E. coli* as intein fusions can undergo at least partial maturation prior to association, resulting in rapid fluorescence complementation (Demidov *et al.*, 2006). The same study demonstrated a decrease in the fluorescence intensity when the conditions were adjusted to disfavor the interaction, suggesting that bimolecular fluorescent complex formation can be reversible (dashed arrow). Unassociated fluorescent protein fragments as well as fragments that are present in complexes that do not contain complementary fragments undergo irreversible misfolding *in vitro* (Hu *et al.*, 2002) (complexes V).

The nonproductive folding of the fluorescent protein fragments that are not in a complex containing a complementary fragment is critical for the specificity of the BiFC assay. Bimolecular fluorescent complex formation is likely to be energetically favorable even when the fluorescent protein fragments are not fused to proteins that interact with each other. The stimulation of fluorescence complementation by an interaction between proteins fused to the fragments is likely to be determined by kinetic rather than thermodynamic factors. The efficiency of bimolecular fluorescent complex formation is determined by the frequency of productive collisions between the fluorescent protein fragments relative to the rate of nonproductive folding. Fusion of the fragments to interaction partners can increase the rate of potentially productive collisions relative to the rates of nonproductive collisions with other proteins in the cell. The spectral characteristics of the bimolecular fluorescent complex and the intact fluorescent protein are indistinguishable, indicating that the β-barrel structure and the tripeptide fluorophore are likely to be identical in the bimolecular complex and the intact protein.

An essential requirement for fusion proteins to be used for BiFC analysis is that the fluorescent protein fragments do not associate with each other efficiently in the absence of an interaction between the proteins fused to the fragments. Spontaneous association between the fluorescent protein fragments can be affected by the characteristics of the proteins fused to the fragments. It is therefore essential to test the requirement for a specific interaction interface for complementation by each combination of interaction partners to be studied using the BiFC approach

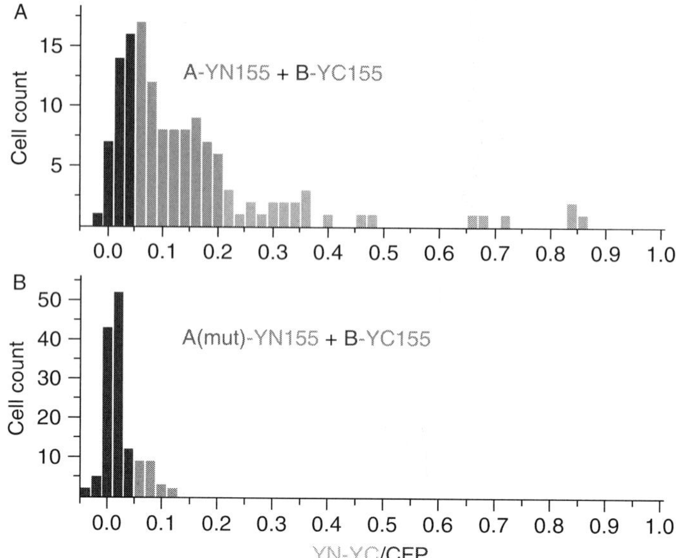

Fig. 2 Determination of the specificity of bimolecular fluorescence complementation (BiFC) by mutational analysis. The specificity of BiFC should be tested by examining the effects of mutations that prevent the association of the interaction partners [data adapted from Hu *et al.* (2002)]. The wild-type (A) and mutated (B) interaction partners should be expressed at the same concentrations. The plasmids should be cotransfected together with an internal reference (i.e., CFP). The fluorescence intensities produced by BiFC (YN-YC) and the internal reference (CFP) are measured in individual cells. The distribution of ratios between the fluorescence intensities in individual cells are plotted in each histogram. (See Plate no. 55 in the Color Plate Section.)

(Fig. 2). Fragments of some fluorescent proteins can form fluorescent complexes in the absence of fusions to specific interaction partners (Cabantous *et al.*, 2005). Likewise, high-level expression of many fusion proteins containing the fragments used for BiFC analysis can result in fluorescent complex formation independent of specific interactions. The fluorescence produced by spontaneous association of the fragments used for BiFC analysis is frequently reduced when the fragments are fused to proteins that do not interact with each other. It is therefore essential to test the effects of mutations that eliminate the protein interaction on fluorescence complementation and to express the fusion proteins at levels comparable to their endogenous counterparts (Fig. 2).

B. Design of Fusion Proteins for BiFC Analysis

The association of the fluorescent protein fragments can be enhanced when they are tethered in the same macromolecular complex. However, the association of the fragments is not determined by molecular proximity alone. Steric constraints can influence the probability of association between the fragments when the fusion proteins form a complex. Fusion proteins that produce optimal signal must generally be identified empirically by testing several combinations of proteins with the fragments fused to different positions. The fluorescence signal produced by BiFC varies widely for interactions between different partners and for different fusions to the same partners. For true interaction partners, it is virtually always possible to find a combination of fusion proteins that produces a detectable signal. Also, since autofluorescence in the visible range is extremely low in most cells, the signal from BiFC is often orders of magnitude higher than cellular background fluorescence.

C. Instruments That Can be Used for BiFC Analysis

Fluorescence complementation can be detected using the same instruments that are used to detect intact fluorescent proteins. The fluorescence intensity of cells expressing fluorescent protein fragment fusions that can form bimolecular fluorescent complexes is generally less than 10% of the fluorescence intensity of cells expressing comparable levels of intact fluorescent proteins. It is likely that only a subset of the fragments associate with each other since the fluorescence intensity of BiFC complexes produced *in vitro* is comparable to that of intact fluorescent proteins (Hu *et al.*, 2002). It is important to optimize the instrument for detection of the bimolecular fluorescent complex(es) in order to maximize the signal to background ratio of the assay.

Bimolecular fluorescent complex formation *in vitro* can be monitored using a fluorometer. In such experiments, the signal produced through complex formation between specific interaction partners can be several orders of magnitude greater than the signal produced through complex formation between proteins in which the interaction interface has been mutated (Hu and Kerppola, 2003; Hu *et al.*, 2002). Bimolecular fluorescent complex formation in cell populations can also be monitored using a fluorometer, but this requires a strong signal and efficient rejection of signal resulting from scatter of the excitation beam.

Microscopy provides the greatest spatial resolution and often the greatest sensitivity for detection of bimolecular fluorescent complexes. Virtually any fluorescence microscope can be used to determine the subcellular localization of BiFC complexes. In many cases, the signal in cells that express fusions to proteins that interact with each other is more than tenfold higher than the signal in cells that express noninteracting fusions (Hu *et al.*, 2002). Microscopy also allows analysis of the variation in fluorescence intensities and subcellular distributions of complexes among different cells in the population. A large number of cells must be analyzed, and strategies to avoid experimental bias must be implemented in order to obtain results that are representative for the cell population. It is generally straightforward to visualize multiple BiFC complexes with different spectral characteristics through the use of excitation and emission filters. This enables comparison of interactions among several proteins in the same cell using multicolor BiFC analysis (see Section IX).

Flow cytometry enables determination of the fluorescence intensities of a large number of individual cells. An instrument with sensitive and stable detectors and accurate correction for scattered excitation light is necessary to obtain quantitative data. Flow analysis is less susceptible to experimental bias but requires careful attention to identical treatment of samples in order to allow comparison of data from different cell populations. Multiwavelength detector systems enable normalization of fluorescence intensities using internal controls and simultaneous analysis of multiple BiFC complexes in each cell.

D. Effects of Fluorescent Protein Fragments on Fusion Protein Properties

Similar to other approaches that make use of fusion proteins, it is necessary to examine the possibility that the fluorescent protein fragments fused to the interaction partners alter their functions. Ideally, the fusion proteins should be tested by substituting them for their endogenous counterparts. This is practical only in prokaryotes and yeast, so alternative assays for the functions of the fusion proteins must be used in other eukaryotes. It is also important to examine potential consequences of the stabilization of the interaction between the fusion proteins by bimolecular fluorescent complex formation (see Box 3 on the dynamics of BiFC complexes). Changes in the dynamics of the interaction caused by association of the fluorescent protein fragments can alter the properties of the complex.

IV. Experimental Strategies for BiFC Analysis

The strategies for investigation of a specific protein interaction must be designed with the purpose of the experiment in mind. However, there are some general strategies that can be useful for the study of many interactions.

A. Design of Plasmid Vectors for Fusion Protein Expression

Plasmid vectors must be designed for fusion of the proteins of interest to the N- and C-terminal fragments of a fluorescent protein. In most cases, fusions to both the N- and C-terminal ends of the proteins of interest should be tested. Schematic diagrams of the different permutations of fusion proteins that can be examined are shown in Fig. 3. The following general guidelines for the construction of expression vectors for BiFC analysis should be considered.

1. Choice of Fluorescent Protein Fragments for Fusions

Several combinations of fluorescent protein fragments that support BiFC have been identified (Hu and Kerppola, 2003; Hu *et al.*, 2002; Shyu *et al.*, 2006). Some combinations of fluorescent protein fragments recommended for BiFC analysis are listed in Table II. For most purposes, fragments of YFP truncated at residue 155

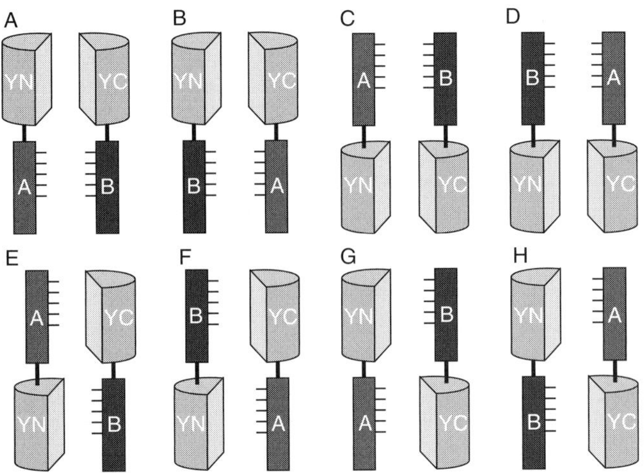

Fig. 3 Recommended combinations of fusion proteins to be tested for bimolecular fluorescence complementation (BiFC). Multiple combinations of fusion proteins should be tested for BiFC. N- and C-terminal fusions can be used to test eight distinct combinations (A–H). Although it may appear that combinations e–h would not be favorable for bimolecular complex formation, this will depend on the precise structures and flexibilities of the fusion proteins, which are difficult to predict. (See Plate no. 56 in the Color Plate Section.)

Table II
Combinations of Fluorescent Protein Fragments Recommended for BiFC Analysis

Fusions[a]	Purpose	Excitation filter (s) (nm)	Emission filter(s)	References
A-YN155 B-YC155	A–B interaction	500/20	535/30 nm	(Hu et al., 2002)
A-YN173 B-YC173	A–B interaction	500/20	535/30 nm	(Hu and Kerppola, 2003)
A-CN155 B-CC155	A–B interaction	436/10	470/30 nm	(Hu and Kerppola, 2003)
A-YN155 B-CN155 Z-CC155	Competition between A and B for interaction with Z	500/20 and 436/10	535/30 and 470/30	(Hu and Kerppola, 2003)

[a]YN155 corresponds to residues 1–154 of EYFP. YC155 corresponds to residues 155–238 of EYFP (Zhang et al., 2002). YN173 corresponds to residues 1–172 of EYFP. YC173 corresponds to residues 173–238 of EYFP. VN155 corresponds to residues 1–154 of Venus (Nagai et al., 2002). VC155 corresponds to residues 155–238 of Venus. VN173 corresponds to residues 1–172 of Venus. VC173 corresponds to residues 173–238 of Venus. CN155 corresponds to residues 1–154 of ECFP (Zhang et al., 2002). CC155 corresponds to residues 155–238 of ECFP.

(YN155, N-terminal residues 1–154; and YC155, C-terminal residues 155–238) are recommended, as they exhibit a relatively high complementation efficiency when fused to many interaction partners, yet produce low fluorescence when fused to proteins that do not interact with each other. Fragments of YFP truncated at residue 173 (YN173, N-terminal residues 1–172; and YC173, C-terminal residues 172–238) can also be used. Complementation between fragments of Venus truncated at either residue 155 or 173 (VN155, VC155; or VN173, VC173) often produces brighter fluorescence, but can result in higher nonspecific signal due to complementation between fragments fused to proteins that do not interact specifically in the absence of the fluorescent protein fusions.

2. Choice of Positions to which the Fragments Are Fused

The fluorescent protein fragments can support fluorescence complementation when they are fused to either the N- or C-terminal ends of the proteins to be investigated. Ideally, all eight different combinations of fusion proteins should be tested for complementation (Fig. 3). The positions of the fusions should be determined empirically based on the following criteria:

i. The fusions must allow the fragments of the fluorescent proteins to associate with each other if the fusion proteins interact. Information about the structure and location of the interaction interface may be useful since fusions near that interaction interface can in some cases produce more efficient complementation than fusions far from the interface. However, structural information is not necessary since multiple combinations of fusion proteins can be tested by screening for fluorescence complementation. A simple strategy for the identification of fusion proteins that support BiFC is to fuse each of the fluorescent protein fragments to the N- and C-terminal end of each interaction partner and to test all eight combinations of fusion proteins that contain both fragments for complementation (Fig. 3).

ii. The fusions must not affect the subcellular distributions or the stabilities of the proteins. The localization and stability of each fusion protein should be compared with those of the endogenous or wild-type proteins lacking the fusions.

iii. The fusions must not affect the functions of the proteins to be investigated. Ideally, the functions of the proteins should be tested using assays that evaluate all of the known functions of the endogenous proteins, and these assays should be performed under the conditions used to visualize the protein interactions.

iv. In general, the fragments should be fused to the full-length proteins whose properties are to be investigated. In many cases, comparisons with truncated proteins will be important to test the roles of specific regions of the proteins in the interaction. In these cases, it is essential that the fluorescent proteins fragments are fused to the same positions of the full-length and truncated proteins.

3. Choice of Peptide Linkers to Connect the Proteins to the Fluorescent Protein Fragments

Peptide linkers are recommended to connect the fragments of the fluorescent proteins to the proteins of interest. These linkers should provide flexibility for independent motion of the fluorescent protein fragments and the interaction partners, allowing the fluorescent protein fragments to associate when the proteins interact. We have used the RSIAT and RPACKIPNDLKQKVMNH (single amino acid code) linker sequences in many fusion constructs used for BiFC analysis (Hu and Kerppola, 2003; Hu *et al.*, 2002). These linkers have been used for the visualization of interactions between many structurally unrelated proteins. Linker sequences encoding multiple copies of a GGGS sequence have also been successfully used in many BiFC experiments. Although these linker sequences worked well for the proteins we have examined, it is possible that other linkers are optimal for complementation between other proteins.

B. Strategies for Fusion Protein Expression

The strategy for expression of the fusion proteins should be based on the purpose of the experiment. To determine whether a pair of proteins can interact in cells and to determine the subcellular location of the complex, a transient expression system can be used. However, the levels of protein expression in different cells in a transiently transfected population are likely to vary over a wide range. Protein overexpression can result in mislocalization of the proteins and formation of nonnative complexes. This problem can be ameliorated by the use of plasmids with weak promoters and plasmid vectors that do not replicate in mammalian cells. Additionally, cells can be transfected using small amounts of plasmid DNA, and they can be observed soon after transfection, before the protein expression level becomes too high.

To ensure that the observed fluorescence signal reflects native interactions, the fusion proteins should be expressed at levels comparable to the endogenous proteins. The levels of protein expression can be controlled by using inducible expression vectors integrated into the genome of stable cell line. Such cell lines allow control of protein expression at relatively uniform levels in the cell population.

C. Controls Required for Interpretation of BiFC Experiments

To interpret results from BiFC analysis, it is necessary to include negative controls in each experiment (Fig. 2). This is essential since the fluorescent protein fragments are able to form fluorescent complexes with low efficiency in the absence of a specific interaction. This nonspecific complementation is generally reduced when the fragments are fused to proteins that do not interact with each other. Thus, expression of the fluorescent protein fragments alone frequently produces more fluorescence than expression of fusions to proteins that do not specifically interact with each other. The validity of BiFC results must be confirmed by testing complementation by fusion proteins in which the interaction interface has been mutated (Grinberg *et al.*, 2004; Hu and Kerppola, 2003; Hu *et al.*, 2002). The mutated proteins should be fused to the fluorescent protein fragments in the same way as the wild-type protein. The level of expression and the localization of the mutated proteins should be compared with the wild-type proteins by Western blot and indirect immunofluorescence analyses. The efficiencies of fluorescence complementation by the wild-type and mutant proteins should be quantified as described in Section IV.D.

If the interaction interface has not been previously characterized, it is possible to screen for mutations that alter the efficiency of fluorescence complementation using the BiFC assay. If such mutations selectively alter complementation with a particular interaction partner, and they do not affect the levels of expression or subcellular locations of the proteins, it is likely that complementation by the wild-type fusion proteins reflects a specific interaction. The BiFC assay can therefore be used to determine whether two proteins interact in cells without prior knowledge of the location or the structural nature of the interaction interface.

D. Quantification of the Efficiency of BiFC

The efficiency of fluorescence complementation is defined as the fluorescence intensity produced by bimolecular fluorescent complex formation, relative to the levels of fusion proteins present in the cell. The efficiencies of BiFC by structurally unrelated fusion proteins cannot be used to compare the efficiencies of complex formation since the efficiency of BiFC is influenced by many factors in addition to the efficiency of complex formation. For example, the levels of expression of the fusion proteins and the spatial arrangement of the fragments of the fluorescent protein affect the efficiency of fluorescence complementation. However, in situations where these factors are constant, differences in the efficiency of BiFC can provide information about the relative efficiencies of complex formation. Thus, the effects of single amino acid substitutions on complex formation can be examined by testing their effects on complementation efficiency as long as the substitutions do not affect protein expression or localization (Hu and Kerppola, 2003; Hu *et al.*, 2002).

To quantify the efficiency of fluorescence complementation, it is necessary to include an internal control in the experiment to normalize for differences in transfection efficiency and the level of protein expression. For this purpose, cells are cotransfected with plasmids encoding the two fusion proteins (e.g., fused to YN155 and YC155), together with a plasmid encoding a full-length fluorescent protein with distinct spectral characteristics (e.g., CFP). The fluorescence intensities produced by both BiFC (e.g., YN-YC) and the internal control (e.g., CFP) are measured in individual cells. The ratio of YN-YC to CFP emission is calculated after subtraction of background signal (Fig. 2). The ratio of YN-YC to CFP fluorescence is a measure of the efficiency of BiFC. The ratios for different structurally related fusion proteins can reflect the relative efficiencies of complex formation in living cells.

E. Interpretation of Results from BiFC Analysis

i. Fluorescence complementation detected

If fluorescence is detected when wild-type proteins fused to the fluorescent protein fragments are expressed and this signal is eliminated or significantly reduced by mutations that do not affect the expression or localization of the protein, it is likely that the fluorescence reflects a specific interaction between the proteins fused to the fluorescent protein fragments. If mutations that are known to eliminate the interaction of the wild-type proteins do not eliminate the fluorescence, then the bimolecular complementation is due to nonspecific interactions between the chimeric fusion proteins. Other combination of fusion proteins or linkers can be tested to determine if they produce a specific signal.

ii. No fluorescence complementation detected

The lack of fluorescence complementation does not prove the absence of an interaction, even if one of the fusion proteins displays complementation with fusions to other interaction partners. The spatial arrangement of the fluorescent protein fragments can sterically prevent bimolecular complex formation. In addition, fusion of the fluorescent protein fragments can alter protein structure, which could selectively eliminate interactions with some proteins. Only in cases where fluorescence complementation can be induced by an external signal, the lack of fluorescence complementation in the absence of the signal tentatively considered to reflect the absence of an interaction or a change in protein complex architecture. If no complementation is observed between fusions to proteins that are known to interact based on other assays, additional fusion proteins containing different linker sequences or fluorescent protein fragments should be tested.

V. Examples of Protein Interactions That Have Been Visualized Using the BiFC Assay

BiFC analysis has been used to visualize interactions among a wide range of proteins in many different cell types and organisms (Table III). The results have validated interactions between many putative interaction partners and have identified several new complexes. Discovery of the subcellular locations of many protein complexes has provided new insights into their functions.

A. BiFC Analysis of Nuclear Proteins

The BiFC assay was originally developed using transcription regulatory proteins as a model and has been used to visualize interactions among many different classes of transcription factors (Deppmann *et al.*, 2003; Diaz *et al.*, 2005; Farina *et al.*, 2004; Grinberg *et al.*, 2004; Hu and Kerppola, 2003; Hu *et al.*, 2002; Jang *et al.*, 2005; Kanno *et al.*, 2004; Laricchia-Robbio *et al.*, 2005; Rajaram and Kerppola, 2004; Zhu *et al.*, 2004). These studies have provided insights into the regulation of subcellular localization by protein interactions. In many cases, the localization of transcription factor complexes differs from those of the

Table III
Examples of Protein Interactions That Have Been Visualized Using the BiFC Assay

Category	Class	Proteins[a]	Organism	References
Peptides	Coiled coil	Anti-parallel NZ-CZ	*E. coli*	(Ghosh *et al.*, 2000)
	Heat shock	Hsc70, Hsp90-TPR1, TPR2A, TPR2B	*E. coli*	(Magliery *et al.*, 2005)
Nuclear proteins	bZIP	Fos-Jun-ATF2; BATF-Jun; Maf-Sox	Mammalian cells	(Hu *et al.*, 2002; Deppmann *et al.*, 2003; Rajaram and Kerppola, 2004; Liu *et al.*, 2006)
	Rel	p50, IkB-p65	Mammalian cells	(Hu *et al.*, 2002)
	bHLHZIP	Myc, Mxi1, Mad3, Mad4-Max, Mist-Mist	Mammalian cells	(Grinberg *et al.*, 2004; Zhu *et al.*, 2004)
	Bromodomain	AcH4-Brd2; SPA-1, P-TEFb-Brd4	Mammalian cells	(Kanno *et al.*, 2004; Farina *et al.*, 2004; Jang *et al.*, 2005)
	Smad	PKB/Akt, Smad4-Smad3	Mammalian cells	(Remy *et al.*, 2004)
	IRF-Ets	IRF8-PU.1	Mammalian cells	(Laricchia-Robbio *et al.*, 2005)
	Winged helix	AcFKH1-CPCR1	*A. chrysogenum*	(Hoff and Kuck, 2005)
Ubiquitination	E3 ligase-substrate	Skp2-Myc	Mammalian cells	(von der Lehr *et al.*, 2003)
		Grr1-Hof1	*S. cerevisiae*	(Blondel *et al.*, 2005)
		EID1-ASK1,2,4,9,13,14,15, SSK1 AtCUL1-ASK1, EID1, P0CA-ASK1	*S. alba, P. crispum*	(Stolpe *et al.*, 2005; Marrocco *et al.*, 2006; Pazhouhandeh *et al.*, 2006)
Peptide conjugates	Jun-Ub, Jun-SUMO1		Mammalian cells	(Fang and Kerppola, 2004)
Plant	Type IV secretion	VirE2-VirD4	*A. tumefaciens*	(Atmakuri *et al.*, 2003; Cascales *et al.*, 2005)
pathogens	Host-pathogen	VirE2, VirF, H2A-AtVIP, VirE3	*Nicotiana tabacum*	(Tzfira *et al.*, 2004; Loyter *et al.*, 2005; Lacroix *et al.*, 2005)

Signaling	MAP kinase network	MEKK3-IκBα; MEKK2-IκBβ; ERK1-p65; ERK2-p65	Mammalian cells	(Schmidt *et al.*, 2003)
	PKB-PDK kinases	PKB/Akt, PDK1-hFt1	Mammalian cells	(Remy and Michnick, 2004)
	Heterotrimeric G-proteins	Gβ1-Gγ7	*Dictyostelium discoideum*, Mammalian cells	(Hynes *et al.*, 2004a,b)
	Phospholipases	PLCβ2-PLCδ1	Mammalian cells	(Guo *et al.*, 2005)
	Apoptosis	Bif1-Bax, TRAF6-Src	Mammalian cells	(Takahashi *et al.*, 2005; Wang *et al.*, 2006)
	Photosensitivity	FpsA-FpsA	*Aspergillus nidulans*	(Blumenstein *et al.*, 2005)
Enzyme complexes	ACCS	ACS1, ACS4–ACS6, ACS7, ACS8	*E. coli*	(Tsuchisaka and Theologis, 2004)
	P450	P450C2, P450E1-P450 reductase; P4502C2-BAP31	Mammalian cells	(Ozalp *et al.*, 2005; Szczesna-Skorupa and Kemper, 2006)
Membrane proteins	Integrin signaling	Integrin αIIbβ3, Syk-Src	Mammalian cells	(de Virgilio *et al.*, 2004)
	Arf GTPases	Arf1, Arf3, Arf4, Arf5-GBF1	Mammalian cells	(Niu *et al.*, 2005)
	Lectin-glycoprotein	MCFD2, Cathepsin-ERGIC53	Mammalian cells	(Nyfeler *et al.*, 2005)
	Cytokine receptors	gp130—LIFR, gp130	Mammalian cells	(Giese *et al.*, 2005)
	APP processing	APP-Notch2, APP	Mammalian cells	(Chen *et al.*, 2006)
Nucleic acid binding	RNA binding	IMP, FMRP, hStau1, IRP1, PTB1-RNA; Nef-Nef; NXF1-Y14	Mammalian cells	(Rackham and Brown, 2004; Ye *et al.*, 2004; Schmidt *et al.*, 2006)
	DNA binding	Zif268, PBSII-DNA	*In vitro*	(Stains *et al.*, 2005)
Plant proteins	Transcription factors	FIE-MEA; bZIP63-bZIP63; LSD1-LSD1; bHLH1-OFP1; SAD, BPBF-GAMYB; LIP19-OBF19; GRP23-RBP36B	*N. benthamiana, N. tabacum, A. thaliana, Allium* sp.	(Bracha-Drori *et al.*, 2004; Walter *et al.*, 2004; Hackbusch *et al.*, 2005; Diaz *et al.*, 2005; Shimizu *et al.*, 2005; Ding *et al.*, 2006)
	Protein modification	PFTα-PFTβ; T143c-T143c	*N. benthamiana, A. thaliana*	(Bracha-Drori *et al.*, 2004; Walter *et al.*, 2004)
	Flowering	FD-FT	*N. benthamiana*	(Abe *et al.*, 2005)
	Plastid division	MinD1-MinE1; FtsZ1,ARC6-FtsZ2	*N. tabacum*	(Maple *et al.*, 2005)
	Enzyme complex	AtSufE-AtSufS, AtNifS	*N. tabacum*	(Xu and Moller, 2006)

[a]Protein pairs that have been tested are separated by a dash. In cases where several protein pairs have been tested, the alternative partners are separated by a comma. Different combinations of proteins that have been tested are separated by semicolons.

individual proteins (Grinberg *et al.*, 2004; Hu *et al.*, 2002; Rajaram and Kerppola, 2004; Shyu *et al.*, 2006). Interactions among transcription factors can therefore regulate their subcellular and subnuclear localization. The ATF2 transcription regulatory protein is localized to the cytoplasm when expressed alone but is translocated into the nucleus upon dimerization with Jun family transcription factors (Hu *et al.*, 2002; Liu *et al.*, 2006). Complexes formed by Max with different Myc and Mad family members are localized to different subnuclear locations (Grinberg *et al.*, 2004). Complexes between the exon junction complex components Y14 and NXF1 are formed only during ongoing transcription and are localized to nuclear splicing speckles (Schmidt *et al.*, 2006). Further studies of the mechanisms regulating the localization of protein complexes will increase our understanding of the roles of subnuclear compartmentalization in regulating nuclear protein function.

B. BiFC Analysis of Enzyme–Substrate Complexes

Interactions between several enzymes and their protein substrates have been visualized using the BiFC assay. Complementation between ubiquitin ligases, kinases, and guanine nucleotide exchange factors and potential substrate proteins has been used to investigate substrate recognition in living cells (Blondel *et al.*, 2005; de Virgilio *et al.*, 2004; Niu *et al.*, 2005; Remy *et al.*, 2004; Stolpe *et al.*, 2005; von der Lehr *et al.*, 2003). Determination of the substrate specificities and subcellular sites of action of these enzymes in living cells has yielded new hypotheses for their functions. The *Saccharomyces cerevisiae* ubiquitin E3 ligase Grr1 interacts with the Hof1 regulator of cytokinesis within the bud neck (Blondel *et al.*, 2005). This interaction is restricted to the M phase of the cell cycle, presumably because of rapid degradation of Hof1 following cytokinesis. The substrate specificities and sites of action of ubiquitin ligases in many other organisms have also been visualized using the BiFC assay (Blondel *et al.*, 2005; Marrocco *et al.*, 2006; Stolpe *et al.*, 2005; von der Lehr *et al.*, 2003).

C. BiFC Analysis of Signal Transduction Pathways

Interactions among many signaling proteins have been visualized using the BiFC assay. The study of membrane protein interactions presents unique challenges because of the role of the membrane environment in determination of interaction specificity. BiFC analysis has therefore been particularly valuable for the visualization of membrane protein interactions (de Virgilio *et al.*, 2004; Giese *et al.*, 2005; Guo *et al.*, 2005; Hynes *et al.*, 2004; Ozalp *et al.*, 2005). Contrary to expectation, the constrained mobilities of membrane proteins apparently neither prevent the association of the fluorescent protein fragments nor eliminate the requirement for specific protein interactions. The integrin αIIβ3 receptor exhibited a specific interaction with the Src kinase upon interaction with extracellular fibrin (de Virgilio *et al.*, 2004). These complexes were localized to membrane ruffles, focal adhesions, and focal complexes. These results demonstrate spatially restricted assembly of signaling complexes. Further studies of interactions among additional signal transduction proteins will enable elucidation of the spatial organization of signal transduction networks.

D. BiFC Analysis of Complex Relocalization

The BiFC assay is ideally suited for visualization of the subcellular localization of protein complexes. However, it is important to establish that the association of the fluorescent protein fragments does not result in mislocalization of the complex. In the case of the differences in subnuclear localization by BiFC complexes formed by Max with different Myc/Max/Mad family proteins, a similar relocalization of Max was observed in cells that overexpressed different Myc/Max/Mad family proteins in the absence of fluorescence complementation (Grinberg *et al.*, 2004). Likewise, ATF2 was localized to different subcellular compartments in cells that expressed different levels of the Jun dimerization partner (Shyu *et al.*, 2006). Upon isopreterenol stimulation, the β and γ subunits of the heterotrimeric G-protein internalized as a complex separate from the β-adrenergic receptor (Hynes *et al.*, 2004a). A complex formed by the guanine nucleotide exchange factors GBF1 and the small GTPase Arf1 was recruited to the Golgi in cells treated with brefeldin A (Niu *et al.*, 2005). Similarly, a complex formed by the Bcl-2 family proteins Bif-1 and Bax was relocated to mitochondria in cells induced to undergo apoptosis (Takahashi *et al.*, 2005). BiFC enables visualization of complex localization in many different subcellular compartments and does not appear to interfere with the translocation of protein complexes between these compartments.

E. BiFC Analysis of Interactions Induced by Posttranslational Modifications

Many interactions are thought to require specific posttranslational modifications. The requirement for these modifications in living cells can be tested using the BiFC assay. The bromodomain protein Brd2 binds selectively to acetylated histones. Complementation between Brd2 and H4 required both the bromodomain of Brd2 and the H4 tail that contains the acetylation site (Kanno *et al.*, 2004). Fluorescence complementation by the ERGIC53 receptor and cathepsins catZ and catC required the lectin-binding domain of ERGIC53, suggesting that these interactions require ligand glycosylation (Nyfeler *et al.*, 2005). BiFC can therefore be used to detect posttranslational modifications that alter protein interactions in cells.

F. BiFC Analysis of Interactions on Molecular Scaffolds

The interaction that brings together the fluorescent protein fragments need not be direct. Fusion proteins that are brought together by assembly in a macromolecular complex can produce bimolecular fluorescent complexes in the absence of direct contact between the proteins that are fused to the fragments. Similarly, fusion proteins that bind to the same molecular scaffold can support fluorescence complementation even if the fusion partners do not contact each other directly. Using this principle, RNA binding by the human zipcode-binding protein 1 ortholog (IMP1), the iron regulatory protein (IRP1), the fragile X mental retardation protein (FMRP), and the human Staufen homologue (hStau1) was visualized in living cells (Rackham and Brown, 2004). Likewise, co-occupancy by zinc-finger DNA binding proteins on oligonucleotides has been detected *in vitro* (Stains *et al.*, 2005). Complexes between the exon junction complex components Y14 and NXF1 were also detected only in the presence of newly synthesized transcripts (Schmidt *et al.*, 2006). Therefore, although the association of fluorescent protein fragments in the BiFC assay is bimolecular, this assay is not limited to the visualization of binary interactions.

VI. BiFC Analysis of Interactions in Different Organisms

The BiFC assay has been used to visualize interactions in a variety of species from many different phyla (see column 4 of Table III). In *E. coli*, many endogenous and heterologous proteins as well as peptides have been shown to interact with each other using BiFC analysis (Ghosh *et al.*, 2000; Hu *et al.*, 2002; Magliery *et al.*, 2005; Tsuchisaka and Theologis, 2004). In *Agrobacterium tumefaciens*, interactions between components of the type IV secretion machinery have been visualized some interaction among these components are blocked by inhibitors of transformation (Atmakuri *et al.*, 2003; Cascales *et al.*, 2005). In *S. cerevisiae*, interactions between the Grr1 ubiquitin E3 ligase and the Hof1 regulator of cytokinesis have been shown to be regulated during the cell cycle (Blondel *et al.*, 2005). In the slime mold *Dictyostelium discoides,* interactions between the β and γ subunits of heterotrimeric G-proteins have been visualized (Hynes *et al.*, 2004b). In the filamentous fungus *Acremonium chrysogenum*, interactions between transcription factors have been visualized in the nucleus (Hoff and Kuck, 2005). In tobacco, onion, and *Arabidopsis thaliana*, interactions between many different types of proteins have been visualized by introducing expression vectors encoding the fusion proteins using *Agrobacterium* infiltration or particle bombardment (Abe *et al.*, 2005; Bracha-Drori *et al.*, 2004; Diaz *et al.*, 2005; Hackbusch *et al.*, 2005; Lacroix *et al.*, 2005; Li *et al.*, 2005; Loyter *et al.*, 2005; Maple *et al.*, 2005; Shimizu *et al.*, 2005; Tzfira *et al.*, 2004; Walter *et al.*, 2004). In *Caenorhabditis elegans*, BiFC has been used to mark cells in which specific promoters are transcribed, but the role of specific protein interactions was not examined (Zhang *et al.*, 2004). The BiFC assay is therefore likely to be generally applicable for visualization of protein interactions in virtually every cell type and organism that can be genetically modified to express proteins that are fused to the fluorescent protein fragments.

VII. Screens Using the BiFC Approach

Complementation assays, and in particular the yeast two-hybrid transcription activation assay, have been used to identify new interaction partners for many proteins. The Ft1 protein was identified as an interaction partner of PKB/Akt using a BiFC-based library screening approach (Remy and Michnick, 2004). The advantage of BiFC-based screens is that the interactions can be detected in their normal environment, and the effects of stimuli on the interaction can be directly tested. One limitation of BiFC-based screens is that differences in protein expression levels are likely to influence the partners that can be identified. Nevertheless, BiFC analysis has the potential to identify partners that interact with a protein of interest under specific cellular conditions. BiFC analysis can also be used to identify synthetic molecules or cellular factors that can modulate protein interactions.

VIII. Analysis of Complex Dynamics Using the BiFC Approach

Interactions between many proteins are regulated in response to extracellular stimuli or developmental programs. Given the time required for fluorophore maturation and the stabilization of protein interactions by association of the fluorescent protein fragments (Hu *et al.*, 2002; Magliery *et al.*, 2005) (see Box 3), it is unclear how closely bimolecular fluorescent complex fluorescence reflects the dynamics of complex formation and dissociation. However, some fluorescent protein fragments purified as intein fusions from *E. coli* can undergo at least partial maturation prior to association (Demidov *et al.*, 2006). Moreover, the addition of high concentrations of protein competitors (Guo *et al.*, 2005) or adjusting conditions to destabilize complexes formed by nucleic acid hybridization (Demidov *et al.*, 2006) can reduce the fluorescence of bimolecular fluorescent complexes *in vitro*.

Fluorescence complementation by fusions to MEKK and IκB was enhanced within 2 min of TNFα treatment of HEK293 cells, and returned to basal level within 15 min of removal of the stimulus (Schmidt *et al.*, 2003). MEKK3 exhibited selective complementation with IκBα whereas MEKK2 exhibited selective fluorescence complementation with IκBβ. Similarly, insulin enhanced and TGF-β treatment reduced fluorescence complementation by PKB/Akt and Smad3 within 30 min of stimulation (Remy *et al.*, 2004). Complementation between fluorescent protein fragments fused to PLCβ2 and PLCδ1 was reduced by $\sim$50% within 5 min of acetylcholine or carbachol treatment of HEK293 cells (Guo *et al.*, 2005). The fluorescence of bimolecular fluorescent complexes can therefore be dynamically modulated in response to regulatory signals. It is unclear if the rapid changes in fluorescence intensity observed in these experiments reflect the rates of formation or dissociation of protein complexes. It is also important to consider the possibility that changes in fluorescence may reflect changes in protein synthesis or degradation.

IX. Simultaneous Visualization of Several Protein Complexes

The discovery of GFP has transformed cell biology. This transformation has been accelerated by the development of numerous variants with altered spectral and photophysical characteristics (Zhang *et al.*, 2002). We have developed a multicolor BiFC assay that enables simultaneous visualization of multiple protein complexes in the same cell (Hu and Kerppola, 2003). The multicolor BiFC assay is based on complementation between fragments of different fluorescent proteins that produce bimolecular fluorescent complexes with distinct spectra (Fig. 4; Hu and Kerppola, 2003). The fragments are fused to alternative interaction partners such that complexes formed between different partners can be visualized independently in the same cell, using different excitation and emission wavelengths.

A. Comparison of the Distributions of Different Complexes in the Same Cell

Direct comparison of the distributions of multiple complexes in the same cell eliminates the need to identify markers that colocalize with the complexes as is necessary for comparison of complex distributions between different cells. Comparison of complexes in the same cell also allows determination whether differences in complex localization reflect intrinsic differences in their localization signals. When the distributions of the complexes are compared between different cells, it is possible that indirect effects of the expression of different fusion proteins in different cells alter complex localization. Similarities in the distributions of two or more protein complexes suggest that the complexes have related functions, especially if their distributions are coordinately regulated.

B. Competition Among Mutually Exclusive Interaction Partners for Complex Formation

Interactions with different structurally related interaction partners are often mediated by the same contact interface. For some intensely studied proteins, scores of putative interaction partners have been identified using *in vitro* assays and genetic screens in yeast. It is physically impossible for one protein to simultaneously associate with all of these partners. Interactions with many of these partners are therefore likely to be mutually exclusive. This results in competition for interactions among alternative partners in the cell. Competition among mutually exclusive interaction partners is likely to be an important determinant of the specificity of protein interactions in the cell.

Interactions with different partners can occur in distinct subcellular locations. The selectivity of protein interactions in the cell is determined by many factors including the relative binding affinities of alternative interaction partners and the local concentrations of each protein. It is difficult to predict the selectivity of protein interaction in cells based on *in vitro* studies, since many factors, including covalent modifications, differences in subcellular distributions, and interactions with other cellular proteins, can affect complex formation.

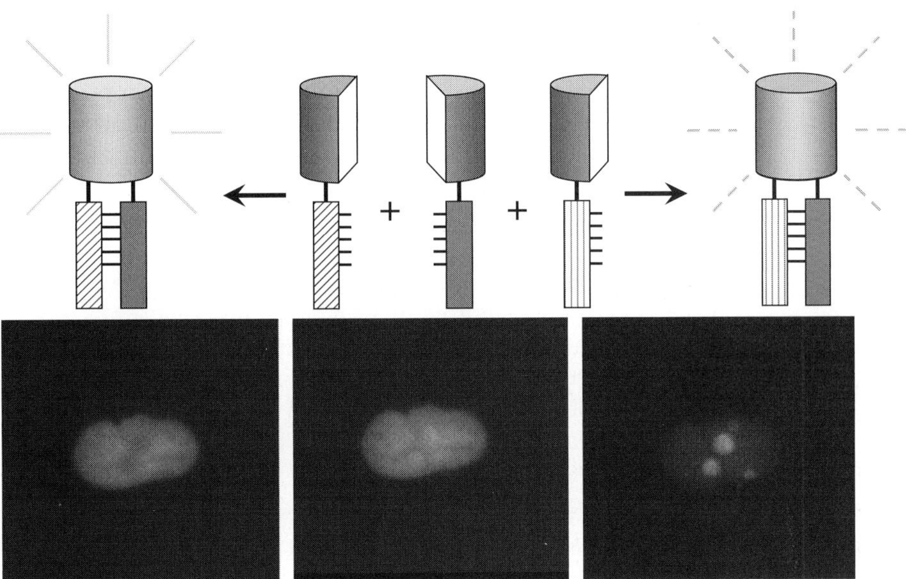

Fig. 4 Simultaneous visualization of multiple protein complexes in the same cell using multicolor bimolecular fluorescence complementation (BiFC) analysis. Multicolor BiFC analysis is based on the enhanced association of different fluorescent protein fragments through interactions between different proteins fused to the fragments. The bimolecular fluorescent complexes formed by fragments of different fluorescent proteins have distinct spectra and can be distinguished using interference filters. Since bimolecular fluorescent complex formation can stabilize protein interactions at least *in vitro*, the relative efficiencies of complex formation do not necessarily reflect the equilibrium binding affinities of the interaction partners in the cell. However, the rate of association between the fluorescent protein fragments in a complex ($t_{1/2} \approx 1$ min) is slower than the rate of dissociation for the majority of protein interactions. It is therefore likely that the relative fluorescence intensities observed in the multicolor BiFC assay reflect the ratio of complexes formed with each interaction partner shortly after synthesis. Quantitative comparison of the efficiencies of complex formation between alternative interaction partners requires that the fluorescent protein fragments can associate with the same efficiency within each complex. This is likely to be true only when the structures of the alternative interaction partners are closely related and should be verified by using fusion proteins with different linker sequences. In cases where quantitative comparison of the efficiencies of complex formation is not possible using multicolor BiFC analysis, this assay can be used for qualitative comparison of the distributions of complexes formed with different interaction partners. The images below the diagrams display competition between full-length Jun and the bZIP domain of Jun for dimerization with the bZIP domain of Fos (the fluorescence of these complexes are shown in green and red, respectively). Image acquired by Changdeng Hu (Hu and Kerppola, 2003) reproduced with permission from Nature Publishing group © 2003. (See Plate no. 57 in the Color Plate Section.)

The multicolor BiFC assay can be used to investigate the competition between mutually exclusive interaction partners for complex formation with a common partner (Grinberg *et al.*, 2004; Hu and Kerppola, 2003). When two mutually exclusive interaction partners fused to fragments of different fluorescent proteins are expressed with a limiting amount of a shared interaction partner fused to a complementary fragment, the proportion of bimolecular fluorescent complexes formed with each interaction partner reflects the relative efficiencies of complex formation with each interaction partner in the cell. To use multicolor BiFC analysis for investigation of the relative efficiencies of complex formation, it is necessary to design fusion proteins that exhibit equal efficiencies of association between the fluorescent protein fragments upon formation of each of the complexes to be compared (Grinberg *et al.*, 2004; Hu and Kerppola, 2003). Differences in the fluorescence intensities of bimolecular fluorescent complexes formed by different combinations of fluorescent protein fragments and possible differences in the intrinsic efficiencies of association of fragments from different fluorescent proteins can be normalized by using proteins in which these fragments are fused to the same interaction partners as described (Grinberg *et al.*, 2004; Hu and Kerppola, 2003). Thus, multicolor BiFC analysis enables comparison of the relative efficiencies of complex formation by alternative interaction partners in the normal cellular environment.

X. Experimental Strategies for Multicolor BiFC Analysis

The strategies for multicolor BiFC analysis of several complexes in the same cell are for the most part identical to those used for BiFC analysis of individual complexes described in Section IV. In addition to these, it is necessary to consider the requirements for separation of the signals from two different BiFC complexes in the same cell.

A. Design of Plasmid Vectors for Multicolor BiFC Analysis

The basic principles for the design of plasmids for BiFC analysis (Section IV.A) also apply to the design of plasmids for multicolor BiFC analysis. The main difference is that it is important to select fluorescent protein fragments that provide maximal spectral separation of the fluorescence signals from different bimolecular complexes. There are several combinations of fragments that can be used for multicolor BiFC analysis (Hu and Kerppola, 2003). Complex formation by CC155 (C-terminal fragment of CFP) with YN155 versus CN155 (N-terminal fragment of CFP) results in complexes with good spectral separation and complementation efficiency (Table II). These combinations are therefore appropriate for the simultaneous analysis of two protein interactions. For the simultaneous analysis of more than two interactions, more selective interference filters and more complex spectral separation algorithms are required.

If quantitative comparison of the efficiencies of complex formation is contemplated, the fluorescent protein fragments should be fused to the alternative interaction partners in the same manner. It is essential that steric constraints to the association between the fluorescent protein fragments are identical in each complex. One way to test this is to determine if the fragments fused to the alternative interaction partners associate with the same efficiency with the shared interaction partner. This can be accomplished by comparing the fluorescence intensities of cells expressing the proteins fused to the same fluorescent protein fragments in the absence of competitors.

B. Strategies for Coexpression of Proteins for Multicolor BiFC Analysis

To investigate the competition between two alternative interaction partners (e.g., A and B) for a shared partner (e.g., Z), the three proteins should be fused to fluorescent protein fragments that can form spectrally distinct bimolecular fluorescent complexes (i.e., A-YN155, B-CN155, and Z-CC155; Fig. 5; for definitions of fusion proteins, see Table II). Cell can be transiently cotransfected with plasmid vectors expressing each of the fusion proteins. Alternatively, cell lines that express the fusion proteins can be developed. Ideally, the proteins should be expressed at levels comparable to their endogenous counterparts. Because of the large differences in excitation and emission spectra, the fluorescence signals from the two complexes can be separated with less than 2% cross talk.

C. Quantitation of Competition Between Alternative Interaction Partners Using Multicolor BiFC Analysis

We have developed two methods for quantification of the efficiencies of complex formation using the multicolor BiFC assay (Grinberg *et al.*, 2004; Hu and Kerppola, 2003). Both methods can provide information about the relative

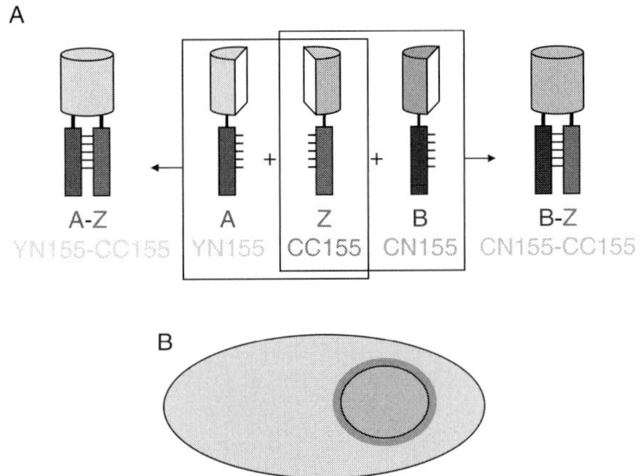

Fig. 5 Simultaneous visualization of multiple protein complexes using multicolor fluorescence complementation analysis. (A) Two alternative interaction partners, A and B, are fused to fragments of different fluorescent protein fragments (YN155 and CN155, respectively). These fusions are coexpressed in cells with a shared interaction partner, Z, fused to a complementary fragment (CC155). Complexes formed by A-YN155 and Z-CC155 can be distinguished from complexes formed by B-CN155 and Z-CC155 based on their fluorescence spectra. (B) Schematic representation of the visualization of multiple protein complexes in the same cell (A-YN155-Z-CC155, cytoplasmic and perinuclear; B-CN155-Z-CC155, nuclear and perinuclear). (See Plate no. 58 in the Color Plate Section.)

efficiencies of complex formation by mutually exclusive interaction partners in living cells. The absolute fluorescence intensities of bimolecular fluorescent complexes formed by different interaction partners cannot be used to compare their efficiencies of complex formation since many factors unrelated to the efficiency of complex formation affect the fluorescence intensities. However, the relative efficiencies of fluorescence complementation between different interaction partners can be compared in the same cell. As long as the steric constraints to fragment association and other factors affecting fluorescence complementation are equivalent for the alternative complexes, this strategy can be used to determine the relative efficiencies of complex formation by alternative interaction partners.

XI. Limitations of the Multicolor BiFC Assay for Analysis of the Efficiencies of Protein Interactions in Cells

The multicolor BiFC assay of the relative efficiencies of complex formation does not provide information about the binding affinities of the interaction partners, but about their relative efficiencies of complex formation. The efficiencies of complex formation do not necessarily reflect binding affinities since the complexes are not at thermodynamic equilibrium. Association of the fluorescent protein fragments is relatively slow ($t_{1/2} \approx 60$ s) and can stabilize the protein interactions at least *in vitro* (Hu *et al.*, 2002). Under these conditions, the relative efficiencies of complex formation reflect competition between alternative interaction partners prior to association of the fluorescent protein fragments. In this case, the multicolor BiFC assay is predicted to give a valid estimate of the relative efficiencies of complex formation for proteins with rapid exchange rates, but interaction partners with slow rates of association may not reach equilibrium prior to association of the fluorescent protein fragments.

Since the multicolor BiFC approach compares the relative amounts of different bimolecular fluorescent complexes, differences between the levels of expression and rates of degradation of the fusion proteins can affect the ratio of complexes that are formed. It is essential to compare the levels of expression of the alternative interaction partners and to take any differences in their expression levels into account when interpreting the data.

XII. Interaction Partners Whose Competition Has Been Visualized Using the Multicolor BiFC Assay

The multicolor BiFC assay has been used to visualize the relative efficiencies of dimerization among the bZIP domains of Fos, Jun, and ATF2. The results show that the bZIP domains of Fos-Jun heterodimers form more efficiently than either Fos-ATF2 or Jun-ATF2 heterodimers in living cells (Hu and Kerppola, 2003).

Jun-ATF2 heterodimers are thought to regulate the expression of many genes. Heterodimer formation may therefore be regulated by regions outside the bZIP domains of these proteins. Alternatively, Jun-ATF2 heterodimers may form only in cells or in subcellular locations where Jun is present in excess relative to the amount of Fos.

Multicolor BiFC analysis has also been used to compare the relative efficiencies of Max interactions with the bHLHZip domain of Myc versus Mad family proteins (Grinberg *et al.*, 2004). Max favored heterodimer formation with bMyc over homodimerization, consistent with the intrinsic binding affinities of these complexes measured *in vitro*. This preference was reversed by point mutations in the leucine zipper of Max. The dimerization preferences of these bHLHZIP proteins in living cells reflected their relative binding affinities *in vitro*.

XIII. Visualization of Ubiquitin Family Peptide Conjugates in Cells

Fluorescence complementation assays have been adapted for the visualization of covalent protein modifications in living cells. Conjugation of ubiquitin family peptides to many substrate proteins modulates their functions and stabilities. Ubiquitin conjugates have been visualized in cells using a ubiquitin-mediated fluorescence complementation (UbFC) assay (Fang and Kerppola, 2004). In this approach, the fluorescent protein fragments used for BiFC analysis are fused to ubiquitin and to a putative substrate protein. The covalent attachment of ubiquitin to a substrate can facilitate association between the fragments, enabling selective visualization of the ubiquitin conjugate (Fig. 6). The UbFC assay shares many of the characteristics of the BiFC assay. Of particular significance is the ability to visualize a small subpopulation of modified proteins in the presence of an excess of unmodified proteins. This is critical for studies of ubiquitin family peptide conjugation since the proportion of most proteins that is modified by a ubiquitin family peptide at any one time is small.

A. Limitations of the UbFC Assay for the Detection of Ubiquitin Family Peptide Conjugates

One well-established function of ubiquitin conjugation is to induce degradation of the modified protein. It is therefore possible that some conjugates are not detected before they are degraded. It is also possible that only conjugates modified at specific lysine residues or conjugates with a specific linkage between the ubiquitin monomers are detected. Such effects of the steric arrangement of fluorescent protein fragments on fluorescence complementation can be detected by comparing UbFC using substrates in which the fluorescent protein fragments are fused to different ends of the protein. If fusions to opposite ends of the substrate protein produce comparable results, it is unlikely that steric factors influence the detection of the conjugates.

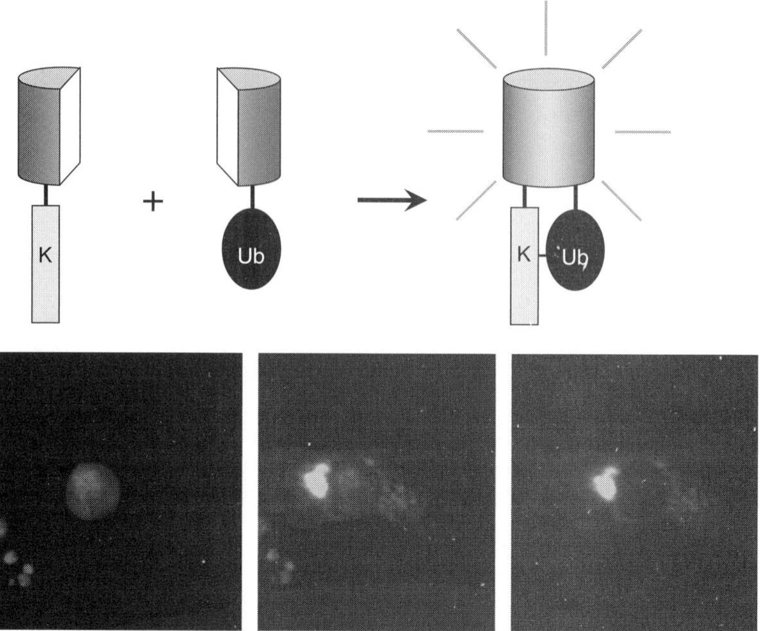

Fig. 6 Visualization of ubiquitin family peptide conjugates using the ubiquitin-mediated fluorescence complementation (UbFC) assay. The UbFC assay is based on the association between nonfluorescent fragments of a fluorescent protein when they are brought together by covalent conjugation of ubiquitin fused to one fragment to a substrate fused to the complementary fragment. The fluorescent protein fragment must be fused to the N-terminal ends of ubiquitin family peptides for them to retain their abilities to be conjugated to substrate proteins at their C-terminal ends. Fusions to the N-terminus of ubiquitin and SUMO1 do not interfere with their conjugation to many substrates (Fang and Kerppola, 2004). It is essential to determine if the conjugates containing the fluorescent protein fragments retain the biological functions of the unmodified conjugates. Since ubiquitin can be conjugated in different monomeric and polymeric configurations to substrates, it is important to establish that the stoichiometry and configuration of the ubiquitin conjugate are not altered by the fusions. The images below the diagrams display the total pool of Jun and the Jun-ubiquitin conjugate in blue and green, respectively. Images acquired by Deyu Fang (Fang and Kerppola, 2004) reproduced with permission © (2004) the National Academy of Sciences. (See Plate no. 59 in the Color Plate Section.)

XIV. Ubiquitin Family Peptide Conjugates That Have Been Visualized Using the UbFC Assay

The UbFC assay has been used to visualize conjugates formed by different ubiquitin family peptides with the Jun protein. Surprisingly, ubiquitinated Jun is exported from the nucleus and is translocated to lysosomes for degradation. In contrast, SUMO1-modified Jun is localized to nuclear foci. Multicolor analysis of ubiquitination and SUMO1-conjugation confirmed that the conjugates are localized to different subcellular compartments in the same cell. Thus, different peptide modifications can induce translocation to different subcellular locations, with different consequences for protein function. Recent studies have revealed a broad range of functions for ubiquitin family peptide modifications and more will undoubtedly be discovered. The direct visualization of these conjugates using the UbFC assay will be a valuable tool for the investigation of these modifications in living cells.

XV. Comparison of BiFC Analysis with Other Methods for the Visualization of Protein Interactions in Living Cells

Several methods have been developed to study protein interactions in living cells. One of the most commonly employed methods is FRET (Jares-Erijman and Jovin, 2003; Miyawaki, 2003; Schmid and Neumeier, 2005; Zal and Gascoigne, 2004; Zhang *et al.*, 2002). To investigate protein interactions using FRET, two different fluorophores are either chemically linked or genetically fused to the two proteins whose interaction is to be examined. The interaction is detected based on a change in the intensity or lifetime of the donor fluorophore or the intensity of the acceptor. Compared to the BiFC assay, FRET analysis in living cells generally requires higher levels of protein expression for reliable detection of energy transfer. In addition, structural information, or a great deal of luck, is required to place the two fluorophores within 100 Å of each other. This is the maximum distance over which significant energy transfer between fluorescent proteins can be detected. The fraction of proteins that interact with each other must also be sufficiently high to produce a change in the donor and acceptor fluorescence intensities that is greater than changes caused by fluctuations in fluorescence intensity due to other effects. To exclude alternative interpretations of the results, numerous controls must be performed and the fluorescence intensities must be measured with high quantitative accuracy. Despite these limitations, FRET has been successfully used for the analysis of many protein interactions in living cells (Hink *et al.*, 2003; Larson *et al.*, 2003; Li *et al.*, 2001; Majoul *et al.*, 2002; Miyawaki, 2003; Sorkin *et al.*, 2000; Tsien, 2003). A great advantage of FRET relative to BiFC analysis is the real-time detection of complex formation and dissociation, which potentially allows analysis of the interaction under equilibrium conditions. FRET is therefore potentially superior to BiFC analysis in studies of the kinetics of protein association and dissociation.

Several characteristics of the BiFC assay make it useful for studying protein interactions. First, it enables direct visualization of protein interactions and does not rely on their secondary effects. Second, the interactions can be visualized in living cells, eliminating potential artifacts associated with cell lysis or fixation. Third, the proteins are expressed in a relevant biological context, ideally at levels comparable to their endogenous counterparts. Thus, they are predicted to reflect the properties of native proteins, including any posttranslational modifications. Fourth, the BiFC assay does not require stoichiometric complex formation but can detect interactions between a subpopulation of each protein. Fifth, multicolor BiFC analysis allows simultaneous visualization of multiple protein complexes in the same cell and enables analysis of the competition between alternative interaction partners for complex formation. Finally, BiFC does not require specialized equipment, apart from an inverted fluorescence microscope equipped with objectives that allow imaging of fluorescence in cells. The simple detection of bimolecular complex formation requires no postacquisition image processing for interpretation of the data. Consequently, BiFC is a powerful tool for scientists seeking to understand protein interactions in living cells.

XVI. Future Opportunities and Challenges

The visualization of molecular interactions in living cells and organisms has a great potential for elucidating fundamental biological mechanisms. There do not appear to be any limitations on use of the BiFC approach in essentially any aerobic organism and cell type except for those that encode endogenous fluorescent or luminescent proteins. Likewise, the BiFC assay appears to be applicable for the visualization of interactions among a wide variety of structurally and functionally diverse proteins. Future applications of the BiFC assay will likely include studies in transgenic animals and plants where the developmental and tissue-specific regulation of protein interactions can be investigated in the context of the living organism. Studies in cultured cells will likely make greater use of regulated expression systems to control fusion protein expression at more uniform levels in the cell population. The native transcription regulatory regions may be used to control expression of the fusion proteins in studies where the timing and the cell-type specificity of protein expression are critical but gene replacement through homologous recombination is impractical.

Multicolor BiFC analysis of the subcellular distributions of complexes formed with different partners is likely to provide insight into functional differences between complexes formed with different partners. The multicolor BiFC approach will also allow the incorporation of internal standards into experiments and elimination of many artifactual sources of variation in fluorescence intensity through ratiometric imaging. These advantages will prove particularly valuable in adaptation of the BiFC approach for high-throughput analysis and library screens. Many biological processes occur asynchronously in different cells in a population. Simultaneous comparison of several molecular interactions in the same cell can be used to detect heterogeneity in cell populations.

The BiFC approach can be used for many purposes in addition to the study of protein interactions. The wide range of spectral variants that are produced by different combinations of fluorescent protein fragments lends itself to combinatorial tagging of cells in studies of cell lineages and migration. The UbFC approach can be applied to the analysis of conjugates formed by other ubiquitin family peptides with a wide range of different substrates. These studies will lead to a better understanding of the relationships between different ubiquitin family peptide modifications of the same substrate.

There are several limitations of present versions of the BiFC approach that provide opportunities for future improvements. Chief among these limitations is the stabilization of the interaction by association of the fluorescent protein fragments under some conditions. A better understanding of the dynamics of bimolecular fluorescent complex formation in living cells and the development of complementing fragments that minimally perturb the dynamic exchange of interaction partners would be valuable contributions. Another limitation is the intrinsic ability of fluorescent protein fragments to associate independent of an interaction between proteins fused to the fragments. This major source of background signal in the BiFC assay varies depending on the fusion proteins and their levels of expression. Development of fluorescent protein fragments with a reduced tendency for intrinsic association, but without a reduction in their ability to associate when brought together by a protein interaction would make the assay less sensitive to the levels of protein expression. Finally, a variety of fluorescent protein variants with useful photophysical characteristics including photoactivation, photoconversion, and time-dependent spectral transformation have been identified (Ando *et al.*, 2002; Chudakov *et al.*, 2003; Patterson and Lippincott-Schwartz, 2002; Terskikh *et al.*, 2000). Fragments that form bimolecular fluorescent complexes with photophysical properties appropriate for specific experimental purposes would be a useful addition to the present selection of spectral variants. Screening of combinatorial libraries of fluorescent protein fragments also provides a strategy for the identification of novel characteristics that may not be found within the sequence space that can be explored using mutated variants of full-length fluorescent proteins.

Acknowledgments

I thank C.-D. Hu for his participation in the design and implementation of the BiFC assay in mammalian cells and all members of the Kerppola laboratory for their contributions to the improvement and application of the BiFC approach.

References

Abe, M., Kobayashi, Y., Yamamoto, S., Daimon, Y., Yamaguchi, A., Ikeda, Y., Ichinoki, H., Notaguchi, M., Goto, K., and Araki, T. (2005). FD, a bZIP protein mediating signals from the floral pathway integrator FT at the shoot apex. *Science* **309,** 1052–1056.

Ando, R., Hama, H., Yamamoto-Hino, M., Mizuno, H., and Miyawaki, A. (2002). An optical marker based on the UV-induced green-to-red photoconversion of a fluorescent protein. *Proc. Natl. Acad. Sci. USA* **99,** 12651–12656.

Atmakuri, K., Ding, Z. Y., and Christie, P. J. (2003). VirE2, a type IV secretion substrate, interacts with the VirD4 transfer protein at cell poles of Agrobacterium tumefaciens. *Mol. Microbiol.* **49,** 1699–1713.

Blondel, M., Bach, S., Bamps, S., Dobbelaere, J., Wiget, P., Longaretti, C., Barral, Y., Meijer, L., and Peter, M. (2005). Degradation of Hof1 by SCFGrr1 is important for actomyosin contraction during cytokinesis in yeast. *EMBO J.* **24,** 1440–1452.

Blumenstein, A., Vienken, K., Tasler, R., Purschwitz, J., Veith, D., Frankenberg-Dinkel, N., and Fischer, R. (2005). The Aspergillus nidulans phytochrome FphA represses sexual development in red light. *Curr. Biol.* **15,** 1833–1838.

Bracha-Drori, K., Shichrur, K., Katz, A., Oliva, M., Angelovici, R., Yalovsky, S., and Ohad, N. (2004). Detection of protein-protein interactions in plants using bimolecular fluorescence complementation. *Plant J.* **40,** 419–427.

Brock, R., and Jovin, T. M. (1998). Fluorescence correlation microscopy (FCM)-fluorescence correlation spectroscopy (FCS) taken into the cell. *Cell. Mol. Biol. (Noisy-le-grand)* **44,** 847–856.

Cabantous, S., Terwilliger, T. C., and Waldo, G. S. (2005). Protein tagging and detection with engineered self-assembling fragments of green fluorescent protein. *Nat. Biotechnol.* **23,** 102–107.

Cascales, E., Atmakuri, K., Liu, Z., Binns, A. N., and Christie, P. J. (2005). Agrobacterium tumefaciens oncogenic suppressors inhibit T-DNA and VirE2 protein substrate binding to the VirD4 coupling protein. *Mol. Microbiol.* **58,** 565–579.

Chen, C. D., Oh, S. Y., Hinman, J. D., and Abraham, C. R. (2006). Visualization of APP dimerization and APP-Notch2 heterodimerization in living cells using bimolecular fluorescence complementation. *J. Neurochem.* **97,** 30–43.

Chudakov, D. M., Belousov, V. V., Zaraisky, A. G., Novoselov, V. V., Staroverov, D. B., Zorov, D. B., Lukyanov, S., and Lukyanov, K. A. (2003). Kindling fluorescent proteins for precise *in vivo* photolabeling. *Nat. Biotechnol.* **21,** 191–194.

de Virgilio, M., Kiosses, W. B., and Shattil, S. J. (2004). Proximal, selective, and dynamic interactions between integrin alpha II beta 3 and protein tyrosine kinases in living cells. *J. Cell Biol.* **165,** 305–311.

Demidov, V. V., Dokholyan, N. V., Witte-Hoffmann, C., Chalasani, P., Yiu, H. W., Ding, F., Yu, Y., Cantor, C. R., and Broude, N. E. (2006). Fast complementation of split fluorescent protein triggered by DNA hybridization. *Proc. Natl. Acad. Sci. USA* **103,** 2052–2056.

Deppmann, C. D., Thornton, T. M., Utama, F. E., and Taparowsky, E. J. (2003). Phosphorylation of BATF regulates DNA binding: A novel mechanism for AP-1 (activator protein-1) regulation. *Biochem. J.* **374,** 423–431.

Diaz, I., Martinez, M., Isabel-LaMoneda, I., Rubio-Somoza, I., and Carbonero, P. (2005). The DOF protein, SAD, interacts with GAMYB in plant nuclei and activates transcription of endosperm-specific genes during barley seed development. *Plant J.* **42,** 652–662.

Ding, Y. H., Liu, N. Y., Tang, Z. S., Liu, J., and Yang, W. C. (2006). Arabidopsis GLUTAMINE-RICH PROTEIN23 is essential for early embryogenesis and encodes a novel nuclear PPR motif protein that interacts with RNA polymerase II subunit III. *Plant Cell* **18,** 815–830.

Fang, D. Y., and Kerppola, T. K. (2004). Ubiquitin-mediated fluorescence complementation reveals that Jun ubiquitinated by Itch/AIP4 is localized to lysosomes. *Proc. Natl. Acad. Sci. USA* **101,** 14782–14787.

Farina, A., Hattori, M., Qin, J., Nakatani, Y., Minato, N., and Ozato, K. (2004). Bromodomain protein Brd4 binds to GTPase-activating SPA-1, modulating its activity and subcellular localization. *Mol. Cell. Biol.* **24,** 9059–9069.

Galarneau, A., Primeau, M., Trudeau, L. E., and Michnick, S. W. (2002). Beta-lactamase protein fragment complementation assays as *in vivo* and *in vitro* sensors of protein protein interactions. *Nat. Biotechnol.* **20,** 619–622.

Gavin, A. C., Bosche, M., Krause, R., Grandi, P., Marzioch, M., Bauer, A., Schultz, J., Rick, J. M., Michon, A. M., Cruciat, C. M., Remor, M., Hofert, C., *et al.* (2002). Functional organization of the yeast proteome by systematic analysis of protein complexes. *Nature* **415,** 141–147.

Ghosh, I., Hamilton, A. D., and Regan, L. (2000). Antiparallel leucine zipper-directed protein reassembly: Application to the green fluorescent protein. *J. Am. Chem. Soc.* **122,** 5658–5659.

Giese, B., Roderburg, C., Sommerauer, M., Wortmann, S. B., Metz, S., Heinrich, P. C., and Muller-Newen, G. (2005). Dimerization of the cytokine receptors gp130 and LIFR analysed in single cells. *J. Cell Sci.* **118,** 5129–5140.

Giot, L., Bader, J. S., Brouwer, C., Chaudhuri, A., Kuang, B., Li, Y., Hao, Y. L., Ooi, C. E., Godwin, B., Vitols, E., Vijayadamodar, G., Pochart, P., *et al.* (2003). A protein interaction map of *Drosophila* melanogaster. *Science* **302,** 1727–1736.

Grinberg, A. V., Hu, C. D., and Kerppola, T. K. (2004). Visualization of Myc/Max/Mad family dimers and the competition for dimerization in living cells. *Mol. Cell. Biol.* **24,** 4294–4308.

Guo, Y. J., Rebecchi, M., and Scarlata, S. (2005). Phospholipase C beta(2) binds to and inhibits phospholipase C delta(1). *J. Biol. Chem.* **280,** 1438–1447.

Hackbusch, J., Richter, K., Muller, J., Salamini, F., and Uhrig, J. F. (2005). A central role of Arabidopsis thaliana ovate family proteins in networking and subcellular localization of 3-aa loop extension homeodomain proteins. *Proc. Natl. Acad. Sci. USA* **102,** 4908–4912.

Hink, M. A., Borst, J. W., and Visser, A. J. (2003). Fluorescence correlation spectroscopy of GFP fusion proteins in living plant cells. *Meth. Enzymol.* **361,** 93–112.

Hoff, B., and Kuck, U. (2005). Use of bimolecular fluorescence complementation to demonstrate transcription factor interaction in nuclei of living cells from the filamentous fungus Acremonium chrysogenum. *Curr. Genet.* **47,** 132–138.

Hu, C. D., Chinenov, Y., and Kerppola, T. K. (2002). Visualization of interactions among bZIP and Rel family proteins in living cells using bimolecular fluorescence complementation. *Mol. Cell* **9,** 789–798.

Hu, C. D., and Kerppola, T. K. (2003). Simultaneous visualization of multiple protein interactions in living cells using multicolor fluorescence complementation analysis. *Nat. Biotechnol.* **21,** 539–545.

Hynes, T. R., Mervine, S. M., Yost, E. A., Sabo, J. L., and Berlot, C. H. (2004a). Live cell imaging of G(s) and the beta(2)-adrenergic receptor demonstrates that both alpha(s) and beta(1)gamma(7) internalize upon stimulation and exhibit similar trafficking patterns that differ from that of the beta(2)-adrenergic receptor. *J. Biol. Chem.* **279,** 44101–44112.

Hynes, T. R., Tang, L. N., Mervine, S. M., Sabo, J. L., Yost, E. A., Devreotes, P. N., and Berlot, C. H. (2004b). Visualization of g protein beta gamma dimers using bimolecular fluorescence complementation demonstrates roles for both beta and gamma in subcellular targeting. *J. Biol. Chem.* **279,** 30279–30286.

Ito, T., Chiba, T., Ozawa, R., Yoshida, M., Hattori, M., and Sakaki, Y. (2001). A comprehensive two-hybrid analysis to explore the yeast protein interactome. *Proc. Natl. Acad. Sci. USA* **98,** 4569–4574.

Jang, M. K., Mochizuki, K., Zhou, M. S., Jeong, H. S., Brady, J. N., and Ozato, K. (2005). The bromodomain protein Brd4 is a positive regulatory component of P-TEFb and stimulates RNA polymerase II-dependent transcription. *Mol. Cell* **19,** 523–534.

Jares-Erijman, E. A., and Jovin, T. M. (2003). FRET imaging. *Nat. Biotechnol.* **21,** 1387–1395.

Johnsson, N., and Varshavsky, A. (1994). Split ubiquitin as a sensor of protein interactions *in vivo*. *Proc. Natl. Acad. Sci. USA* **91,** 10340–10344.

Kanno, T., Kanno, Y., Siegel, R. M., Jang, M. K., Lenardo, M. J., and Ozato, K. (2004). Selective recognition of acetylated histones by bromodomain proteins visualized in living cells. *Mol. Cell* **13,** 33–43.

Kerppola, T. K. (2006a). Complementary methods for studies of protein interactions in living cells. *Nat. Methods* **3,** 969–971.

Kerppola, T. K. (2006b). Visualization of molecular interactions by fluorescence complementation. *Nat. Rev. Mol. Cell Biol.* **7,** 449–456.

Lacroix, B., Vaidya, M., Tzfira, T., and Citovsky, V. (2005). The VirE3 protein of Agrobacterium mimics a host cell function required for plant genetic transformation. *EMBO J.* **24,** 428–437.

Laricchia-Robbio, L., Tamura, T., Karpova, T., Sprague, B. L., McNally, J. G., and Ozato, K. (2005). Partner-regulated interaction of IFN regulatory factor 8 with chromatin visualized in live macrophages. *PNAS* **102,** 14368–14373.

Larson, D. R., Ma, Y. M., Vogt, V. M., and Webb, W. W. (2003). Direct measurement of Gag-Gag interaction during retrovirus assembly with FRET and fluorescence correlation spectroscopy. *J. Cell Biol.* **162,** 1233–1244.

Li, H. Y., Ng, E. K., Lee, S. M., Kotaka, M., Tsui, S. K., Lee, C. Y., Fung, K. P., and Waye, M. M. (2001). Protein-protein interaction of FHL3 with FHL2 and visualization of their interaction by green fluorescent proteins (GFP) two-fusion fluorescence resonance energy transfer (FRET). *J. Cell. Biochem.* **80,** 293–303.

Li, J. X., Krichevsky, A., Vaidya, M., Tzfira, T., and Citovsky, V. (2005). Uncoupling of the functions of the Arabidopsis VIN protein in transient and stable plant genetic transformation by Agrobacterium. *Proc. Natl. Acad. Sci. USA* **102,** 5733–5738.

Li, S., Armstrong, C. M., Bertin, N., Ge, H., Milstein, S., Boxem, M., Vidalain, P. O., Han, J. D., Chesneau, A., Hao, T., Goldberg, D. S., Li, N., *et al.* (2004). A map of the interactome network of the metazoan C. elegans. *Science* **303,** 540–543.

Liu, H., Dengl, X.,Shyu,Y. J., Li, J. J., Taparowsky, E. J., and Hu, C.-D. (2006). Mutual regulation of c-Jun and ATF2 by transcriptional activation and subcellular localization. *EMBO J.* **25,** 1058–1069.

Loyter, A., Rosenbluh, J., Zakai, N., Li, J. X., Kozlovsky, S. V., Tzfira, T., and Citovsky, V. (2005). The plant VirE2 interacting protein 1. A molecular link between the Agrobacterium T-complex and the host cell chromatin? *Plant Physiol.* **138,** 1318–1321.

Magliery, T. J., Wilson, C. G. M., Pan, W. L., Mishler, D., Ghosh, I., Hamilton, A. D., and Regan, L. (2005). Detecting protein-protein interactions with a green fluorescent protein fragment reassembly trap: Scope and mechanism. *J. Am. Chem. Soc.* **127,** 146–157.

Majoul, I., Straub, M., Duden, R., Hell, S. W., and Soling, H. D. (2002). Fluorescence resonance energy transfer analysis of protein-protein interactions in single living cells by multifocal multiphoton microscopy. *J. Biotechnol* **82,** 267–277.

Maple, J., Aldridge, C., and Moller, S. G. (2005). Plastid division is mediated by combinatorial assembly of plastid division proteins. *Plant J.* **43,** 811–823.

Marrocco, K., Zhou, Y. C., Bury, E., Dieterle, M., Funk, M., Genschik, P., Krenz, M., Stolpe, T., and Kretsch, T. (2006). Functional analysis of EID1, an F-box protein involved in phytochrome A-dependent light signal transduction. *Plant J.* **45,** 423–438.

Miyawaki, A. (2003). Visualization of the spatial and temporal dynamics of intracellular signaling. *Dev. Cell* **4,** 295–305.

Nagai, T., Ibata, K., Park, E. S., Kubota, M., Mikoshiba, K., and Miyawaki, A. (2002). A variant of yellow fluorescent protein with fast and efficient maturation for cell-biological applications. *Nat. Biotechnol.* **20,** 87–90.

Nagai, T., Sawano, A., Park, E. S., and Miyawaki, A. (2001). Circularly permuted green fluorescent proteins engineered to sense Ca^{2+}. *Proc. Natl. Acad. Sci. USA* **98,** 3197–3202.

Niu, T. K., Pfeifer, A. C., Lippincott-Schwartz, J., and Jackson, C. L. (2005). Dynamics of GBF1, a brefeldin A-sensitive Arf1 exchange factor at the Golgi. *Mol. Biol. Cell* **16,** 1213–1222.

Nyfeler, B., Michnick, S. W., and Hauri, H. P. (2005). Capturing protein interactions in the secretory pathway of living cells. *Proc. Natl. Acad. Sci. USA* **102,** 6350–6355.

Ozalp, C., Szczesna-Skorupa, E., and Kemper, B. (2005). Bimolecular fluorescence complementation analysis of cytochrome P4502C2, 2E1, and NADPH-cytochrome P450 reductase molecular interactions in living cells. *Drug Metab. Dispos.* **33,** 1382–1390.

Ozawa, T., Kaihara, A., Sato, M., Tachihara, K., and Umezawa, Y. (2001). Split Luciferase as an optical probe for detecting protein-protein interactions in mammalian cells based on protein splicing. *Anal. Chem.* **73,** 2516–2521.

Patterson, G. H., and Lippincott-Schwartz, J. (2002). A photoactivatable GFP for selective photolabeling of proteins and cells. *Science* **297,** 1873–1877.

Paulmurugan, R., and Gambhir, S. S. (2003). Monitoring protein-protein interactions using split synthetic renilla luciferase protein-fragment-assisted complementation. *Anal. Chem.* **75,** 1584–1589.

Paulmurugan, R., Umezawa, Y., and Gambhir, S. S. (2002). Noninvasive imaging of protein-protein interactions in living subjects by using reporter protein complementation and reconstitution strategies. *Proc. Natl. Acad. Sci. USA* **99,** 15608–15613.

Pazhouhandeh, M., Dieterle, M., Marrocco, K., Lechner, E., Berry, B., Brault, V., Hemmer, O., Kretsch, T., Richards, K. E., Genschik, P., and Ziegler-Graff, V. (2006). F-box-like domain in the polerovirus protein P0 is required for silencing suppressor function. *Proc. Natl. Acad. Sci. USA* **103,** 1994–1999.

Pelletier, J. N., Campbell-Valois, F. X., and Michnick, S. W. (1998). Oligomerization domain-directed reassembly of active dihydrofolate reductase from rationally designed fragments. *Proc. Natl. Acad. Sci. USA* **95,** 12141–12146.

Petersen, N. O., Hoddelius, P. L., Wiseman, P. W., Seger, O., and Magnusson, K. E. (1993). Quantitation of membrane receptor distributions by image correlation spectroscopy: Concept and application. *Biophys. J.* **65,** 1135–1146.

Pfleger, K. D. G., and Eidne, K. A. (2006). Illuminating insights into protein-protein interactions using bioluminescence resonance energy transfer (BRET). *Nat. Methods* **3,** 165–174.

Rackham, O., and Brown, C. M. (2004). Visualization of RNA-protein interactions in living cells: FMRP and IMP1 interact on mRNAs. *EMBO J.* **23,** 3346–3355.

Rajaram, N., and Kerppola, T. K. (2004). Synergistic transcription activation by Maf and Sox and their subnuclear localization are disrupted by a mutation in Maf that causes cataract. *Mol. Cell. Biol.* **24,** 5694–5709.

Remy, I., and Michnick, S. W. (2004). Regulation of apoptosis by the Ft1 protein, a new modulator of protein kinase B/Akt. *Mol. Cell. Biol.* **24,** 1493–1504.

Remy, I., and Michnick, S. W. (2006). A highly sensitive protein-protein interaction assay based on Gaussia Luciferase. *Nat. Methods* **3,** 977–979.

Remy, I., Montmarquette, A., and Michnick, S. W. (2004). PKB/Akt modulates TGF-beta signalling through a direct interaction with Smad3. *Nat. Cell Biol.* **6,** 358–365.

Richards, F. M. (1958). On the enzymic activity of subtilisin-modified ribonuclease. *Proc. Natl. Acad. Sci. USA* **44,** 162–166.

Rossi, F., Charlton, C. A., and Blau, H. M. (1997). Monitoring protein-protein interactions in intact eukaryotic cells by beta-galactosidase complementation. *Proc. Natl. Acad. Sci. USA* **94,** 8405–8410.

Schmid, J. A., and Neumeier, H. (2005). Evolutions in science triggered by green fluorescent protein (GFP). *Chembiochem* **6,** 1149–1156.

Schmidt, C., Peng, B. L., Li, Z. K., Sclabas, G. M., Fujioka, S., Niu, J. G., Schmidt-Supprian, M., Evans, D. B., Abbruzzese, J. L., and Chiao, P. J. (2003). Mechanisms of proinflammatory cytokine-induced biphasic NF-kappa B activation. *Mol. Cell* **12,** 1287–1300.

Schmidt, U., Richter, K., Berger, A. B., and Lichter, P. (2006). *In vivo* BiFC analysis of Y14 and NXF1 mRNA export complexes: Preferential localization within and around SC35 domains. *J. Cell Biol.* **172,** 373–381.

Shimizu, H., Sato, K., Berberich, T., Miyazaki, A., Ozaki, R., Imai, R., and Kusano, T. (2005). LIP19, a basic region leucine zipper protein, is a fos-like molecular switch in the cold signaling of rice plants. *Plant Cell Physiol.* **46,** 1623–1634.

Shyu, Y. J., Liu, H., Deng, X., and Hu, C. D. (2006). Identification of new fluorescent protein fragments for bimolecular fluorescence complementation analysis under physiological conditions. *Biotechniques* **40,** 61–66.

Sorkin, A., McClure, M., Huang, F., and Carter, R. (2000). Interaction of EGF receptor and grb2 in living cells visualized by fluorescence resonance energy transfer (FRET) microscopy. *Curr. Biol.* **10,** 1395–1398.

Spotts, J. M., Dolmetsch, R. E., and Greenberg, M. E. (2002). Time-lapse imaging of a dynamic phosphorylation-dependent protein-protein interaction in mammalian cells. *Proc. Natl. Acad. Sci. USA* **99,** 15142–15147.

Stains, C. I., Porter, J. R., Ooi, A. T., Segal, D. J., and Ghosh, I. (2005). DNA sequence-enabled reassembly of the green fluorescent protein. *J. Am. Chem. Soc.* **127,** 10782–10783.

Stelzl, U., Worm, U., Lalowski, M., Haenig, C., Brembeck, F. H., Goehler, H., Stroedicke, M., Zenkner, M., Schoenherr, A., Koeppen, S., Timm, J., Mintzlaff, S., *et al.* (2005). A human protein-protein interaction network: A resource for annotating the proteome. *Cell* **122,** 957–968.

Stolpe, T., Süsslin, C., Marrocco, K., Nick, P., Kretsch, T., and Kircher, S. (2005). In planta analysis of protein-protein interactions related to light signaling by bimolecular fluorescence complementation. *Protoplasma* **226,** 137–146.

Szczesna-Skorupa, E., and Kemper, B. (2006). BAP31 is involved in the retention of cytochrome P4502C2 in the endoplasmic reticulum. *J. Biol. Chem.* **281,** 4142–4148.

Takahashi, Y., Karbowski, M., Yamaguchi, H., Kazi, A., Wu, J., Sebti, S. M., Youle, R. J., and Wang, H. G. (2005). Loss of Bif-1 suppresses Bax/Bak conformational change and mitochondrial apoptosis. *Mol. Cell. Biol.* **25,** 9369–9382.

Terskikh, A., Fradkov, A., Ermakova, G., Zaraisky, A., Tan, P., Kajava, A. V., Zhao, X., Lukyanov, S., Matz, M., Kim, S., Weissman, I., and Siebert, P. (2000). "Fluorescent timer": Protein that changes color with time. *Science* **290,** 1585–1588.

Tsien, R. Y. (2003). Imagining imaging's future. *Nat. Rev. Mol. Cell. Biol.* (Suppl.), SS16–SS21.

Tsuchisaka, A., and Theologis, A. (2004). Heterodimeric interactions among the 1-amino-cyclopropane-1-carboxylate synthase polypeptides encoded by the Arabidopsis gene family. *Proc. Natl. Acad. Sci. USA* **101,** 2275–2280.

Tzfira, T., Vaidya, M., and Citovsky, V. (2004). Involvement of targeted proteolysis in plant genetic transformation by Agrobacterium. *Nature* **431,** 87–92.

von der Lehr, N., Johansson, S., Wu, S. Q., Bahram, F., Castell, A., Cetinkaya, C., Hydbring, P., Weidung, I., Nakayama, K., Nakayama, K. I., Soderberg, O., Kerppola, T. K., *et al.* (2003). The F-Box protein Skp2 participates in c-Myc proteosomal degradation and acts as a cofactor for c-Myc-regulated transcription. *Mol. Cell* **11,** 1189–1200.

Walter, M., Chaban, C., Schutze, K., Batistic, O., Weckermann, K., Nake, C., Blazevic, D., Grefen, C., Schumacher, K., Oecking, C., Harter, K., and Kudla, J. (2004). Visualization of protein interactions in living plant cells using bimolecular fluorescence complementation. *Plant J.* **40,** 428–438.

Wang, K. Z. Q., Wara-Aswapati, N., Boch, J. A., Yoshida, Y., Hu, C. D., Galson, D. L., and Auron, P. E. (2006). TRAF6 activation of PI 3-kinase-dependent cytoskeletal changes is cooperative with Ras and is mediated by an interaction with cytoplasmic Src. *J. Cell Sci.* **119,** 1579–1591.

Wehr, M. C., Laage, R., Bolz, U., Fischer, T. M., Grünewald, S., Scheek, S., Bach, A., Nave, K.-A., and Rossner, M. (2006). Monitoring regulated protein-protein interactions using Split-TEV. *Nat. Methods* **3,** 985–993.

Wehrman, T., Kleaveland, B., Her, J. H., Balint, R. F., and Blau, H. M. (2002). Protein-protein interactions monitored in mammalian cells via complementation of beta-lactamase enzyme fragments. *Proc. Natl. Acad. Sci. USA* **99,** 3469–3474.

Xu, X. M., and Moller, S. G. (2006). AtSufE is an essential activator of plastidic and mitochondrial desulfurases in Arabidopsis. *EMBO J.* **25,** 900–909.

Ye, H. H., Choi, H. J., Poe, J., and Smithgall, T. E. (2004). Oligomerization is required for HIV-1 nef-induced activation of the Src family protein-tyrosine kinase, Hck. *Biochemistry* **43,** 15775–15784.

Zal, T., and Gascoigne, N. R. (2004). Using live FRET imaging to reveal early protein-protein interactions during T cell activation. *Curr. Opin. Immunol.* **16,** 418–427.

Zhang, J., Campbell, R. E., Ting, A. Y., and Tsien, R. Y. (2002). Creating new fluorescent probes for cell biology. *Nat. Rev. Mol. Cell Biol.* **3,** 906–918.

Zhang, S. F., Ma, C., and Chalfie, M. (2004). Combinatorial marking of cells and organelles with reconstituted fluorescent proteins. *Cell* **119,** 137–144.

Zhu, L. Q., Tran, T., Rukstalis, J. M., Sun, P. C., Damsz, B., and Konieczny, S. F. (2004). Inhibition of Mist1 homodimer formation induces pancreatic acinar-to-ductal metaplasia. *Mol. Cell. Biol.* **24,** 2673–2681.

CHAPTER 20

Protein–Protein Interactions Determined by Fluorescence Correlation Spectroscopy

J. Langowski
German Cancer Research Center
Division Biophysics of Macromolecules
Im Neuenheimer Feld 580
D-69120 Heidelberg, Germany

Abstract

Fluorescence correlation spectroscopy (FCS) is an emerging technique where the interaction between biomolecules is detected through their correlated motion. It offers the advantage of high (single-molecule) sensitivity; independence of molecular orientation or distance; and simultaneous measurement of molecular interactions, concentrations, and mobilities. Here we introduce the principle of the technique and review some recent examples from the literature where FCS has been used with autofluorescent proteins for measuring protein–protein interactions and mobilities in living cells.

0091-679X/08 $35.00
DOI: 10.1016/S0091-679X(08)85020-0

I. Introduction

The quantitative characterization of biomolecular interactions is of fundamental importance for our understanding of cellular mechanisms. In recent years, two developments have revolutionized the way such interactions can be measured: the discovery of autofluorescent proteins (which this book deals with) and the possibility to detect such molecules with single-molecule sensitivity. Confocal optics, as schematized in Fig. 1, are a typical experimental setup that allow single-molecule detection. The fluorescent molecules are excited in a very small detection volume ($\approx$1 fl) with a laser beam focused through a microscope lens. The emitted fluorescence is detected through the same optics and a pinhole whose image in the sample overlaps with the laser focus; excitation and emission wavelengths are separated by dichroic mirrors and filters. This way, a few hundred photons can be detected from a molecule during its random motion through the laser focus.

The diffusion of the fluorescent particles in and out of the detection volume causes the fluorescence intensity in the detectors to fluctuate randomly. Fluorescence correlation spectroscopy (FCS) measures and analyzes these fluctuations, thus yielding information about the motion of the molecules through the laser beam.

A particular intriguing application of FCS is the measurement of biomolecular interactions. When a fluorescent ligand binds to a macromolecule, its mobility will be restricted by the presence of the large interaction partner. The extreme case occurs when the fluorophore is bound to large cellular structures, such as membranes, the cytoskeleton, or chromatin: it will then be immobilized, the amount of residual mobility being determined by the binding kinetics to the structure and any slow motion of the structure itself. Interactions of this type have often been studied with techniques such as fluorescence recovery after photobleaching (see Chapter 14 by McNally, this volume) or related methods, such as continuous microphotolysis (Wachsmuth *et al.*, 2003).

If the fluorescent ligand binds to a mobile macromolecule, the mobility of the complex will be determined by that of the largest interaction partner. This mobility is characterized by the diffusion coefficient D in cell biology often given in units of $\mu m^2 s^{-1}$. Measuring D will reveal eventual interactions of the fluorescent molecule with larger macromolecules, as long as D changes significantly. Since D is in first approximation proportional to the largest linear dimension of the macromolecule, its dependence on the molecular mass M of a globular protein is not very strong; for D to double, M would have to increase by $2^3 = 8$. An association between two ligands of equal size would decrease D by only 26%, a change that can easily be undetectable in the presence of noise.

A much more sensitive way to use the random motion of macromolecules to detect their interaction is two-color fluorescence cross-correlation spectroscopy (FCCS) (Ricka and Binkert, 1989; Schwille *et al.*, 1997). Here both interaction partners are labeled with fluorophores that can be spectrally distinguished. If they form a complex, they will always enter and exit the laser focus at the same time;

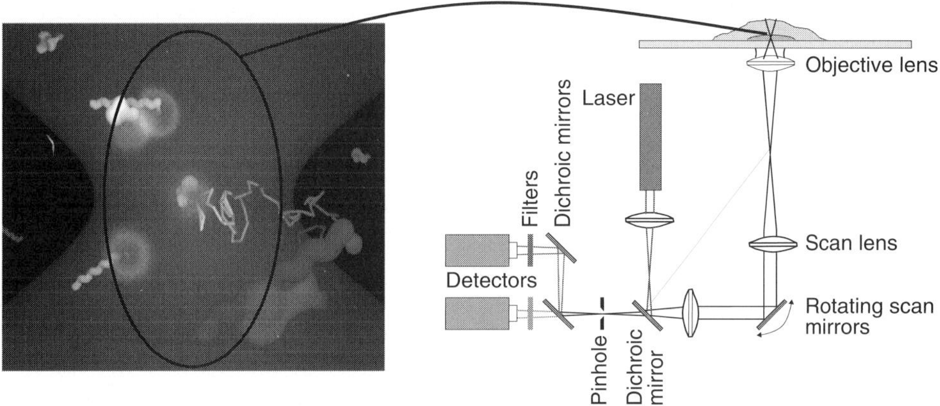

Fig. 1 Schematic principle of the FCS method. (See Plate no. 61 in the Color Plate Section.)

if they do not interact, their respective movements (and fluorescence fluctuations) will be uncorrelated. In practice, the fluorophores are excited by either one or two laser lines focused into the same spot; emitted light from the focal volume is detected at two wavelengths, and particles that fluoresce at both wavelengths will give simultaneous bursts of intensity in the two channels. This correlated emission is detected by computing the cross-correlation function. FCCS is a convenient means to show binding between two ligands labeled with different fluorophores because the complex will show correlated fluorescence at the two wavelengths. The examples given in this chapter will concentrate on the application of FCCS to *in vivo* biomolecular interactions.

FCS and its theoretical foundations have been described some time ago (Elson and Magde, 1974; Magde *et al.*, 1974; Webb, 1976), but in the early work its application was severely limited by sensitivity issues. Meanwhile, FCS has undergone an amazing development that now allows routine measurements of biomolecular interactions. The recent improvements which made this possible (Qian and Elson, 1991; Rigler *et al.*, 1993) are mostly the use of confocal optics for excitation and detection, and avalanche photodiode detectors that offer a quantum efficiency >50% in the red range of the visible spectrum (a factor of 10 over most photomultipliers).

Currently, many manufacturers of confocal microscopes offer avalanche photodiode-based FCS accessories that substitute for the detection photomultipliers and directly detect and analyze the confocally collected fluorescence emission from the sample (Zeiss Confocor 3, Leica FCS accessory for the TCS SP2 AOBS; individual solutions exist for Olympus and Nikon confocal microscopes). Very small concentrations (<1 pM) may be detected because individual fluorescent particles will give clearly distinguishable bursts of fluorescence intensity above the background arising from detector noise, Raman scattering, and optical imperfections. At somewhat higher concentrations, in the nanomolar range, typically several molecules are present simultaneously in the focus. However, the instantaneous number of molecules will fluctuate: at a concentration c, the fluctuation of the number of solute molecules N in a given volume element V is $\langle \partial N^2 \rangle = \langle N \rangle$, where $\langle N \rangle = cV$ is the average number of molecules in V and $\langle \partial N^2 \rangle = \langle (N - \langle N \rangle)^2 \rangle$ is the mean squared fluctuation (as an example, see Table I). The time dependence of the fluctuations is directly related to the diffusion coefficient of the molecule (see below). By observing the concentration fluctuation of a solute in a very small volume of known size, one can thus determine its concentration and its diffusion coefficient.

Due to the small focus of the laser beam, measurements inside living cells become possible. A typical high-resolution microscope lens has a focal spot of 300 nm diameter and 1.5 μm length, such that diffusion processes inside cells or organelles can be probed in a position-dependent manner. FCS has for instance been used to probe chromatin in the cell nucleus (Sorscher *et al.*, 1980) or to assess anomalous diffusion of proteins (Wachsmuth *et al.*, 2000; Weiss *et al.*, 2003).

The primary data obtained in an FCS measurement is the time-dependent fluorescence intensity $F(t)$, which is proportional to the number of particles in the observation volume at time t. The timescale of these fluctuations is determined by the speed with which the molecules move through the laser focus (Fig. 2).

Table I
Number Fluctuations in a 1-nM Solution as a Function of Volume

Size (mm)	Volume (liter)	No. of particles	ΔN	$\Delta N/N$ (%)
10	10^{-3}	6.023×10^{11}	7,76,080	0.00013
1	10^{-6}	6.023×10^{8}	24,541	0.0041
0.1	10^{-9}	6.023×10^{5}	776	0.129
0.01	10^{-12}	602.3	24.5	4.075
0.001	10^{-15}	0.6023	0.776	128.9

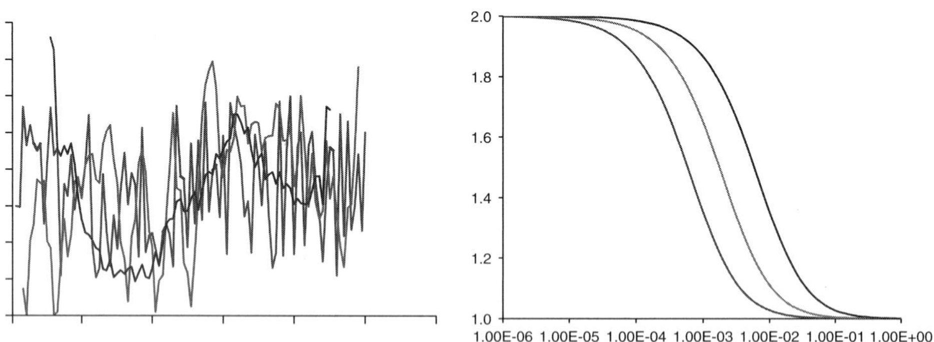

Fig. 2 Example of fluctuating fluorescence intensity for fast (red), medium (green), and slow (blue) moving particles, and corresponding autocorrelation functions $G(\tau)$. (See Plate no. 62 in the Color Plate Section.)

II. FCS Theory

Let us assume that we measure the fluorescence of a 10^{-9} M rhodamine solution in volume elements of various sizes. Table I shows the absolute and relative number fluctuations for this case. In a classical fluorescence spectrometer, the typical observation volume is of the order of 1 ml. It is easily seen that at this sample size no observable fluctuation is expected. If one, however, measures the fluorescence of the same solution in a smaller volume, the fluctuations become increasingly important until they reach the size of the fluorescence signal itself at a sample size of 1 fl; here, less than one molecule is present in the observation volume on average. The characteristics of the fluorescence fluctuations and their relation to molecular properties are summarized in the following sections.

A. Autocorrelation Function for One Species

The fluorescence fluctuations are characterized by their *autocorrelation function* $G(\tau)$, which describes the random motion of the fluorophores (e.g., see Figs. 2 and 3). It is defined as

$$G(\tau) = \frac{\langle F(t)F(t+\tau)\rangle}{\langle F(t)\rangle^2} \quad (1)$$

For obtaining quantities such as diffusion coefficients, concentrations, or reaction rate constants, one has to fit a theoretical correlation function to the measured $G(\tau)$ which is based on a model that contains these quantities as free parameters. For a solution of a single fluorescent species with diffusion coefficient D and molar

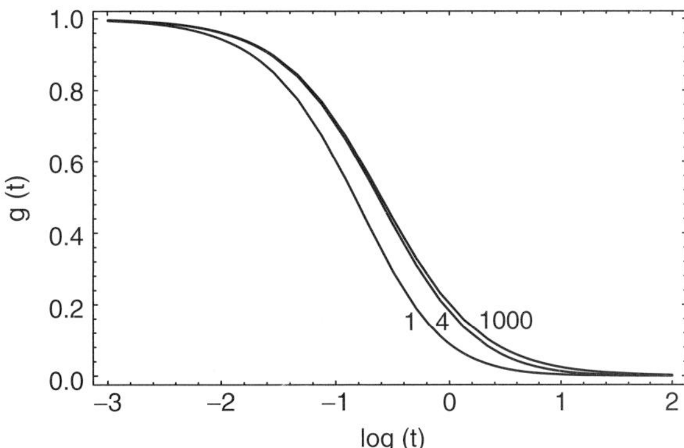

Fig. 3 Influence of the structure factor κ on the FCS autocorrelation function. Three curves are displayed for the same diffusion time τ and $\kappa = 1, 4, 1000$. Since $\kappa \geq 4$ for typical confocal optics, the relevant range in practice is between the rightmost two curves.

concentration c and for Gaussian profiles for the excitation intensity and detection efficiency, $G(\tau)$ evaluates to (Rigler *et al.*, 1993):

$$G(\tau) = \frac{1}{cV_{\text{eff}}}\left(1+\frac{4D\tau}{w_0^2}\right)^{-1}\left(1+\frac{4D\tau}{z_0^2}\right)^{-1/2}+1 \quad (2)$$

Here V_{eff} is the effective observation volume which depends on the geometry of the focus for excitation and emission, w_0 and z_0 are the half-widths of the focus in the x-y plane (the observation plane of the lens) and in the z-direction, respectively.

V_{eff}, w_0, and z_0 can be measured independently by calibration with a solution of a fluorophore of known concentration and diffusion coefficient. If only relative changes are of interest, one can use the average particle number $N = cV_{\text{eff}}$ and an effective diffusion time $\tau_{\text{diff}} = {w_0}^2/4D$ as parameters:

$$G(\tau) = \frac{1}{N}\left(1+\frac{\tau}{\tau_{\text{diff}}}\right)^{-1}\left(1+\frac{\tau}{\tau_{\text{diff}}\kappa^2}\right)^{-1/2}+1 \quad (3)$$

κ (also called the structure factor) is the axial ratio of the observation volume z_0/w_0 (see Fig. 3).

The intercept of the FCS autocorrelation function $G(\tau)$ is inversely proportional to the number of particles in the focal volume, and thus to their concentration. In practice, deviations from this ideal behavior are found at very high and very low concentrations. At low concentrations these deviations are due to the background which becomes comparable to the fluorescence signal, and which is caused by incomplete suppression of the excitation light, detector dark counts, and background fluorescence. At a particle concentration c, the measured particle number N in the presence of background is then

$$N = cV_{\text{eff}}\left(1+\frac{v}{c}\right)^2 \quad (4)$$

where $v = \langle U\rangle/\widetilde{F}$ is the ratio of the background signal to the normalized fluorescence intensity of the fluorophor.

B. Multiple Species

In a mixture of molecules with different diffusion coefficients, the fluorescence intensity autocorrelation function is a sum of the contributions of the individual species. The general form of $G(\tau)$ for a mixture m of different fluorescent species with diffusion times $\tau_{\text{diff},i}$ is then given by the following expression:

$$G(\tau) = \frac{1}{N}\sum_{i=1}^{m}\rho_i g_i(\tau)+1 \quad (5)$$

with $g_i(\tau) = \left(1+\frac{\tau}{\tau_{\text{diff},i}}\right)^{-1}\left(1+\frac{\tau}{\tau_{\text{diff},i}\kappa^2}\right)^{-1/2}$.

The ρ_i are the relative amplitudes corresponding to molecules with distinct diffusion coefficients; they are related to their concentrations c_i by the following expression:

$$\rho_i = \frac{\phi_i^2 c_i}{\sum_{i=1}^{m} \phi_i^2 c_i} \tag{6}$$

where ϕ_i is the quantum yield of species i.

C. Triplet Contribution

Up to now only number fluctuations in the detection volume have been considered to contribute to the fluctuations of the light intensity at the detector, under the simplifying assumption that an excited fluorophor will emit a constant light flux. Because of the quantum nature of light and the photophysics of fluorescent molecules, this is not the case. The most important effect that has to be considered is a transition of the excited molecule into the triplet state. This will "interrupt" the stream of photons for approximately the triplet lifetime of the fluorophor and add another contribution to the autocorrelation function, which in good approximation is then (Widengren *et al.*, 1995):

$$G(\tau) = (1 + \beta e^{-\lambda\tau}) \left(\frac{1}{N} \sum_{i=1}^{m} \rho_i g_i(\tau) \right) + 1 \tag{7}$$

The amplitude of the triplet term β and its relaxation time λ increase with the excitation light intensity up to a limit given by the excitation, emission, and intersystem crossing probabilities of the fluorophore. Practically, β can reach amplitudes higher than the number correlation function itself. Since relaxation time of the triplet term is of the same order as the diffusion times of small molecules (some microseconds), it is important to conduct the FCS experiment with a laser intensity that keeps β as small as possible.

III. Two-Color Cross-Correlation

The detection of specific binding between biomolecules by FCS depends on a change in molecular size: when the diffusion coefficient D changes sufficiently on binding, the complex can be distinguished in $G(\tau)$ as a second species and its concentration determined [Eqs. (5) and (6)]. However, in cases when D changes only very slightly or not at all, that is, when a nonfluorescent ligand binds to a larger fluorescent particle (see above), this approach is not practicable anymore.

Schwille *et al.* (1997) were the first to show the feasibility of two-color FCCS. In this method the fluorescence is detected simultaneously at two distinct wavelengths in the same detection volume. The signals from the two detectors are analyzed by computing their cross-correlation function. It is easily seen that in a mixture of two fluorescent molecules emitting at the two wavelengths but not interacting with each other the particles will diffuse independently and the amplitude of the cross-correlation function will be zero. On the contrary, when the particle is labeled with two dyes and emits simultaneously at the two detection wavelengths, the cross-correlation function is equal to the autocorrelation function for single-color FCS (assuming equal detection efficiencies and exact overlap of the detection volumes for the two channels). This latter case occurs when the two fluorescent species form a complex.

In FCCS, therefore, the amount of complex formation between two fluorescently labeled biomolecules can be obtained simply by measuring the cross-correlation amplitude. A comprehensive summary of the theory of FCCS of associating species has been given by Weidemann *et al.* (2002); a recent review summarizes some of the applications of FCCS *in vivo* (Bacia *et al.*, 2006).

IV. Protein–Protein Interactions Using FCCS and Nongenetic Labels

While the possibility of determining biomolecular interactions *in vivo* by FCCS has been evoked in the earlier FCS literature [see, e.g., Hink *et al.* (2002) for a review], only recently have attempts been successful to show these interactions in living cells. The first results were obtained using chemically labeled proteins. Bacia *et al.* (2002) followed the fate of cholera toxin labeled with Cy2 on one subunit and Cy5 on the other during endocytosis and could show the separation of the two subunits in the Golgi apparatus. Furthermore, they could demonstrate cross-correlation after endocytosis of a mixture of Cy2- and Cy5-labeled holotoxins, showing the tight association between different protein molecules in the same endocytic pathway. For analyzing the motion of the proteins bound to the membrane, photobleaching was a problem (as it is in most FCS studies of slowly moving biomolecules); in cases of strong bleaching, data was only taken after the strongly immobilized molecules had been bleached out and the fluorescence was approximately constant. Proteins that were free to move in the cytosol (during movement from the plasma membrane to the Golgi apparatus) did not show this strong bleaching effect.

Chemical labeling of viral capsid proteins with Alexa 568, together with labeling of the viral genome through a GFP-histone fusion protein was used by Bernacchi *et al.* (2004) to investigate the infectious pathway of *Simian virus 40*. Through a mobility analysis of the two fluorescent components it could be shown that the virus disintegrates close to the nucleus, before being transported through the nuclear membrane. Simultaneously, cross-correlation was lost on virus disassembly. FCS also showed that some nonpermissive cells incorporate the virus, but do not transport it into the nucleus.

In a recent FCCS study, Oyama *et al.* (Oyama and Yanagawa, 2004; Oyama *et al.*, 2006) used an *in vitro* translation system (Doi *et al.*, 2002) to attach a fluorescent puromycin derivative to the C-terminus of the protein. They measured *in vitro* the diffusion coefficients and association constants of c-Fos and c-Jun proteins, calmodulin and calmodulin-binding proteins, and various members of the Polycomb group protein family. Another intriguing new possibility for specific labeling of proteins in FCCS was recently presented by Becker *et al.* (2006). They used proteins recombinantly expressed in *Escherichia coli* and modified at the C-terminal end with a thioester which were then chemically ligated to a peptide containing the fluorescent dye. Although both these latter strategies have not been applied for *in vivo* studies yet, they constitute a promising route for characterizing protein–protein interactions by FCCS and other single-molecule fluorescence techniques.

V. Protein–Protein Interactions *In Vivo* Using FCCS and Autofluorescent Proteins

At the time of this writing, only few studies have been published where protein–protein interactions have been observed *in vivo* by FCCS using two autofluorescent protein-labeled interaction partners. Baudendistel *et al.* (2005) have used the Fos-Jun system to demonstrate protein–protein interaction. Fos and Jun are two components of the AP-1 general transcription activator; they are known to exert their function as a dimer and can therefore serve as a reference for dimer formation (Allegretto *et al.*, 1990; Kerppola and Curran, 1991; Leonard *et al.*, 1997). FCCS was measured in cells expressing equal levels of Fos-EGFP and Jun-mRFP1 fusion proteins. As a control, the deletion mutants FosΔdimΔDNA-EGFP and JunΔdimΔDNA-mRFP1 were used in which the dimerization (Δdim) as well as the DNA-binding (ΔDNA) bZip domains were removed to abolish the dimerization and DNA-binding reactions.

Figure 4 summarizes the results. The negative control, in which free EGFP and mRFP1 were expressed at equimolar amounts in HeLa cells, shows autocorrelation decays with diffusion coefficients corresponding to intracellular diffusion of typical small proteins. The cross-correlation function (in blue) shows only a small background amplitude due to spectral cross talk, with a decay on the same timescale as the single species autocorrelation functions. The positive control for maximum cross-correlation, using a fusion protein of EGFP and mRFP1 expressed in HeLa cells, is shown in Fig. 4B. Again, the decay curves for the red and green channel autocorrelation functions give a diffusion coefficient in the range expected for a small protein of this size; the amplitude of the

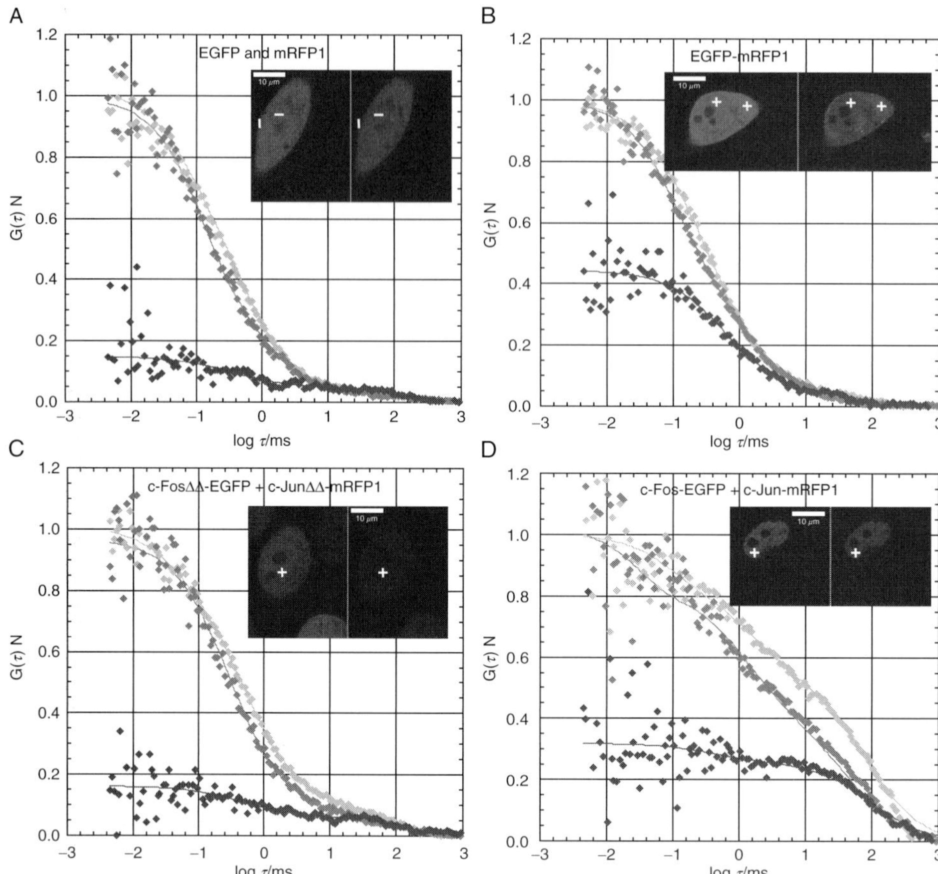

Fig. 4 A, B: *In vivo* FCCS controls for independent (A) and covalently tethered (B) fluorophores. Normalized autocorrelation amplitudes G(τ)N in the red (red line) and the green (green line) channels measured in HeLa cells expressing (A) EGFP and mRFP1 separately; (B) a two-color fusion construct of EGFP and mRFP1. Cross-correlation functions are shown in blue. Rhombi represent measured data, lines fitted curves. Inset: confocal images of cells in the EGFP (green) and the mRFP1 (red) channels. The white crosses indicate the measurement points. C, D: Protein–protein interactions in the AP-1 system. Normalized autocorrelation amplitudes G(τ)N in the red (red line) and the green (green line) channels measured in HeLa cells expressing (C) the AP-1 deletion mutants c-FosΔdimΔDNA-EGFP and c-JunΔdimΔDNA-mRFP1; (D) the AP-1 wildtype proteins c-Fos-EGFP and c-Jun-mRFP1. (See Plate no. 63 in the Color Plate Section.)

cross-correlation function is almost half of the maximum value of 100% expected in the ideal case. Due to imperfect overlap of the focus volumes for excitation and detection of the two chromophores used, this maximum value is practically never reached in an FCCS experiments and the 45% obtained here were taken as the reference for complete binding of the two proteins.

Intracellular FCCS measurements with the Fos and Jun deletion mutants that lacked DNA binding and dimerization domains gave a cross-correlation amplitude not significantly exceeding the background level (Fig. 4C). Again, the diffusion coefficients are in the range expected for freely diffusing proteins without DNA binding.

The demonstration of the direct interaction and DNA binding of c-Fos and c-Jun *in vivo* finally followed by expressing the AP-1 wild-type fusion proteins c-Fos-EGFP and c-Jun-mRFP1 together in the cells. These cells showed fluorescence only in the nuclei (Fig. 4D) due to the nuclear localization sites present in full-length AP-1 proteins. The cross-correlation amplitude in these cells is strongly increased over the background level and clearly indicates the Fos-Jun dimerization (Fig. 4D). The autocorrelation functions showed a fast component corresponding to freely diffusing c-Fos in the green and c-Jun protein in the red channel. Additionally, a second slow component of about equal amplitude as the fast one, but a diffusion coefficient two orders of magnitude smaller, was detected in both the red and the green channel. This slow component had a much higher amplitude for the wild-type proteins than for the nonbinding controls, indicating a strong retention or immobilization of the AP-1 proteins. For the cross-correlation function, only about 25% could be attributed to faster diffusive processes, corresponding to free Fos-Jun dimers. The main decay component of the cross-correlation curve was slow, with a diffusion coefficient equal to the slow component of the autocorrelation functions and a relative amplitude of 75%. This result showed directly that binding of c-Fos and c-Jun to DNA strongly favors the formation of the heterodimer and that a significant amount of the free proteins exists as monomer *in vivo*, showing no cross-correlation.

In a more recent study, Muto *et al.* (2006) showed the interaction between *Arabidopsis thaliana* auxin response factors MP/ARF5 or NPH4/ARF7 and their repressive regulator, MSG2/IAA19, using FCCS with autofluorescent labeling. They transiently expressed GFP and mRFP1 fusions of these proteins in HeLa cells, measured FCS auto- and cross-correlation functions *in vivo* and compared the FCCS amplitudes with positive and negative controls similar to those used by Baudendistel *et al.* (2005) The different interaction pairs investigated fell into two classes, with either about 80% or about 20% of the proteins bound in a complex. However, these data were not corrected for differences in the expression level of the two proteins.

The cross-correlation amplitudes in this work were significantly higher than those obtained in the work by Baudendistel *et al.* (2005); however, both studies show about a factor of 3 in amplitude of the positive control over the background. This seems therefore to define the maximum dynamic range that can be obtained with FCCS for quantitative binding studies *in vivo*.

Some studies have looked at the interaction of membrane-bound proteins by FCCS. Vamosi *et al.* (2004) studied the membrane localization and interaction of subunits of the IL-2 and IL-15 interleukin receptors and major histocompatibility complex (MHC) class I by fluorescence resonance energy transfer (FRET) and FCCS. In this case, the proteins were labeled chemically or with fluorescent antibodies. A strong decrease by about three orders of magnitude of the diffusional mobility of the Alexa 488-labeled antibody was detected on binding to the IL-2 α subunit. Fluorescence cross-correlation on the same timescale as the immobilized antibody could then be shown between Cy5-labeled MHC class I and the Alexa 488-labeled antibody, indicating that MHC and IL-2 diffuse in the same membrane subdomain (raft).

Autofluorescent protein labeling was employed in the work of Larson *et al.* (2005) who studied the interaction of the IgE receptor FcϵRI with Lyn kinase, the enzyme responsible for initiating cell signaling by phosphorylating the IgE-associated FcεRI. Here IgE was chemically labeled with Alexa 546 and the Lyn kinase fused to EGFP. In an approach similar to the work by Vamosi *et al.* (2004), the authors showed a slowing down of the diffusion of the protein components after antigen stimulation, indicating association in the membrane. This association could then be proved unequivocally by the appearance of a cross-correlation signal 6 min after antigen stimulation, which decreased again after 30 min. Thus, the association between Lyn kinase and IgE receptor is transient.

These examples, albeit few, demonstrate that FCCS is a very powerful method to characterize association between fluorescent proteins in living cells. Both in the papers by Baudendistel *et al.* (2005) and Muto *et al.* (2006) the analogy between yeast two-hybrid techniques (YTH) and FCCS with double autofluorescent fusion proteins ("two-hybrid FCCS," THFCCS) has been evoked. In both cases, a reporter protein is genetically linked to the protein whose association is to be measured: a signal is generated by the association between the reporter groups that are brought together by the binding of the proteins under study. While YTH does not require advanced (and expensive) optical equipment, optical techniques have several advantages: first, since YTH uses transcriptional activators for generating a signal, it is difficult to use for measuring interactions between transcription factors themselves. Also, for *in vivo* interactions in higher eukaryotes, evidently other cell lines must be available. Alternatives to THFCCS are FRET and fluorescent protein complementation, both of which require close spatial proximity of the interaction partners. THFCCS has the great advantage that binding is detected through the correlated motion of the two proteins, which does not restrict the technique to interactions within the range of the Förster radius.

It is conceivable that THFCCS combined with a scanning system may be applied to screen protein–protein interaction in a large number of cells and/or in a position-dependent manner. Thus, automatic detection of such interactions and of the cellular compartment in which they occur becomes feasible, making this effectively a high-throughput technique usable in automated microscopy setups where biomolecular interactions and cell morphology are monitored at the same time.

References

Allegretto, E. A., Smeal, T., Angel, P., Spiegelman, B. M., and Karin, M. (1990). DNA-binding activity of Jun is increased through its interaction with Fos. *J. Cell. Biochem.* **42,** 193–206.

Bacia, K., Kim, S. A., and Schwille, P. (2006). Fluorescence cross-correlation spectroscopy in living cells. *Nat. Methods* **3,** 83–89.

Bacia, K., Majoul, I. V., and Schwille, P. (2002). Probing the endocytic pathway in live cells using dual-color fluorescence cross-correlation analysis. *Biophys. J.* **83,** 1184–1193.

Baudendistel, N., Muller, G., Waldeck, W., Angel, P., and Langowski, J. (2005). Two-hybrid fluorescence cross-correlation spectroscopy detects protein-protein interactions *in vivo*. *Chemphyschem* **6,** 984–990.

Becker, C. F., Seidel, R., Jahnz, M., Bacia, K., Niederhausen, T., Alexandrov, K., Schwille, P., Goody, R. S., and Engelhard, M. (2006). C-terminal fluorescence labeling of proteins for interaction studies on the single-molecule level. *Chembiochem* **7,** 891–895.

Bernacchi, S., Mueller, G., Langowski, J., and Waldeck, W. (2004). Characterization of simian virus 40 on its infectious entry pathway in cells using fluorescence correlation spectroscopy. *Biochem. Soc. Trans.* **32,** 746–749.

Doi, N., Takashima, H., Kinjo, M., Sakata, K., Kawahashi, Y., Oishi, Y., Oyama, R., Miyamoto-Sato, E., Sawasaki, T., Endo, Y., and Yanagawa, H. (2002). Novel fluorescence labeling and high-throughput assay technologies for *in vitro* analysis of protein interactions. *Genome Res.* **12,** 487–492.

Elson, E. L., and Magde, D. (1974). Fluorescence correlation spectroscopy. I. Conceptual basis and theory. *Biopolymers* **13,** 1–27.

Hink, M. A., Bisselin, T., and Visser, A. J. (2002). Imaging protein–protein interactions in living cells. *Plant Mol. Biol.* **50,** 871–883.

Kerppola, T. K., and Curran, T. (1991). Fos-Jun heterodimers and Jun homodimers bend DNA in opposite orientations: Implications for transcription factor cooperativity. *Cell* **66,** 317–326.

Larson, D. R., Gosse, J. A., Holowka, D. A., Baird, B. A., and Webb, W. W. (2005). Temporally resolved interactions between antigen-stimulated IgE receptors and Lyn kinase on living cells. *J. Cell Biol.* **171,** 527–536.

Leonard, D. A., Rajaram, N., and Kerppola, T. K. (1997). Structural basis of DNA bending and oriented heterodimer binding by the basic leucine zipper domains of Fos and Jun. *Proc. Natl. Acad. Sci. USA* **94,** 4913–4918.

Magde, D., Elson, E. L., and Webb, W. W. (1974). Fluorescence correlation spectroscopy. II. An experimental realization. *Biopolymers* **13,** 29–61.

Muto, H., Nagao, I., Demura, T., Fukuda, H., Kinjo, M., and Yamamoto, K. T. (2006). Fluorescence cross-correlation analyses of the molecular interaction between an Aux/IAA protein, MSG2/IAA19, and protein–protein interaction domains of auxin response factors of arabidopsis expressed in HeLa cells. *Plant Cell Physiol.* **47,** 1095–1101.

Oyama, R., Takashima, H., Yonezawa, M., Doi, N., Miyamoto-Sato, E., Kinjo, M., and Yanagawa, H. (2006). Protein–protein interaction analysis by C-terminally specific fluorescence labeling and fluorescence cross-correlation spectroscopy. *Nucleic Acids Res.* **34,** e102.

Oyama, R., and Yanagawa, H. (2004). Protein–protein interaction assay by dual-color fluorescence cross correlation spectroscopy. *Tanpakushitsu Kakusan Koso* **49,** 2775–2779.

Qian, H., and Elson, E. L. (1991). Analysis of confocal laser-microscope optics for 3-D fluorescence correlation spectroscopy. *Appl. Opt.* **30,** 1185–1195.

Ricka, J., and Binkert, T. (1989). Direct measurement of a distinct correlation function by fluorescence cross correlation. *Phys. Rev. A* **39,** 2646–2652.

Rigler, R., Mets, Ü, Widengren, J., and Kask, P. (1993). Fluorescence correlation spectroscopy with high count rate and low background: Analysis of translational diffusion. *Eur. Biophys. J.* **22,** 169–175.

Schwille, P., Meyer-Almes, F. J., and Rigler, R. (1997). Dual-color fluorescence cross-correlation spectroscopy for multicomponent diffusional analysis in solution. *Biophys. J.* **72,** 1878–1886.

Sorscher, S. M., Bartholomew, J. C., and Klein, M. P. (1980). The use of fluorescence correlation spectroscopy to probe chromatin in the cell nucleus. *Biochim. Biophys. Acta* **610,** 28–46.

Vamosi, G., Bodnar, A., Vereb, G., Jenei, A., Goldman, C. K., Langowski, J., Toth, K., Matyus, L., Szollosi, J., Waldmann, T. A., and Damjanovich, S. (2004). IL-2 and IL-15 receptor {alpha}-subunits are coexpressed in a supramolecular receptor cluster in lipid rafts of T cells. *Proc. Natl. Acad. Sci. USA* **101,** 11082–11087.

Wachsmuth, M., Waldeck, W., and Langowski, J. (2000). Anomalous diffusion of fluorescent probes inside living cell nuclei investigated by spatially-resolved fluorescence correlation spectroscopy. *J. Mol. Biol.* **298,** 677–689.

Wachsmuth, M., Weidemann, T., Muller, G., Hoffmann-Rohrer, U. W., Knoch, T. A., Waldeck, W., and Langowski, J. (2003). Analyzing intracellular binding and diffusion with continuous fluorescence photobleaching. *Biophys. J.* **84,** 3353–3363.

Webb, W. W. (1976). Applications of fluorescence correlation spectroscopy. *Q. Rev. Biophys.* **9,** 49–68.

Weidemann, T., Wachsmuth, M., Tewes, M., Rippe, K., and Langowski, J. (2002). Analysis of ligand binding by two-colour fluorescence cross-correlation spectroscopy. *Single Mol.* **3,** 49–61.

Weiss, M., Hashimoto, H., and Nilsson, T. (2003). Anomalous protein diffusion in living cells as seen by fluorescence correlation spectroscopy. *Biophys. J.* **84,** 4043–4052.

Widengren, J., Mets, Ü., and Rigler, R. (1995). Fluorescence correlation spectroscopy of triplet states in solution: A theoretical and experimental study. *J. Phys. Chem.* **99,** 13368–13379.

CHAPTER 21

Recent Advances on *In Vivo* Imaging with Fluorescent Proteins

Robert M. Hoffman

AntiCancer, Inc.,
San Diego, California 92111

Department of Surgery
University of California
San Diego, California 92103

Abstract

In vivo imaging with green fluorescent protein (GFP) and other fluorescent proteins is revolutionizing cancer biology and other fields of *in vivo* biology (Hoffman, 2005; Hoffman and Yang, 2006a,b,c). Our laboratory pioneered the use of GFP for *in vivo* imaging in 1997 (Chishima *et al.*, 1997). This chapter highlights recent developments from our laboratory on both macro and micro *in vivo* imaging by using fluorescent proteins.

METHODS IN CELL BIOLOGY, VOL. 85

0091-679X/08 $35.00
DOI: 10.1016/S0091-679X(08)85021-2

I. Macroimaging with Fluorescent Proteins

We have recently reported that a blue LED flashlight (LDP LLC, Woodcliff Lake, NJ; www.maxmax.com/OpticalProducts.htm), with an excitation filter (mid-point wavelength peak of 470 nm) and an emission D470/40 filter (Chroma Technology, Brattleboro, VT) for viewing, could be used for whole-body imaging of mice with green fluorescent protein (GFP) and red fluorescent protein (RFP)-expressing tumors growing in or on internal organs (Hoffman and Yang, 2006a). Whole-body imaging of two tumors, one expressing GFP and the other expressing RFP implanted in the brain of a single mouse, was carried out. The GFP and RFP tumors were simultaneously excited with the blue LED flashlight and readily visualized. A GFP-expressing tumor implanted in the bone was noninvasively imaged with the blue LED flashlight. A GFP-expressing tumor implanted on the colon was also noninvasively imaged with the blue LED flashlight. An RFP tumor implanted on the liver and GFP tumor implanted on the pancreas of a single nude mouse were simultaneously imaged with the blue LED flashlight. After excitation with the blue LED flashlight of an RFP-expressing tumor implanted on the pancreas that has metastasized, an image was readily captured with a Hamamatsu C5810 three-chip cooled color charge-coupled device (CCD) camera (Hamamatsu Photonics, Hamamatsu City, Japan). However, a much simpler digital camera could be used with acceptable results. The images were readily seen by the naked eye with no anesthesia, substrate, or restraint of the animal needed (Yang *et al.*, 2005).

The size and intensity of images of the tumor implanted on the colon was visualized by both whole-body and open imaging using the LED flashlight. The size of the imaged tumor was comparable for both the whole-body and open images. Even more striking is that the intensity of the whole-body image is 70% of the open image. Although some information is lost with whole-body imaging due to light scattering, a remarkable amount of information is obtained, even with such simple instrumentation.

It has been previously shown that whole-body imaging correlates with actual tumor volume (Katz *et al.*, 2003). Low-fluorescence tumors, may require more sophisticated equipment for whole-body images. The data described above resulted in the following conclusions: (i) very strong signals emit from GFP- and RFP-expressing tumors inside the animal; (ii) the images are readily quantifiable; (iii) there is negligible interference from autofluorescence; and (iv) very simple and low-cost instruments can be used for GFP and RFP whole-body macroimaging. These data show the great potential of fluorescent protein-based whole-body imaging for high-throughput *in vivo* screening of drug efficacy and other applications (Yang *et al.*, 2005). The data also correct serious misconceptions in the literature stating "limits" of *in vivo* fluorescent protein imaging (Choy *et al.*, 2003; Gross and Piwnica-Worms, 2005; Ntziachristos *et al.*, 2005; Weissleder and Ntziachristos, 2003).

We simultaneously compared single mice with a highly fluorescent, RFP-expressing orthotopic pancreatic cancer xenografts with both fluorescent protein imaging and high resolution magnetic resonance imaging (MRI). Images were acquired at multiple time points after tumor implantation in the pancreas. Indwelling pancreatic primary tumors and metastatic foci were detected by both fluorescence and MRI. Moreover, a strong correlation existed between images taken with these two technologies. The use of fluorescent proteins in *in vivo* imaging permitted rapid, high-throughput imaging without the need for either anesthesia or contrast agents. Both fluorescence and MRI enabled accurate imaging of tumor growth and metastasis, though MRI enabled tissue structure to be visualized as well. Imaging using fluorescent proteins has high resolution and is exceedingly rapid with instant image capture. The data suggest a complimentary role for these two imaging modalities (Bouvet *et al.*, 2005).

II. Single-Cell *In Vivo* Imaging with Fluorescent Proteins

Cancer cell deformation and migration in narrow vessels is a critical step in metastasis. In order to visualize the cytoplasmic and nuclear dynamics of cells migrating in capillaries, RFP was expressed in the cytoplasm, and GFP, linked to histone H2B, was expressed in the nucleus of cancer cells. Immediately after the cells were injected in the heart of nude mice, a skin flap on the abdomen was made. With a color CCD camera, we could observe highly elongated cancer cells and nuclei in capillaries in the skin flap in living mice. The migration velocities of the cancer cells in the capillaries were measured by capturing images of the dual-color fluorescent cells over time. The cells and nuclei in the capillaries were elongated to fit the width of these vessels. The average length of the major axis of the cancer cells in the capillaries increased to approximately four times their normal length. The nuclei increased their length 1.6 times in the capillaries. Cancer cells in capillaries over 8 μm in diameter could migrate up to 48.3 μm/h. The data suggest that the minimum diameter of capillaries where cancer cells are able to migrate is ~8 μm (Yamauchi *et al.*, 2005).

Recently, the Olympus OV100 small animal imaging system with a sensitive CCD camera and five objective lenses, parcentered and parfocal, enabling imaging from macrocellular to subcellular, has become available. With this new instrument, the nuclear and cytoplasmic behavior of cancer cells was imaged in real time in blood vessels, as they moved by various means or adhered to the vessel surface in the abdominal skin flap. During extravasation, real-time dual-color imaging showed that cytoplasmic processes of the cancer cells exited the vessels first, with nuclei following along the cytoplasmic projections. Both cytoplasm and nuclei underwent deformation during extravasation. Different cancer cell lines seemed to strongly vary in their ability to extravasate. With the dual-color cancer cells and the highly sensitive whole-mouse imaging system described here, the subcellular dynamics of cancer metastasis can now be observed in live mice in real time. This imaging technology will enable further understanding of the critical steps of metastasis and provide visible targets for antimetastasis drug development (Yamauchi *et al.*, 2006).

III. Imaging Dual-Color Angiogenesis and Tumors with Fluorescent Properties

We have recently shown that the neural-stem cell marker nestin is expressed in hair follicle stem cells and the blood vessel network interconnecting hair follicles in the skin of transgenic mice with nestin regulatory element-driven green fluorescent protein (ND-GFP) (Amoh *et al.*, 2004). The hair follicles were shown to give rise to the nestin-expressing blood vessels in the skin. In a subsequent study, we visualized tumor angiogenesis by dual-color fluorescence imaging in ND-GFP transgenic mice after transplantation of the murine melanoma cell line B16F10 expressing red fluorescent protein. ND-GFP was highly expressed in proliferating endothelial cells and nascent blood vessels in the growing tumor. Results of immunohistochemical staining showed that the blood vessel-specific antigen CD31 was expressed in ND-GFP-expressing nascent blood vessels. ND-GFP expression was diminished in the vessels with increased blood flow. Progressive angiogenesis during tumor growth was readily visualized during tumor growth by GFP expression. Doxorubicin inhibited the nascent tumor angiogenesis as well as tumor growth in the ND-GFP mice transplanted with B16F10-RFP (Amoh *et al.*, 2005a).

The nestin ND-GFP gene was crossed into nude mice on the C57/B6 background to obtain ND-GFP nude mice. ND-GFP was expressed in the brain, spinal cord, pancreas, stomach, esophagus, heart, lung, blood vessels of glomeruli, blood vessels of skeletal muscle, testes, hair follicles, and blood vessel network in the skin of ND-GFP nude mice. Human lung cancer, pancreatic cancer, and colon cancer cell lines as well as a murine melanoma cell line and a breast cancer tumor cell line expressing red fluorescent protein were implanted orthotopically, and a red fluorescent protein-expressing human fibrosarcoma was implanted s.c. in the ND-GFP nude mice. These tumors grew extensively in the ND-GFP mice. ND-GFP was highly expressed in proliferating endothelial cells and nascent blood vessels in the growing tumors, visualized by dual-color fluorescence imaging. Results of immunohistochemical staining showed that CD31 was expressed in the ND-GFP-expressing nascent blood vessels. The ND-GFP transgenic nude mouse model enables the visualization of nascent angiogenesis in human as well as mouse tumor progression (Amoh *et al.*, 2005b).

Dual-color fluorescence imaging tumor visualized angiogenesis in the ND-GFP transgenic nude mice after orthotopic transplantation of the MIA PaCa-2 human pancreatic cancer line expressing red fluorescent protein. ND-GFP was highly expressed in proliferating endothelial cells and nascent blood vessels in the growing tumor. The density of nascent blood vessels in the tumor was readily quantitated.

Gemcitabine significantly decreased the mean nascent blood vessel density in the tumor as well as decreased tumor volume. These results demonstrated for the first time that gemcitabine is an inhibitor of angiogenesis as well as tumor growth in pancreatic cancer (Amoh *et al.*, 2006).

In another study, nascent angiogenesis of pancreatic cancer liver metastasis in the ND-GFP transgenic nude mice was imaged. Human pancreatic cancer cells were visualized by RFP. ND-GFP was highly expressed in proliferating endothelial cells and nascent blood vessels in the growing liver metastasis. Immunohistochemical staining showed that CD31 colocalized in ND-GFP-expressing nascent blood vessels. The density of nascent blood vessels in the tumor was readily quantitated. Gemcitabine significantly decreased the mean nascent blood vessel density in the pancreatic liver metastases (Amoh *et al.*, 2006).

Angiogenesis of sarcoma formed by the HT-1080 human fibrosarcoma cell line expressing RFP was also imaged in the ND-GFP mice. Tumor cells were injected into either the muscle or the bone. Nestin was highly expressed in proliferating endothelial cells and nascent blood vessels in the growing tumors, including the surrounding tissues. Immunohistochemical staining showed that CD31 colocalized with ND-GFP in nascent blood vessels. The density of nascent blood vessels in the tumor was readily quantitated. The mice were given daily i.p. injections of 5 mg/kg doxorubicin after implantation of tumor cells. Doxorubicin significantly decreased the mean nascent blood vessel density in the tumors as well as decreased tumor volume. Thus, the dual-color model of the ND-GFP nude mouse and RFP sarcoma cells is also useful for the visualization and quantitation of bone and soft tissue tumor angiogenesis and evaluation of angiogenetic inhibitors for such tumors (Hayashi *et al.*, 2007).

IV. Imaging Tumor–Host Interaction with Fluorescent Proteins

We used dual-color *in vivo* cellular imaging to visualize trafficking, nuclear-cytoplasmic dynamics, and the viability of cancer cells after their injection into the portal vein (PV) of mice. For these studies, we used the dual-color fluorescent cancer cells that express GFP linked to histone H2B in the nucleus and retroviral RFP in the cytoplasm. Human HCT-116-GFP-RFP colon cancer and mouse mammary tumor (MMT) cells were injected in the PV of nude mice. The cells were observed intravitally in the liver at the single-cell level using the Olympus OV100 small animal imaging system. Most HCT-116-GFP-RFP cells remained in sinusoids near peripheral PVs. Only a small fraction of the cancer cells invaded the lobular area. Extensive clasmocytosis (destruction of the cytoplasm) of the HCT-116-GFP-RFP cells occurred within 6 h. The number of apoptotic cells rapidly increased within the PV within 12 h of injection. Apoptosis was readily visualized in the dual-color cells by their altered nuclear morphology. The data suggest rapid death of HCT-116-GFP-RFP cells in the PV. In contrast, dual-color

MMT-GFP-RFP cells injected into the PV mostly survived in the liver of nude mice within 24 h of injection. Many surviving MMT-GFP-RFP cells showed invasive figures with cytoplasmic protrusions. The cells grew aggressively and formed colonies in the liver. However, when the host mice were pretreated with cyclophosphamide, the HCT-116-GFP-RFP cells also survived and formed colonies in the liver after PV injection. These results suggest that a cyclophosphamide-sensitive host cellular system attacked the HCT-116-GFP-RFP cells but could not effectively kill the MMT-GFP-RFP cells (Tsuji *et al.*, 2006).

Intrasplenic injection of tumor cells has long been known as an effective method of developing liver metastases in nude mice, whereas PV injection of tumor cells can result in rapid death of the tumor cells as described above. After splenic injection of tumor cells, splenocytes comigrated with the tumor cells to the liver and facilitated metastatic colony formation. HCT-116-GFP-RFP cells were injected in either the PV or spleen of nude mice and imaged at the subcellular level *in vivo*. Extensive clasmocytosis of the cancer cells occurred within 6 h after PV injection and essentially all the cancer cells died. In contrast, splenic injection of these tumor cells resulted in the aggressive colonization of liver and distant metastasis. GFP spleen cells were found in the liver metastases that resulted from intrasplenic injection of the tumor cells in transgenic nude mice ubiquitously expressing GFP. When GFP spleen cells and the RFP cancer cells were coinjected in the PV, liver metastasis resulted that contained GFP spleen cells. These results suggest a novel tumor–host interaction that enables efficient formation of liver metastasis via intrasplenic injection (Bouvet *et al.*, 2006).

In order to noninvasively image cancer-cell–stromal-cell interaction in the tumor microenvironment and drug response at the cellular level in live animals in real time, we developed a new imageable three-color animal model. The model consists of GFP-expressing mice transplanted with the dual-color tumor cells labeled with GFP in the nucleus and RFP in the cytoplasm. The Olympus IV100 Laser Scanning Microscope with ultra-thin microscope objectives ("stick objectives"), was used for three-color whole-body imaging of the two-color cancer cells interacting with the GFP-expressing stromal cells. In this model, drug response of both cancer and stromal cells in the intact live animal was also imaged in real time. Various *in vivo* phenomena of tumor–host interaction and cellular dynamics were imaged including mitotic and apoptotic tumor cells, stromal cells interacting with the tumor cells, tumor vasculature, and tumor blood flow. This new model and technology system enabled the first cellular and subcellular images of unperturbed tumors in the live, intact animal. New visible real-time targets for novel anticancer agents are provided in this model, including the color-coded interacting cancer and stromal cells, tumor vasculature, and blood flow. This imageable model should lead to a new paradigm of *in vivo* cancer cell biology and to new visible real-time targets for cancer drug discovery (Yang, *et al.*, 2007).

V. New Applications for Fluorescent Proteins *In Vivo*: The Development of Effective Bacterial Therapy of Cancer

We developed a genetically modified bacterial strain, *Salmonella typhimurium* A1, selected for anticancer activity *in vivo*. The strain grows in tumor xenografts. In sharp contrast, normal tissue is cleared of these bacteria even in immunodeficient athymic mice. *S. typhimurium* A1 is auxotrophic (Leu/Arg-dependent) but apparently receives sufficient support from the neoplastic tissue to grow locally. Whether additional genetic lesions are present is not known. The A1 mutant was labeled with GFP. *In vitro*, the GFP-expressing bacteria grew in the cytoplasm of PC-3 human prostate cancer cells and caused nuclear destruction. These effects were visualized in cells labeled with GFP in the nucleus and RFP in the cytoplasm. *In vivo*, the bacteria caused tumor inhibition and regression of xenografts visualized by whole-body imaging. The bacteria, introduced i.v. or intratumorally, invaded and replicated intracellularly in PC-3 prostate cancer cells labeled with RFP and grafted into nude mice. By day 15, *S. typhimurium* A1 was undetectable in the liver, lung, spleen, and kidney, but it continued to proliferate in the PC-3 tumor, which stopped growing. When the bacteria were injected intratumorally, the tumor completely regressed by day 20. There were no obvious adverse effects on the host when the bacteria were injected by either route. The *S. typhimurium* A1 strain grew throughout the tumor, including viable malignant tissue. This result is in marked contrast to bacteria previously tried for cancer therapy that were confined to necrotic areas of the tumor, which may account, in part, for the strain's unique antitumor efficacy (Zhao *et al.*, 2005).

In order to increase tumor-targeting capability of A1, the strain was reisolated after infection of a human colon tumor growing in nude mice. The tumor-isolated strain, termed A1-R, had increased targeting for tumor cells *in vivo* as well as *in vitro* compared with A1. Treatment with A1-R resulted in highly effective tumor targeting, including viable tumor tissue and significant tumor shrinkage in mice with s.c. or orthotopic human breast cancer xerographs. Survival of the treated animals was significantly prolonged. Forty percent of treated mice were cured completely and survived as long as non-tumor-bearing mice did. These results suggest that amino-acid-auxotrophic virulent bacteria, which selectively infect and attack viable tumor tissue, are a promising approach to cancer therapy (Zhao *et al.*, 2006).

We then determined the efficacy of A1-R on metastatic PC-3 human prostate tumors orthotopically implanted in nude mice. Twenty mice were implanted orthotopically with PC-3 tumors expressing GFP. Ten of the 20 mice served as untreated controls and were followed until death. Ten mice were given the *S. typhimurium* A1-R weekly intravenously, beginning at 2 weeks when the tumor GFP was first externally visible. Survival time was compared to the untreated mice. Of the 10 mice with the PC-3 tumors injected weekly with *S. typhimurium* A1, 7 were alive and well at the time the last untreated mouse died. Four A1-R-treated mice remain alive and well for 6 months following implantation. Ten additional non-tumor-bearing mice were used to determine toxicity of *S. typhimurium* A1-R, which were injected weekly. No toxic effects were observed. Since human prostate cancer eventually becomes non-androgen-dependent similar to PC-3 tumors, genetically altered bacteria represent a novel potential treatment for this disease. The approach described here, where bacterial monotherapy could cure metastatic prostate tumors, is a significant improvement over other bacterial tumor-therapy strategies that require combination with toxic chemotherapy (Zhao, *et al.,* 2007).

VI. Conclusions

This chapter demonstrates the power of fluorescent protein for *in vivo* imaging. Macro imaging can be effected with the simplest of equipment—an LED flashlight, an excitation filter, and an emission filter. The excitation filter should be of narrow bandwidth to avoid autofluorescence.

Fluorescent protein imaging is much more powerful than the photon-counting technique using luciferase. No image can be formed with luciferase, since there are insufficient numbers of photons emitted at any one time. A summary comparison between fluorescent protein and luciferase imaging follows.

The GFP approach has several important advantages over other optical approaches to imaging tumor growth *in vivo*. In comparison with the luciferase reporter, GFP has a much stronger signal, and therefore can be used to image

Table I
Comparison of Optical Imaging Methods for Tumor Growth and Metastasis

Parameters	GFP/RFP (references)	Luciferase (references)
Strength of signal	6.6×10^9/pixels/sec/cm^2/steradian (Ray *et al.*, 2004)	7.4×10^6/pixels/sec/cm^2/steradian (Ray *et al.*, 2004)
Minimum number of cells imageable *in vitro*	1 (Yang *et al.*, 2003)	300 (Yang *et al.*, 2005)
Minimum number of cells imageable *in vivo*	1 (Chaudhuri *et al.*, 2002; Yamauchi *et al.*, 2006; Yang *et al.*, 2002, 2003)	3000 (Hoffman, 2002)
Need for substrate	No (Yang *et al.*, 2002, 2003)	Yes (Hoffman, 2002)
Need for anesthesia	No (Yang *et al.*, 2002)	Yes (Hoffman, 2002)
Method of visualization	Direct imaging (Yang *et al.*, 2000)	Photon-counting (pseudo-image) (Hoffman, 2002)
Multicolor imaging	Yes (Yang *et al.*, 2002)	No
Stability of signal	Yes (Hoffman, 2005)	No (Burgos *et al.*, 2003)
Need for excitation light	Yes (Hoffman, 2005)	No (Hoffman, 2002)

Source: Hoffman and Yang (2006b).

unrestrained animals—irradiation with nondamaging blue light is the only step needed. Images can be captured with fairly simple apparatus, and there is no need for total darkness. The fluorescence intensity of GFP is strong (Cormack *et al.*, 1996; Crameri *et al.*, 1996; Delagrave *et al.*, 1995; Heim *et al.*, 1995), and the protein sequence of GFP has also been "humanized," which enables it to be highly expressed in mammalian cells (Zolotukhin *et al.*, 1996). Importantly, unlike luciferase, fluorescent proteins come in a multitude of colors (Shaner *et al.*, 2004), allowing for multiple events to be imaged. In addition, GFP fluorescence is relatively unaffected by the external environment, as the chromaphore is protected by the three-dimensional structure of the protein (Cody *et al.*, 1993). A triple fusion reporter vector harboring a *Renilla* luciferase reporter gene, a reporter gene encoding a monomeric RFP, and a mutant herpes simplex virus type thymidine kinase were tested *in vivo*. A highly sensitive cooled CCD camera that is compatible with both luciferase and fluorescence imaging compared these two signals from the fused reporter gene using a lentivirus vector in 293T cells implanted in nude mice. The signal from RFP was found to be ~1000 times stronger than the signal from luciferase (Ray *et al.*, 2004). The weak signal from luciferase necessitates photon counting, with the construction of a pseudo-image *in vivo* rather than true imaging, therefore greatly reducing resolution and precluding the *in vivo* cellular imaging that is an important feature of GFP imaging. In addition, the rapid clearance of the injected luciferase results in an unstable signal that makes comparison of data difficult (Burgos *et al.*, 2003). The stronger signals from fluorescent proteins allow much more cost-efficient instrumentation. To overcome limits on fluorescent protein imaging imposed by the skin, reversible skin-flap window models have been developed that allow single-cell imaging on most organs of the mouse (Yang *et al.*, 2002). The main advantage of luciferase-based imaging is that no excitation light is required, unlike fluorescence imaging. This feature of luciferase-based imaging could allow deeper imaging than with fluorescence, because excitation light is scattered and attenuated as it goes through skin, body tissues and fluids. However, new bright red-shifted proteins have great potential for deep-tissue imaging (Shcherbo *et al.*, 2007, Merzlyak *et al.*, 2007). See Table I for a comparison of GFP and luciferase imaging (Hoffman and Yang, 2006b).

References

Amoh, Y., Li, L., Tsuji, K., Moossa, A. R., Katsuoka, K., Hoffman, R. M., and Bouvet, M. (2006). Dual-color imaging of nascent blood vessels vascularizing pancreatic cancer in an orthotopic model demonstrates antiangiogenesis efficacy of gemcitabine. *J. Surg. Res.* **132,** 164–169.

Amoh, Y., Li, L., Yang, M., Jiang, P., Moossa, A. R., Katsuoka, K., and Hoffman, R. M. (2005a). Hair follicle-derived blood vessels vascularize tumors in skin and are inhibited by doxorubicin. *Cancer Res.* **65,** 2337–2343.

Amoh, Y., Li, L., Yang, M., Moossa, A. R., Katsuoka, K., Penman, S., and Hoffman, R. M. (2004). Nascent blood vessels in the skin arise from nestin-expressing hair follicle cells. *Proc. Natl. Acad. Sci. USA* **101,** 13291–13295.

Amoh, Y., Li, L., Tsuji, K., Moossa, A. R., Katsuoka, K., Hoffman, R. M., and Bouvet, M. (2006). Dual-color imaging of nascent blood vessels vascularizing pancreatic cancer in an orthotopic model demonstrates antiangiogenesis efficacy of gemcitabine. *J. Surg. Res.* **132,** 164–169.

Amoh, Y., Yang, M., Li, L., Reynoso, J., Bouvet, M., Moossa, A. R., Katsuoka, K., and Hoffman, R. M. (2005b). Nestin-linked green fluorescent protein transgenic nude mouse for imaging human tumor angiogenesis. *Cancer Res.* **65,** 5352–5357.

Bouvet, M., Spernyak, J., Katz, M. H., Mazurchuk, R. V., Takimoto, S., Bernacki, R., Rustum, Y. M., Moossa, A. R., and Hoffman, R. M. (2005). High correlation of whole-body red fluorescent protein imaging and magnetic resonance imaging on an orthotopic model of pancreatic cancer. *Cancer Res.* **65,** 9829–9833.

Bouvet, M., Tsuji, K., Yang, M., Jiang, P., Moossa, A. R., and Hoffman, R. M. (2006). *In vivo* color-coded imaging of the interaction of colon cancer cells and splenocytes in the formation of liver metastases. *Cancer Res.* **66,** 11293–11297.

Burgos, J. S., Rosol, M., Moats, R. A., Khankaldyyan, V., Kohn, D. B., Nelson, M. D., Jr., and Laug, W. E. (2003). Time course of bioluminescent signal in orthotopic and heterotopic brain tumors in nude mice. *Biotechniques* **34,** 1184–1188.

Chaudhuri, T. R., Cao, Z., Rodriguez-Burford, C., LoBuglio, A. F., and Zinn, K. R. (2002). A noninvasive approach for monitoring breast tumor cells during therapeutic intervention. *Cancer Biother. Radiopharm.* **17,** 205–212.

Chishima, T., Miyagi, Y., Wang, X., Yamaoka, H., Shimada, H., Moossa, A. R., and Hoffman, R. M. (1997). Cancer invasion and micrometastasis visualized in live tissue by green fluorescent protein expression. *Cancer Res.* **57,** 2042–2047.

Choy, G., O'Connor, S., Diehn, F. E., Costouros, N., Alexander, H. R., Choyke, P., and Libutti, S. K. (2003). Comparison of noninvasive fluorescent and bioluminescent small animal optical imaging. *Biotechniques* **35,** 1022–1026, 1028–1030.

Cody, C. W., Prasher, D. C., Westler, V. M., Prendergast, F. G., and Ward, W. W. (1993). Chemical structure of the hexapeptide chromophore of the Aequorea green fluorescent protein. *Biochemistry* **32,** 1212–1218.

Cormack, B., Valdivia, R., and Falkow, S. (1996). FACS-optimized mutants of the green fluorescent protein (GFP). *Gene* **173,** 33–38.

Crameri, A., Whitehorn, E. A., Tate, E., and Stemmer, W. P. C. (1996). Improved green fluorescent protein by molecular evolution using DNA shuffling. *Nat. Biotechnol.* **14,** 315–319.

Delagrave, S., Hawtin, R. E., Silva, C. M., Yang, M. M., and Youvan, D. C. (1995). Red-shifted excitation mutants of the green fluorescent protein. *Biotechnology* **13,** 151–154.

Gross, S., and Piwnica-Worms, D. (2005). Spying on cancer: Molecular imaging *in vivo* with genetically encoded reporters. *Cancer Cell* **7,** 5–15.

Hayashi, K., Yamauchi, K., Yamamoto, N., Tsuchiya, H., Tomita, K., Amoh, Y., Hoffman, R. M., and Bouvet, M. (2007). Dual-color imaging of angiogenesis and its inhibition in bone and soft tissue sarcoma. *J. Surg. Res.* **140**(2), 165–170.

Heim, R., Cubitt, A. B., and Tsien, R. Y. (1995). Improved green fluorescence. *Nature* **373,** 663–664.

Hoffman, R. M. (2002). Green fluorescent protein imaging of tumour growth, metastasis, and angiogenesis in mouse models. *Lancet Oncol.* **3,** 546–556.

Hoffman, R. M. (2005). The multiple uses of fluorescent proteins to visualize cancer *in vivo*. *Nat. Rev. Cancer* **5,** 796–806.

Hoffman, R. M., and Yang, M. (2006a). Subcellular imaging in the live mouse. *Nat. Protoc.* **1,** 775–782.

Hoffman, R. M., and Yang, M. (2006b). Whole-body imaging with fluorescent proteins. *Nat. Protoc.* **1,** 1429–1438.

Hoffman, R. M., and Yang, M. (2006c). Color-coded fluorescence imaging of tumor–host interactions. *Nat. Protoc.* **1,** 928–935.

Katz, M. H., Takimoto, S., Spivack, D., Moossa, A. R., Hoffman, R. M., and Bouvet, M. (2003). A novel red fluorescent protein orthotopic pancreatic cancer model for the preclinical evaluation of chemotherapeutics. *J. Surg. Res.* **113,** 151–160.

Merzlyak, E. M., Goedhart, J., Shcherbo, D., Bulina, M. E., Shcheglov, A. S., Fradkov, A. F., Gaintzeva, A., Lukyanov, K. A., Lukyanov, S., Gadella, T. W., and Chudakov, D. M. (2007). Bright monomeric red Fluorescent protein with an extended fluorescence lifetime. *Nat. Meth.* **4,** 555–557.

Ntziachristos, V., Ripoll, J., Wang, L. V., and Weissleder, R. (2005). Looking and listening to light: The evolution of whole-body photonic imaging. *Nat. Biotechnol.* **23,** 313–320.

Ray, P., De, A., Min, J.-J., Tsien, R. Y., and Gambhir, S. S. (2004). Imaging tri-fusion multimodality reporter gene expression in living subjects. *Cancer Res.* **64,** 1323–1330.

Shaner, N. C., Campbell, R. E., Steinbach, P. A., Giepmans, B. N., Palmer, A. E., and Tsien, R. Y. (2004). Improved monomeric red, orange and yellow fluorescent proteins derived from *Discosoma sp.* red fluorescent protein. *Nat. Biotechol.* **22,** 1567–1572.

Shcherbo, D., Merzlyak, E. M., Chepurnykh, T. V., Fradkov, A. F., Ermakova, G. V., Solovieva, E. A., Lukyanov, K. A., Bogdanova, E. A., Zaraisky, A. G., Lukyanov, S., and Chudakov, D. M. (2007). Bright far-red fluorescent protein for whole-body imaging. *Nat. Meth.* **4,** 741–746.

Tsuji, K., Yamauchi, K., Yang, M., Jiang, P., Bouvet, M., Endo, H., Kanai, Y., Yamashita, K., Moossa, A. R., and Hoffman, R. M. (2006). Dual-color imaging of nuclear-cytoplasmic dynamics, viability, and proliferation of cancer cells in the portal vein area. *Cancer Res.* **66,** 303–306.

Weissleder, R., and Ntziachristos, V. (2003). Shedding light onto live molecular targets. *Nat. Med.* **9,** 123–128.

Yamauchi, K., Yang, M., Jiang, P., Xu, M., Yamamoto, N., Tsuchiya, H., Tomita, K., Moossa, A. R., Bouvet, M., and Hoffman, R. M. (2006). Development of real-time subcellular dynamic multicolor imaging of cancer cell trafficking in live mice with a variable-magnification whole-mouse imaging system. *Cancer Res.* **66,** 4208–4214.

Yamauchi, K., Yang, M., Jiang, P., Yamamoto, N., Xu, M., Amoh, Y., Tsuji, K., Bouvet, M., Tsuchiya, H., Tomita, K., Moossa, A. R., and Hoffman, R. M. (2005). Real-time *in vivo* dual-color imaging of intracapillary cancer cell and nucleus deformation and migration. *Cancer Res.* **65,** 4246–4252.

Yang, M., Baranov, E., Jiang, P., Sun, F.-X., Li, X.-M., Li, L., Hasegawa, S., Bouvet, M., Al-Tuwaijri, M., Chishima, T., Shimada, H., Moossa, A. R., *et al.* (2000). Whole-body optical imaging of green fluorescent protein-expressing tumors and metastases. *Proc. Natl. Acad. Sci. USA* **97,** 1206–1211.

Yang, M., Baranov, E., Wang, J.-W., Jiang, P., Wang, X., Sun, F.-X., Bouvet, M., Moossa, A. R., Penman, S., and Hoffman, R. M. (2002). Direct external imaging of nascent cancer, tumor progression, angiogenesis, and metastasis on internal organs in the fluorescent orthotopic model. *Proc. Natl. Acad. Sci. USA* **99,** 3824–3829.

Yang, M., Jiang, P., and Hoffman, R. M. (2007). Whole-body subcellular multicolor imaging of tumor-host interaction and drug response in real time. *Cancer Res.* **67,** 5195–5200.

Yang, M., Li, L., Jiang, P., Moossa, A. R., Penman, S., and Hoffman, R. M. (2003). Dual-color fluorescence imaging distinguishes tumor cells from induced host angiogenic vessels and stromal cells. *Proc. Natl. Acad. Sci. USA* **100,** 14259–14262.

Yang, M., Luiken, G., Baranov, E., and Hoffman, R. M. (2005). Facile whole-body imaging of internal fluorescent tumors in mice with an LED flashlight. *Biotechniques* **39,** 170–172.

Zhao, M., Geller, J., Ma, H., Yang, M., Penman, S., and Hoffman, R. M. (2007). Monotherapy with a tumor-targeting mutant of *Salmonella typhimurium* cures orthotopic metastatic mouse models of human prostate cancer. *Proc. Natl. Acad. Sci. USA* **104,** 10170–10174.

Zhao, M., Yang, M., Li, X.-M., Jiang, P., Li, S., Xu, M., and Hoffman, R. M. (2005). Tumor-targeting bacterial therapy with amino acid auxotrophs of GFP-expressing *Salmonella typhimurium. Proc. Natl. Acad. Sci. USA* **102,** 755–760.

Zhao, M., Yang, M., Ma, H., Li, X., Tan, X., Li, S., Yang, Z., and Hoffman, R. M. (2006). Targeted therapy with a *Salmonella typhimurium* leucine-arginine auxotroph cures orthotopic human breast tumors in nude mice. *Cancer Res.* **66,** 7647–7652.

Zolotukhin, S., Potter, M., Hauswirth, W. W., Guy, J., and Muzyczka, N. (1996). A 'humanized' green fluorescent protein cDNA adapted for high-level expression in mammalian cells. *J. Virol.* **70,** 4646–4654.

CHAPTER 22

Computational Processing and Analysis of Dynamic Fluorescence Image Data

Jonas F. Dorn, Gaudenz Danuser, and Ge Yang

Laboratory for Computational Cell Biology
Department of Cell Biology, CB167
The Scripps Research Institute
La Jolla, California 92037

0091-679X/08 $35.00
DOI: 10.1016/S0091-679X(08)85022-4

Abstract

With the many modes of live cell fluorescence imaging made possible by the rapid advances of fluorescent protein technology, researchers begin to face a new challenge: How to transform the vast amounts of unstructured image data into quantitative information for the discovery of new cell behaviors and the rigorous testing of mechanistic hypotheses? Although manual and semiautomatic computer-assisted image analysis are still used extensively, the demand for more reproducible and complete image measurements of complex cellular dynamics increases the need for fully automatic computational image processing approaches for both mechanistic studies and screening applications in cell biology. This chapter provides an overview of the issues that arise with the use of computational algorithms in live cell imaging studies, with particular emphasis on the close coordination of sample preparation, image acquisition, and computational image analysis. It also aims to introduce the terminology and central concepts of computer vision to facilitate the communication between cell biologists and computer scientists in collaborative imaging projects.

I. Introduction

Fluorescence live cell microscopy has become the tool of choice for the visualization and analysis of spatial and temporal dynamics of molecular assemblies as well as single molecules in living cells. Driven by the advances in automated microscopy and the green fluorescent protein (GFP) revolution, and by the need to capture the full range of protein dynamics to decipher protein functions in cellular processes, it is now common in cell biological studies to generate large volumes of time-resolved fluorescence image data in two and three dimensions. This is especially the case for high-throughput imaging approaches in functional genomics, proteomics, and drug screening (Abraham *et al.*, 2004; Drubin *et al.*, 2006; Neumann *et al.*, 2006; Yarrow *et al.*, 2004). However, it remains a fundamental challenge to obtain accurate and complete measurements from these image data for statistical classification and quantitative modeling of cellular processes based on mechanistic hypotheses. The goal of computational processing and analysis of dynamic fluorescence images is to acquire and analyze measurement data in a highly automated manner.

The importance of obtaining quantitative information from fluorescence image data for the testing of biological hypotheses is well recognized (Eils and Athale, 2003; Swedlow *et al.*, 2003). However, in practice, investigators often encounter obstacles that limit the broader application of computational image processing and analysis in cell biology.

First, it is not always clear which measurements are necessary to characterize a molecular system, and whether these measurements are sufficient to characterize the cellular process investigated. While there are no universal solutions to these problems, this chapter will review a few essential image-based representations of molecular system dynamics in live cells. Second, even if the requirements for measurements are well defined, it is often difficult to find a software tool to extract these data. It is even more challenging to find software tools that can answer specific questions that are raised by the mechanistic hypotheses underlying the experiments.

One solution is for investigators to develop their own software tools. This is feasible for some applications with the assistance of commercial and open source software packages that support the assembly and integration of custom-designed algorithms, even for users with limited computational expertise. Another solution is for investigators to develop interdisciplinary collaboration with computer scientists. Such collaboration may require close interaction between the computer scientists and experimental biologists to jointly optimize the data acquisition and analysis procedures. Thus, a common language between these fields must be established.

This chapter aims to serve the different approaches to overcome the obstacles by introducing to readers, especially microscopists and cell biologists, the basic terminology, concepts, and methods of computational processing and analysis of dynamic fluorescence images.

II. Rationale

The use of computational image analysis methods in cell biology has been boosted by the recent application of light microscopy to large-scale screen for genes or small molecule compounds affecting specific cell functions (Goshima *et al.*, 2005; Neumann *et al.*, 2006; Perlman *et al.*, 2004; Shay *et al.*, 2004; Yarrow *et al.*, 2005; Ziauddin and Sabatini, 2001). In these applications, computers are required to handle large volume of experimental data. Besides the aspect of *efficiency* in data processing, this section discusses two additional issues, which make computational image analysis an invaluable tool also for small-scale, image-based studies of cell biological mechanisms: *data consistency* and *completeness*.

A. Why Use Computational Techniques for Quantitative Testing of Biological Hypotheses?

1. Efficiency

Efficient extraction of quantitative measurements is currently the main motivation for the application of computational image analysis of live cell data. With the development of microscopes for live cell genome-wide screens (Smith and Eisenstein, 2005), it is possible to acquire vast amounts of data in ever shorter time. For example, even at low spatiotemporal sampling, a live cell siRNA screen of 49 mitotic genes generated over 100 GB of image data (Neumann *et al.*, 2006). Such quantities of movies make data management challenging, as discussed in Chapter 24 by Moore *et al.*, this volume, and they make manual data analysis unrealistic. Instead, these types of experiments require computational image analysis to extract image features for the classification of cell behavior in response to perturbations. Mostly, simple algorithms are applied, which are robust enough to produce meaningful features without need for manual validation of image analysis results (Abraham *et al.*, 2004). Alternatively, robustness is achieved by manually training the computer to recognize a small number of phenotypes (Chen *et al.*, 2006; Conrad *et al.*, 2004). Therefore, image-based screens currently are limited to the extraction of relatively coarse phenotypes with limited mechanistic insight of cell behavior.

2. Consistency

Computational image analyses yield consistent data, that is, different experiments are processed based on the same criteria. This eliminates measurement uncertainty associated with subjective interpretations of image contents by one investigator or among investigators. Furthermore, computational image analysis permits the quantification of measurement uncertainty that originates from noise in the raw imagery (Dorn *et al.*, 2005). High consistency and known uncertainty are particularly useful when the study of a certain cell function requires distinction of weak yet significant phenotypes.

3. Completeness

Computational image analyses yield complete data, that is, every image event that fulfills an objective set of criteria is considered. Human observers are skilled at recognizing the "odd one out." At the same time, they have a tendency—by nature or necessity—of concentrating on the apparently interesting events. This biases the analysis and increases the risk of overlooking rare events associated with weaker phenotypes. Together, consistent and complete image measurements permit the statistical selection of obvious and less obvious, for example, highly transient, events. The recovery of rapid transient events is particularly relevant to reconstituting the relationships between the dominant states of a molecular system. Thus, computational image analysis is a critical tool to systematically decipher the mechanism of molecular dynamics in cell function.

In this chapter, we will provide examples where computational image analysis was applied to detailed mechanistic studies of cell functions based on weak and transient phenotypes. Application of computational image analysis in the context of large-scale screens is discussed in the following chapter (Chapter 23).

B. Role of User Input in Computational Processing and Analysis of Image Data

Maximum efficiency, consistency, and completeness imply minimal user input. Yet, it is impossible to design image analysis algorithms that are universally robust to achieve complete understanding of image contents in all applications. Human–machine interaction provides an important way to augment automated computational image analysis by drawing on the application-specific knowledge of users and the power of human visual cognition system. There are three levels of user input: adjustment of parameters controlling the image processing algorithms, correction of outputs generated by image processing algorithms, and manual execution of entire image analysis steps. The higher the level of user input, the more difficult it becomes to ensure efficiency, consistency, and completeness.

Adjustments of control parameters represent the lowest level of user input and are necessary as a way to supply the algorithm with application-specific prior knowledge. However, the amount of information provided and the constraints it imposes on the outcome of the computational image analysis step has to be carefully traded against the risk of biasing the data. Therefore, image analysis algorithms must be equipped with objective diagnostic tools that indicate data distortions induced by inappropriate parameter settings.

Corrections of outputs generated by image processing algorithms allow the elimination of obviously false measurements. In many cases, such outlier data can be detected automatically and thus objectively. However, it may be necessary to achieve a measurement success rate of 100%. For example, the analysis of the turnover kinetics of a molecular structure via measurement of the lifetime of its fluorescent components requires 100% success rate in computational tracking of the structure. If the image of the structure is lost, the statistics will be systematically distorted to fast kinetics. Such perfection in tracking may be difficult to achieve with computational image analysis. Thus, after tracking, the investigator may be given the opportunity to validate and modify marker trajectories. Similar to the adjustment of control parameters, manual data interventions bear the risk of "tailoring" data to the investigator's specific hypothesis. It is critical that corrections are limited to measurements where diagnostic tools indicate high uncertainty in the automated data acquisition.

Manual execution of an entire image analysis step, for example, feature selection, cell delineation, or tracking of fluorescent tags, is not recommended. However, developing an image analysis package for one specific project can be very time-consuming. In some cases, it may be justified to trade consistency, accuracy, and completeness of data against lower expense of manual measurements.

Again, investigators need to be aware of the potential bias of manual measurements and suitable controls must be designed and performed.

III. Image Features and Representation of Dynamic Events

A. Overview

Most of the concepts, theories, and methods for computational processing and analysis of fluorescence images originated from the fields of *image processing* and *computer vision*. Usually, image processing refers to low-level image-to-image transformations such as filtering, compression, enhancement, and restoration (Gonzalez and Woods, 2002). While these transformations enhance certain image properties, they do not generate knowledge. Image transformations generate new images whose content still has to be interpreted by a human observer. In contrast, the output of a computer vision method is the numerical representation of image content (Trucco and Verri, 1998), for example, the type of objects found in an image or their trajectories. There is overlap between the two fields. In particular, most computer vision techniques require image processing to convert images into forms suitable for subsequent analysis.

The computational analysis of fluorescence images discussed in this chapter mostly falls within the domain of computer vision. Thus, this section starts with introducing computer vision terminology that is essential to understanding the primary literature and to communicating with computer vision specialists. Then, as this volume focuses on live cell imaging, different representations of dynamic image events that can be extracted by computer vision methods are discussed.

B. Image Features

Image features are local parts of an image that contain meaningful information and are detectable (Trucco and Verri, 1998). The following are a few basic two-dimensional (2D) image features relevant for the computational analysis of live cell fluorescence images.

1. Point

Also referred to as *particle*, *dot*, or *marker*. A point feature describes an object whose shape and intensity distribution can be ignored. In addition to its coordinate, a point feature can also possess properties such as intensity and wavelength. A diffraction-limited fluorescent speckle, which represents the image of a cluster of fluorophores, gives an example of a point feature [Fig. 1A (I)]. Other examples include fluorescently tagged single molecules or small vesicles whose size is at or below the resolution limit of the microscope optics so that their shapes are not measurable.

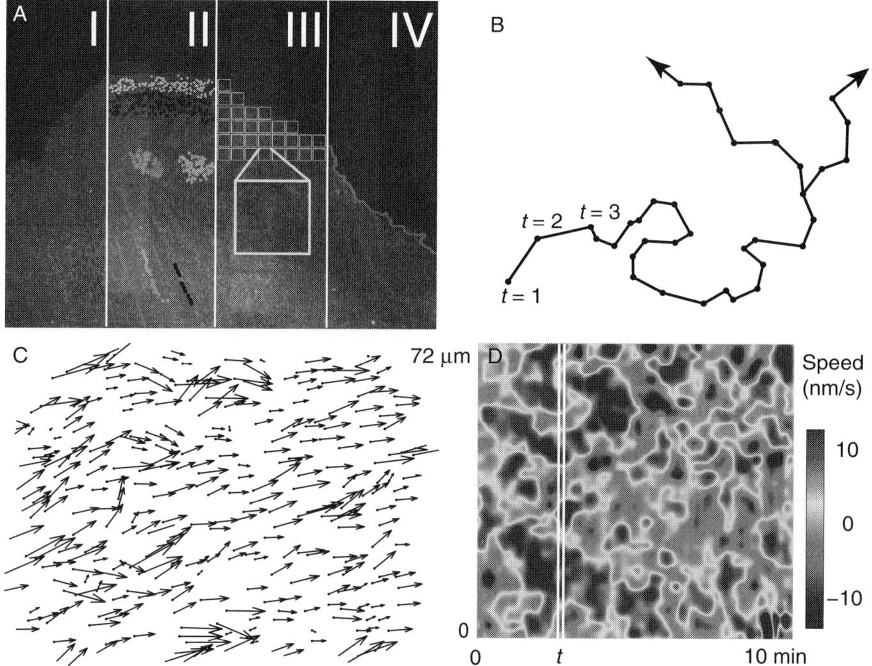

Fig. 1 Image features and representation of dynamic data. (A) Image features. I, Particles: fluorescent speckles detected as significant local intensity maxima. II, Groups of objects: groups of fluorescent speckles with similar behavior. III, Region: fluorescence distribution creating a textured template. IV, Shape: cell edge. (B)–(D) Representation of dynamic data. (B) Trajectory of the centroid of a single moving cell; the split indicating a cell division. (C) Vector field of speckle flux. (D) Morphological profile of the cell edge. The vertical slice shows displacement speed along the cell edge at time *t*; colors encode protrusion (positive speeds: red) or retraction (negative speeds: blue). (See Plate no. 64 in the Color Plate Section.)

2. Point Group

Also referred to as *cloud* or *swarm*. It is composed of multiple point features with similar dynamic behavior. An example of point group features is shown in Fig. 1A (II). Individual speckles are aggregated in units of consistent behavior.

3. Region

A region is a continuous part of an image. A region in which the image intensity can be considered uniform is called a *blob*. A *template* is an artificially defined region of interest, often a rectangular window. An example of templates is shown in Fig. 1A (III). In contrast to the examples of point features or point group features, individual speckles have no meaning within the context of a region. Regional features are described by their *shape* and enclosed intensity distribution, also referred to as a *pattern*. When the pattern within a region is organized according to specific rules, the region is said to have a *texture*. Another related feature concept useful in shape description is the *ridge* or *skeleton* of a region, which represents objects by their medial axes (Haralick, 1983).

4. Edge

An edge represents a group of connected pixels along which the image exhibits an abrupt change in any of its properties, for example, intensity, color, or texture. There are several related image feature concepts. A proximal pair of edges delineating a local change in image properties corresponds to a *line* or a *curve*. The edge circumscribing a region feature or part of a region feature is also referred to as a *boundary* or a *contour*. An example of a boundary is shown in Fig. 1A (IV), delineating a section of the cell edge as perceived by the transition from the speckled foreground to the nonspeckled background. Properties of edges are described by parameters such as length or curvature; or by functions, such as splines, which parameterize the positions of the edge-forming pixel sequence.

Some of these feature concepts differ in 3D. For example, edges must be replaced by surfaces. See Nixon and Aguado (2002) and Trucco and Verri (1998) for more detailed description of feature concepts in 3D.

C. Representation of Dynamic Events

One of the main goals of live cell imaging is to analyze how image features that describe cellular and subcellular components evolve over time. This is achieved by computational detection and tracking of image features. Depending on the feature, various representations of these dynamics exist. Refer to Chapter 23 by Harder *et al.*, this volume for a more detailed discussion of data types capturing feature dynamics in microscopy. Here, we limit the discussion to three representations of dynamic events used throughout the rest of the chapter.

1. Trajectory

A trajectory defines the sequence of image feature properties over time. Its most common application is in describing the positional change of a point, a point group, a region, or an edge feature (Fig. 1B). However, the evolution of nonpositional properties, for example, the intensity and size of a region feature, can be represented in the same fashion. This representation is usually referred to as a *phase space* trajectory.

2. Vector Field

A vector field represents the spatial organization of the instantaneous motion of the image features being tracked in terms of velocity magnitude and direction (Fig. 1C). It is usually used in two cases. The first case is when stable and distinct point features are not available to provide trajectory representations. The second case is when the primary interest is in the motion of the ensemble of image features. A term related to the representation of motion of multiple image features is *optical flow*, which refers to the apparent motion of image intensity patterns (Trucco and Verri, 1998).

3. Morphodynamic Profile

Edge features evolve morphologically, that is, their shape changes over time. One way to visualize such changes is by plotting the local displacement of the edge as a function of the position along the tracked edge segment (Fig. 1D vertical axis) and time (Fig. 1D horizontal axis). The displacement magnitude can be color coded so that spatiotemporal relationships between displacements are revealed as patterns. In Machacek and Danuser (2006), we referred to this representation as a *morphodynamic profile*. When applied to visualization of the dynamics of cell edges, morphodynamic profiles display coordinated displacement patterns that turned out to be indicative of the spatiotemporal regulation of cell edge protrusion (Fig. 1D). Other image properties measured along edge features can be visualized in the same way, for which we use the more general term *activity map*.

IV. Methods

Over the past 50 years, research in image processing and computer vision has generated a broad range of techniques serving a large variety of applications. This section will review methods for image filtering, feature detection, and feature tracking as related to computational analysis of live cell images. Obviously, it is beyond the scope of this chapter to provide a comprehensive summary of the relevant literature. The key message of this section is that meaningful measurements by computational image analysis require sample preparation, microscope settings, and choice of algorithms to be closely coordinated in an integrated experimental design.

A. Definition of Informative Image Measurements

Ideally, an imaging experiment starts with a simulation of synthetic images that recapitulate the expected cellular dynamics according to existing knowledge of the studied mechanism. These simulations serve two main purposes.

1. Sensitivity Analysis

Simulations can reveal prior to expensive experimentation whether changes in the dynamics of the studied molecular system yield measurable changes in the image data, given the spatial and temporal sampling of the images, the noise level, and the available image analysis algorithms. If the answer is negative, simulations will indicate which component of the experimental pipeline has to be enhanced. If no further adjustments can be made, the simulations provide an initial guess of the magnitude of molecular effects that can be resolved by the imaging experiment. With this information, the number of control experiments testing the validity of image data can be substantially reduced. An example of a sensitivity analysis by simulation is given in Dorn *et al.* (2005), where the strategies of fluorescent tagging and image sampling were optimized to faithfully report modulations of the dynamic instability of yeast spindle microtubules (MTs).

2. Interpretation of Image Measurements

In many cases, the relationship between an event at the molecular scale and its effect on the image measurement is not well defined. Simulations can indicate how changes in the image measurement can be interpreted in terms of modulated dynamics of the molecular system (Gardner *et al.*, 2005; Jaqaman *et al.*, 2006; Marshall *et al.*, 1997; Sbalzarini *et al.*, 2005; Sprague *et al.*, 2003). Furthermore, simulations will reveal nonlinear and nonmonotonic behavior, such as two qualitatively distinct molecular events yield the same image measurement. In this case, the measurements must be complemented by additional image parameters. For example, simulations of speckle formation in polymer networks in Ponti *et al.* (2003) revealed that speckle appearance was not an unambiguous measure of local network assembly but that speckles also appeared because of network disassembly. A refined algorithm was then designed, accounting for fluctuations in both foreground and background intensity to properly identify the kinetics of monomer association and dissociation.

B. Acquisition of Optimized Fluorescent Images

Accurate capture of molecular dynamics in live cells requires sufficient resolution in space and time. Image analysis algorithms require a minimum level of signal-to-noise ratio (SNR) and the length of the movies must cover enough time to reflect all possible states of the probed molecular system. Spatiotemporal resolution, SNR, and observation length are tightly interdependent and partially conflicting imaging parameters (Fig. 2). High spatiotemporal resolution implies fast sampling at high magnification, which results in fewer photons reaching the imaging sensor and thus low SNR. SNR could be improved by prolonging the exposures or by increasing the power of the illumination. This, in turn, will increase photobleaching and phototoxicity, limiting the number of possible exposures and hence observation length. To design an imaging experiment, the minimum requirements in all three variables need to be defined and their compatibility with the available specimen and microscope hardware tested. In each of the three imaging parameters, specimen and microscope hardware define a maximum performance point. These points can be derived from the microscope specifications

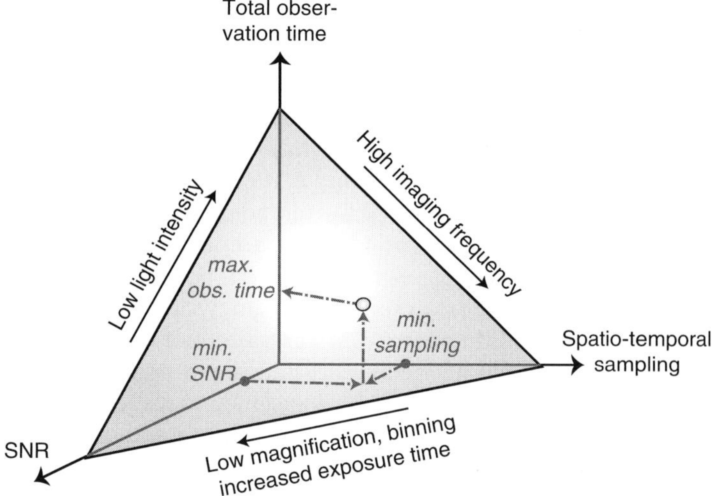

Fig. 2 The iron triangle of imaging. An ideal microscopy movie requires high sampling in space and time, high signal-to-noise ratio (SNR), and long observation time. In practice, these three requirements cannot be satisfied simultaneously. For example, pixel binning, lower magnification, and longer exposure increase SNR at the cost of lower spatiotemporal sampling. A minimum spatiotemporal sampling is required to resolve fluorescent objects or dynamic processes, and a minimum SNR is required for image analysis algorithms to work properly. After these minimum requirements are determined, the maximum possible observation length can be determined.

(fastest acquisition rate of the camera, highest magnification) or be determined experimentally (acquisition time before bleaching or phototoxic damaging of the specimen; SNR obtained under very long exposures, e.g., in fixed specimens). Given the mutual interdependence between the parameters, the joint performance can be conceptualized by the plane through the three maximum performance points (Fig. 2). If the minimum requirements of a specific experiment combine in a point beyond the performance plane, it is impossible to acquire the necessary image data. In this case, the experimental setup has to be redesigned, or the plane has to be shifted, either by investing in better microscopy hardware or by improving the stability and efficiency of the fluorescent probes (see Chapter 1 by Dobie *et al.*, this volume).

In the following, we provide a few general guidelines on how to determine the minimal requirements for a specific experiment.

1. Sampling

To allow any computational image analysis of the spatiotemporal dynamics of a live cell, specimen must be sampled at least three times finer than the highest spatial and temporal frequency of interest (Stelzer, 2000). For the sampling in space, this means that the magnification of the microscope must be selected so that at least three pixels and three z-slices cover the lateral and axial extension of the point-spread function (PSF), respectively (Inoue and Spring, 1997). For the sampling in time, the characteristic time scale of the probed dynamics must be assumed *a priori*. Generally, this is difficult to achieve and educated guesses based on kinetic data and simulations of the molecular processes must be made. However, for any confined motion, for example, the diffusion of a particle inside a cellular compartment, the requirements for temporal sampling can be determined by plotting the measured velocities at different sampling intervals. If the trajectory is oversampled, multiple feature positions will be observed before the confinement boundary influences the trajectory. Accordingly, a plot of velocity versus time interval will be constant. In contrast, when trajectories are undersampled, the observed feature positions are dictated by the number of boundary-induced motion reversals in the trajectory. The expected displacement between any two observations only depends on the shape of the confinement region, and not the actual velocity of the feature. Therefore, the apparent velocity that is measured is inversely proportional to the time interval. Accordingly, a plot of the velocity versus the inverse of the time interval will result in a straight line with nonzero slope. The characteristic time scale of the probed dynamics is indicated by the time interval at which the two lines intersect. In practice, the intersection point can be determined by first acquiring image data at the maximum sampling rate affordable, ignoring the limitations too fast sampling imposes on the observation length. Then, the image sequence can be downsampled artificially, and the sampling interval be defined at which the plot velocity versus inverse of the time interval deviates from a horizontal line. This sets the minimal requirement for the time sampling.

2. Signal-to-Noise Ratio

SNR requirements are determined entirely by the image analysis algorithm. Given a high SNR image of the specimen, the breakdown of algorithms can be identified by applying them to images with increasing levels of noise added. It should be noted that once the required SNR level is known, the imaging parameters have to be set such that these conditions are fulfilled at the end of the time-lapse image sequence where the effect of photobleaching is strongest.

3. Observation Length

The observation length required to capture all possible states of a dynamic molecular system is the most difficult criterion to determine *a priori*. Instead, individual measurements have to be repeated until their distribution does no longer change with additional data. This is the number of data points that represents the entire dynamics of the system. Often, it is impossible to acquire a single movie that contains sufficient time points. In this case, measurements from multiple movies have to be pooled (Dorn *et al.*, 2005).

C. Filtering of Images

The first data processing step in computational image analysis is the filtering of images for the purposes of both noise reduction and selective enhancement of image features. Users need to be aware of how different filtering schemes transform the images.

1. Image Noise Reduction

Assuming that the noise in adjacent pixels is statistically independent, noise-induced intensity fluctuations can be suppressed by spatially averaging the image information (low-pass filter) at the expense of spatial resolution. As shown in Fig. 3A, a Gaussian filter of scale 2 pixels effectively reduces noise, best visible in the background region. When increasing the scale to 10 or 20 pixels, noise is further reduced but the structural details of the stress fibers are lost. How filtering compromises image resolution is illustrated in a simpler example in Fig. 3B. As a rule of thumb, the resolution of a filtered image expressed in pixels corresponds to the scale of the filter. To preserve the optical resolution of the microscope, only filters with a scale equal to the extension of the PSF in the image space can be used.

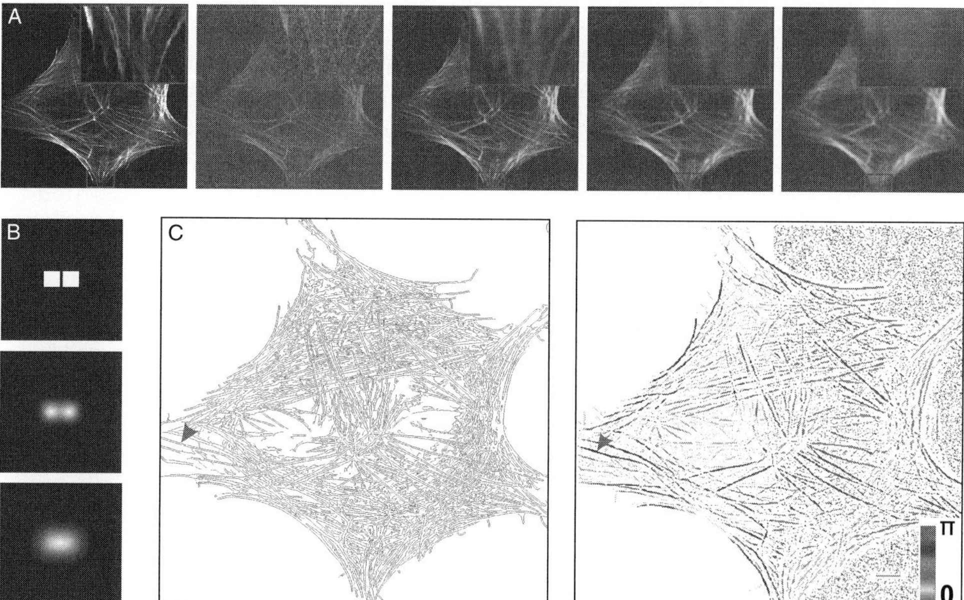

Fig. 3 Filtering of images. (A) From left to right: original image; original image contaminated by noise; noise-contaminated image filtered using a Gaussian low-pass filter at a scale of 2, 10, and 20 pixels, respectively. (B) From top to bottom: original image; original image filtered by a Gaussian low-pass filter at a scale of 10 and 20 pixels, respectively. (C) Left: detected edge pixels. Right: detected line pixels. The right half-plane uses color coding to represent orientation of line pixels. (See Plate no. 65 in the Color Plate Section.)

Thus, to allow noise reduction by low-pass filtering in the pixel domain, the spatial sampling must be substantially finer than the width of the PSF (at least three times, see the earlier description).

2. Selective Feature Enhancement

Image filtering can be used to selectively enhance features of interest while suppressing irrelevant features. This is essential to subsequent feature detection. Importantly, with the choice of a filter, the user makes the choice of the feature type, as illustrated in Fig. 3C for edge and line filtering. In the left panel, an edge filter (Canny, 1986) highlights the boundaries of the stress fibers. In the right panel, a line filter (Koller *et al.*, 1995) highlights the axes of the stress fibers. Unfortunately, there is no "magic filter," which can highlight both feature types consistently. If lines and edges should eventually be extracted for the same image, the image has to be filtered separately with both a line and an edge filter. Subsequently, the two filter responses have to be merged in an additional processing step. The design of optimal image filters and the combination of several filters in multifunctional filter benches is an area of active research in the image processing community. Introductory chapters on these aspects are provided in Gonzalez and Woods (2002) and Trucco and Verri (1998).

Feature enhancement by filtering does not detect features, as falsely suggested by some image processing packages. In the right panel of Fig. 3C, it is still up to the observer to visually link pixels that belong to one stress fiber. No quantitative information can yet be extracted of, for example, the position, length, and curvature of stress fibers. All filtering provides is an unstructured selection of pixels, which have a higher probability to represent a line-like image signal. Moreover, filtering can generate additional information, such as the local orientation of a potentially line-like image signal (color-coded map in the right half of Fig. 3C, right panel). To detect actual image features and to measure their parameters, higher-level processing of these filtering responses is required, referred to as feature detection or feature extraction.

D. Detection of Image Features

The goal of image feature detection is to obtain numerical representations of the location and properties of image features (Nixon and Aguado, 2002). Returning to the task of line detection introduced earlier, this requires primarily the grouping of line pixels belonging to the same stress fiber (Davies, 2004; Sonka, 1999). A large number of feature detection techniques have been developed, many of them involving application-specific rules. The goal of this section is to introduce the basic concepts and ideas common to these techniques.

1. Point Feature Detection

Point features are local intensity minima or maxima whose intensity level is significantly different from the neighborhood. Consequently, point detection techniques must define a meaning of "neighborhood" for the computation of a representative background intensity distribution; and a meaning of "significantly different." The most rigorous way is to cast the comparison of foreground and background intensity as a statistical test. This is particularly important for low SNR fluorescence images. For example, Ponti *et al.* (2003) describe a statistical hypothesis test for the detection of speckles in which local intensity maxima are compared to its three neighboring local minima using a *t*-test that incorporates a precalibrated model of camera noise (Fig. 4A). The selectivity of the point detector is controlled via the confidence level required for the *t*-test to be accepted. As for any statistical test, this choice implies the finding of a trade-off between false positives (yellow circle), whose number increases with lowering the confidence level, and false negatives, whose number increases with increasing the confidence level (see also Section VI). We refer to Nixon and Aguado (2002) and Starck *et al.* (2000) for a variety of other point detection approaches.

2. Edge, Line, and Curve Feature Detection

The detection of curvilinear features involves the grouping of edge or line pixels into larger entities. There are local and global strategies to achieve this goal. In local grouping strategies, the decision whether two pixels belong to the same curvilinear feature is made based on neighborhood relationships, such as the

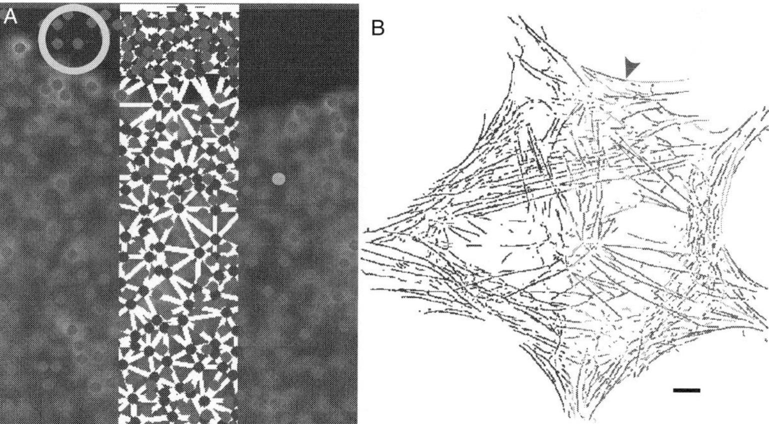

Fig. 4 Detection of image features. (A) Fluorescent speckle detection. The central band shows neighborhood definition by Delaunay triangulation (white). (B) Line detection by applying to the line filter response (left half-plane) shown in Fig. 3 a connected component labeling algorithm (right half-plane). (See Plate no. 66 in the Color Plate Section.)

spatial proximity, line or edge orientation, or magnitude of the filter response. The simplest of these local algorithms is the *connected component labeling*, which links immediately neighboring pixels (Fig. 4D right half-plane). Local grouping algorithms are very sensitive to image noise, which perturbs neighborhood relationships between pixels, leading to line or edge fragments. To avoid fragmentation global grouping approaches can be applied. These algorithms employ some form of clustering to extract collinear pixel arrays by trading two opposing criteria (Duda *et al.*, 2001; Höppner *et al.*, 1999): The number of extracted curvilinear features should be minimized while maximizing the overall proximity of pixels to the line they are assigned to. Many algorithms have been developed to solve this problem. They vary in how the number of extracted line or edge features is determined, how the proximity of a pixel to the feature is defined, and how they deal with the assignment of pixels in regions with intersecting features. A discussion of these important details is beyond the scope of this section. After obtaining a general overview of clustering and data classification methods in Duda *et al.* (2001), Höppner *et al.* (1999), and Jain *et al.* (1999), the interested reader is referred to, for example, Stanford and Raftery (2000).

3. Region Detection

The techniques for image segmentation into regions of similar intensity or texture properties (see Section III) can be classified from two perspectives.

Contour-based versus region-based techniques: Contour-based segmentation techniques start with the detection of edges, followed by a region filling. The critical issue is how to restrict the filling process in areas with incomplete edges that leave open contours. Region-based methods start with initial seed regions and perform segmentation through seed growth (Snyder and Qi, 2004; Sonka, 1999). Adjoining regions can be merged if they exhibit a high similarity in intensity or texture. Region growth is ceased if the image area to be added significantly deviates from the region properties.

Local versus global segmentation techniques: Early developments of contour-based and region-based segmentation usually relied on empirical rules to make local decisions on edge detection or region growing. Such decisions are often incompatible among neighboring contours or regions. This is remedied in more recently developed techniques such as active contours (Kass *et al.*, 1987) or level sets (Osher, 2003), which solve the segmentation problem globally, that is, simultaneously for all regions. Global segmentation techniques often provide robust results but tend to omit local details. This is particularly relevant for the segmentation in fluorescence image data. The generally low SNR of the raw images requires the use of global approaches for segmentation, yet the many salient and faint structures in the cell image require a local reconstruction of the morphology. Thus, hybrid segmentation methods combining global and local approaches will have to be developed to solve the problem of region segmentation in live cell fluorescence microscopy.

4. Detection of Other Features

The detection of the various other features mentioned in Section III, such as corners, ridges, surfaces, and textures (patterns), has also been studied in computer vision (Davies, 2004; Nixon and Aguado, 2002; Trucco and Verri, 1998). These features may also be encountered in computational processing and analysis of fluorescence images.

E. Tracking of Features

The most powerful asset of live cell fluorescence microscopy is the ability to visualize the dynamics of molecular structures. Correspondingly, computational feature tracking techniques must be developed to capture the full information of live cell images (Meijering *et al.*, 2006). Dependent on the feature, three types of tracking tasks are distinguished.

1. Point/Particle Tracking

Here, the input data consists of the coordinates and possibly additional properties of fluorescent point features in a time-lapse sequence. The tracking algorithm should establish point-to-point correspondences between features in consecutive time points. In live cell fluorescence imaging, point tracking is the predominant tracking task. Point features originate from detection of subresolution fluorophore clusters associated with small molecular structures, or from blobs representing vesicles or more extended organelles. Blob shapes are not tracked explicitly but may be used as an attribute for identifying the same blob in consecutive time points. In the biophysical and cell biological literature, point-to-point tracking is often referred to as particle tracking. Whereas much discussion has been devoted to the question of how much precision the displacement of a single particle can be tracked (Cheezum *et al.*, 2001), the critical issue in particle tracking is how to find corresponding particle images in consecutive frames. Most particle tracking techniques apply simple nearest neighbor (NN) assignment. Stepping through the list of particles in one frame, particles are linked to the closest particle in the next frame. This is sufficient under the condition that the ratio between particle displacement and the mean distance between particles is $\ll 0.5$ [see Ponti *et al.* (2005) for a more formal discussion of the breakdown of NN tracking]. The breakdown of NN is illustrated in Fig. 5A. Both P_1 and P_2 compete for the same NN, that is, the image of particle P_2. The outcome of a NN algorithm in this situation depends on the order by which the assignments are made. In general, this is arbitrary. In some cases, simple heuristics may be sufficient to remedy (Ponti *et al.*, 2003); however, in general, a global solution is required to achieve satisfactory tracking results.

Point tracking techniques with global criteria have been developed in the field of radar tracking and computer vision (Bar-Shalom *et al.*, 2001; Blackman and

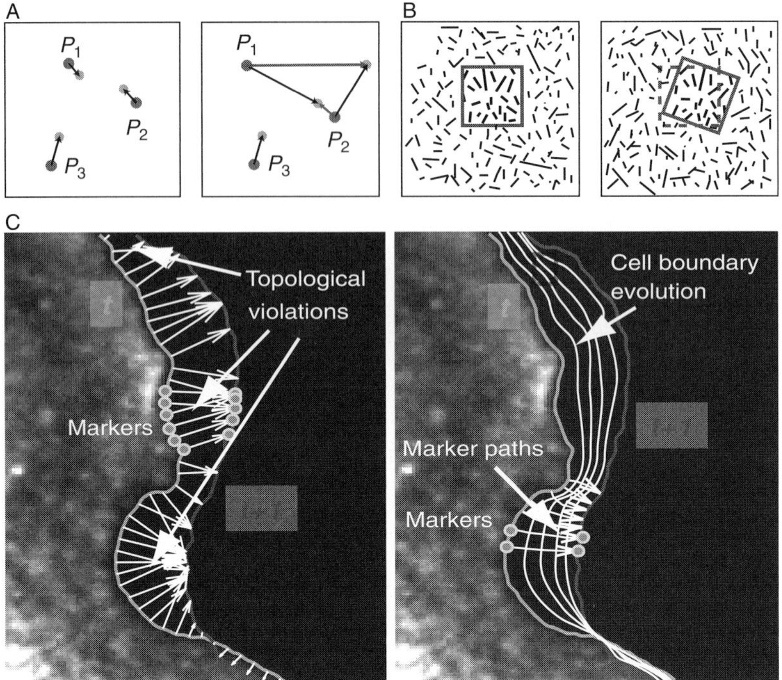

Fig. 5 Tracking of image features. (A) Point tracking by local (left) and global (right) nearest neighbor (NN) assignment. (B) Template tracking by search for the matching image pattern between two time points. (C) Contour tracking by Level Set Method. [Adapted from Machacek and Danuser (2006) with permission.] (See Plate no. 67 in the Color Plate Section.)

Popoli, 1999; Cox, 1993; Cox and Hingorani, 1996; Reid, 1979; Shafique and Shah, 2005; Veenman *et al.*, 2001), of which some have recently been applied to cell biological studies (Bonneau *et al.*, 2005; Genovesio *et al.*, 2006; Sage *et al.*, 2005; Shafique and Shah, 2005; Vallotton *et al.*, 2003). The simplest global extension of NN assignment is the global NN (GNN) assignment. This algorithm makes the assignment of particle pairs in two consecutive frames by searching the combination of pairings that minimizes the total displacement of all particles. Thus, while locally some assignments will sacrifice the NN rule, as is the case for P_1 in Fig. 5A, the solution to the problem will be optimal from a global perspective. Very efficient numerical algorithms have been developed to solve this problem of combinatorial optimization for millions of particles simultaneously (Burkard and Cela, 1999). However, the disadvantage of GNN tracking is that the decision on the assignment is made between frame pairs, without consideration of the particle motion in the past. This provides still suboptimal solutions, particularly, when the average displacement of particles between frames exceeds half the mean distance between particles (Ponti *et al.*, 2005). Two algorithmic frameworks exist, which delay the decision on point assignments until evidence from multiple frames is available:

joint probabilistic data association filter and multiple hypothesis testing (Blackman and Popoli, 1999); but for most tracking applications in cell biology the computational cost of these algorithms is too high. Alternatively, the robustness of GNN assignment can be increased by propagating the particle motion from the past into the future, that is, the assignment is no longer made between particles in consecutive frames t and $t + 1$, but between the particles in t projected into $t + 1$ and the particles in $t + 1$. The type of projection is application specific. Possible approaches to project particles between frames are to estimate the global organization of particle motion iteratively from the available particle assignments (Ponti *et al.*, 2005) or other tracking methods (Ji and Danuser, 2005; Yang *et al.*, 2005), or to formulate explicit motion models for each particle, whose parameters are inferred based on the already tracked particle paths (Genovesio *et al.*, 2006).

2. Template/Pattern Tracking

Here, the input data consist of patterns of specific image regions. The tracking algorithm should establish a transformation model, including the translocation of pattern, between consecutive time points. Image regions are delimited either by object boundaries segmented in a previous step or by an arbitrarily selected window frame, often referred to as a template. Pattern tracking is applied to image data where the detection of stable features is difficult, for example, because of low SNR or fast turnover of the fluorescently labeled structure. Therefore, unlike with point tracking, it is usually irrelevant to determine the trajectory of a template. Instead, template tracking provides a powerful means to follow the instantaneous average motion of a fuzzy and unstable image structure (Ji and Danuser, 2005; Miyamoto *et al.*, 2004). Very often, many templates are selected to cover a larger image region. Each template indicates the displacement of a small subregion, allowing dense mapping of the organization of the flow of extended molecular structures (Ji and Danuser, 2005).

As illustrated in Fig. 5B, the principle of pattern tracking is to estimate transformation parameters for a template region, which maximizes the similarity between the corresponding template patterns in frames t and $t + 1$. Besides the translocation of the pattern the transformation model may include rotation, scaling, and higher-order deformations of the template. The similarity of the patterns can be formulated either by the signal cross-correlation or by the sum of the pixel-wise intensity difference, defining a maximization or a minimization problem of parameter estimation. A detailed comparison of the two approaches and methods to solve these optimization problems are given in Danuser *et al.* (2000).

3. Object/Contour Tracking

Here, the input data consist of a sequence of contours. The tracking algorithm should establish a warping model between consecutive time points. In the simpler case, the contour lines can be parameterized and the warping defined between explicit shape models such as splines. In the more challenging case, an explicit correspondence between boundary points is sought. As illustrated in the left panel of Fig. 5C, this is a nontrivial problem. For example, when following boundary points simply along the direction locally normal to the boundary in time point t until the normals intersect with the boundary in time point $t + 1$, the order of points might get disrupted (topological violations).

Several techniques are available to establish the correspondence between contours at t and $t + 1$. One of the most powerful techniques for cell biological applications, where complex shapes and large object deformations are the norm, is the Level Set Method (Osher, 2003). As discussed in Machacek and Danuser (2006), the Level Set Method permits the interpolation of the boundary evolution between t and $t + 1$ on a nearly continuous time scale. It is then straightforward to track auxiliary boundary points (edge markers) continuously along the direction normal to the interpolated boundaries (Fig. 5C right panel). The result is a set of topologically consistent (nonintersecting) edge marker paths whose lengths define local boundary displacements. Level Set Methods have also been used in an approach where the detection and tracking of cell edges are performed simultaneously (Dufour *et al.*, 2005).

F. Data Analysis

The previous sections have summarized methods for extracting from the unstructured information of a time-lapse image sequence lists of objects with attributes, coordinates, velocities, etc. However, this data still needs to be converted into a format that allows accepting or rejecting a mechanistic hypothesis based on rigorous statistical grounds. As this procedure is highly application dependent, we limit this section to a few remarks on the potential of computationally processed image measurements to reveal molecular mechanisms and on common pitfalls in the data analysis.

1. Identification and Interpretation of Unimodal and Multimodal Distributions

The analysis of image measurements within one cell or multiple cells is often limited to testing changes in the average of the measurements under different experimental conditions. This disregards that in some cases, the most significant aspect of a molecular perturbation may be a change in the distribution of a measurement that is not recognized as a shift in the average. In contrast, variations in the distributions of measurements among cells or among different experiments under supposedly equal conditions indicate data heterogeneity that requires caution with the pooling of measurements. Thus, before pooling data and testing mechanistic hypotheses based on it, the experimental variation of the distribution of measurements should be identified.

For qualitative examination of one-dimensional measurements, the data can be visualized in histograms. Critically, the appearance of the distribution depends on the bin size. As a rule of thumb a bin width of $2 \times \mathrm{IQD}/N^{(1/3)}$ provides a satisfactory representation of the distributions, where inter-quantile distance (IQD) is the difference between the 25th and the 75th percentile, and N the number of observations (Scott, 1992). An improved, more elaborate method for selecting the bin width is presented in Shimazaki and Shinomoto (2007).

Not all data is normally distributed! Representing skewed or multimodal distributions with a mean and standard deviation could cause information loss. Before reducing measurements into a description with mean and standard deviation, the underlying assumption of normality or at least the symmetry of data must be tested, for example, with the Lilliefors test (Sheskin, 2004). If normality is confirmed, the distributions of different conditions can be represented graphically as bar graphs.

To identify alternative distributions that reflect the measurements or to evaluate the significance of distribution variation among experiments one can apply the nonparametric Kolmogorov–Smirnov test (Sheskin, 2004). Asymmetric distributions should be represented graphically either by the full histogram or in the more condensed format of boxplots, which indicate the position of the distribution quantiles. Importantly, histograms or boxplots should indicate the number, the distribution, and the location of outliers. Outliers may be due to errors in the image analysis, or reflect a rare but mechanistically relevant event, especially when some experimental conditions exhibit outliers while others do not.

Multimodal distributions of measurements indicate that the system adopts different states. These states may be separable spatially, that is, the dynamics of a molecular machinery in one cellular location differs from the dynamics of the same machinery in another location (Ponti *et al.*, 2004); or temporally, that is, the dynamics undergoes switches between discrete states over the time course of an experiment (Genovesio *et al.*, 2006; Huet *et al.*, 2006); or completely overlap in time and space. Very often the multimodality of data is not obvious from the histogram (Fig. 6A). A vast number of algorithms have been developed to partition one-dimensional and multidimensional data sets into statistically significant modes. For an introduction to the fields of statistical learning and data mining, we refer the reader to the following classic textbooks (Hastie *et al.*, 2001; Höppner *et al.*, 1999; Jain and Dubes, 1988). As with the grouping of pixels into curvilinear features, the issues resolved by these algorithms are the identification of the number of modes and the estimation of parameters characterizing each one of the modes, which must be performed simultaneously.

Multimodal data analysis is particularly relevant in situations where a molecular perturbation selectively oppresses or adds one mode (Fig. 6B) as opposed to an overall change of the parameter distribution (Fig. 6C). Selective elimination or addition of a mode in response to specific perturbation of a molecule reflects the contribution of this molecule to the ensemble behavior of a multicomponent molecular system.

Figure 6D demonstrates that unless the data is explicitly tested for multimodality the differential response of modes is often buried in the bulk of the

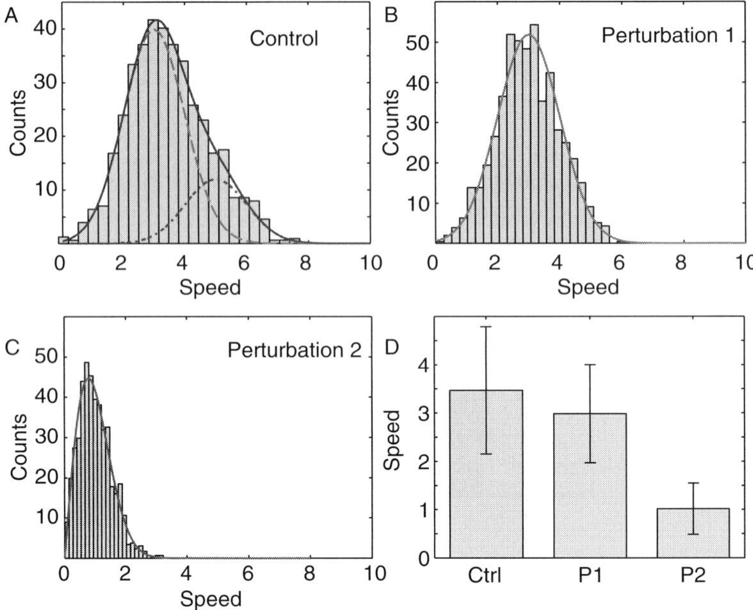

Fig. 6 Learning mechanisms by analysis of data distributions. (A) The speed distribution in wild type is the mixture of a slow (dashed line) and a fast (dash-dotted line) dynamic state. (B) Perturbation 1 eliminates the fast moving state, but leaves the slow moving state intact. (C) Perturbation 2 completely disrupts the system; only random positional fluctuations of the tracked objects are left. (D) Bar graph representation of the mean and standard deviation of the distributions in (A)–(C). In this display perturbation 1 has apparently no phenotype despite the specific disruption of a significant speed population in (A).

distribution. Although the differences in the distributions of control and perturbed condition P_1 are highly significant by multimode analysis, a statistical test that assumes unimodality and/or normality of the data will not distinguish the two conditions. Importantly, multimodality can falsely result from incomplete sampling of unimodal distributions. Thus, revealing multiple modes relies on large and unbiased data sets as obtained by computational image processing. A simple way to test the stability of a distribution for a given amount of data points is by bootstrapping (Efron, 1979), where subsets of data points are randomly selected and their distribution tested for equality with the distribution of the full data set (Jaqaman *et al.*, 2006).

2. Data Regression to Probe Properties of Molecular Systems Indirectly

In many cases, primary image measurements cannot report the actual properties of interest of a molecular system. Here, empirical statistics have to be complemented with mathematical models that link the properties of interest to one or several of the directly observable image parameters. Subsequently, the distributions of derived properties can be estimated by regression of the mathematical model to the distributions of the primary data. There are several aspects to be considered in data regression, for example, reviewed in a recent User's Guide to Systems Biology (Jaqaman and Danuser, 2006). The most important ones are; *uniqueness* of the model, that is, whether there are any two different distributions of primary image measurements, which could generate the same distribution for the derived properties; *determinability*, that is, whether the available set of primary image measurements is sufficient to estimate derived properties; and *sensitivity* of image measurements, that is, whether changes in the properties of the molecular system induced, for example, by experimental perturbation, yield a measurable change in one or several of the primary image measurements. If any of these aspects is not fulfilled, the regression will fail. As summarized in Jaqaman and Danuser (2006), well-defined protocols exist to validate regression results in these terms.

3. Analysis of Motion by Diffusion Measures

One of the main goals of fluorescence live cell imaging is to probe the motion of molecular structures. Thus, we conclude this section on data analysis with a brief summary of the most commonly used approaches to statistically characterize the motion properties as captured by a population of feature trajectories. The simplest statistical measure of motion is the mean square displacement (MSQD) as a function of the elapsed time Δt:

$$\mathrm{MSQD}(\Delta t = n \times \delta t) = \frac{1}{M}\sum_{i=1}^{M}\frac{1}{N_i - n}\sum_{k=1}^{N_i - n}|\mathrm{x}_i((k+n)\delta t) - \mathrm{x}_i(k\delta t)|^2,$$

$\mathrm{x}(t)$ denotes the coordinate of the feature at time point t, N_i is the total number of time points in trajectory i, M the number of trajectories analyzed, and δt the time interval at which trajectories are sampled. The larger Δt becomes, the fewer displacements are being averaged, decreasing the reliability of the MSQD. Therefore, MSQD values should be tested only up to $\Delta t = N/4(\delta t)$ (Saxton, 1997). At short Δt, the MSQD values are dominated by the type of motion (Saxton and Jacobson, 1997). In a log–log plot of MSQD(Δt), a slope of 1 means that the features undergo random Brownian motion. A slope larger than 1 indicates that the diffusion is superimposed by a directed, active motion component. A slope smaller than 1 indicates subdiffusion, where the Brownian motion of features is geometrically limited in one or multiple spatial directions (Sbalzarini *et al.*, 2005). Recent, more advanced diffusion analyses have applied the general *moments of displacements* (Ferrari *et al.*, 2001) to classify virus trajectories (Ewers *et al.*, 2005) and windowing schemes to extract MSQD statistics from trajectories with multiple overlapping modes of motion (Huet *et al.*, 2006).

In the case of geometrically confined motion, MSQD values reach a plateau at long-time intervals. The relation between the plateau level of MSQD values and the size of the confinement region can be explicitly calculated for many geometries of the confinement region (Dorn *et al.*, 2005; Kusumi *et al.*, 1993; Rosa *et al.*, 2006; Saxton, 1993). In cases where there is no analytical solution for this relation, it can be determined via numerical simulations (Marshall *et al.*, 1997).

V. Two Case Studies of Image Analysis Applied to Mechanistic Cell Biology

This section presents two case studies to illustrate in two different ways how the use of computational image analysis provides information on the function of molecular systems, not accessible by visual inspection and manual measurement. The first case describes applications of 3D particle tracking methods to measuring single MT length dynamics in the spindle of the budding yeast *Saccharomyces cerevisiae*. Computational image analysis made it possible to resolve MT lengths less than the diffraction limit of the light microscope. The study also shows how image acquisition and analysis must be tightly coupled and a novel approach for the classification of stochastic system behavior. In combination with the powerful yeast genetics, these data begin to reveal information of the mechanism of MT regulation by the kinetochore.

The second case illustrates computational image analysis of MT flux patterns in mammalian and vertebrate spindles by quantitative Fluorescent Speckle Microscopy. Here, the particle tracking methods were critical to following hundreds of thousands of stochastic markers of tubulin flow, providing a high-resolution map of the spatial organization of the flow. The high density of vector information makes it possible to probe the mechanisms of force generation and the dynamic organization of MTs across the spindle.

A. Measurement of Kinetochore MT Length Dynamics in Budding Yeast to Study the Function of Kinetochore Proteins

Budding yeast is an important model system for the study of mitosis, in particular also for mathematical modeling of the mechanisms that yield symmetric chromosome segregation (Mogilner *et al.*, 2006). It contains all the key architectural elements of a mitotic spindle (Winey and O'Toole, 2001): MTs radiate from two opposing spindle poles to establish bipolar attachment to 32 replicated sister chromatids via kinetochores. Moreover, overlapping interpolar MTs connect to the two half-spindles via binding of MT-associated motors. However, the budding yeast spindle is mechanically much simpler than mammalian spindles and powerful genetics allows the easy introduction of proteins mutations and fusions to GFP. Of particular importance to the present study is that only one kinetochore-MT (kMT) binds to each kinetochore. Thus, we can directly address the question of if and how kinetochore proteins regulate kMT dynamics by correlating kinetochore protein mutations to the changes in the MT length dynamics they induce relative to wild type kMT dynamics.

To this end, we need to tackle three methodological challenges:

Measuring kMT length in yeast: The length of the yeast spindle ranges from 1 to 2 μm. Thus, the 32 kMTs and kinetochores are separated mostly by distances below the resolution limit of the microscope.

Describing MT dynamics in different mutants: MTs switch randomly between growth and shrinkage. Thus, direct comparison of kMT trajectories among mutants is impossible. Furthermore, most mutations in kinetochore proteins induce weak effects on kMT dynamics, requiring a very sensitive parameterization of the dynamics.

Modeling MT dynamics: Parametric descriptions of kMT dynamics are invaluable to identifying phenotypes among mutants. However, to understand the function of kinetochore proteins as regulators of kMT dynamics, we will need a mechanochemical model of the interaction between kinetochore proteins and kMTs and procedures to calibrate the parameters of such model against the pool of kMT length trajectories in wild type and mutants. In the following, we describe how the methods in Section IV were applied to tackle the first two challenges. The mechanochemical modeling of kinetochore—kMT interaction based on image measurements is beyond the scope of this chapter.

1. Experimental Design

kMT dynamics in yeast cannot be measured directly because of the small size of the spindle. GFP labeling of tubulin or a kinetochore protein results in a fluorescent blob representing the entire spindle but individual kinetochores or MTs can no longer be distinguished. Elegant fluorescence recover after photobleaching experiments have been designed to infer MT dynamics under these circumstances (Maddox *et al.*, 2000; Pearson *et al.*, 2006) and models that relate temporal variation of the blob intensity to the kinetics of MT growth and shrinkage (Sprague *et al.*, 2003). However, these approaches define only the average behavior of kMTs.

To study the dynamics of individual kMTs, we adopted the approach by Belmont and Straight (1998) to GFP-label a single chromosome (chromosomes IV or V) next to the centromere (CEN) (He *et al.*, 2000), and to measure the motion of the sister chromatids relative to the position of the GFP-labeled spindle pole body (SPB) protein Spc42p. The distance between CEN and SPB tags reports the length of an individual kMT (Fig. 7A).

To address the concern that diffusive motion of the CEN tag around the kinetochore might mask kMT dynamics, we investigated the sensitivity and

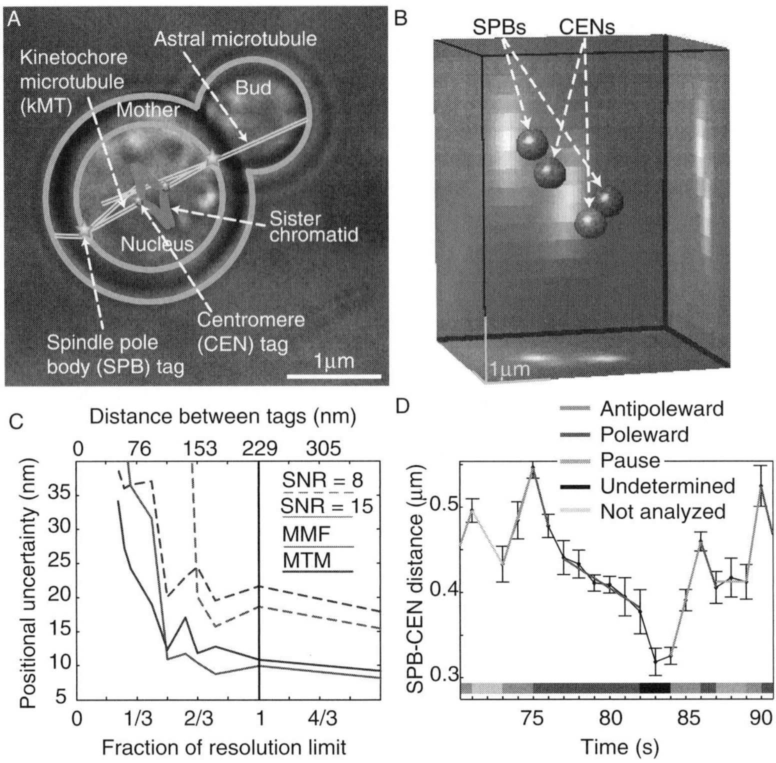

Fig. 7 Measurement and analysis of chromosome dynamics in budding yeast. (A) Phase contrast image of metaphase budding yeast. Representative positions of 1 of 16 chromosome pairs (red) and microtubules (MTs) (yellow). Green stars: location of fluorescent tags next to the centromere (CEN), and on the spindle pole body (SPB). (B) 3D fluorescent image of a metaphase spindle with SPB and CEN tags extracted. Due to the short extension of the spindle tag signals strongly overlap. (C) Resolution performance of computational tracking on simulated data. Both mixture model fitting (MMF) and multitemplate matching (MTM) algorithms resolve tags separated by a distance below the diffraction limit. (D) Classification of chromosome dynamics in yeast. The SPB–CEN length trajectories are statistically grouped into phases of antipoleward (green) motion, poleward (red) motion, and pause (blue). The algorithm also detects significant changes in speed throughout one phase of motion (e.g., between 75 and 83 sec). (See Plate no. 68 in the Color Plate Section.)

accuracy of our assay with a Monte Carlo simulation of kMT dynamics (Dorn *et al.*, 2005). Parameters of kMT kinetics were selected to produce qualitative agreement between simulates and experimental kMT length trajectories. Using the model the influence of CEN tag motion was predicted for a worst-case scenario that the chromosome could rotate freely around the MT plus end. The simulations indicated that while the offset of the CEN tag from the kMT tip did influence the absolute values of growth and shrinkage speeds, and the frequencies of directional switching, it did not affect the sensitivity of the readouts. Furthermore, given the fast dynamics of kMTs in yeast and the sampling rate achievable with state-of-the-art light microscopy (see later description), the four parameters of MT dynamics were found to be sensitive, that is, mutations can be distinguished, but not unique. Thus, novel ways to describe MT dynamics are required (Jaqaman *et al.*, 2006). This example demonstrates how simple simulations at the outset of an experiment direct the planning of an experimental setup and data analysis strategy that are eventually compatible.

2. Acquisition of Optimized Fluorescent Images

During the optimization of the image acquisition, we had to decide whether 3D image stacks were required, and how many time points were needed to fully capture kinetochore MT dynamics (Dorn *et al.*, 2005). Imaging in 3D substantially reduces the temporal resolution and expedites photobleaching of the specimen. As an alternative, 2D images could be acquired and time points analyzed only for which all fluorescent tags are located in the same plane of focus (Sprague *et al.*, 2003). However, the depth-of-field of a single optical slice [~800 nm for our setup (Inoue and Spring, 1997)] is similar to the length of a half spindle. Therefore, even with both CEN and SPB tags detectable in the same plane, kMTs could be significantly tilted relative to the horizontal plane, resulting in systematic underestimation of the kMT length. We compared the apparent kMT dynamics of 2D movies to the full kMT dynamics measurable in 3D stacks and found that 3D image acquisition is required to distinguish potential phenotype changes between mutants (Dorn *et al.*, 2005).

Finally, we determined the number of time points, and thus, movies, required to achieve convergence of the descriptors of kMT dynamics. We found that at ~2500 data points, or ~25 movies, the descriptors became stable. Together these initial investigations defined an experimental design *a priori*, which ensures meaningful measurement of kMT dynamics for all mutants.

3. Image Analysis

Tracking of kMT length trajectories via localization of SPB and CEN tags is accomplished in three steps: First, the images of the fluorescent tags are detected frame-by-frame using a mixture-model-fitting (MMF). One or multiple PSFs are fitted to each one of the 3D local intensity maxima of the image stack. The number of PSFs necessary to fit the intensity distribution of a local maximum is decided by a statistical information criterion that weighs the residuals of the fit against the number of free parameters of the fit. If several PSF are fitted tightly to one local maximum, several tags colocalize at distances below the diffraction-limited resolution (Fig. 7B; Thomann *et al.*, 2002). The result of a MMF could also be the rejection of even a single PSF fit. In this case, the SNR of the local maximum is too weak to provide evidence for a tag in this location.

Second, corresponding tag images must be assigned between frames. We apply a GNN algorithim (Section IV.E.1) to track tags by globally minimizing the overall tag displacement and intensity change. Since the subsequent analyses require error-free tag links and the spindle geometry is occasionally ambiguous, manual user input is allowed to support this step. Users can delete frames, for example, when tags move outside of the field of view, or delete tags whose images passed the statistical tests of the MMF as false positives. However, to minimize operator-bias in the final data, no new tags or manual adjustments of the positions of existing tags are permitted.

Third, exploiting the prior knowledge that tag signals change little between consecutive frames, the tag locations of the initial detection step are enhanced by a multitemplate matching (MTM) algorithm (Thomann *et al.*, 2003). This is a variant of the template-based tracking methods discussed in Section IV.E.2 , which accounts for the superposition of templates representing multiple tags as the tags come closer. In contrast to MMF, MTM is a relative measurement, that is, no explicit model of the image of a tag is assumed. Thus, MTM adds accuracy to the tag tracking algorithm, as well as increases resolution (Thomann *et al.*, 2003). Simulations have shown that MTM resolves tags at distances 30% of the Rayleigh limit (Fig. 7C).

4. Data Analysis

Diffusion analysis alone was sufficient to demonstrate that chromosomes are attached to MTs in the G1 phase of the yeast cell cycle (Dorn *et al.*, 2005). However, this technique was too global, hence not sensitive to distinguish between different kMT dynamics of different mutations of kinetochore proteins and other components of the spindle. Thus, kMT length trajectories were characterized by local descriptors of the distribution of kMT growth and shrinkage speeds, and catastrophe and rescue frequency. An algorithm was developed for the automatic classification of the length trajectory into phases of growth, shrinkage, and pause (Fig. 7D). Given an estimate of the uncertainty of the kMT length measurement in each time point, the statistical significance of length changes between two time-points or over multiple time-points was determined. The length uncertainties were estimated by error propagation in the MMF and MTM modules, which converts noise-induced intensity fluctuations in the raw image stack into positional fluctuations. An important finding of this local classification was that the speeds of, for example, kMT growth varies significantly between different growth phases of one trajectory, as well as within a single growth phase. We have increasing evidence from our mutant database that this heterogeneity in growth and shrinkage speed is the signature of regulated kinetics of kMT polymerization and depolymerization by kinetochore proteins (Dorn *et al.*, 2005; Jaqaman *et al.*, 2006). In particular, we found that mutations in kinetochore proteins primarily affect the shape of the growth and shrinkage speed distribution, whereas the mean values remain unchanged (Dorn *et al.*, 2005). This illustrates that the mechanistically most informative part of live cell image measurements can often be extracted from the distribution of the data.

B. Tracking of Fluorescent Speckles for the Study of Mechanisms Driving MT Flux in Metaphase Spindles

A remarkable feature of metaphase spindles in higher eukaryotic cells is the constant poleward movement of MTs, a phenomenon referred to as MT flux (Sawin and Mitchison, 1991). Fluorescent speckle microscopy (FSM) provides a powerful tool to visualize this dynamic process (Waterman-Storer *et al.*, 1998). Initially, however, manual speckle tracking was the only tool available to obtain quantitative measurements from FSM images (Maddox *et al.*, 2003), which significantly limited the application of FSM to the quantitative study of spatial and temporal organization of MT flux. Computational speckle detection and tracking played a critical role in overcoming this limitation, as demonstrated in the following two application examples.

1. Tracking MT Flux in PtK1 Spindles

Initial FSM studies in combination with drug perturbation established a central role of tetrameric kinesin-5 (Eg5) in driving MT flux through MT sliding in *Xenopus* egg extracts (Miyamoto *et al.*, 2004). However, it remained an open question how much Eg5 activity contributes to the MT flux in mammalian cells and whether in these cells MT depolymerization at the poles can be a mechanism sufficient to drive MT flux independent of kinesin motors. This question can be answered by monitoring MT flux in a monopolar spindle, which can be generated from bipolar spindles by application of Eg5-inhibitor monastrol (Kapoor *et al.*, 2000). We used the PtK1 mammalian cell line due to optimal properties (large size, flat, low scattering) for light microscopy. However, the small size (typically ~10 μm) of the spindles and the short lifetime of the majority of MTs (~30 sec) in combination with the relatively slow flux rate (600 nm/min) make it difficult to detect and track individual speckles by manual analysis. Furthermore, PtK1 spindles often undergo rapid drifts and rotation during image acquisition, which precludes flux analysis by kymographs. To overcome these obstacles, computational techniques were developed and applied for image alignment, speckle detection and speckle tracking (Cameron *et al.*, 2006; Ponti *et al.*, 2003). Thanks to the complete analysis of every speckle attained by computational speckle tracking, sufficient trajectories could be collected to confirm consistent poleward MT flux in bipolar and monopolar spindles (Fig. 8A). The measured flux rate matched well with measurements from photoactivation experiments. Furthermore, the large number of computationally extracted trajectories allowed us to establish a statistically significant decrease of ~25% in flux rate in monopolar spindles compared to bipolar spindles, despite the substantial heterogeneity of speeds. Together these results suggested that in PtK1 cells the contribution of Eg5 to MT flux is minor

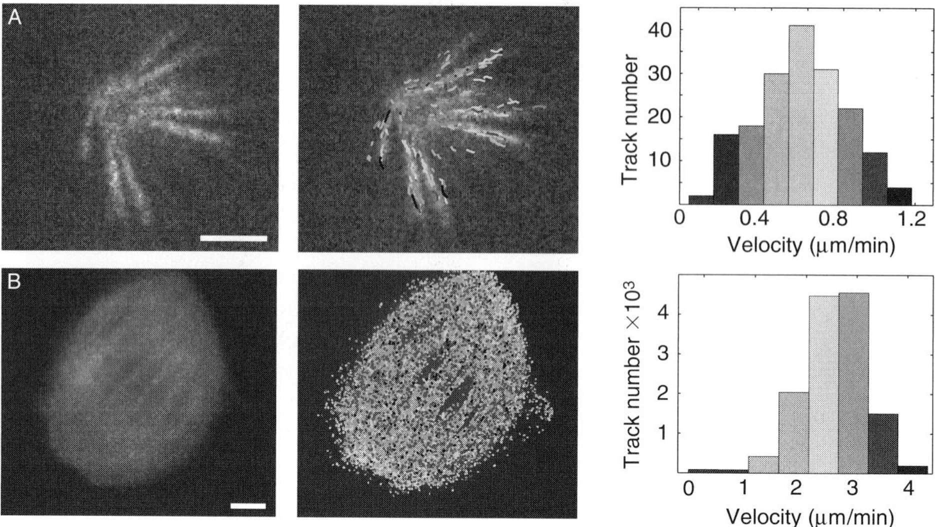

Fig. 8 Tracking of fluorescent speckles. (A) Speckle tracking in a PtK1 cell monopolar spindle. Left: original image, Middle: image with track overlaid. Each track is color coded based on its average velocity following the color scheme shown in the histogram (right). Total number of tracks: 176. Total number of tracked speckle positions: 851. (Adapted from Cameron *et al.*, 2006 with permission.) (B) Speckle tracking in a *Xenopus* egg extract spindle. Total number of tracks: 13,337. Total number of tracked speckle positions: 70,396. (See Plate no. 69 in the Color Plate Section.)

(Cameron *et al.*, 2006). Instead a "pulling-in" mechanism was proposed based on this data in which a depolymerase localized in the spindle pole at the MT minus ends is the critical driver of poleward flux.

2. Tracking MT Flux in *Xenopus laevis* Extract Spindles

Although the study of MT flux in PtK1 spindles revealed the coexistence of two mechanisms driving MT flux *in vivo*, it remained unclear how these mechanisms are regulated spatially and temporally and whether there is cross-talk between them. For example, Eg5-driven sliding of overlapping MTs in the spindle midzone could promote MT depolymerization at the poles. The number of trackable speckles in PtK1 spindles is typically too small (<100 per frame) to answer this question with sufficient spatial resolution. Instead, *Xenopus laevis* extract spindles, whose length typically ranges between 30 and 80 μm, provide a suitable assay. Although speckle detection and spindle drift remain two issues to be resolved also with this application, the primary challenge comes with the large number of speckles within each frame, typically between 2000 and 10,000. For example, analysis of an image series of 50 frames requires tracking of half a million speckles. Extension of FSM into 3D will further increase the scale and complexity of the tracking problem.

To address this challenge, the first choice to be made was to choose an algorithm with low computational complexity, especially since the selected algorithm is expected to be scalable to 3D speckle tracking applications. Consequently, joint probabilistic data association filter and multiple hypothesis testing (Section IV.E.1) were excluded. Instead, we drew on the low computational complexity of GNN and improved its performance by developing a new algorithm that incorporates information of the global vector field of MT flux to constrain the speckle assignment of individual speckles (Yang *et al.*, 2005). Speckle tracking results revealed that spatial distribution of MT flux is regionally heterogeneous but organized (Fig. 8B). Specifically, MT flux rate near the poles is reduced by ~20% compared to that near the equator, which not only suggests the coexistence of different driving mechanisms but provides a tool to dissect the organization of these mechanisms in space and time through various perturbations to this system. For example, we found that inhibition of dynein, which bundles MTs into poles, significantly reduces the flux heterogeneity. This result confirms the conclusion drawn from PtK1 cells that MT flux is critically controlled by a pole-located depolymerase.

VI. Performance Evaluation for Quality Control

Like any other experimental methods, computational techniques have to be evaluated by controls to ensure correctness, reliability, and efficiency. Development of general methods for the performance evaluation of algorithms is actively studied in computer science in areas such as *algorithm analysis* (Levitin, 2006) and *software engineering* (Sommerville, 2006). Most users of computational image analysis techniques will not need to understand the technical details of general performance evaluation. However, their knowledge and skills are indispensable in defining meaningful test cases and in providing application-specific understanding of image data for the evaluation process. This section is intended to prepare software users for this purpose. After a brief discussion of the relations between algorithm and software, methods for verification of the correctness of computational techniques are reviewed, and commonly used terms in algorithm efficiency evaluation are introduced.

A. Algorithms and Software

The core of a computational technique is its algorithms. An *algorithm* refers to a method, normally in the form of a set of mathematically defined rules or procedures that can be implemented as a computer program to solve a problem. The conceptual abstraction of algorithms allows them to be implemented on any suitable computer platform. In contrast, *software* is the implementation of algorithms in specific computer languages on specific computer systems.

The *performance* of computational image analysis is fundamentally determined by correctness of the underlying algorithms. Software implementation itself cannot change the correctness, but it does play a critical role in achieving efficiency. Thus, different programs using the same algorithm may have completely different performance characteristics. Evaluation of an image analysis technique usually starts with mathematical analysis of its algorithms and continues with empirical performance evaluation through testing the software implementation. This section will focus on the second of these aspects, where experimentalists play an important role.

B. Verification of Correctness

1. Test Data Preparation

The test data required for performance evaluation consists of two parts: input data and its corresponding correct output required as the reference for testing (often called *ground truth*). Input data comes essentially from two sources: experiments and computer simulation. Actual experimental data allows direct testing of software performance on actual applications. However, there are several issues to be addressed.

a. Correct Output Corresponding to Input Data Is Unknown

Several strategies may be adopted to address this issue. First, results from manual image analysis may be used as the ground truth. Second, ground truth can be generated by reducing the complexity of the experiment to the level where the output is known (Gross *et al.*, 2002). Third, perturbations with known effects on the specimen behavior can generate useful test data (Ponti *et al.*, 2003). Fourth, for some applications it is sufficient to check the internal consistency of the measurements without requirement for an explicit ground truth (Christensen and Phillips, 2002).

b. Acquisition of Test Data Is Too Costly

If the error rate of an image analysis method is on the order a few percentage points, hundreds of experiments and ground truth measurements are required to accurately validate the algorithmic performance. Often, this is not practical.

c. Available Experimental Data May Not Represent the Full Range of Possible Input

The test data and ground truth measurements may represent only a subset of the input data the algorithms are expected to process. Qualitatively, this can be addressed with the assistance of the experimentalist familiar with the variability of the data. However, on a quantitative level, the intrinsic heterogeneity of the molecular process being probed is generally unknown, so that it may be impossible to select a representative sample of the data.

Computer simulation of input images provides an important remedy to these problems (Hu *et al.*, 2007; Ponti *et al.*, 2003; Vallotton *et al.*, 2004). First, ground truth for input image data is directly available, making it possible to rigorously define the relation between input image data and the correct output. Second, simulation experiments can be repeated many times so that the performance of algorithms in detecting and processing rare events can be effectively tested. Third, stochastic simulations can generate a large variety of different image scenes based on assumptions of the statistical distribution of the mechanistic parameters characterizing the probed process.

However, computer simulations also have limitations. The complexity of the molecular system and the lack of mechanistic knowledge of the probed process make it difficult to reproduce the full image dynamics that are to be expected for an experiment. To address this limitation, an iterative cycle of real-world data acquisition, inference of the molecular mechanism of the dynamic process based on existing algorithms, image data simulation and performance evaluation, and improvement of the design of the algorithms have to be implemented. In computer vision, this procedure is referred to as *realistic simulation* (Boyer and Philips, 1998; Woolfson and Pert, 1999). While realistic simulations are clearly the most precise approach to prepare test data, the risk remains that image data simulation and image analysis algorithm are derived from the same set of assumptions and thus cannot reflect the true performance. Thus, all of the above-mentioned experimental strategies, in particular the approach of perturbation experiments, are required to complement performance evaluation based on simulated data.

2. The Role of Manual Analysis in Performance Evaluation

Human vision is particularly powerful for tasks such as image segmentation and object recognition. Thus, manual image analysis can be effective in generating ground truth data. However, it is important to recognize that manual analysis can also be biased due to factors such as the knowledge and training of the investigator. These human factors must be controlled as demonstrated in, for example, Heath *et al.* (1997).

There are also tasks where human vision is *a priori* inferior to computer vision. In these cases, manual generation of ground truth data is not feasible. Empirically, we found that one such example is the tracking of dense particles. We performed an experiment with a total of 10 users who were given the same FSM movie for manual tracking. The selection of speckles to be tracked was made by a random number generator, programmed so that users had to repeat ~30% of their own measurements and ~50% of the measurements by other users. The result of this test was that the reproducibility of trajectory measurements by individual users was <70%, while reproducibility between users was <50%. Clearly, these values are insufficient to generate a ground truth data for the performance evaluation of a tracking algorithm. In contrast, simulation of speckle data with a similar characteristics as the experimental data provided more consistent evaluation results (Cameron *et al.*, 2006). This example illustrates that the performance of manual image analysis must be carefully verified before it can be used as ground truth to evaluate computational image analysis methods.

3. Test Protocol Development

After generation of test data, the performance test must be executed in a consistent and systematic manner, addressing the following issues (Boyer and Philips, 1998; Christensen and Phillips, 2002; Liu and Dori, 1997).

a. Parameter Setting

Changes in control parameter settings can significantly influence performance of an algorithm. There are two commonly used approaches to address this issue. First, the parameter setting is optimized for the test data set. Here, the limitation is that if the test data does not encompass the full range of actual experimental data, the performance will be overestimated. Second, key control parameters can be identified so that the change in performance can be characterized under different settings. While this approach effectively evaluates the robustness of an algorithm, the procedure can become tedious as the number of control parameters increases. Thus, for automated processing and analysis of large volume of image data, it is essential to design algorithms with data-driven self-adjustment of control parameters.

b. Quantification of Correctness

The term "correctness" itself needs to be precisely defined in the testing process. This can be illustrated in a simple application in event testing. For example, the event could be whether a cell is undergoing mitosis, or whether a specific protein has been expressed. If the test answer is yes, it is called a positive event, otherwise a negative event. Table I summarizes the four possible cases of software output. The four classes (True Positive, True Negative, False Negative, and False Positive) of readout all provide measurement of the correctness from a somewhat different perspective. Therefore the meaning of "correctness" must be precisely defined before a scoring method can be developed. Here, we use the mitosis test as an example to demonstrate these concepts. When a mitotic cell is identified correctly by the algorithm as undergoing mitosis, a true positive is obtained. The ratio of the total number of true positives to the total number of actual mitotic cells in the test image is defined as the *sensitivity* (or *true positive rate*) of the detection algorithm. Similarly, the ratio of the total number of false positives (nonmitotic cells that are falsely detected as mitotic) to the total number of nonmitotic cells is defined as

Table I
Different Types of Output of a Simple Event Test

Classification	Definition
True positive (TP, true acceptance, true match)	A positive event is correctly identified as positive
True negative (TN, true rejection, true nonmatch)	A negative event is correctly identified as negative
False negative (FN, false rejection, false nonmatch, type I error)	A positive event is incorrectly identified as negative
False positive (FP, false acceptance, false match, type II error)	A negative event is incorrectly identified as positive

the *false positive rate*. The *specificity* of the detection algorithm is defined as one over false positive rate. Plotting the true positive rate versus the false positive rate gives the receiver operating characteristic curve, a widely applied performance evaluation and validation tool in biostatistics (Lasko *et al.*, 2005). Different applications usually place different requirements on sensitivity and specificity. For example, if the goal is to identify mitotic cells in a screen for mitosis inhibitors, detection algorithms with higher specificity would be more desirable since it is more important to exclude nonmitotic cells than to include mitotic cells. In contrast, if the goal is to compute the percentage of mitotic cells (the mitotic rate), algorithms with higher sensitivity will be more desirable. Thus, correctness may have to be defined in different ways for different applications.

4. Test Administration

Most software testing is performed by individual laboratories for individual research projects. To allow objective comparison of different algorithms and to avoid duplication of effort, it will be particularly important to establish publicly available data sets with high-quality fluorescence microscopy image data and standardized test protocols for software performance evaluation. It should become routine to indicate the performance of an image analysis technique on one of these test data sets before it is applied to an experiment for which no ground truth is accessible. The significance of standard data sets for the development of computational methods has been demonstrated in other application areas of computer vision, including image alignment in medical imaging (West *et al.*, 1997) and face recognition (Phillips *et al.*, 2000). For the benefit of a more quantitative cell biology, the implementation of a centralized image test data base should be initiated.

C. Efficiency Evaluation

Efficiency of software is measured by the time and memory it takes to complete a task. In practice, these two factors are measured mostly by running the software on a given input data set on a given computer system. However, this approach has limited power to predict the efficiency of the software on different experimental data. As a useful complement, the efficiency of software can also be evaluated by analyzing the time and space complexity of its algorithms. Such analysis is particularly useful for large-scale problems where software runs for evaluation purposes are too costly.

Terminology commonly used by software developers to describe the time complexity of algorithms is listed in Table II. N represents the primary number that determines the algorithm's running time, usually proportional to the number of pixels or objects in an image series.

A related and widely used term is the O-notation. For example, calling an algorithm $O(N \times \log N)$ or $O(N^2)$ means that its worst running time is proportional to $N \cdot \log N$ or N^2. These definitions are useful as they provide direct predictions of change in computation time as the scale of input data changes (Sedgewick, 1992).

Table II
Commonly Used Time Complexity Terms

If running time of an algorithm is proportional to	Its time complexity is called	Increase in running time when N is increased by 10
1	Constant	0
$\log N$	Logarithmic	2.303
N	Linear	10
$N \log N$	$N \log N$ or linearithmic	23.03
$N^3, N^3, \ldots$	Polynomial	100, 1000, ...
2^N	Exponential	1024

Exponential algorithms are generally not usable, except for very small-scale problems. A related concept is *NP-hard* (nondeterministic polynomial-time hard): When a problem is NP-hard, no algorithm exists to compute its exact optimal solution in at most polynomial time. Instead, approximate solutions must be accepted to keep computation time acceptable (Vazirani, 2003). Unfortunately, many of the problems encountered in computer vision, such as tracking of multiple targets, are NP-hard.

VII. Summary

This chapter provides a selective review of basic terminology, concepts, and techniques of computational processing and analysis of dynamic fluorescence images. It also introduces several related statistical data analysis techniques and quality control methods. In general, computational image processing shares many common aspects with running wet bench experiments. First of all, the selection of techniques and their strengths and limitations must be taken into account during experiment design. Furthermore, appropriate controls must be designed and performed. Finally, the procedures have to be optimized to achieve the best performance for a specific application. When properly applied, computational image analysis provides unique insights into the dynamic organization and regulation of molecular systems in living cells.

The critical importance of computational image processing and analysis for cell biology studies resides in the reliability and efficiency with which quantitative measurements of molecular system dynamics can be extracted from fluorescence images, in the ability to achieve high spatial and temporal accuracy and to improve resolution, and in the minimization of investigator bias and information loss throughout the data analysis. Although significant progress has been made in developing computational image analysis in other fields of research, many challenges remain to render these tools an everyday instrument for cell biologists. First, the mathematical nature of these techniques limits their broader accessibility to the cell biology community. The selection of the right algorithms and the tuning of control parameters require special training and significant expertise. We are still far away from "black-box" solutions to the analysis of live cell image data. Second, the complexity of structure and dynamics inherent to cellular systems and the often low SNR achievable in fluorescent images pose significant challenges even to the most advanced algorithms in computer vision. Thus, in parallel to the ever growing number of new imaging approaches, the field needs novel algorithms that are tailored to live cell images. Third, the requirement for customized algorithms to address specific questions remains an obstacle in practice despite the constant growth of commercial software in number and in functionality. Although for some applications it is feasible for investigators to develop their own computational image analysis tools, effective interdisciplinary collaboration seems to be the most promising route to perform advanced quantitative fluorescence microscopy. Most importantly, in the design of an imaging experiment, the molecular techniques applied to the preparation and manipulation of the specimen, the selection of microscopy protocols, and the choice of algorithms and control parameters must all be considered from the very beginning, and then iteratively optimized. Only by considering image analysis as an integral part of an experiment will the full potential of these methods be realized.

Important advances are being made in light microscopy, especially in overcoming resolution limits and extensions to three dimensions. Many of these advances depend heavily on computational techniques, for example, image reconstruction, image segmentation, and tracking, which increasingly overwhelm human visual inspection and manual data analysis. Together with the drive to study ever larger molecular systems under live conditions, these needs will undoubtedly expedite the development of computational fluorescence image analysis techniques to the benefit of cell biology studies.

Acknowledgments

Funded by NIH grants R01 GM60678, GM68956 to G.D., a Roche Research Foundation fellowship to J.D., and a Burroughs-Wellcome Fund La Jolla Interfaces in Sciences fellowship to G.Y.

References

Abraham, V. C., Taylor, D. L., and Haskins, J. R. (2004). High content screening applied to large-scale cell biology. *Trends Biotechnol.* **22,** 15–22.

Bar-Shalom, Y., Li, R. X., and Kirubarajan, T. (2001). "Estimation with Applications to Tracking and Navigation." John Wiley & Sons, New York.

Belmont, A. S., and Straight, A. F. (1998). *In vivo* visualization of chromosomes using lac operator-repressor binding. *Trends Cell Biol.* **8,** 121–124.

Blackman, S. S., and Popoli, R. (1999). "Design and Analysis of Modern Tracking Systems." Artech House, Norwood, MA.

Bonneau, S., Dahan, M., and Cohen, L. D. (2005). Single quantum dot tracking based on perceptual grouping using minimal paths in a spatiotemporal volume. *IEEE Trans. Image Process.* **14,** 1384–1395.

Boyer, K., and Philips, P. (1998). "Empirical Evaluation Techniques in Computer Vision." Wiley—IEEE Computer Society Press, Los Alamitos.

Burkard, K. E., and Cela, E. (1999). Linear assignment problems and extensions. *In* "Handbook of Combinatorial Optimization" (D. Z. Du and P. M. Pardalos, eds.), Supp. A. pp. 75–149. Kluwer Academic Publishers, Dordrecht, NL.

Cameron, L. A., Yang, G., Cimini, D., Canman, J. C., Evgenieva, O. K., Khodjakov, A., Danuser, G., and Salmon, E. D. (2006). Kinesin 5-independent poleward flux of kinetochore microtubules in PtK1 cells. *J. Cell Biol.* **173,** 173–179.

Canny, J. (1986). A computational approach to edge-detection. *IEEE Trans. Pattern Anal. Mach. Intell.* **8,** 679–698.

Cheezum, M. K., Walker, W. F., and Guilford, W. H. (2001). Quantitative comparison of algorithms for tracking single fluorescent particles. *Biophys. J.* **81,** 2378–2388.

Chen, X., Murphy, R. F., and Kwang, W. J. (2006). Automated interpretation of protein subcellular location patterns. *Int. Rev. Cytol.* **249,** 193–227.

Christensen, H. I., and Phillips, P. J. (2002). "Empirical Evaluation Methods in Computer Vision." World Scientific, River Edge, NJ.

Conrad, C., Erfle, H., Warnat, P., Daigle, N., Lorch, T., Ellenberg, J., Pepperkok, R., and Eils, R. (2004). Automatic identification of subcellular phenotypes on human cell arrays. *Genome Res.* **14,** 1130–1136.

Cox, I. J. (1993). A review of statistical-data association techniques for motion correspondence. *Int. J. Comput. Vis.* **10,** 53–66.

Cox, I. J., and Hingorani, S. L. (1996). An efficient implementation of Reid's multiple hypothesis tracking algorithm and its evaluation for the purpose of visual tracking. *IEEE Trans. Pattern Anal. Mach. Intell.* **18,** 138–150.

Danuser, G., Tran, P. T., and Salmon, E. D. (2000). Tracking differential interference contrast diffraction line images with nanometre sensitivity. *J. Microsc.* **198,** 34–53.

Davies, E. R. (2004). "Machine Vision: Theory, Algorithms, Practicalities." Elsevier Science & Technology Books, San Francisco.

Dorn, J. F., Jaqaman, K., Rines, D. R., Jelson, G. S., Sorger, P. K., and Danuser, G. (2005). Interphase kinetochore microtubule dynamics in yeast analyzed by high-resolution microscopy. *Biophys. J.* **89,** 2835–2854.

Drubin, D. A., Garakani, A. M., and Silver, P. A. (2006). Motion as a phenotype: The use of live-cell imaging and machine visual screening to characterize transcription-dependent chromosome dynamics. *BMC Cell Biol.* **7,** 19.

Duda, R. O., Hart, P. E., and Stork, D. G. (2001). "Pattern Classification." John Wiley & Sons, New York.

Dufour, A., Shinin, V., Tajbakhsh, S., Guillen-Aghion, N., Olivo-Marin, J.-C., and Zimmer, C. (2005). Segmenting and tracking fluorescent cells in dynamic 3-D microscopy with coupled active surfaces. *IEEE Trans. Image Process.* **14,** 1396–1410.

Efron, B. (1979). Bootstrap methods: Another look at the Jackknife. *Ann. Statist.* **7,** 1–26.

Eils, R., and Athale, C. (2003). Computational imaging in cell biology. *J. Cell Biol.* **161,** 477–481.

Ewers, H., Smith, A. E., Sbalzarini, I. F., Lilie, H., Koumoutsakos, P., and Helenius, A. (2005). Single-particle tracking of murine polyoma virus-like particles on live cells and artificial membranes. *PNAS* **102,** 15110–15115.

Ferrari, R., Manfroi, A. J., and Young, W. R. (2001). Strongly and weakly self-similar diffusion. *Physica D.* **154,** 111–137.

Gardner, M. K., Pearson, C. G., Sprague, B. L., Zarzar, T. R., Bloom, K., Salmon, E. D., and Odde, D. J. (2005). Tension-dependent regulation of microtubule dynamics at kinetochores can explain metaphase congression in yeast. *Mol. Biol. Cell* **16,** 3764–3775.

Genovesio, A., Liedl, T., Emiliani, V., Parak, W. J., Coppey-Moisan, M., and Olivo-Marin, J. C. (2006). Multiple particle tracking in 3-D+t microscopy: Method and application to the tracking of endocytosed quantum dots. *IEEE Trans. Image Process.* **15,** 1062–1070.

Gonzalez, R., and Woods, R. (2002). "Digital Image Processing." Prentice Hall, Upper Saddle River, NJ.

Goshima, G., Wollman, R., Stuurman, N., Scholey, J. M., and Vale, R. D. (2005). Length control of the metaphase spindle. *Curr. Biol.* **15,** 1979–1988.

Gross, S. P., Tuma, M. C., Deacon, S. W., Serpinskaya, A. S., Reilein, A. R., and Gelfand, V. I. (2002). Interactions and regulation of molecular motors in Xenopus melanophores. *J. Cell Biol.* **156,** 855–865.

Haralick, R. M. (1983). Ridges and valleys on digital images. *Comput. Vision Graphics Image Process.* **22,** 28–38.

Hastie, T., Tibshirani, R., and Friedman, J. (2001). "The Elements of Statistical Learning—Data Mining, Inference and Prediction." Springer, New York.

He, X., Asthana, S., and Sorger, P. (2000). Transient sister chromatid separation and elastic deformation of chromosomes during mitosis in budding yeast. *Cell* **101,** 763–775.

Heath, M. D., Sarkar, S., Sanocki, T., and Bowyer, K. W. (1997). A robust visual method for assessing the relative performance of edge-detection algorithms. *IEEE Trans. Pattern Anal. Mach. Intell.* **19,** 1338–1359.

Höppner, F., Klawonn, F., Kruse, R., and Runkler, T. (1999). "Fuzzy Cluster Analysis." John Wiley and Sons, Chichester, UK.

Hu, K., Ji, L., Applegate, K., Danuser, G., and Waterman-Storer, C. M. (2007). Differential transmission of actin motion within focal adhesions. *Science* **315,** 111–115.

Huet, S., Karatekin, E., Tran, V. S., Fanget, I., Cribier, S., and Henry, J. P. (2006). Analysis of transient behavior in complex trajectories: Application to secretory vesicle dynamics. *Biophys. J.* **91,** 3542–59.

Inoue, S., and Spring, K. R. (1997). "Video Microscopy: The Fundamentals." Plenum, New York and London.

Jain, A. K., and Dubes, R. C. (1988). "Algorithms for Clustering Data." Prentice Hall, Englewood Cliffs, NJ.

Jain, A. K., Murty, M. N., and Flynn, P. J. (1999). Data clustering: A review. *ACM Comput. Surv.* **31,** 264–323.

Jaqaman, K., and Danuser, G. (2006). Linking data to models: Data regression. *Nat. Rev. Mol. Cell Biol.* **7,** 813–819.

Jaqaman, K., Dorn, J. F., Jelson, G. S., Tytell, J. D., Sorger, P. K., and Danuser, G. (2006). Comparative autoregressive moving average analysis of kinetochore microtubule dynamics in yeast. *Biophys. J.* **91,** 2312–2325.

Ji, L., and Danuser, G. (2005). Tracking quasi-stationary flow of weak fluorescent signals by adaptive multi-frame correlation. *J. Microsc.* **220,** 150–167.

Kapoor, T. M., Mayer, T. U., Coughlin, M. L., and Mitchison, T. J. (2000). Probing spindle assembly mechanisms with monastrol, a small molecule inhibitor of the mitotic kinesin Eg5. *J. Cell Biol.* **150,** 975–988.

Kass, M., Witkin, A., and Terzopoulos, D. (1987). Snakes—active contour models. *Int. J. Comput. Vis.* **1,** 321–331.

Koller, T., Gerig, G., Szekely, G., and Dettwiler, D. (1995). Multiscale detection of curvilinear structure in 2-D and 3-D image data. *In* "Fifth International Conference on Computer Vision ICCV'95," pp. 864–869. IEEE Computer Society Press, Cambridge, MA.

Kusumi, A., Sako, Y., and Yamamoto, M. (1993). Confined lateral diffusion of membrane-receptors as studied by single-particle tracking (Nanovid microscopy)—effects of calcium-induced differentiation in cultured epithelial-cells. *Biophys. J.* **65,** 2021–2040.

Lasko, T. A., Bhagwat, J. G., Zou, K. H., and Ohno-Machado, L. (2005). The use of receiver operating characteristic curves in biomedical informatics. *J. Biomed. Inform.* **38,** 404–415.

Levitin, A. (2006). "Introduction to the Design and Analysis of Algorithms." Addison-Wesley, Reading, MA.

Liu, W. Y., and Dori, D. (1997). A protocol for performance evaluation of line detection algorithms. *Mach. Vis. Appl.* **9,** 240–250.

Machacek, M., and Danuser, G. (2006). Morphodynamic profiling of protrusion phenotypes. *Biophys. J.* **90,** 1439–1452.

Maddox, P., Straight, A., Coughlin, P., Mitchison, T. J., and Salmon, E. D. (2003). Direct observation of microtubule dynamics at kinetochores in Xenopus extract spindles: Implications for spindle mechanics. *J. Cell Biol.* **162,** 377–382.

Maddox, P. S., Bloom, K. S., and Salmon, E. D. (2000). The polarity and dynamics of microtubule assembly in the budding yeast Saccharomyces cerevisiae. *Nat. Cell Biol.* **2,** 36–41.

Marshall, W., Straight, A. F., Marko, J. F., Swedlow, J., Dernburg, A., Belmont, A. S., Murray, A., Agard, D., and Sedat, J. W. (1997). Interphase chromosomes undergo constrained diffusional motion in living cells. *Curr. Biol.* **7,** 930–939.

Meijering, E., Smal, I., and Danuser, G. (2006). Tracking in molecular bioimaging. *IEEE Signal Process. Mag.* **23,** 46–53.

Miyamoto, D. T., Perlman, Z. E., Burbank, K. S., Groen, A. C., and Mitchison, T. J. (2004). The kinesin Eg5 drives poleward microtubule flux in Xenopus laevis egg extract spindles. *J. Cell Biol.* **167,** 813–818.

Mogilner, M., Wollman, R., Civelekoglu-Scholey, G., and Scholey, J. (2006). Modeling mitosis. *Trends Cell Biol.* **16,** 88–96.

Neumann, B., Held, M., Liebel, U., Erfle, H., Rogers, P., Pepperkok, R., and Ellenberg, J. (2006). High-throughput RNAi screening by time-lapse imaging of live human cells. *Nat. Methods* **3,** 385–390.

Nixon, M., and Aguado, A. (2002). "Feature Extraction in Computer Vision and Image Processing." Butterworth-Heinemann/Newnes, Oxford.

Osher, S. (2003). "Geometric Level Set Methods in Imaging, Vision, and Graphics." Springer, New York, NY.

Pearson, C. G., Gardner, M. K., Paliulis, L. V., Salmon, E. D., Odde, D. J., and Bloom, K. (2006). Measuring nanometer scale gradients in spindle microtubule dynamics using model convolution microscopy. *Mol. Biol. Cell* **17,** 4069–4079.

Perlman, Z. E., Slack, M. D., Feng, Y., Mitchison, T. J., Wu, L. F., and Altschuler, S. J. (2004). Multidimensional drug profiling by automated microscopy. *Science* **306,** 1194–1198.

Phillips, P. J., Moon, H., Rizvi, S. A., and Rauss, P. J. (2000). The FERET evaluation methodology for face-recognition algorithms. *IEEE Trans. Pattern Anal. Mach. Intell.* **22,** 1090–1104.

Ponti, A., Machacek, M., Gupton, S. L., Waterman-Storer, C. M., and Danuser, G. (2004). Two distinct actin networks drive the protrusion of migrating cells. *Science* **305,** 1782–1786.

Ponti, A., Matov, A., Adams, M., Gupton, S., Waterman-Storer, C. M., and Danuser, G. (2005). Periodic patterns of actin turnover in lamellipodia and lamellae of migrating epithelial cells analyzed by Quantitative Fluorescent Speckle Microscopy. *Biophys. J.* **89,** 3456–3469.

Ponti, A., Vallotton, P., Salmon, W. C., Waterman-Storer, C. M., and Danuser, G. (2003). Computational analysis of F-actin turnover in cortical actin meshworks using fluorescent speckle microscopy. *Biophys. J.* **84,** 3336–3352.

Reid, D. B. (1979). An algorithm for tracking multiple targets. *IEEE Trans. Automatic Contr.* **24,** 843–854.

Rosa, A., Neumann, F. R., Gasser, S. M., and Stasiak, A. (2006). Long time behaviour of diffusing particles in constrained geometries; application to chromatin motion. *J. Phys. Condens. Matter* **18,** S235–S243.

Sage, D., Neumann, F. R., Hediger, F., Gasser, S. M., and Unser, M. (2005). Automatic tracking of individual fluorescence particles: Application to the study of chromosome dynamics. *IEEE Trans. Image Process.* **14,** 1372–1383.

Sawin, K. E., and Mitchison, T. J. (1991). Poleward microtubule flux in mitotic spindles assembled invitro. *J. Cell Biol.* **112,** 941–954.

Saxton, M. J. (1993). Lateral diffusion in an archipelago—single-particle diffusion. *Biophys. J.* **64,** 1766–1780.

Saxton, M. J. (1997). Single-particle tracking: The distribution of diffusion coefficients. *Biophys. J.* **72,** 1744–1753.

Saxton, M. J., and Jacobson, K. (1997). Single-particle tracking: Application to membrane dynamics. *Annu. Rev. Biophys. Biomol. Struct.* **26,** 373–399.

Sbalzarini, I. F., Mezzacasa, A., Helenius, A., and Koumoutsakos, P. (2005). Effects of organelle shape on fluorescence recovery after photobleaching. *Biophys. J.* **89,** 1482–1492.

Scott, D. W. (1992). "Multivariate Density Estimation: Theory, Practice, and Visualization." Wiley, New York, NY.

Sedgewick, R. (1992). "Algorithms in C++." Addison-Wesley Pub. Co., Reading, MA.

Shafique, K., and Shah, M. (2005). A noniterative greedy algorithm for multiframe point correspondence. *IEEE Trans. Pattern Anal. Mach. Intell.* **27,** 51–65.

Shay, T., Naffar-Abu-Amara, S., Paran, Y., Zamir, E., Liron, Y., Geiger, B., and Kam, Z. (2004). Cell-based screening for function. "IEEE International Symposium on Biomedical Imaging: Macro to Nano," Vol. 2, pp. 1243–1246. IEEE Press, Piscataway, NJ.

Sheskin, D. J. (2004). "Handbook of Parametric and Nonparametric Statistical Procedures." Chapman and Hall, Boca Raton.

Shimazaki, H., and Shinomoto, S. (2007). A method for selecting the bin size of a time histogram. *Neural Comput.* **19**(6), 1503–1527.

Smith, C., and Eisenstein, M. (2005). Automated imaging: Data as far as the eye can see. *Nat. Meth.* **2,** 547–555.

Snyder, W. E., and Qi, H. (2004). "Machine Vision." Cambridge University Press, Cambridge.

Sommerville, I. (2006). "Software Engineering." Addison-Wesley Pub. Co., London.

Sonka, M. (1999). "Image Processing, Analysis, and Machine Vision." PWS Pub., Pacific Grove, CA.

Sprague, B. L., Pearson, C. G., Maddox, P. S., Bloom, K. S., Salmon, E. D., and Odde, D. J. (2003). Mechanisms of microtubule-based kinetochore positioning in the yeast metaphase spindle. *Biophys. J.* **84,** 3529–3546.

Stanford, D., and Raftery, A. E. (2000). Finding curvilinear features in spatial point patterns: Principal curve clustering with noise. *IEEE Trans. Pattern Anal. Mach. Intell.* **22,** 601–609.

Starck, J. L., Murtagh, F., and Bijaoui, A. (2000). "Image Processing and Data Analysis: The Multiscale Approach." Cambridge University Press, Cambridge.

Stelzer, E. H. K. (2000). Practical limits to resolution in fluorescence light microscopy. *In* "Imaging Neurons" (R. Yuste, F. Lanni, and A. Konnerth, eds.), pp. 12.1–12.9. Cold Spring Harbor Press, Cold Spring Harbor, NY.

Swedlow, J. R., Goldberg, I., Brauner, E., and Sorger, P. K. (2003). Informatics and quantitative analysis in biological imaging. *Science* **300,** 100–102.

Thomann, D., Dorn, J., Sorger, P. K., and Danuser, G. (2003). Automatic fluorescent tag localization II: Improvement in super-resolution by relative tracking. *J. Microsc.* **211,** 230–248.

Thomann, D., Rines, D. R., Sorger, P. K., and Danuser, G. (2002). Automatic fluorescent tag detection in 3D with super-resolution: Application to the analysis of chromosome movement. *J. Microsc.* **208,** 49–64.

Trucco, E., and Verri, A. (1998). "Introductory Techniques for 3-D Computer Vision." Prentice Hall, Upper Saddle River, NJ.

Vallotton, P., Gupton, S. L., Waterman-Storer, C. M., and Danuser, G. (2004). Simultaneous mapping of filamentous actin flow and turnover in migrating cells by quantitative fluorescent speckle microscopy. *Proc. Natl. Acad. Sci. USA* **101,** 9660–9665.

Vallotton, P., Ponti, A., Waterman-Storer, C. M., Salmon, E. D., and Danuser, G. (2003). Recovery, visualization, and analysis of actin and tubulin polymer flow in live cells: A Fluorescence Speckle Microscopy Study. *Biophys. J.* **85,** 1289–1306.

Vazirani, V. V. (2003). "Approximation Algorithms." Springer, Berlin.

Veenman, C. J., Reinders, M. J. T., and Backer, E. (2001). Resolving motion correspondence for densely moving points. *IEEE Trans. Pattern Anal. Mach. Intell.* **23,** 54–72.

Waterman-Storer, C. M., Desai, A., Bulinski, J. C., and Salmon, E. D. (1998). Fluorescent speckle microscopy, a method to visualize the dynamics of protein assemblies in living cells. *Curr. Biol.* **8,** 1227–1230.

West, J., Fitzpatrick, J. M., Wang, M. Y., Dawant, B. M., Maurer, C. R., Kessler, R. M., Maciunas, R. J., Barillot, C., Lemoine, D., Collignon, A., Maes, F., Suetens, P., *et al.* (1997). Comparison and evaluation of retrospective intermodality brain image registration techniques. *J. Comput. Assist. Tomogr.* **21,** 554–566.

Winey, M., and O'Toole, E. T. (2001). The spindle cycle in budding yeast. *Nat. Cell Biol.* **3,** E23–E27.

Woolfson, M. M., and Pert, G. J. (1999). "An Introduction to Computer Simulation." Oxford University Press, Oxford, UK.

Yang, G., Matov, A., and Danuser, G. (2005). Reliable tracking of large scale dense antiparallel particle motion for fluorescence live cell imaging. *In* "Proc. IEEE Computer Vision and Pattern Recognition." Vol. 3, 138. IEEE Press, San Diego.

Yarrow, J., Perlman, Z., Westwood, N., and Mitchison, T. (2004). A high-throughput cell migration assay using scratch wound healing, a comparison of image-based readout methods. *BMC Biotechnol.* **4,** 21.

Yarrow, J. C., Totsukawa, G., Charras, G. T., and Mitchison, T. J. (2005). Screening for cell migration inhibitors via automated microscopy reveals a rho-kinase inhibitor. *Chem. Biol.* **12,** 385–395.

Ziauddin, J., and Sabatini, D. M. (2001). Microarrays of cells expressing defined cDNAs. *Nature* **411,** 107–110.

CHAPTER 23

Automated Classification of Mitotic Phenotypes of Human Cells Using Fluorescent Proteins

N. Harder, R. Eils, and K. Rohr

Department of Bioinformatics and Functional Genomics
German Cancer Research Center (DKFZ)
University of Heidelberg, IPMB
Im Neuenheimer Feld 267, D-69120 Heidelberg, Germany

METHODS IN CELL BIOLOGY, VOL. 85

0091-679X/08 $35.00
DOI: 10.1016/S0091-679X(08)85023-6

Abstract

High-throughput screens of the gene function provide rapidly increasing amounts of data. In particular, the analysis of image data acquired in genome-wide cell phenotype screens constitutes a substantial bottleneck in the evaluation process and motivates the development of automated image analysis tools for large-scale experiments. In this chapter, we present a computational scheme to process multicell time-lapse images as they are produced in high-throughput screens. We describe an approach to automatically segment and classify cell nuclei into different mitotic phenotypes. This enables automated identification of cell cultures that show an abnormal mitotic behavior. Our scheme proves high classification accuracy, suggesting a promising future for automating the evaluation of high-throughput experiments.

I. Introduction

The technology of RNA interference (RNAi) is an effective method to identify the biological function of genes in the field of functional genomics. Together with the availability of complete genome sequences from several organisms, RNAi enables genome-wide high-throughput screening of gene function (Friedman and Perriman, 2004). In high-throughput RNAi screens, all known genes of the observed organism are systematically silenced one after the other by inhibiting their expression and then the resulting morphological changes are analyzed. However, such large-scale knockdown experiments produce huge amounts of image data that require tools for automated evaluation.

Previous work on automated analysis of cell images has been published in different fields of application. Based on fluorescence microscopy, complete cells as well as single subcellular structures have been studied. Classification of complete cells has been performed, for example, to investigate the influence of drugs on cellular proteins (Lindblad *et al.*, 2003). There, cells are classified based on morphological characteristics of the plasma membrane which depend on the level of protein activation. Another application of complete cell classification has been considered in the work by Gallardo *et al.* (2004), where the detection of mitotic cells in images from automated microscope systems has been studied. Furthermore, work has been done on automated recognition of subcellular structures, which is a major task in location proteomics (e.g., Boland and Murphy, 2001; Conrad *et al.*, 2004; Danckaert *et al.*, 2002). In this field, the subcellular location of proteins is investigated in order to understand their function. Automated analysis of cell images also plays an increasing role in cytopathology, where computational methods have been developed to segment and classify different cell types in bright-field microscopy images of cell smears for early cancer detection (Würflinger *et al.*, 2004). Image processing workflows for the automated analysis of high-throughput screening experiments have been described, for example, by Perlman *et al.* (2004); Zhou *et al.* (2005); and Neumann *et al.* (2006). In the work by Perlman *et al.* (2004), computational methods have been used for analyzing screens to profile dose-dependent phenotypic effects of drugs in human cells. An approach for the automated analysis of RNAi screens to study cell signaling in *Drosophila* embryonic cells has been proposed by Zhou *et al.* (2005). Neumann *et al.* (2006) present an approach for analyzing images of human cells from RNAi screens. There a similar workflow is used as in our case, however, a different and simpler adaptive thresholding scheme is applied, and a smaller number of features is used for classification.

In this chapter, we present a fully automated approach to analyze multicell images from large-scale RNAi screens (for a preliminary version see Harder *et al.*, 2006). We analyze cell array images that include a large number of cell nuclei in different mitotic stages and classify the nuclei into four different phases of the cell life cycle. This enables an evaluation of the mitotic behavior of the considered cell culture over time. Our automated image analysis workflow comprises three steps: (1) segmentation of multicell images, (2) image feature extraction, and (3) classification. In order to find the most appropriate algorithm for fast and accurate segmentation, we compared three different thresholding techniques according to their segmentation quality and computation time. Our approach allows minimization of the manual evaluation effort that still slows down such large-scale experiments considerably. Furthermore, the results of an automated evaluation are objective and reproducible. Our approach has been tested using real multicell images from high-throughput screens in the framework of the EU project MitoCheck (Neumann *et al.*, 2006). This project aims to elucidate the coordination of mitotic processes in human cells at a molecular level to provide a better understanding of the mechanisms of cancer development. To identify the genes that are involved in cell division, genome-wide RNAi screens are performed. The effect of a silenced gene on mitosis is studied based on fluorescence microscopy time-lapse images of the treated cell culture. Thus for each known human gene, image sequences are acquired leading to an enormous amount of image data that has to be analyzed. An automated evaluation of the resulting image sequences requires classification and quantification of different mitotic patterns.

This chapter is organized as follows. Below, we first describe our segmentation algorithm that is based on region adaptive thresholding. Then we detail the feature extraction and classification schemes. Finally, experimental results are presented and conclusions are drawn.

II. Segmentation of Multicell Images

The high-throughput screening images in our application typically contain multiple cells (see Fig. 1). Consequently, segmentation and labeling of single objects is a crucial step. Various advanced segmentation algorithms have been described in the literature, but as computation time plays an important role when

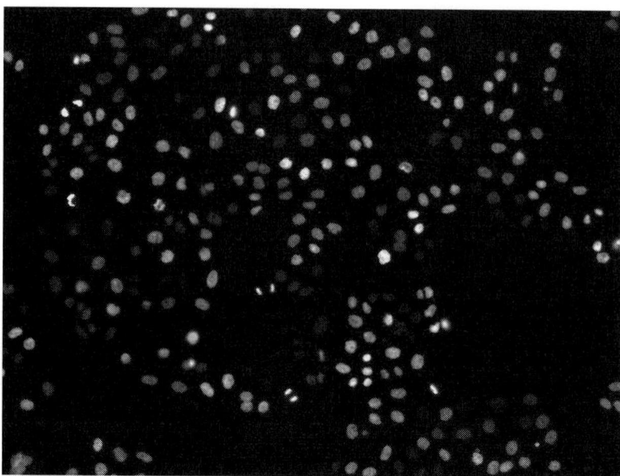

Fig. 1 Multicell image from a high-throughput experiment.

dealing with huge amounts of data, segmentation accuracy as well as speed of the algorithm are decisive criteria. We investigated three different thresholding techniques and evaluated the segmentation accuracy by manually counting correctly and incorrectly segmented cells. This manual evaluation has been performed for each tested algorithm for four different images that included in total 761 cell nuclei. The first technique (GLOBAL) is a global thresholding scheme, where the threshold is automatically calculated using Otsu's threshold selection method (Otsu, 1979) which is based on the image histogram. With this scheme it is presumed that the histogram can be divided into two distributions representing background and foreground. The optimal threshold to separate these two classes is chosen in such a way that the weighted sum of the within-class variances is minimal. However, from our experiments it turned out that a global threshold did not yield satisfying results, even after applying background correction. The reason is that the gray value intervals of background and objects overlap. As can be seen in Fig. 2A, B, using this approach the low-contrast cell nuclei are not or only partly segmented since the corresponding gray values are below the computed threshold (see the image regions marked by the ellipses). The evaluation of the segmentation accuracy resulted in only 55.9% correctly segmented cell nuclei. Note, that in all experiments we set the threshold for the minimum contrast of considered cells to about 2% of the maximum gray value of the image.

The second and the third techniques are two versions of an adaptive thresholding algorithm. This algorithm uses a quadratic sliding window to calculate local thresholds for different regions of an image. A local threshold is only calculated if the variance within the window reaches a user-defined threshold, else a global threshold is used (Gonzalez and Woods, 2002). This ensures that only for regions that have a high variance and thus contain a certain amount of information

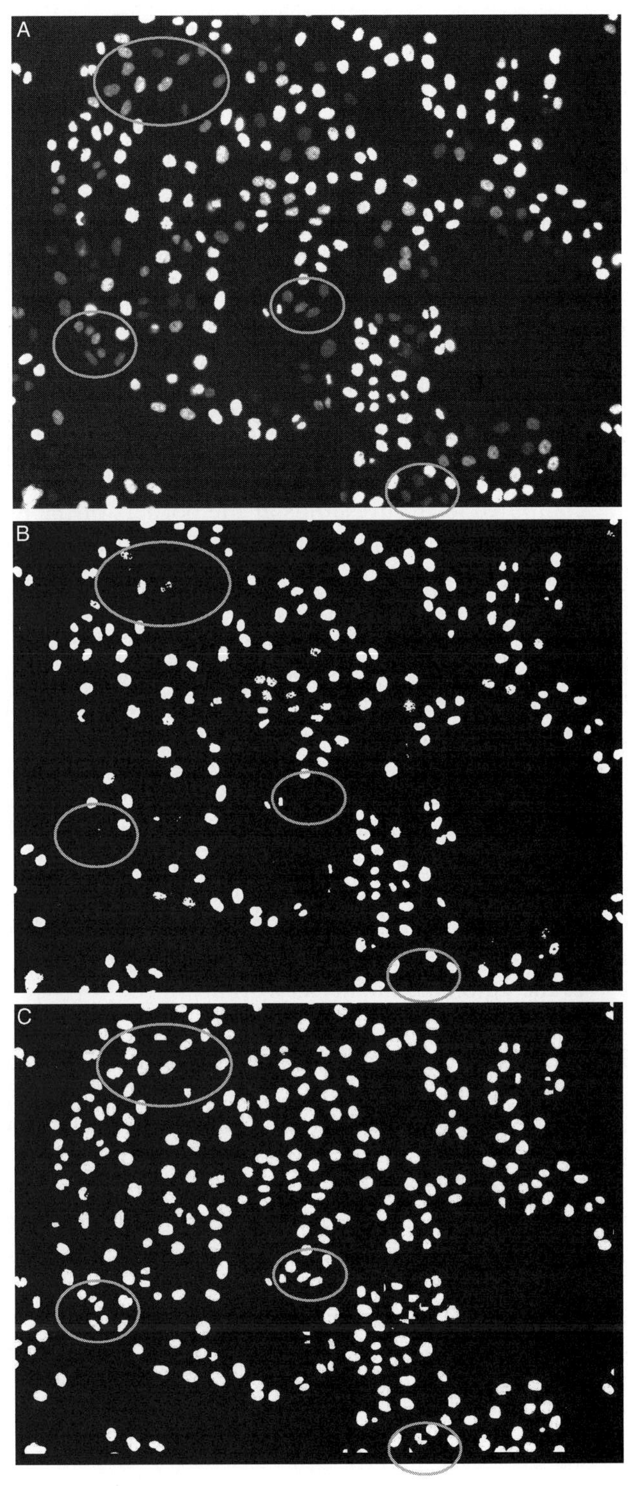
A
B
C

(e.g., regions which contain a boundary between an object and the background) the local threshold is calculated, which reduces the computation time. Global as well as local gray value thresholds are automatically calculated using Otsu's threshold selection method. In the first version of the algorithm (ADAPT-1), we apply the gray value threshold only to the central pixel of the sliding window and shift the window one pixel per step (see Fig. 3A), which leads to a high segmentation accuracy of 98.0%. Since here the sliding windows are heavily overlapping and the variance calculation has to be carried out for each pixel of the image, the computation time is rather high. In the second version of the algorithm (ADAPT-2), the threshold is applied to every pixel of the sliding window and the window is shifted one window width per step (see Fig. 3B). Thus, the single windows do not overlap and the number of variance and local threshold calculations is decreased enormously. In our experiments, this strategy reduced the computation time by a factor of 330 and still yielded 92.1% correctly segmented nuclei. The loss in segmentation accuracy is due to artifacts that result from the rapidly changing computed local thresholds at the borders of the nonoverlapping region windows. Thus, cell nuclei of all classes are affected by this decrease of accuracy to the same extent. The result of the ADAPT-2 algorithm for the previously shown example image is displayed in Fig. 2C. It can be seen that most of the low-contrast nuclei are now segmented correctly. The above-mentioned artifacts can be observed for some single cell nuclei which are only partly segmented. The absolute time consumption of each algorithm on an AMD Opteron processor machine with a clock frequency of 1.8 GHz is given in Table I. Here, we used a window width in accordance with the average nucleus diameter (e.g., 30 pixels). Since the speed of the segmentation algorithm is a major criterion in our application and an accuracy of 92.1% in this context is acceptable, we chose to adopt the second version ADAPT-2.

After image segmentation, an object labeling step is performed, holes within objects are filled, and bounding boxes are calculated. The labeled bounding boxes together with the segmented image can be used to locate and access each single nucleus of the multicell image separately. This is necessary for the following steps of the automated analysis workflow.

III. Extraction of Image Features

For analyzing the mitotic behavior of a cell culture, the number of cells in the different stages of the cell life cycle per image frame is an important information. For example, an increased number of dying cells or a higher frequency of mitotic cells gives an indication that mitosis is affected. To extract this information from the images we automatically classify the cells in each image frame using machine learning methods.

Fig. 2 (A) Original example image with marked regions of low-contrast cells (for visualization purposes the brightness and contrast has been adjusted); (B) image segmentation result using global Otsu thresholding; and (C) segmentation result using the local adaptive thresholding algorithm ADAPT-2.

To derive a quantitative description of single cell nuclei as input for a classifier, a set of robust image features has to be extracted. In our approach, we adopt a large feature set previously used for the classification of subcellular phenotypes (Conrad *et al.*, 2004) and apply it to cell nuclei from multicell array images. The feature set includes granularity features, object- and edge-related features, tree-structured wavelet features, Haralick texture features, gray scale invariants, and Zernike moments. A list of the number of features per type is given in Table II. In total, we compute 353 features per cell object.

IV. Image Features

A. Object- and Edge-Related Features

The object-related features comprise basic attributes like the area (number of pixels) of an object, the object's mean gray value, and the standard deviation of the gray values. As edge-related features we compute the contour length of an object after an edge detection step. For edge detection, Laplace and Sobel filters are used.

B. Haralick Texture Features

The Haralick texture features (Haralick, 1979) are computed on the basis of so-called co-occurrence matrices of the image. In a co-occurrence matrix, all occurring gray value pairs of pixels at a given distance and angle are counted.

Table I
Average Time Consumption in Seconds on an AMD Opteron Processor (1.8 GHz) for One Image (1344 × 1024 pixels × 12 bits) and Obtained Accuracies for the Different Thresholding Techniques

Segmentation algorithm	Global	Adapt-1	Adapt-2
Time (s)	0.0875	60.1281	0.1812
Accuracy (%)	55.9	98.0	92.1

Table II
Feature Types and Numbers of Extracted Features

Feature type	Number
Object-related	8
Edge-related	3
Haralick texture	260
Granularity	21
Tree-structured wavelets	2
Gray scale invariants	10
Zernike moments	49

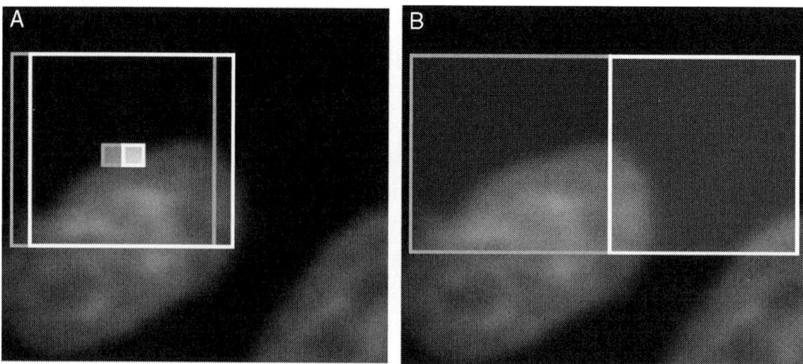

Fig. 3 Region adaptive thresholding. (A) the threshold is applied only to the central pixel of the region leading to overlapping regions (ADAPT-1). (B) the threshold is applied to every pixel of the region leading to adjacent regions (ADAPT-2).

Here, we compute co-occurrence matrices for angles of 0°, 45°, 90°, and 135° and the distances of 1–5 pixels. For each co-occurrence matrix, 13 second-order statistics (e.g., angular second moment, contrast, correlation, variance, entropy, and so on) are computed leading to 260 image features that describe the texture of an image.

C. Granularity Features

Granularity features also depend on the relation of neighboring pixel pairs. Here, the gray value differences of the center pixel to all pixels within a given distance in eight directions (0°, 45°, 90°, 135°, 180°, 225°, 270°, and 315°) are computed and the maximum difference in each direction is stored. As features, we compute the mean and the standard deviation of the maxima. The granularity features are computed for distances of 1–10 pixels.

D. Tree-Structured Wavelet Features

A wavelet transform decomposes a signal into different frequency channels. Given an image, this results in a set of subimages, where each subimage comprises a part of the whole frequency bandwidth. The resulting subimages have only half the resolution of the input image. These subimages can be further decomposed using the wavelet transform. The idea of the tree-structured wavelet transform (Chang and Kuo, 1993) is to decompose further only the significant frequency channels that contain the most information. The criterion to decide which subimages represent the most significant frequency channels is the image energy. Here, the averaged l_1-norm of the gray values is used as image energy function, that is, the sum over the magnitudes of the gray values of all image pixels divided by the number of pixels. Depending on the image size the input image is decomposed recursively several times. As features, we compute in each decomposition step the product of the highest energy value and the decomposition depth of the subimage.

E. Gray Scale Invariants

Gray scale invariant features are computed by combining a pair of neighboring pixels using a simple nonlinear kernel function (Burkhardt and Siggelkow, 2001). This function is computed for a pixel and its neighbors in all possible directions at a given distance and the results are integrated. This scheme is applied to all pixels of an image, and again the results are integrated. The applied kernel functions are the following: (1) the product of the gray values and (2) the product of the square roots of the gray values. Different gray scale invariant features can be computed using different kernel scales (i.e., different distances between center and neighboring pixels).

F. Zernike Moments

Moments are commonly used in statistics to characterize distributions and can similarly be used in image processing by considering an image region as a two-dimensional density distribution function. Moment sets of different orders and with different basis functions can be used to describe the information contained in an image region (Prokop and Reeves, 1992). Zernike moments, which use Zernike polynomials (Zernike, 1934) as moment basis set, enable the computation of image features that are invariant to translation and rotation. We calculate the Zernike moments up to degree 12 and use the moment's magnitudes as image features as proposed, for example, by Boland *et al.* (1998).

V. Classification of Mitotic Patterns

In previous work, different types of classifiers have been used for the classification of cellular and subcellular patterns. One can distinguish between unsupervised (clustering) methods and supervised methods. For unsupervised methods, a priori knowledge of the classes is not required but the results are generally of lower accuracy. Supervised methods can achieve high accuracies, but they require a learning step based on training data which associates predefined classes with image features. Since in our application we are able to define the classes, and since we also need a high classification accuracy, we use a supervised classification method. The most common supervised methods are Support Vector Machines (SVMs) (e.g., Conrad *et al.*, 2004; Huang and Murphy, 2004), Artificial Neural Networks (ANN; e.g., Boland and Murphy, 2001; Danckaert *et al.*, 2002), and Bayesian classifiers (e.g., Lindblad *et al.*, 2003; Würflinger *et al.*, 2004).

In our work we use SVMs because this classifier generally proves to have a high classification accuracy. Recently, in the context of our application we have performed a detailed comparison of six different classifiers comprising K-means clustering, Hard Competitive Learning, and Neural Gas as well as Hierarchical Clustering, SVMs, and Random Forests. In this study, it turned out that SVMs and Random Forests yielded the highest accuracy while SVMs require a significantly lower computation time (Kovalev *et al.*, 2006).

An SVM (Vapnik, 1995, 1998) is a binary classifier that finds for a set of training samples in the feature space the hyperplane which separates the two classes with the widest margin (i.e., the training samples of both classes that define the hyperplane have the greatest possible distance). If the two classes are not linearly separable, the data is mapped from the input feature space into a higher dimensional space where linear separation is possible. To this end, different kernel functions can be applied. In our case, we use a Gaussian Radial Basis Function (RBF) as kernel function. SVMs are mathematically well-founded and they are particularly suited for high-dimensional data compared to other classification methods. This property allows us to work with a high number of features and we can skip the feature selection step as it is not crucial. The binary classifier can be easily extended to a multiclass classification problem by applying multiple SVMs. For this extension different common strategies exist: "one-against-one," "one-against-all," and "Directed Acyclic Graph Support Vector Machine (DAGSVM)" (for a comparison, see e.g., Hsu and Lin, 2002). In our application we use the "one-against-one" approach. For k classes, this approach constructs $k(k-1)/2$ binary classifiers and each of them is trained with data from two classes. Then the class with most votes is assigned to the test sample. To optimize the penalty parameter C and the kernel parameter γ for the RBF, we perform a threefold cross-validation with varying values of C and γ on the training set (model selection) prior to the actual training of the classifier. The goal is to find the combination (C, γ) that leads to the highest classification accuracy (i.e., the highest percentage of correctly classified data), which is determined for each combination using cross-validation on the training set. For the threefold cross-validation, the training set is split into three equally sized subsets. The classifier is trained sequentially with two subsets and tested with the third subset for all combinations of the subsets. This way, each training sample is used once for testing. After parameter optimization, an SVM classifier with the resulting optimal parameters is trained using the complete training set and tested on a so far unseen test set. An overview of this procedure is given in Fig. 4. In our case, we use the package libSVM (Chang and Lin, 2001) for SVM classification.

VI. Experimental Results

A. Image Data

In our study, we use images that have been acquired using a wide-field microscopy screening system in the framework of the EU project MitoCheck at the European Molecular Biology Laboratory (EMBL), Heidelberg. The images are from the pilot RNAi screen (Neumann *et al.*, 2006). Chromosome morphology is visualized using a HeLa (Kyoto) cell line stably expressing the chromosomal

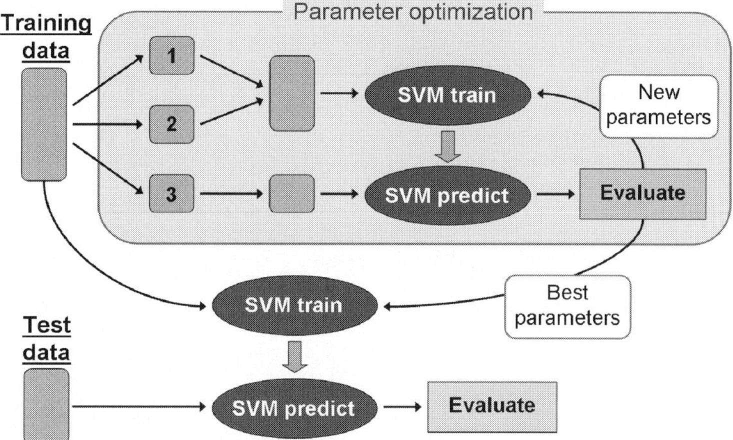

Fig. 4 Overview of the procedure for training and testing the SVM classifier.

marker histone 2B-EGFP. The cells have been imaged every 30 min for 44 h starting 20 h after seeding. In one image there are about 200–400 nuclei with an average diameter of ca.30 pixels. All images have a gray value depth of 12 bits and a resolution of 1344 × 1024 pixels. An example image can be seen in Fig. 1.

A representative set of single cell nuclei from different images taken from different experiments has been manually classified by experts at EMBL. The objects have been assigned to four classes: (1) *interphase*, (2) *mitosis*, (3) *apoptosis* (several cell death phenotypes with supercompacted or fragmented nuclei), and (4) *shape* (clustered nuclei and nuclei of abnormal shape including a high percentage of binucleated cells). Figure 5 shows example images for each class. The total number of manually classified cell objects is 637 and the number of objects per class is given in Table III.

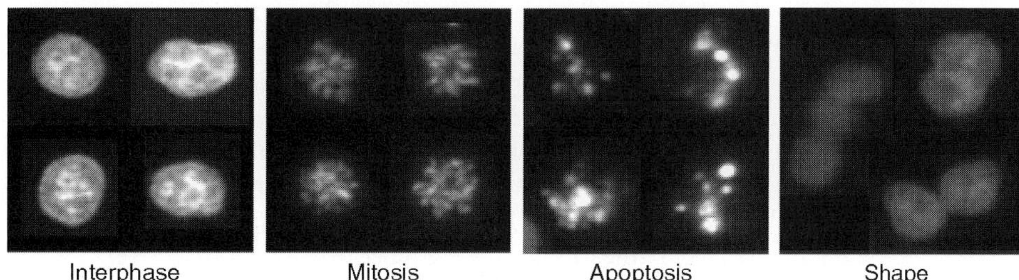

Fig. 5 Example images representing the four classes.

Table III
Numbers of Samples per Class and Set Sizes

Classes	Training set	Test set	Total
Interphase	258	64	322
Mitosis	88	22	110
Apoptosis	48	12	60
Shape	116	29	145
Total	510	127	637

B. Classification Results

We tested our approach using the above-described manually classified image data. The cell nuclei were extracted automatically from the multicell images using the segmentation approach described above. For each of the 637 cell images, we extracted 353 features as detailed above. We split the available samples for each class randomly in training data and test data at the ratio of 4:1, resulting in a training set size of 510 and a test set size of 127. Table III shows the number of training and testing samples per class. For the training set, we standardized each feature to a mean value of zero and a standard deviation of one. In the test set, the feature values were linearly transformed based on the transformation parameters from the training set. Then, we trained an SVM classifier with a Gaussian RBF kernel based on the training data set as described above. The samples of the test set were classified into the four classes, that is, *interphase*, *mitosis*, *apoptosis,* and *shape* (Fig. 5). An evaluation of the classification result yielded an overall classification accuracy of 96.9%. Thereby, 123 out of the 127 test set samples were correctly classified.

Since our test set was relatively large, the result obtained from this classification can already be considered to be significant. In order to check the reliability of the result, we repeated the classification step applying a tenfold outer cross-validation on the whole data set of 637 images. As already described above for the threefold cross-validation, in each loop of the tenfold cross-validation we train the classifier with nine subsets and test it with the remaining one (for an overview of the tenfold cross-validation procedure see Fig. 6). Thus, the training set is virtually enlarged to comprise the whole dataset which makes the results more general and more reliable. Based on this procedure the classification yielded an overall accuracy of 96.1% (i.e., 612 out of the 637 samples were correctly classified). Thus, both classification results correspond very well and we can draw the conclusion that we can rely on an overall classification accuracy of around 96%.

We have also determined the confusion matrices for both classification experiments (see Tables IV and V). The confusion matrices represent the frequencies of confusions between the different classes. The true classes are listed in the rows of each matrix and the output of the classifier is given in the columns. Thus, the

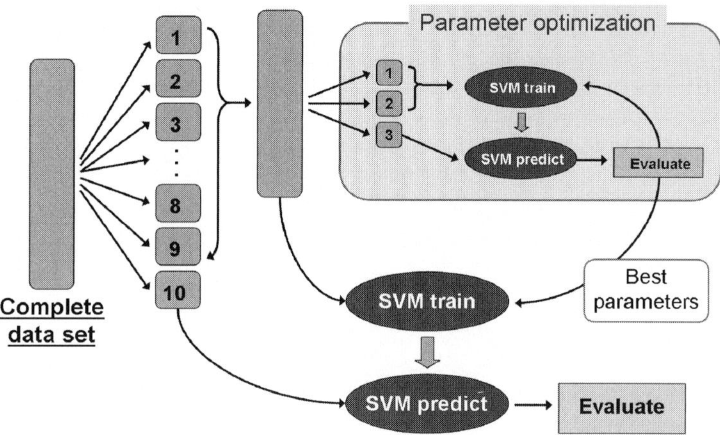

Fig. 6 Overview of the tenfold outer cross-validation procedure to determine the classification accuracy.

Table IV
Confusion Matrix for 510 Training and 127 Test Samples and Classification Accuracies per Class

	Classifier output				
True class	Interphase	Mitosis	Apoptosis	Shape	Accuracy
Interphase	64	0	0	0	100%
Mitosis	0	20	2	0	91%
Apoptosis	0	2	10	0	83%
Shape	0	0	0	29	100%

Overall accuracy: 96.9%.

diagonal of the matrix represents the correct classified samples. In Table V, for example, we can see that 10 samples of the class *mitosis* have been wrongly assigned to the class *apoptosis*. By contrast, 12 samples of the class *apoptosis* have been wrongly classified as *mitosis*. Besides that there are only very few other misclassifications. The same tendency can be observed in Table IV. Here, misclassifications appeared only between the classes *mitosis* and *apoptosis*. Taking a closer look at the samples of these two classes indicates a reason for the relatively high confusion probability: The samples from these two classes often look very similar. Sometimes the samples seem to be nearly indistinguishable even for a trained human observer as illustrated by Fig. 7. In this case, a classification on single images does not seem to be possible; however, a human observer can exploit the information from the previous and consecutive time frames to determine the correct class.

Table V
Confusion Matrix for Tenfold Cross-Validation on 637 Samples and Classification Accuracies per Class

	Classifier output				
True class	Interphase	Mitosis	Apoptosis	Shape	Accuracy
Interphase	320	1	0	1	99.38%
Mitosis	0	100	10	0	90.91%
Apoptosis	1	12	47	0	78.33%
Shape	0	0	0	145	100%

Overall accuracy: 96.1%.

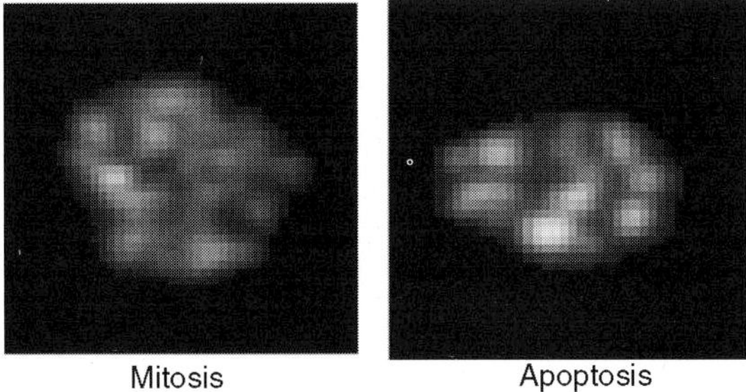

Fig. 7 Example images of the classes *mitosis* and *apoptosis* which are very similar (classification based on manual classification by an expert).

Another reason for the generally lower classification accuracy of the *apoptosis* class is probably the comparatively small number of samples for this class (60 samples, see Table III).

VIII. Conclusion

We have presented an approach for automated recognition of mitotic patterns which can handle multicell images and classifies the segmented cell nuclei with a high accuracy into four classes. Applying this approach for subsequent images of image sequences of cell cultures provides useful information about the mitotic behavior throughout the sequences. Our approach enables fast and reliable automatic analysis of large-scale high-throughput RNAi screens.

From our experimental results, it turned out that we reach an overall classification accuracy of more than 96% and that the classes *interphase* and *shape* (clustered nuclei and multinuclear cells) are recognized with an accuracy of nearly 100%. It also turned out that the classes *mitosis* and *apoptosis* are sometimes hard to distinguish on the basis of single images. In future work this problem could be solved in an analogous manner as a human expert by incorporating information from adjacent time frames. To this end a tracking scheme is required which determines the temporal connections between nuclei in consecutive images. Another issue for further study is the application of feature selection methods to determine those features with the highest discriminative power and to use only those for classification. Even though this step is not crucial for the classification accuracy when using SVMs, it is a possibility to further reduce the computation time.

Acknowledgments

This work has been supported by the EU project MitoCheck. We thank B. Neumann, M. Held, U. Liebel, H. Erfle, and J. Ellenberg (European Molecular Laboratory, EMBL, Heidelberg) for providing the image data and the ground truth for the classification as well as for discussions.

References

Boland, M. V., Markey, M., and Murphy, R. F. (1998). Automated recognition of patterns characteristic of subcellular structures in fluorescence microscopy images. *Cytometry* **33,** 366–375.

Boland, M. V., and Murphy, R. F. (2001). A neural network classifier capable of recognizing the patterns of all major subcellular structures in fluorescence microscope images of HeLa cells. *Bioinformatics* **17**(12), 1213–1223.

Burkhardt, H., and Siggelkow, S. (2001). "Invariant Features in Pattern Recognition—Fundamentals and Applications." John Wiley & Sons, New York.

Chang, C.-C., and Lin, C.-J. (2001). "LIBSVM: A Library for Support Vector Machines." Software available athttp://www.csie.ntu.edu.tw/cjlin/libsvm.

Chang, T., and Kuo, C.-C. J. (1993). Texture analysis and classification with tree-structured wavelet transform. *IEEE Trans. Image Process.* **2**(4), 429–441.

Conrad, C., Erfle, H., Warnat, P., Daigle, N., Lörch, T., Ellenberg, J., Pepperkok, R., and Eils, R. (2004). Automatic identification of subcellular phenotypes on human cell arrays. *Genome Res.* **14,** 1130–1136.

Danckaert, A., Gonzalez-Couto, E., Bollondi, L., Thompson, N., and Hayes, B. (2002). Automated recognition of intracellular organelles in confocal microscope images. *Traffic* **3,** 66–73.

Friedman, A., and Perriman, N. (2004). Genome-wide high-throughput screens in functional genomics. *Curr. Opin. Genet. Dev.* **14,** 470–476.

Gallardo, G., Yang, F., Ianzini, F., Mackey, M., and Sonka, M. (2004). Mitotic cell recognition with hidden Markov models. *In* "Medical Imaging 2004: Visualization, Image-Guided Procedures, and Display, Proc SPIE" (J. R. L. Galloway, ed.), Vol. 5367, pp. 661–668, SPIE, Bellingham, WA.

Gonzalez, R. C., and Woods, R. E. (2002). "Digital Image Processing." Prentice Hall, 2nd edn. New Jersey, NJ.

Haralick, R. M. (1979). Statistical and structural approaches to texture. *Proc IEEE* **67,** 786–804.

Harder, N., Neumann, B., Held, M., Liebel, U., Erfle, H., Ellenberg, J., Eils, R., and Rohr, K. (2006). Automated recognition of mitotic patterns in fluorescence microscopy images of human cells. *In* "Proceedings of the IEEE International Symposium on Biomedical Imaging: From Nano to Macro (ISBI'06)" (J. Kovačević and E. Meijering, eds.), pp. 1016–1019. Arlington, VA, April 2006, IEEE, New York.

Hsu, C.-W., and Lin, C.-J. (2002). A comparison of methods for multiclass support vector machines. *IEEE Trans. Neural Netw.* **13,** 415–425.

Huang, K., and Murphy, R. F. (2004). Automated classification of subcellular patterns in multicell images without segmentation into single cells. *In* "Proceedings of the IEEE International Symposium on Biomedical Imaging: From Nano to Macro (ISBI '04)" (R. M. Leahy and C. Roux, eds.), pp. 1139–1142. Arlington, VA, April 2004, IEEE, New York.

Kovalev, V., Harder, N., Neumann, B., Held, M., Liebel, U., Erfle, H., Ellenberg, J., Eils, R., and Rohr, K. (2006). Feature selection for evaluating fluorescence microscopy images in genome-wide cell screens. *In* "Proceedings of the IEEE Computer Society Conference on Computer Vision and Pattern Recognition (CVPR'06)" (C. Schmid, S. Soatto, and C. Tomasi, eds.), pp. 276–283. New York, NY, June 2006, IEEE, New York.

Lindblad, J., Wählby, C., Bengtsson, E., and Zaltsman, A. (2003). Image analysis for automatic segmentation of cytoplasms and classification of Rac1 activation. *Cytometry A* **57A,** 22–33.

Neumann, B., Held, M., Liebel, U., Erfle, H., Rogers, P., Pepperkok, R., and Ellenberg, J. (2006). High-throughput RNAi screening by time-lapse imaging of live human cells. *Nat. Methods* **3,** 385–390.

Otsu, N. (1979). A threshold selection method from grey level histograms. *IEEE Trans. Syst. Man Cybern.* **9,** 62–66.

Perlman, Z., Slack, M., Feng, Y., Mitchison, T., Wu, L., and Altschuler, S. (2004). Multidimensional drug profiling by automated microscopy. *Science* **306,** 1194–1198.

Prokop, R. J., and Reeves, A. P. (1992). A survey of moment-based techniques for unoccluded object representation and recognition. *CVGIP: Graph. Models Image Process.* **54**(5), 438–460.

Vapnik, V. (1995). "The Nature of Statistical Learning Theory." Springer-Verlag, New York.

Vapnik, V. (1998). "Statistical Learning Theory." John Wiley and Sons, New York.

Würflinger, T., Stockhausen, J., Meyer-Ebrecht, D., and Böcking, A. (2004). Robust automatic coregistration, segmentation, and classification of cell nuclei in multimodal cytopathological microscopic images. *Comput. Med. Imaging Graph.* **28,** 87–98.

Zernike, F. (1934). Beugungstheorie des Schneidenverfahrens und seiner verbesserten Form, der Phasenkontrastmethode. *Physika* **1,** 689–704.

Zhou, X., Liu, K.-Y., Bradley, P., Perrimon, N., and Wong, S. T. (2005). Towards automated cellular image segmentation for RNAi genome-wide screening. *In* "Proceedings of the Eighth International Conference on Medical Image Computing and Computer-Assisted Intervention (MICCAI'2005)" (J. Duncan and G. Gering, eds.), Vol. 3749 of LNCS, (Palm Springs), CA pp. 302–309 Springer-Verlag, Berlin, Heidelberg, Oct. 2005.

CHAPTER 24

Open Tools for Storage and Management of Quantitative Image Data

Joshua Moore, Chris Allan, Jean-Marie Burel, Brian Loranger, Donald MacDonald, Jonathan Monk, and Jason R. Swedlow

Division of Gene Regulation and Expression
College of Life Sciences
Wellcome Trust Biocentre
University of Dundee, Dundee
Scotland DD1 5EH
United Kingdom

Abstract

The explosion in quantitative imaging has driven the need to develop tools for storing, managing, analyzing, and viewing large sets of data. In this chapter, we discuss tools we have built for storing large data sets for the lifetime of a typical

0091-679X/08 $35.00
DOI: 10.1016/S0091-679X(08)85024-8

research project. As part of the Open Microscopy Environment (OME) Consortium, we have built a series of open-source tools that support the manipulation and visualization of large sets of complex image data. Images from a number of proprietary file formats can be imported and then accessed from a single server running in a laboratory or imaging facility. We discuss the capabilities of the OME Server, a Perl-based data management system that is designed for large-scale analysis of image data using a web browser-based user interface. In addition, we have recently released a lighter weight Java-based OME Remote Objects Server that supports remote applications for managing and viewing image data. Together these systems provide a suite of tools for large-scale quantitative imaging that is now commonly used throughout cell and developmental biology.

I. Introduction

Quantitative digital imaging assays have now become universal in cell and developmental biology and are one of the fundamental techniques for the growing field of systems biology. As new imaging methods extend the resolution of spatial, temporal, and spectral dimensions and applications expand into large-scale genome-wide and chemical library screens, there is a growing requirement for tools that support the storage, archiving, analysis, visualization, and querying of large multidimensional sets of data. The new age of quantitative biology requires the informatics tools to provide methods of accessing the quantitative data at its foundation (Swedlow *et al.*, 2006).

In some cases, the success of the genomics projects provides a good template for the management of quantitative imaging data. Central repositories for shared data sets using defined formats for data transfer have worked extremely well for projects where a single reference set of data is to be shared by a large community (e.g., the Model Organism Databases—http://www.ncbi.nlm.nih.gov/entrez/query.fcgi?db=Genome). Indeed, similar image-based resources that are intended to be community-wide resources are now being released (Christiansen *et al.*, 2006; Lein *et al.*, 2007). However, experimental, hypothesis-driven cell and developmental biology also make increasing use of multidimensional imaging, yet it is often not clear initially which primary data represents a "reference" data set that will be used for a publication or what combination of data is appropriate for sharing. To complicate matters, the range and complexity of data collected for a single project—the links between original and processed images, analytic results, experimental descriptions, and final summarized results are only maintained manually by experimental biologists (usually dedicated but itinerant Ph.D. and postdoctoral students). The fact that such data is often physically stored on-line on CD or DVD media (with limited life spans) only adds to the difficulty of the problem.

In this chapter, we discuss methods we have built and used for physically storing quantitative image data in a safe, secure, and accessible method, and then discuss the much harder problem of building software tools for managing, visualizing, and querying image.

II. Secure, Archived and Available Storage for Biological Image Data

Many of the debates about image data management center around standardized file formats and image databases. However, in most cases, even if these tools were in place, most laboratories and imaging facilities do not have the means to store the volume of data generated by their microscopes in a manageable and affordable way. The requirements for such a storage system are not simple to satisfy. Data must be stored securely and available for ~2–5 years, the lifetime of most projects. Image data files are often multiple gigabytes, so the system must be able to store and deliver these very large files. The entire system including the physical media, the data servers, networking, and data transfer facilities must be easily expandable to meet the needs of new experiments. In addition, most microscopy facilities will require nightly backups so that reasonable disaster recovery is possible.

At the Wellcome Trust Biocentre, University of Dundee, we have installed a Storage Area Network (SAN) based on a hierarchical file system composed of pools of high-performance disk arrays and redundant tape libraries for data archiving and backup. The layout of the system is shown in Fig. 1. The overall cost of the system is minimized by extensive use of Serial ATA drives within fiber channel enclosures. The disk arrays are connected by fiber channel to Linux x86 servers that provide access and manage all the data in the system, ensuring 25% of the available disk space is free at all times for storage of new data. Additional servers provide connections to two tape robots that keep redundant, physically distinct copies of the whole file system as backups. As new data are stored on the SAN system, older, less accessed data are moved through the storage hierarchy to the one of the tape libraries. This not only ensures that all data from a project is stored and available but also ensures that the most accessed data is kept available for user access. This hierarchical strategy is cost efficient and eliminates the need for users to archive their own data using CDs and DVDs. The SAN supports access from any computer in the center, regardless of location or operating system. In addition, the installation of a high-speed redundant network allows data to be recorded directly onto the SAN without any compromise in data acquisition performance. Perhaps most importantly, the installation of a truly enterprise solution ensures that expansion of the storage capacity of the system is easily accomplished by simply buying more physical storage media.

In our experience, this type of storage system that is normally found in commercial data centers is critical for multidimensional imaging and quantitative analysis.

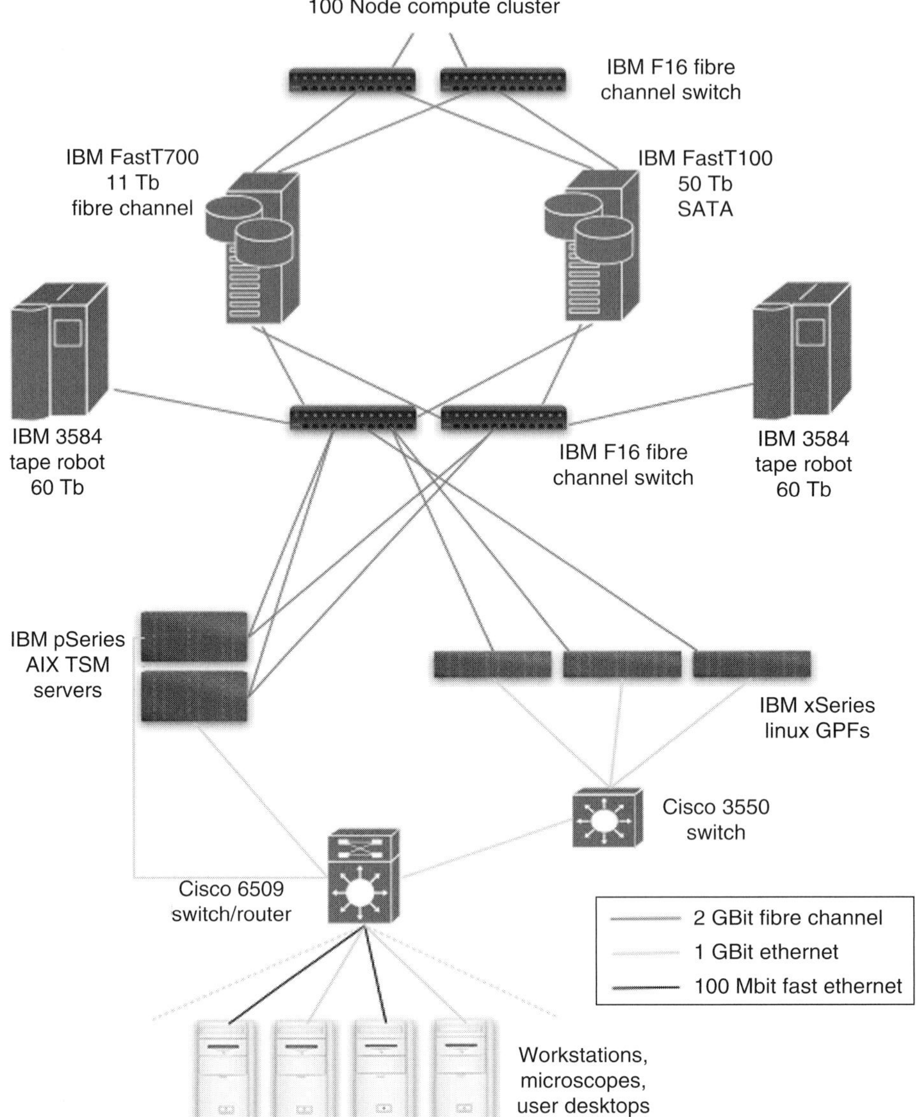

Fig. 1 Simple architecture diagram for the SAN used for image storage at Dundee. A hierarchical file system, run using Tivoli Storage Manager installed on two IBM pSeries servers, manages all backup and archiving. All data is stored live either on the fiber channel or on the SATA disk systems, and then automatically migrated to tape in archive form if not accessed for a significant period. Data is accessed from user workstations and desktops via a high-performance network and the SAN is connected to a compute cluster for computationally demanding processing. (See Plate no. 70 in the Color Plate Section.)

Moreover, once installed, this system can be used to support other data intensive activities in a research center like proteomics and bioinformatics.

III. The Open Microscopy Environment: Data Management Tools for Biological Research

The Open Microscopy Environment (OME) is a multisite software development consortium (http://openmicroscopy.org/about/) dedicated to building open, freely available software tools for biological image data management. Originally founded in 2000 by Peter Sorger, Jason Swedlow, and Ilya Goldberg, the project now has grown to include over 20 developers and designers in the United States and the United Kingdom (http://openmicroscopy.org/about/). The project is attempting to provide a complete set of specifications and tools for biological imaging research, which is a significant task. As detailed on the project website (http://openmicroscopy.org), OME is releasing file format specifications, software tools to support these specifications, and software tools for data management. All tools are open and available and the project maintains mailing lists (http://openmicroscopy.org/getting-involved/) to ensure that it can capture feedback from users and external developers.

In the past six years, the OME Consortium has released a series of software tools, specifications, and documents:

- A white paper summarizing the goals and the concepts of the project (Swedlow *et al.*, 2003);
- A description of the data model (http://openmicroscopy.org/XMLschemas/OME/latest/ome_xsd/) used in the OME project;
- A specification for a metadata format, OME-XML, that can be used for sharing image data and metadata between laboratories and collaborators (Goldberg *et al.*, 2005);
- A new file format, OME-TIFF, that serves as a high-performance data file format for image data acquisition and visualization (http://www.loci.wisc.edu/ome/ome-tiff.html (Eliceiri and Rueden, 2005);
- A software tool, Bio-Formats, that converts over 40 proprietary file formats into OME-TIFF (http://www.loci.wisc.edu/ome/formats.html);
- A package of software tools, OME Server (versions 2.0–2.6), that provide data management and analysis tools for biological imaging (Johnston *et al.*, 2006; Schiffmann *et al.*, 2006; Swedlow *et al.*, 2003). The OME Server (latest version is 2.6.0 as of this writing) is available at http://cvs.openmicroscopy.org.uk; and
- A new Java-based server that supports remote client applications providing a range of image visualization and management tools.

The interested reader can consult online and published resources for all aspects of the OME project. In this chapter, we focus briefly on the OME Server and then describe in depth the latest technology released by the OME Consortium, the OME Remote Objects (OMERO) Server, and its client applications.

A. The OME Server

The OME Server is a Perl-based application running on top of a PostgreSQL database system (http://www.postgresql.org) that supports upload of image data from a variety of proprietary file formats; organization of image data into different types of hierarchies; annotation of individual images, and visualization of all image data and metadata in a web browser-based user interface (http://openmicroscopy.org/getting-started/web-client.html (Johnston *et al.*, 2006; Schiffmann *et al.*, 2006; Swedlow *et al.*, 2003). Most importantly for the purposes of the discussion that follows, the OME Server includes facilities for running and managing image analysis modules. Analysis modules written outside of OME and without any awareness of the OME Server can be accessed by the OME Server using customized code (technically, this is called "wrapping" the application code). The OME Server contains facilities for updating its data model to store new data types, so new analyses can use the OME database to store any generated results. Most recently, this OME Server has been used to define and measure components of the human kinetochore (Schiffmann *et al.*, 2006) and has been modified to support the commercial data analysis tool MATLAB, allowing user-created analysis tools to be easily integrated into the OME Server (http://www.openmicroscopy.org/howto/quantitative-image-analysis-MATLAB.html).

Integration of analysis modules into an OME Server provides two major advantages. For analyses that require a series of sequential steps, the data and parameter inputs and outputs for each step are stored in an OME database that facilitates the provision of output from one module as input for the next. In addition, the results from any analysis are available in a relational database and can be queried, either directly using SQL or using database querying tools like Microsoft Excel or Spotfire. In fact, an Excel-based file for querying and downloading analysis data from an OME Server is available (OME-Excel; http://cvs.openmicroscopy.org.uk/horde/chora/browse.php?f=OME%2Fsrc%2FExcel%2FOME-Excel.xls). Combining image data import, a relational database for metadata and analytics, and analysis module management system and an external querying tool like OME-Excel completes a standard workflow for image visualization and quantitative analysis. A step-by-step guide for this approach is available (http://www.openmicroscopy.org/howto/FindSpots-v2.pdf). Together, these tools provide a complete solution for management and automated analysis for large sets of image data.

IV. The OMERO Server: A New Server Application for Data Management

A. Rationale for a New Server

Developing new tools for new applications is always a challenge; however, the OME Consortium is developing its tools to enhance its own work and simultaneously releasing tools to support the world-wide community of imaging laboratories. Many laboratories generate software that they use themselves but releasing software for a large community requires much more robust testing, documentation, and consideration of diverse sets of requirements. It is also critical that released software is updated to adapt to changes in operating systems and the software libraries upon which it depends. In the case of the OME Server, there are a number of custom-built functions that are both large and have many external library dependencies. An example is our custom-built object-relational mapping (ORM) code ("DBObject") that provides communication between the underlying database with the OME Server. Over the last two years, as new Linux and Perl versions were released, DBObject became incompatible with new versions of various Perl modules. This problem is by no means insurmountable, and DBObject was appropriately updated, but this is a significant code maintenance burden. Although DBObject worked well, there are simply other development teams who specialize in ORM and, for instance, have dedicated their efforts to providing a tool that can easily switch between different relational database management systems. One of the most successful is Hibernate (http://www.hibernate.org) that is an open-source library written in Java. Using Hibernate would significantly reduce our code maintenance burden and also provide substantial new flexibility for choosing different database systems including MySQL (http://www.mysql.com) and Oracle (http://www.oracle.com). This combination of flexibility and reduced code maintenance is critical for a small, open-source effort—code maintenance is a critical activity to ensure that users can continue to use released software. As the burden of maintenance grows, more and more precious developer time is spent on this activity, and less is spent on developing new tools for users.

A major goal of OME is the support and provision of remote client applications, allowing users to manage, view, and analyze images in a location and platform-independent manner. The OME Server, written in Perl, required an interface called OME-JAVA, written and maintained by the OME Consortium. OME-JAVA supports the Java-based Shoola client application, communicating to an OME Server. This communication occurs via a data transport mechanism (XML-RPC) that matches a common standard in the community, but unfortunately causes severe performance problems when large data graphs are passed from the OME Server to the Shoola client. Moreover, OME-JAVA provides an interface for Java clients but provides no support for other languages—C++, .NET, Python, etc., that were required by the community (Fig. 2).

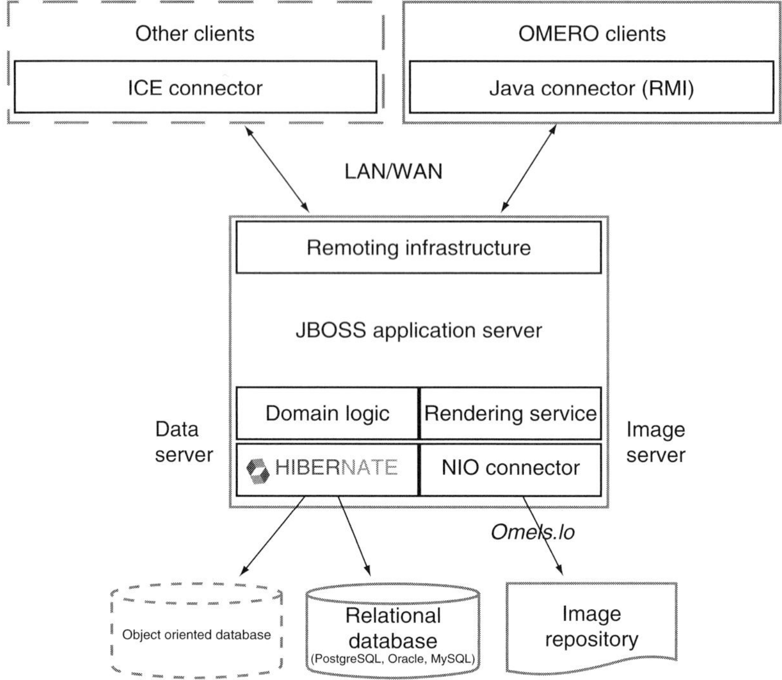

Fig. 2 Architectural diagram of the OMERO system. In the Beta1 release (http://trac.openmicroscopy.org.uk/omero/MilestoneDownloads), the OMERO Server is deployed within a JBOSS Application Server (http://www.jboss.org), using Hibernate (http://www.hibernate.org) for ORM, and PostgreSQL http://www.postgresql.org) as a relational database management system. The image server and the Rendering Engine provide remote clients with either quantitative or rendered versions of pixel data. In the latter case, this substantially reduces the bandwidth required for transfer to a remote client and the computational requirements on the client computer. Remote communication in OMERO3.0-Beta1 is achieved using Java Remote Method invocation (RMI; http://www.java.co/rmi) to support Java-based clients. In Beta2 (scheduled for release in May 2007), a remoting interface based on ICE (http://www.zeroc.com) will be added to support a diverse set of remote client environments (C++, .NET, Python, CORBA, etc.).

Why are remote clients important? Given the complexity of multidimensional image data and imaging experiments, visualization, management annotation, and analysis all require complex software tools to support user's requirements. There are many powerful existing software packages that OME's tools should at least try to support. Flexible support for a variety of application environments with good performance for complex sets of data is thus a critical requirement for an image data management system.

B. Design Criteria for OMERO

For all of these reasons, we decided to design and build a new server, the OMERO Server that would port much of the functionality in the Perl-based OME Server to Java. In taking on this project, it was clear that the code maintenance burden must be substantially reduced, the system must be simple to install, and the performance of the remoting system must be significantly improved. A major design goal was the reduction of self-written code through the reuse of existing middleware and ORM tools where possible. Numerous successful middleware projects have grown up since the first years of OME and it only made sense to use them. Rather than writing and maintaining our ORM or remoting framework, we can integrate off-the-shelf components and, where necessary and/or possible, contribute back to those projects. After extensive research, it became clear that a number of facilities were available for Java that matched well with the requirements of a server for image data and metadata. Moreover, no other main stream programming language has the same level of widely used specifications and high-performance libraries available for the tasks at hand.

We chose the JavaEE5 specification because of its simplified programming model (and therefore increased maintainability) despite the fact that it was still in the specification stage. After the relative success of JavaEE4, we expected that JavaEE5 would become equally widespread. In addition, the choice of Java for the server was fairly straightforward. No other main stream programming language had the same level of widely used specifications and high-performance libraries available for the tasks at hand. While Java is not the fastest environment, performance of the database system, disk access, and network bandwidth will limit the speed of a Java-based image data server. Given the amount of work in JavaEE middleware, if a bottleneck is discovered in one of the Java components, a host of other solutions are available, simply because of the standardized specifications in the JavaEE environment.

Our goal in building a new server was not to simply replicate the functionality in the existing OME Server, but instead to generate a parallel project that would allow complementary technology. This strategy would certainly create potential difficulties in the short term for users, as desired functionality was split across the two server systems. For example, one of the major features of the OME Server is the use of a dynamic data model, with associated software that supports on-the-fly updates of the database, the application programming interface (API), and the web browser-based user interface included in the OME Server. This feature allows individual scientists to customize the OME Server system to their own needs. This type of functionality is much harder to support in a system designed to support remote applications that could include different applications written in different programming environments. For ease of entry into a system that nevertheless provides significant data management and visualization, the loss of dynamicity and flexibility, which are the hallmarks of OME Server, is an acceptable compromise, at least for the short term. Loss of this facility makes the OMERO Server simpler, and many of the currently missing features, like running chains of analysis modules, can be added back in a controlled way using any one of a wide variety of workflow tools available in Java (http://java-source.net/open-source/workflow-engines). This strategy allows the stepwise development of a new application that can eventually be integrated with the functionality in the OME Server.

Perhaps most importantly from the user's perspective, the installation and use of the OMERO system must be relatively easy. Users are now accustomed to automatic installation systems, where all dependencies are included and installation proceeds reliably, without fail or perturbation of the user's computer. However, installation of a database system, application server, server applications and clients is somewhat more complex than installing a single standalone application. With the growing maturation of Unix-based operating systems (e.g., OS X and Linux), providing the required level of functionality is relatively easy, and we have taken advantage of many of these facilities in building the OMERO distributions (see http://trac.openmicroscopy.org.uk/omero/MilestoneDownloads). As a result, the major burdens for the system administrator are those of installing a relational database like PostgreSQL (http://www.postgresql.org), creating directories, and dealing with user administration.

C. Technical Details of the OMERO Server

Having described the rationale and design of the OMERO Server, the following sections outline the technical aspects of the different components of the OMERO Server.

1. OMERO Metadata Services

The OMERO Server consists of services implemented as EJB3 session beans (http://java.sun.com/products/ejb) that can operate in any JavaEE5 application server. These services make use of Hibernate (http://www.hibernate.org), a high-performance ORM solution, to retrieve metadata from the database. The OMERO services form a layered interface for controlled interaction with the database. The most basic layer consists of ome.api.IQuery and ome.api.IUpdate that provide object-oriented versions of the established SQL verbs SELECT, INSERT, UPDATE, and DELETE with additional logic for security and auditing (http://cvs.openmicroscopy.org.uk/tiki/tiki-index.php?page=Omero API). IUpdate accepts arbitrary graphs of entities defined by the OME specification and guarantees transitive persistence of the entire graphs. All new entities associated with the graph as well as all changes to existing entities are detected by the backend and saved to the database if allowed. The security checks take place on a row basis and function similarly to the Unix file system. Each entity has an owner, a group, and a permissions settings, which defaults to RWR-R- ("read/write for user, read for group, read for world"). Once authorized, all changes to the database are logged in Event table.

The central feature of IQuery is the storage of Hibernate Query Language (HQL) queries on the server that can then be referenced by name, permitting code reuse, and sharing. SELECT security throws an exception if an unreadable entity is directly requested, and otherwise inserts statements into the SQL for collections to filter unreadable objects. Reads are currently not logged for performance reasons. Except for a few methods in the administration interface, all other interactions with the OMERO Server are built from these simple premises, but can become arbitrarily complex. See ome.api.IPojos (http://trac.openmicroscopy.org.uk/omero/browser/trunk/components/common/src/ome/api/IPojos.java) for an example of a well-rounded API that guarantees that the entity graphs returned do not contain nulls. A unique benefit of the OMERO design is that these metadata services sit in the same process as the binary data services, which can make use of preloaded metadata to efficiently parse data from disk.

2. OMERO Image Rendering Engine

A major requirement for any image data application is the ability to display images. In most applications, this is achieved by reading pixel data from a file system and then mapping the pixel data to the 256 gray level available on most computer display monitors. In fluorescence imaging, it is common to record and display multiple channels at once. Typically three, four, or even five separate images must be mapped, and then presented as a color image for painting on a monitor. Because these operations can require many thousands of operations and must be displayed rapidly to support the display of time-lapse movies, most image display software applications use a high-speed graphics CPU and dedicated hardware for image rendering and display. This requirement limits the deployment of these applications to high-powered workstations.

In both the OME and OMERO Servers, an image server—a software application that delivers rendered images to a client—is included that ensures that client applications can display image data. In both cases, this image server reads binary pixel data (sets of "Pixels") from the server's file system, renders them based on defined settings, and then transfers the rendered image, ready for display, to the client. Rendering is defined as the process of mapping the raw, multidimensional pixel data from a biological imaging system (e.g., a microscope) to a format that is suitable for visualization on a personal computer monitor or laptop display. The main goal of any visualization system is to provide data for display at speeds that meet or exceed real time.

The OMERO Rendering Engine (OMERO-RE) has been designed to minimize the amount of data transferred to the client and thus removes the requirement for specific graphics CPU, allowing high-performance image viewing on standard laptop computers. The OMERO-RE achieves this by limiting data transfer times because of being close to the data, using highly efficient network transfer protocols, utilizing modern multiprocessor and multicore machines and providing the data to clients in a format that is efficient to display as possible.

When displaying large sets of data, an image visualization system must ensure that an initial viewing of an image and a calculated thumbnail represents a reasonable or "pretty good" image. For this purpose, the OMERO-RE uses statistical parameters that are calculated during image import and stored in the OMERO database as the basis for initial rendering settings of all images. The details of this statistic-based image rendering will be discussed in a separate publication. Using this strategy, RE and OMERO clients provide high-performance image visualization and thumbnailing in a remote client (Fig. 3C).

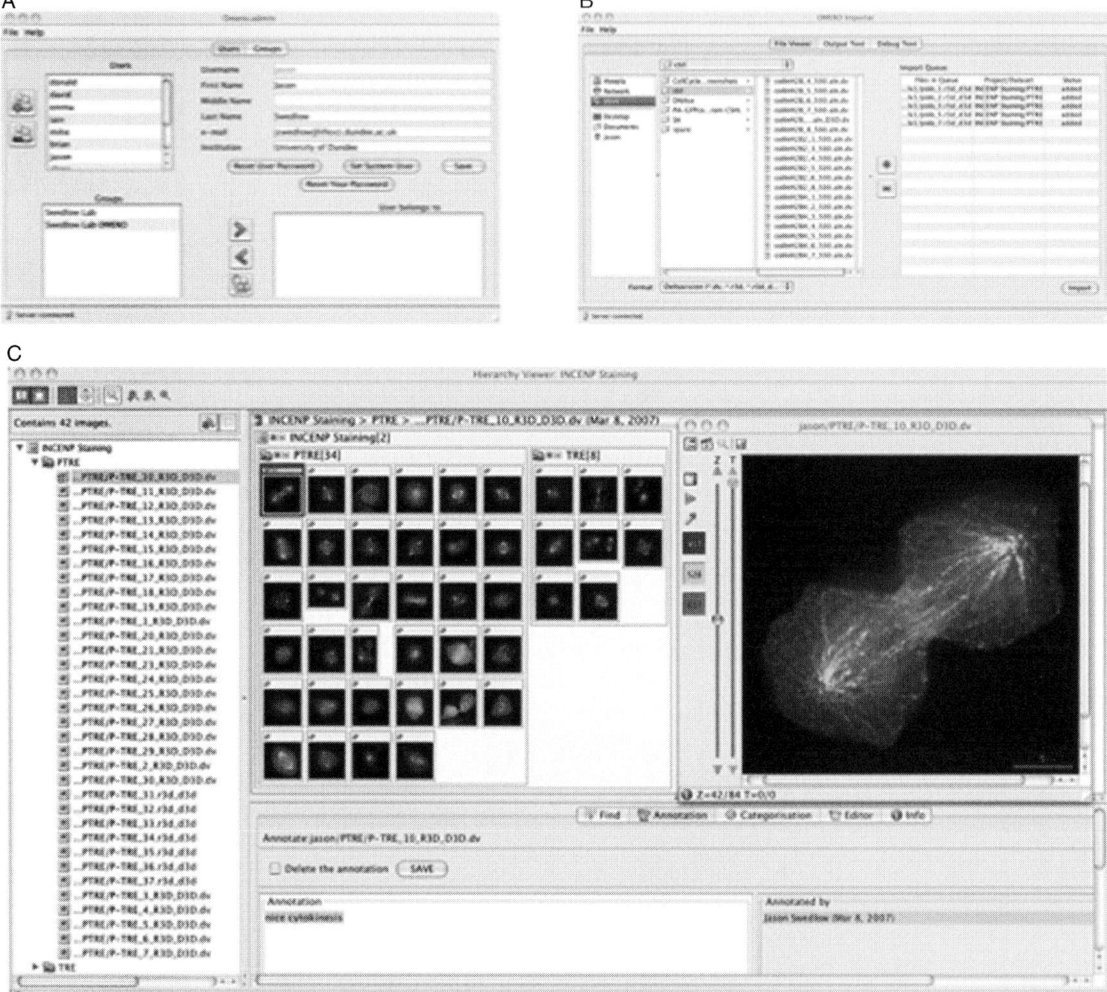

Fig. 3 Screen shots of the OMERO clients, connected to an OMERO Server. (A) The OMERO. Admin tool, for managing users, passwords, and groups. (B) The OMERO.Importer tool, shown during setup for an import of proprietary image files (DeltaVision format) into an OMERO Server. (C) The OMERO.Insight tool, shown with the HierarchyViewer and the ImageViewer. For more examples of this tools, see http://www.openmicroscopy.org/getting-started/shoola.html. (See Plate no. 71 in the Color Plate Section.)

3. A Simple Interface for Image Data Management

As mentioned, ease of use was a central design goal of the OMERO system. This was of importance not only for server installation and administration but also for the access of data from a client. With all the metadata and rendering logic on the server, clients can begin working with an OMERO Server extremely quickly. With a Java client, getting started with a running server amounts to:

```
ServiceFactory sf = new ServiceFactory (
   new Server ("myhost"),
   new Login ("name","password"));
sf.getQueryService ().findAll (Image.class);
```

The use of the Internet Communication Engine (ICE) from http://zeroc.com extends this ease of use to other languages like C++, C#, Python, and Ruby. Interacting with data in an OMERO Server can be as simple as starting the Python interpreter and interacting with OMERO via the console. For example, for using OMERO with languages other than Java, see https://trac.openmicroscopy.org.uk/omero/wiki/OmeroBlitz.

D. A Flexible Server for Binary Data Applications

In general, OMERO can be of immediate benefit to anyone who has metadata information to attach to binary image data that falls within the boundaries of the OME data specification. However, another feature that makes OMERO unique is the use of a Model-driven architecture with extensive code generation from a domain-specific language. This allows not only for the simple addition of new types into the system (conceivably through the use of UML or other simple domain description languages) but also for the application of the entire system to other domains and data intensive applications.

Adapting OMERO for use with other data models (e.g., mass spectrometry) is straightforward. The OMERO developers would suggest some collaboration in order to support the most possible code sharing. Briefly, the files that describe the data model need to be rewritten for each new domain. Some types, however, such as user administration and image files can be shared between instances, and should then be moved to a common package. Although some OME services only make sense for the OME data model, most of the current services could be reused without change. Further services can be added by simply dropping the Java class file into the.ear application file. For more information, see http://trac.openmicroscopy.org.uk/omero/wiki/HowToCreateAService.

E. Clients for the OMERO Server: Leading by Design

The facilities built into the OMERO Server allow the development of new types of data management tools for quantitative analysis of biological data. Most laboratories use a number of different imaging platforms and thus require tools to manage, visualize, and analyze heterogeneous sets of image data recorded in a range of file formats. Ideally, a single set of applications, running on a user's laptop or workstation, could access all sets of data and provide easy-to-use access to this data.

We have designed and built three OMERO client applications for use with the OMERO Server. All are written in Java and require Java 1.5 to be installed on the user's computer (automatically installed on up-to-date OS X and Windows XP systems). OMERO.admin is a simple tool that helps manage an OMERO Server, allowing management of users, groups, and passwords (Fig. 3A). OMERO.importer is a standalone application that allows a user to import proprietary image data files from a file system accessed from the user's computer to a running OMERO Server. This tool uses a standard file browser to help the user find and specify files for import into the server and then uploads the files to an OMERO Server (Fig. 3B). OMERO.importer uses an open library known as Bio-Formats (http://www.loci.wisc.edu/ome/formats.html) that translates proprietary file formats in preparation for upload to an OMERO Server. Finally, OMERO.insight provides a number of tools for accessing and using data in an OMERO Server (Fig. 3C). OMERO.insight contains three tools for viewing and managing data:

- DataManager, a traditional tree-based view of the data hierarchies in an OMERO Server.
- ImageViewer, for visualization of 5D images (space, channel, time). The ImageViewer makes use of the OMERO-RE, and can provide high-performance viewing of multidimensional images on standard workstations, without requiring installation of high-powered graphics cards. Most importantly, image viewing at remote locations is now possible.
- HierarchyViewer, for viewing, annotating, and manipulating large sets of image data.

These documents will be described in detail in a separate manuscript. A tutorial that shows the use of these tools is available at http://openmicroscopy.org/getting-started/

A key aspect of these tools is the way they are designed, tested, and updated according to user feedback. Initially, we used users in our own laboratories and directly asked them to review our plans and software. Interestingly, this did not produce the most useful feedback—often users clearly did not like what they received, but getting clear statements of what specifically was required seemed very hard. In the last year, we have introduced a more professional process, using a team of design and human computer interaction experts to interview users, record, and analyze their reactions and provide detailed documentation of users' reactions and requests. This project, known as Usable Image (http://www.usableimage.org), will be fully detailed in a future publication. Currently, most of our major design decisions and new software releases are presented to users by this team and all reactions and comments recorded and acted upon before full public release. This approach ensures that developers get honest unfiltered feedback from users and provides a mechanism for fully fleshing out user suggestions and ideas before any requested changes or features before changes are made to the software codebase.

F. Current Capabilities

The OMERO Suite of Server and Clients will go through a series of beta releases in 2007, culminating with a "release" of a software package to support a "complete" workflow:

- Import acquired data from a microscope workstation from a variety of proprietary file formats
- thumbnail views of image in the system
- visualize imported data using a remote application
- annotate, organize, and share data in hierarchies
- export image data for publication (TIFF, JPEG)
- manually measure distances and image signal intensities in defined regions of interest
- export image data in OME-TIFF format (http://www.loci.wisc.edu/ome/ome-tiff.html)

The Beta1 release supports the first four of these items, with Beta2 (scheduled for May 2007) filling in some of the missing functionality. In our experience, users most value the flexibility of viewing images at remote locations (office, library, café, and home) and the provision of a data organization tool. A data management tool to users who have struggled to keep track of their data across the 3–5 years of a project appears to be valuable.

V. Future Directions

The OMERO suite of tools provides a strong foundation for users and developers of image management software. Through 2007, the project seeks to expand the functionality of the OMERO suite, to eventually provide complete support for a "standard" workflow—image import, annotation, simple processing and analysis, visualization, and export for publication. A major goal is to provide tools that handle sets of images as a single group to reflect the experimental relationships between them. To this end, OMERO development will focus on user interfaces and server functionality that allow visualization and analysis of complex sets of data, but that always use the power of a data and image server to reduce the data to a representation that is easy to view and comprehend.

Acknowledgments

The authors gratefully acknowledge helpful discussions with our academic and commercial partners (http://openmicroscopy.org/about/). Work in the authors' laboratory on OME and OMERO is supported by grants from the Wellcome Trust (Ref. 068046, 077128, and 080087) and the BBSRC (BB/D00151X/1). J.R.S. is a Wellcome Trust Senior Research Fellow and declares a potential conflict as a founder of Glencoe Software, Inc., that develops image data management tools based on the OMERO Server and Clients.

References

Christiansen, J. H., Yang, Y., Venkataraman, S., Richardson, L., Stevenson, P., Burton, N., Baldock, R. A., and Davidson, D. R. (2006). EMAGE a spatial database of gene expression patterns during mouse embryo development. *Nucleic Acids Res.* **34,** 637–641.

Eliceiri, K. W., and Rueden, C. (2005). Tools for visualizing multidimensional images from living specimens. *Photochem. Photobiol.* **81,** 1116–1122.

Goldberg, I. G., Allan, C., Burel, J.-M., Creager, D., Falconi, A., Hochheiser, H. S., Johnston, J., Mellen, J., Sorger, P. K., and Swedlow, J. R. (2005). The Open Microscopy Environment (OME) Data Model and XML file: Open tools for informatics and quantitative analysis in biological imaging. *Genome Biol.* **6,** R47.

Johnston, J., Nagaraja, A., Hochheiser, H., and Goldberg, U. (2006). A flexible framework for web interfaces to image databases: Supporting user-defined ontologies and links to external databases. *Third IEEE International Symposium on Biomedical Imaging: Macro to Nano,* pp.1380–1383.

Lein, E. S., Hawrylycz, M. J., Ao, N., Ayres, M., Bensinger, A., Bernard, A., Boe, A. F., Boguski, M. S., Brockway, K. S., Byrnes, E. J., Chen, L., Chen, Li, *et al.* (2007). Genome-wide atlas of gene expression in the adult mouse brain. *Nature* **445,** 168–176.

Schiffmann, D. A., Dikovskaya, D., Appleton, P. L., Newton, I. P., Creager, D. A., Allan, C., Nathke, I. S., and Goldberg, I. G. (2006). Open microscopy environment and findspots: Integrating image informatics with quantitative multidimensional image analysis. *Biotechniques* **41,** 199–208.

Swedlow, J. R., Goldberg, I., Brauner, E., and Sorger, P. K. (2003). Informatics and quantitative analysis in biological imaging. *Science* **300,** 100–102.

Swedlow, J. R., Lewis, S. E., and Goldberg, I. G. (2006). Data models across labs, genomes, space, and time. *Nat. Cell Biol.* **8,** 1190–1194.

INDEX

D

G

Q

R

Z

VOLUMES IN SERIES

Founding Series Editor
DAVID M. PRESCOTT

Volume 1 (1964)
Methods in Cell Physiology
Edited by David M. Prescott

Volume 2 (1966)
Methods in Cell Physiology
Edited by David M. Prescott

Volume 3 (1968)
Methods in Cell Physiology
Edited by David M. Prescott

Volume 4 (1970)
Methods in Cell Physiology
Edited by David M. Prescott

Volume 5 (1972)
Methods in Cell Physiology
Edited by David M. Prescott

Volume 6 (1973)
Methods in Cell Physiology
Edited by David M. Prescott

Volume 7 (1973)
Methods in Cell Biology
Edited by David M. Prescott

Volume 8 (1974)
Methods in Cell Biology
Edited by David M. Prescott

Volume 9 (1975)
Methods in Cell Biology
Edited by David M. Prescott

Volume 10 (1975)
Methods in Cell Biology
Edited by David M. Prescott

Volume 11 (1975)
Yeast Cells
Edited by David M. Prescott

Volume 12 (1975)
Yeast Cells
Edited by David M. Prescott

Volume 13 (1976)
Methods in Cell Biology
Edited by David M. Prescott

Volume 14 (1976)
Methods in Cell Biology
Edited by David M. Prescott

Volume 15 (1977)
Methods in Cell Biology
Edited by David M. Prescott

Volume 16 (1977)
Chromatin and Chromosomal Protein Research I
Edited by Gary Stein, Janet Stein, and Lewis J. Kleinsmith

Volume 17 (1978)
Chromatin and Chromosomal Protein Research II
Edited by Gary Stein, Janet Stein, and Lewis J. Kleinsmith

Volume 18 (1978)
Chromatin and Chromosomal Protein Research III
Edited by Gary Stein, Janet Stein, and Lewis J. Kleinsmith

Volume 19 (1978)
Chromatin and Chromosomal Protein Research IV
Edited by Gary Stein, Janet Stein, and Lewis J. Kleinsmith

Volume 20 (1978)
Methods in Cell Biology
Edited by David M. Prescott

Advisory Board Chairman
KEITH R. PORTER

Volume 21A (1980)
Normal Human Tissue and Cell Culture, Part A: Respiratory, Cardiovascular, and Integumentary Systems
Edited by Curtis C. Harris, Benjamin F. Trump, and Gary D. Stoner

Volume 21B (1980)
Normal Human Tissue and Cell Culture, Part B: Endocrine, Urogenital, and Gastrointestinal Systems
Edited by Curtis C. Harris, Benjamin F. Trump, and Gray D. Stoner

Volume 22 (1981)
Three-Dimensional Ultrastructure in Biology
Edited by James N. Turner

Volume 23 (1981)
Basic Mechanisms of Cellular Secretion
Edited by Arthur R. Hand and Constance Oliver

Volume 24 (1982)
The Cytoskeleton, Part A: Cytoskeletal Proteins, Isolation and Characterization
Edited by Leslie Wilson

Volume 25 (1982)
The Cytoskeleton, Part B: Biological *Systems* and *In Vitro* Models
Edited by Leslie Wilson

Volume 26 (1982)
Prenatal Diagnosis: Cell Biological Approaches
Edited by Samuel A. Latt and Gretchen J. Darlington

Series Editor
LESLIE WILSON

Volume 27 (1986)
Echinoderm Gametes and Embryos
Edited by Thomas E. Schroeder

Volume 28 (1987)
***Dictyostelium discoideum:* Molecular Approaches to Cell Biology**
Edited by James A. Spudich

Volume 29 (1989)
Fluorescence Microscopy of Living Cells in Culture, Part A: Fluorescent Analogs, Labeling Cells, and Basic Microscopy
Edited by Yu-Li Wang and D. Lansing Taylor

Volume 30 (1989)
Fluorescence Microscopy of Living Cells in Culture, Part B: Quantitative Fluorescence Microscopy—Imaging and Spectroscopy
Edited by D. Lansing Taylor and Yu-Li Wang

Volume 31 (1989)
Vesicular Transport, Part A
Edited by Alan M. Tartakoff

Volume 32 (1989)
Vesicular Transport, Part B
Edited by Alan M. Tartakoff

Volume 33 (1990)
Flow Cytometry
Edited by Zbigniew Darzynkiewicz and Harry A. Crissman

Volume 34 (1991)
Vectorial Transport of Proteins into and across Membranes
Edited by Alan M. Tartakoff

Selected from Volumes 31, 32, and 34 (1991)
Laboratory Methods for Vesicular and Vectorial Transport
Edited by Alan M. Tartakoff

Volume 35 (1991)
Functional Organization of the Nucleus: A Laboratory Guide
Edited by Barbara A. Hamkalo and Sarah C. R. Elgin

Volume 36 (1991)
***Xenopus laevis:* Practical Uses in Cell and Molecular Biology**
Edited by Brian K. Kay and H. Benjamin Peng

Series Editors
LESLIE WILSON AND PAUL MATSUDAIRA

Volume 37 (1993)
Antibodies in Cell Biology
Edited by David J. Asai

Volume 50 (1995)
Methods in Plant Cell Biology, Part B
Edited by David W. Galbraith, Don P. Bourque, and Hans J. Bohnert

Volume 51 (1996)
Methods in Avian Embryology
Edited by Marianne Bronner-Fraser

Volume 52 (1997)
Methods in Muscle Biology
Edited by Charles P. Emerson, Jr. and H. Lee Sweeney

Volume 53 (1997)
Nuclear Structure and Function
Edited by Miguel Berrios

Volume 54 (1997)
Cumulative Index

Volume 55 (1997)
Laser Tweezers in Cell Biology
Edited by Michael P. Sheetz

Volume 56 (1998)
Video Microscopy
Edited by Greenfield Sluder and David E. Wolf

Volume 57 (1998)
Animal Cell Culture Methods
Edited by Jennie P. Mather and David Barnes

Volume 58 (1998)
Green Fluorescent Protein
Edited by Kevin F. Sullivan and Steve A. Kay

Volume 59 (1998)
The Zebrafish: Biology
Edited by H. William Detrich III, Monte Westerfield, and Leonard I. Zon

Volume 60 (1998)
The Zebrafish: Genetics and Genomics
Edited by H. William Detrich III, Monte Westerfield, and Leonard I. Zon

Volume 61 (1998)
Mitosis and Meiosis
Edited by Conly L. Rieder

Volume 62 (1999)
Tetrahymena thermophila
Edited by David J. Asai and James D. Forney

Volume 63 (2000)
Cytometry, Third Edition, Part A
Edited by Zbigniew Darzynkiewicz, J. Paul Robinson, and Harry Crissman

Volume 64 (2000)
Cytometry, Third Edition, Part B
Edited by Zbigniew Darzynkiewicz, J. Paul Robinson, and Harry Crissman

Volume 65 (2001)
Mitochondria
Edited by Liza A. Pon and Eric A. Schon

Volume 66 (2001)
Apoptosis
Edited by Lawrence M. Schwartz and Jonathan D. Ashwell

Volume 67 (2001)
Centrosomes and Spindle Pole Bodies
Edited by Robert E. Palazzo and Trisha N. Davis

Volume 68 (2002)
Atomic Force Microscopy in Cell Biology
Edited by Bhanu P. Jena and J. K. Heinrich Hörber

Volume 69 (2002)
Methods in Cell–Matrix Adhesion
Edited by Josephine C. Adams

Volume 70 (2002)
Cell Biological Applications of Confocal Microscopy
Edited by Brian Matsumoto

Volume 71 (2003)
Neurons: Methods and Applications for Cell Biologist
Edited by Peter J. Hollenbeck and James R. Bamburg

Volume 72 (2003)
Digital Microscopy: A Second Edition of Video Microscopy
Edited by Greenfield Sluder and David E. Wolf

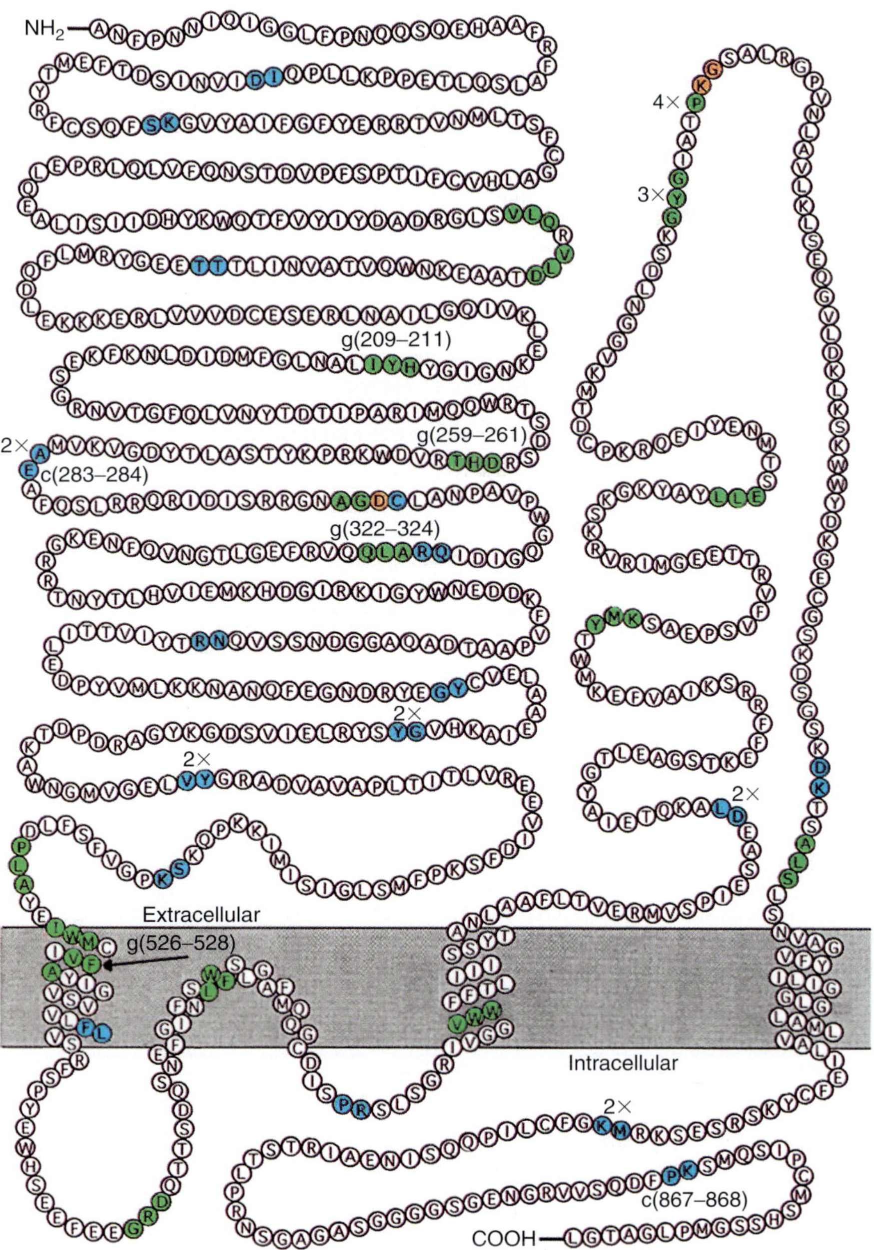

Plate 1 (Figure 2.1 on page 26 of this volume)

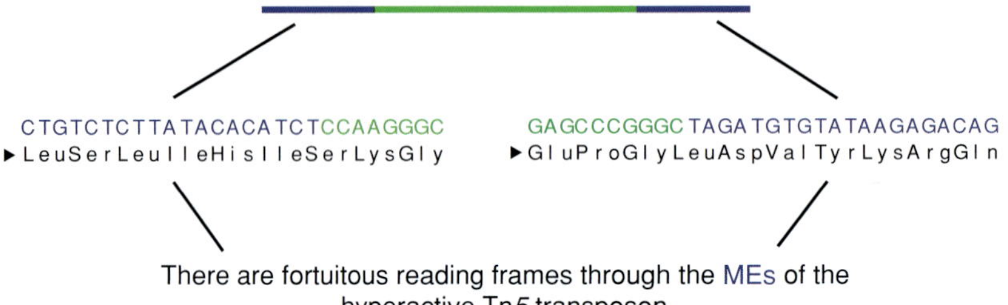

Plate 2 (Figure 2.2 on page 28 of this volume)

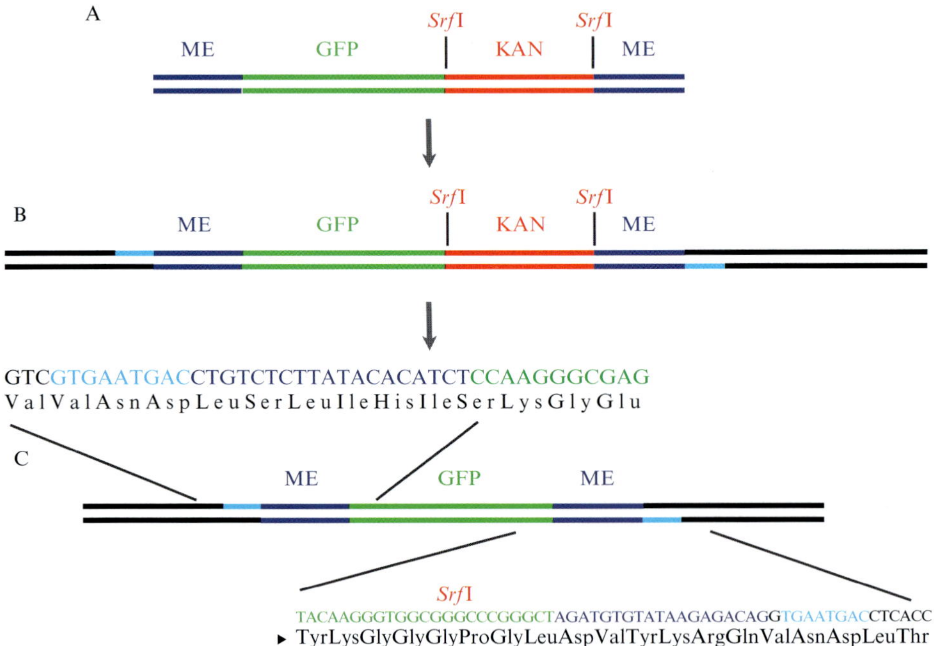

Plate 3 (Figure 2.3 on page 29 of this volume)

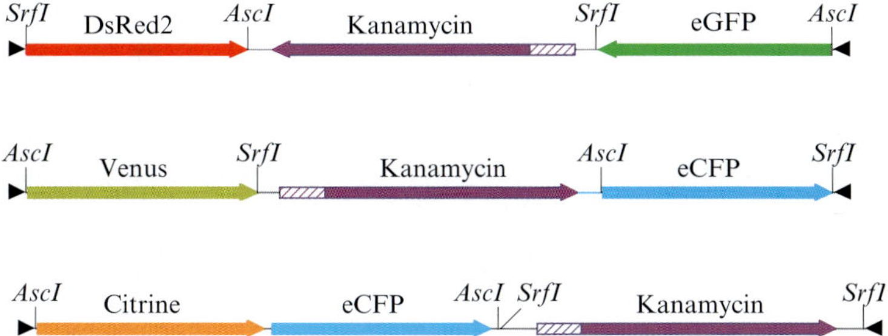

Plate 4 (Figure 2.4 on page 30 of this volume)

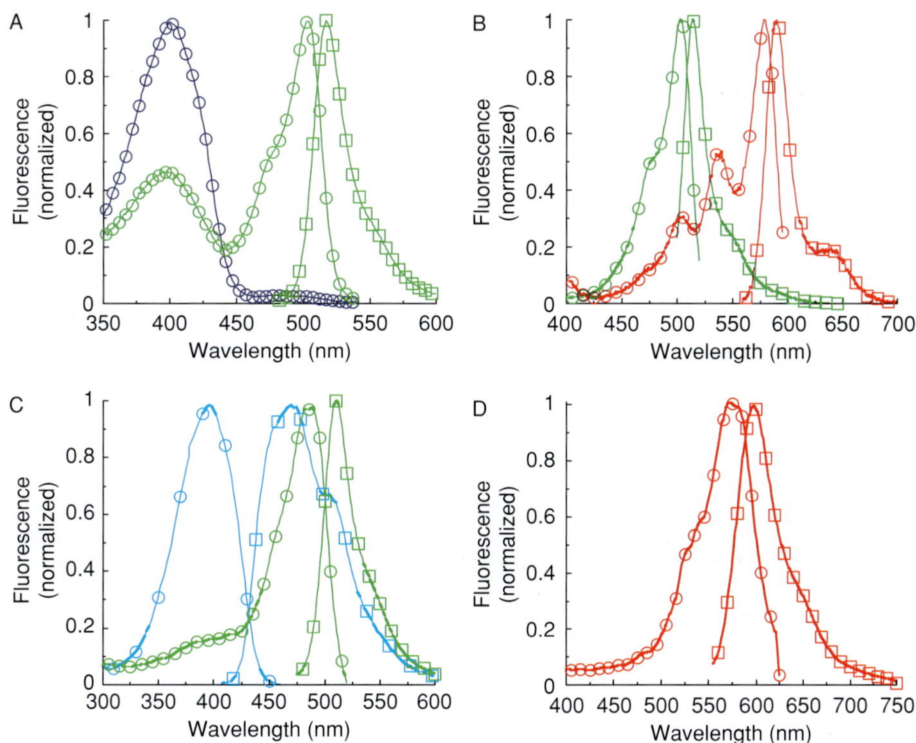

Plate 5 (Figure 3.1 on page 49 of this volume)

Plate 6 (Figure 3.2 on page 53 of this volume)

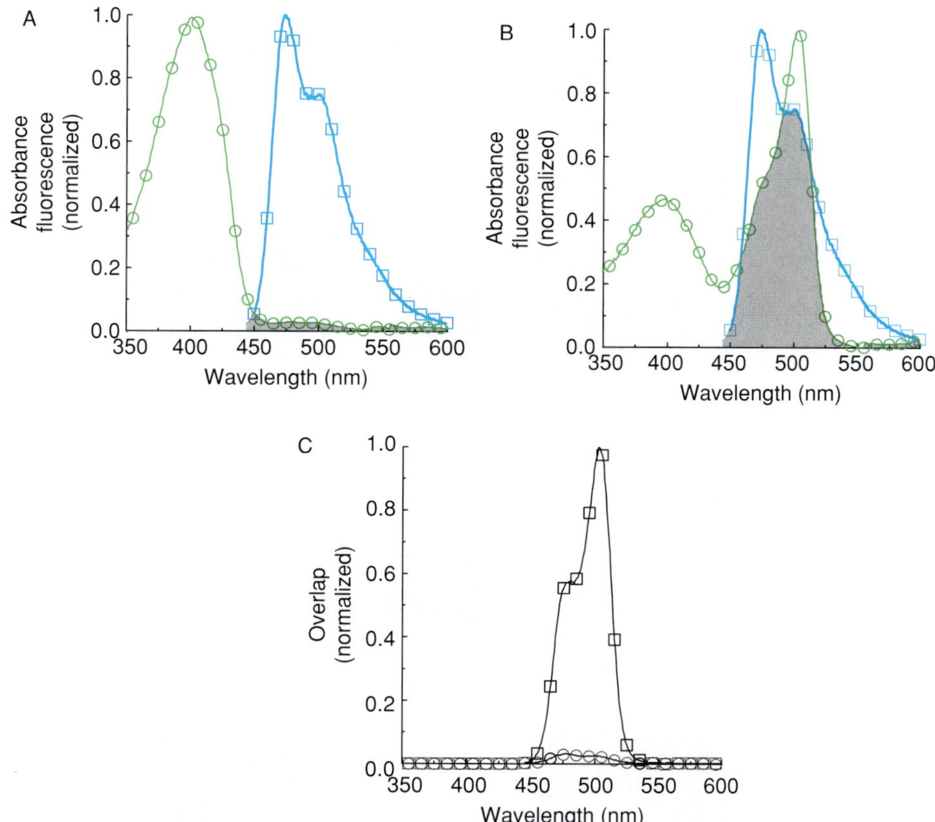

Plate 7 (Figure 3.3 on page 56 of this volume)

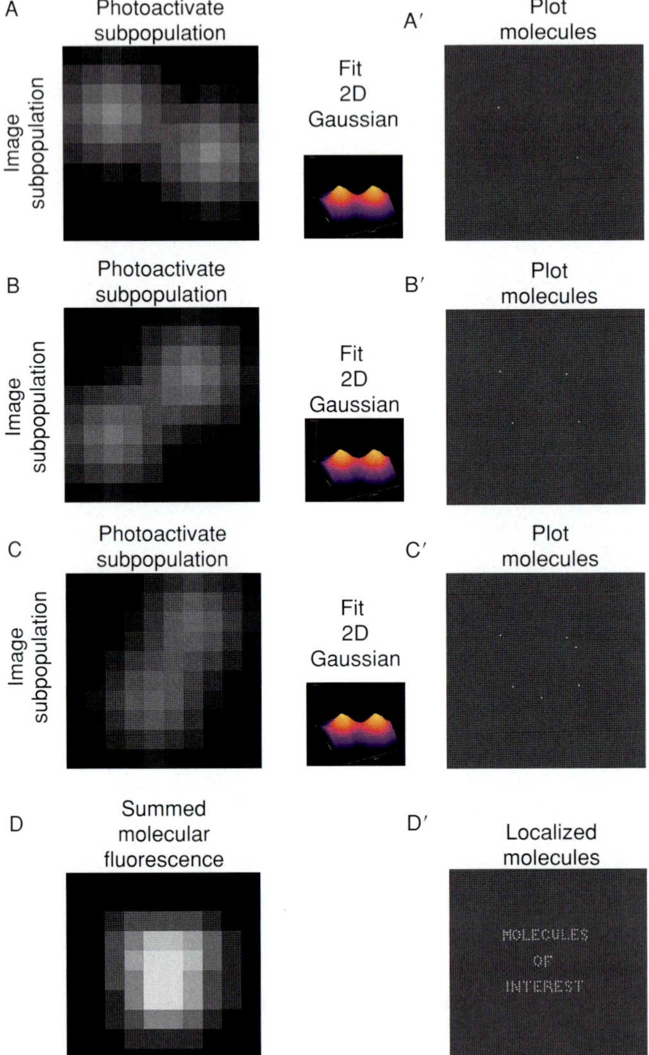

Plate 8 (Figure 3.4 on page 57 of this volume)

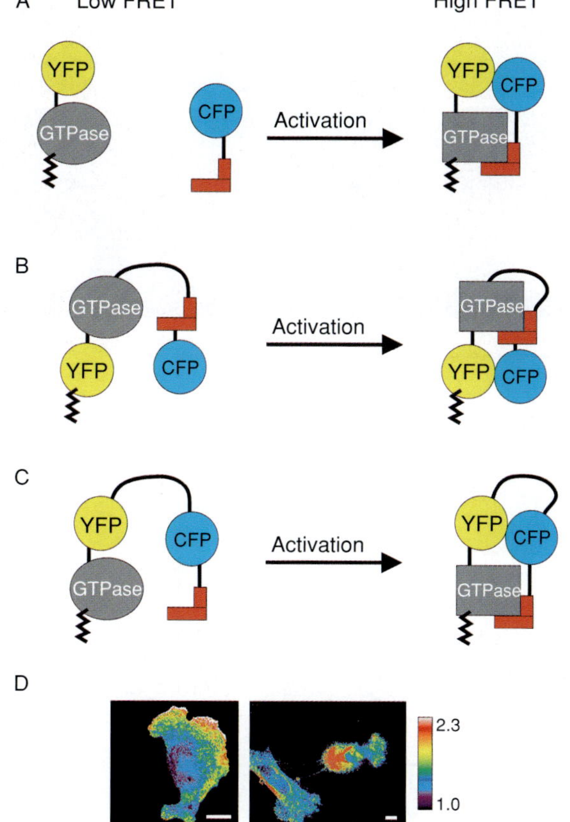

Plate 9 (Figure 4.1 on page 65 of this volume)

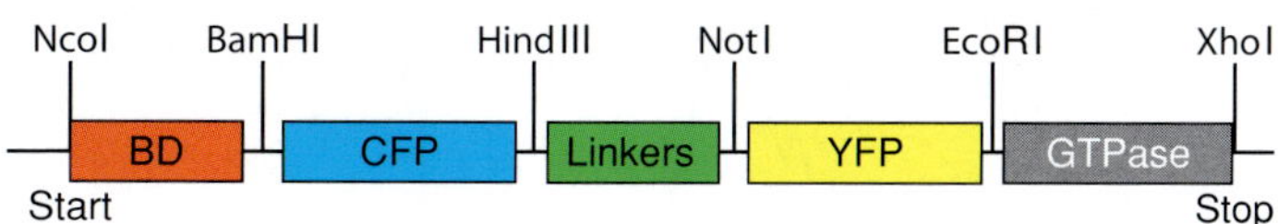

Plate 10 (Figure 4.2 on page 69 of this volume)

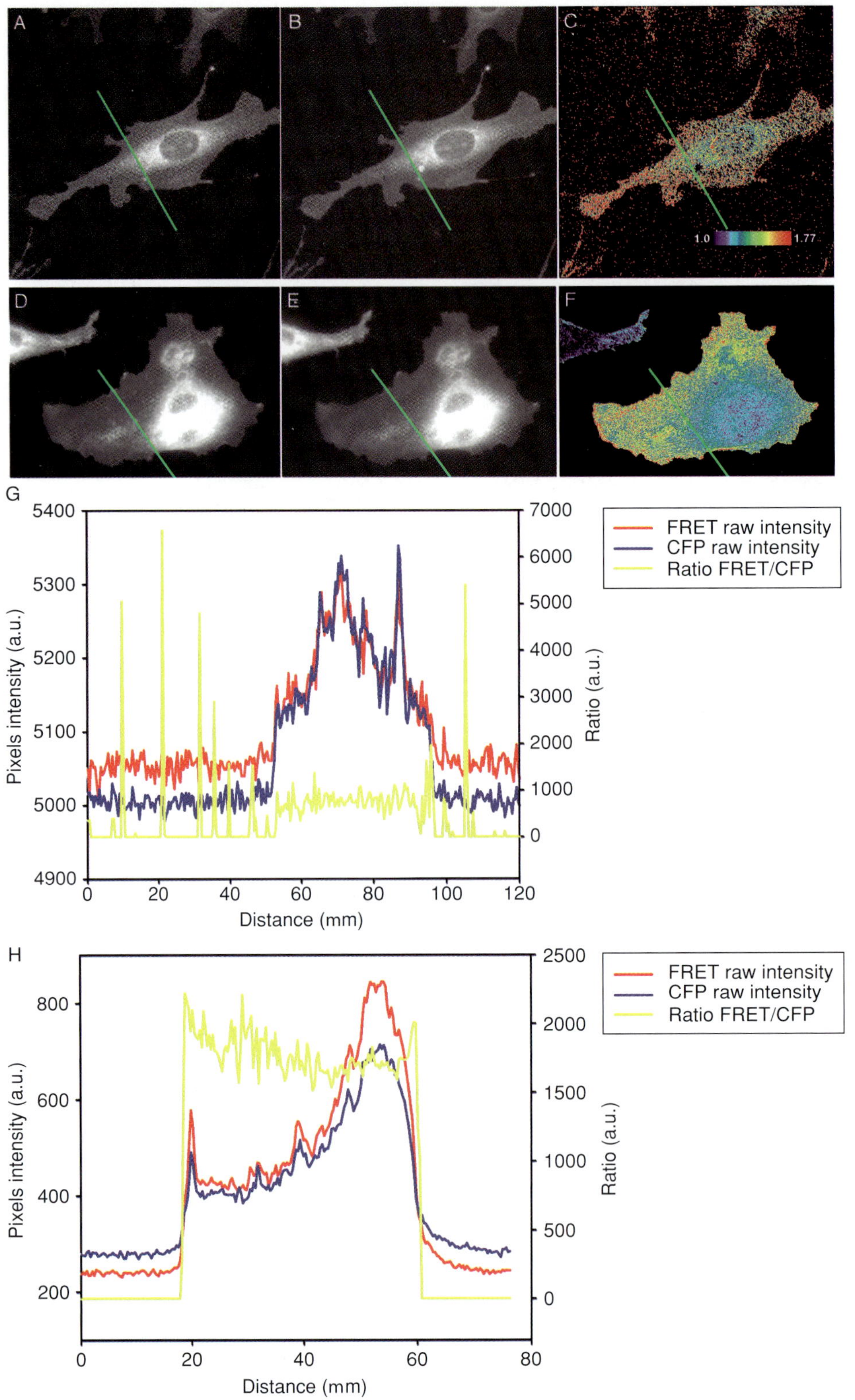

Plate 11 (Figure 4.4 on page 74,75 of this volume)

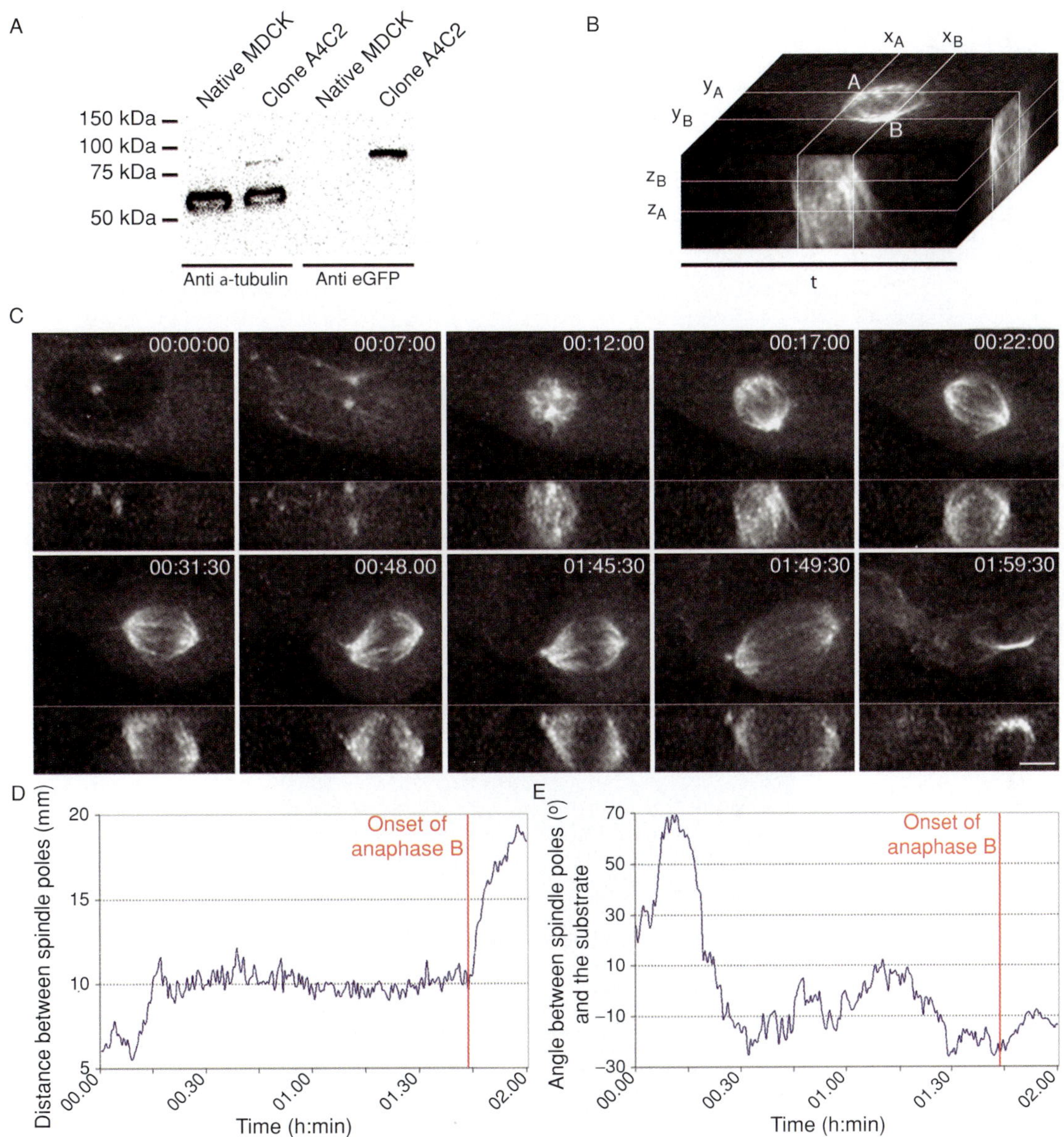

Plate 12 (Figure 5.2 on page 88 of this volume)

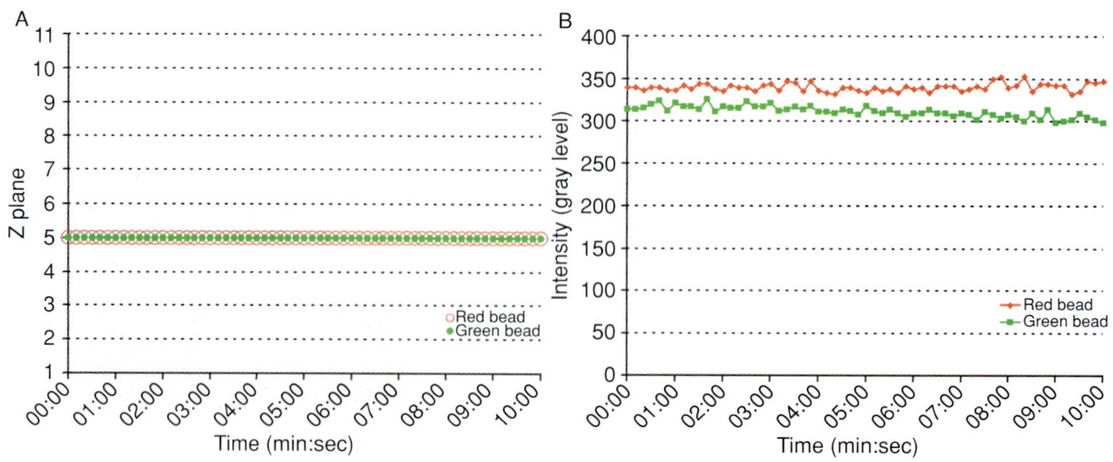

Plate 13 (Figure 5.3 on page 94 of this volume)

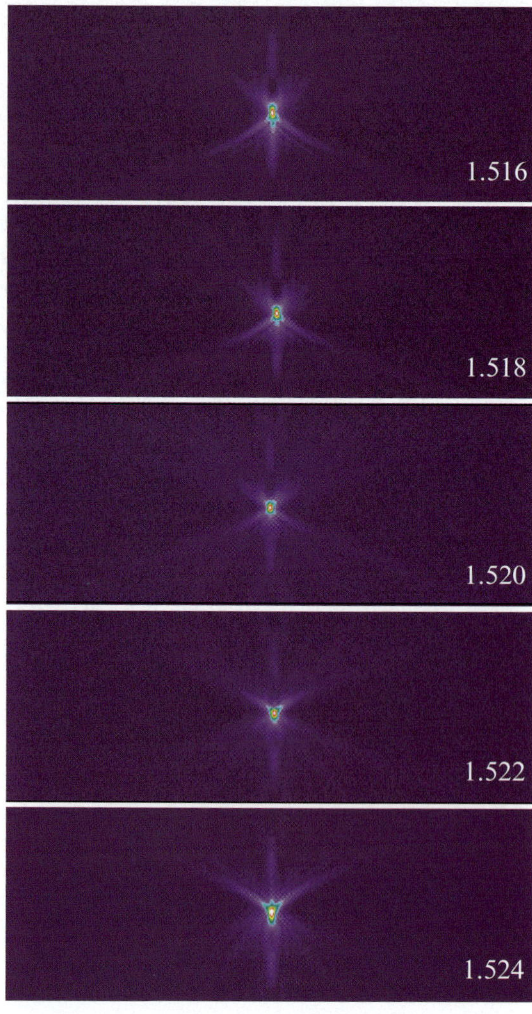

Plate 14 (Figure 5.6 on page 104 of this volume)

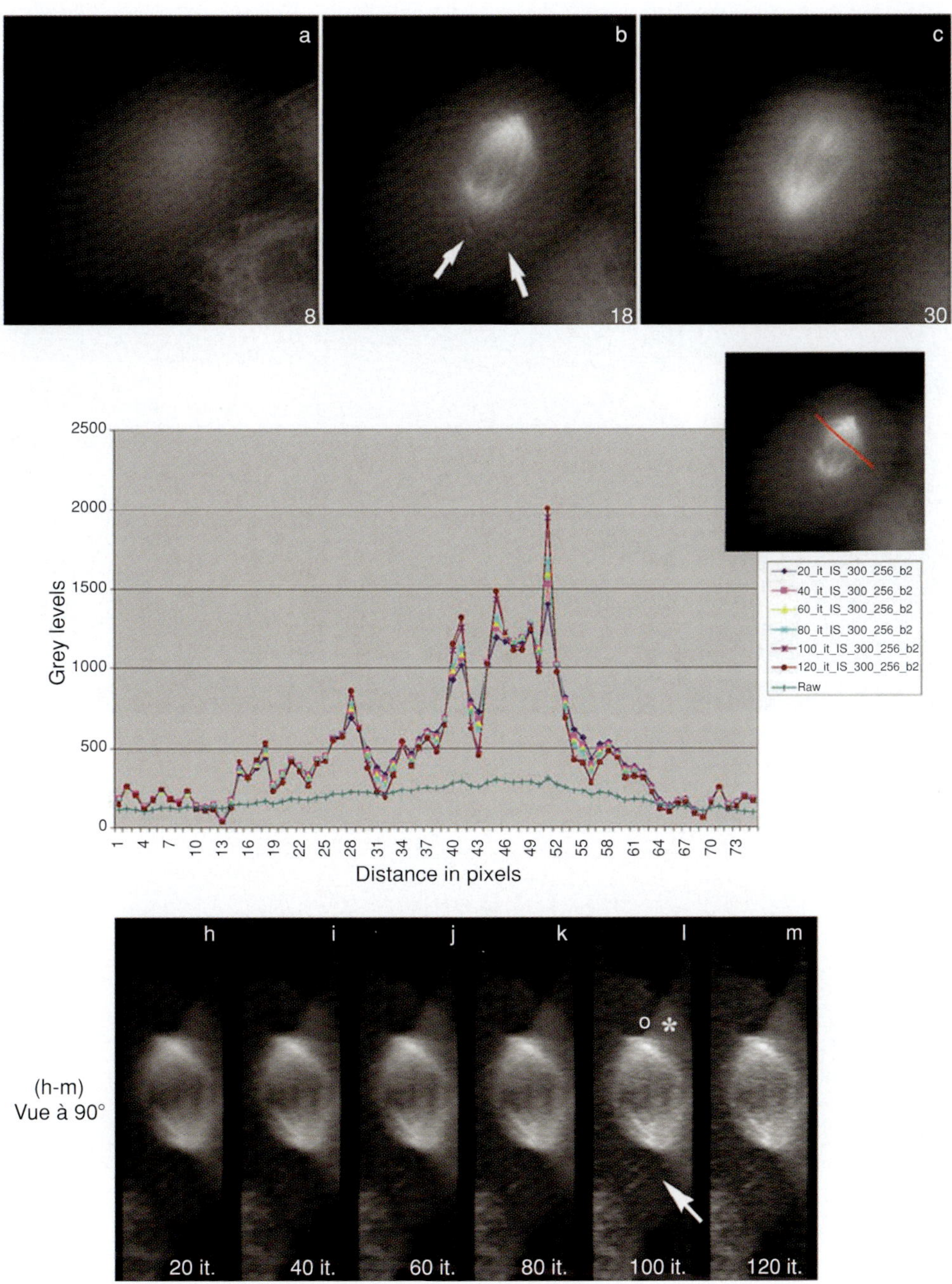

Plate 15 (Figure 5.9 on page 108 of this volume)

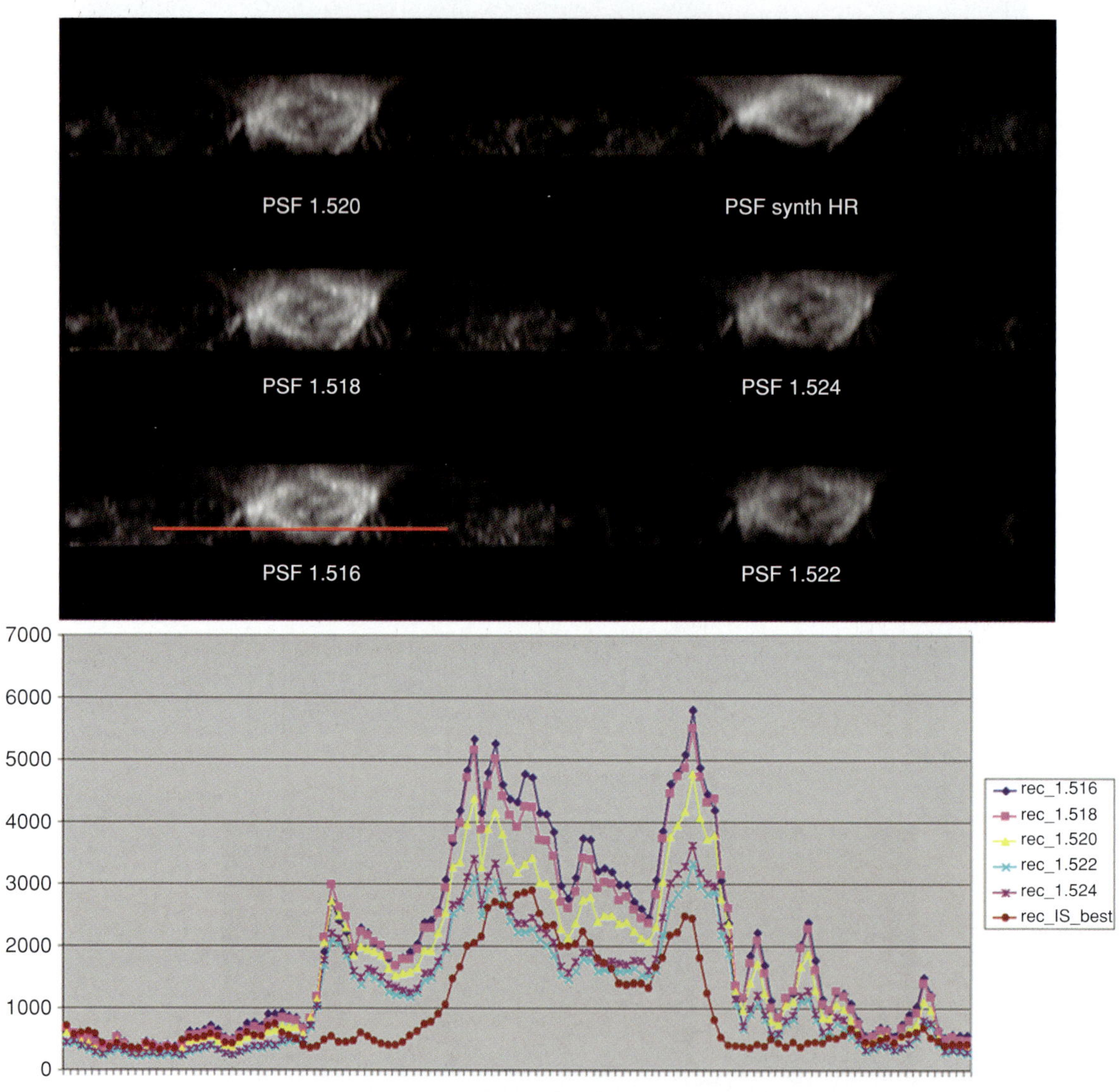

Plate 16 (Figure 5.10 on page 109 of this volume)

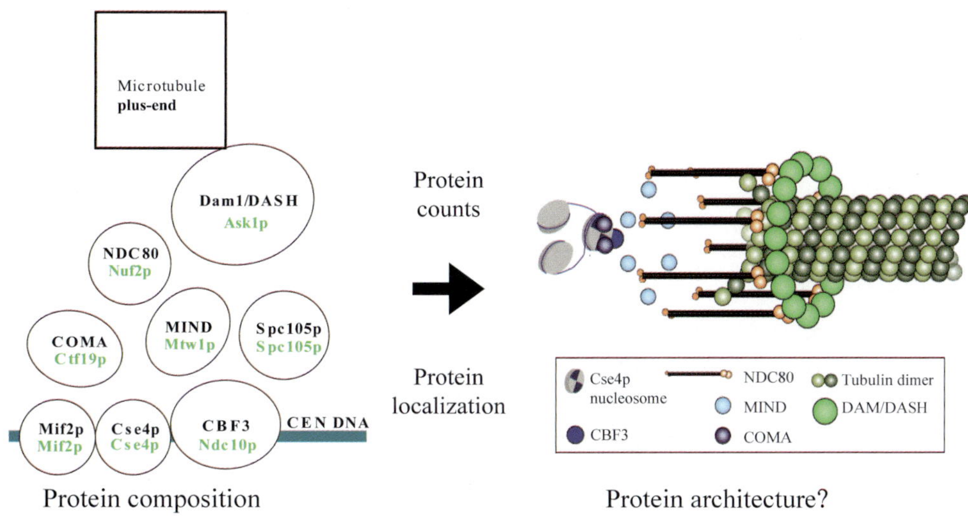

Plate 17 (Figure 7.1 on page 131 of this volume)

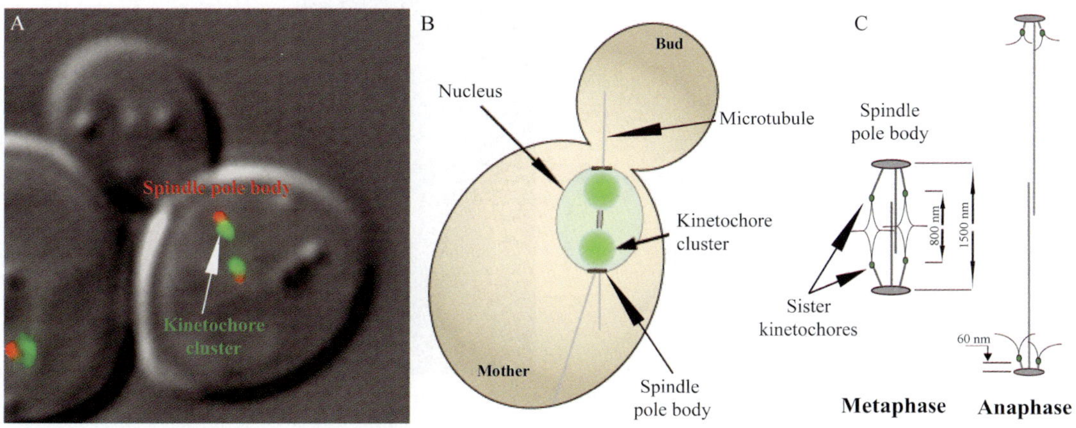

Plate 18 (Figure 7.2 on page 132 of this volume)

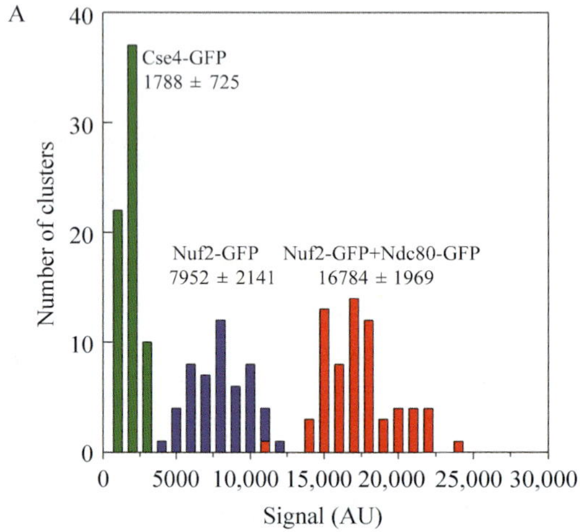

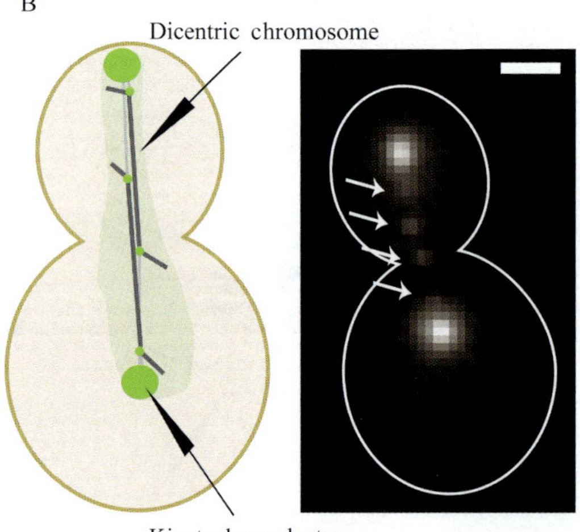

Plate 19 (Figure 7.5 on page 141 of this volume)

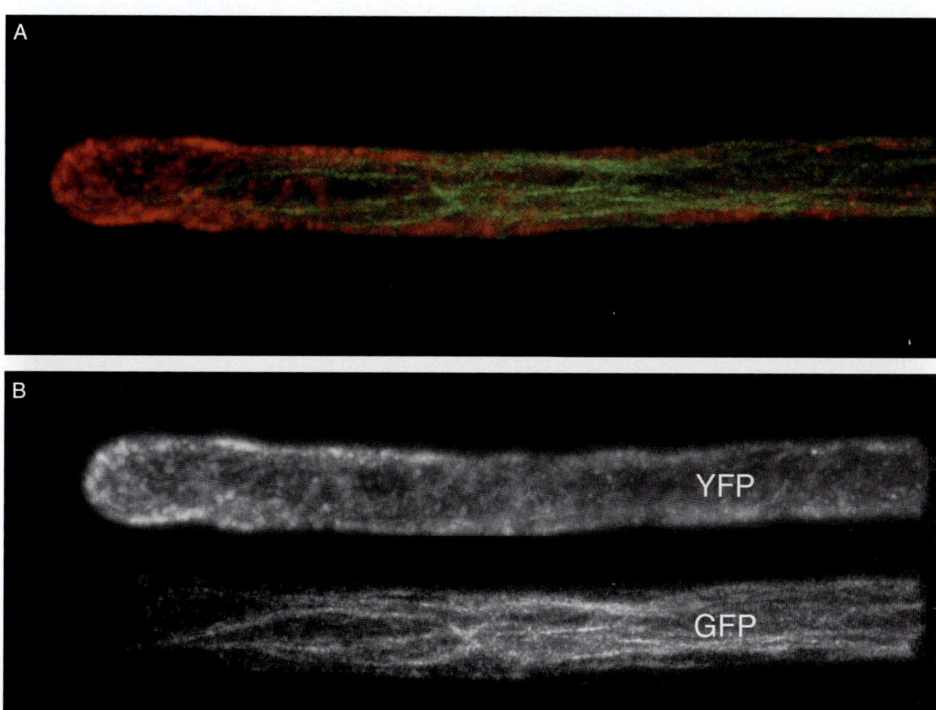

Plate 20 (Figure 8.11 on page 171 of this volume)

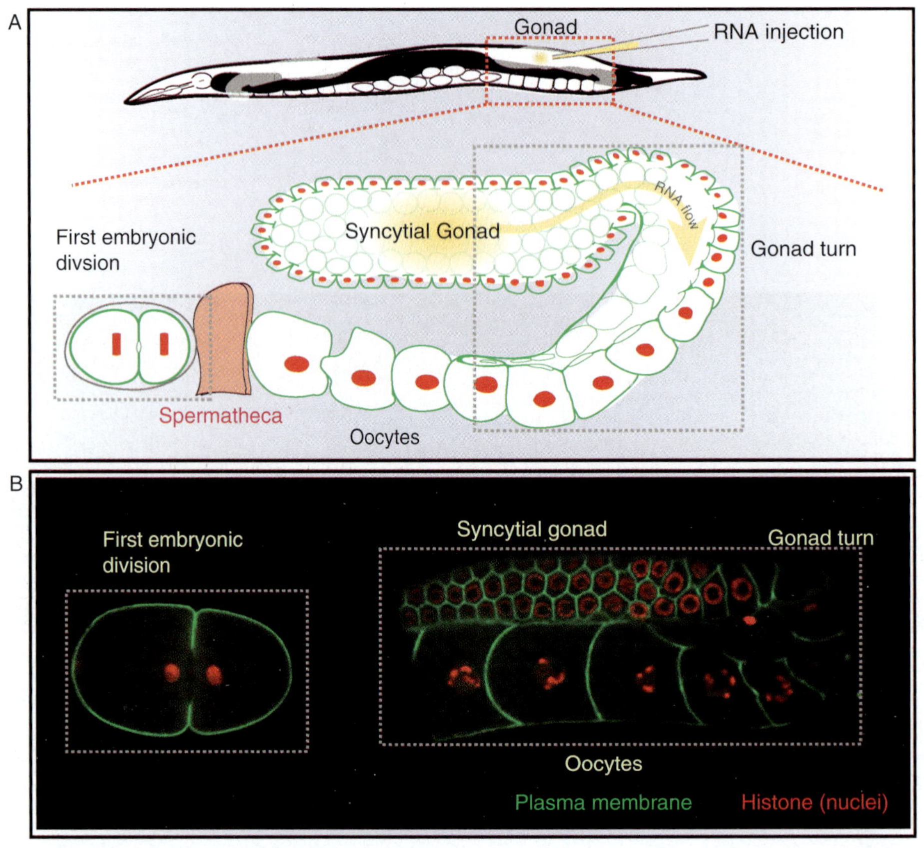

Plate 21 (Figure 9.1 on page 181 of this volume)

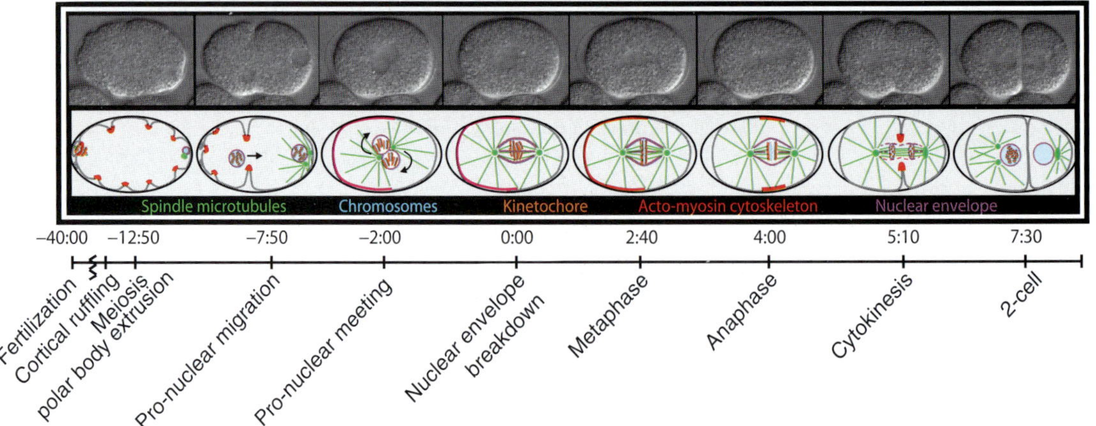

Plate 22 (Figure 9.2 on page 183 of this volume)

The codons for each amino acid are listed in order by frequency. Codons are color-coded to indicate whether their frequency is ***increased****, ***decreased,*** or ***unaltered*** in highly-expressed genes.

ALA	ARG	ASN	ASP
GCU*	CGU*	AAC*	GAU
GCC*	AGA	AAU	GAC*
GCA	CGC*		
GCG	CGA		
	CGG		
	AGG		
CYS	**GLN**	**GLU**	**GLY**
UGC*	CAA	GAG*	GGA*
UGU	CAG*	GAA	GGU
			GGC
			GGG
HIS	**ILE**	**LEU**	**LYS**
CAC*	AUC*	CUC*	AAG*
CAU	AUU	CUU*	AAA
	AUA	UUG	
		CUG	
		UUA	
		CUA	
MET	**PRO**	**PHE**	**SER**
AUG*	CCA*	UUC*	UCC*
	CCG	UUU	UCU
	CCU		UCA
	CCC		UCG
			AGC
			AGU
STOP	**THR**	**TYR**	**VAL**
UAA*	ACC*	UAC*	GUC*
UAG	ACU	UAU	GUU
UGA	ACA		GUG
	ACG		GUA

Adapted from Duret and Mouchiroud (1999)

Plate 23 (Figure 9.3 on page 185 of this volume)

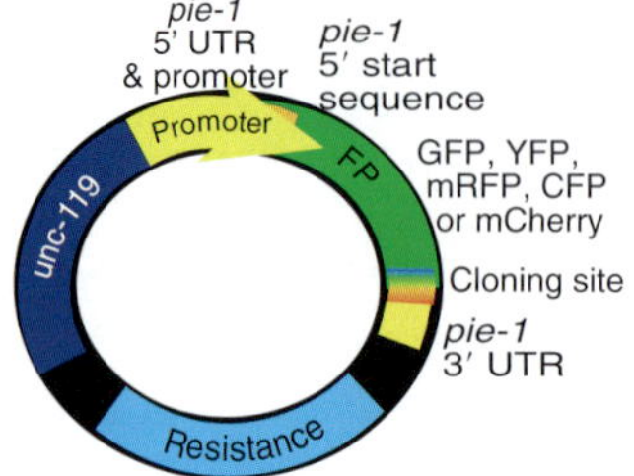

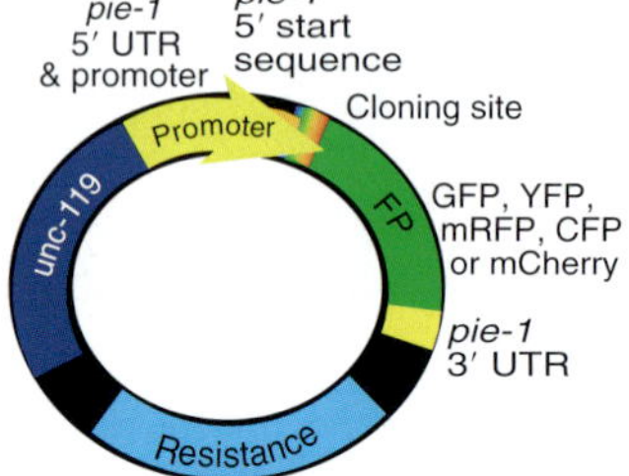

	Plasmid name	Promoter	Fluorescent protein	Location	Resist-ance	Selection marker	Additional features	Source
Parental vectors	pPD95.75		GFP					Fire lab
	pJH4.52	*pie-1*	GFP	N-terminus	Amp			Seydoux lab
	TH317-NoTag-AMP	*pie-1*	No tag	N-terminus	Amp	*unc-119*	short *unc-119*	A. Pozniakovsky; Hyman lab
	TH318-NoTag-CM	*pie-1*	No tag	N-terminus	Cm	*unc-119*	short *unc-119*	A. Pozniakovsky; Hyman lab
	TH319-NoTag-KAN	*pie-1*	No tag	N-terminus	Kan	*unc-119*	short *unc-119*	A. Pozniakovsky; Hyman lab
GFP vectors	pAZ132 (N-GFP)	*pie-1*	GFP	N-terminus	Amp	*unc-119*		Fire lab
	pIC26 (N-LAP)	*pie-1*	GFP	N-terminus	Amp	*unc-119*	LAP	I Cheeseman; Desai lab
	TH304-GFP (C) AMP	*pie-1*	GFP	C-terminus	Amp	*unc-119*	short *unc-119*, MCS	A. Pozniakovsky; Hyman lab
	TH303-GFP (N) AMP	*pie-1*	GFP	N-terminus	Amp	*unc-119*	short *unc-119*, MCS	A. Pozniakovsky; Hyman lab
	TH305-GFP (C) CM	*pie-1*	GFP	C-terminus	Cm	*unc-119*	short *unc-119*, MCS	A. Pozniakovsky; Hyman lab
	TH314-LAP (N) AMP	*pie-1*	GFP	N-terminus	Amp	*unc-119*	short *unc-119*, MCS, LAP	A. Pozniakovsky; Hyman lab
	TH315-LAP(N)GW-AMP	*pie-1*	GFP	N-terminus	Amp/Cm	*unc-119*	short *unc-119*, MCS, GateWay, LAP	A. Pozniakovsky; Hyman lab
mCherry vectors	pAA65(LAP)	*pie-1*	mCherry	N-terminus	Amp	*unc-119*	LAP, codon optim.	A. Audhya; Oegema lab
	pAA64	*pie-1*	mCherry	N-terminus	Amp	*unc-119*	codon optim.	A. Audhya; Oegema lab
	TH313-Cherry (C) AMP	*pie-1*	mCherry	C-terminus	Amp	*unc-119*	short *unc-119*, MCS codon optim.	A. Pozniakovsky; Hyman lab
	TH312-Cherry (N) AMP	*pie-1*	mCherry	N-terminus	Amp	*unc-119*	short *unc-119*, MCS codon optim.	A. Pozniakovsky; Hyman lab
mRFP vectors	TH309-mRFP (N) AMP	*pie-1*	mRFP	N-terminus	Amp	*unc-119*	short *unc-119*, MCS	A. Pozniakovsky; Hyman lab
	TH310-mRFP (C) AMP	*pie-1*	mRFP	C-terminus	Amp	*unc-119*	short *unc-119*, MCS	A. Pozniakovsky; Hyman lab
	TH311-mRFP (N) CM	*pie-1*	mRFP	N-terminus	Cm	*unc-119*	short *unc-119*, MCS	A. Pozniakovsky; Hyman lab
CFP vector	TH306-CFP(N) AMP	*pie-1*	CFP	N-terminus	Amp	*unc-119*	short *unc-119*, MCS	A. Pozniakovsky; Hyman lab
YFP vectors	TH307-YFP(N) AMP	*pie-1*	YFP	N-terminus	Amp	*unc-119*	short *unc-119*, MCS	A. Pozniakovsky; Hyman lab
	TH308-3xYFP(N) AMP	*pie-1*	Triple YFP	N-terminus	Cm	*unc-119*	short *unc-119*, MCS	A. Pozniakovsky; Hyman lab

Plate 24 (Figure 9.4 on page 189 of this volume)

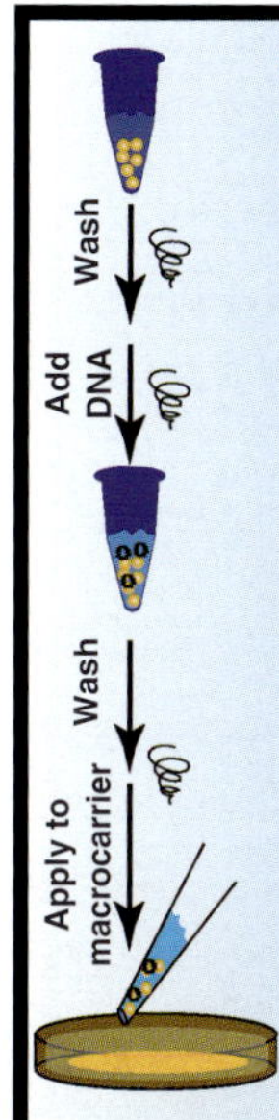

Bead preparation for 10 bombardment plates

1. Weigh 6 mg of **gold beads** into **siliconized eppendorf tube.**
2. Add 1 ml **70% ethanol** and vortex for 5 min.
3. Allow beads to settle (1 min).
4. Briefly spin and remove supernatant.
5. Add 1 ml water and vortex for 1 min.
6. Allow beads to settle (1 min).
7. Briefly spin and remove supernatant.
8. Repeat steps 5–7 twice more.
9. Resuspend beads in 100 μl of **50% glycerol.**
10. Vortex beads for 5 min.
11. Continue vortexing and add the following in order:
    (a)10 μl **DNA** (~1μg), vortex 30–60 sec;
    (b)100 μl **2.5 M $CaCl_2$,** vortex for 30–60 sec;
    (c) 40 μl **0.1 M spermidine**, vortex for at least 3 min.
12. Allow beads to settle 1 min.
13. Briefly spin and then remove the supernatnatant.
14. Add 300 μl of 70% Ethanol, vortex/pipet to resuspend thoroughly.
15. Allow beads to settle 1 min.
16. Briefly spin and then remove the supernatnatant.
17. Resuspend beads in 100 μl of **95% Ethanol**.
18. Vortex for at least 3 min to resuspend.
19. Apply 10 μL of beads to sterile **macrocarrier** with wide-bore pipet for bombardment.

Worm preparation for bombardment

1. From slightly starved stock ***unc-119*** plates seed 10–12 **c600/peptone plates.**
2. Allow to grow one week at 20 °C.
3. Wash the worms off 10 plates with sterile **M9** and collect in a 50 ml falcon tube.
4. Allow worms to settle for about 5 min.
5. Remove supernatant, leave about 1–2 mL worm slurry.
6. Resuspend the worms gently, use a **wide-bore 200 μl tip.**
7. Aliquot ~100 μl of worm suspension onto the center of a round **OP50 bacterial lawn** on a 60 mm **NGM plate**. Repeat for 10 plates total.
8. Allow excess liquid to soak into the plate (about 5–10 min).
9. The worm patch is ready to bombard when it has a matte-like appearance.

Bombardment protocol

1. Spray down the chamber with 70% EOH.
2. Place a **rupture disk** into into the upper assembly. Screw the holder into place tightly.
3. Place a **stopping screen** in to the bottom of the macrocarrier holder (bottom assembly).
4. Invert a macrocarrier holder into the assembly and screw the holding ring onto the top of the assembly.
5. Slide the assembly into the uppermost slot in the chamber.
6. Place the specimen platform is in the second position in the chamber. Place the worm plate in the center of the specimen platform.
7. Close the chamber door and draw a vacuum.
8. When the vacuum reaches ~27 mm/Hg, depress and hold the fire button. When the pressure gauge reaches 1350 psi, the rupture disk will break, and the chamber should be immediately vented.
9. Remove the worms from the chamber and allow to recover for 1–2 hours.
10. Distribute the worms from each small bombardment plate onto 2 large NGM seeded plates and return to the 20 °C incubator.
11. Check for moving worms after two weeks

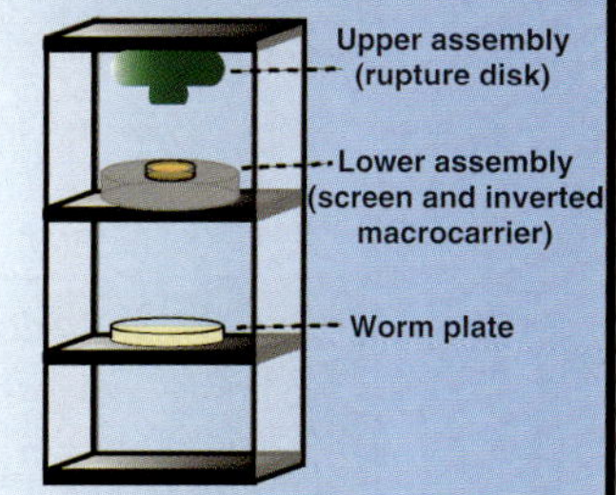

Plate 25 (Figure 9.5 on page 191 of this volume)

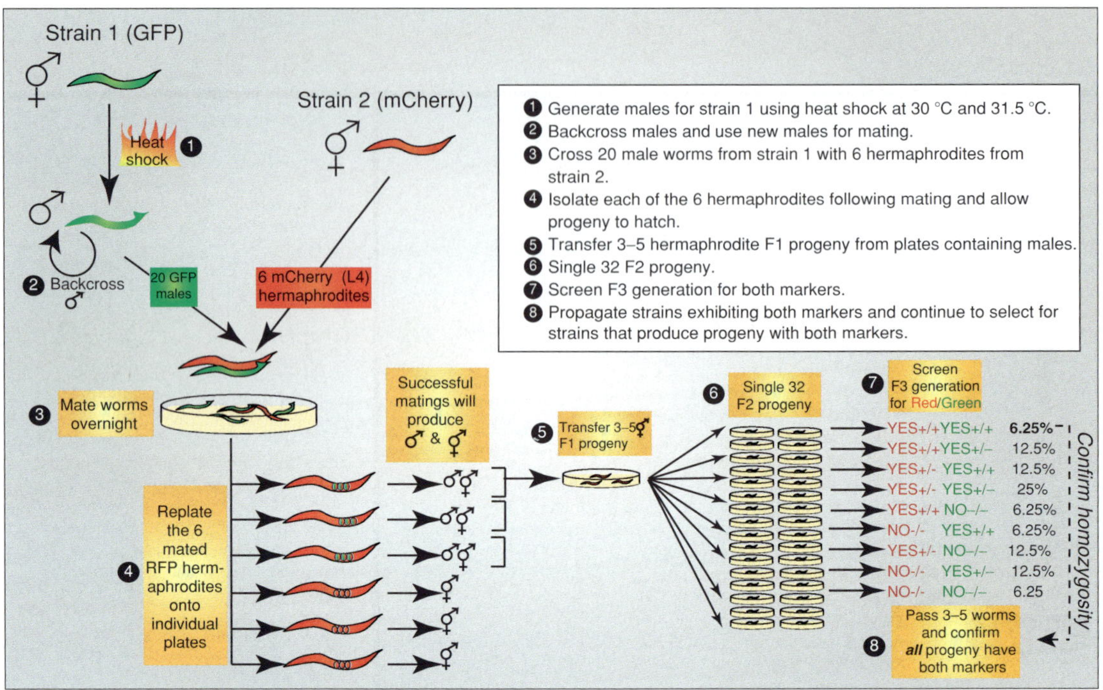

Plate 26 (Figure 9.6 on page 199 of this volume)

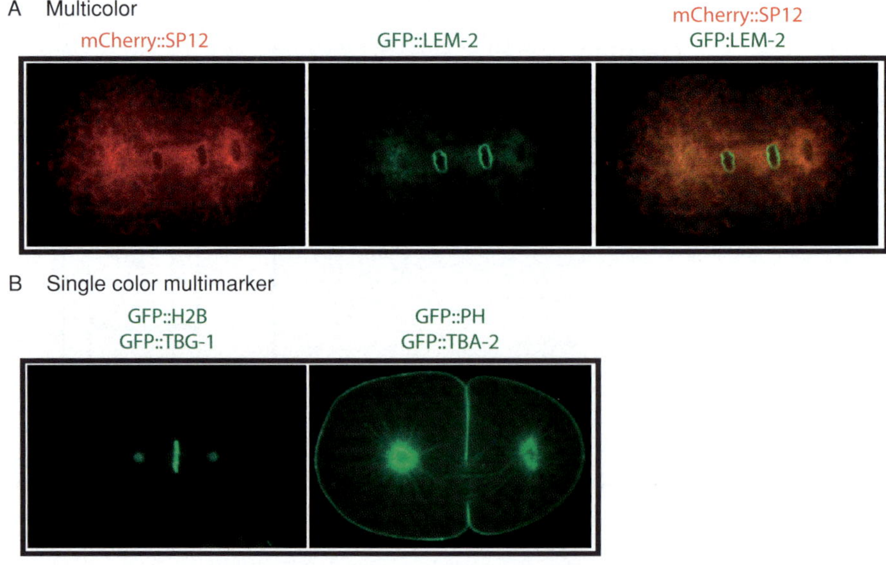

Plate 27 (Figure 9.7 on page 202 of this volume)

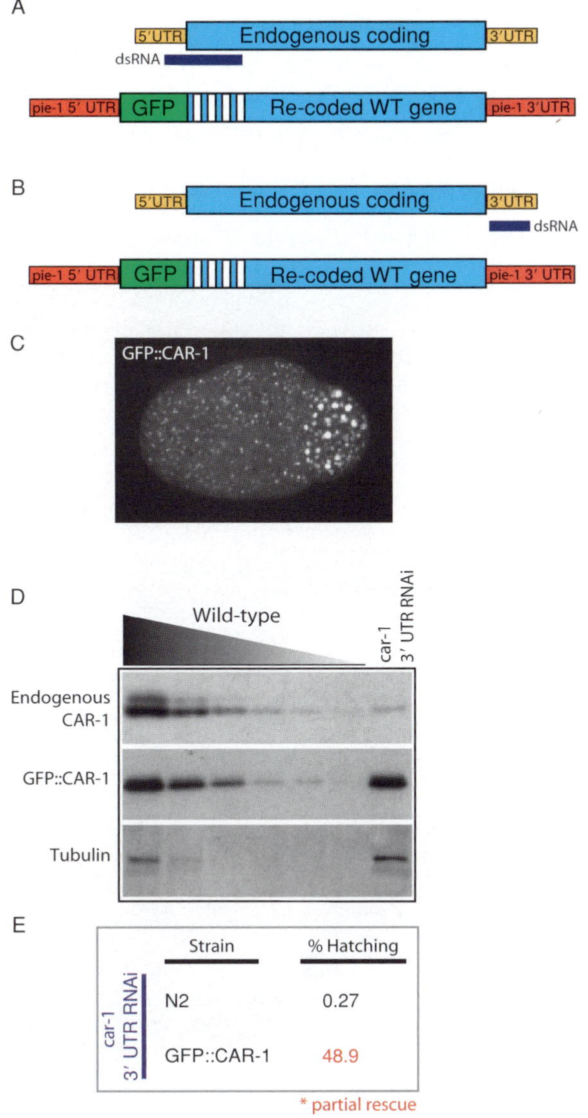

Plate 28 (Figure 9.8 on page 204 of this volume)

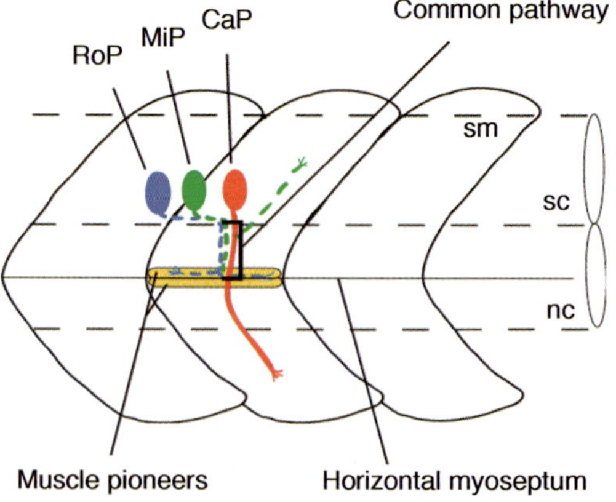

Plate 29 (Figure 10.3 on page 227 of this volume)

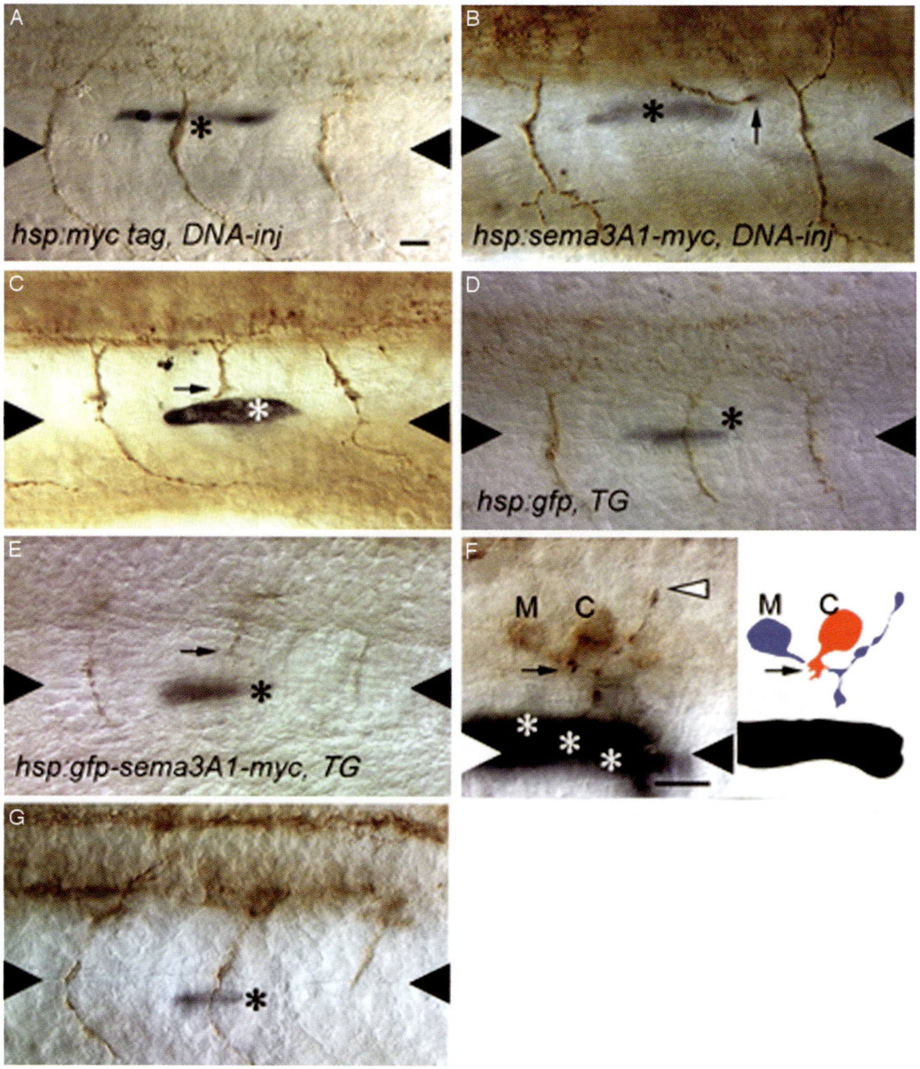

Plate 30 (Figure 10.4 on page 230 of this volume)

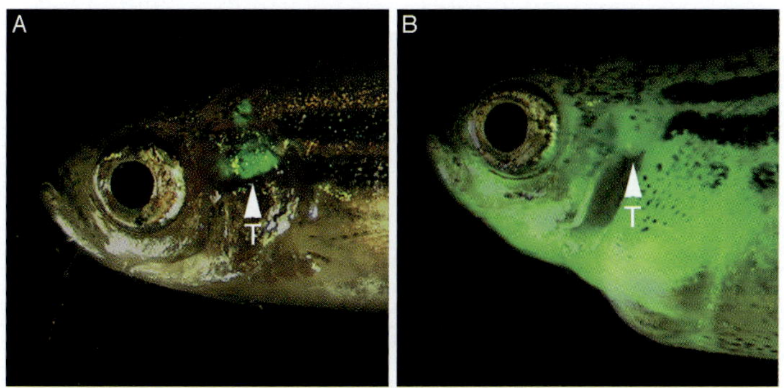

Plate 31 (Figure 10.5 on page 232 of this volume)

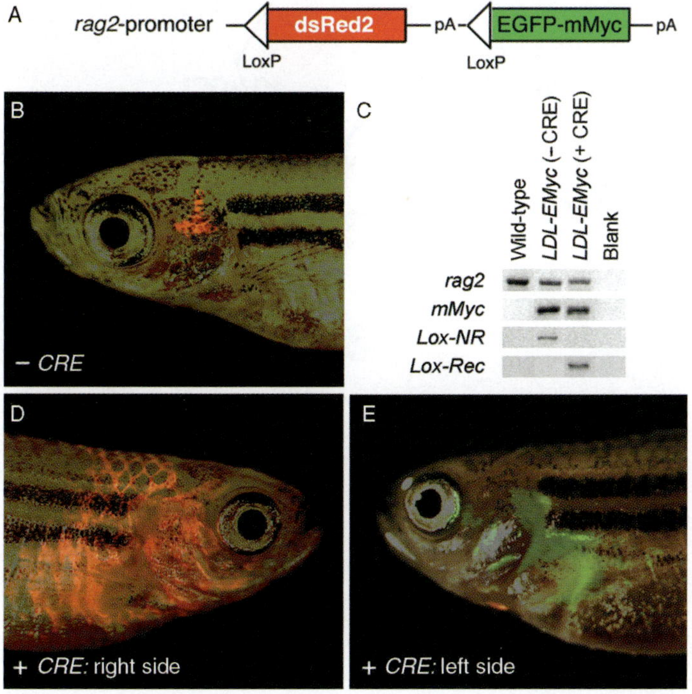

Plate 32 (Figure 10.6 on page 234 of this volume)

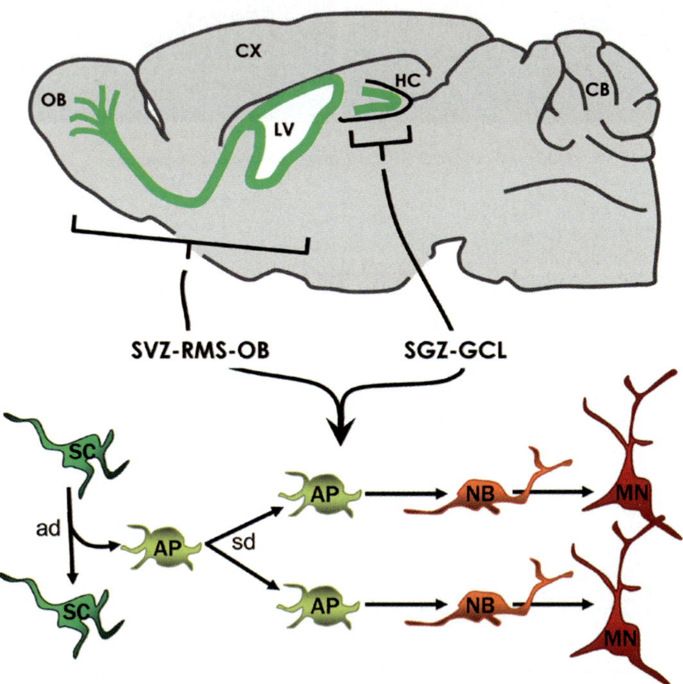

Plate 33 (Figure 11.1 on page 245 of this volume)

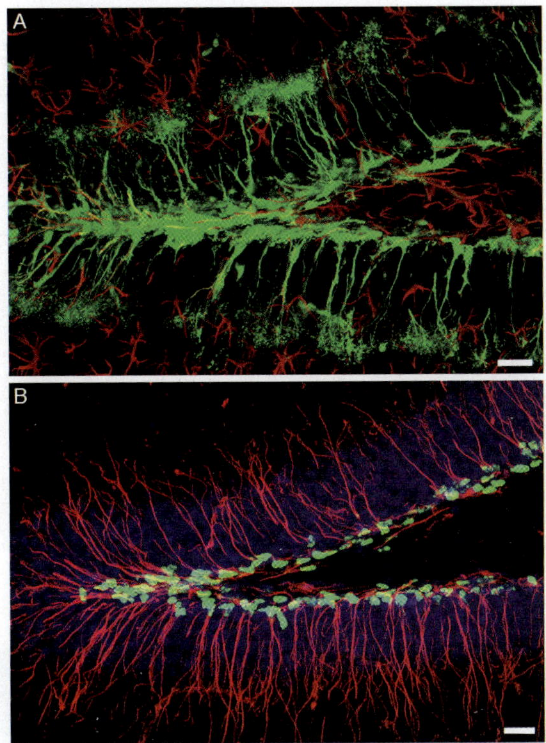

Plate 34 (Figure 11.2 on page 249 of this volume)

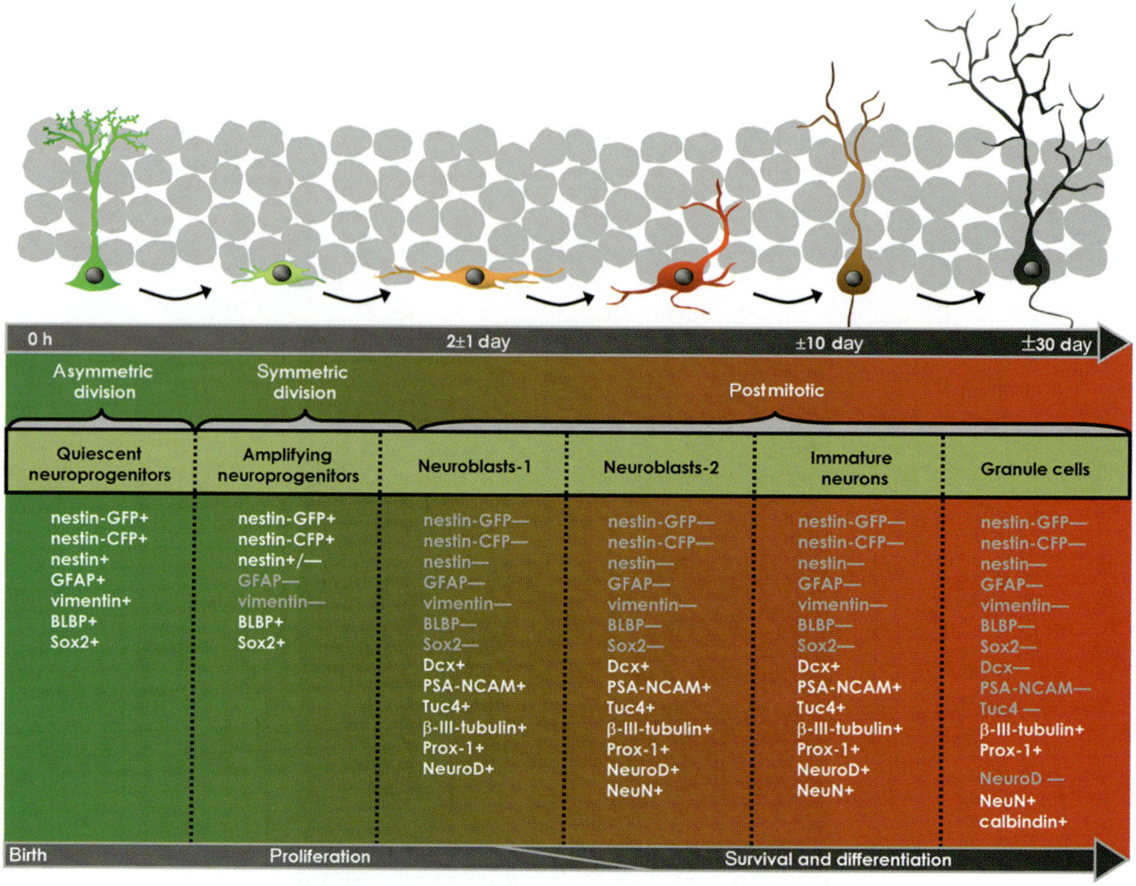

0 h | 2±1 day | ±10 day | ±30 day

Asymmetric division | Symmetric division | Postmitotic

Quiescent neuroprogenitors	Amplifying neuroprogenitors	Neuroblasts-1	Neuroblasts-2	Immature neurons	Granule cells
nestin-GFP+	nestin-GFP+	nestin-GFP—	nestin-GFP—	nestin-GFP—	nestin-GFP—
nestin-CFP+	nestin-CFP+	nestin-CFP—	nestin-CFP—	nestin-CFP—	nestin-CFP—
nestin+	nestin+/—	nestin—	nestin—	nestin—	nestin—
GFAP+	GFAP—	GFAP—	GFAP—	GFAP—	GFAP—
vimentin+	vimentin—	vimentin—	vimentin—	vimentin—	vimentin—
BLBP+	BLBP+	BLBP—	BLBP—	BLBP—	BLBP—
Sox2+	Sox2+	Sox2—	Sox2—	Sox2—	Sox2—
		Dcx+	Dcx+	Dcx+	Dcx—
		PSA-NCAM+	PSA-NCAM+	PSA-NCAM+	PSA-NCAM—
		Tuc4+	Tuc4+	Tuc4+	Tuc4 —
		β-III-tubulin+	β-III-tubulin+	β-III-tubulin+	β-III-tubulin+
		Prox-1+	Prox-1+	Prox-1+	Prox-1+
		NeuroD+	NeuroD+	NeuroD+	NeuroD —
			NeuN+	NeuN+	NeuN+
					calbindin+

Birth | Proliferation | Survival and differentiation

Plate 35 (Figure 11.3 on page 253 of this volume)

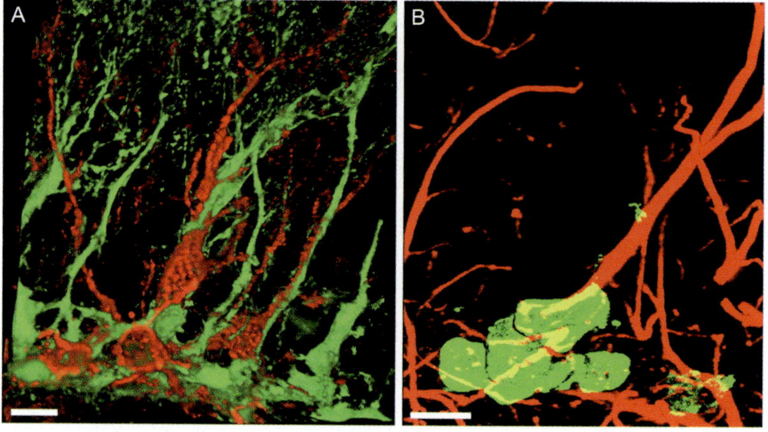

Plate 36 (Figure 11.4 on page 254 of this volume)

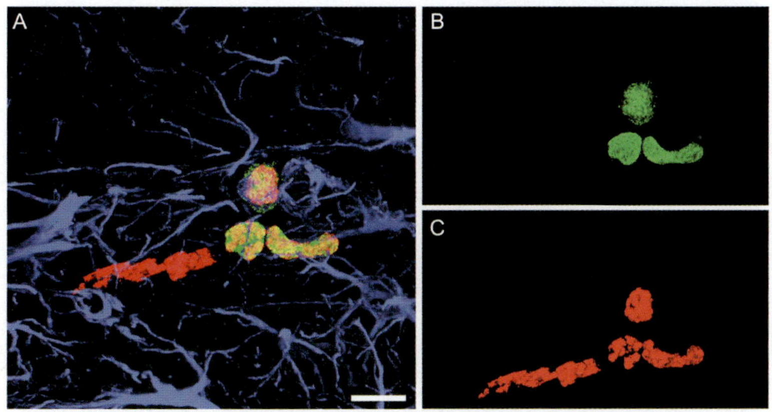

Plate 37 (Figure 11.5 on page 258 of this volume)

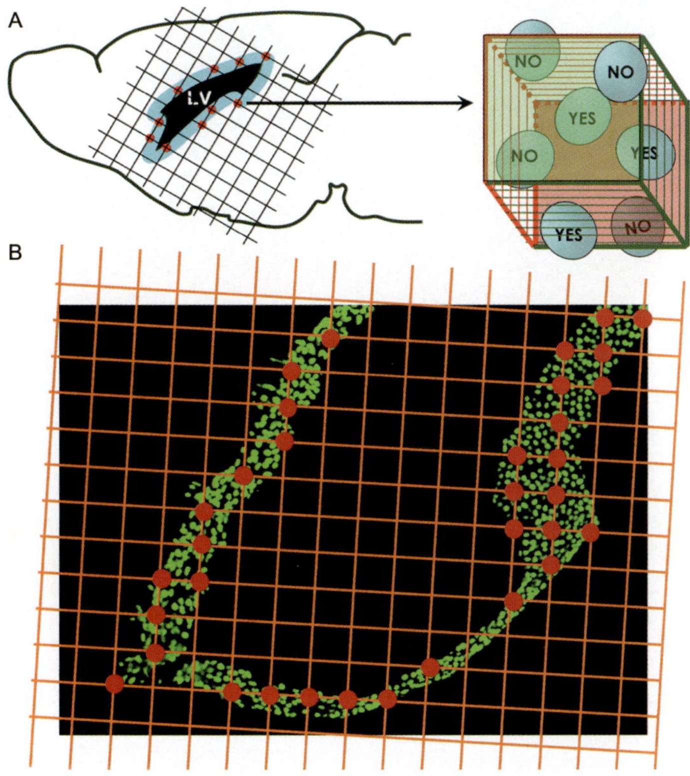

Plate 38 (Figure 11.7 on page 263 of this volume)

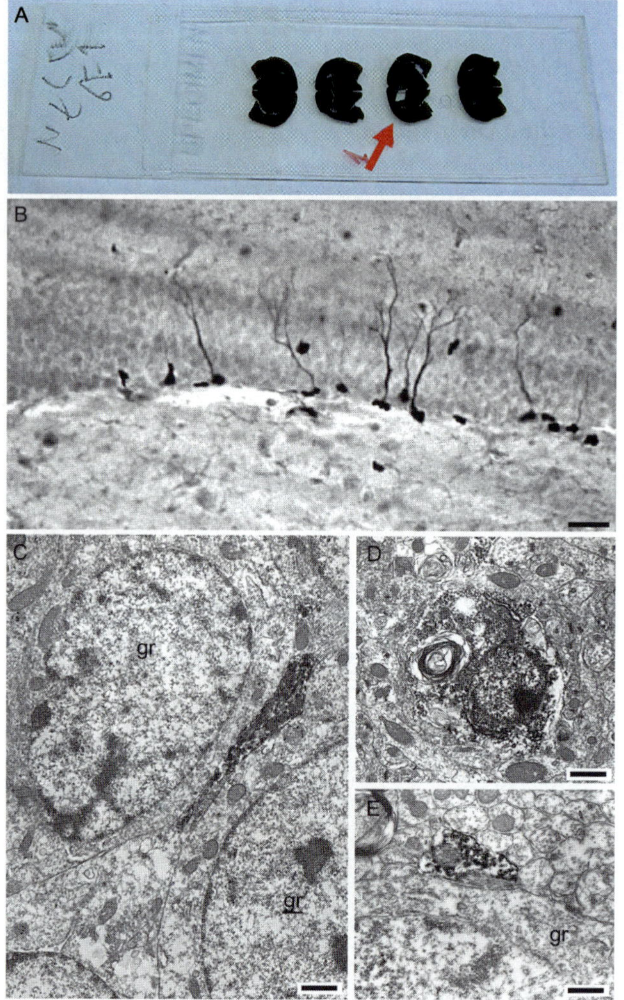

Plate 39 (Figure 11.8 on page 266 of this volume)

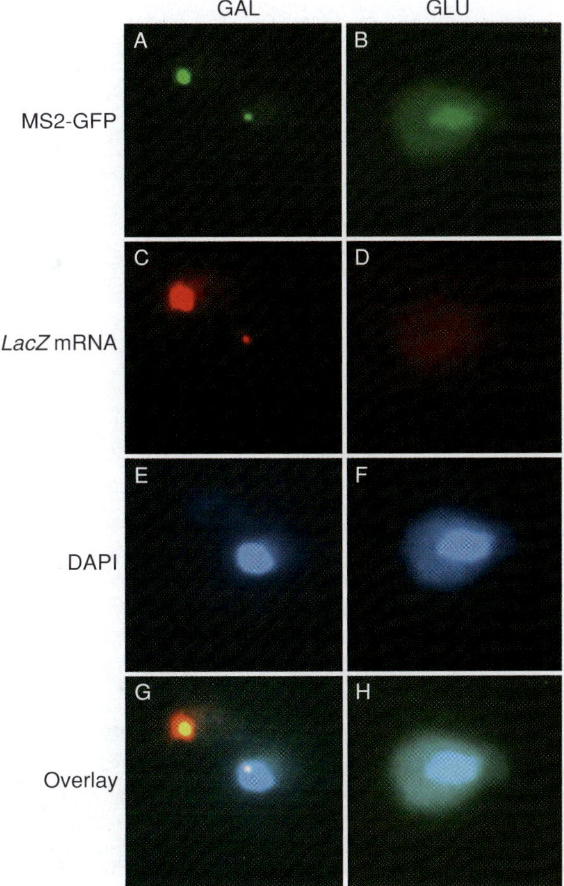

Plate 40 (Figure 12.4 on page 281 of this volume)

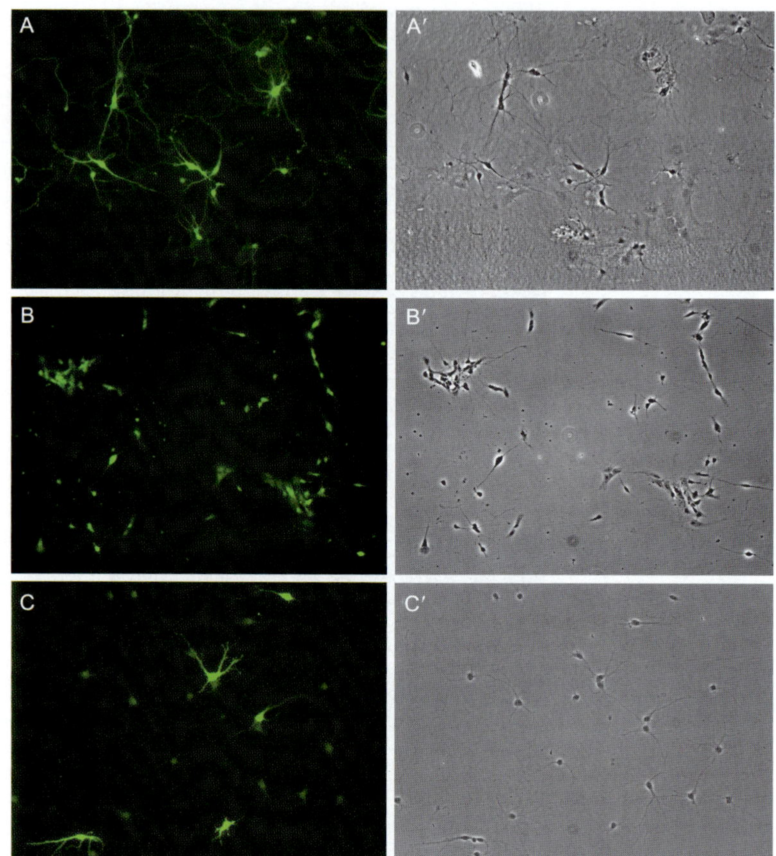

Plate 41 (Figure 13.2 on page 305 of this volume)

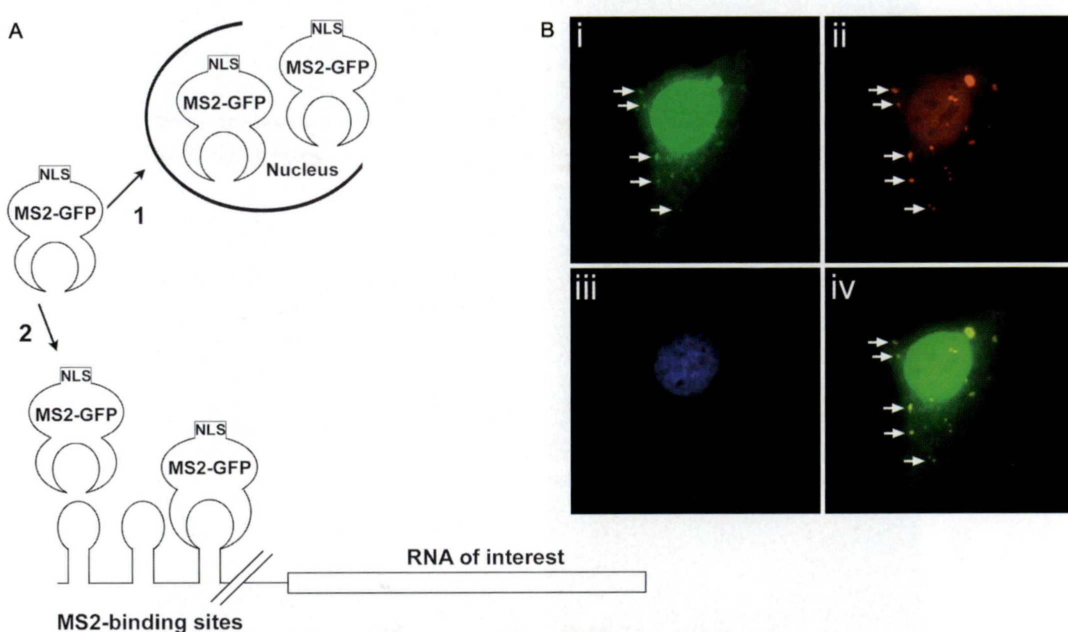

Plate 42 (Figure 13.3 on page 310 of this volume)

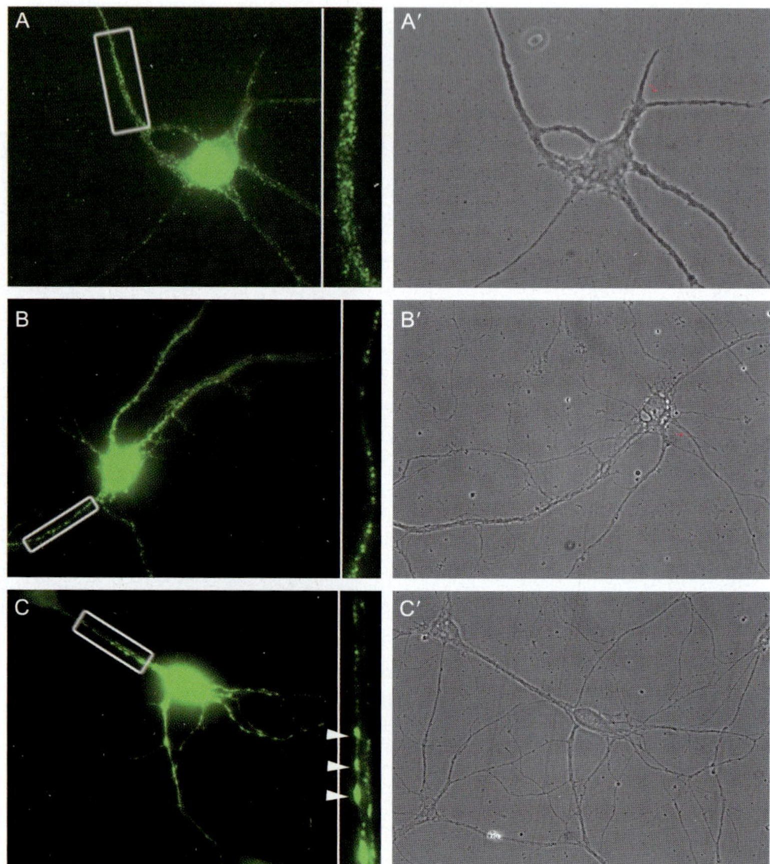

Plate 43 (Figure 13.4 on page 313 of this volume)

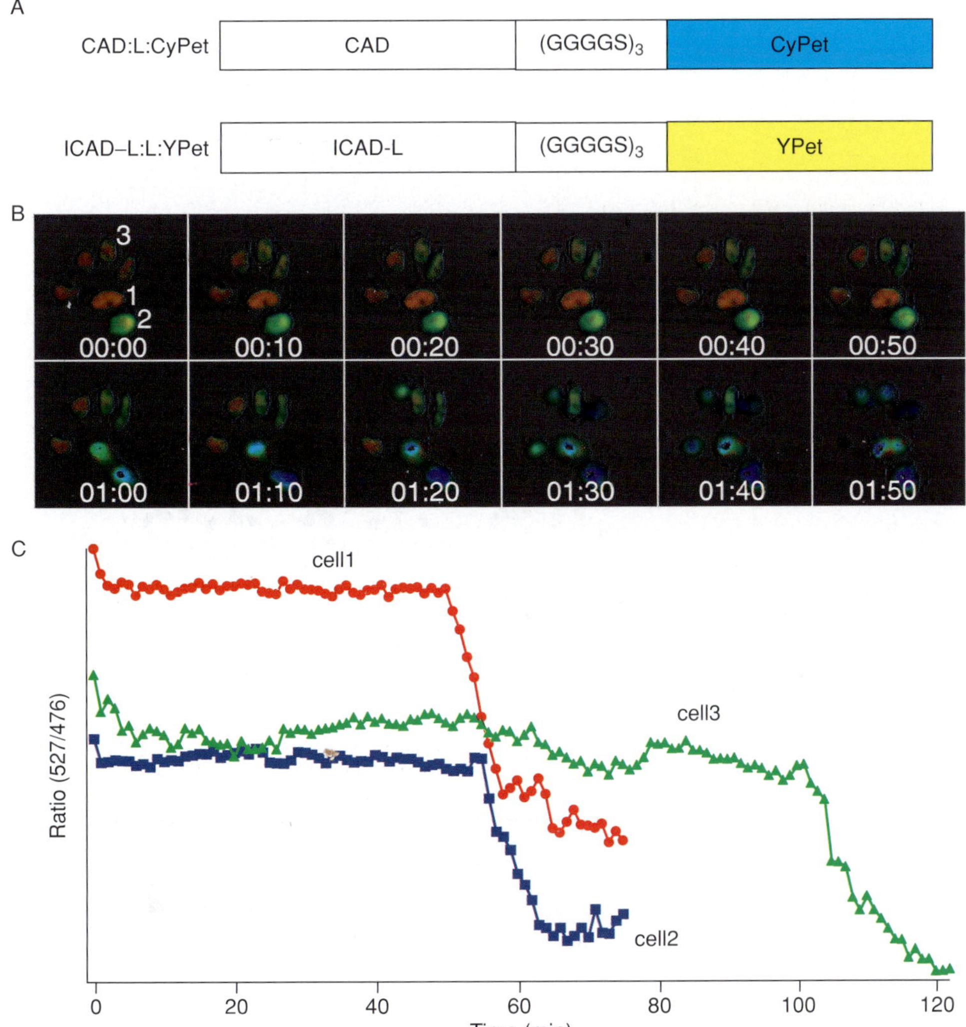

Plate 44 (Figure 16.3 on page 392 of this volume)

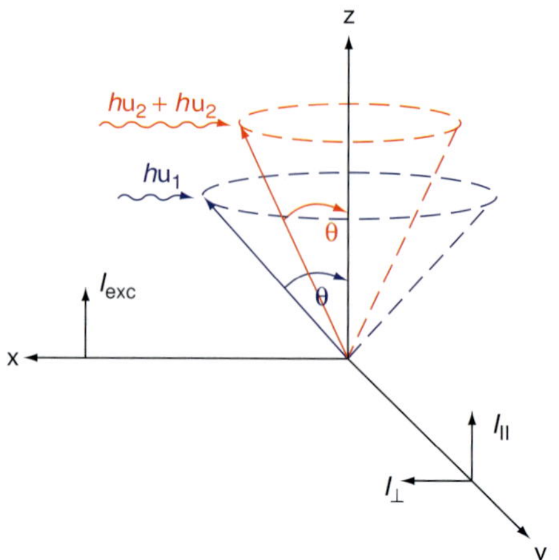

Plate 45 (Figure 17.1 on page 398 of this volume)

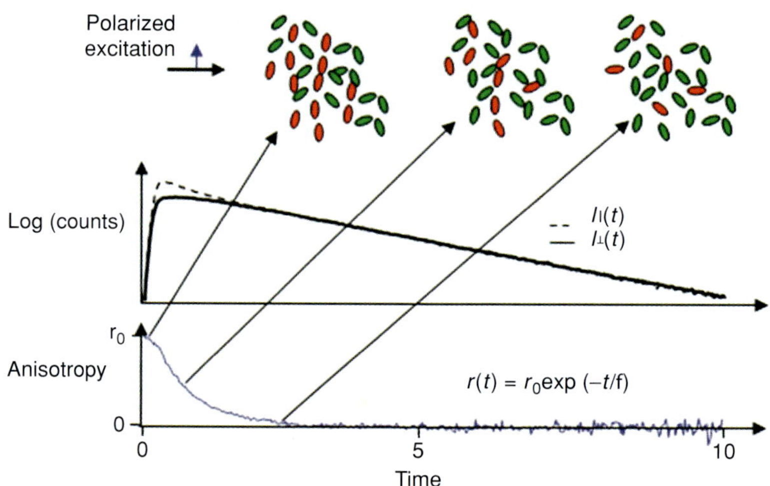

Plate 46 (Figure 17.2 on page 399 of this volume)

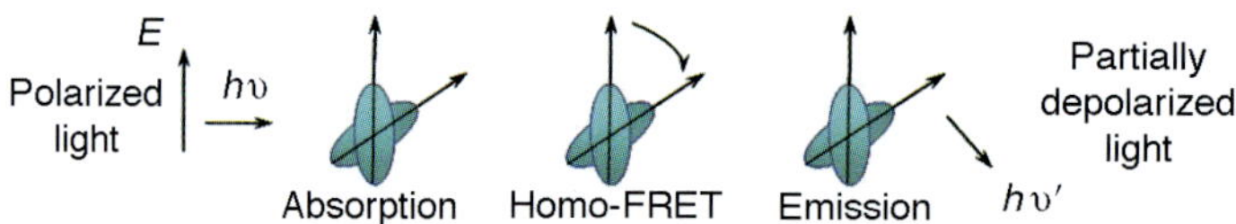

Plate 47 (Figure 17.7 on page 404 of this volume)

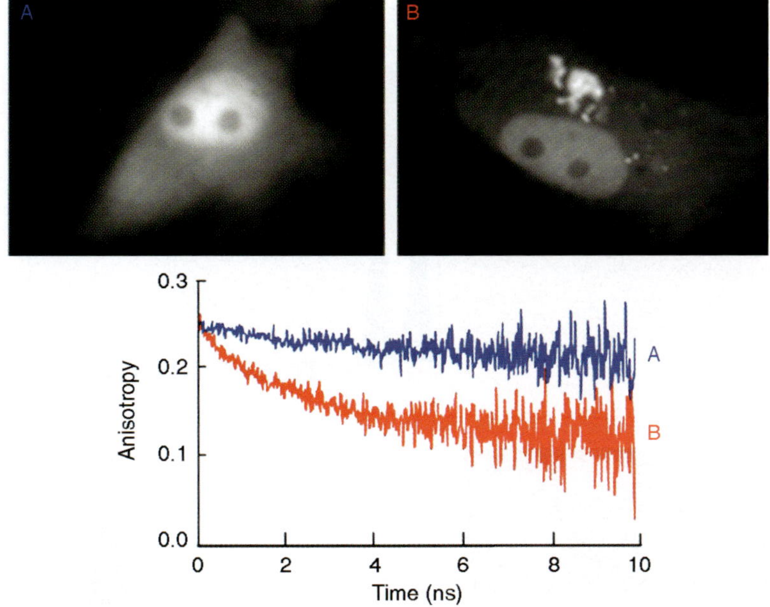

Plate 48 (Figure 17.8 on page 405 of this volume)

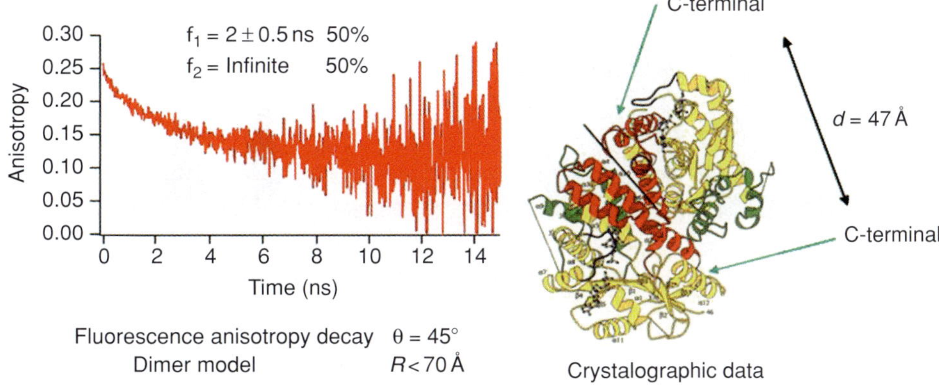

Plate 49 (Figure 17.9 on page 406 of this volume)

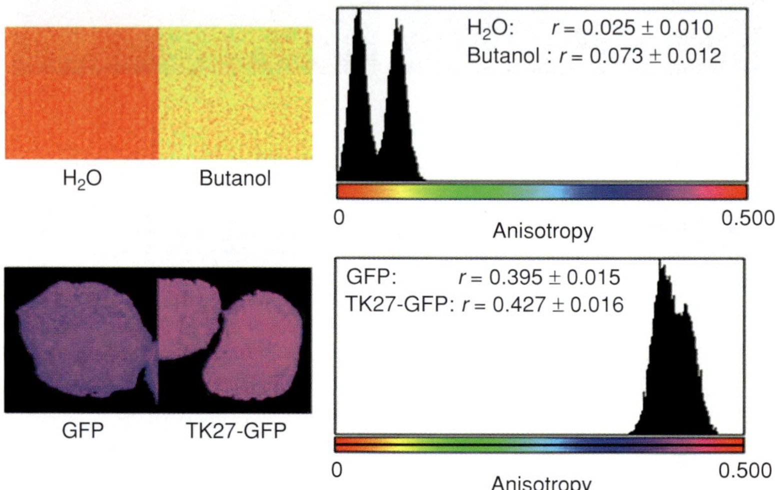

Plate 50 (Figure 17.11 on page 408 of this volume)

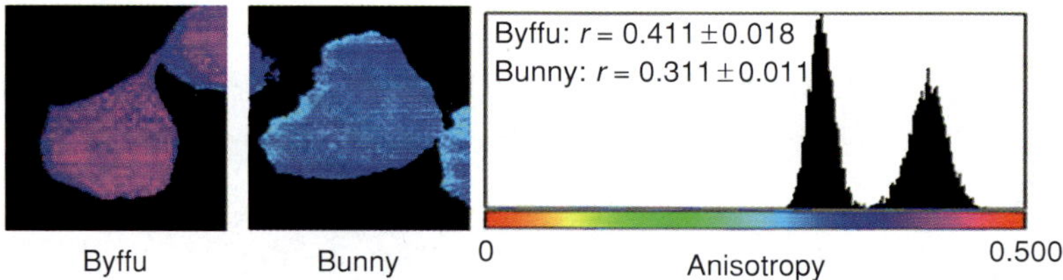

Plate 51 (Figure 17.12 on page 410 of this volume)

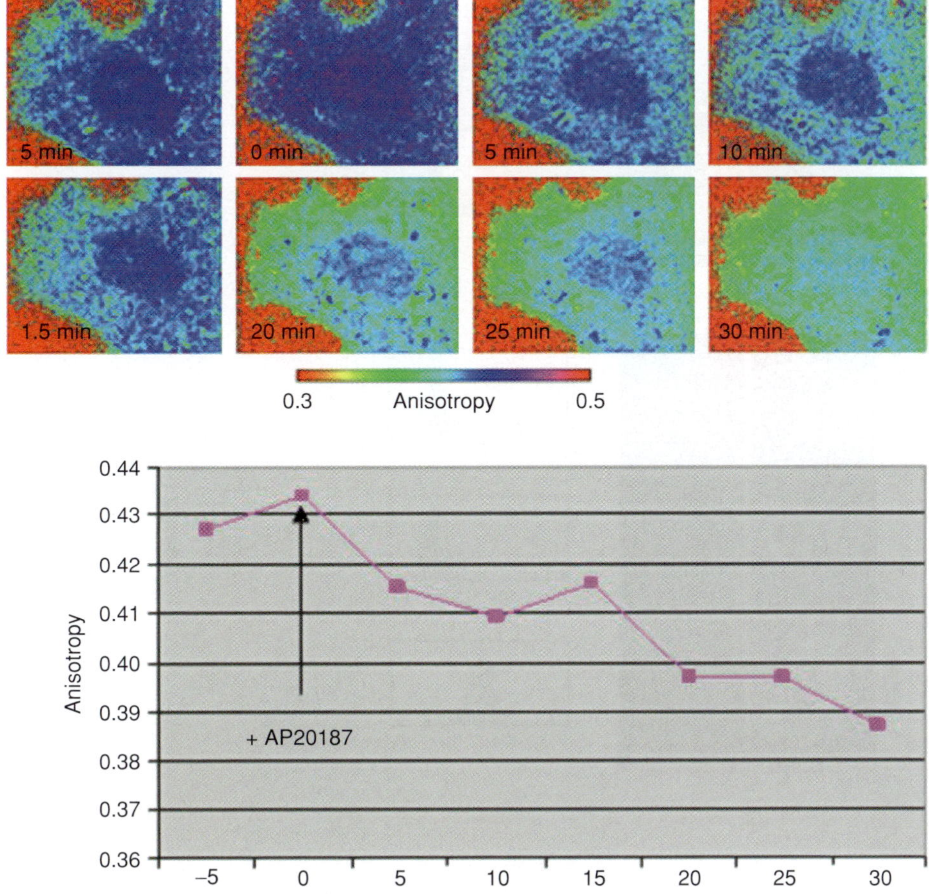

Plate 52 (Figure 17.13 on page 411 of this volume)

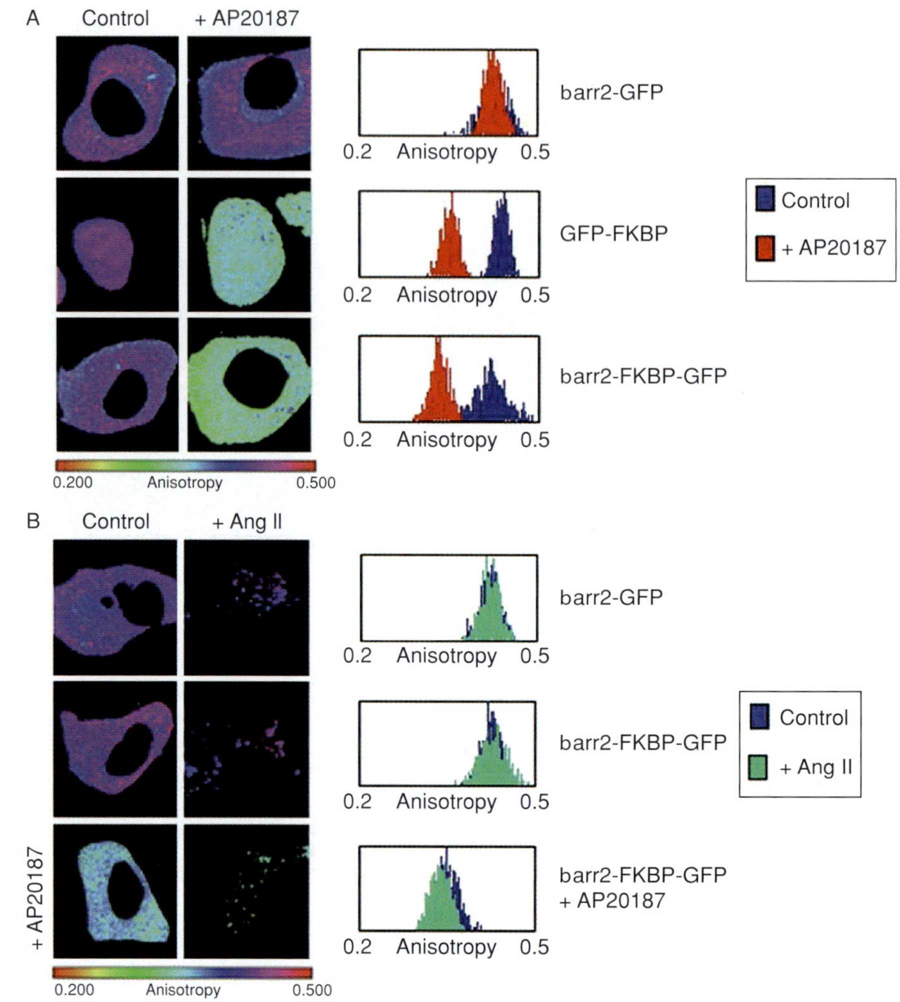

A Control + AP20187

barr2-GFP

0.2 Anisotropy 0.5

GFP-FKBP

0.2 Anisotropy 0.5

barr2-FKBP-GFP

0.2 Anisotropy 0.5

Control

+ AP20187

0.200 Anisotropy 0.500

B Control + Ang II

barr2-GFP

0.2 Anisotropy 0.5

barr2-FKBP-GFP

0.2 Anisotropy 0.5

Control

+ Ang II

+ AP20187

barr2-FKBP-GFP
+ AP20187

0.2 Anisotropy 0.5

0.200 Anisotropy 0.500

Plate 53 (Figure 17.14 on page 412 of this volume)

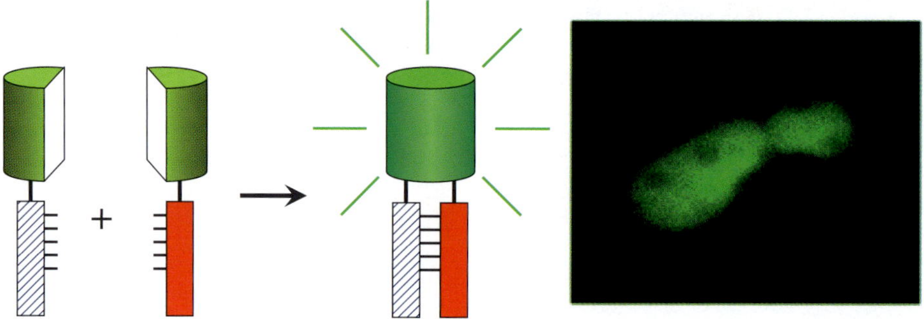

Plate 54 (Figure 19.1 on page 439 of this volume)

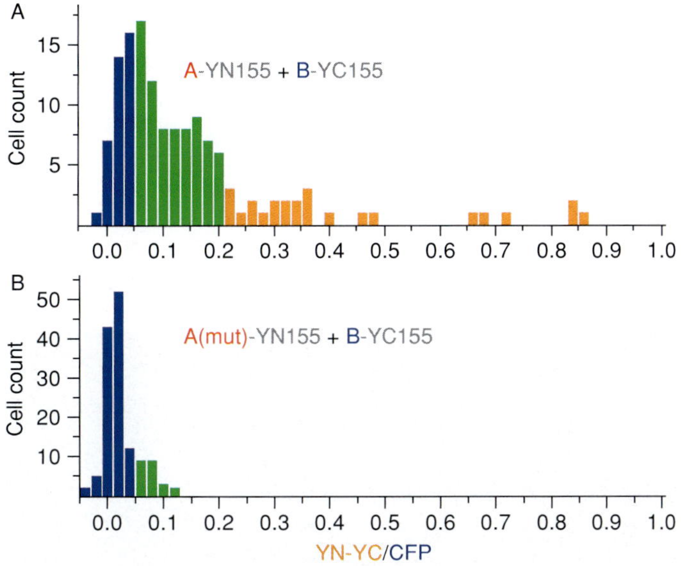

Plate 55 (Figure 19.2 on page 442 of this volume)

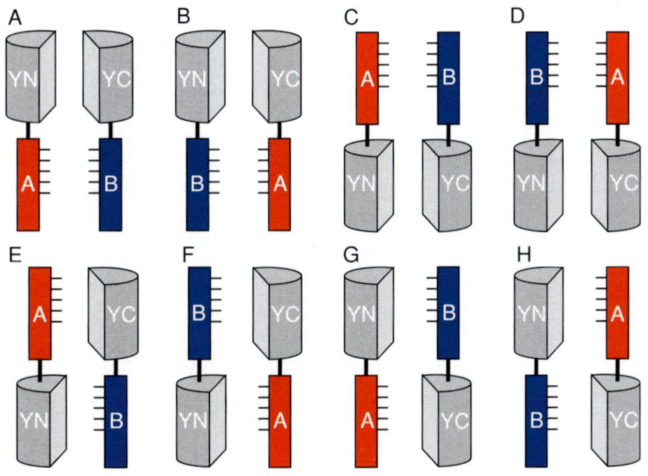

Plate 56 (Figure 19.3 on page 445 of this volume)

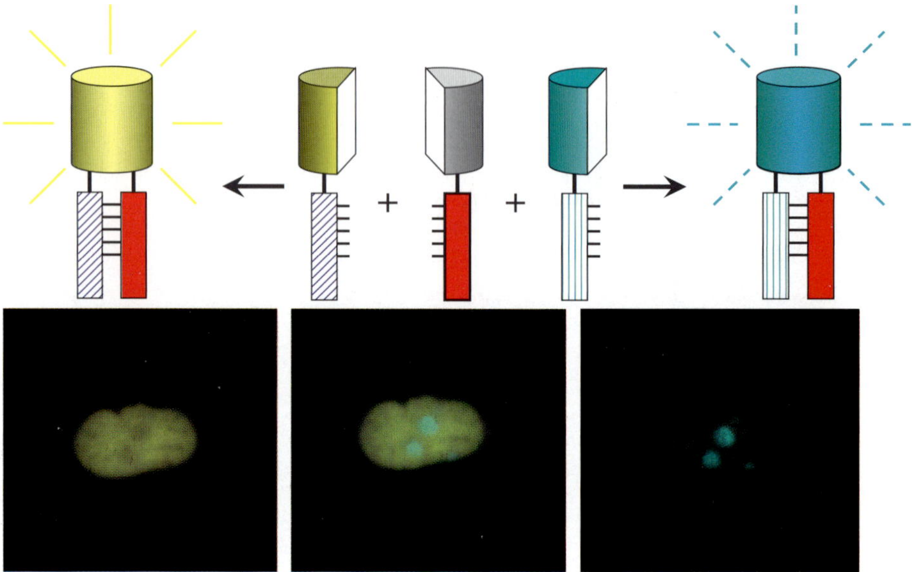

Plate 57 (Figure 19.4 on page 457 of this volume)

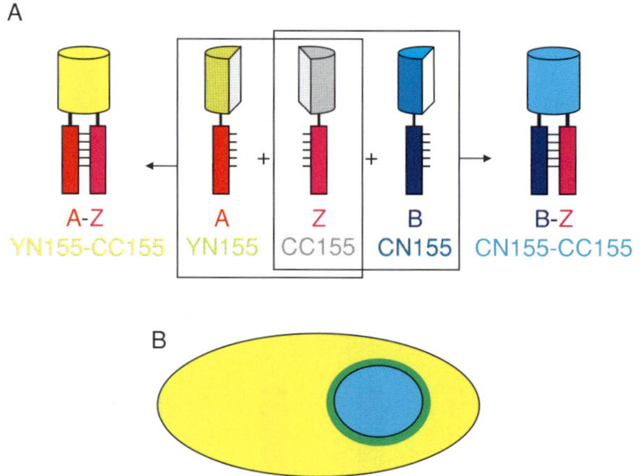

Plate 58 (Figure 19.5 on page 459 of this volume)

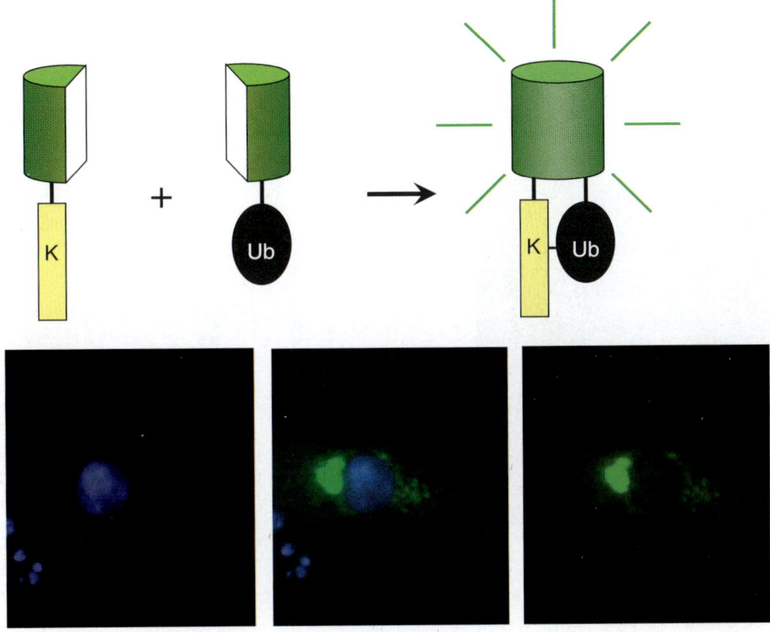

Plate 59 (Figure 19.6 on page 462 of this volume)

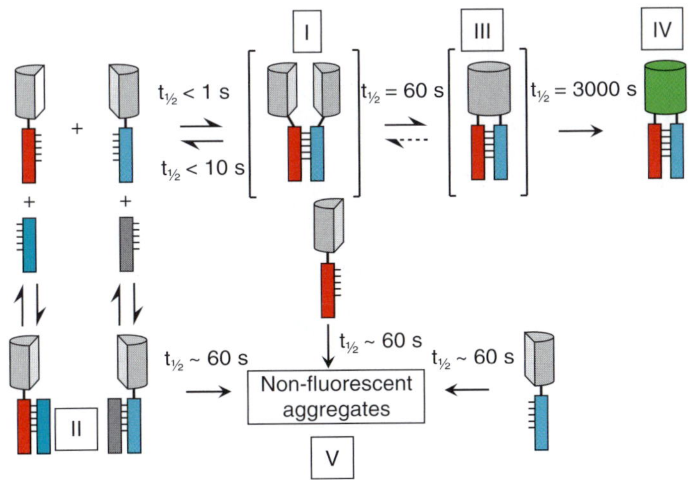

Plate 60 (Box 3 on page 440 of this volume)

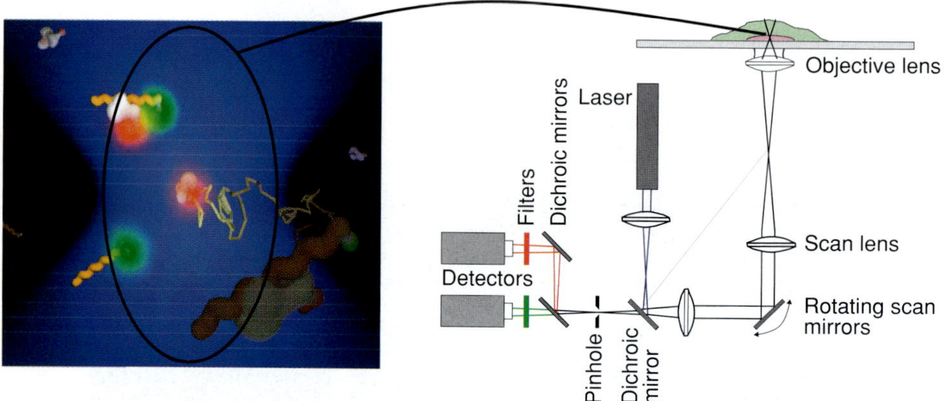

Plate 61 (Figure 20.1 on page 473 of this volume)

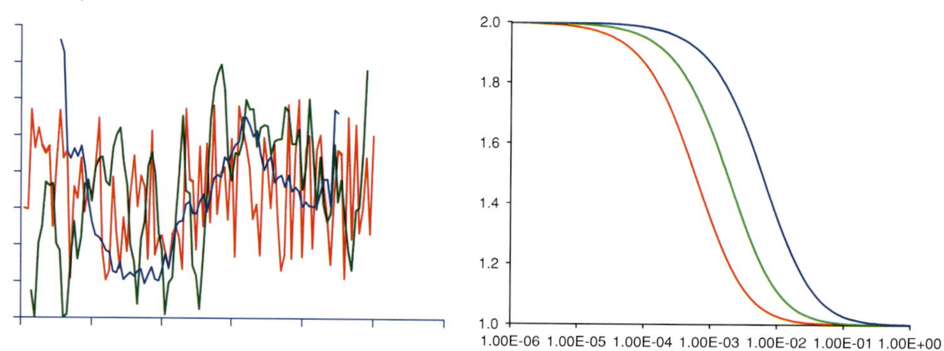

Plate 62 (Figure 20.2 on page 474 of this volume)

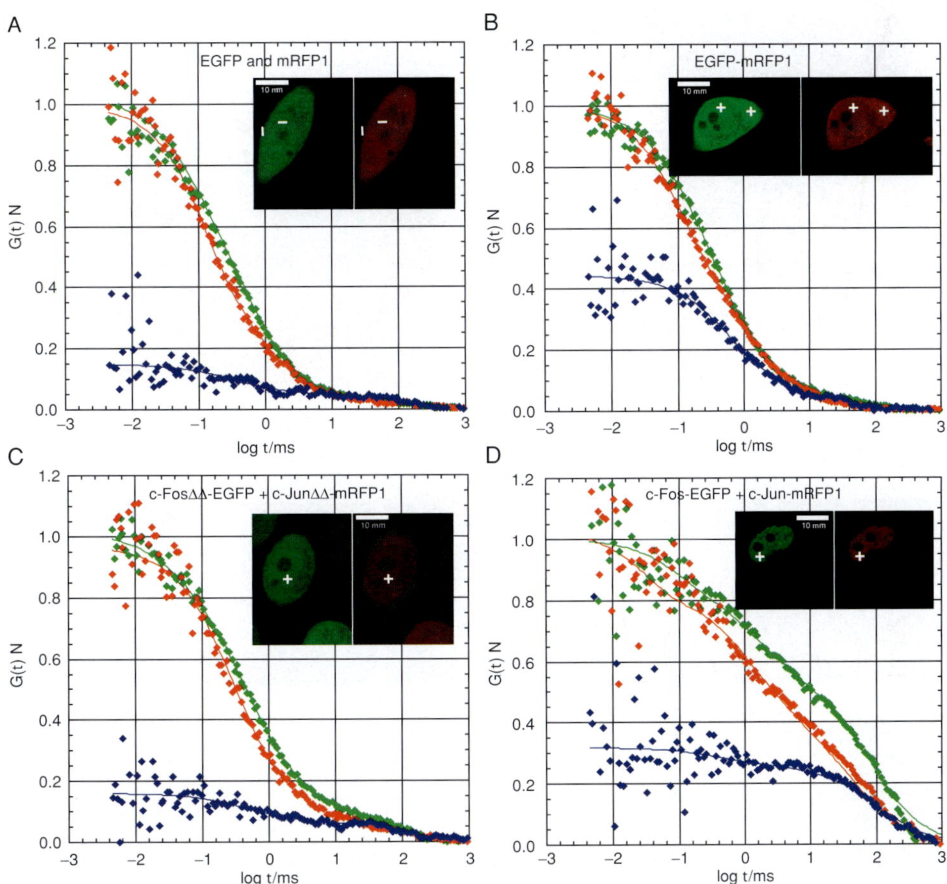

Plate 63 (Figure 20.4 on page 480 of this volume)

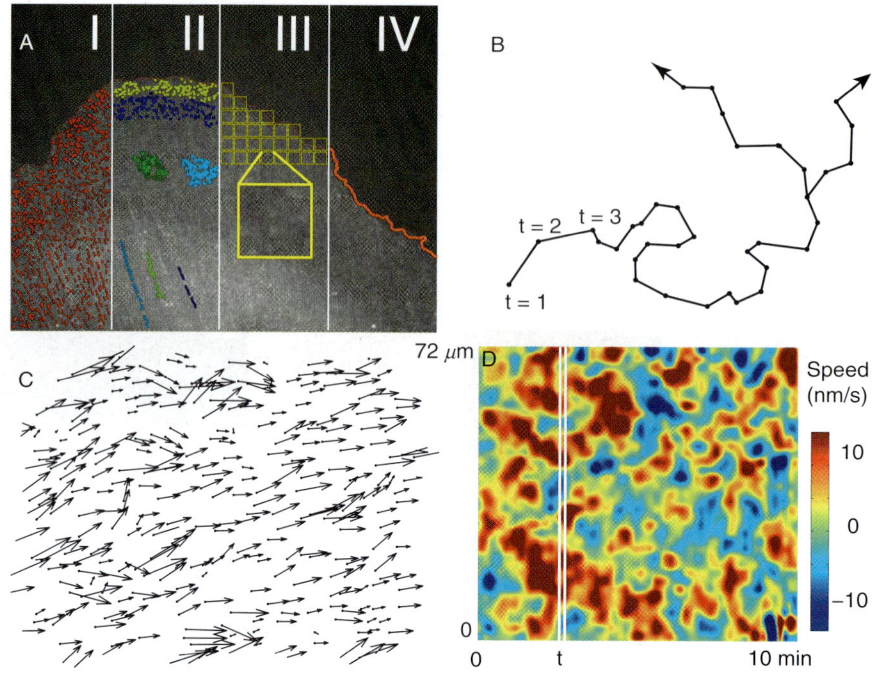

Plate 64 (Figure 22.1 on page 503 of this volume)

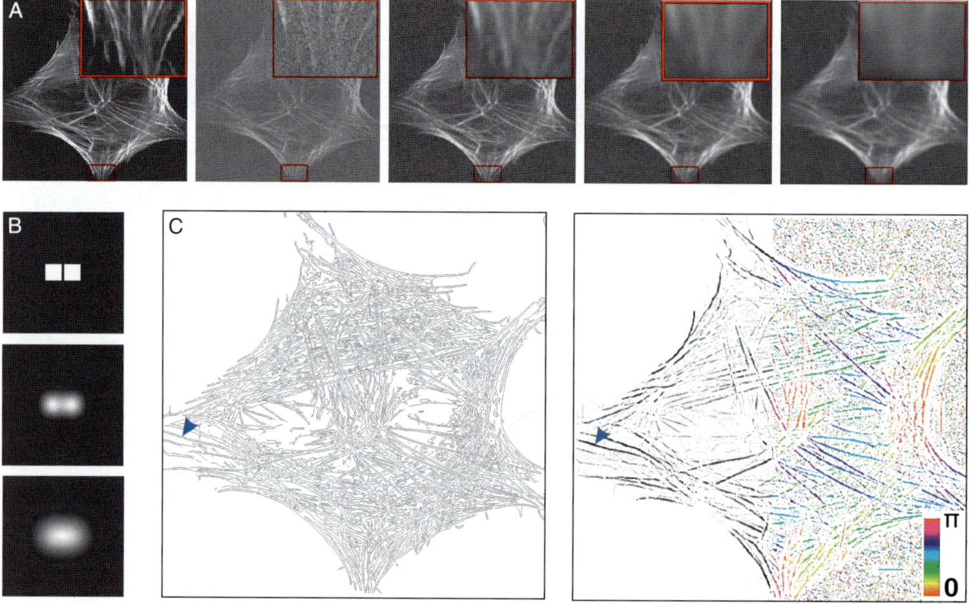

Plate 65 (Figure 22.3 on page 509 of this volume)

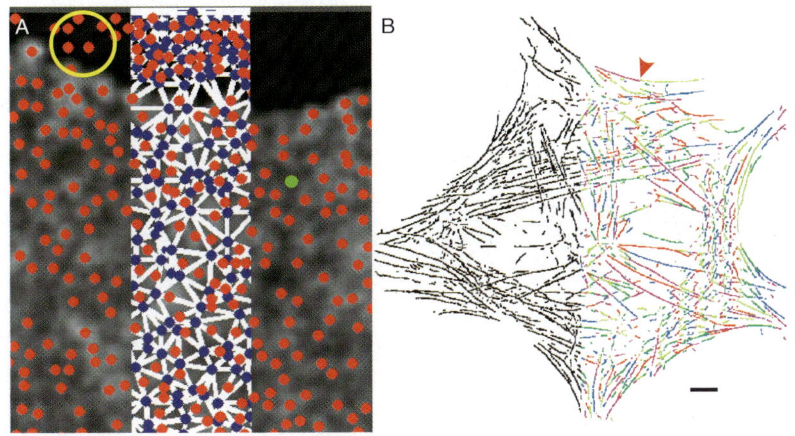

Plate 66 (Figure 22.4 on page 511 of this volume)

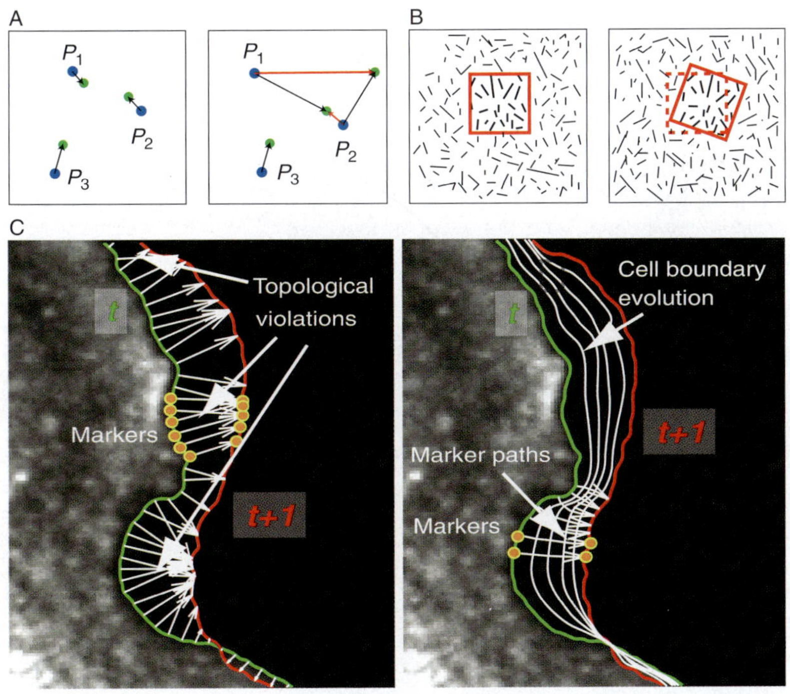

Plate 67 (Figure 22.5 on page 514 of this volume)

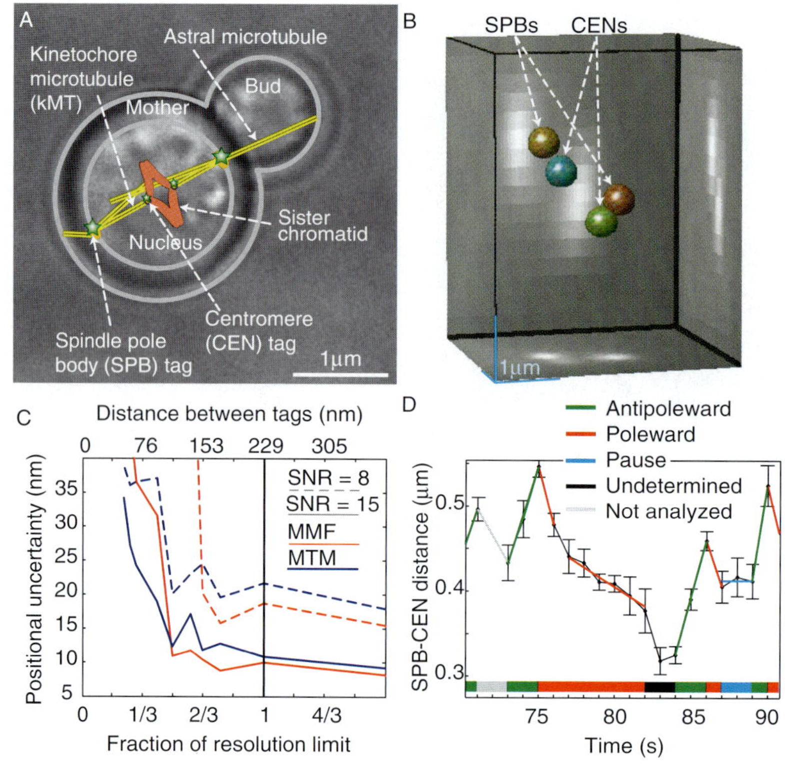

Plate 68 (Figure 22.7 on page 522 of this volume)

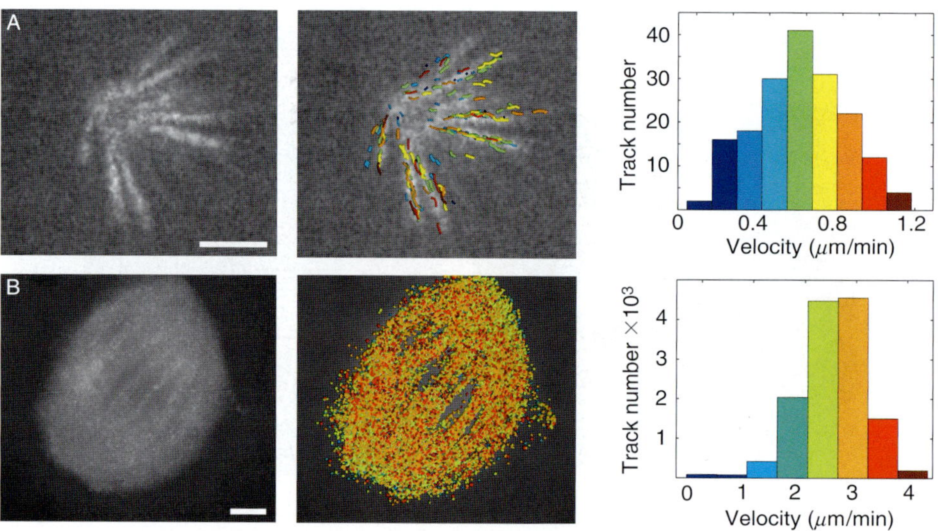

Plate 69 (Figure 22.8 on page 526 of this volume)

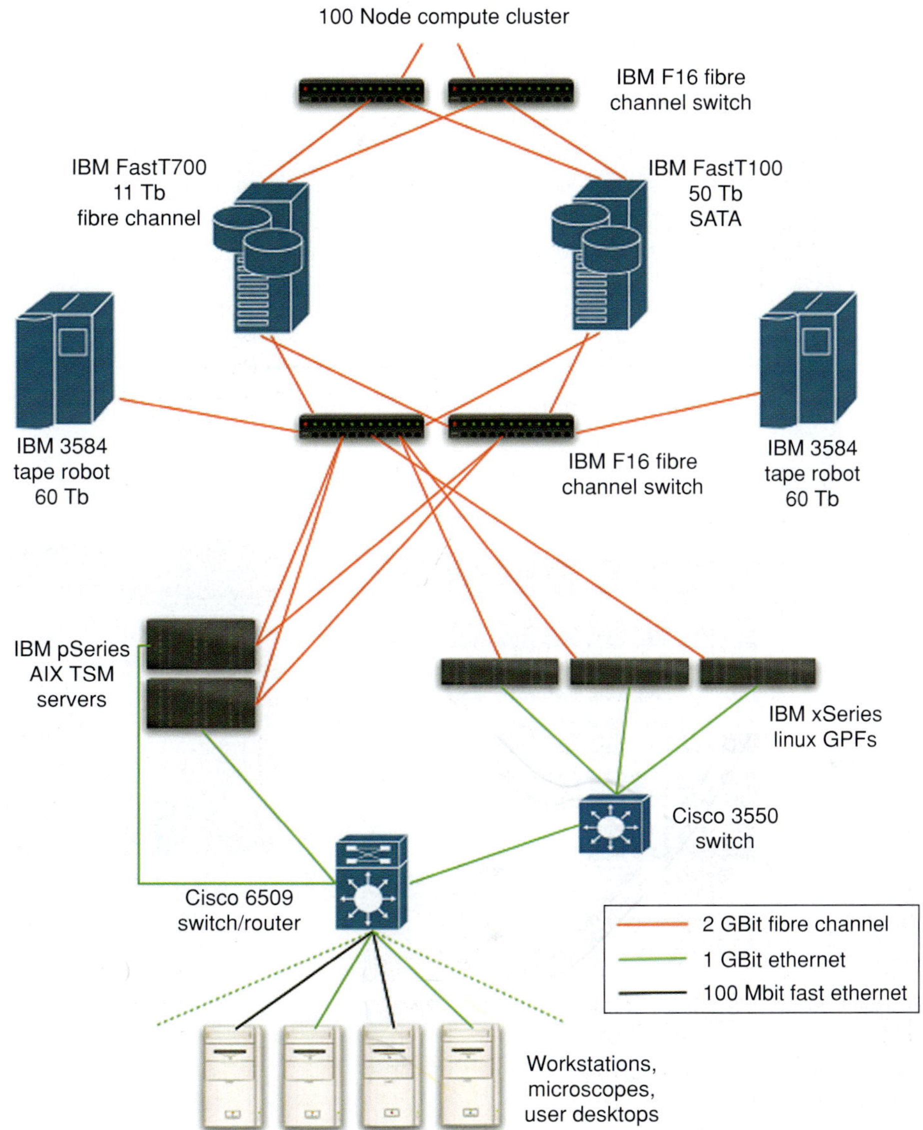

Plate 70 (Figure 24.1 on page 558 of this volume)

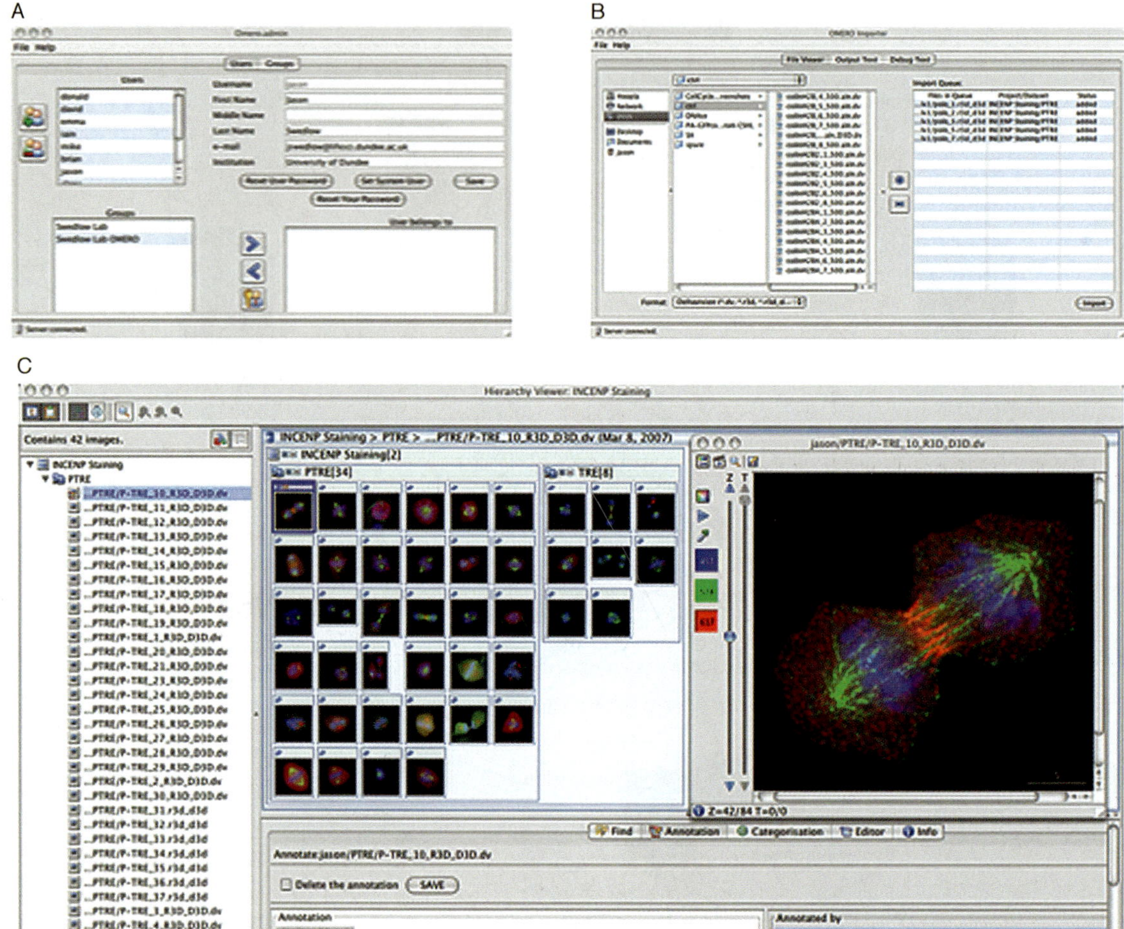

Plate 71 (Figure 24.3 on page 566 of this volume)